Water-Rock Interactions, Ore Deposits, and Environmental Geochemistry

A Tribute to David A. Crerar

SPECIAL PUBLICATION SERIES

No. 1. *Magmatic Processes: Physicochemical Principles.* B. O. Mysen (Editor), 1987.
No. 2. *Fluid-Mineral Interactions: A Tribute to H. P. Eugster.* R. J. Spencer and I-Ming Chou (Editors), 1990.
No. 3. *Stable Isotope Geochemistry: A Tribute to Samuel Epstein.* H. P. Taylor, Jr., J. R. O'Neil and I. R. Kaplan (Editors), 1991.
No. 4. *Victor Moritz Goldschmidt: Father of Modern Geochemistry.* Brian Mason, 1992.
No. 5. *Mineral Spectroscopy: A Tribute to Roger G. Burns.* M. D. Dyar, C. McCammon and M. W. Schaefer (Editors), 1996.
No. 6. *Mantle Petrology: Field Observations and High-Pressure Experimentation: A Tribute to Francis R. (Joe) Boyd.* Y. Fei, C. M. Bertka and B. O. Mysen (Editors), 1999.
No. 7. *Water-Rock Interactions, Ore Deposits, and Environmental Geochemistry: A Tribute to David A. Crerar.* Roland Hellmann and Scott A. Wood (Editors), 2002.

Special Publication Series Editor: Scott A. Wood
Department of Geological Sciences
University of Idaho
Box 443022
Moscow, ID 83844-3022
USA

Water-Rock Interactions, Ore Deposits, and Environmental Geochemistry

A Tribute to David A. Crerar

Edited by:

Roland Hellmann[1] and Scott A. Wood[2]

[1]Crustal Fluids and Dynamics Group, LGIT, CNRS UMR C5559, Observatoire de Grenoble, Université J. Fourier, BP 53X, 38041 Grenoble Cedex 9, FRANCE

[2]Department of Geological Sciences, University of Idaho, Box 443022, Moscow, ID 83844-3022 USA

Special Publication No. 7

THE GEOCHEMICAL SOCIETY

Published by the Geochemical Society
Department of Earth & Planetary Sciences
Washington University
One Brookings Drive
St. Louis, MO 63130-4899 USA

Printed in United States of America

Library of Congress Catalog Card Number: 2001132168

ISBN 0-941809-06-4

TABLE OF CONTENTS

Preface vii

DAVID A. CRERAR: Some vignettes of his life and career ix

Part A. Water-Rock Interactions and Fluid-Mineral Interfaces

William J. Ullman and Susan A. Welch: Organic ligands and feldspar dissolution 3

Art F. White, Alex E. Blum, Marjorie S. Schulz, Thomas G. Huntington, Norman E. Peters, and David A. Stonestrom: Chemical weathering of the Panola granite: Solute and regolith elemental fluxes and the weathering rate of biotite 37

Sherry D. Samson and Carrick M. Eggleston: Nonsteady-state dissolution of goethite and hematite in response to pH jumps: the role of adsorbed Fe(III) 61

Clarence D. Chang, Cynthia T.-W. Chu, and John L. Schlenker: Corrosion forms and twinning in zeolite ZSM-5 crystals 75

Nathan P. Mellott, Susan L. Brantley, and Carlo G. Pantano: Topography of polished plates of albite crystal and glass during dissolution 83

Dirk Bosbach: Linking molecular-scale barite precipitation mechanisms with macroscopic crystal growth rates 97

Steven R. Higgins, Andrew G. Stack, Kevin G. Knauss, Carrick M. Eggleston, and Guntram Jordan: Probing molecular-scale adsorption and dissolution-growth processes using nonlinear optical and scanning probe methods suitable for hydrothermal applications 111

Roland Hellmann, Peter J.N. Renders, Jean-Pierre Gratier, and Robert Guiguet: Experimental pressure solution compaction of chalk in aqueous solutions Part 1. Deformation behavior and chemistry 129

Roland Hellmann, Patrick Gaviglio, Peter J.N. Renders, Jean-Pierre Gratier, Samir Békri, and Pierre Adler: Experimental pressure solution compaction of chalk in aqueous solutions Part 2. Deformation examined by SEM, porosimetry, synthetic permeability, and X-ray computerized tomography 153

Part B. Thermodynamics, Hydrothermal Geochemistry, and Ore Deposits

Greg M. Anderson: Stable and metastable equilibrium: The third constraint. 181

Christopher H. Gammons, Scott A. Wood, and Youning Li: Complexation of the rare earth elements with aqueous chloride at 200 °C and 300 °C and saturated water vapor pressure. 191

Ru Ding and Scott A. Wood: The aqueous geochemistry of the rare earth elements and yttrium. Part X. Potentiometric determination of stability constants of acetate complexes of La^{3+}, Nd^{3+}, Gd^{3+} and Yb^{3+} at 25-70 °C and 1 bar. 209

Scott A. Wood, Donald A. Palmer, David J. Wesolowski, and Pascale Bénézeth: The aqueous geochemistry of the rare earth elements and yttrium. Part XI. The solubility of $Nd(OH)_3$ and hydrolysis of Nd^{3+} from 30 to 290 °C at saturated water vapor pressure with in-situ pH_m measurement. 229

William E. Seyfried, Jr., Kang Ding, and Bin Rao: Experimental calibration of metastable plagioclase-epidote-fluid equilibria at elevated temperatures and pressures: applications to the chemistry of hydrothermal fluids at mid-ocean ridges. 257

Anthony E. Williams-Jones, Artashes A. Migdisov, Sandy M. Archibald, and Zhifeng Xiao: Vapor-transport of ore metals. 279

Christoph Wahrenberger, Terry M. Seward, and Volker Dietrich: Volatile trace element transport in high-temperature gases from Kudriavy volcano (Iturup, Kurile Islands, Russia). 307

Brenda A. MacMurray and Greg M. Anderson: A fluid-inclusion study of the drusy quartz of the Potosi Dolostone, Southeast Missouri. 329

Part C. Environmental Geochemistry

Christopher J. Tadanier, John C. Little, Duane F. Berry, and Michael F. Hochella, Jr.: Microbial acquisition of nutrients from mineral surfaces 339

Jeremy B. Fein: The effects of ternary surface complexes on the adsorption of metal cations and organic acids onto mineral surfaces 365

Peter R. Jaffé, Sookyun Wang, Peter L. Kallin, and Sherwood L. Smith: The dynamics of arsenic in saturated porous media: fate and transport modeling for deep aquatic sediments, wetland sediments, and groundwater environments 379

David M. Tuck: Primary drainage of NAPL governed by time-dependent interfacial properties. 399

Uta Gabriel, Laurent Charlet, and Carl W. Schläpfer: In situ speciation of uranium(VI) at the silica-water interface: A combined TRLIFS and surface complexation study 423

Colleen M. Hansel, Matthew J. La Force, Stephen S. Sutton, and Scott Fendorf,: Ecosystem dynamics of zinc and manganese within a mine-waste impacted wetland. 441

Index. 455

PREFACE

Water-Rock Interactions, Ore Deposits, and Environmental Geochemistry: A Tribute to David A. Crerar was organized as a tribute to David Crerar's exceptional contributions to various fields of geochemistry. This volume is a collection of 23 articles written mostly by former students, friends, and colleagues who are currently involved in fields of study that defined David's research interests: water-rock interactions, thermodynamics, hydrothermal geochemistry, ore deposit formation, and environmental geochemistry. David's diverse research activities, both during his student days at the University of Toronto and Pennsylvania State University, and as a professor of geochemistry in the Department of Geological and Geophysical Sciences, Princeton University, attest to his broad scientific intellect and curiosity. The development and enlargement of his scientific research would surely have continued, but his life was tragically cut short at the age of 49 by an incurable illness.

The first theme of this volume, water-rock interactions, concerns mainly non-equilibrium interactions between minerals and aqueous solutions, interfacial adsorption, rates of dissolution and precipitation reactions, the mechanisms of these reactions at a molecular scale, and the behavior of fluid-mineral interfaces. David's foresight into the importance of this field of research made him and his students and colleagues among the early leaders in this relatively new area of research. The water-rock interaction topics explored in this volume are representative of very recent advances in experimental and applied field studies. The studies we have included vary enormously in scale and methodology, ranging from chemical weathering and denudation rates at the watershed scale, to experimental approaches investigating macroscopic dissolution rates as well as fluid-assisted deformation via pressure solution, to molecular scale studies of dissolution and precipitation rates and mechanisms, based on some of the newest techniques in microscopy and spectroscopy available today.

The second part of the volume is devoted to thermodynamics, hydrothermal geochemistry, and ore deposit formation. These are essentially the subjects that propelled David into the world of geochemistry as a young student. Thermodynamics was a subject that David mastered, but in addition, he developed a reputation for his abilities to convey his knowledge to others in a pedagogical manner. Both of these qualities led, perhaps naturally, to the recently published preeminent text, *Thermodynamics in Geochemistry*, co-authored by Greg Anderson and David. Thus, it is fitting that the first contribution is a thermodynamic examination of equilibrium, written by Greg (who was also David's undergraduate and Master's advisor). The next group of papers examines complexation of rare earth elements, stability constants of organic-metal complexes, and the solubility and hydrolysis constants of neodymium at hydrothermal conditions. The application of thermodynamics and hydrothermal geochemistry to natural systems is treated by the last group of papers, which are based on fluid inclusions, hydrothermal fluids at mid-ocean ridges, and vapor-phase metal transport and deposition in active volcanic systems.

The third and last part of this volume deals with environmental geochemistry. This domain of research is arguably the most pertinent of David's research activities to modern-day problems associated with ground and water pollution and their environmental impacts. Environmental geochemistry is in large part based on the interface between the worlds of biology, geology, and chemistry, and today, biogeochemistry is a field which has literally exploded in importance. The papers in this last part of the volume reflect the broad scope of what defines environmental geochemistry today: microbe-mineral systems; interactions between metals, organics, and minerals; theoretical arsenic dynamics in natural porous media; modeling NAPL migration; uranium speciation at silica surfaces; and zinc and manganese behavior in a mine-waste impacted wetland.

We thank the Geochemical Society for providing the necessary resources and funding that made this volume possible. The Department of Geological and Geophysical Sciences, Princeton University, also provided a valuable financial contribution. As editors, we thank the many referees who provided detailed evaluations and critiques of the individual chapters. In addition, we also acknowledge the essential contributions provided by Brian Evans (MIT) and Tom Giordano (New Mexico State University) with the editorial handling of the chapters that R.H. and S.A.W., respectively, submitted to this volume. Peggy Cross, Bob Key, and Bill Bonini provided much assistance with retrieving archived information, as well as obtaining the photographs of David and David's former graduate students that appear in this volume. Many thanks.

ROLAND HELLMANN
SCOTT A. WOOD
April, 2001

UNKNOWN PHOTOGRAPHER

DAVID A. CRERAR

Shell Professor of Geochemistry, Princeton University

PHOTO BY BOB KEY

Former graduate students and research associates of David at Princeton University in May 1993. Back row, from left to right: Richard Murnane, Roland Hellmann, Jeffrey Means, Bear McPhail, David Tuck. Second row, from left to right: Rémy Hennet, Scott Wood, Maria Borcsik, Susan Brantley, Nicholas Susak, Lusheng Yan. Front row, from left to right: Lynn Vogel-Koplitz, Ann Maest, Jennifer Snow Boles, Mary Yang, Lori Williams, Robert Kleinmann. Not shown: Ellen Axtmann, Michael So Chyi, Patricia Dove, Kevin Costello, Jessica Posey Dowty, Carl Yuretich.

DAVID ALEXANDER CRERAR

Some vignettes of his life and career

David Crerar died on September 6, 1994 at the Queen Elizabeth Hospital in Toronto, from complications arising from a degenerative neurological disease. He was 49 years old.

David was born in Toronto. His father was a professional musician, and David displayed precocious musical and artistic talents at an early age. He played violin and piano, and was a member of the Canadian Youth Orchestra. Photographs, silk screen prints, and other works made by David still hang in the homes of his friends. Another of his interests was rocks and minerals, perhaps one of the reasons for David having followed the undergraduate geology program at the University of Toronto.

I was a newly appointed professor at U of T. in 1966 when David and his friend Doug Macdougall (now at Scripps) approached me about teaching a course in thermodynamics. The three of us had a great time that year. They let nothing pass unquestioned. I think I learned more than they did. The following year David became my first graduate student. It was an exciting time, and his M.Sc. thesis of 59 pages was beautiful. Following his Ph.D. at Penn State with Hu Barnes, he joined the faculty at Princeton. Sheldon Judson relates that David's application arrived quite late in the search process, but after his visit their decision to hire him was the fastest on record.

A few years later, we began to write a book on thermodynamics together. To finish this up, we arranged that we would spend a sabbatical leave together in Toulouse, France. David was at first a bit reluctant about this, as he hated flying, but he came to love living in France, so much so that he talked seriously about working for the CNRS. Another of David's interests was the preparation and the organic chemistry of food, and Dave and his wife Scotia visited just about every restaurant and vineyard within driving distance of Toulouse, and even got to know a few of the chefs. Dinner with the Crerars was always a delightful occasion. We managed to get some writing done, though I think some of our best ideas occurred at Le Baril, a café around the corner where the owners knew us quite well.

However, what I remember best about David has nothing to do with his scientific career, but how much fun it was to have him around; how much more interesting life was when I was with him. David was forever recounting hilariously funny things that had happened to him, and I used to wonder why so many interesting things happened to him, and not to the rest of us. Of course, they happen to everyone, but only those blessed with a heightened awareness, sensitivity, sense of humour, and a great love of life can recognize them. David was a very special person, and those who knew him are grateful for the memories.

Greg Anderson
Toronto
February, 2001

DAVID A. CRERAR AS A GRADUATE STUDENT

David (never "Dave") came to Penn State already recognized during his undergraduate and M.Sc. studies at the University of Toronto as a uniquely gifted scientist. He had received there during his M.Sc. studies multiple, simultaneous awards carrying an income extraordinary for a graduate student. His unusual three-quarter time assistantship at Penn State continued to support his enduring penchant to acquire art, leading sometimes to a near impecunious condition. That hobby, together with his avid violin-playing, balanced his passion for science.

Academically, he was an outlier off normal grade scales. That was a consequence, not of his disinterest in field work, but of his superb practical facility with theory. That characteristic, and his active imagination and corresponding innovative flair, plus an adequate experimental ability and surplus energy, allowed him to carry out several research projects simultaneously while excelling in his courses. Those efforts were extraordinarily productive, resulting in the publication of six major papers from his Penn State investigations. Among these papers was the intellectually demanding investigation of aqueous complexing in the Cu-Fe-S system, a fundamental study that has been used extensively in modeling of ore genesis.

David was also sensitive to his environment. Around his desk, he built a gazebo with vivid colors-to the mixed reactions of his office mates. He was artfully involved in reciprocal practical jokes with, or on them, such as the exquisite nylon mono-filament spider web that appeared surrounding the desk and chair of one who left for a few days for an interview.

When near completion of his doctoral program, David made no visible effort to find a more permanent position. At the same time, Princeton had underway a search for a new professor in geochemistry with over 200 applications in hand. At a lunch with the chairman of the search committee, I mentioned that a sterling character was about to graduate from our group and that he seemed to match their wishes nicely. The chairman expressed interest but warned that the search was about closed and asked that an application be submitted quickly. Relaying that message had little impact on David. Only after two weeks could I persuade him that he might well be the winning candidate. His application catalyzed a prompt invitation to visit Princeton and present the normally required colloquium. His Friday lecture and interactions with the faculty generated sufficient enthusiasm that on Monday he had by phone an informal offer of the appointment, an inconceivably rapid reaction by any university and an astute compliment to his abilities.

In honor of the phenomenal volume and quality of contributions to geochemistry by our alumnus, the David A. Crerar Memorial Seminar has been established in our Geosciences Department as a permanent, continuing, weekly series of seminars on geochemistry in the broad sense, befitting the variety in David's research.

H. L. Barnes
Department of Geosciences,
Pennsylvania State University
April, 2001

DAVID A. CRERAR: A SCIENTIFIC HISTORY AT GUYOT HALL, PRINCETON UNIVERSITY

David and his wife Scotia MacRae moved to Princeton, N.J. in early 1974, as David was invited to join the Department of Geological and Geophysical Sciences of Princeton University as a lecturer in February of that year, and officially became an assistant professor shortly thereafter in September. David inherited the geochemical labs in Guyot Hall left behind by Professor H.D. Holland, who had recently left Princeton to accept an appointment at Harvard University. David very quickly established a strong research and teaching program in geochemistry, and because of his approachability and wide-ranging interests, attracted a relatively large group of undergraduate and graduate students who worked with him on a variety of projects in the areas of environmental, sedimentary, and hydrothermal geochemistry, thermodynamics, and water-rock interactions. David's career rapidly advanced at Princeton, with him becoming a tenured associate professor in 1980. In 1980, in recognition of excellence in scientific research, David was awarded a Shell Distinguished Chair in Geochemistry by the Shell Foundation. In 1982 he was awarded the Lindgren Award by the Society of Economic Geologists. In the same year, David was promoted again, this time to full professor. David was also affiliated with the Princeton Water Resources Program, a part of the civil engineering department. David served on the editorial boards of *Economic Geology* and *Chemical Geology*; he was also a member of advisory groups associated with various environmental commissions.

In addition to his students, David always could depend on the expert help of permanent staff member Maria Borcsik (a true Guyot Hall institution), who was an integral part of nearly all of the David's research projects. Maria's expertise in analytical chemistry was invaluable for running the numerous analytical instruments, as well for supervising generations of undergraduate and graduate students in the trials and tribulations of experimental and analytical geochemistry. Maria was also the social glue of the A-floor (where geochemistry was mainly housed); morning coffee at 10 a.m. sharp was *de rigueur*, as was partaking in her homemade Hungarian gulash during lunch-time parties. Another important person in David's group was Ted Forseman; he was instrumental in the design and machining of numerous high-temperature/high-pressure autoclaves and optical cells that were a key component of the hydrothermal geochemistry studies that were carried out over the nearly two decade-long career of David's in Guyot Hall.

Describing the evolution of David's scientific endeavors at Princeton was not an easy task to accomplish, given both the diverse and continuously evolving nature of his research, as well as the large number of graduate students and colleagues that were involved. Of course, many undergraduates were also involved in these research projects, but because of space limitations, their names have not been included below. In the following narrative essay I have attempted to describe in relatively succinct fashion the evolution of David's geochemical research program, starting from his early days in Guyot Hall in the mid-seventies, and ending approximately in 1990-1991, when his abilities started becoming increasingly impaired due to his illness. Many apologies are presented in advance for inadvertent omissions and inaccuracies in the following essay.

Right from the early stages of David's career at Princeton, one of his principal fields of interest concerned environmental geochemistry. Perhaps the most pressing topic in this field was radioactive waste containment and migration. David, graduate students Jeffrey Means and Ann Maest, and collaborator J. Duguid (Oak Ridge N.L.) investigated chemical transport, complexation, and adsorption of radionuclides on iron and manganese oxides in soils and sediments. The adsorption of specific radionuclides, which were used as proxy's for high-level radioactive waste, onto potential repository materials was studied to temperatures of 250 °C; this included examination of the effect of organics on adsorption rates. Their research also showed that certain chelating agents, both synthetic and naturally occurring (such as humic acids), effectively retarded adsorption, and thus their presence could dramatically increase the mobility of radionuclides escaping from underground radwaste storage sites. Related to the above topic was a study of the degradation kinetics of organic chelating agents, such as EDTA and similar compounds, at temperatures up to 300 °C. This research, primarily applicable to the decontamination of nuclear reactors, was carried out by graduate students Jeffrey Means and Jennifer Snow, research associate Tonalee Key, and David.

The acidification of streams and ground waters due to mine drainage was the subject of Robert Kleinmann's doctoral thesis. Perhaps the most important result of this research was the recognition of the interrelationship between chemical and biological processes responsible for acid-mine drainage. Robert and David discovered that certain bacteria actually had a deleterious effect by enhancing acid production. This thesis also resulted in the development of a possible method for controlling acid formation in pyritic coals in contact with water.

The nearby New Jersey coastal plain, especially the Pine Barrens, an area of relatively pollution-free natural waters, offered a suitable natural laboratory for geochemical studies of its aquifers, rivers, and estuary systems. David, graduate students Richard Yuretich and Jeffrey Means, and David Kinsman (the Princeton faculty member who initiated these studies) studied geochemical mass balances, the trace metal content of stream and ground waters, as well as transport and deposition mechanisms of metals, such as aluminum and iron, dissolved organic matter, and selected trace elements.

A few years later on, David and graduate students Ann Maest, Susan Brantley, Lusheng Yan, Kevin Costello, Lois Hernandez, faculty Robert Stallard and Bob Key, and research associate Rich Rotter started working extensively on trace-metal contamination in polluted streams, estuaries, bays, and coastal waters, with particular emphasis given to the industrialized areas of northern and central New Jersey (where polluted waters are not hard to find....). This group examined in detail river-estuary-ocean mixing processes, the distribution of trace metals in coastal waters, and sinks of trace metals in coastal environments, with the ultimate goal of better understanding the chemical and physical processes leading to trace metal deposition. These studies showed that the speciation of metals changes during transport from riverine to estuarine environments, with metals often being converted to biologically available forms within estuaries. Perhaps most importantly, this research provided evidence that the geochemical behavior of heavily polluted river-estuarine systems differs significantly from their non-polluted analogs. A somewhat related project, led by Kevin, Rich, Bob, and David, studied eutrophication processes and nutrient balances in Barnegat Bay, a large near-shore marine environment between the N.J. mainland and the offshore barrier islands on the Atlantic.

Another important aspect of environmental geochemistry research involved collaboration with the Water Resources Program in the civil engineering dept., and particularly with faculty and research staff members P. Jaffé, R. Ferrara, G. Pinder, W. Gray, and V. Nguyen. One of the first projects, a collaboration between David, P. Jaffé, and R. Ferrara, studied the adsorption kinetics of toxic organic substances, such as chlorinated hydrocarbons, onto clays and sediments. Their results revealed that the adsorption and desorption behavior were more complicated than expected, which demonstrated the need for more complex numerical models for simulating toxic waste mobilities. Another project involving David, in collaboration with G. Pinder, W. Gray, and V. Nguyen, was a study of fluid flow in reactive porous media, with particular application to solution mining. David's main contribution was integration of chemical kinetic parameters for water-rock interactions into the numerical codes. The developed models were applied to uranium solution mining and the high-temperature dissolution of silica. Graduate student David Tuck, supervised by David and P. Jaffé, developed a research project pertaining to the cleanup of toxic hydrocarbons in porous media. David Tuck's doctoral thesis specifically examined the physical and chemical aspects of non-aqueous phase liquid (NAPL) "blob" movement in porous media. One important outcome of this research was a method involving flushing of surfactant solutions into contaminated sediments in order to reduce intergranular capillary forces, thereby making physical recovery of NAPL blobs economically feasible.

Sedimentary geochemistry was also an area that did not escape David's interests, and this field was an important part of his group's research efforts throughout his Guyot Hall days. Among the initial projects was a study of ophiolites in the Californian Franciscan assemblage. Graduate students Michael Chyi, Jay Namson, Mark Feigenson, and Loretta Williams, in association with R. Carlson of the Carnegie Institution, made detailed measurements of the mineralogy and chemistry of the assemblage, including major, minor and trace elements, rare earth elements, and Sr, O, Pb, and Nd isotopic compositions. According to David and his collaborators, the data could be used to infer how hydrothermal circulation cells operated within the sea floor spreading center, as well as the alternation between cycles of pelagic sedimentation and hydrothermal deposition of minerals.

Loretta Williams, G. Parks (Stanford), and David investigated silica diagenesis in marine sediments. The specific problem they looked at was the transformation of amorphous silica through opal-CT to microcrystalline quartz. Their experimental and theoretical data suggested that these transformations are a function of the relative solubilities, structures, and colloidal properties of the silica phases. Closely related to this project was a study by David, graduate student Ellen Axtmann, and R. Axtmann and B. Fleming (Dept. of Chemical Engineering, P.U.) on the growth and ripening of silica polymers in solution.

Other projects in the field of sedimentary geochemistry were based on diverse subjects. One of the earlier ones was with research associate Deba Bhattacharyya on the diagenesis of common marine clays to produce iron-rich minerals such as chamosite or glauconite. One of the goals of this research was to understand the differences between Precambrian iron formations and Phanerozoic ironstones. Another project, involving Rob Hargraves and E. Dowty (faculty), Jessica Posey and Bruce Moskowitz (research associates), and David, was based on synthesizing iron oxides with the same chemical and magnetic characteristics as those iron oxides thought to be present on Mars' surface. The major question addressed was understanding the differences in ferric iron oxide precipitates on Mars and Earth. Their main finding was that the basaltic environment of Mars probably favors the precipitation of lepidocrocite, a precursor to magnetic maghemite, whereas on Earth, with its more siliceous crust, the formation of non-magnetic iron oxides, such as goethite and hematite, is favored. A final example is a project that involved graduate student Susan Brantley and David, in collaboration with N. Møller-Weare and J. Weare (both from UCSD), on the chemical evolution of a modern, natural evaporite system in Peru. Using the chemical data collected in the field by Susan, the aqueous speciation and the sequence of salt precipitation within the evaporite were successfully predicted using a numerical model.

Since the early nineties, the field of geomicrobiology has literally exploded in importance, as more and more examples are found of biological systems that are a part of, and in many cases control, geochemical processes. David was truly among the early pioneers in this field. Early on, he realized that microbes can play key roles in mineral dissolution, precipitation, and aqueous geochemistry. The doctoral thesis of Robert Kleinmann determined that *Thiobacillus ferrooxidans* had a dramatic effect on acid mine drainage by enhancing acid production in water in contact with pyritic coal mine spoils. Understanding the role of organisms with respect to acid formation led Robert and David to propose a method for controlling acid formation. David's interest in the geochemistry of the Pine Barrens also led to research with Jeffrey Means on the biogeochemical precipitation of bog iron ore. This was part of more general research in the laboratory and the field involving iron- and manganese-oxidizing bacteria responsible for precipitation of Fe and Mn oxides in fresh waters. Al Fischer's (faculty) discovery of the bacterium *Metallogenium personatum* in Cretaceous-Paleocene cherts resulted in collaborative work with David concerning the biogeochemistry of manganese and its biologically-controlled precipitation throughout Earth's history. In a project with a completely different aim, graduate student Joseph Kirschvink and David investigated magnetotactic bacteria. In this case, the goal was understanding how these organisms were able to adapt and evolve with changing magnetic fields, and in particular, polarity reversals.

David's extensive research program in thermodynamics, hydrothermal geochemistry, and ore deposit formation in Guyot Hall could be considered a direct offshoot of his doctoral research activities at Penn State on the hydrothermal solubilities of chalcocite and chalcopyrite. Aside from directly working with his own graduate students, David established close ties with various faculty members of the chemistry department, most notably A. Bocarsly, J. Schwartz, and D. McClure. One of David's first graduate students who worked in hydrothermal geochemistry was Nick Susak, who determined the solubility of the buffer assemblage pyrite + pyrrhotite + magnetite in NaCl solutions at 200-350 °C. Graduate students Scott Wood and Nick, along with David, carried out a series of solubility and solvation experiments on gold and a variety of metal sulfide minerals in aqueous NaCl solutions, at controlled pH and fixed oxygen and sulfur fugacities, from 200-350 °C. This work has led to a better understanding of ore genesis and zoning. Scott and David also developed an improved method for the derivation of stability constants of complex ions from solubility data. David, with graduate students Scott and Susan Brantley, also worked on theoretical studies concerning ore genesis and zoning, where the stoichiometry and coordination of metal complexes were shown to be important parameters.

Scott and Ted Forseman also designed and built new hydrothermal equipment with increased temperature and pressure capabilities, such that metal solubilities could be studied close to magmatic conditions. Graduate student Bear McPhail continued in this direction with experiments investigating the behavior of iron in high temperature chloride brines at 350 °C and vapor-saturated pressure, and 500 °C and 1 kbar, using the assemblage magnetite + K-feldspar + biotite + quartz. This work suggested $FeCl_4$-type complexes at supercritical conditions, and $FeCl_2$ complexes at subcritical conditions. Using differential pressure gauges and a specially designed high-temperature oil bath, David and graduate students Scott and Susan measured the mean molal activity coefficients and related non-ideal thermodynamic parameters of a series of single and double-salt aqueous solutions to 350 °C.

Despite the already extensive body of research dedicated to Mississippi Valley-Type Pb-Zn sulfide deposits, their origin was still unclear (which is still the case today), and this provided the stimulus for further investigations into their genesis at Princeton. Graduate student Rémy Hennet and David, with assistance from T. Giordano (New Mexico State U.), experimentally determined the stability constants of Pb and Zn in the presence of water-soluble organic acids, based on analytical measurements at elevated temperatures using specific ion electrodes, NMR, and polarographic techniques. Their results indicated that Pb- and Zn-acetate complexes predominate over chloride complexes in oil field brines and similar fluids in deep sedimentary basins, suggesting that metal-organic complexes may have played an important role in the genesis of Pb-Zn MVT deposits. In a related study, Rémy, David, and J. Schwartz found that CO_2 increased sphalerite solubilities, which led them to postulate the existence of pressure-sensitive thiocarbonate complexes that incorporate both reduced sulfur and metal in the same molecule. The thermal stability of natural organic acids at hydrothermal conditions was a project investigated by graduate student Jennifer Snow, research associate Tonalee Key, and David. Jennifer completed a doctoral thesis on the degradation kinetics of simple carboxylic and phenolic aromatic acids. An important finding was that the activation energies of these reactions were very low, suggesting water as a natural catalyst. In another project related to organics, the presence of organic acids in deep oil field brines drilled in Mississippi was measured by Rémy and David, in collaboration with A. Maest and Y. Kharaka (USGS-Menlo Park).

With the technical assistance of Ted Forseman, Nick Susak also developed a high-pressure/high-temperature spectrophotometric cell that was used to measure in situ the ligand field molecular spectra of Fe, Ni, and Co complex ions in aqueous NaCl solutions to 300 °C. Chemistry graduate student Lynn Vogel, with assistance from D. McClure, David, and M. Buback (U. Göttingen, Germany), continued working in this domain, measuring the electronic absorption of aqueous Fe(II), Fe(III), and Mn(II) complexes in solutions to 300 °C. Visiting professor K. Kim (U. Korea, Seoul) did similar studies on electronic spectroscopic determinations of aqueous Mn(II) chloride to 300 °C; the spectra showed the step-wise change in Cl ligation number from 1 to 4, as well as a change in geometry from octahedral to tetrahedral with increasing temperature and Cl activity. Work on Mn(II) was continued by graduate student Al Henry. Based on the spectroscopy measurements by Lynn, K. Kim, and Al, the limiting complex at higher temperatures and high chloride concentrations was determined to be tetra-chloro for metal ions with electron configurations less than d^{10}. K. Kim, Bear, and David developed a method for measuring the pH of hydrothermal solutions in situ, using optical absorbance measurements of quinoline and acridine, two organic indicators. A new hydrothermal Raman spectrophotometer was employed by graduate student Mary Yang, who, working with David and D. Irish (U. Waterloo, Canada), used it to measure the speciation of Zn and Cd bromide complexes to 300 °C, and Pb and Zn acetate complexes to 200 °C, yielding estimates of complex stability constants and ligation numbers. The information garnered from these various types of spectroscopic measurements provided fundamental information on the molecular chemistry (bond strengths, symmetries, stoichiometry) and thermodynamics that are crucial to determining the behavior of metal ions in hydrothermal solutions.

David's expertise in thermodynamics was well known, as was his ability to convey his immense knowledge with enthusiasm and clarity, which was equally appreciated by undergraduate and graduate students alike. David was fascinated by thermodynamics, so this almost inevitably led to a long collaborative effort over several years with his former mentor and advisor, Greg Anderson from the University of Toronto, with the goal of writing a new book on chemical thermodynamics applied to geochemistry. They spent an entire sabbatical year together from 1986 to 1987 at the geochemistry laboratory, U. Paul Sabatier (Toulouse, France), developing ideas and completing the remaining chapters of their future book. Life in southern France also had a pronounced effect: David's *bon vivant* side flourished as he and Scotia indulged in a year-long culinary romance with *la gastronomie française*. David's intimate knowledge of every good restaurant in southern France made many people wonder whether he was moonlighting as an official tester for the Michelin *Guide Rouge*. The fruits of Greg and David's long travail resulted in the pre-eminent text "Thermodynamics in Geochemistry *The Equilibrium Model*", which was published in 1993. Greg and David note in the preface of their book that far from simply presenting applications of thermodynamics to geochemistry, the defining hallmark of their work is the conveyance of the fundamentals required to understand thermodynamics itself.

While the previously described hydrothermal solubility and spectroscopic studies were based on chemical equilibrium conditions, David recognized the vast scientific potential of studying the kinetics of rock-water interactions at elevated temperatures and pressures, a field which was largely unexplored in geochemistry, even in the early eighties. David initiated collaboration with friend and colleague C. Chang (Mobil R&D, Princeton), whose expertise with the design and application of flow systems was very valuable to the beginning efforts of David's water-rock interaction group. Research associate Jessica Posey built a prototype hydrothermal circulation apparatus, based on a tubular reactor design, and proceeded to measure the dissolution kinetics of fluorite and calcite at elevated temperatures and pressures. Jessica, graduate student Ellen Axtmann, and David also carried out a project investigating the dissolution rates of uraninite and uranium roll-front ores.

With the arrival of graduate student Roland Hellmann, a set of 4 independent tubular-reactor circulation systems was built by him (with technical assistance from Ted Forseman) and used to measure fluorite and albite feldspar dissolution kinetics to 300 °C over a wide range of pH. The initial non-stoichiometric dissolution of albite suggested that the formation of Na- and Al-leached layers was an integral part of the hydrolysis mechanism of feldspars; this was verified by XPS measurements that were carried out in collaboration with C. Eggleston and M. Hochella at Stanford. The XPS depth profiles proved the existence of leached layers up to ≈1000 Å thick, which showed that with certain minerals, such as feldspars, dissolution is a 3-dimensional process. Assistance with some of the kinetic studies of Roland and Jessica was provided by R. Zhang, a frequent visiting scientist from the Chinese Academy of Sciences, Beijing.

The study of mineral dissolution kinetics was continued by graduate student Patricia Dove, who used mixed-flow reactors to investigate the hydrothermal dissolution kinetics of quartz in cation-chloride solutions. The kinetic rate data were elegantly fit to a rate equation incorporating a Langmuir adsorption isotherm. According to a hypothesis put forth by Patricia and David, the observed increases in dissolution rates (compared to pure water) are a function of the identity and concentration of the cation, where the role of cations is adsorbing on and disrupting the structure of the mineral-solution interface. Patricia also led a study on the dissolution of three isostructural sulfate minerals: barite, celestite, and anglesite. This permitted insights on the influence of crystal chemistry on dissolution rates.

Susan Brantley and David, in collaboration with C. Willaime, J. Schott, and C. Guy (U. Rennes and Toulouse, France), looked into the effect of strain energy on dissolution rates. By artificially inducing high dislocation densities in calcite and then measuring the dissolution rate, they were able to show that the increase in dissolution rates (2-3x) was small compared to the three order of magnitude increase in dislocation density. Another research project by Susan, Roland, David, and Robert Stallard (faculty) examined the effect of chemical affinity on the development of etch pits and on the dissolution rate of quartz. Results showed that at affinities below a certain critical value, etch pits do not form; this has important implications with respect to dissolution mechanisms.

Water-mineral interactions at still higher temperatures and pressures were the focus of a research project by graduate student Susan, Brian Evans (faculty), and David. Brian and Susan designed and built equipment for studying rock-water interactions and rock mechanics to temperatures and pressure of 1100 °C and 10 kbars, conditions approaching those of the lower continental crust. This equipment had the advantage of permitting rock-water reactions at non-hydrostatic stresses, such as those involved in pressure solution. The major investigations carried out involved rock sintering, porosity reduction of quartz aggregates, and crack healing.

David's excellent research record in so many fields of geochemistry earned him enormous respect from his peers and from the Earth sciences community. His presence at Guyot Hall is sorely missed, even today. During David's almost two decade-long career at Princeton, 22 doctoral candidates were mentored by him. A large majority of them have continued on in various fields of geochemistry, either in academia, industry, government, or private consulting. Most of his former graduate students, as well as many colleagues and friends, were able to return to Guyot Hall in May, 1993 to attend a 2 day-long international symposium honoring David's contributions to geochemistry and to celebrate David's life. David's last few years were spent in a hospital in Toronto, where he passed away on 6 September, 1994, at the age of 49. He is survived by his wife Scotia and daughter Aurora of Princeton, as well as by his mother and brother in Toronto.

It is hard to say where David's career would have led had his life not been tragically cut short by his illness, but it assuredly would have been full of intellectual fervor and diversity.

Roland Hellmann
Grenoble
February, 2001

ODE TO A GEOCHEMIST

There once was a Canadian student named Dave,
About whom all the Toronto profs did rave,
    His mentor Greg told him his fate,
    Was to go south to Penn State,
So off to America for a Ph.D. did go Dave.

At Penn State everything was bright and new.
And Dave quickly took a liking to venerable Doctor Hu.
    And using Hu's advice as a guide
    He studied metals and bisulfide
And the data sets he collected soon grew.

However, after studying his data he saw
That his numbers, while still completely raw,
    Showed little complexes of bisulfide
    And lots and lots by chloride
So he tried to develop a chloride-complexing law.

Well, we all know the stubbornness of Hu.
David tried to convince him until he turned blue.
    But it had to be bisulfide
    It could not be chloride
So David despaired of publishing a theory so new.

At that point in Dave's life entered Wayne
Who said, oh Dave, let's be sane!
    If chloride it must be
    We'll get you your Ph.D.
And so Wayne helped David the data to Hu to explain.

But even more important than Wayne in David's life,
Was his meeting of the woman who became his wife.
    He found her so precocia
    The woman we all know as Scotia
That he married her and loved her all his life.

At that point, a university named P.U.
Offered David a job and a new venue.
    He stopped being a grad
    And he started his own lab,
And began training students to do what he could do.

Once at Princeton, David soon came to befriend,
Fischer, Hargraves, Spera, and Dahlen.
    With students Means, Kleinmann and Maest,
    He studied water, mines, and radioactive waste.
But it was Maria who was the glue that kept us all on trend.

He built bombs with students Sue and Scott,
He read spectra in waters cold and hot.
    He worked with Mary and Lynn,
    And Michael, Bear, and Lusheng.
And he became friends with the Frenchman Jacques Schott.

He studied how water and its solutes move with Dave Tuck
He looked at metal complexes with Nick Susak.
    Roland dissolved minerals of all sorts
    While Trish strictly worked with quartz,
And Jennifer studied complexing by gooey organic muck.

While Ellen worried about the intricacies of solution mining
Scott pursued large expensive malleable gold lining.
    And Remy blew up a bomb
    With great Swiss aplomb
And generally instructed the rest of us about fine wining and dining.

Although conferences David generally did shun
The Gordon Conferences actually convinced him to run one.
    But in New Hampshire of course it was Hal
    Whom the town wanted to throw into jail
For swimming with practically nothing on under the sun.

But there came a day when David looked at all he had
And he still wanted more—he was still sad
    So he and Scotia did confide
    And to make a family they did decide
And nine months later little baby Aurora is what they had.

For men and women both David was a good role model,
Especially when he brought in baby Aurora with her bottle.
    He loved his baby girl,
    And he showed her to all the world,
And in a laboratory the little girl soon learned to toddle.

But even with Scotia and Aurora, David still had a great fear
Of airplanes flying in the sky wide and clear.
    He bravely took flying class
    And finally he did pass
So that the Crerar family came to fly to France one year.

Now France was a place David did adore.
The wine, the women, the food—he always wanted more.
    So in France he did stay
    For a year of work and for play
And he even hosted a trip for the entire A-floor.

So David excelled at the game of publish or perish.
And he taught and did research with a flarish.
    His students did excel
    And his grant proposals did well
And his office became more and more garish.

But, why, in essence, was David so great?
For one thing, he published at a prodigious rate.
    He did science with the best
    He instructed the rest
And, unlike other chemists, enemies he just did not create.

But he cared more about his students more than his fame
And he cared more about teaching well than playing the talk circuit game.
    He loved music and life
    And especially his daughter and wife
And he encouraged his students and peers to do the same.

So his career, while it ended too soon,
Was filled with success and happiness and gazing at the moon.

And so to David I toast
He was a mentor of whom I boast
And may his spirit live on in every friend in this room!

Susan Brantley,
Department of Geosciences,
Pennsylvania State University

Written in May, 1993 and presented at Princeton during a 2-day long symposium in honor of David's life and career.

Part A.
Water-Rock Interactions and Fluid-Mineral Interfaces

Water-Rock Interactions, Ore Deposits, and Environmental Geochemistry: A Tribute to David A. Crerar

Editors: Roland Hellmann and Scott A. Wood

Organic ligands and feldspar dissolution

William J. Ullman[1*] and Susan A. Welch[1,2]

[1] College of Marine Studies, University of Delaware, Lewes, DE 19958

[2] Department of Geology and Geophysics, University of Wisconsin, Madison, WI 53706

* corresponding author: ullman@udel.edu

Abstract—Organic compounds can both enhance and inhibit the rates of feldspar dissolution in the laboratory and presumably in field settings where organic matter is abundant. Organic acids can directly enhance dissolution by either a proton- or a ligand-mediated mechanism. They can also indirectly enhance dissolution by the formation of complexes in solution with reaction products and as a consequence increase the chemical affinity for the overall dissolution reaction. The former mechanism is likely to be important at high chemical affinities and the latter at lower chemical affinities. The direct ligand-mediated mechanism involves the adsorption of ligands to Al sites on the feldspar surface, the formation of Al-organic complexes at the surface site, and the subsequent leaching of Al into solution. The degree of dissolution enhancement by the direct ligand-mediated mechanism is dependent on the structure and concentration of the ligand, the composition of the feldspar, the composition of the solution, and the temperature. However, overall dissolution rates are not a simple function of these factors. Organic polymers can either inhibit dissolution by irreversibly binding to the mineral surface and reducing the effective reactive surface area (i.e. the number of reactive sites) of the mineral or enhance dissolution by a ligand-mediated mechanism. The experimental determinations of the impact of a wide variety of organic ligands on feldspar dissolution rates provide evidence of multiple mechanisms of mineral ligand interaction that can impact the dissolution process. This literature is reviewed in light of conceptual models of the dissolution process.

1. INTRODUCTION

The dissolution of primary rock-forming minerals, the release of ions to solution, and the formation of secondary minerals take place in all earth-surface environments where rocks and water coexist. Weathering is an important process in the creation of soils and the maintenance of soil productivity, the evolution of ground- and surface-water compositions, the denudation of continents, the flux of continental solids and solutes to the ocean, the maintenance of seawater composition and productivity, and the regulation of atmospheric composition and global climate. Secondary porosity generated by mineral dissolution contributes to the success of petroleum migration and recovery, and may also impact the transport of solutions and solutes through aquifers. Weathering reactions involving feldspars are particularly important since they are the most abundant rock-forming mineral at the earth's surface and represent a potentially important source of non-carbonate calcium that controls the long-term concentrations of atmospheric carbon dioxide, and therefore regulates the global heat balance (Berner, 1991; Brady, 1991).

Traditionally, silicate and alumino-silicate weathering has been viewed by geochemists as an abiotic, acid-mediated, hydrolysis process initiated by the reaction of primary minerals with water and dilute inorganic acids, typically carbonic acid (H_2CO_3), but also nitric (HNO_3) and sulfuric acid (H_2SO_4) in some geological settings (Garrels and Mackenzie, 1971). Increasingly, however, it is recognized that the weathering process can be further mediated by organisms and their metabolites. Respiration by plant roots and microbial degradation of organic matter can elevate carbonic acid concentrations in soils and groundwater, leading to an increase in the weathering rates of minerals (Chapelle et al., 1987; Hendry et al., 1992; Keller and Wood, 1993). In addition to carbonic acid (carbon dioxide and water), plants, bacteria, and other microorganisms produce other compounds (such as other inorganic acids, alkaline compounds, low- and high-molecular-weight organic compounds, oxidants, and reductants) which can affect mineral weathering reactions (Huang and Schnitzer, 1986; Ehrlich, 1990; Drever and Vance, 1994; Barker et al., 1997; Stone, 1997). Ligand-mediated dissolution represents one of the additional mechanisms by which minerals dissolve (hydroxyl-mediated and redox-mediated dissolution are two others). Ligands can affect feldspar weathering rates by complexing with ions at the mineral surface or in solution. Ligands may also inhibit feldspar dissolution rates under some conditions (Welch and Vandevivere, 1994; Welch et al., 1999). The impact of these ligand-mediated reactions on the rates of mineral weathering must be considered, particularly in near-surface environments where microbes are present and where microbially produced metabolites and/or organic contaminants may be abundant. They also may be important in oilfield reservoir systems or at contaminated sites where partial oxidation of petroleum compounds can also lead to high levels of organic ligand production (Kharaka et al., 1986; Fisher, 1987; Bennett and Siegel, 1987; Cozzarelli et al., 1994).

In this review we consider primarily the effect of naturally occurring organic ligands on feldspar dissolution rates. Increasingly, however, organic contaminants and their decomposition products are found in natural waters and many of these may have the capacity to influence silicate and alumino-silicate dissolution (e.g., Bennett and Siegel, 1987). Less is known, however, about these compounds and their impact on mineral weathering reactions. For the purpose of this review, we also presume that the principal effect of organisms on feldspar dissolution rates is due to the production and excretion of compounds to the solution. There is, however, increasing evidence that direct contact between bacteria and minerals may be important in mineral alteration reactions, as microbial surfaces can also complex with metal ions (Ferris et al., 1988; Daughney et al., 1998; Fein et al., 1997). Microorganisms attached to mineral surfaces can also create micro-environments where concentrations of ligands, acidity, and redox activity can be substantially elevated compared to the bulk solution, thus affecting mineral exchange reactions (Hiebert and Bennett, 1992; Barker and Banfield, 1996, 1998; Rogers et al., 1998a,b; Barker et al., 1998). The impact of direct contact between organisms and mineral surfaces on mineral weathering reactions is not addressed here.

It is clear from a large number of studies that feldspar dissolution rates are a strong function of pH (Blum and Lasaga, 1988, 1991; Brady and Walther, 1989; Drever and Vance, 1994; Blum and Stillings, 1995). Rates are lowest in near-neutral solutions (pH ≈ 5 to 8) and then increase dramatically at low and high pH due to proton-mediated and hydroxyl-mediated mechanisms, respectively. The direct effects of ligands on the dissolution process are of secondary importance except when protons and hydroxyl concentrations are at their minimum. However, at these near-neutral pHs, which correspond to the pHs of many natural environments, ligands may significantly enhance dissolution rates by up to an order of magnitude (Wogelius and Walther, 1991; Welch and Ullman, 1993,1996; Stillings et al., 1998). The mechanism by which ligands interact with the dissolving mineral, the degree of dissolution enhancement or inhibition, the effect of ligand structure and concentration on the degree of enhancement, and the interactions between ligand-mediated and other dissolution mechanisms are the important questions that need to be addressed in order to make reasonable predictions of chemical weathering rates on a local, regional, or global scale.

The objective of this review is to summarize the existing experimental results concerning the role of organic ligands in the feldspar weathering process, both as dissolution enhancers and inhibitors, as compared to the rates of dissolution due to proton- or hydroxyl-mediated feldspar dissolution alone. The dependence of ligand-mediated dissolution rates on mineral chemistry, solution compositions, and physical factors, principally temperature, are also considered. We consider primarily feldspars because of their abundance in crustal rocks. However, results from other mineral systems are also useful in interpreting feldspar weathering rates, and a sampling of such studies are included in this review. We also evaluate the simple empirical and thermokinetic models of the dissolution process that have been found to reproduce the experimental results and discuss the extrapolation of laboratory results to field settings.

The overall weathering process, however, is complicated. It is not possible to consider all of the possible combinations and permutations of dissolution and precipitation reactions involving all of the phases and components that may occur in field settings. It is also not entirely clear how the various rate-controlling factors interact to obtain the overall rates of mineral dissolution observed in the laboratory or in the field. We therefore focus only on the general trends in the primary dissolution of feldspars, a single, but important group of minerals. These trends, however, provide a useful blueprint for the conditions under which ligand-mediated dissolution enhancement and inhibition may be significant, and therefore this review should prove useful for those interested in interpreting or predicting the rates of mineral alteration in field settings where organic ligands are important participants in the weathering process.

2. A CONCEPTUAL MODEL OF FELDSPAR DISSOLUTION

Feldspars are tectosilicate minerals in which a regular three-dimensional network of Si and Al tetrahedra provides the structural framework, and alkali and alkaline earth metals in interstitial positions balance the net charge of this structure (Smith and Brown, 1988). The charge-balancing cations are known to easily exchange with cations in solution (especially with H^+), resulting in a leached near-surface layer that is amorphous and stable (Petrovič, 1973; Chou and Wollast, 1985; Casey and Bunker, 1990; Stillings and Brantley, 1995). For this reason, most studies of feldspar dissolution, including this one, are focused on the release rates of the structural cations, Al and Si.

2.1. Dissolution in Ligand-Free Aqueous Solution

In a ligand-free aqueous solution, the stoichiometry of a generalized overall feldspar dissolution reaction can be described as a sum of three types of reactions. In neutral solutions, the feldspar hydrolysis (HD) reaction predominates and is given by:

$$(K_xNa_y)Ca_{(1-x-y)}Al_{(2-x-y)}Si_{(2+x+y)}O_8 + 8H_2O \rightarrow (2+x+y)Si(OH)_4^o + (2-x-y)Al(OH)_3^o + (2-x-y)OH^- + xK^+ + yNa^+ + (1-x-y)Ca^{2+} \quad (1)$$

The principal feldspar dissolution reaction in acidic solutions is an acid hydrolysis (proton-mediated: PMD) reaction and can be given as:

$$(K_xNa_y)Ca_{(1-x-y)}Al_{(2-x-y)}Si_{(2+x+y)}O_8 + 4(x+y)H_2O + 4(2-x-y)H^+ \rightarrow (2+x+y)Si(OH)_4^o + (2-x-y)Al^{3+} + xK^+ + yNa^+ + (1-x-y)Ca^{2+} \quad (2)$$

In basic solution, the feldspar hydroxylation (hydroxyl-mediated: HMD) reaction can be written as:

$$(K_xNa_y)Ca_{(1-x-y)}Al_{(2-x-y)}Si_{(2+x+y)}O_8 + (6-x-y)H_2O + (2+x+y)OH^- \rightarrow (2+x+y)SiO(OH)_3^- + (2-x-y)Al(OH)_4^- + xK^+ + yNa^+ + (1-x-y)Ca^{2+} \quad (3)$$

The above reactions and stoichiometries reflect a conceptualization and simplification of the overall reactions that actually take place through a series of elementary reaction steps involving, at the molecular scale, direct electronic interaction between reactants and formation and/or destruction of molecular bonds (Lasaga, 1981; Brezonik, 1994).

Much of the interesting weathering behavior of feldspar is not reflected in these conceptual stoichiometries. For example, initial rates of feldspar dissolution are usually not stoichiometric and non-stoichiometric release of component ions may persist even after steady dissolution rates are achieved in longer experiments. Casey and Bunker (1990) suggest that this behavior is the result of multipathway leaching during which each element is subjected to a different set of solubilization reactions. Also, some of the reaction products may react with each other or with some of the reactants to form different solute phases or secondary solids. Secondary clays are a common product of feldspar weathering reactions (Garrels and Mackenzie, 1971) and the nature of these secondary reactions and phases plays an important role in soil development and fertility.

The three reactions given by Eqns. 1-3 have different stoichiometric dependencies with respect to the concentrations of H^+ and OH^- and therefore predominate at different ranges of pH. The rates of hydrolysis in neutral solution are known to be exceedingly slow, accounting for the long-term geological stability of feldspars in many near-neutral environments. The net effect of these inorganic reactions on feldspar dissolution rates, R_{inorg}, can be described as a linear sum of the three mechanisms:

$$R_{inorg} = R_{PMD} + R_{HD} + R_{HMD} \quad (4)$$

where R_{HD}, R_{PMD}, and R_{HMD} are the contributions due to the hydrolytic (Eqn. 1), proton-mediated (Eqn. 2), and hydroxyl-mediated (Eqn. 3) dissolution mechanisms to the overall surface-normalized dissolution process in the absence of ligands. The pH dependence of the three mechanisms and overall dissolution behavior in the absence of ligands, R_{inorg}, is shown schematically in Fig. 1.

The acid hydrolysis and hydroxylation mechanisms of feldspar dissolution are far more rapid than the hydrolysis reaction and are often described using catalytic terminology (proton-promoted and hydroxyl-promoted dissolution), in spite of the consumption of protons and hydroxyls by the overall reaction (Eqns. 2 and 3). Protons and hydroxyls may have catalytic effects on specific elementary reactions in the silicate dissolution process (Xiao and Lasaga, 1994,1996). We have, however, adopted a less controversial terminology (proton-mediated dissolution, PMD, and hydroxyl-mediated dissolution, HMD) to describe the increase in the rate of dissolution above the hydrolytic dissolution rate (HD), but this terminology is synonymous with the catalytic terminology used in most of the papers that we have reviewed.

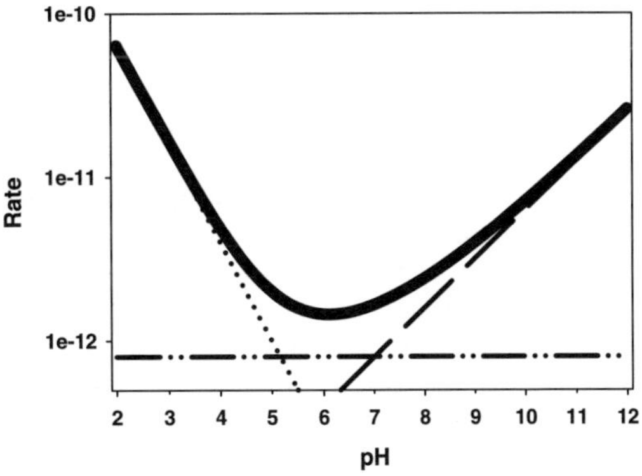

Fig. 1. A schematic view of the relative importance of proton-mediated feldspar dissolution (PMD; ·······), hydrolytic dissolution (HD; — ·· —), and hydroxyl-mediated dissolution (HMD; —— ——). The net effect of these processes as a function of pH (Eqn. 10) is given by (———). At near neutral pH, there is a region where the dissolution rate is at a minimum and it is within this region where ligand-mediated dissolution (LMD) influences rates most dramatically (Eqn. 18). Rate in moles/m^2/sec.

2.2. Dissolution in Ligand-Containing Aqueous Solutions

This review concerns the importance of a fourth reaction, the ligand-mediated dissolution (LMD) reaction that can be written for a single hypothetical monovalent ligand, L^-, in neutral solution as:

$$(K_xNa_y)Ca_{(1-x-y)}Al_{(2-x-y)}Si_{(2+x+y)}O_8 + 8H_2O + (2-x-y)L^- \rightarrow (2+x+y)Si(OH)_4^o + (2-x-y)AlL^{2+} + 4(2-x-y)OH^- + xK^+ + yNa^+ + (1-x-y)Ca^{2+} \quad (5)$$

There are a wide variety of possible ligands with a wide range of chemical reactivities that may either enhance feldspar dissolution rates or inhibit dissolution. In nature, it is likely that many ligands will be present in solution and each will react with feldspars according to Eqn. 5. Under these conditions, the overall rates of LMD will be some function of the relative concentrations of the contributing ligands and the impact of each ligand individually on dissolution. There are only few controlled experiments involving multiple ligands and feldspar dissolution, and therefore the nature of this function, whether additive, multiplicative, or more complicated, is, as yet, uncertain. Some inferences concerning ligand-ligand and other ligand-solute interactions and their effect on LMD are discussed below.

Ligands can be either inorganic or organic, but this review focuses primarily on the effect of organic compounds on the dissolution process. Polyvalent ligands may have slightly different stoichiometries than given above. An analogous reaction can be written for LMD in acidic or basic solutions, but there is little experimental evidence that this process is important at higher pH.

The stoichiometry of Eqn. 5 does not require that ligands be involved in any of the fundamental steps in the dissolution process. Equation 5 could be described as a stepwise reaction involving first HD and subsequent reaction to form ligand complexes with the reaction products ($Al(OH)_3^o$) of the HD reaction subsequent to dissolution. There is, however, sufficient experimental evidence to indicate that the overall rate of feldspar dissolution into solution containing ligands, R_{total}, given conceptually by

$$R_{total} = R_{inorg} + R_{LMD} = R_{HD} + R_{PMD} + R_{HMD} + R_{LMD} \quad (6)$$

increases with ligand concentration when the reaction conditions and composition of the reacting solutions are held constant. This suggests that LMD, like PMD and HMD, involves an important rate-controlling process in feldspar dissolution and represents a mechanism that is independent of, or at least, partially independent of the other contributing reactions. Recent experimental work and *ab initio* modeling suggest that direct interaction of ligands and silicate surfaces does occur and that organic ligands can disrupt the aluminum-silicate structure leading to dissolution (Kubicki et al., 1997, 1999)

3. SOURCES, FUNCTIONS, AND CONCENTRATIONS ORGANIC LIGANDS

Although mineral weathering reactions can occur in a wide variety of geochemical environments, the importance of ligand-mediated reactions will certainly depend on the amounts and types of organic compounds that are present. It is therefore instructive to review the processes responsible for organic ligand production, the types of ligands produced, and the ranges of concentrations that may be achieved in contrasting geochemical settings.

3.1. Sources and Functions

Low-molecular-weight organic anions are commonly found in natural waters (Table 1). Microbes, plants, and animals produce organic acids as a result of their metabolism due to the anabolic synthesis or catabolic breakdown of larger organic compounds (Brock and Madigan, 1988). In the absence of other electron acceptors, anaerobic and facultative anaerobic microorganisms ferment the available organic carbon compounds and produce ligands such as formate, acetate, propionate, lactate, and butyrate (Gottschalk, 1986; Maeda and Kuwaii, 1988). Under extremely reducing conditions, microbes can produce organic anions (primarily acetate) from inorganic carbon (Brock and Madigan, 1988). Oxidative microbial degradation of macromolecules (such as polysaccharides and proteins) can result in the production of low-molecular-weight organic acids (Thurman, 1985). Bacteria can also produce and excrete high concentrations of organic acids when grown under conditions of nutrient stress, relatively high carbon availability, but limited availability of other nutrients, such as N and P (Neijssel and Tempest, 1975; Tempest and Neijssel 1992; Linton 1990). A number of ligands, notably oxalate, but also pyruvate, citrate, succinate, malate, gluconate, lactate, fumarate and formatel, have been detected in soils and on weathered rock colonized by fungi and bacteria (e.g., Fox and Comerford, 1990; Palmer et al., 1991; Baziramakenga et al., 1995; Krzyszowska et al., 1996). Ectomyccorrhizal organisms are known to produce soluble organic compounds that affect mineral weathering reactions (Jones 1998; Paris et al., 1995, 1996; Leyval et al., 1990). Plants produce and excrete organic ligands through their roots as well (Ochs et al., 1993; Laheurte et al., 1990). It is difficult to differentiate between compounds produced directly by the plants and those produced by their ectomyccorhizal communities (Laheurte et al., 1990; Paris et al., 1995).

Bacteria, fungi, and plants also produce other soluble compounds that are capable of interacting with mineral surfaces and altering mineral reaction rates. Under conditions of Fe limitation, organisms produce a specific class of soluble ligand, siderophores, that bind very strongly to Fe^{3+}, solubilizing normally insoluble iron compounds and making the iron available to the organism. Siderophores can also bind with other metals, particularly Al^{3+}, and can therefore, by the LMD mechanism, affect alumino-silicate mineral solubility and dissolution rates (Ochs et al., 1993; Bennett et al., 1998; Rogers et al., 1998a,b).

Although low-molecular-weight ligands are thought to dominate the interaction between rocks and solution in biologically mediated rock weathering, organisms produce other higher molecular-weight compounds, such as cell surfaces and extracellular polymers, that can affect the rates of mineral/water exchange reactions. Biopolymers can function as attachment or protective structures, food-storage reservoirs, or colonization substrates for other potential food organisms (Christensen and Characklis, 1990). Microbial extracellular polymers are typically comprised of sugars (including oxidized sugars such as gluconate or uronic acids) or amino acids. These compounds create microenvironments around attached organisms on mineral surfaces where they may play a role in mineral-water exchange reactions. All cell surfaces also have reactive functionality (carboxylic acids, teichoic acids, etc.) that can similarly bind either to metals at mineral surface or to metals released from the minerals to solution, and thus potentially affect mineral diagenesis and mineral-water exchange (Fein et al., 1997; Daughney et al, 1998).

3.2. Concentrations

Acetate is typically the most abundant organic anion in natural systems (Sansone and Martens, 1982; Sansone, 1986; Surdam et al., 1984; Fisher, 1987; Talbot et al., 1990; Martens, 1990; Kawamura and Nissenbaum, 1992; Table 1). Acetate concentrations greater than 10,000 ppm (160 mM) have been measured and acetate has been detected at temperatures up to 200 °C in sedimentary basins (Surdam et al., 1984; Kharaka, et al., 1986). Acetate can be produced by both abiotic and biological processes (Shock, 1988). Oxalate is also abundant in soils where it is often found to precipitate as metal-oxalate salts. Hydrated Ca-oxalate ($Ca(C_2O_4) \cdot H_2O$, whewellite) and its non-crystalline form are common constituents of soil litter (Graustein et al. 1977; Lapeyrie et al., 1990).

The total concentration of low-molecular-weight organic anions in natural waters associated with petroleum can exceed 1 g/L. Short chained aliphatic organic ligands (acetate, propionate, butyrate, and valerate) are typically the most abundant single ligands detected in these waters, although difunctional ligands (such as oxalate, malonate, and succinate) and aromatic compounds have also been observed in appreciable quantities (Peltzer and Bada, 1981; Kharaka et al., 1986; Fisher, 1987; MacGowan and Surdam, 1988; Cozzarelli, et al., 1994; see Table 1). Maximum organic ligand concentrations tend to be found in the 70 to 120 °C range, above the temperature (~70 °C) where microbial uptake commonly depletes organic concentrations (although low levels of microbial activity have been found in deep subsurface environments at elevated temperatures, Onstott et al., 1998), and below the temperatures where thermal decarboxylation consumes these compounds (Kharaka et al, 1986; Fisher, 1987; MacGowan and Surdam, 1988). Because of thermal decarboxylation of potential ligands, ligand-mediated dissolution is likely to be less important at hydrothermal temperatures (> 100 °C; Stoessel and Pitman, 1990).

The total concentration of organic carbon (dissolved + particulate) in soils typically ranges up to 6% (Brady and Weil, 1996). Most of this is the relatively refractory, high-molecular-weight residue of the degradation of plant and animal material (Thurman, 1985). However, microbial degradation of this material produces a myriad of soluble low-molecular-weight organic ligands that can potentially react with minerals. The low-molecular-weight ligands that are extremely common in relatively organic-poor environments, such as acetate, oxalate, etc., are also common in soils. Aromatic organic compounds have also been detected in natural waters where concentrations can range from μM to mM levels (see Table 1). In the rhizosphere, the microenvironments surrounding plants roots that include the ectomycorrhizal community, organic acid concentrations can be up to 1 to 2 orders of magnitude higher than in bulk soils. Grierson (1992) measured mg/L levels of maleic, fumaric, malic, acontitic, and citric acids in leachates from the rhizosphere of *Banksia integrifolia*. The concentrations of these compounds were approximately an order of magnitude lower in leachates of the surrounding soil and leaf litter. In a similar study, Fox and Comerford (1990) found up to 1mM of oxalate in the soil rhizosphere. Similar concentrations of oxalate, 10 to 100 μM of leachate, were measured at an Antarctic lichen community (Johnston and Vestal, 1993).

Table 1. Low molecular weight organic ligands in natural settings and some of their thermodynamic properties. All thermodynamic data from Smith and Martell (1998).

Ligand (Acid Formula)	Concentrations	pK_a*	Al Complexes**
	Aliphatic Monocarboxylic Acids		
Formate (HCOOH)	Few hundred nmol/g in stone (a); 5 to 174 μM in soil (b); 10 to 20 μM in beech forest soil (d); Few hundred nmol/g in soil extract (e); 10 to 50 μM in groundwater (f); 0.2 to 1.4 μM in forest floor solution (g); 50 to 300 μmol/kg in soil leachate (k)	3.75	$\log \beta_1 = 1.36$ $\log \beta_2 = 2.02$ other complexes I = 1
Acetate (CH_3COOH)	10 to 20 μM in beech forest soil (d); 140-1000 nmol/g in soil extract (e); 10 to 50 μM in groundwater (f); 0 to 50 μM in forest floor solution (g); 50 μM to 250 mM in formation water (h,j); 5 to 800 μmol/kg in soil leachate (k)	4.76	$\log \beta_1 = 1.51$ other complexes I = 1
Propionate (CH_3CH_2COOH)	Few μM in beech forest soil (d); 30 μM to 24 mM in formation water (h,j); 5 to 800 μmol/kg in soil leachate (k)	4.87	$\log \beta_1 = 1.69$ other complexes I = 1
Butyrate ($CH_3(CH_2)_2COOH$)	Up to 169 μM in formation water (h,j); up to 10 μmol/kg in soil leachate (k)	4.82	$\log \beta_1 = 1.58$ (I = 1)
Valarate ($CH_3(CH_2)_3COOH$)	Up to 7 μM in formation waters (j)	4.83	No data
	Aliphatic Dicarboxylic Acids		
Oxalate (HOOCCOOH)	Few tens nmol/g in stone (a); 25 to 1000 μM in soil (b); 20 to 100 μmol/g dry sediment in reducing marine sediment (c); Few μM in beech forest soil (d); 40-100 nmol/g in soil extract (e); 1-5 μM in forest floor solution (g); up to 5 mM in formation water (j); up to 200 μmol/kg in soil leachate (k)	1.25, 4.27	$\log \beta_1 = 5.97$ $\log \beta_2 = 10.93$ $\log \beta_3 = 14.88$ (I = 0.5)
Malonate ($HOOCCH_2COOH$)	Up to 25 mM in formation water (j); up to 200 μmol/kg in soil leachate (k)	2.85, 5.70	$\log \beta_1 = 6.26$ $\log \beta_2 = 11.11$ $\log \beta_3 = 13.3$ (I = 0.15)
Succinate ($HOOC(CH_2)_2COOH$)	Few tens nmol/g in stone (a); Few tens μmol/g dry sediment in reducing marine sediment (c); 2 to 500 μM in formation water (h); up to 200 μmol/kg in soil leachate (k)	4.21, 5.64	$\log \beta_1 = 3.20$ other complexes (I = 0.5)
Fumarate (HOOC-CH =CH-COOH)	Few tens nmol/g in stone (a); up to 30 μmol/kg in soil leachate (k)	3.02, 4.48	No data
Malate ($HOOCCH_2CHOHCOOH$)	Few μM in beech forest soil (d); Few nmol/g in soil extract (e); 8 to 30 nmol/ml in soils and root leachate (I); up to 30 μmol/kg in soil leachate (k)	3.46, 5.10	$\log \beta_1 = 4.60$ $\log \beta_2 = 7.62$ other complexes (I = 0.15)
	Aliphatic Tricarboxylic Acids		
Citrate ($COOH-CH_2-C(OH)COOH-CH_2-COOH$)	Few nmol/g in stone (a); trace in soils (b); Few tens (5 to 25) nmol/g in soil extract (e); 0-0.13 μM in forest floor solution (g); 70 to 600 nmol/ml in soils and root leachate (I); up to 30 μmol/kg in soil leachate (k)	3.13, 4.76, 6.40	$\log \beta_1 = 7.14$ $\log \beta_2 = 12.9$ other complexes (I = 0.1)
Isocitrate ($COOH-CH_2-CH(COOH)$ – CHOH-COOH)	Few μM in beech forest soil (d)	3.05, 4.30, 5.74 (I = 0.1)	No data
Aconitic ($HOOC-CH=C-(COOH)CH_2-COOH$)	Few μM in beech forest soil (d); Few nmol/g in soil extract (e); 1 to 20 nmol/ml in soils and root leachate (I)	2.8 4.5	No data

	α-Hydroxy- and α-Keto-carboxylic Acids		
Glycolate (CH_2OH-COOH)	<1 μmol/g dry sediment in reducing marine sediment [(c)]; up to 250 μmol/kg in leachate [(k)]	3.83	No data
Pyruvate ($CH_3CHOCOOH$)	Few tens nmol/g in stone [(a)]; Few μM in beech forest soil [(d)]; up to 40 μM/kg in soil leachate [(k)]	2.48	No data
Lactate ($CH_3CHOHCOOH$)	Few tens nmol/g in stone [(a)]; <1 μmol/g dry sediment in reducing marine sediment [(c)]; to 400 μmol/kg in soil leachate [(k)]	3.86	$\log \beta_1 = 2.36$ $\log \beta_2 = 4.42$ $\log \beta_3 = 5.8$ other complexes (I = 0.5)
Gluconate ($HOCH_2(CHOH)_4$ COOH)	Very reactive. Not reported in nature.	3.46, I = 0.1	$\log \beta_1 = 1.9$ (I = 0.5)
α-Ketoglutarate (HOOC-$(CH_2)_2$-(C = O)-COOH)	Very reactive. Not reported in nature.	1.9, 4.44, I = 0.5	No data
	Aromatic Carboxylic Acids		
Salicylic Acid (2-Hydroxybenzoic Acid)	2 - 5 nmol/g in rhizosphere [(e)]; 0 to 0.5 mM in formation waters [(j)]	13.7	$\log \beta_1 = 13.7$ (I ≤ 0.1) $\log \beta_2 = 24.8$ (I ≤ 0.5)
p-Hydroxybenzoic Acid	5 - 60 nmol/g in rhizosphere [(e)]	4.58, 9.46	No data
Protocatechuic Acid (3,4-Dihydroxybenzoic Acid)	1 – 5 nmol/g in rhizosphere [(e)]	4.49, 8.75, 13.0	$\log \beta_1 = 16.8$ $\log \beta_2 = 29.8$ $\log \beta_3 = 38.5$ other complexes (I = 0.1)
Caffeic Acid (3-(3,4-Dihydroxyphenyl) Propanoic Acid)	No concentration data found.	4.62, 9.07, 12.5, I = 0.1	No data
Gallic Acid (3,4,5-Trihydroxybenzoic Acid) (Note: Dissociation and stability constants refer to the singly protonated ligand. Full deprotonation is not observed)	No concentration data found.	4.4 9.11 11.4 (I ≤ 0.1)	$\log \beta^*_1 = 6.4$ $\log \beta^*_2 = 9.0$ $\log \beta^*_3 = 7.4$ (I ≤ 0.1)
Vanillic Acid (4-Hydroxy, 3-Methoxy-benzoic Acid)	1-14 nmol/g in rhizosphere [(e)]	No data	No data
	Other Aromatic Compounds		
Catechol (1,2-Dihydroxybenzene) (Note: Also forms strong complexes with Si; Smith and Martell, 1998)	No concentration data found.	9.45, 13.3	$\log \beta_1 = 16.7$ $\log \beta_2 = 30.35$ $\log \beta_3 = 39.4$ other complexes (I ≤ 0.1)

• Acid dissociation constants at 25 °C and I = ionic strength = 0 unless noted

** Association constants for complexes of the form AlL_n where $\beta_n = [AlL_n] / [Al^{3+}][L]^n$. The stabilities of complexes of differing form (other complexes) are described in Smith and Martell (1998).

Concentrations reported by:

a) Palmer et al., 1991
b) Fox and Comerford, 1990
c) Peltzer and Bada, 1981
d) Shen et al., 1996
e) Baziramakenga et al., 1995
f) McMahon and Chapelle, 1991
g) Krzyszowska et al., 1996
h) Kharaka et al., 1986
i) Grierson, 1992
j) MacGowan and Surdam, 1988, 1990
k) Cordt and Kussmaul, 1992

Below the soil zone, dissolved organic carbon concentrations on the order of a few to few 10's of mg/l are found in sediments and groundwater. Most of this material consists of relatively refractory high-molecular-weight humic and fulvic material (Thurman, 1985). In shallow anoxic ground water, up to tens of µM of fermentation products (acetate, butyrate, formate, and lactate) have been measured (Welch, unpublished). McMahon and Chapelle (1991) also measured low concentrations (tens of µM) of acetate and formate in an oxic aquifer. They hypothesized, however, that these organic acids were produced by microbial fermentation in the adjacent anoxic aquitard and that microbial uptake of these compounds was rapidly consuming them in the oxic groundwater solution.

4. INTERACTION OF ORGANICS WITH MINERAL SURFACES

The first step in the interaction of organic ligands and the mineral surface, by analogy with PMD and HMD, must involve adsorption of the ligand to the surface (Kubicki et al., 1997). Adsorption will depend on ligand structure, speciation, and charge, as well as the speciation and charge of the surface and the concentrations of other cations and anions competing for the ligand or reactive sites at the mineral surface. Soluble ions in solution may impact the activity of the ligand in solution and may compete with ligands for adsorption sites at the mineral surface. Adsorption and solution equilibria will have additional dependencies on temperature.

A number of investigators have studied the pH dependence of surface charge and the availability of surface exchange sites on alumino-silicate minerals. Blum and Lasaga (1988, 1991) concluded on the basis of acid and base titrations of albite and a simple Langmuir model that, under acidic conditions, feldspar surface charge is primarily controlled by the Al sites on the mineral surface and that the isoelectric point of the mineral was therefore similar to that of tetrahedral surface Al (Parks, 1967; Brady and Walther, 1992). Brady and Walther (1989, 1992) and Blum and Lasaga (1991) proposed that surface Si sites, although three times more abundant than surface aluminum sites based on the bulk albite stoichiometry, play a negligible role in PMD of this mineral. Oxburgh et al. (1994) further demonstrated that proton adsorption increases with An content of the feldspar from An_{12} to An_{75}. Several experimental studies have demonstrated that feldspar dissolution rates under acidic conditions increase dramatically with increasing Al content (Manley and Evans, 1986; Casey et al., 1991, Welch and Ullman 1992; 1996). This is an additional indication that, at least for PMD, Al content of the mineral and the relative abundance of Al-sites at the mineral surface play an important role in the first step in the dissolution process.

Brady and Walther (1989, 1992) suggest, based on the similarity of the pH dependence of HMD for a number of classes of minerals, that deprotonation of silica sites may play an important role in silicate dissolution in basic solutions, as well. Blum and Lasaga (1991), however, prefer a model dependent on the protonation and deprotonation of Al sites that also reproduces the observed pH dependence. More complex models involving multiple site adsorption (Stillings et al., 1992) and constant capacitance, double- and triple layer models (Sverjensky and Sahai, 1996) do not substantially alter the qualitative interpretations based on these simple models.

The one problem with the direct application of the Blum and Lasaga (1991) model to ligand adsorption is that it predicts that more proton adsorption occurs on feldspar than there are Al-sites on the surface. They propose that this is due to more than one proton or hydroxyl adsorbing to a given Al site (see also Stillings et al., 1998), a condition that may be realistic for the extremely fine-grained, and presumably stressed, surface of their experimental mineral. Alternatively, proton and hydroxyl adsorption may be occurring at some depth below the surface layer (Brantley and Stillings, 1996). While small protons may be able to penetrate into the feldspar surface to react with the bulk mineral, it is unlikely that larger ions or molecules, such as complexing organic ligands, will be able to do the same. As a result of this size constraint, ligand adsorption and therefore ligand-mediated reactions, are likely limited to the surface-most layer of the dissolving aluminum silicate and to one or fewer ligands per reactive surface site. Stillings et al. (1998), however, demonstrated that a Freundlich isotherm model that does not presume a fixed number of adsorption sites does a better job of describing oxalate adsorption to feldspar surfaces and LMD rates than a Langmuir isotherm that limits adsorption to a fixed number of surface sites. Both of these models suggest that one or more ligands adsorb at each Al site, if adsorption at Si sites is presumed negligible.

The condition of the surface is not the only factor influencing the adsorption of organic ligands to the mineral surface, and therefore LMD rates. The speciation of the attacking ligand is also important (Kubicki et al., 1997). Ligand speciation is controlled by pH and the concentration of other ions that form stable complexes in solution. If the charge of the dominant ligand species in solution is neutral or of the same sign as the surface, adsorption will be inhibited due to electrostatic repulsion (Morel and Hering, 1993). Thus, the expectation is that ligand adsorption and LMD rates should be greatest in those regions of pH where surface and ligand charges are predominantly of opposite sign and where there are few ions that form stable complexes with the ligand in solution. Figure 2 schematically shows the speciation of oxalic acid and its dissociated anions as a function of pH. Below pH = 1.25, neutral oxalic acid (H_2Ox^o = $H_2(COO)_2^o$) predominates in solution and above pH = 4.27 , oxalate ion (Ox^{2-} = $(COO)_2^{2-}$) is dominant. Over this same range of pH, a typical feldspar surface changes from a net positive charge to a net negative charge. Measurements of proton adsorption indicate that the pH (or point) of zero net proton charge (PZNPC) for feldspar is somewhere in the region of 5.2 to 7.5 (see discussions in Brantley and Stillings, 1996; Chen and Brantley, 1997). Above the PZNPC it is unlikely, because of charge similarities, that HOx^- (= $H(COO)_2^-$) or Ox^{2-}, the dominant ionic forms of oxalate in this range of pH, will significantly adsorb to the feldspar surface, and if adsorption is an important step in the LMD dissolution process, oxalate will have little or no impact on overall dissolution rates.

The presence of organic ligands in solution is only one of the requirements for LMD of feldspars. PMD and HMD predominate as dissolution mechanisms where protons and hydroxyls are in substantial concentration and where they compete with ligands for adsorption sites. Although ligands can affect the dissolution reaction over a wide range of pH, their major positive impact on feldspar dissolution occurs in the near-neutral pH range where proton and hydroxyl concentrations are at a joint minimum. The composition of the bulk feldspar and the feldspar surface also affects the number and type of reactive sites available to attacking protons, hydroxyls, and ligands, and therefore influence dissolution rates. Temperature may also influence the strength of interactions between ions in solution and on surfaces, and therefore potentially affects both the mechanism and the rates of mineral/solute interactions. It is therefore necessary to consider the impact of all of these processes on the LMD mechanism in order to identify the general trends that may be useful in modeling the stability of feldspar or other minerals under the wide range of conditions that may be found on the Earth's surface.

5. EXPERIMENTAL RESULTS

The results of experimental work summarized in Table 2 represent the best information available to date from which to infer the effects of ligands in solution and on surfaces on the rates of feldspar dissolution. For each experiment, we give the range of rates and some of the experimental conditions under which the rates were determined.

There is a great deal of variability in the rates that have been determined, although most of the rates at a given temperature and pH for the same mineral are within an order of magnitude of each other (Welch and Ullman, 1996). This variability is probably due to differences in experimental methods and materials. This variation reflects: (1) the choice of experimental minerals; (2) the preparation of the minerals for experiments; (3) the presence of trace contaminants and compositional intergrowths in the experimental minerals; (4) the assignment of specific surface areas to the reacting phases; (5) differences in the types of reactors used in the experiments; (6) the length of the experiment. As a result of these differences in experimental protocols, it has often been difficult to compare the results from one laboratory to those of another, and such comparisons, when made, are often not reliable in detail. Intra-laboratory comparisons, with common experimental constraints, are better suited for examining trends with respect to solution composition, mineral chemistry, and other parameters, and these general trends are often confirmed from laboratory to laboratory. These trends provide the basis of our understanding of the effect of ligands on feldspar dissolution rates and the mechanism of LMD reactions.

In Eqn. 6, we conceptually described the overall dissolution rate, R_{total}, as a sum of contributions from HD, PMD, HMD, and LMD, even though it is not clear how the factors controlling these reactions interact. Each of these mechanisms involve the same limited mineral surface and therefore it is likely that a simple sum of contributions will not adequately reproduce observations in an experimental situation where more than one mechanism contributes to dissolution. A more general equation describing the overall reaction rates as a function of the solution and mineral surface composition, the physical conditions of the reaction, and which allows for the possibility of interaction between the various dissolution mechanisms, can be given as:

$$R_{total} = f(L, S, M, T) \quad (7)$$

where L describes the ligand or ligands, their chemical characteristics, and concentration; S describes the composition of other components of the solution and includes the effect of protons and hydroxyls on dissolution, as well as the saturation state of the solution with respect to any primary or secondary minerals (note that L could conceptually be included in this term, but we have chosen not to do so because of our focus on the ligand-mediated mechanism of dissolution); M is a parameter describing the chemical, mineralogical, and textural characteristics of the dissolving mineral and mineral surface; and *T* describes the physical conditions of the experiment, principally temperature. In typical laboratory experiments, one of these parameters is varied while the other three remain approximately fixed. However, these factors are not entirely independent as, for example, temperature can alter equilibria, or solution composition can affect mineral surface characteristics. We will, nonetheless, attempt to discuss these variables one at a time.

5.1. Ligand Composition and Concentration

Table 2 indicates that a wide variety of organic ligands have been found to affect feldspar dissolution rates. These results universally demonstrate that ligands can enhance dissolution rates compared to controlled experiments in the absence of ligands. Generally, it has been found that low-molecular-weight ligands capable of forming strong complexes with Al^{3+} in solution (Table 1), and therefore, by analogy, with Al-sites on the mineral surface, have the greatest positive impact of dissolution rates (Furrer and Stumm 1986; Welch and Ullman, 1993). The best example of this effect is oxalate ion. Al^{3+} forms strong bidentate inner-sphere complexes with many multifunctional organic ligands (such as oxalate) in solution, and many of these ligands dramatically affect feldspar dissolution rates (Martell and Motekaitis, 1989; Bennett and Casey, 1994). In a survey of aliphatic ligands, Welch and Ullman (1993) found that equimolar amounts of the polyfunctional ligands that form strong, stable complexes with Al^{3+} (and possibly with $Si(OH)_4$; Bennett 1991) in solution, such as oxalate, citrate, succinate, α-ketoglutarate, and pyruvate, enhanced dissolution rates to a larger extent than the monofunctional acids acetate and propionate, although both classes of ligands enhanced dissolution over the rates found in inorganic solutions. This order of effectiveness of dissolution enhancement follows, approximately, the order of the strength of the first aluminum/ligand complex in solution (log β_1 in Table 1). However, the strength of this first association cannot be the only controlling factor. Many of the aromatic ligands studied, which form extremely strong complexes with Al in solution, are not as effective at promoting LMD as oxalate and citrate (Manley and Evans, 1986). Apparently other characteristics of the association of the ligand with the Al sites on the feldspar surface are also important (Bennett and Casey, 1994).

The observation that LMD rates appear to be higher in the presence of ligands that strongly complex Al in solution further suggests that the LMD mechanism is acting primarily at the Al sites or at Al-O-Si bonds

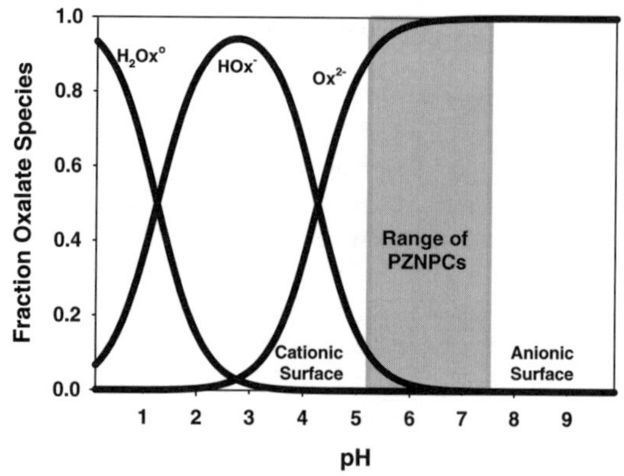

Fig. 2. The importance of ligand and surface charge on dissolution rates. In the pH region below the point of zero net proton charge (PZNPC) the average charge on the feldspar surface is positive. In this range of pH, the adsorption of anionic organic ligands (oxalate is given as an example), which is the first step in the ligand-mediated dissolution (LMD) reaction mechanism, is favored. Above the PZNPC, the anionic surface will electrostatically inhibit ligand adsorption, and therefore LMD as well. The PZNPC is given as a range of values (stippled) in this diagram.

(Kubicki et al., 1997) at the feldspar surface, in support of other lines of evidence. For example, in complexing organic acid solutions at near neutral pH and low temperature, feldspar dissolution is more rapid than HD alone and approximately stoichiometric (Fig. 3; Welch and Ullman, 1993; 1996; Stillings et al., 1996; Schweda, 1989; Mast and Drever, 1987; Schenk et al., 1989). In contrast, in similar inorganic solutions, where overall dissolution rates are low, Si is often released preferentially to Al and Al may not be detected in solution (Fig. 3; Welch and Ullman, 1993; Manley and Evans, 1986). However, Bennett et al. (1988) observed LMD of quartz, suggesting that, in the absence of Al sites on the mineral surface, ligands can interact with Si sites in much the same way in which they interact with Al sites on feldspar surfaces. Inorganic ligands may also contribute to LMD; experiments have shown that fluoride can enhance feldspar dissolution by a LMD mechanism as well (Amrhein and Suarez, 1988; Shotyk and Nesbitt, 1992).

Manley and Evans (1986) found that salicylic, protocatechuic, gallic, p-hydroxybenzoic, vanillic, and caffeic acids enhanced feldspar dissolution and that the degree of enhancement increased with the Al content of the feldspar and with the strength of the organic acid (pK_a, Table 1). However, they could not distinguish between the impact of LMD and PMD in their experimental design. A number of authors have also demonstrated that complex organic solutions

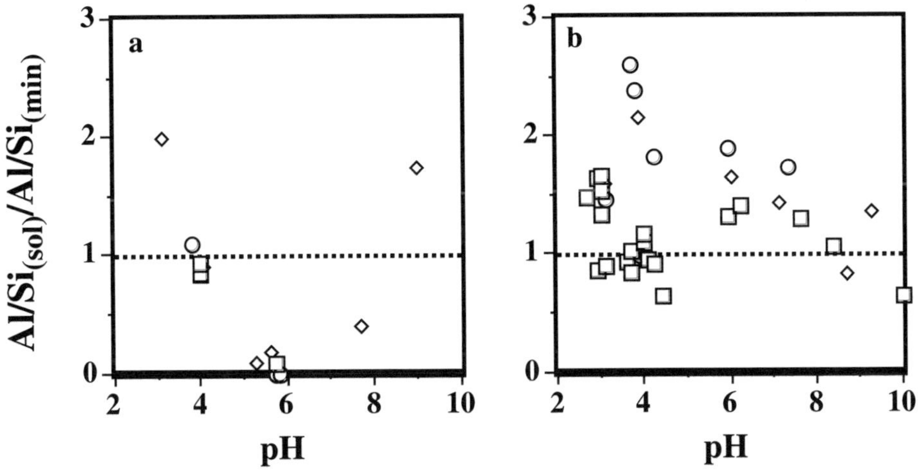

Fig. 3. The ratio of Al/Si released to solution during dissolution normalized to the Al/Si ratio of the dissolving feldspar. (Bytownite = ❒; Labradorite = ◊, ο ; data from Welch and Ullman, 1993). (a) The Al/Si release ratio in the absence of organic ligands is a strong function of pH, reflecting the solubility of Al(III) species in solution. (b) The addition of ligands that form strong and stable complexes with Al(III) in solution, and therefore, presumably at the mineral surface as well, leads to a dramatic increase in the Al/Si release ratio and the rates of mineral dissolution (Welch and Ullman, 1993). This is strong evidence of the direct interaction of ligands with Al sites or perhaps Al-O-Si bonds at the mineral surface leading to LMD.

such as stream water and peat, soil, and leaf extracts can also enhance the dissolution rates of feldspars by a LMD mechanism (Tan, 1980; Lundström and Öhman, 1990; Heyes and Moore, 1992). Although these experiments with complex organic materials are difficult to interpret, Lundström and Öhman (1990) suggest that the impact of their mixtures on dissolution rates is due to an easily metabolizable fraction of the mixtures, suggesting that lower molecular weight and/or multifunctional components of the mixture are responsible for the observed LMD enhancement.

At higher temperatures, Bevan and Savage (1989) and Blake and Walter (1996, 1999) demonstrated that both acetate and oxalate are effective at enhancing dissolution rates compared to inorganic controls at 70 - 100 °C. At 100 to 160 °C, acetate and oxalate also enhance dissolution rates (Surdam et al., 1984; Stossel and Pittman, 1990; Manning et al., 1992; Gestsdóttir and Manning, 1992; Hajash et al, 1992, 1998; Huang and Longo, 1992; Franklin et al., 1994). This is in spite of the suggestion, based on low-temperature studies, that the relative importance of LMD compared to PMD may decrease with increasing temperatures (Welch and Ullman, 2000). These observations suggest that there may be some unexplored details of the LMD mechanism that change as a function of temperature. Many of the qualitative observations of LMD at low temperature, however, appear to be consistent with the higher temperature observations.

There are a number of experiments that demonstrate that the ligand-mediated rate of dissolution, where and when observed, increases monotonically with ligand concentration (Fig. 4). Welch and Ullman (1992) demonstrated that the oxalate-mediated dissolution rate of bytownite feldspar (An_{75}) was linearly proportional to total oxalate concentration up to 1 mM, but that the constant of proportionality was a function of pH. The absolute enhancement of dissolution was greatest at pH = 3 and decreased with increasing pH, suggesting that oxalate-mediated dissolution was linearly dependent on HOx^- concentration, or that there is some other synergy between LMD and PMD mechanisms. Stillings et al. (1996) found that the oxalate-mediated dissolution rate of andesine (An_{47}) was approximately linear with total oxalate concentration up to 8 mM at pH = 3 and pH = 4 (Fig. 4c), but that oxalate apparently inhibited dissolution at high total oxalate concentrations (greater than 2 mM) at neutral pH. These results were used to suggest that the effect of oxalate on LMD is dependent on the changing speciation of oxalate in solution (see Fig. 2), and that dissolution rates should be best described in terms of the activities of the ionic oxalate species HOx^- and Ox^{2-}. Stillings et al. (1996) suggest that an alternative non-linear, but nonetheless monotonically increasing function of the total free ionic oxalate activity ($\{HOx^-\} + \{Ox^{2-}\}$) better fits the data at these high total oxalate concentrations. This empiri-

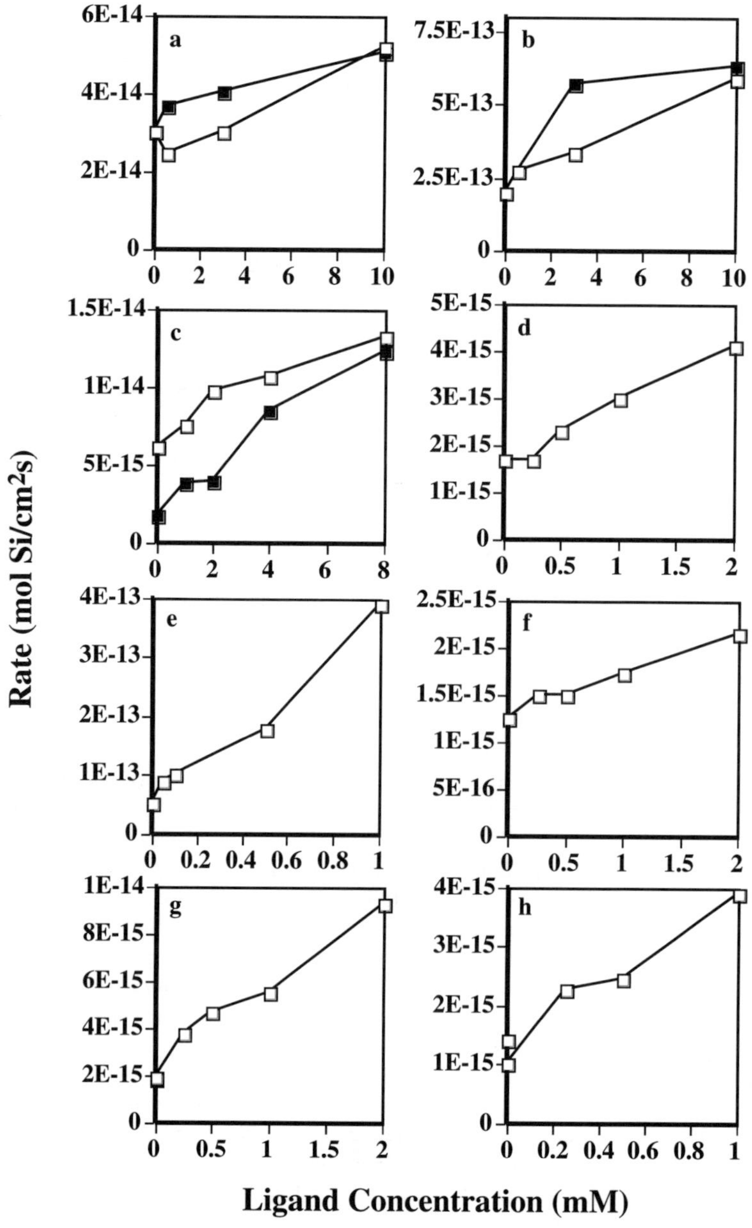

Fig. 4. Feldspar dissolution rates (as quantified by silica release rates) in the presence of low-molecular weight ligands are a monotonic function of ligand concentrations when other factors are held constant. The same ligands that enhance dissolution at lower concentrations may inhibit dissolution rates at high concentrations (>1-2 mM under some experimental conditions; Stillings et al., 1996; Ullman and Welch, unpublished). (a) Orthoclase dissolution in batch reactors at pH 6 and 80 °C (oxalate = ❒; citrate = ■; data from Blake and Walter, 1999); (b) Labradorite dissolution in batch reactors at pH 6 and 80 °C (oxalate = ❒; citrate = ■; data from Blake and Walter, 1999); (c) oxalate-mediated dissolution of andesine in a flow-through reactor (pH 3 = ❒; pH 4 = ■; data from Stillings et al., 1996); (d) gluconate-mediated dissolution of bytownite in a flow-through reactor at neutral pH (Vandevivere et al., 1994); (e) oxalate-mediated dissolution of bytownite in a flow-through reactor at pH ≈ 4.5; (f) oxalate-mediated dissolution of albite in a flow-through reactor at pH ≈ 4.5; (g) oxalate-mediated dissolution of andesine in a flow-through reactor at pH ≈ 4.5; (h) oxalate-mediated dissolution of albite in a flow-through reactor at pH ≈ 6.0 (e - h, data from Ullman and Welch, unpublished).

cal relationship may be the result of the non-linearity of adsorption behavior at high total oxalate concentration and the ultimate dependence of dissolution rate on this adsorption step (Amrhein and Suarez, 1988). Vandevivere et al. (1994) found that the gluconate-mediated rate of bytownite (An_{75}) dissolution was also linearly proportional to total ligand concentration at pH = 6 (Fig. 4d). Blake and Walter (1999) indicate that LMD rates at 80 °C increase monotonically but non-linearly with oxalate and citrate concentration and similarly conclude from these results that adsorption of the ligand is an important step in the LMD process (Figs. 4a and 4b). Knauss and Copenhaver (1995) also observed a monotonic but non-linear rate increase due to malonate-mediated dissolution.

Mixtures of ligands also tend to enhance dissolution compared to PMD alone. Franklin et al. (1994) found that the addition of oxalate to a solution of acetate at 100 °C enhanced LMD rates by comparison with the acetate solution alone. However, Huang and Longo (1992) note that mixtures of ligands may also lead to dissolution inhibition. The addition of EDTA or oxalate to solutions of catechol lead to overall LMD rates less than rates measured in catechol solution alone. This observation suggests that there are competitive reactions between dissolution enhancing species that may be important in field settings where many different ligands with different dissolution enhancing properties may play a role.

Not all ligands enhance dissolution rates. Welch and Vandevivere (1994) and Welch et al. (1999) demonstrated that several acidic polysaccharides (alginate, gum xanthan, and pectin) inhibited feldspar dissolution rates in near-neutral solutions. Inhibition was a function of the concentration of the polymer in solution and the abundance and orientation of carboxylic acid functional groups on the polymer chain. In addition, a high-molecular-weight fraction of polyaspartate (26kD) inhibited dissolution, while a lower molecular-weight fraction (6kD) had no effect on dissolution rates. A number of other polysaccharides had no impact on dissolution rates. However, under mildly acidic conditions, several of the acidic polysaccharides greatly enhanced feldspar dissolution compared to the controls. The maximum enhancement occurred near the pH of the pK_a of the carboxylate groups of the polymers and not at the pH of maximum Al-organic complexation in solution (Welch et al., 1999). On the basis of these results, Welch et al. (1999) suggest that the enhancement of Al leaching from the feldspar surface, and therefore the overall feldspar dissolution rate, is related not only to the strength of the Al-L complex, but also to the solubility and stability of this complex once formed.

Some bacterially produced polysaccharides enhanced dissolution compared to buffered controls, but inhibited dissolution rates in the presence of gluconate, a dissolution-enhancing ligand (Welch and Vandevivere, 1994). This dual nature of polymers can be explained by considering the competition between polymer complexation in solution and polymer adsorption and binding to the mineral surface. At near-neutral pH, the carboxylic groups of the polymers are deprotonated and bind irreversibly to multiple sites on the mineral surface, thereby inhibiting the dissolution reaction by a reduction in the number of reactive sites or reactivity of the adsorption sites. Under mildly acidic conditions, only a fraction of the carboxyls are deprotonated. Under these conditions, the polymers can bind to the surfaces, but less strongly, and protons are also available to react with the mineral surface and release ions by the PMD reaction. Once reaction products have been released to solution, these can react with the polymers to further reduce the activity of the carboxylic sites on the polymer and therefore the strength of surface adsorption and binding, thus further enhancing the PMD reaction. It may be difficult to distinguish between this combined PMD and LMD mechanism and a mechanism where the primary effect of protons is to change the solubility of a multiply-bound ligand-aluminum surface complex allowing dissolution by a primarily LMD mechanism.

Low-molecular-weight ligands can also have this dual dissolution-enhancing and inhibiting nature. In feldspar dissolution experiments, pyrocatechol violet initially increased feldspar dissolution compared to controls (Schenk et al., 1989; Petersen et al., 1992). However, long-term dissolution rates were inhibited by this ligand due to irreversible adsorption to the surface. Subsequent experiments with these mineral samples showed that this inhibition could be reversed by pyrolytic removal of the adsorbed organic layer.

What is clear from these results is that the impact of ligands on the feldspar dissolution process will require consideration of the thermodynamic and steric competition between protons and ligands for Al sites at the mineral surface and between Al and protons (and perhaps other cations) for the charged functional groups of the ligand. Some of the models, discussed below, attempt to integrate these competitive processes into estimates of overall dissolution rates.

5.2. *Effect of Overall Solution Composition*

Mineral/water exchange is a function of solution and surface speciation, both of which are strong functions of solution pH, and to a lesser extent a function of redox potential, ionic strength, and the ionic composition of the solution. In addition, solution composition may affect the saturation states with respect to dissolving primary phases, reactive intermediates, and secondary precipitating phases which can also affect mineral dissolution rates.

5.2.1. pH dependence of dissolution

The pH dependence of mineral weathering has been studied extensively in the last several decades since acidification of natural waters is a major environmental problem and chemical weathering of minerals can neutralize this environmental acidity (see Eqns. 1 and 2). As shown in Fig. 1, silicate mineral dissolution in inorganic solutions is not a simple function of pH (Chou and Wollast, 1985; Blum and Lasaga, 1988; Welch and Ullman, 1993; 1996; Brady and Walther, 1989; Carroll and Walther, 1990; Wogelius and Walther, 1992; Hellmann, 1994; Tole et al., 1986; Oxburgh et al., 1994). The form of the PMD dissolution curve suggests that the reaction rate can be described by:

$$R_{PMD} = k_{PMD}\ [H^+]^n \qquad (8)$$

where R_{PMD} is the surface area-normalized PMD rate, k_{PMD} is a kinetic rate constant, $[H^+]$ is proton activity, and n is a factor related to proton adsorption on the mineral surface (Blum and Lasaga, 1988). Both k_{PMD} and n increase with increasing Al content of the feldspar mineral (Welch and Ullman, 1996). k_{PMD} increases by four orders of magnitude and n varies from approximately 0.2 to 1 with increasing An content along the plagioclase series in the absence of ligands (Brady and Walther, 1989; Oxburgh et al., 1994; Welch and Ullman, 1996), indicating that in acidic solutions, proton adsorption to Al sites at or near the feldspar surface is catalyzing the dissolution reaction. Welch and Ullman (1996) found that the pH dependence of n at low temperature was the same in oxalate solutions as in inorganic control experiments from An_3 to An_{49}, but was somewhat lower for An_{77}. This difference in An_{77} behavior, however, may be an experimental artifact resulting from transport limitation of the dissolution rate of this reactive member of the feldspar series at high dissolution rates. These observations suggest that the effect of ligands on dissolution rates in highly acidic solution is to increase the rates of the PMD reaction, by perhaps buffering the reactive solution at acidic pH, rather than to control a completely independent dissolution pathway, as suggested by Eqn. 5. This is consistent with the observations that dissolution rates are independent of H_2Ox° in solution (Welch and Ullman, 1992; Stillings et al., 1998).

In basic solutions above pH ≈ 8 to 9, mineral dissolution rates follow a rate law of the form:

$$R_{HMD} = k_{HMD}\ [OH^-]^m \qquad (9)$$

where R_{PMD} is the surface area-normalized HMD rate, k_{HMD} is a kinetic rate constant, $[OH^-]$ is hydroxyl activity, and m is a factor related to the adsorption of hydroxyl ions at aluminum (Blum and Lasaga, 1991) or Si sites (Brady and Walther, 1989). Surprisingly, Brady and Walther (1989) have found that m is relatively insensitive to mineral composition and has a value close to 0.3 for many silicate and alumino-silicate minerals. There are insufficient experimental determinations of LMD rates in basic solution to determine whether ligands affect the overall pH dependence of feldspar dissolution at basic pH. In one experimental study, Knauss and Copenhaver (1995) found no significant effect of malonate on alkali feldspar dissolution compared to the HMD rate in slightly alkaline solutions, in spite of a relatively large effect of malonate over the PMD rate in mildly acidic solutions. This may be the result of the inhibition of anionic ligand adsorption onto the negatively charged feldspar surface at basic pH (Fig. 2). The direct impact of ligands on dissolution rates may be negligible under basic conditions. However, ligands may still enhance dissolution rates indirectly at high pH by complexing with reaction products in solution and lowering the solution saturation state with respect to the dissolving mineral, as they do at lower pH (see discussion further on).

Equations 8 and 9 are often combined with a rate constant, k_{HD}, describing the slow rate of hydrolysis in neutral solutions, R_{HD}, to give a general empirical expression for the dissolution of feldspars in inorganic solutions, R_{inorg}:

$$R_{inorg} = R_{PPD} + R_{HD} + R_{HPD} =$$
$$k_{HPD}\ [H^+]^n + k_{HD} + k_{HPD}\ [OH^-]^m \qquad (10)$$

(Wollast and Chou, 1988; Drever and Vance, 1994). This is the function plotted in Fig. 1 describing the pH dependence of feldspar dissolution rates in ligand-free solutions.

5.2.2. Other ions

There is good evidence that other ions in solution can directly affect the PMD rates of feldspar (Stillings and Brantley, 1995; Brantley and Stillings, 1996; Blake and Walter, 1999). In the case of the PMD rates, the effect of other cations in solution is to substitute for protons at the mineral surface, thus reducing the specific surface activity of the proton

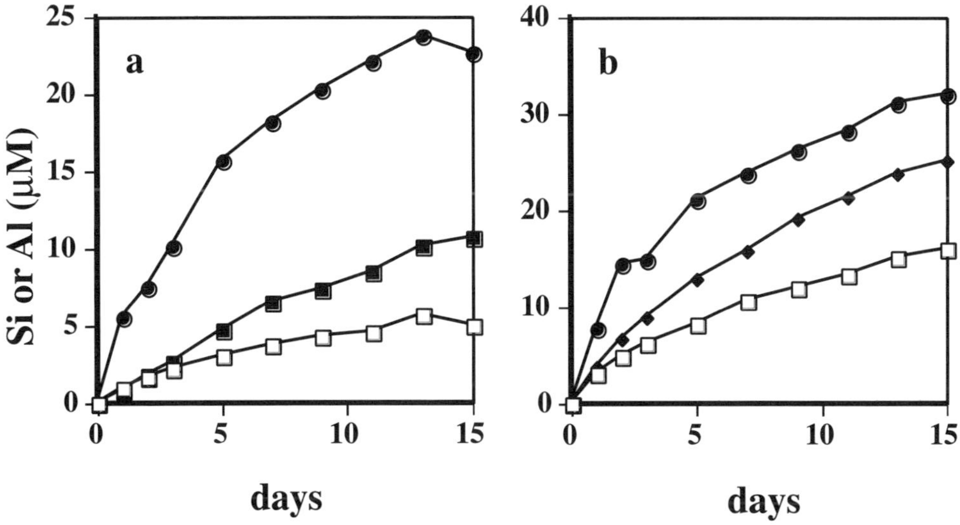

Fig. 5. The release of Si and Al to solution as a result of PMD and LMD reactions in a batch reactor at (a) 20 °C and (b) 5 °C (pH = 6-6.5; LMD Si = ■, ◆; LMD Al = ●; PMD Si control = ❒; no dissolved Al detected in the control; data taken from Welch and Ullman, 2000). Both initial rates and subsequent, more steady rates of Si release are higher than the controls. The initially high release rates of both Si and Al from the surface at high chemical affinities is consistent with the LMD mechanism involving surface complexation at Al sites on the mineral surface.

and therefore the PMD rate (Stillings and Brantley, 1995) This process has been described as an example of competitive cation adsorption for a finite number of reactive surface sites. In one LMD study, Blake and Walter (1999) observed that the oxalate-mediated enhancement of plagioclase dissolution could be offset by the addition of NaCl. This effect could be due to competition between Na^+ and protons for adsorption sites at or near the mineral surface as in the case of PMD, or alternatively, to the reduction of oxalate activity by the formation of Na-Ox complexes. However, Na-Ox complexes are fairly weak ($\log K_1(NaOx^-) \approx 1$; Smith and Martell, 1998), and therefore the former competitive adsorption mechanism, analogous to the proton/Na^+ mechanism, but involving cation/anion competition, appears to be the most reasonable. However, other ions that form stronger complexes with ligands, including for example some of the products of feldspar weathering, Ca^{2+} and Al^{3+}, may be able to inhibit the rates of LMD by the solution complexation mechanism. Inhibition of dissolution by Al^{3+} has been proposed as a dominant rate-controlling process in feldspar dissolution (Oelkers et al., 1994). These authors propose that the effect is due to Al^{3+} adsorption and the poisoning of reactive sites. However, this effect may equally well be related to the formation of solution complexes and the reduction of free ligand activity (see below).

The indirect impact of solution processes on ligand activity may be a dominant process in some field settings. Any ion that can form stable complexes with ligands in solution will reduce the ligand activity, and therefore reduce the adsorption of ligands to dissolving mineral surfaces. This effect of ligand speciation and activity on LMD rates has been demonstrated by Welch and Ullman (1992) and Stillings et al. (1998). This effect may also be responsible, in part, for the rapid decrease in dissolution rates with time observed in batch experiments where the ligand activity may be significantly reduced as a result of the initial Al^{3+} and Ca^{2+} release to solution from the dissolving mineral (Gestsdóttir and Manning, 1992; Manning et al., 1992). The release of Ca^{2+} from feldspars during dissolution may also lead to the formation of stable mineral phases such as whewellite in experiments involving oxalate (Stillings et al., 1998).

5.2.3. *Effect of chemical affinity*

The overall dissolution rate of a mineral may be affected by the saturation state with respect to dissolving phase. When solutions are significantly undersaturated with respect to the dissolving phase, reaction rates are essentially independent of saturation state (Nagy and Lasaga, 1992; Nagy et al., 1991; Burch et al., 1993; Taylor et al., 2000). However, as the solution approaches saturation, the net dissolution

Table 2. A summary of experimental observations concerning ligand-mediated dissolution of feldspars.

Study	Organic Ligand(s)	Mineral(s) [Reactor Type]	pH [T°C]	Log(Rate) (mol/m²/sec)	Notes
Amrhein and Suarez (1988)	Oxalate (also Fluoride)	Anorthite (An_{93}) [Batch]	5 to 6.5 [25]	−11.2 to −10.6	Rates dependent on adsorbed ligand concentration; oxalate and fluoride have similar effects
Barker et al. (1998)	Microbial cultures	Bytownite (Also Biotite) [Batch]	3.5 to neutral [22]	No Rate Data	Microbial metabolism enhances dissolution by a factor of 10 to 100 by a combination of PMD and LMD. SEM evidence of extensive feldspar dissolution.
Barman et al. (1992)	Oxalate, Citrate, Glycine, Salicylate	Microcline (Other minerals) [Batch]	0.5 to 5.6 [Room]	No Rate Data	Al released preferentially compared to Si in 0.5 M oxalic acid, citric acid, and glycine, and in 10 mM salicylic acid.
Bennett et al. (1988)	Citrate, Oxalate, Salicylate, Acetate	Quartz [Batch]	3 to 7 [25]	−10.1 to −8.5	Dissolution enhancement was: Citrate > Oxalate > Salicylate > Acetate = Water. Rates in citrate solution increase with concentration.
Bennett (1991)	Citrate, Oxalate	Quartz [Batch]	Neutral [25–70]	−12.7 to −9.9	Strong evidence is presented that organic-Si complexes form and play a role in mineral dissolution. Ligands significantly lower the apparent activation energy of dissolution.
Bennett et al. (1991)		[Field Study]	4.6 –7.7	No Rate Data	The weathering of silicates in a peat bog is controlled by silica-organic associations.
Bennett et al. (1996)		Microcline, Anorthoclase, Albite, Oligoclase (and other silicates) [Field Study]		No Rate Data	Silicate surfaces are colonized by bacteria in an anoxic, contaminated aquifer setting. Etching of silicates is correlated with bacterial settlement.
Bevan and Savage (1989)	Oxalate	K-feldspar (Or_{83}, Perthitic) [Batch and Flow-through]	1, 4, 9 [70 and 95]	−12 to −9.1	At pH 1, oxalate inhibits dissolution Under other conditions oxalate enhanced rates by a factor of 1.5 to 4. Temperature dependence of dissolution (E_a) is from 14.4 to 57.7 kJ/mol.
Blake and Walter (1996)	Oxalate, Citrate	Orthoclase (Or_{70}) (also Quartz) [Batch]	6 [80]	No Rate Data	Ligands enhance Si release by a factor of 3 compared to buffered controls. Al release is not detected in controls, but is approximately stoichiometric with ligands.
Blake and Walter (1999)	Oxalate, Citrate	Albite (An_1), Orthoclase (Or_{69}), (also Quartz), Labradorite (An_{69}), [Batch]	6 [70-80]	−12.0 to −8.8	Dissolution rates increase monotonically but not linearly with ligand concentration. Enhancement increases with An content. The addition of Na^+ to solution inhibits LMD.

Brady and Carroll (1994)	Acetate	Labradorite (An_{60}) [Flow-through]	4 [21, 35, 60]	No Rate Data	Determined E_a of labradorite dissolution of 48 kJ/mol at two different levels of pCO_2
Franklin et al. (1994)	Acetate, Oxalate (and combinations)	Albite (An_2), also Quartz [Flow-through column, 345 bars]	2.5 to 7 [100]	−10.8 to −8.9	Mixtures of acetate and oxalate enhance rates over acetate alone. Adsorption isotherm model gives good prediction of rates.
Gestsdóttir and Manning (1992)	Acetate, Oxalate	Albite [Batch, 500 bars]	6.75 – 8 [150]	No Rates per Surface Area	The results indicate that acetate increases the rate of dissolution but not overall albite solubility. Both the solubility and dissolution rate increase with increasing oxalate concentration.
Hajash et al. (1992)	Acetate, (and mixtures with Oxalate)	Albite (An_3) [Batch, 345 bars]	2.8 – 7 [100]	−11.0 to −9.0	Acetate increases dissolution rates. Mixtures of oxalate with acetate increase dissolution rates relative to acetate alone above pH = 3. Dissolution rates proportional to total oxalate concentration.
Hajash et al. (1998)	Acetate, Citrate (and in combination)	Albite [Flow–through Column, 1 and 345 bars]	4.4 – 5.7 [100, 160]	−10.9 to −9.0	Rates in organic solution are approximately 2 – 10x that of inorganic controls. Rate uncertainties due to surface area uncertainty.
Heyes and Moore (1992)	Leaf and Peat Extracts	Microcline (and other minerals) [Batch]	5.5-7.4 [25]	No Rate Data	Leaf leachates containing 50 and 300 mg/L DOC were more effective weathering agents than peat extract, HCl, or water.
Huang and Keller (1970)	Acetate, Salicylate, Tartrate, Aspartate	Microcline, Labradorite (other minerals) [Batch]	Mildly acidic	No Rate Data	Organic ligands leach Al from minerals preferentially to Si.
Huang and Kiang (1972)	Acetate, Salicylate, Aspartate, Citrate	K-spar, Albite, Oligoclase, Labradorite, Bytownite, Anorthite [Batch]	Mildly acidic	No Rate Data	The net release of feldspar into solutions of organic acids is proportional to An content. Al was preferentially leached compared to Si in strongly complexing organic solutions, Citrate > Salicylate > Aspartate > Acetate.
Huang and Longo (1992)	Catechol, Acetate, Oxalate, and EDTA (and in combination)	Anorthite (An_{98}) K-spar (Or_{73}), Oligoclase (An_{28}) [Batch]	2.9 to 8.5 [95, 100]	−11.6 (pH = 6) to −10.7 (pH = 3.5, K–spar) to −10.4 (for An_{98})	Catechol and EDTA enhance dissolution to a larger extent than oxalate. Mixtures of ligands show some inhibition.
Jorgensen (1976)	Catechol	Microcline Oligoclase (also Quartz and other Minerals) [Batch]	9.4 – 9.7 [25]	No Rate Data	Minerals dissolve incongruently in catechol solutions, leaving a residual surface enriched in Al. Five times more quartz dissolution in catechol compared to inorganic controls. No inorganic controls for feldspars.

Knauss and Copenhaver (1995)	Malonate	Albite (An_1), Microcline (Or_{95}) (and Quartz) [Flow-through]	4 – 10 [70]	–11.1 to –10.9	Feldspar dissolution rates increase monotonically by a factor of 2 – 10 in malonate solutions. The predominant impact is at mildly acidic to neutral pH.
Lundström and Öhman (1990)	Citrate, (also Streamwater and Organic Extracts from forest soils and peat)	K-spar (Or_{64}), Oligoclase (An_{19}) [Batch]	5.1 [20]	–12 to –11.5	Citrate, streamwater and extracts increase rates by a factor of 1.7 to 2.7. Microbial inoculation of extracts decreased dissolution rates to that of controls, suggesting that dissolution enhancing compounds are easily metabolizable.
MacGowan and Surdam (1988)	Acetate Oxalate Malonate	Andesine [Batch]	5.8 – 8.6 [100]	No Rate Data	Dissolution experiments with formation waters and organic acids showed extensive dissolution of mineral fragments by SEM correlated with increase of Al in solution.
Manley and Evans (1986)	Citrate, Oxalate, (and Aromatic Acids, see notes)	Labradorite (An_{69}), Albite (An_6), Microcline (Or_{67}) [Batch]	No Data	No Rate Data	The relative effectiveness of organic acids in dissolution enhancement was Oxalic ≈ citric > salicylic > protocatechuic ≈ gallic > p-hydroxybenzoic > vanillic ≈ caffeic. Labradorite dissolves more readily than microcline, which dissolves faster than albite.
Manning et al. (1992)	Acetate, Oxalate, Citrate	Albite [Batch 500 bars]	3 – 9 [150]	No Rates per Surface Area	Dissolution rates increase with increasing acetate concentration, not due to increases in overall mineral solubility. Oxalate and citrate increase dissolution rate and overall solubility. However, the effect may be related to PMD rather than LMD.
Mast and Drever (1987)	Oxalate	Oligoclase (An_{13}) (Also Tremolite) [Flow-through]	4 – 9 [Room]	–11.7	Neither pH nor oxalate concentration had any effect on Si release rates.
Panov et al. (1983)	Citrate, Tartrate	Oligoclase (An_{13}) Microcline (Or_{85}) [Batch]	Acidic [65]	No Rate Data	Trace elements are leached from feldspars by ligands.
Petersen et al. (1992)	Salicylate, Tartarate, Citrate, Pyrocatechol-Violet	Feldspar-rich Pegmatite Sand [Batch and Flow-through]	4.5 – 6	Initially: –11.4 to –8.6 Long–term: –12 to –10.5	Dissolution is approximately stoichiometric at pH 4.5 and 6 with ligands. The organic ligands increased Si release by at most a factor of 2.5. (See also Schenck et al., 1989). Pyrocatechol violet both increased and inhibited under differing conditions.

Reed and Hajash (1992)	Acetate (Acetate + Oxalate)	Granitic Sand (63% Feldspar) [Batch, 345 bars]	4.2 – 4.5 [100]	No Rate Data	Organic ligands increased release of feldspar dissolution products. Oxalate + acetate was more effective than acetate alone.
Schenk et al. (1989)	Salicylate, Tartarate, Citrate, Pyrocatechol-Violet	Feldspar-rich Pegmatite sands [Batch and Flow-through]	pH 4.5 to 6	−11.4 to −8.6	Dissolution is approximately stoichiometric at pH 4.5 and 6 with ligands, Al is undetectable in DI. Organic ligands increased Si release by at most a factor of 2.5. Pyrocatechol-violet inhibited feldspar dissolution by irreversibly adsorbing to the surface. (see also Petersen et al., 1992)
Schweda (1989)	Citrate (Also Cations)	Microcline, (Also Sanidine) [Flow-through]	4, 5.7 [25]	−13 to −10.5	At pH 4, citrate (0.1 to 1 mM) had no effect on sanidine dissolution rates. At pH 5.7 in citrate, sanidine and microcline rates approximately double. Al release in citrate solutions is approximately stoichiometric.
Shotyk and Nesbitt (1992)	Oxalate, Fluoride	Labradorite, Anorthite [Flow-through]	4.0 – 4.1 [not specified]	No Rate Data	Based on SIMS profiling labradorite dissolved incongruently and anorthite dissolved congruently. LMD produced thicker leached layers than PMD. Ligands accelerate dissolution rates.
Song and Huang (1988)	Oxalate Citrate	Microcline, Orthoclase [Batch]	Neutral and Acidic	No Rate Data	K release from minerals enhanced by citrate and oxalate. Al released preferentially to Si in acidic oxalate solution.
Stillings et al. (1996)	Oxalate	Microcline (Or_{78}), Albite (An_2), Oligoclase (An_{22}), Andesine (An_{43}), Bytownite (An_{76}) [Flow through]	3 – 7 [Room]	−11.7 to −9.4	Experiments to look at effect of mineral composition and, for one mineral, effect of oxalate concentration over a range of pH. Ligand dependence of dissolution rates best described in terms of ionic activities of ligand and adsorption model.
Stoessel and Pittman (1990)	Acetate, Propionate, Oxalate, Malonate	Microcline [Batch, 300 bars]	1.5 – 8.3 [100]	No Rate Data No Rate Data	Acetate and propionate had limited effect on dissolution rates. Oxalate and malonate enhanced dissolution. Cannot distinguish between PMD and LMD effects.
Surdam et al. (1984)	Acetate Oxalate	Microcline, Albite, Andesine, Labradorite [Batch]	2.9-3.4 [100]	No Rate Data	Acetate in formation water can increase Al solubility by an order of magnitude, Oxalate can increase Al solubility by 3 orders of magnitude.

Surdam and MacGowan (1987)	Acetate, Malonate, Salicylate	Andesine [Batch]	5-5.2 [100]	No Rate Data	Organic ligands enhance Al release and solubility.
Tan (1980)	Humic and Fulvic Acid Extracts	Microcline	2.5 – 7 [24 ± 2]	No Rate Data	At pH=7, humic acids released 2–10 times greater concentrations of Si and Al than fulvic acids. Al preferentially released to Si. At pH=2.5, fluvic acids released 25% more Si and Al than inorganic control.
Vandevivere et al. (1994)	Live Bacteria and Gluconate	Bytownite (An_{77}) and labradorite [Batch and flow through]	3 to neutral [25]	−11.6 (abiotic) to −9.2 (bacteria); −11.1 to −10.8 (from 0 to 2 mM gluconate)	Microorganisms increase Si release from feldspars at neutral pH by producing organic acid. Not clear whether LMD or PMD is more important. Dissolution rate linearly proportional to gluconate concentration.
Varadachari et al. (1994)	Oxalate, Citrate, Glycine, Salicylate	Microcline (and other minerals) [Batch]	not specified [27]	No Rate Data	SEM and XRD characterization of reacted minerals.
Welch, Barker, and Banfield (1999)	Poly-saccharides	Bytownite An_{76} [Batch]	3 – 7 [25]	−11.1 (pH=6, polymer) to −10.9 (pH=6 control) to −9.5 (pH=3, polymer)	The same polysaccharides that increased Si release in acidic solutions by a factor of 10 to 100 (compared to a control) inhibited dissolution by a factor of 2 in neutral solutions.
Welch and Ullman (1992)	Oxalate	An_3, An_{49}, An_{74}, An_{77}, and Quartz) [Flow-through]	3 – 10 [Room]	−11.0 (Quartz) to −8.3 (An_{77})	Dissolution rates dependent on Al content of feldspar. Linear dependence of dissolution rates on oxalate concentration varies with pH. LMD increases overall dissolution by a factor of 10 at neutral pH.
Welch and Ullman (1993)	Acetate, Propionate, Oxalate, Citrate, Pyruvate, Succinate, α-Ketoglutarate	Bytownite (An_{77}), Labradorite (An_{49}) [Flow-through]	3 – 10 [Room]	−11.0 to −8.2 (An_{77}) −11.2 to −9.5 (An_{49})	Polyfunctional organic ligands have a greater impact on dissolution that monofunctional ligands.
Welch and Ullman (1996)	Oxalate, Catechol	An_3, An_{49}, An_{77}, and Quartz) [Flow-through]	3, 6 [Room]	−11.3 (An_3) to −8.6 (An_{77})	PMD rate increases with Al content of mineral from quartz to An_{77}. LMD rate has pH dependence.
Welch and Ullman (1999)	Gluconate producing bacteria	Bytownite (An_{75}) [Batch]	5.7 – 7 [5 – 35]	No Rate Data	Glucose metabolism produces gluconic acid. In the presence of gluconate, dissolution rates increase by either PMD or LMD, or both. The temperature dependence of metabolism leads to higher rates of dissolution at lower temperatures.

Welch and Ullman (2000)	Acetate, Oxalate, Gluconate	Bytownite (An_{75}) [Batch]	5.7 – 7 [5-35]	–11.9 to –11.0	KNO_3 and acetate solutions have the same apparent activation energy as distilled water (≈ 43 kJ/mol). Gluconate and oxalate have lower activation energy (≈ 29 kJ/mol)
Welch and Vandevivere (1994)	Gluconate, Various Polysaccharides	Bytownite (An_{76}) and labradorite (An_{50}) [Batch]	Neutral [25]	No Rate Data	Si release from feldspar inhibited by acidic polysaccharides. Bacterial exopolymers enhanced dissolution compared to control but inhibited dissolution in gluconate solutions.

rate approaches zero. This dependence can be described by the function (Lasaga, 1995, 1998):

$$\left(1 - \frac{r_-}{r_+}\right) = 1 - \exp(-A/\sigma RT) \tag{11}$$

where (r_- / r_+) is the ratio of the rate of the reverse (precipitation) reaction over the forward (dissolution) reaction involving the primary phase, R is the universal gas constant, T is the absolute temperature, σ is a term describing the reaction stoichiometry (e.g. Aagaard and Helgeson, 1982), and A is the chemical affinity of the overall reaction given by:

$$A = RT \ln\left(\frac{K}{Q}\right) \tag{12}$$

where K and Q are the equilibrium constant and activity product of the overall dissolution reaction.

In the experiments of Lasaga and coworkers (Nagy and Lasaga, 1992; Nagy et al., 1991; Burch et al., 1993; Taylor et al., 2000) it is clear that at chemical affinities above some critical value, there is a plateau where dissolution rates are, as expected, independent of saturation state. For plagioclases this critical value of A is approximately 40 kJ/mol (Burch et al., 1993; Taylor et al., 2000). Both the dissolution rate and mechanism may depend on chemical affinity (Lasaga and Blum, 1986). Although, the direct inhibitory effect of chemical affinity has not been demonstrated experimentally for LMD because most experiments are conducted above this critical value, it is undoubtedly an important effect closer to saturation. Typically, undersaturation in LMD experiments is aided, in part, by the formation of ligand complexes that reduce the activity of reaction products, and therefore, future studies will require some additional certainty concerning the nature and stability of ligand complexes in solution in order to determine the effect of chemical affinity on LMD. While the complexation of ligands by reaction products may increase the chemical affinity and therefore tend to increase dissolution rates, the concomitant reduction of ligand activity may also reduce dissolution rates if free ligand activity and ligand adsorption are responsible for the LMD enhancement mechanism (Stillings et al., 1998). The experimental separation of such competing processes may be extremely difficult.

Ligand-mediated enhancement of dissolution, however, cannot be due to the solution complexation mechanism alone. If this were so, there would be no dissolution enhancement at high chemical affinities, in contrast to the observations cited above. This is clearly at odds with the experimental evidence of LMD enhancement and the concentration dependence of LMD rates under such conditions.

It is clear in some LMD experiments that there is still an apparent rate dependence on chemical affinity above the critical value (Stillings et al., 1998). Oelkers et al. (1994) also observed that the PMD rate of albite had a dependence on chemical affinity well above the critical value measured by Burch et al. (1993). Oelkers et al. (1994) and others (Chou and Wollast, 1985; Stillings et al., 1998) attribute this inhibitory effect to another process, Al^{3+} adsorption and stabilization of the dissolving surface. However, this mechanism does not appear to be important for LMD. If the release of Al^{3+} to solution and subsequent adsorption is the mechanism inhibiting the rates of dissolution in the inorganic controls, the initial rates of both LMD and analogous PMD control experiments should initially be the same, with a subsequent decline in the PMD rates with time. In batch experiments, however, ligands enhance both the initial rates of dissolution compared to inorganic controls, as well as the subsequent steady state rates (Fig. 5; Welch and Ullman, 2000). Al^{3+} adsorption is likely to have a lesser impact on LMD because the formation of Al-L complexes will limit the activity of free Al^{3+} in solution and available for adsorption (see below). Further experiments will be needed to determine the relative importance of these two co-varying mechanisms on LMD.

5.3. Effect of Feldspar Composition

Several experimental studies have shown that the PMD rate of feldspar is dependent on mineral composition (Brady and Walther, 1989; Casey, et al., 1991; Welch and Ullman, 1992; 1993; Oxburgh et al., 1994). The pH dependence of dissolution rates also increases with increasing Al content from albite to anorthite along the plagioclase series. Similarly, since organic ligands are interacting primarily with Al sites, ligands tend to have a far more pronounced effect on more Ca, Al-rich plagioclases (e.g., Manley and Evans, 1986; Welch and Ullman, 1992, 1993, 1996; Oxburgh et al., 1994; Blake and Walter 1999). Welch and Ullman (1996) found that LMD rates have a similar first- or higher-order dependence on the An content of plagioclase feldspar at pH = 3 and 6. They argue that this dependence further supports the idea that the reactivity of surface Al sites toward oxalate controls the overall feldspar dissolution rate.

There are no direct determinations of the effect of feldspar heterogeneities (intergrowths and zonation) on LMD rates. Heterogeneous dissolution and the formation of periodic dissolution textures have been observed for PMD by transmission electron microscopy (Inskeep et al., 1991). Similar evidence that differences in mineral chemistry can lead to the development of periodic textures on feldspar surfaces has been observed by scanning electron microscopy for LMD (Welch, 1991). At elevated chemical affinities, the overall dissolution of heterogeneous feldspars should be predictable on the basis of the relative amounts of the chemically distinct intergrowths and their An (or Al) content. Ordering of the Al and Si tetrahedra could also be a factor in LMD and PMD, as Al-sites linked to other Al sites in a more disordered feldspar are more susceptible to attack (Oxburgh et al., 1994). The effects of chemical zonation and structural ordering on overall feldspar dissolution rates have not been experimentally demonstrated.

5.4. Effect of Temperature

The rates of feldspar dissolution increase with increasing temperature. The temperature dependence for feldspar dissolution can be described using the Arrhenius relationship:

$$k = A_o e^{\frac{-E_a}{RT}} \tag{13}$$

where k is the rate constant, A_o is the preexponential factor, and E_a is the activation energy of the dissolution reaction. This form of the temperature dependence of a reaction is most appropriately applied to a single elementary step in a reaction mechanism. However, for a large number of complex geochemical reactions that proceed through a number of elementary steps, but where a single process controls the overall dissolution rate (such as feldspar dissolution), it is a useful approximation for describing the overall dependence of reaction rate on temperature (Lasaga, 1998). Under these conditions, E_a is more appropriately called the apparent activation energy. The magnitude of E_a, however, will depend on the kinetic expression used. In most cases where the temperature dependence of feldspar dissolution rate has been measured, all parameters except temperature are held constant and therefore, the temperature dependence of the overall reaction rate will be the same as that for the measured rate constant:

$$R_{total} = A_0^* e^{\frac{-E_a}{RT}} \tag{14}$$

The apparent activation energy for PMD has been estimated previously in a number of laboratory and field studies (Lasaga et al., 1994, and references therein; Velbel, 1990,1993; Brady, 1991; Casey and Sposito, 1992; Hellmann, 1994; Brady and Carroll, 1994; Dorn and Brady, 1995). Knauss and Wolery (1986) calculated an E_a of 54 kJ/mole for albite dissolved in a flow-through reactor at neutral pH over a temperature range of 25 to 70 °C. Bevan and Savage (1989) found E_a values for both PMD and LMD at pH=1 to be similar, 59 and 50 kJ/mole, respectively, and for LMD at pH = 3, E_a was 59 kJ/mole. However, in the absence of oxalate, the apparent activation energy of PMD at pH =3 was significantly lower (15 kJ/mole). Hellmann (1994) determined an apparent activation energy for HD of 69 kJ/mole for albite at neutral pH and temperatures between 100 and 300 °C, although E_a for albite dissolution was higher by approximately 17 to 20 kJ/mole under basic and acidic conditions, respectively, due to the enthalpy of adsorption of hydroxyls and protons to the feldspar surface (Casey and Sposito, 1992). Brady and Carroll (1994) calculated an E_a of 48 kJ/mole for PMD from labradorite feldspar in a flow-through reactor in slightly acidic solutions (pH 4) from 21- 60 °C. A tabulation of additional estimates of E_a for feldspar dissolution is given by Blum and Stillings (1995). In general, the E_a for silicate minerals dissolved at low temperatures over a range of experimental conditions varies from 40-80 kJ/mole (Lasaga et al., 1994; Blum and Stillings, 1995).

Welch and Ullman (2000) found that the apparent activation energy for bytownite (An_{75}) dissolution at near-neutral pH at low temperature and in the absence of complexing ligands was 43.0±1.2 kJ/mole. The addition of acetate had no impact on the apparent activation energy, consistent with the predictions of Kubicki et al. (1997). However, the addition of oxalate or gluconate, both ions that form strong and stable complexes with Al^{3+} in solution, and presumably at the mineral surface as well, reduced E_a by about 14 kJ/mole. Whether this decrease in E_a to 28.7± 1.2 kJ/mole reflects a change in reaction mechanism or the catalysis of PMD and/or HD by the organic acids cannot be determined on the basis of these experiments alone. This decrease in E_a for feldspars dissolving in organic acid solutions is consistent with a field study on biological weathering; Brady et al. (1999) estimated E_a = 55.6 kJ/mole for lichen-mediated weathering, compared to 96.7 kJ/mole for bare rock surfaces.

Several researchers have noted an increase in Al-organic stability constants in solution with increasing temperature (Fein and Hestrin, 1994; Tait et al., 1992; Thyne et al., 1992). In apparent contrast, Welch and Ullman (2000) report that organic ligands have a relatively greater impact on feldspar dissolution rates at lower temperatures. Although the factors controlling this apparent discrepancy are not certain, it may reflect the change in organic adsorption characteristics with temperature. Since surface adsorption is an exothermic reaction, adsorption coefficients should decrease with increasing temperature (Thyne et al., 1992; Tait et al., 1992; Piatt et al., 1996). If ligand adsorption to a mineral surface is an important contribution to the LMD mechanism, the decrease in adsorption with temperature may decrease the relative importance of LMD dissolution compared to other dissolution mechanisms at higher temperature. Alternatively, the effect of stronger ligand complexing in solution may be sufficient to reduce free ligand activity and therefore the amount of ligand available for adsorption (Thyne et al., 1992; Fein and Hestrin, 1994). Further experiments will be needed to confirm these observations and determine the mechanism of this relative change in rates.

5.5. Dissolution Stoichiometry

Feldspar dissolution in inorganic solution is often non-stoichiometric; the charge-balancing cations, Na, K, and Ca are generally released in preference to the structural cations, Al and Si. In addition, there may be preferential release of either Si or Al from the feldspar framework depending on experimental conditions. Non-stoichiometric dissolution results from the formation of leached layers during the initial stages of dissolution (Chou and Wollast, 1984; Pačes, 1973), precipitation of secondary mineral phases (Holdren and Adams, 1982), and preferential dissolution of exsolved phases (Gardner, 1983; Shotyk and Nesbitt, 1992; Inskeep et al., 1991).

At near-neutral conditions, low temperature, and in the absence of ligands, Si is usually preferentially released to solution, leaving a thin residual surface enriched in Al (10 – 20 nm, Chou and Wollast, 1995); often Al is not detectable in solution (Fig. 3; Welch and Ullman, 1993, 1996; Blake and Walter, 1996, 1999). As pH decreases, Al release increases with respect to Si release (Welch and Ullman, 1993). In mildly acidic solutions (pH ≈ 4) the release of the feldspar mineral framework ions, Si and Al, is approximately stoichiometric (Fig. 3). In more acidic solutions (pH < 4) Al is preferentially leached, leaving a surface enriched in Si. At higher temperatures (≥ 100 °C) Al is preferentially released with respect to Si under acidic, near-neutral, and extremely basic conditions; under mildly basic conditions, dissolution is congruent or slightly favors Si (Hellmann, 1995). The feldspar dissolution stoichiometry is often much closer to that of the primary mineral phase for longer experiments and at both low and high temperature (Stillings and Brantley, 1995; Hellmann, 1995). In many dissolution experiments, however, non-stoichiometric release persists up to thousands of hours. The non-stoichiometric dissolution of feldspars is consistent with the observations of Casey and Bunker (1990) who suggest that the dissolution of mixed oxide minerals is a multipathway process, with each component of the mixed oxide being leached from the mineral surface by different mechanisms, at different rates, and having differing controlling factors. Because feldspar dissolution is often non-stoichiometric, care must be taken in the extrapolation of the release rates of individual ions into solution to the overall mineral dissolution rate. Nonetheless, a comparison of release rates under different experimental conditions can still yield useful information concerning the factors controlling mineral weathering in nature.

Dissolution is similarly non-stoichiometric in many solutions containing organic ligands. However, structural ions are either released stoichiometrically from neutral to acidic pH or, in some cases, Al is released preferentially to Si (Fig. 3). Al release to solution is often greater in solutions containing complexing ligands than in inorganic solutions, particularly at near-neutral pH (Huang and Kiang, 1972; Manley and Evans, 1986; Mast and Drever, 1987; Huang and Longo, 1992; Welch and Ullman, 1993; Blake and Walter, 1996, 1999), consistent with the implication that the LMD mechanism is acting primarily at Al sites at the feldspar surface.

6. MODELS OF LIGAND-MEDIATED DISSOLUTION AND RATE PREDICTION

One of the principal reasons for laboratory studies of feldspar dissolution is to estimate the rates of chemical weathering in field situations. In order to accomplish this, the above results, integrating the effects of PMD, HD, HMD, and LMD, need to be included in models of dissolution and rates need to be scaled to the field. A number of different models and approaches have been used to integrate experimental results, and these are discussed below. The question of scaling laboratory measurements of dissolution rates to field conditions and settings has not been resolved and remains an important research direction. Some of the work in this area is briefly reviewed.

6.1. Empirical Models

The simplest empirical models of feldspar dissolution that incorporate LMD are based on the linear relationship between LMD rates and ligand concentration that have been observed by a number of investigators (Fig. 4; Welch and Ullman, 1992; Vandevivere et al., 1994):

$$R_{LMD} = k_{LMD}\,[L] \qquad (15)$$

This model appears to be most useful at low ligand concentrations. By analogy to the PMD and HMD reactions, as well as to experimental evidence that indicates that the dependence on ligand concentration may not be linear over the whole range of concentrations (Knauss and Copenhaver, 1995; Stillings et al., 1996; Blake and Walter, 1999), a non-linear form is often preferred:

$$R_{LMD} = k_{LMD}\,[L]^{\,p} \qquad (16)$$

This form is appropriate for a single ligand and can be expanded to include a number of similar terms for each reactive ligand:

$$\Sigma\, R_{LMD} = \Sigma k^{i}_{LMD}\,[L_i]^{\,p_i} \qquad (17)$$

The equation above can be added to that for R_{inorg} (Eqn. 4) to give:

$$\begin{aligned} R_{total} &= R_{inorg} + \Sigma\, R_{LMD} = R_{PMD} + R_{HD} + R_{HMD} + \Sigma\, R_{LMD} \\ &= k_{PMD}\,[H^+]^{\,n} + k_{HD} + k_{HMD}\,[OH^-]^{\,m} + \Sigma k^{i}_{LMD}\,[L_i]^{\,p_i} \end{aligned} \qquad (18)$$

For single ligands, this empirical equation does a good job of reproducing several experimental observations. It reproduces the observation that the pH dependence of overall dissolution reaction with ligands is the same as PMD alone (Welch and Ullman, 1996). This model also reproduces the observation that the major impact of ligands on dissolution is largely focused in the near-neutral pH range. The observed effect of feldspar composition on dissolution rates can be largely incorporated into the k_{PMD} and k_{LMD} terms; the impact of composition on k_{HD} and k_{HMD} has not been investigated. Empirical equations for k_{PMD} and n are given in Welch and Ullman (1996). This empirical model (Eqn. 18), however, fails to fully reproduce the observation that the dependence of oxalate-mediated dissolution is also a function of pH (Welch and Ullman, 1992).

A number of other empirical forms have been suggested for incorporating the pH-dependence of R_{total}. These have involved incorporating an additional, pH-dependent term into Eqn. 16 (Welch and Ullman, 1992), or by adding a ligand-dependent term into Eqn. 8 (Welch and Ullman, 1996).

However, none of the empirical models take into account the apparent inhibition of dissolution that has been observed for mixtures of ligands and the possibility that competition between ligands, protons, and other ions for adsorption sites at the dissolving feldspar surface controls overall dissolution rates (see above and Table 2). A more mechanistic model of the dissolution process is needed to incorporate and model these observations.

6.2. Adsorption Models

The success of the empirical LMD model (Eqn. 16) may result from the fact that the form of the empirical equation may be derived from a consideration of ligand adsorption to the dissolving feldspar surface (Stillings et al., 1998). Stillings et al. (1998) proposed two adsorption models for LMD reactions based on the assumption (similar to that previously proposed for PMD and HMD; Blum and Lasaga, 1988, 1991; Brady and Walter, 1989; Brantley and Stillings, 1996) that the rate of LMD is controlled by the adsorption of ligands to sites on the mineral surface. These models specifically incorporate the speciation of oxalate in solution and the adsorption coefficients for both oxalate anions, HOx^- and Ox^{2-}, into a dissolution model that treats adsorption as the rate-controlling step in the feldspar dissolution process. The dissolution rate can then be expressed as a sum of linear kinetic terms dependent on the concentrations of adsorbed protons ($[H^+]_{ads}$), HOx^- ($[HOx^-]_{ads}$) and Ox^{2-} ($[Ox^{2-}]_{ads}$) on the andesine surface:

$$R_{total} = R_{PMD} + R_{LMD} = k_{H^+}\,[H^+]_{ads} + k_{HOx}\,[HOx^-]_{ads} + k_{Ox}\,[Ox^{2-}]_{ads} \qquad (19)$$

This adsorption model does an excellent job of reproducing the observed oxalate-mediated dissolution of andesine (An_{47}) up to 8 mM total oxalate from pH = 3 to 4.

The two adsorption models proposed by Stillings et al. (1998) differ in the type of adsorption isotherm used. In the proposed Langmuir model, the total number of adsorption sites is held constant and the number of protonated sites is determined by the difference between the total number of adsorption sites (determined by curve fitting) and the number occupied by HOx^- and Ox^{2-}. These model parameters were determined by the best fit to adsorption experiments. The total number of adsorption sites determined by the best fit procedure is greater than the expected number of Al-sites on the andesine surface, but falls within the range of adsorption sites determined by proton adsorption. This observation would suggest that more than one ligand or proton may adsorb at a single Al site at the mineral surface, or that there are other sites, perhaps bridging Al - O - Si sites near the mineral surface, that adsorb protons and ligands. This latter observation is consistent with the predictions of *ab initio* calculations (Xiao and Lasaga, 1994; Kubicki et al., 1997).

This competitive Langmuir adsorption model has been used to describe other aspects of mineral dissolution and has the potential for incorporating the observations of both enhancement and inhibition of dissolution by ligands. Stillings and Brantley (1995) and Brantley and Stillings (1996) successfully used this model to describe the effect of other cations in solution on PMD. Although best suited for examining cation-cation competition, it also appears to adequately describe anionic ligand-cation interactions (Stillings et al., 1998; Ullman and Welch, 1998).

The alternative formulation given by Stillings et al (1998) uses the Freundlich isotherm to estimate $[HOx^-]_{ads}$ and $[Ox^{2-}]_{ads}$. This model does not limit the total amount of ligand adsorption to the mineral surface to any fixed number and reproduces the experimental trends for andesine dissolution somewhat better than the Langmuir model at high total oxalate concentrations. An intriguing aspect of these models is that the rate constants, k_{H+}, k_{HOx}, and k_{Ox} given in Eqn. 19, are essentially identical and appear to be largely independent of the adsorption model used. Since these models are based only on the number of adsorption sites (Langmuir model) and the amount of adsorption (both models), and not the specific characteristics of the sites, these rate coefficients developed for andesine may be useful to model dissolution over the whole feldspar series. These results may also make it possible to generalize this model to a wide range of different adsorbed ions on the feldspar surface to determine the overall impact of a complex mixture of ligands on feldspar dissolution, given some knowledge of the adsorption characteristics of the individual ligands. However, further tests of this model will be needed to verify that the various adsorption coefficients (protons and ligands) and reaction coefficients for each type of adsorption site are constant over the range of possible conditions.

6.3. *Transition-State Models*

In recent years, there has been a great deal of interest in applying Transition-State Theory (TST) to the dissolution of minerals (Nagy et al., 1991; Nagy and Lasaga, 1992; Burch et al., 1993; Gérard et al., 1998; Lasaga, 1998). The application of this theory implies that the rate of dissolution has a first-order dependence on the concentration of a transition-state (or activated) complex that is formed at the mineral surface in the course of the dissolution reaction. A term incorporating the chemical affinity of the overall reaction may be included to account for the possibility of the reverse (precipitation) reaction at low chemical affinities. Although there has been some discussion concerning the nature of the transition-state complex that is responsible for feldspar dissolution (and there may be more than one dissolution pathway and therefore more than one such complex), the approach has been very useful for describing the rates of PMD for a number of aluminum silicates.

Both Oclkers and Schott (1998) and Stillings et al. (1998) have applied this theory to ligand-mediated dissolution of feldspar. The transition-state model appears to reproduce some of the features of LMD behavior. However, the model as proposed does not include any direct dependence on ligand concentration or activity in solution that the experimental evidence requires. In contrast to a great deal of experimental evidence (see section 6.4 below), Oelkers and Schott (1998) conclude on the basis of this observation that the only impact of ligands on feldspar dissolution is on the reduction of aluminum activity in solution due to complex formation, and that there is no direct impact of ligand adsorption on the reaction with the mineral surface.

Oelkers and Schott (1998) suggest by analogy to quartz dissolution, the transition-state complex is an aluminum-depleted feldspar surface created by the elementary reaction involving the exchange of protons for Al at the surface:

$$f\,(>Al) + 3f\,H^+ \rightarrow [P^*] + f\,Al^{3+} \quad (20)$$

where $[P^*]$ represents the activated complex for silica release, $(>Al)$ represents a surface aluminum site, and f is a stoichiometric coefficient. Let us, as an alternative, propose another stoichiometry that leads to the formation of this same activated complex, but that includes an additional monovalent ligand, L^-, as a reactant:

$$f(>Al) + 3f\,H^+ + f\,L^- \rightarrow [P^*] + f\,AlL^{2+} \quad (21)$$

In Eqn. 21, both ligands and protons participate in the elementary reaction involving the removal of Al from the surface. Under these conditions, the rate of reaction at high chemical affinities and when

$$K^\circ\left(\frac{a^3_{H^+}\,a_{L^-}}{a_{AlL^{2+}}}\right)^f \gg 1$$

(where K^o is the equilibrium constant for the elementary reaction represented by Eqn. 21), instead of being dependent on

$$\left(\frac{a^3_{H^+}}{a_{Al^{3+}}}\right)^f$$

(Oelkers and Schott, 1998), is dependent on

$$\left(\frac{a^3_{H^+}\,a_{L^-}}{a_{AlL^{2+}}}\right)^f$$

(where a_i refers to the activity of ion i in solution):

$$R^* = k^*\left(\frac{a^3_{H^+}\,a_{L^-}}{a_{AlL^{2+}}}\right)^f \quad (22)$$

However, the activity of the aluminum-ligand complex in solution, $a_{AlL^{2+}}$, is dependent on its formation constant, $K_{AlL^{2+}}$, and the activities of Al^{3+} and L^- in solution:

$$K_{AlL^{2+}} = \frac{a_{AlL^{2+}}}{a_{Al^{3+}}\,a_{L^-}} \quad (23)$$

Substituting this form into Eqn. 22 gives exactly the same functional dependence on

$$\left(\frac{a^3_{H^+}}{a_{Al^{3+}}}\right)^f$$

that Oelkers and Schott (1998) used to conclude that ligands do not participate in the formation of the transition state complex for feldspar dissolution:

$$R^* = k^{**}\left(\frac{a^3_{H^+}}{a_{Al^{3+}}}\right)^f \quad (24)$$

where k^{**} incorporates the formation constant $K_{AlL^{2+}}$. The fact that the same form of the transition-state model applies equally well to PMD and LMD is therefore not an indication that ligands do not adsorb and react with the mineral surface as part of the LMD process. Also, some of the variation in f observed by Oelkers and Schott (1998) and Stillings et al. (1998) may be related to slightly different, but reasonable, alternative stoichiometries for Eqn. 21. The identical forms of the transition state expressions, with and without ligands, provide a very useful tool for incorporating LMD into reactive-transport models without having to determine adsorption constants for each ligand species. Only the nature and stability of aqueous complexes will be relevant to this model, although some independent knowledge of reaction stoichiometry f will be required.

For the moment, all that can be concluded is that a number of models can predict equally well LMD dissolution rates as measured in experiments. Some of these models have mechanistic components whereas others are purely empirical. No best model, however, can be identified because of the covariance of the many chemical and thermodynamic parameters that occur in these models (Stillings et al., 1998). The choice of model, therefore, will depend on the conditions to be modeled and the availability of data to apply the chosen model.

6.4. Adsorption vs. Solution Control of Feldspar Dissolution

It is apparent from the previous discussions of the dissolution models that there is considerable debate whether complexation of Al in solution or at the mineral surface is the primary factor influencing the rates of LMD. Oelkers, Schott, and coworkers (Oelkers et al., 1994) demonstrate that Al^{3+} added to solution inhibits feldspar dissolution by adsorbing to mineral surfaces and inhibiting the formation of the activated complex for silica release (P^* in Eqns. 20 and/or 21). From this they conclude that the major impact of organic ligands on dissolution is to decrease the activity of Al^{3+} in solution by the formation Al-organic complexes (Oelkers and Schott, 1998). In contrast to this, Burch et al. (1993) and Taylor et al. (2000) have demonstrated that for feldspar dissolution in the absence of complexing ligands and at high chemical affinities, the rate is independent of solution composition, including the Al^{3+} released from the feldspar. Therefore, formation of Al-organic complexes in solution should have negligible effects on LMD. The model of Chen and Brantley (1997) suggests that the discrepancy between the low-temperature experiments of Burch et al. (1993) and Taylor et al. (2000) and the high-temperature experiments of Oelkers et al. (1994) may be due to an increase in Al^{3+} adsorption with increasing temperature.

Some degree of organic complexation at Al-sites on mineral surfaces does occur (Furrer and Stumm, 1986; Fein and Brady, 1995; Kubicki et al., 1999), but it is not known how important this effect is for mineral dissolution kinetics. There are a number of lines of experimental evidence showing that Al complexing at the mineral surface and removal of Al from the feldspar plays an important role in LMD. It is possible that Al release from the feldspar surface, leading to the formation of the activated complex for silica release (P^* in Eqns. 20 and 21), involves the initial formation of an Al-ligand activated complex for Al release. This is consistent with Casey and Bunker's (1990) view of feldspar dissolution as a multipathway leaching process.

There is clear experimental evidence that LMD dissolution rates are directly proportional to total ligand concentrations under a wide range of conditions (see Fig. 4 and Section 5.1), including at neutral pH where free Al^{3+} is undetectable in the presence or absence of complexing ligands. Also, there is clear evidence of the compositional dependence of feldspar dissolution with the more Ca, Al-rich members of the plagioclase series dissolving significantly more rapidly than the Na, Si-rich members, under comparable conditions. If dissolution was inhibited by Al^{3+} in solution, the more aluminous feldspars, which release more Al to solution, should have lower overall dissolution rates in contrast to the experimental observations (Casey et al., 1991; Welch and Ullman, 1996).

The adsorption of organics is greater at lower temperatures (see Section 5.4 above) and this leads to a larger relative enhancement of dissolution rates due to LMD at lower temperatures compared to inorganic controls. This also leads to a lower apparent activation energy for LMD compared to PMD alone (Welch and Ullman, 2000). At higher temperatures, where more total aluminum is available in solution, the relative impact of ligands on feldspar dissolution is less.

In batch reactors where Al^{3+} should and does increase with time, the initial rates of dissolution at high chemical affinities are higher in ligand-containing solutions than in non-complexing media, and remain higher throughout most experiments (Fig. 5; Welch and Ullman, 1996, 2000; Blake and Walter, 1996, 1999). Under conditions of initially high chemical affinity, the initial rates should be identical only if Al inhibition is important.

It is likely that both solution and surface complexation affect reaction rates (Stillings et al., 1998). Surface complexation should be more important at: (1) lower temperatures where ligand adsorption is stronger; (2) under conditions of high chemical affinity; (3) at neutral pH. Solution complexation is likely to be more important at: (1) low chemical affinity; (2) at higher temperatures (as the stability of aqueous Al-ligand complexes in solution increase, ligand adsorption to surfaces decreases and Al^{3+} adsorption increases); (3) for ligands (such as polymers) that cannot adsorb to the reacting surface due to steric hindrances.

6.5. *Laboratory vs. Field Measurements*

It is well documented that apparent dissolution rates in field settings are significantly lower than rates measured in the laboratory (Velbel, 1990, 1993; Brantley, 1992). There are several reasons for this discrepancy. Laboratory experiments are often performed in acidic solutions (pH < 3), where mineral dissolution rates are elevated due to PMD. However, even mineral dissolution experiments in neutral solutions yield dissolution rates greater than rates estimated from field studies (Velbel, 1986, 1990, 1993; Swoboda-Colberg and Drever, 1992). Most laboratory experiments are run at temperatures that are warmer than conditions in soils and groundwater and would therefore result in higher mineral dissolution rates (Velbel, 1990). Furthermore, minerals used commonly in laboratory experiments have fresh, clean, and very reactive surfaces which should weather faster than natural surfaces. Natural mineral surfaces have already had the more reactive components of their surfaces removed, have developed leached layers, and have been covered by secondary precipitates or other adsorbed compounds that inhibit dissolution. Laboratory experiments with naturally weathered mineral samples can better reproduce natural dissolution rates (Suarez and Wood, 1996).

This latter issue plays an important role in the potential use of laboratory estimates of LMD rates in field settings. Because feldspars used in the laboratory are so reactive, very high concentrations of ligands must be used to demonstrate a significant LMD effect over the PMD and HD effects. It is not likely that millimolar concentrations of any single dissolution-enhancing ligand are going to be found in bulk groundwater or soil solutions. However, there may be micro-environments isolated from the bulk solution, where higher ligand concentrations can be created and sustained. Recent field evidence of microbial-mediated dissolution suggests that such concentrations do occur and that LMD may be an important process in a range of field settings (Bennett and Siegel, 1987; Bennett et al., 1988, 1996; Thorseth et al., 1992; Staudigel et al., 1995; Barker and Banfield, 1996, 1998). For the moment, the LMD process can only be inferred from field observations and rates cannot be directly measured. However, experimental and observational advances may make it possible to do both simultaneously, or almost simultaneously, in the future.

Feldspar dissolution rates in field settings are also likely to be depressed by the elevated saturation state of solutions relative to the dissolving phase. In contrast to most laboratory experiments, natural waters are often closer to equilibrium saturation with respect to the dissolving phases (low chemical affinities), resulting in a net decrease in dissolution rates according to Eqn. 11. Mineral dissolution rates are also a function of the physical rate of exchange of the solution in contact with the mineral surfaces, which may impact the chemical affinity near the reacting surface (Berner, 1978). Even in flowing, open hydrological systems, there may be isolated microenvironments where solutions exchange only slowly with the bulk flow and therefore flow has minimal impact. Laboratory experiments do not adequately reproduce the limited interaction of the mineral surface with the bulk solution in soils and aquifers due to transport limitations. Although laboratory experiments are not very effective for predicting the absolute rates of mineral weathering in nature, they are nonetheless useful for determining the differences in mineral weathering rates under different experimental conditions and for determining which factors may play an important role in mineral weathering in nature. Furthermore, laboratory dissolution experiments are useful for determining mineral dissolution reaction mechanisms.

7. SUMMARY

Organic ligands participate in the overall feldspar dissolution process. They may enhance or inhibit dissolution and the release of reaction products to solution by forming complexes in solution or at mineral surfaces. The indirect effects of the formation of metal-ligand complexes in solution may inhibit dissolution by reducing the activity of free reactive ligands and therefore the adsorption of ligands to the sites (primarily Al-sites) on the mineral surface. However, when solutions are nearly saturated with respect to the dissolving mineral, the formation of ligand-metal complexes with reaction products may enhance the overall dissolution rate by increasing the chemical affinity of the overall reaction.

Experiments show that a wide range of organic ligands enhance feldspar dissolution rates. Polyfunctional aliphatic anions which form stable and soluble bidentate complexes with the ionic components of the mineral, principally Al, appear to enhance dissolution rates to a larger extent than mono-functional anions, although it is not clear to what extent this is due to the formation of stable complexes in solution or at the mineral surface. Polyfunctional aromatic ligands may also enhance dissolution rates, but not to the same extent as polyfunctional aliphatic ions. Organic polymers may either enhance or inhibit dissolution depending on their affinity for adsorption to mineral surfaces. The inhibitory or enhancement effect of ligands and polymers is a strong function of pH in acid solution, and this dependence needs to be incorporated into models of the dissolution process.

Ligands affect not only the rate of the dissolution reaction but may affect the mechanism and stoichiometry of the reaction. In solutions containing complexing organic ligands, the concentrations of Al and the Al/Si ratio released to solution are commonly greater than in the absence of complex-forming ligands. This effect is most pronounced at near-neutral pH where the solubilities of Al-containing secondary products are at their minimum. Since the major effect of organic ligands is interaction with Al at the mineral surface and in solution, the effects of ligand-mediated dissolution are most pronounced for the more aluminous members of the feldspar group.

The inhibition of ligand-mediated dissolution by mixtures of organic ligands and by other cations in solution suggests that there is competition between these solutes and organic ligands for adsorption on surface sites at mineral surfaces. There is also competition between cations (some of which may be products of the dissolution reaction) for the formation of cation-ligand complexes in solution. Models of the dissolution process are only beginning to take these competitive reactions into account.

Empirical, adsorption, and transition-state models are all useful for predicting rates of dissolution in the laboratory. However, none of these models is applicable to all experimental conditions. These models do not appear to be useful, at present, for estimating the rates of feldspar dissolution in field settings. The scaling of laboratory models to field conditions is an important future direction for feldspar weathering studies.

Acknowledgements—Both authors have received support from the Subsurface Science Program of the U.S. Department of Energy for work on ligand-mediated and microbially-mediated mineral dissolution. S.A.W. has also been supported since 1997 by grants to Prof. J.F. Banfield from the Office of Basic Energy Sciences of the U.S. Department of Energy and by the Geology and Paleontology Program of the National Science Foundation. We thank our funding agencies and our colleagues for the many fruitful discussions that are summarized in this review. Critical comments and suggestions on an earlier draft of this manuscript by J.B. Fein, E. Valsami-Jones, D.L. Suarez, and R. Hellmann are gratefully acknowledged. Their suggestions greatly improved the final version of this review.

Editorial handling: R. Hellmann

REFERENCES

Aagaard P. and Helgeson H. C. (1982) Thermodynamic and kinetic constraints on reaction rates among minerals and

aqueous solutions. I. Theoretical considerations. *Amer. J. of Sci.* **282**, 237-285.
Amrhein C. and Suarez D. L. (1988) The use of a surface complexation model to describe the kinetics of ligand-promoted dissolution of anorthite. *Geochim. Cosmochim. Acta* **52**, 2785-2793.
Barker W. W. and Banfield J. F. (1996) Biologically versus inorganically-mediated weathering reactions: Relationships between minerals and extracellular microbial polymers in lithobiotic communities. *Chem. Geol.* **132**, 55-69
Barker W. W. and Banfield J. F. (1998) Zones of chemical and physical interaction at interfaces between microbial communities and minerals. *Geomicrobiol. J.* **15**, 223-244.
Barker W. W., Welch S. A., and Banfield J. F. (1997) Geomicrobiology of silicate mineral weathering. In *Geomicrobiology: Interactions between Microbes and Minerals* (eds. J. F. Banfield and K. H. Nealson). Reviews in Mineralogy **35**, Mineralogical Society of America, Washington, D.C. pp. 391-428.
Barker W. W., Welch S. A., Chu S., and Banfield J. F. (1998) Experimental observations of the effects of bacteria on aluminosilicate weathering. *Amer. Mineral.* **83**, 1551-1563.
Barman A. K., Varadachari C., and Ghosh K. (1992) Weathering of silicate minerals by organic acids. I. Nature of cation solubilisation. *Geoderma* **53**, 45-63.
Baziramakenga R., Simard R. R., and Leroux G. D. (1995) Determination of organic acids in soil extracts by ion chromatography. *Soil Biol. Biochem.* **27**, 349-356.
Bennett P. C. (1991) Quartz dissolution in organic-rich aqueous systems. *Geochim. Cosmochim. Acta* **55**, 1781-1797.
Bennett P. C. and Casey W. H. (1994) Chemistry and mechanisms of low temperature dissolution of silicates by organic acids. In *Organic Acids in Geological Processes* (eds. E. D. Pittman and M. D. Lewan), Springer Verlag, Berlin. pp. 162-200.
Bennett P.C. and Siegel D. I. (1987) Increased solubility of quartz in water due to complexation by dissolved organic compounds. *Nature* **326**, 684-687.
Bennett P. C., Melcer M. E., Siegel D. I., and Hassett J. P. (1988) The dissolution of quartz in dilute aqueous solutions of organic acids at 25 °C. *Geochim. Cosmochim. Acta* **52**, 1521-1530.
Bennett P. C., Siegel D. I., Hill B. M. and Glaser P. H. (1991) Fate of silicate minerals in a peat bog. *Geology* **19**, 328-331.
Bennett P. C., Hiebert F. K., and Choi W. J. (1996) Microbial colonization and weathering of silicates in a petroleum-contaminated groundwater. *Chem. Geol.* **132**, 45-53.
Bennett P. C., Choi W. J., and Rogers, J. R. (1998) Microbial destruction of feldspars. In *Proceedings of the V. M. Goldschmidt Conference 1998 (Toulouse, France). Mineral. Mag.* **62A**, 149-150.
Berner R. A. (1978) Rate control of mineral dissolution under Earth surface conditions. *Amer. J. Sci.* **278**, 1235-1252.
Berner R. A. (1991) Model for atmospheric CO_2 over Phanerozoic time. *Amer. J. Sci.* **291,** 339-376.
Bevan J. and Savage D. (1989) The effect of organic acids on the dissolution of K-feldspar under conditions relevant to burial diagenesis. *Mineral. Mag.* **53**, 415-425.
Blake R. E. and Walter L. M. (1996) Effects of organic acids on the dissolution of orthoclase at 80 °C and pH 6. *Chem. Geol.* **132**, 91-102.
Blake R. E. and Walter L. M. (1999) Kinetics of feldspar and quartz dissolution at 70-80 °C and near neutral pH: Effects of organic acids and NaCl. *Geochim. Cosmochim. Acta* **63**, 2043-2059.
Blum A. and Lasaga A. C. (1988) Role of surface speciation in the low temperature dissolution of minerals. *Nature* **331**, 431-433
Blum A. and Lasaga A. C. (1991) The role of surface speciation in the dissolution of albite. *Geochim. Cosmochim. Acta* **55**, 2193-2201.
Blum A., and Stillings L. L. (1995) Feldspar dissolution kinetics. In *Chemical Weathering Rates of Silicate Minerals* (eds. A. F. White and S. L. Brantley). Reviews in Mineralogy **31**, Mineralogical Society of America, Washington, D.C. pp. 291-351.
Brady P. V. (1991) The effect of silicate weathering on global temperature and atmospheric CO_2. *J. Geophys. Res.* **96**, 18101-18106.
Brady P. V. and Carroll S. A. (1994) Direct effects of CO_2 and temperature on silicate weathering: Possible implications for climate control. *Geochim. Cosmochim. Acta* **58**,1853-1856.
Brady P. V., and Walther J. V. (1989) Controls on silicate dissolution rates in neutral and basic pH solutions at 25 °C. *Geochim. Cosmochim. Acta* **53**, 2823-2830
Brady P. V., and Walther J. V. (1992) Surface chemistry and silicate dissolution at elevated temperatures. *Amer. J. Sci.* **292**, 639-658.
Brady, N. C., and Weil R. R. (1996) *The Nature and Properties of Soils (Eleventh Edition).* Prentice Hall, Englewood Cliffs.
Brady P. V., Dorn R. I., Brazel A. J., Clark J., Moore R. B., and Glidewell T. (1999) Direct measurement of the combined effects of lichen, rainfall, and temperature on silicate weathering. *Geochim. Cosmochim. Acta* **63**, 3293-3300.
Brantley S. L. (1992) Kinetics of dissolution and precipitation- Experimental and field results. In *Water-Rock Interaction Volume 1 Low Temperature Environments. International Symposium on Water-Rock Interaction* (eds. Y. K. Kharaka and A.S. Maest). A.A. Balkema, Rotterdam. pp. 3-6.
Brantley S. L. and Stillings L. L. (1996) Feldspar dissolution at 25 °C and low pH. *Amer. J. Sci.* **296**, 101-127.
Brezonik P. L. (1994) *Chemical Kinetics and Process Dynamics in Aquatic Systems.* Lewis Publishers, Boca Raton.
Brock T. D. and Madigan M. T. (1988) *Biology of Microorganisms.* Prentice Hall, Englewood Cliffs.
Burch T. E., Nagy K. L., and Lasaga A. C. (1993). Free energy dependence of albite dissolution kinetics at 80 °C and pH 8.8. *Chem. Geol.* **105**, 137-162.
Carroll S. A. and Walther J. V. (1990). Kaolinite dissolution at 25°, 60°, and 80 °C. *Amer. J. Sci.* **290**, 797-810
Casey W. H. and Bunker B. (1990) Leaching of mineral and glass surfaces during dissolution. In *Mineral-Water Interface Geochemistry* (eds. M. F. Hochella, Jr. and A. F. White). Reviews in Mineralogy **23**, Mineralogical Society of America, Washington, D.C. pp. 397-426.
Casey W. H. and Sposito G. (1992) On the temperature dependence of mineral dissolution rates. *Geochim. Cosmochim. Acta* **56**, 3825-3830.
Casey W. H., Westrich H. R., and Holdren G. R. (1991) Dissolution rates of plagioclase at pH=2 and 3. *Amer. Mineral.* **76**, 211-217
Chapelle F. H., Zelibor J. L. Jr., Grimes D. Y., and Knobel L. L. (1987) Bacteria in deep coastal plain sediments of Maryland: A possible source of CO_2 to groundwater. *Water Resources Res.* **23**, 1625-1632.
Chen Y. and Brantley S. L. (1997) Temperature and pH dependence of albite dissolution rates at acid pH. *Chem. Geol.* **135**, 275-290.

Chou L. and Wollast R. (1984) Study of the weathering of albite at room temperature and pressure with a fluidized bed reactor. *Geochim. Cosmochim. Acta* **48**, 2205-2217.

Chou L. and Wollast R. (1985) Steady-state kinetics and dissolution mechanisms of albite. *Amer. J. Sci.* **285**, 963-993.

Christensen B. E., and Characklis W. G. (1990) Physical and chemical properties in biofilms. In *Biofilms* (eds. Characklis W.G., and Marshall K. C.). John Wiley & Sons, New York. pp. 93-130.

Cordt T. and Kussmaul H. (1992) Characterization of some organic acids in the subsurface of the Sandhausen Ecosystem. In *Progress in Hydrogeochemistry; Organics; Carbonate Systems; Silicate Systems; Microbiology; Models* (eds. G. Matthess, F.H. Frimmel, H.D. Schulz, and E. Uschowski). Springer Verlag, Berlin. pp. 93-100.

Cozzarelli I. M., Baedecker M. J., Eganhouse R. P., and Goerlitz D. F. (1994) The geochemical evolution of low-molecular-weight organic acids derived from the degradation of petroleum contaminants in groundwater. *Geochim. Cosmochim. Acta* **58**, 863-877.

Daughney C. J., Fein J. B., and Yee N. (1998) A comparison of the thermodynamics of metal adsorption onto two common bacteria. *Chem. Geol.* **144**, 161-176.

Dorn R. I. and Brady P. V., (1995) Rock-based measurement of temperature dependent plagioclase weathering. *Geochim. Cosmochim. Acta* **59**, 2847-2852.

Drever J. I. and Vance G. F. (1994) Role of soil organic acids in mineral weathering processes. In *Organic Acids in Geological Processes* (eds. E. D. Pittman and M. D. Lewan). Springer Verlag. pp. 138-161.

Ehrlich H. L. (1990) *Geomicrobiology, Second Edition.* Marcel Dekker, New York.

Fein, J. B. and Brady P.V. (1995) Mineral surface controls on the diagenetic transport of oxalate and aluminum. *Chemical Geology* **121**, 11-18.

Fein J. B. and Hestrin J. E. (1994) Experimental studies of oxalate complexation at 80 °C — gibbsite, amorphous silica, and quartz solubilities in oxalate-bearing fluids *Geochim. Cosmochim. Acta* **58**, 4817-4829.

Fein J. B., Daughney C. J., and Yee N. (1997) A chemical equilibrium model for metal adsorption onto bacterial surfaces. *Geochim. Cosmochim. Acta* **61**, 3319-3328.

Ferris F. G., Fyfe W. S., and Beveridge T. J. (1988) Metallic ion binding by *Bacillus subtilis*: Implications for the fossilization of microorganisms. *Geology* **16**, 149-152.

Fisher J. B. (1987) Distribution and occurrence of aliphatic acid anions in deep subsurface waters. *Geochim. Cosmochim. Acta* **51**, 2459-2468.

Fox T. R. and Comerford N. B. (1990) Low-molecular-weight organic acids in selected forest soils of the southeastern USA. *Soil Sci. Soc. Am. J.* **54**, 1139-1144.

Franklin S. P., Hajash A., Dewers T. A., and Tieh T. T. (1994) The role of carboxylic-acids in albite and quartz dissolution-an experimental study under diagenetic conditions. *Geochim. Cosmochim. Acta* **58**, 4259-4279.

Furrer G. and Stumm W. (1986) The coordination chemisty of weathering: I. Dissolution kinetics of δ-Al_2O_3 and BeO. *Geochim. Cosmochim. Acta* **53**, 1469-1470.

Gardner L. R. (1983) Mechanisms and kinetics of incongruent feldspar dissolution. *Geology* **11**, 418-421.

Garrels R. M. and Mackenzie F. T. (1971) *Evolution of Sedimentary Rocks.* Norton, New York.

Gérard F., Fritz B., Clément, A. and Crovisier, J. L. (1998) General implications of aluminium speciation-dependent kinetic dissolution rate law in water-rock modeling. *Chem. Geol.* **151**, 247-258.

Gestsdóttir K. and Manning D. A. C. (1992) An experimental study of the dissolution of albite in the presence of organic acids. In *Water-Rock Interaction Volume 1 Low Temperature Environments. International Symposium on Water-Rock Interaction* (eds. Y. K. Kharaka and A.S. Maest). A.A. Balkema. pp. 315-318.

Gottschalk G. (1986) *Bacterial Metabolism.* Springer-Verlag, New York.

Graustein W. C., Cromack K. Jr, and Sollins P. (1977) Calcium oxalate: Occurrence in soils and effect on nutrient and geochemical cycles. *Science* **198**,1252-1254.

Grierson P. F. (1992) Organic acids in the rhizosphere of *Banksia integrifolia* L.f. *Plant and Soil* **144**, 259-265.

Hajash A., Franklin S. P., and Reed C. L. (1992) Experimental feldspar dissolution in acetic and oxalic acids at 100 °C, 345 bars. In *Water-Rock Interaction Volume 1 Low Temperature Environments, Proceedings of the 7th International Symposium on Water-Rock Interaction.* (eds. Y. K. Kharaka and A. S. Maest). A.A. Balkema. pp. 325-328.

Hajash A. Jr., Carpenter T. D., and Dewers T. A. (1998) Dissolution and time-dependent compaction of albite sand; experiments at 100 °C and 160 °C, in pH-buffered organic acids and distilled water. *Tectonophysics* **295**, 93-115.

Hellmann R. (1994) The albite-water system: Part I. The kinetics of dissolution as a function of pH at 100, 200, and 300 °C. *Geochim. Cosmochim. Acta* **58**, 595-611.

Hellmann R. (1995) The albite-water system: Part II. The time-evolution of the stoichiometry of dissolution as a function of pH at 100, 200, and 300 °C. *Geochim. Cosmochim. Acta* **59**, 1669-1697.

Hendry M. J., Lawrence J. R., Kirkland R., and Zanyk B. N. (1992) Microbial production of carbon dioxide in the unsaturated zone of a meso scale model. In *Water-Rock Interaction Volume 1 Low Temperature Environments, Proceedings of the 7th International Symposium on Water-Rock Interaction* (eds. Y. K. Kharaka and A. S. Maest). A.A. Balkema. pp. 287-290.

Heyes A. and Moore T. R. (1992) The influence of dissolved organic carbon and anaerobic conditions on mineral weathering. *Soil Science* **154**, 226-236.

Hiebert F. K. and Bennett P. C. (1992) Microbial control of silicate weathering in organic-rich ground-water. *Science* **258**, 278-281.

Holdren G. R. Jr. and Adams J. E. (1982) Parabolic dissolution kinetics of silicate minerals: An artifact of nonequilibrium precipitation processes? *Geology* **10**, 186-190.

Huang W. H. and Keller W. D. (1970) Dissolution of rock-forming silicate minerals in organic acids: Simulated first stage weathering of fresh mineral surfaces. *Amer. Mineral.* **55**, 2076-2094.

Huang W. H. and Kiang W. C. (1972) Laboratory dissolution of plagioclase in water and organic acids at room temperature. *Amer. Mineral.* **57**, 1849-1859.

Huang W. L. and Longo J. M. (1992) The effect of organics on feldspar dissolution and the development of secondary porosity. *Chem. Geol.* **98**, 271-292.

Huang P. M. and Schnitzer M. (1986) Interactions of Soil Microbes with Natural Organics and Microbes. *Soil Science Society of America, Special Publication No. 17.*

Inskeep W. P., Nater E. A., Bloom P. R., Vandervoort D. R., and Erich M. S. (1991) Characterization of laboratory weathered labradorite surfaces using X-ray photelectron spectroscopy and transmission electron microscopy. *Geochim. Cosmochim. Acta* **55**, 787-800.

Johnston C. G. and Vestal J. R. (1993) Biogeochemistry of oxalate in the Antarctic cryptoendolithic lichen-dominated community. *Microb Ecol.* **25,** 305-319.

Jones D. L. (1998) Organic acids in the rhizosphere, a critical review. *Plant and Soil* **205**, 25-44.

Jorgensen S. S. (1976) Dissolution kinetics of silicate minerals in aqueous catechol solutions. *J. Soil Sci.* **27**, 183-195.

Kawamura K. and Nissenbaum A. (1992) High abundance of low molecular weight organic acids in hypersaline spring water associated with a salt diapir. *Org. Geochem.* **18**, 469-476.

Keller C. K. and Wood B. D. (1993) Possibility of chemical weathering before the advent of vascular land plants. *Nature* **364**, 223-225.

Kharaka Y. K., Law L. L., Carothers W. W., and Goerlitz D. F. (1986) Role of organic species dissolved in formation waters from sedimentary basins in mineral diagenesis. In *Roles of Organic Matter in Sediment Diagenesis* (ed. D.L. Gautier). Society of Economic Paleontologists and Mineralogists **38**, pp. 111-122.

Knauss K. G. and Copenhaver S. A. (1995) The effect of malonate on the dissolution kinetics of albite, quartz, and microcline as a function of pH at 70 °C. *Appl. Geochem.* **10**, 17-33.

Knauss, K. G. and Wolery, T. J. (1986) Dependence of albite dissolution kinetics on pH and time at 20 °C and 70 °C. *Geochim. Cosmochim. Acta* **50**, 2481-2498.

Krzyszowska A. J., Blaylock M. J., Vance G. F., and David M. B. (1996) Ion-chromatographic analysis of low molecular weight organic acids in spodosol forest floor solutions. *Soil Sci. Soc. Am. J.* **60**, 1565-1571.

Kubicki, J. D., Blake G. A., and Apitz S. E. (1997) Molecular orbital calculations for modeling acetate-aluminosilicate adsorption and dissolution reactions. *Geochim. Cosmochim. Acta* **61**, 1031-1046.

Kubicki J. D., Schroeter L. M., Itoh M. J., Nguyen B. N. and Apitz S. E. (1999) Attenuated total reflectance Fourier-transform spectroscopy of carboxylic acids adsorbed onto mineral surfaces. *Geochim. Cosmochim. Acta* **63**, 2709-2725.

Lahuerte F., Leyval C., and Berthelin J. (1990) Root exudates of maize, pine, and beech seedlings influenced by mycorrhizal and bacterial innoculation. *Symbiosis* **9**, 111-116.

Lapeyrie F., Picatto C., Gerard J., and Dexheimer J. (1990) T.E.M. study of intracellular calcium oxalate accumulation by ectomycorrhizal fungi in pure culture or in association with eucalyptus seedlings. *Symbiosis* **9**, 163-166.

Lasaga, A. C. (1981). Transition State Theory. In *Kinetics of Geochemical Processes* (eds. A.C. Lasaga and R.J. Kirkpatrick) Reviews in Mineralogy **8**, Mineralogical Society of America, Washington, D.C. pp. 135-168.

Lasaga A. C. (1995) Fundamental approaches in describing mineral dissolution and precipitation rates. In *Chemical Weathering Rates of Silicate Minerals* (eds. A.F. White and S. L. Brantley). Reviews in Mineralogy **31**, Mineralogical Society of America, Washington, D.C. pp. 23-86.

Lasaga A. C. (1998) *Kinetic Theory in the Earth Sciences.* Princeton University Press, Princeton.

Lasaga A. C. and Blum A. E. (1986) Surface chemistry, etch pits, and mineral-water reactions. *Geochim. Cosmochim. Acta* **50**, 2363-2379.

Lasaga A. C., Soler J. M., Ganor J., Burch T., and Nagy K. L. (1994) Chemical weathering rate laws and global geochemical cycles. *Geochim. Cosmochim. Acta* **58**, 2361-2386.

Leyval C., Laheurte G., Belgy G., and Berthelin J. (1990) Weathering of micas in the rhizospheres of maize, pine and beech seedlings influenced by mycorrhizal and bacterial innoculation. *Symbiosis* **9**, 105-109.

Linton J. D. (1990) The relationship between metabolic production and the growth efficiency of the producing organism. *FEMS Microbiol. Rev.* **75**, 1-18

Lundström U. S. and Öhman L. O. (1990) Dissolution of feldspars in the presence of natural organic solutes. *J. Soil. Sci.* **41**, 359-369.

Maeda H. and Kawai A. (1988) Microflora and bacterial organic acid production in the bottom sediment of Lake Biwa. *Nip. Suis. Gak.* **54**, 1375-1383.

MacGowan D. B. and Surdam R. C. (1988) Difunctional carboxylic acid anions in oilfield waters. *Org. Geochem.* **12**, 245-259.

MacGowan D.B. and Surdam R. C. (1990) Carboxylic acid anions in formation waters, San Joaquin Basin and Louisiana Gulf Coast, U.S.A.- Implications for clastic diagenesis. *Appl. Geochim.* **5**, 687-701.

Manley E. P. and Evans L. J. (1986) Dissolution of feldspars by low-molecular-weight aliphatic and aromatic acids. *Soil Sci.* **141**, 106-112.

Manning D. A. C., Gestsdóttir K., Rae E. I. C. (1992) Feldspar dissolution in the presence of organic acid anions under diagenetic conditions; an experimental study. In *Advances in Organic Geochemistry 1991; part II, Advances and applications in energy and the natural environment; Proceedings of the 15th International Meeting on Organic Geochemistry* (eds. C. B. Eckardt, J.R. Maxwell, S.R. Larter, and D.A.C. Manning) *Organic Geochemistry* **19**, 483-492.

Martell A. E. and Motekaitis R. J. (1989) Coordination chemistry and speciation of Al(III) in solution. In *Environmental Chemistry and Toxicology of Aluminum* (ed. T.E. Lewis). Lewis, Chelsea. pp. 3-18.

Martens C. S. (1990) Generation of short chain organic acid anions in hydrothermally altered sediments of the Guaymas Basin, Gulf of California. *Appl. Geochem.* **5**, 71-76.

Mast M. A. and Drever J. I. (1987) The effect of oxalate on the dissolution rates of oligoclase and tremolite. *Geochim. Cosmochim. Acta* **51**, 2559-2568.

McMahon P. B. and Chapelle F. H. (1991) Microbial organic acid production in aquitard sediments and its role in aquifer geochemistry. *Nature* **349**, 233-235

Morel F. M. M., and Hering J. G. (1993) *Principles and Applications of Aquatic Chemistry.* Wiley, New York.

Nagy K. L., Blum A. E., and Lasaga A. C. (1991) Dissolution and precipitation kinetics of kaolinite at 80 °C and pH 3: The dependence on solution saturation state. *Amer. J. Sci.* **291**, 649-686.

Nagy K. L. and Lasaga A. C., (1992) Dissolution and precipitation kinetics of gibbsite at 80 °C and pH 3: The dependence on solution saturation state. *Geochim. Cosmochim. Acta* **56**, 3903-3111.

Neijssel O. M. and Tempest D. W. (1975) The regulation of carbohydrate metabolism in *Klebsiella aerogennes* NCTC 418 organisms, growing in chemostat culture. *Arch. Microbiol.* **106**, 251-258.

Ochs M., Brunner I., Stumm W., and Bozena C. (1993) Effects of root exudates and humic substances on weathering kinetics. *Water, Air, Soil Poll.* **68**, 213-229.

Oelkers E. H. and Schott J. (1998) Does organic acid adsorption affect alkali-feldspar dissolution rates? *Chem. Geol.* **151**, 235-245.

Oelkers E. H., Schott J., and Devidal J. L., (1994) The effect of aluminum, pH, and chemical affinity on the rates of alumino-silicate dissolution reactions. *Geochim. Cosmochim. Acta* **58**, 661-669.

Onstott T. C., Phelps T. J., Colwell F. S., Ringelburg T. J., White D.C., Boone D. R., McKinley J. P., Stevens T. O.,

Long P. E., Balkwill D. L., Griffen W. T., and Kieft T. (1998) Observations pertaining to the origin and ecology of microorganisms recovered from the deep subsurface of Taylorsville basin, Virginia. *Geomicrobiol. J.* **15**, 353-385.

Oxburgh R., Drever J. I., and Sun Y. T. (1994) Mechanism of plagioclase dissolution in acid solution at 25 °C. *Geochim. Cosmochim. Acta* **58**, 661-669.

Pačes T. (1973) Steady-state kinetics and equilibrium between ground water and granitic rock. *Geochim. Cosmochim. Acta* **37**, 2641-2663.

Palmer R. J. Jr., Siebert J., and Hirsch P. (1991) Biomass and organic acids in sandstone of a weathered building: Production by bacterial and fungal isolates. *Microb. Ecol.* **21**, 253-266.

Panov Y. N., Khokhlov V. V., and Latikainen V. I. (1983) Trace-element mobility when organic acids act on feldspars. *Geochem. Internat.* **20**, 115-121.

Paris F., Bonnaud P., Ranger J., Robert M., and Lapeyrie F. (1995) Weathering of ammonium or calcium-saturated 2:1 phyllosilicates by ectomycorrhizal fungi in vitro. *Soil Biol. Biochem.* **27**, 1237-1244.

Paris F., Botton B., and Lapeyrie F. (1996) *In vitro* weathering of phlogopite by ectomycorrhizal fungi. *Plant and Soil* **179**, 141-150.

Parks G. A. (1967). Aqueous surface chemistry of oxides and complex oxide minerals. In *Equilibrium concepts in natural water systems*, Advances in Chemistry Series **67**. American Chemical Society, Washington, D.C. pp 121-160.

Peltzer E. T. and Bada J. L. (1981) Low molecular weight α-hydroxy carboxylic and di-carboxylic acids in reducing marine sediments. *Geochim. Cosmochim. Acta* **45**, 1847-1854.

Petersen A., Matthess G., and Schenk D. (1992) Experiments on the influence of organic ligands upon kinetics of feldspar weathering. *Progress in Hydrogeochemistry; Organics; Carbonate Systems; Silicate Systems; Microbiology; Models* (eds. G. Matthess, F.H. Frimmel, H.D. Schulz, and E. Uschowski). Springer-Verlag, Berlin. pp 86-92.

Petrovič R. (1973) The effect of coherency stress on the mechanism of the reaction albite + $K^+ \leftrightarrow$ K-feldspar + Na^+ and on the mechanical state of the resulting feldspar. *Contrib. Mineral. Petrol.* **52**, 17-28.

Piatt J. J., Backhus D. A., Capel P. D., and Eisenreich S. J. (1996) Temperature-dependent sorption of naphthalene, phenanthrene, and pyrene to low organic carbon aquifer sediments. *Environ. Sci. Tech.* **30**, 751-760.

Reed C. L. and Hajash A. (1992) Dissolution of grantic sand by pH-buffered carboxylic acids—A flow-through experimental study at 100 °C and 345 bars. *AAPG Bull.* **76**, 1402-1416.

Rogers J. R., Bennett P. C., and Choi W. J. (1998a) Feldspars as a source of nutrients for microorganisms. *Am. Mineral.* **83**, 1532-1540.

Rogers J. R., Bennett P. C., and Ullman W. J. (1998b) Biochemical release of a limiting nutrient from feldspars. In *Proceedings of the V. M. Goldschmidt Conference 1998. Mineral. Mag.* **62A**, 1283-1284.

Sansone F. J. (1986) Depth distribution of short-chain organic acids turnover in Cape Lookout Bight sediments. *Geochim. Cosmochim. Acta* **50**, 99-105.

Sansone F. J. and Martens C. S. (1982) Volatile fatty acid cycling in organic-rich marine sediments. *Geochim. Cosmochim. Acta* **46**, 1575-1589.

Schenk D., Petersen A., and Matthess G. (1989) Acceleration and retardation of silicate weathering due to organic substances. In *Water-Rock Interaction Proceedings of the 6th International Symposium on Water-Rock Interaction* (ed. D.L. Miles). A.A. Balkema. pp. 605-607.

Schweda P. (1989) Kinetics of alkali feldspar dissolution at low temperature. In *Water-Rock Interaction Proceedings of the 6th International Symposium on Water-Rock Interaction* (ed. D.L. Miles). pp. 609-612. Balkema

Shen Y., Strom L., Jan-Ake J., and Tyler G. (1996) Low-molecular organic acids in the rhizosphere soil of solution of beech forest (Fagus sylvatica L.) cambisols determined by ion chromatography using supported liquid membrane enrichment technique. *Soil Biol. Biochem* **28**, 1163-1169.

Shock E. L. (1988) Organic acid metastability in sedimentary basins. *Geology* **16**, 886-890.

Shotyk W. and Nesbitt H. W. (1992) Incongruent and congruent dissolution of plagioclase feldspar: effect of feldspar composition and ligand complexation. *Geoderma* **55**, 55-78.

Smith J. V. and Brown W. L. (1988) *Feldspar Minerals 1. Crystal Structures, Physical, Chemical and Microtextural Properties.* Springer-Verlag, Berlin.

Smith R. M. and Martell A. E. (1988) *NIST Critically Selected Stability Constants of Metal Complexes Database (Version 5.0).* NIST Standard Reference Database 46. U.S. Department of Commerce.

Song S. K. and Huang P. M. (1988) Dynamics of potassium release from potassium-bearing minerals as influenced by oxalic and citric acids. *Soil Sci. Soc. Am. J.* **52**, 383-390.

Staudigel H., Chastain R. A., Yayanos A., and Bourcier W. (1995) Biologically mediated dissolution of glass. *Chem. Geol.* **126**, 147-154.

Stillings L. L. and Brantley S. L. (1995) Feldspar dissolution at 25 °C and pH 3: Reaction stoichiometry and the effect of cations. *Geochim. Cosmochim. Acta* **59,** 1483-1496.

Stillings L. L., Brantley S. L., and Machesky M. L. (1992) Multisite proton adsorption at the feldspar-water interface. In *Water-Rock Interaction Volume 1 Low Temperature Environments, Proceedings of the 7th International Symposium on Water-Rock Interaction* (eds Y. K. Kharaka and A.S. Maest). A.A. Balkema. pp. 67-72.

Stillings L. L., Drever J. I., Brantley S. L., Sun Y., and Oxburgh R. (1996) Rates of feldspar dissolution at pH 3-7 with 0-8 mM oxalic acid. *Chem. Geol.* **132**, 79-89.

Stillings L. L., Drever J. I., and Poulson S. R., (1998) Oxalate adsorption at a plagioclase (An_{47}) surface and models for ligand-promoted dissolution. *Environ. Sci. Technol.* **32**, 2856-2864.

Stoessell R. K. and Pittman E. D. (1990) Secondary porosity revisited; the chemistry of feldspar dissolution by carboxylic acids and anions. *AAPG Bulletin.* **74**, 1795-1805.

Stone A. T. (1997) Reactions of extracellular organic ligands with dissolved metal ions and mineral surfaces. In *Geomicrobiology: Interactions between Microbes and Minerals* (eds. J. F. Banfield and K. H. Nealson). Reviews in Mineralogy **35**, Mineralogical Society of America, Washington, D.C. pp. 309-344.

Suarez D. L and Wood J. D. (1996) Short- and long-term weathering rates of a feldspar fraction isolated from an arid soil zone. *Chem. Geol.* **132**, 143-150.

Surdam R. C. and MacGowan D. B. (1987) Oilfield waters and sandstone diagenesis. *Appl. Geochem.* **2**, 613-619.

Surdam R. C., Boese S. W., and Crossey L. J. (1984) The chemistry of secondary porosity. In *Clastic Diagenesis, American Association of Petroleum Geologist Memoirs* **37** (eds. D. A. McDonald and R. C. Surdam). AAPG, Tulsa. pp. 127-149.

Sverjensky D. A. and Sahai N. (1996) Theoretical prediction of singe-site surface-protonation equilibrium constants for oxides and silicates in water. *Geochim. Cosmochim. Acta* **60,** 3773-3797.

Swoboda-Colberg N. G. and Drever J. I. (1992) Mineral dissolution rates: a comparison of laboratory and field rates. In *Water-Rock Interaction Volume 1 Low Temperature Environments, Proceedings of the 7th International Symposium on Water-Rock Interaction* (eds. Y. K. Kharaka and A.S. Maest). A.A.Balkema. pp. 115-118.

Tait C. D., Janecky D. R., Clark D. L., and Bennett, P. C. (1992) Oxalate complexation with aluminum(III) and iron(III) at moderately elevated temperatures. In *Water-Rock Interaction Volume 1 Low Temperature Environments, Proceedings of the 7th International Symposium on Water-Rock Interaction* (eds. Y. K. Kharaka and A.S. Maest). A.A. Balkema. pp. 349-352.

Talbot R. W., Andreae M. O., Berresheim H., Jacob D. J., and Beecher K. M. (1990) Sources and sinks of formic, acetic, and pyruvic acids over central Amazonia 2. Wet season. *J. Geophys. Res.* **95**, 16799-16811.

Tan K. H. (1980) The release of silicon, aluminum, and potassium during decomposition of soil minerals. *Soil Science* **129**, 4-11.

Taylor A. S., Blum J. D., and Lasaga A. C. (2000) The dependence of plagioclase dissolution and Sr release rates on solution saturation state. *Geochim. Cosmochim. Acta* **64**, 2389-2400.

Tempest D. W. and Neijssel O. M. (1992) Physiological and energetic aspects of bacterial metabolite overproduction. *FEMS Microbiol. Lett.* **100**, 169-176.

Thorseth I. H., Furnes H., and Heldal M. (1992) The importance of microbial activity in the alteration of natural basaltic glass. *Geochim. Cosmochim. Acta* **56**, 845-850.

Thurman E. M. (1985) *Organic Geochemistry of Natural Waters.* Nijhoff/Junk, Dordrecht.

Thyne G. D., Harrison W. J., and Alloway M. D. (1992) Experimental study of the stability of Al-oxalate complexation at 100 °C and calculations of the effects of complexation on clastic diagenesis. In *Water-Rock Interaction Volume 1 Low Temperature Environments, Proceedings of the 7th International Symposium on Water-Rock Interaction* (eds. Y. K. Kharaka and A.S. Maest). A.A. Balkema. pp. 353 -357.

Tole M. P., Lasaga A. C., Pantano C., and White W. B. (1986) The kinetics of dissolution of nepheline ($NaAlSiO_4$). *Geochim. Cosmochim. Acta* **50**, 379-392.

Ullman W. J. and Welch S. W. (1998) Competitive adsorption at surface aluminum sites: An approach to determining feldspar dissolution rates in mixed proton/ligand systems. In *Proceedings of the V. M. Goldschmidt Conference 1998. Mineral. Mag.* **62A**, 1559-1560.

Vandevivere P., Welch S. A., Ullman W. J. , and Kirchman D. L. (1994) Enhanced dissolution of silicate minerals by bacteria at near-neutral pH. *Microb. Ecol.* **27**, 241-251

Varadachari C., Barman A. K., and Ghosh K. (1994) Weathering of silicate minerals by organic acids; II. Nature of residual products. *Geoderma* **61**, 251-268.

Velbel M. A. (1986) Influence of surface area, surface characteristics, and solution composition on feldspar weathering rates. In *Geochemical Processes at Mineral Surfaces* (eds. J. A. Davis, and K. F. Hayes). American Chemical Society Symposium Series **323**, American Chemical Society, Washington, D.C. pp. 615-634.

Velbel M. A. (1990) Influence of temperature and mineral surface characteristics on feldspar weathering rates in natural and artificial systems: A first approximation. *Water Resources Res.* **26**, 3049-3053.

Velbel M. A. (1993) Temperature dependence of silicate weathering in nature: How strong a negative feedback on long-term accumulation of atmospheric CO_2 and global greenhouse warming? *Geology* **21**, 1059-1062.

Welch S. A. (1991) The effect of organic acids on mineral dissolution rates and stoichiometry. M.S. thesis, Univ. of Delaware.

Welch S. A. and Ullman W. J. (1992) Dissolution of feldspars in oxalic acid solutions. In *Water-Rock Interaction Volume 1 Low Temperature Environments, Proceedings of the 7th International Symposium on Water-Rock Interaction* (eds. Y. K. Kharaka and A.S. Maest). A.A. Balkema. pp. 127 -130.

Welch S. A. and Ullman W. J. (1993) The effect of organic acids on plagioclase dissolution rates and stoichiometry. *Geochim. Cosmochim. Acta* **57**, 2725-2736.

Welch S. A. and Ullman W. J. (1996) Feldspar dissolution in acidic and organic solutions: Compositional and pH dependence of dissolution rate. *Geochim. Cosmochim. Acta* **60**, 2939-2948.

Welch S. A. and Ullman W. J. (1999) The effect of microbial glucose metabolism on bytownite feldspar dissolution rates between 5 and 35 °C. *Geochim. Cosmochim. Acta* **63**, 3247-3259.

Welch S. A. and Ullman W. J. (2000) The temperature dependence of feldspar dissolution in organic acids. *Chem. Geol.* **167**, 337-354.

Welch S. A. and Vandevivere P. (1994) Effect of microbial and other naturally occurring polymers on mineral dissolution. *Geomicrobiol. J.* **12**, 227-238

Welch S. A., Barker W. W., and Banfield J. L. (1999) Microbial extracellular polysaccharides and plagioclase dissolution. *Geochim. Cosmochim. Acta* **63**, 1405-1419.

Wogelius R. A. and Walther J. V. (1991) Olivine dissolution at 25 °C: Effects of pH, CO_2, and organic acids. *Geochim. Cosmochim. Acta* **55**, 943-954.

Wogelius R. A. and Walther J. V. (1992) Olivine dissolution kinetics at near-surface conditions. *Chemical Geology* **97**, 101-112.

Wollast R. and Chou L. (1988) Rate control of weathering of silicate minerals at room temperature and pressure. In *Physical and Chemical Weathering in Geochemical Cycles* (eds. A. Lerman and M. Meybeck). Kluwer, Dordrecht. pp. 11-32.

Xiao Y. and Lasaga A. C. (1994) *Ab initio* quantum mechanical studies of the kinetics and mechanisms of silicate dissolution: H^+ (H_3O^+) catalysis. *Geochim. Cosmochim. Acta* **58**, 5379-5400.

Xiao Y. and Lasaga A. C. (1996) *Ab initio* quantum mechanical studies of the kinetics and mechanisms of quartz dissolution: OH^- catalysis. *Geochim. Cosmochim. Acta* **60**, 2283-2295.

Water-Rock Interactions, Ore Deposits, and Environmental Geochemistry: A Tribute to David A. Crerar
© The Geochemical Society, Special Publication No. 7, 2002
Editors: Roland Hellmann and Scott A. Wood

Chemical weathering of the Panola Granite: Solute and regolith elemental fluxes and the weathering rate of biotite

ART F. WHITE[1]*, ALEX E. BLUM[2], MARJORIE S. SCHULZ[1], THOMAS G. HUNTINGTON[3], NORMAN E. PETERS[3] AND DAVID A. STONESTROM[1]

[1] U. S. Geological Survey, Menlo Park, CA 94025 USA
[2] U. S. Geological Survey, Boulder, CO 80303 USA
[3] U. S. Geological Survey, Atlanta, GA 30360 USA

* corresponding author: afwhite@usgs.gov

Abstract—Present-day elemental and mineral weathering rates based on solute fluxes are compared quantitatively to past long-term rates determined from solid-state elemental fractionation in a saprolitic granite regolith at Panola, Georgia, USA. Saturated fluid flow across a low-permeability kaolin duripan controls the rate of steady-state unsaturated flow in the underlying saprolite. Water and Cl mass balances and experimental conductivities produce a minimum fluid flux density of 8×10^{-2} m yr^{-1} and a fluid residence time of 12 years. Solute Si flux, based on pore water concentrations and infiltration rates, is 27 mmoles yr^{-1}, compared to a long-term flux rate of 17 mmoles yr^{-1}, based on regolith Si loss and reported ^{36}Cl dating of the regolith surface. Similarities in short- and long-term fluxes imply that parameters influencing silicate weathering, including precipitation, temperature, and vegetative cover, while not necessarily constant, have not significantly impacted Si leaching rates during the last several hundred thousand years.

Linear decreases in solid-state Mg with decreasing regolith depth permit the calculation of the long-term biotite weathering rate under isovolumetric steady-state weathering conditions. A rate constant of 3×10^{-17} moles m^{-2} s^{-1} is up to 5 orders of magnitude slower than that reported for experimental dissolution of biotite, implying very different reaction kinetics during natural weathering. Short-term biotite weathering fails to produce expected increases in solute Mg and K concentrations with increasing depth and fluid residence times in the regolith. This discrepancy indicates that ion exchange disequilibrium and open-system biologic uptake in an aggrading forest ecosystem are of sufficient magnitudes to overwhelm solute fluxes resulting from biotite weathering.

1. INTRODUCTION

The quantification of natural silicate weathering rates has important implications in a diverse range of geochemical issues, including the neutralization of acid precipitation, the release rates of macronutrients such as K and Ca in forested catchments, and the linkage between weathering rates, CO_2 draw-down, and long-term climate change. In addressing these issues, extensive efforts have been directed at determining silicate weathering mechanisms and rates based on detailed experiments that involve generally well-characterized silicate minerals. In many cases, the success of these efforts has surpassed our present ability to verify their applicability to natural weathering processes.

Most quantitative estimates of natural silicate weathering rates are derived by simple input-output models, which fall conceptually into two categories (White, 1995). Short-term reaction rates are calculated from solute fluxes derived most commonly from rates of regolith infiltration (Swodoba-Colberg and Drever, 1993; Murphy et al., 1998), watershed solute balances (Velbel, 1985; Pačes, 1986), and from solute increases observed along ground water flow paths (White and Claassen, 1978; Kenoyer and Bowser, 1992). The determination of such rates commonly requires an extensive monitoring network characterizing fluid flow and solute compositions. In such studies, it is commonly difficult to document the *in situ* parameters that control silicate reaction rates, including reactive mineral surface areas and hydrologic flow paths and heterogeneities. Such monitoring efforts are commonly short-term, one to several years, which may or may not approach the actual fluid residence times in these systems. Extrapolation of such short-term rates to longer times, therefore, is susceptible to effects of changing climate, vegetation, and anthropogenic influences.

Long-term reaction rates are calculated from differences between elemental or mineral compositions in the initial protolith and the weathered regolith. While extensive data in the soil literature document such compositional differences, relatively few studies have tied these changes to quantitative mineral weathering rates (Sverdrup and Warfvinge, 1995; White et al., 1996; White et al., 2001). These rates are averages for the duration of protolith weathering, commonly tens of thousands to millions of years. Studies confined solely to the solid-state provide minimal infor-

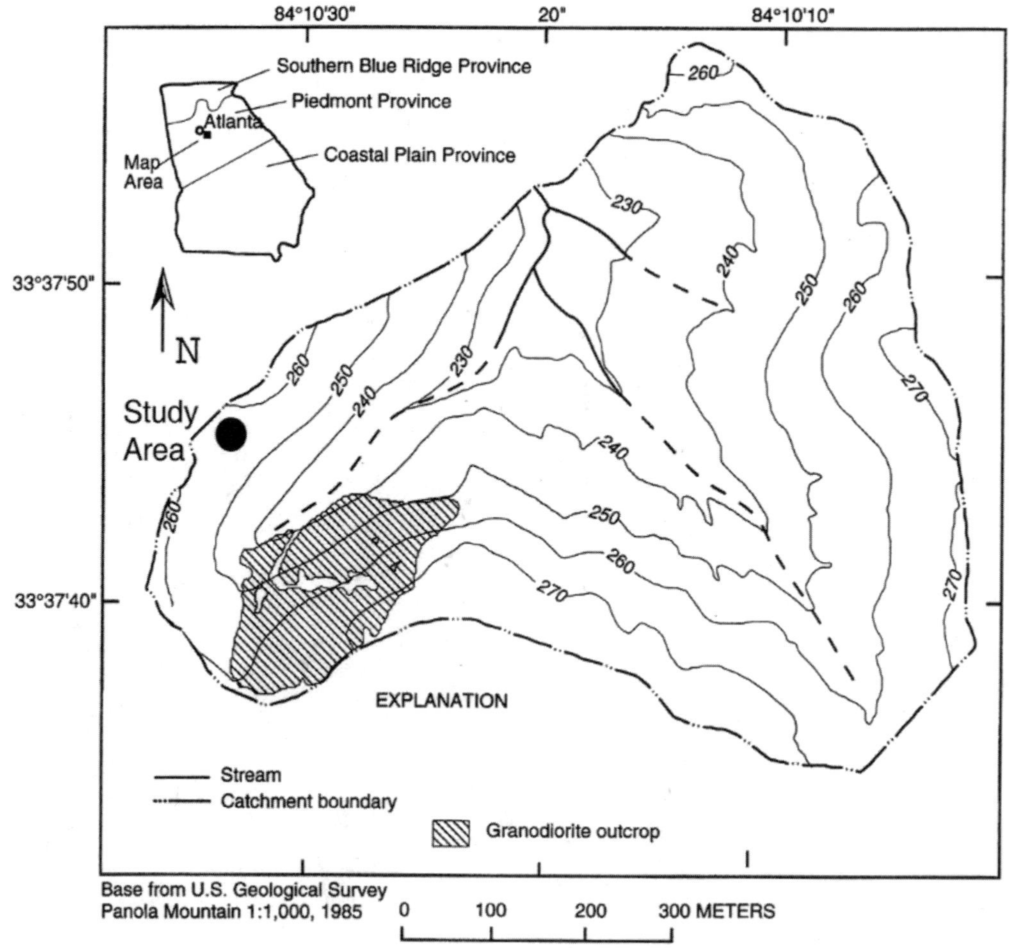

Fig. 1. Location map showing the ridge top sampling site in the Panola watershed. Dashed area represents outcrop areas of the Panola Granodiorite.

mation of long-term hydrochemical processes that have led to observed weathering rates. Weathering rates can change over long times due to the decreasing availability of reactive surfaces and increasing secondary clay and oxide phases which alter permeability, wetted mineral surface areas, and mineral reaction affinities.

Additional insights into natural weathering are derived from direct comparisons of present short-term and past long-term rates for the same regolith. Significant differences imply errors in the assumptions used in calculating weathering rates and/or fundamental changes in the actual rates over time (White, 1995). Current silicate weathering rates, based on watershed solute fluxes in post-glacial watersheds (< 1 x 10^4 yrs) in North America and northern Europe, are estimated be up to a factor of 3 times faster than long-term past weathering rates based on cation and mineral depletions in watershed soils (April et al., 1986; Kirkwood and Nesbitt, 1991; Land et al., 1999). These differences are attributed to the impacts of acidic precipitation on contemporary weathering rates.

In comparing long-term weathering rates for much older saprolites (>10^5 years) with watershed solute discharge in the Piedmont of the eastern USA, Cleaves (1993) concluded that present interglacial weathering rates are 2-5 times faster than for past periglacial periods. This difference is attributed to past periods of lower precipitation, colder temperatures, and lower soil gas CO_2. In contrast, Pavich et al. (1995) concluded that long-term rates of saprolite formation in the Virginia Piedmont are comparable with current-weathering rates based on stream solute fluxes. Rates of saprolite formation are also found similar to current weathering rates based on watershed discharge in the Luquillo Mountains of Puerto Rico, an area which has been subject to only minor long-term climate fluctuations (White et al., 1998).

The present study details methods for deriving both short-term weathering rates based on present-day soil water fluxes and past long-term weathering rates calculated from changes in regolith composition. These approaches are applied to a saprolitic weathering environment in the Georgia Piedmont of the eastern USA. The importance of heterogeneous permeabilities in estimating solute fluxes is investigated. Dissolution rates of biotite, derived from solid-state elemental distributions with depth in the regolith, are compared to experimental rates and rates determined for other natural systems. Finally, the impacts of biologic fluxes and cation exchange reactions on pore water chemistry, used to calculate contemporary weathering rates, are assessed.

2. FIELD SETTING AND METHODOLOGY

Panola Mountain Research Watershed is a 41-hectare watershed within the Panola Mountain State Conservation Park located in the Piedmont province of Georgia, USA (Fig. 1). Chemical weathering in the Panola watershed has been addressed in a number of studies (Grant, 1975; Nixon, 1981; Burns, 1998; Stonestrom et al., 1998; Schroeder and Melear, 1999; White et al., 1999a,b; White et al. 2001). The average annual precipitation is 1.24 m and the mean annual air temperature is 18 °C. Land use records indicate the original forest was cut in the early 1800's and was farmed until the early 1900's (Huntington, 1995). The watershed is currently covered with a mixed deciduous/coniferous forest. Tree root densities are greatest within the upper meter of regolith, with a sparser distribution of roots penetrating to the bedrock surface.

The focus of the present study is a ridge top in the western quadrant of the watershed (Fig. 1), which is underlain by the 320 M yr old Panola Granite, described as a biotite-muscovite oligioclase-quartz-microcline granodiorite (Higgins et al., 1988). The ridge-top is an old erosion surface comparable in age to granodiorite surface exposures in the watershed dated at between 200-500 kyrs by ^{36}Cl (Bierman et al., 1995). The ridge-top soils are highly weathered ultisols developed *in situ* from bedrock residuum and consist of a dense, dark red, unstructured clay-rich A horizon (~0.5 m thick), underlain by a less dense orange-brown B-horizon (~1 m thick). The underlying porous saprolite (~2-3 m thick) retains the original granodiorite texture and grades from friable saprock to competent bedrock over an interval of several cm at a depth of 4.7 m.

Nested porous cup suction water samplers were installed at the bottom of holes augered to depths of 0.1 to 4.6 m in the ridge-top regolith and sampled bi-weekly to monthly from December 1991 to April of 1993. The site was also sampled during a LiBr tracer test in 1995 (Huntington et al., 1994). Water samples from the suction samplers were filtered in the laboratory, pH and alkalinity determined, and chemistry analyzed using ion chromatography and direct current plasma spectroscopy. Porous tip tensiometers with analogue gauges were installed at selected depth intervals at the site. Moisture samples were taken using a soil corer and measured for gravimetric water content. Soil and saprolite cores, taken at depths ranging from 0.78 to 2.35 m, were measured in the laboratory for saturated hydraulic conductivities using a falling head permeameter (Klute and Dirksen, 1986; Stonestrom et al., 1998).

Surface areas of bulk soil and saprolite, in addition to individual mineral separates, were determined by single point BET measurements using N_2. Mineral separates were split and sieved to retain the <2 mm fraction. Clay-size particles (<4 µm) were subsequently removed by centrifugation. Fe oxides were removed from 4 µm-2 mm silt-sand fraction using citrate-bicarbonate-dithionite extractions and organics were removed using hydrogen peroxide (Mehra and Jackson, 1960). K-feldspar was separated from quartz using heavy liquid mixtures of bromoform and acetone. Muscovite was paramagnetically separated from biotite using a Franz magnetic separator. Exchangeable cations and cation exchange capacity measurements were made on soil and saprolite samples using unbuffered 1 NH_4Cl (Huntington et al., 1990). Selected soil and saprolite samples were powdered and analyzed using X-ray fluorescence spectrometry and X-ray diffraction. Thin sections were prepared and characterized optically using an electron microprobe and scanning electron microscopy

3. RESULTS

3.1. Solid State Chemistry and Mineralogy

Average mineralogy of the fresh Panola granodiorite, determined from point counts (Table 1), indicates a preponderance of plagioclase, quartz, K-feldspar, biotite, and muscovite, with much lesser amounts of hornblende (<2%). Accessory minerals included apatite, zircon and calcite. Major mineral stoichiometries based on microprobe data are also included in Table 1. Bulk oxide data indicate strong elemental fractionations in the soils and saprolite compared to the fresh granodiorite composition (Table 2). Na and

Table 1. Mineral compositions and distributions in the unweathered Panola granodiorite. Mineral formulas based on microprobe analyses[1]. Weight percentages based on point counts[1]

Mineral	Formula	Wt %[2]
Quartz	SiO_2	27.5
Plagioclase	$Na_{0.77}Ca_{0.23}Al_{1.23}Si_{2.77}O_8$	31.9
K-feldspar	$Na_{0.03}K_{0.89}Al_{1.02}Si_{2.99}O_8$	20.6
Biotite	$K_{2.00}(Mg_{1.78}Fe(II)_{3.05}Al_{0.51}Ti_{0.30})(Al_{2.40}Si_{5.60})O_{20}(OH)_4$	12.8
Muscovite	$K_{1.62}(Mg_{0.28}Fe(II)_{0.44}Al_{1.04}Ti_{0.10})(Al_{2.94}Si_{5.06})O_{20}(OH)_4$	7.2

[1] Sampled from a depth of 11.55 m at Ridge-top site
[2] Average analyses reported by William Kelly, NY Geological Survey

Table 2. Chemical compositions of the Panola regolith.

Sample	Depth m	Density g cm^{-3}	SiO_2	Al_2O_3	Fe_2O_3	MgO	CaO	Na_2O	K_2O	TiO_2
						Wt%				
"A" soil	0.10	2.34	81.1	8.8	2.74	0.21	0.14	0.44	3.05	1.91
"B"soil	0.30	1.86	77.7	11.8	3.12	0.31	0.12	0.35	3.40	1.50
	0.48	1.86	64.0	19.9	5.83	0.50	0.11	0.31	2.81	1.38
	0.74	1.83	60.4	21.1	6.37	0.39	0.07	0.25	3.05	1.07
	0.84	1.73	59.9	21.7	6.76	0.44	0.06	0.23	2.63	1.21
	1.07	1.81	63.7	20.9	6.60	0.49	0.06	0.20	2.52	1.26
	1.27	1.83	60.3	22.3	6.95	0.48	0.06	0.20	2.61	1.16
	1.52	1.78	58.9	22.1	7.08	0.45	0.06	0.22	2.14	1.08
Saprolite	1.83	1.80	61.3	22.1	7.13	0.39	0.05	0.16	2.65	1.20
	2.16	1.72	61.7	23.1	6.18	0.53	0.06	0.19	3.15	1.21
	2.51	1.99	62.6	22.1	6.58	0.74	0.06	0.14	3.20	1.31
	2.84	2.25	66.0	20.6	6.05	0.81	0.07	0.23	4.41	1.33
	3.23	1.97	69.8	16.7	6.26	1.18	0.06	0.21	4.93	1.49
	3.73	2.00	65.5	21.4	5.31	1.06	0.05	0.20	4.01	1.22
	3.99	1.98	65.7	19.7	5.68	0.98	0.05	0.22	4.18	1.27
	4.29	1.92	63.6	21.0	4.80	0.90	0.05	0.23	3.06	1.00
	4.52	1.98	67.4	19.1	5.39	0.92	0.07	0.21	3.96	1.27
	4.65	2.00	69.4	18.4	5.32	1.09	0.06	0.20	4.13	1.26
	4.79	1.93	60.9	18.7	4.95	1.10	0.08	0.47	5.48	1.08
Granodiorite[1]	11.55	2.65	66.5	14.7	4.34	1.01	2.41	3.16	4.75	0.90

[1] Sample obtained from drill core described by White et al. (2001)

Ca are essentially absent in the saprolite and soil due to the near-complete weathering of plagioclase to kaolinite in the upper 6 m of the bedrock underlying the saprolite (White et al., 2001). K and Mg concentrations decrease upwards in the saprolite. Si concentrations are highest and Al and Fe concentrations are lowest in the shallow soil.

3.2. Weathering Characteristics

The kaolin minerals, kaolinite and halloysite, and amorphous Fe-oxyhydroxides, goethite, and hematite, are the principal secondary minerals in the soil/saprolite profile (Schroeder and Melear, 1999). In the saprolite and soil, both biotite and K-feldspar weather to kaolinite, as evident in backscattered electron SEM images (Fig. 2). As biotite loses Fe, K, and Mg, and forms kaolinite, the mean atomic number is lowered and the image appears darker. Fig. 2A shows two "biotite grains", one of which shows the layered biotite structure becoming replaced with darker zones of kaolinite. A second, more highly weathered "biotite" grain exhibits no residual biotite, consisting of complete pseudomorphic replacement to layered kaolinite with additional box-like kaolinite structures on the grain surface. Similar epitaxial conversion of biotite to kaolinite has been observed in other saprolites (Murphy et al., 1998; Jeong, 1998; Dong et al., 1998). Biotite weathering is often initiated along cleavage planes, causing the expansion, or fanning of

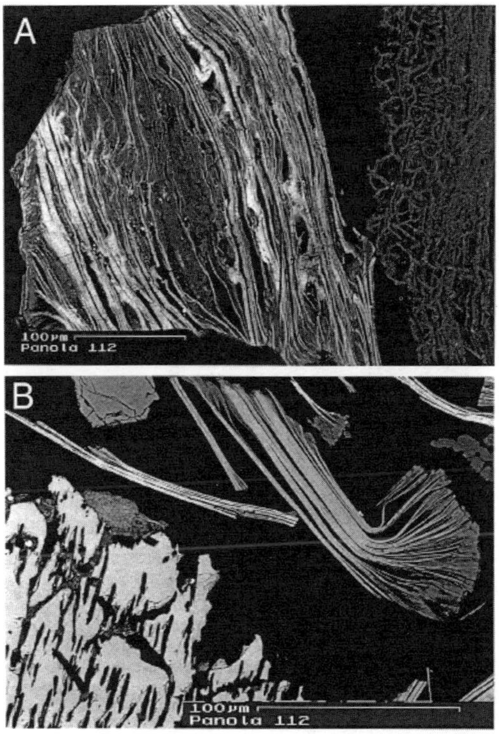

Fig. 2. SEM backscattered photomicrographs of silicate grains at a depth of 2.8 m in the Panola saprolite. (A). Left grain is partially weathered biotite. Bright segments correspond to unaltered biotite and gray areas correspond to epitaxial replacement by kaolinite. Right grain is completely kaolinized biotite with surficial box work structure. (B) Upper right biotite grain exhibits fanning and initial epitaxial replacement with kaolinite. Lower left grain is partially weathered K-feldspar with voids aligned along crystallographic axes.

the biotite grain (Fig. 2B). This morphology is described for other weathered biotites (Kretzschmar et al., 1997; Jeong, 1998; Lee and Parsons, 1999). A highly weathered K-feldspar is also included in Fig. 2B, which exhibits extensive voids and etch pits aligned along selective crystallographic planes.

Muscovite grains from the saprolite and soil appear pristine in thin section. The differences in the intensity of biotite and muscovite weathering are substantiated by compositional differences determined by microprobe analysis (Fig. 3). For single weathered "biotite" grains, the Al/Si ratios for individual dark and light interlayers range between 0.5, representative of fresh biotite, to 1.0, representative of kaolinite (Table 1). The increase in Al/Si ratios correlates with loss of K and Mg from interlayer sites due to the oxidation of octahedral Fe (Fig. 3). Similar composition ranges in epitaxic kaolinite-biotite are reported by Jeong (1998). Muscovites have higher Al/Si ratios than biotite, reflecting less substitution of Mg and Fe into the structure (Fig. 3). Muscovite shows no discernible trend in elemental compositions and has essentially the same compositions as in the fresh granodiorite.

3.3. Pore Water Hydrology

Soil porosity and gravimetric water content measurements are used to calculate hydraulic saturation of the soils and saprolite (Fig. 4A). The extent of saturation varies temporally above the saprolite/soil interface

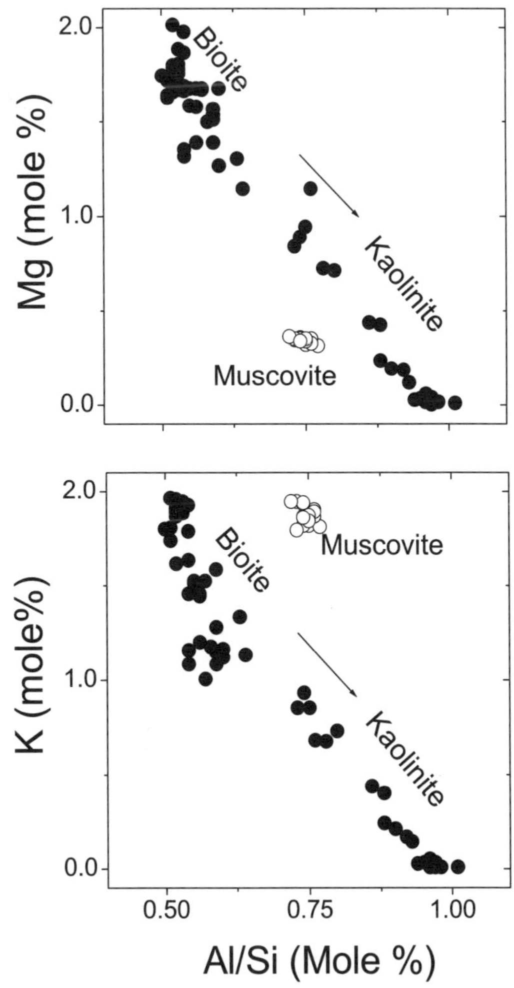

Fig. 3. Changes in Mg and K mole percentages and Al/Si ratios in biotite and muscovite as functions of regolith weathering. Arrows denote composition trends resulting from the progressive epitaxial replacement of biotite by kaolinite.

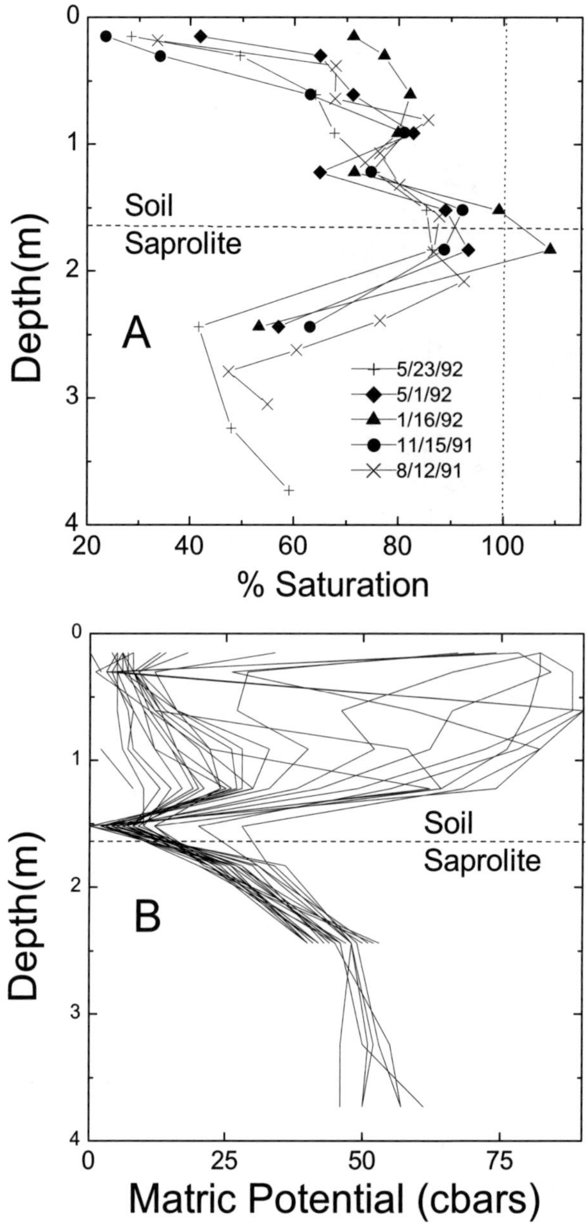

Fig. 4. Soil water hydrology of Panola ridge top. (A) Hydraulic saturation as a function of regolith depth for dates indicated. Values < 100% correspond to unsaturated conditions. (B) Soil tensions as a function of regolith depth. Lines correspond to individual samplings conducted between 8/11/91 and 10/12/93. Zero tension corresponds to hydraulic saturation and progressively higher tensions correspond to increasing undersaturation.

at 1.6 m. In the vicinity of this interface, the regolith is saturated or nearly saturated year-around. In contrast, the underlying saprolite is unsaturated. Field tensiometer readings measure capillary and adsorptive forces produced by the interaction of water with soil mineral surfaces. Tension is zero at fully saturated conditions and increases with decreasing saturation, thus explaining the inverse relationship between saturation and tension shown in Fig. 4. In the soil zone, low moisture tensions occur from November to March (Fig. 5), corresponding to periods of forest dormancy and low evapotranspiration (ET) (Cappellato et al., 1993). High tensions occur during active forest growth and high ET from April to October. Superimposed on this seasonal trend in shallow soil tensions are specific precipitation events such as the intense June 1992 storms shown in the precipitation record in Fig. 5. These events temporarily decreased the matric potential, even during seasonally high rates of transpiration. Moisture tensions below 1.6 m in the saprolite were essentially unaffected by these storm events or longer term seasonal trends in ET.

The soil/saprolite interface (1.6 m) is a zone of consistent pore water saturation and near-zero matric potentials (Fig. 4). These conditions are indicative of "perched water" which overlays a low-permeability duripan consisting of a 10-cm thick layer of relatively competent clay. Duripans serve as aquitards, which restrict the downward percolation of water. Duripans at soil/saprolite boundaries are commonly observed in the southeastern Piedmont of the U.S.A. (Simpson, 1986; Buol and Weed, 1991). From observations in thin sections, O'Brien and Buol (1984) concluded that low hydraulic conductivity is related to the absence of continuous pores due to the formation of thick cutans of clay translocated from the overlying soil. While cyclic wetting and drying occurs in the overlying soil horizons, the underlying saprolite maintains nearly constant moisture contents that are dampened to responses to individual storm events and longer-term trends in seasonal ET (Fig. 5).

3.4. Pore Water Chemistry

Representative pore water solutes obtained from the suction water samplers in soil and saprolite are tabulated in Table 3. Also included are data for open fall and through fall precipitation, surface water, and ground water. Na and Cl are moderately concentrated relative to average precipitation above the duripan and strongly concentrated below the duripan (Fig. 6). The shallow solutes reflect the direct influences of seasonal ET, which concentrates Na and Cl due to selective exclusion by vegetation. Perched water directly above the duripan undergoes significantly greater ET due to longer residence times. Subsequent slow leakage of this water into the upper saprolite explains the high Na and Cl concentrations. This effect may also be exaggerated by sampling bias. During dry periods, the air entry potentials for the porous cup samplers in the shallow soils become very high compared to the potentials in the wetter saprolite (Fig. 4B), thus preventing the sampling of highly evaporated waters. Na/Cl ratios are similar to precipitation (Table 3). The granodiorite is essentially devoid of Cl, and Na, which is associated primarily with plagioclase, is effectively weathered from the underlying bedrock (White et al., 2001).

Solute Si, directly below the duripan, is less concentrated than Na and Cl, suggesting non-conservative behavior related to the incorporation of Si into ongoing kaolinite formation in the duripan. Si increases in the deeper saprolite, particularly near the bedrock interface, thus indicating silicate weathering. Pore water Ca, Mg, and K concentrations exhibit sig-

Table 3. Representative chemical concentrations of pore water, precipitation, stream water, and ground water in Panola watershed ($\mu mole \cdot l^{-1}$).

			Pore Waters[1]			**Wet**[2] **Precip.**	**Thru**[3] **Fall**	**Surface**[4] **Water**	**Ground**[5] **Water**
Depth	0.30	0.91	1.83	3.00	4.62	na	na	na	2.5
pH	5.80	5.54	5.39	5.37	5.58	4.83	4.05	5.09	5.25
Na	24.3	40.0	202	206	163	11.4	10.6	42.1	117.2
K	41.0	36.2	43.7	17.7	12.9	1.8	10.6	8.7	14.5
Ca	55.9	25.5	24.1	12.5	14.5	3.3	32.0	18.4	17.2
Mg	30.4	81.7	66.5	56.4	35.2	1.7	19.2	12.1	19.0
Al	4.6	4.0	0.9	2.3	4.7	na	na	9.5	3.4
SiO_2	75.1	92.9	201	189	372	na	na	76.8	182.2
Alk	86.9	63.4	30.8	25.4	55.6	0.0	0.0	0.0	59.4
Cl	27.6	34.1	251	319	174	14.1	20.6	30.2	67.3
NO_3	1.4	1.1	23.8	2.8	2.1	31.6	19.3	0.0	0.7
SO_4	47.9	43.9	10.3	1.6	6.1	36.8	42.7	51.0	36.5

[1]Sampled 1/21/93

[2]Average of precipitation events 2/82 -7/88

[3]Cappellato et al., 1993

[4]Upper gauge, sampled 1/23/93

[5]Well 700 sampled 1/23/93

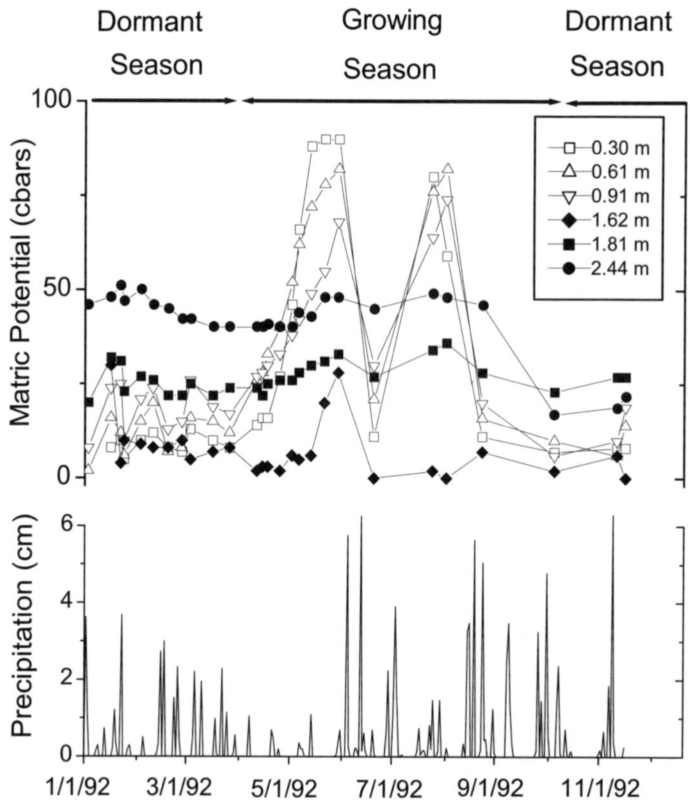

Fig. 5. Seasonal variations in soil tensions and precipitation events for 1992. Also indicated are the approximate periods of forest growth and dormancy. (dates: mo/day/yr)

nificantly greater seasonal variability in the soil than do Na, Cl, and Si. This indicates that, in addition to the effects of ET, these elements are being selectively added and/or removed by biologic cycling by throughfall, leaf fall, and root uptake. In the underlying saprolite, Ca, Mg, and K concentrations are consistently lower than in the soils. The lack of concurrent cation increases along with Si indicates that silicate weathering is not the dominant control on solute Ca, Mg and K in the saprolite.

3.5. Mineral Surface Areas

Bulk regolith BET surface areas increase with increasing depth in the lower soil and upper saprolite (Table 4 and Fig. 7A), corresponding to increasing Al and Fe concentrations and more abundant kaolinite and Fe-oxides (Table 2). These phases decrease in the underlying saprolite, thus accounting for a decrease in bulk BET surface areas. Specific mineral surface areas are also tabulated in Table 4 and plotted versus depth in Fig. 7B. Kaolinite surface areas range between 11.1 and 28.5 $m^2 g^{-1}$, which correspond to a reported range for kaolinite (6 to 39 $m^2 g^{-1}$; Dixon and Weed, 1977). Fe-oxide surface areas calculated from the difference in bulk surface area before and after dithionite extraction range from 10.7 to 38.3 $m^2 g^{-1}$ and approach values reported for goethite (32 to 71 $m^2 g^{-1}$; Buffle, 1988).

The weathered primary silicates exhibit significantly lower specific surface areas than do the secondary clay and Fe oxyhydroxides (Table 4 and Fig. 7B). Mean average silicate grain sizes varied between 0.27 to 0.54 mm with depth in the regolith. Biotite surface areas range between 1.25 and 7.2 $m^2 g^{-1}$, which is similar to that reported for epitaxially-weathered biotites in saprolites in Puerto Rico (Murphy et al., 1998). These values are much higher than surface areas of fresh biotite (0.24 $m^2 g^{-1}$; Acker and Bricker, 1992) and, in part, reflect the kaolinite intergrowths replacing biotite (Fig. 2A). The K-feldspar surface areas are also high (2.3 to 8.1 $m^2 g^{-1}$) and may reflect interstitial

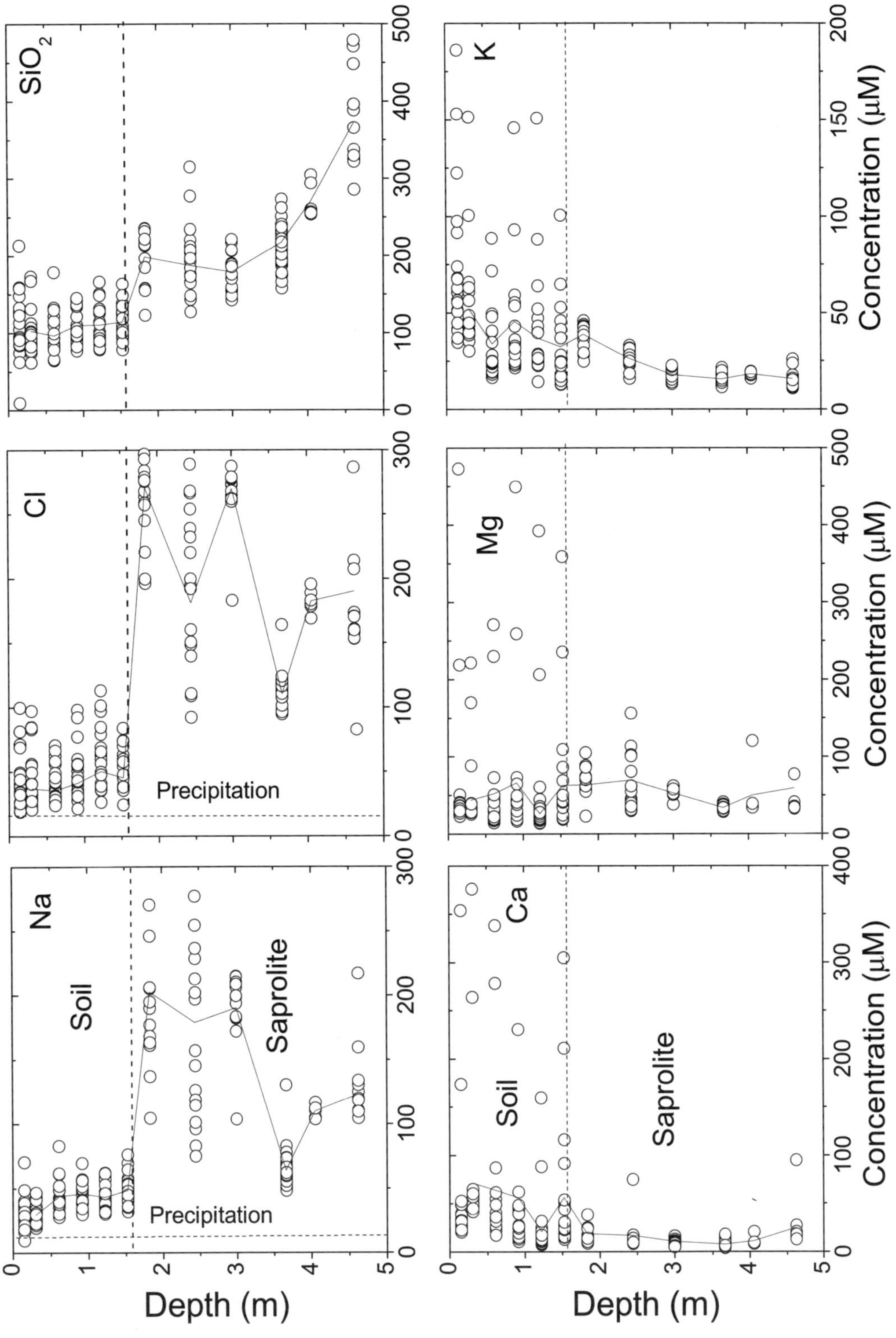

Fig. 6. Pore water solute concentrations with depth in the regolith. Samples collected between 12/91 and 4/93. Vertical dashed lines show average concentrations in open fall precipitation.

Table 4. Surface areas of regolith and specific minerals with depth ($m^2 \cdot g^{-1}$).

Depth m	Bulk Regolith	Kaolinite	Fe-oxide	Qtz	Biotite	K-spar
0.10	2.68	12.9	nd	0.15	nd	2.35
0.20	3.56	11.0	10.7	0.18	nd	4.65
0.38	13.5	20.5	12.6	0.14	1.70	4.64
0.48	20.3	26.7	12.5	0.19	1.25	3.83
0.56	21.4	28.5	43.1	0.20	1.64	3.61
0.74	12.8	15.6	41.7	nd	nd	nd
0.84	16.3	11.1	38.3	0.16	1.47	4.41
1.52	17.9	15.6	22.4	0.11	4.62	8.20
1.83	14.2	11.0	22.2	0.11	2.88	7.10
2.84	7.78	12.6	26.5	nd	nd	nd
3.23	5.64	11.8	32.4	0.22	4.43	1.72
3.53	4.90	13.3	10.9	0.42	7.19	7.23
3.99	7.33	23.3	11.8	0.27	6.58	8.13
4.29	8.28	12.8	11.2	0.23	6.76	nd
4.52	8.10	12.1	16.4	0.24	4.09	nd

nd = not determined.

Table 5. Exchangeable cation concentrations (mM kg^{-1}) and cation exchange capacities (CEC) in mEq kg^{-1}

Depth	Na	K	Mg	Ca	Al	CEC
0.05	0.25	0.91	0.70	2.35	1.77	12.6
0.15	0.09	0.88	0.88	2.82	1.11	11.7
0.25	0.30	1.80	1.55	3.67	2.65	20.5
0.34	0.23	2.59	2.39	5.73	2.26	25.8
0.46	0.38	3.01	4.68	8.36	1.48	33.9
0.52	0.44	2.66	5.09	7.27	3.13	37.2
0.61	0.63	3.15	4.00	4.17	8.52	45.7
0.70	0.53	2.27	4.42	3.30	7.43	40.5
0.79	0.56	2.04	4.09	2.15	8.06	39.3
0.88	0.56	2.16	3.91	1.35	22.36	80.3
1.03	0.54	2.51	3.00	0.71	17.56	63.1
1.22	0.29	2.29	2.27	0.26	5.50	24.1
1.49	0.24	2.15	1.35	0.07	5.56	21.9
1.78	0.12	1.71	1.18	0.06	5.54	20.9
2.11	0.34	1.89	1.13	0.10	8.87	31.3
2.46	0.13	0.51	1.13	0.03	6.09	21.2
2.78	0.21	0.84	0.89	0.09	22.66	71.0
3.16	0.33	0.99	0.82	0.05	14.44	46.4
3.47	0.21	0.75	0.92	0.03	2.63	10.7
3.58	0.26	0.90	1.03	0.06	3.01	12.4
3.68	0.56	1.33	1.08	0.06	9.90	33.9
3.86	0.50	0.84	1.75	0.05	47.83	148
4.05	0.33	0.85	1.98	0.19	53.53	166
4.20	0.25	0.78	2.08	0.18	83.93	257
4.34	0.19	0.57	1.67	0.19	33.45	105
4.46	0.24	0.74	1.14	0.22	39.25	121
4.58	0.25	0.72	0.93	0.20	17.38	55.4

kaolinite. Quartz has the lowest specific surface areas in the regolith (0.11 to 0.42 m^2 g^{-1}), which is consistent with weathered quartz surface areas reported elsewhere (0.10 to 0.23 m^2 g^{-1},White et al., 1996; 0.15 to 0.40 m^2 g^{-1}, Schulz and White, 1999).

The contributions of specific minerals to the bulk regolith surface areas are calculated based on the wt % of each mineral present (Fig. 7C). At the shallowest soil depths, almost all of the bulk surface area is attributed to kaolinite. With increasing soil depth, Fe-hydroxides contribute equally to the bulk surface areas. In the underlying saprolite, increasing percentages of primary minerals dominate the bulk surface area. The line plotted on the far right of Fig. 7C is the sum of the surface areas contributed by the individual minerals. These calculated percentages are only slightly higher than the bulk regolith surface area (100%), implying that the removal of secondary minerals by chemical extraction and sonification does not expose significant additional primary mineral surface areas.

3.6. Cation Exchange

Except in the shallowest soil horizons, Al is the dominant cation produced by NH_4Cl extraction (Table 5 and Fig. 8A). Maximum extractable base cations occur between 0.5 and 1.0 m in the soil; Ca dominates, with lesser amounts of Mg and K, and very low concentrations of Na (Fig. 8B). In the deeper soil (>1.0 m) and in the underlying saprolites, base cations are significantly less abundant, with higher proportions of Mg and K that reflect the diminished role of biological cycling and increased effects of weathering of biotite and K-feldspar.

Cation exchange occurs principally on secondary kaolinite and Fe-oxides in the Panola regolith. Cation exchange also occurs with soil organic matter at shallower depths. The average cation exchange capacity (CEC) for Piedmont kaolinite is 40 meq kg^{-1} (Jardine

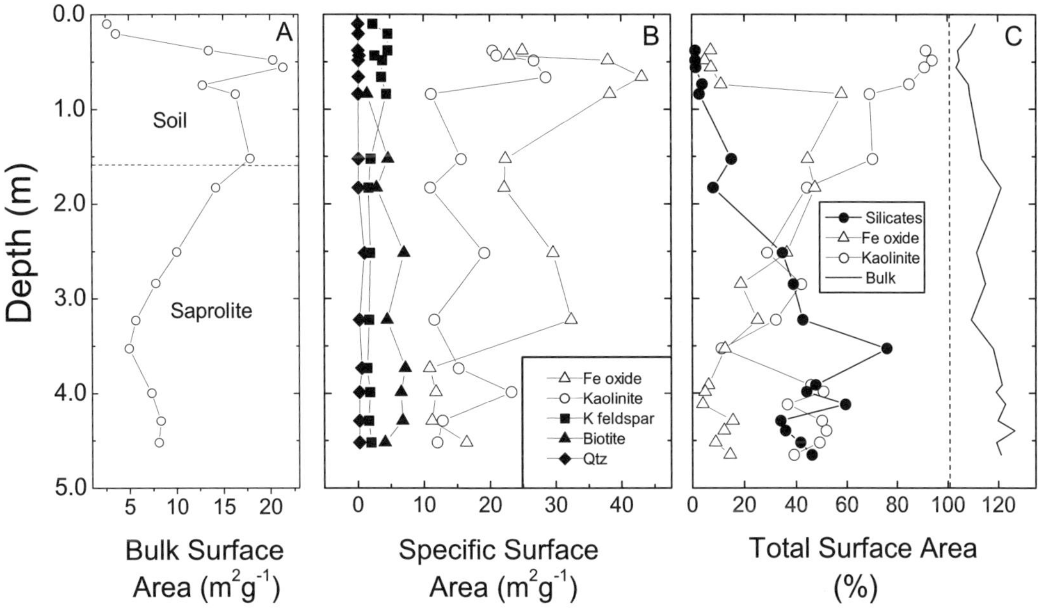

Fig. 7. Surface areas versus regolith depth. (A) Bulk surface areas. (B) Specific mineral surface areas. Solid points are primary minerals and open points are secondary minerals. (C) Percent contribution of specific mineral surface areas to total surface areas. Line on right side of plot is the total percentage contributed by individual minerals compared to bulk surface area (dashed vertical line = 100%).

et al., 1985; Levy et al., 1988; St. Pierre et al., 1992) and the average CEC for goethite is 100 meq kg^{-1} (Al-Kanani and MacKenzie, 1990). The expected CEC for the Panola regolith is calculated based on these average CEC values and wt % kaolinite and goethite present in the Panola regolith. This distribution with depth is then compared to the measured total CEC (Fig. 8D). Kaolinite and Fe-oxide concentrations are low in the shallowest soils (<0.5 m) due to translocation and dissolution by high concentrations of organic acids (Fig. 8C). Consequently, the total cation exchange capacity is low (Fig. 8D). These phases reach maximum concentrations between 0.5 to 1.0 m in the soil, which correspond to the zone of maximum base cation exchange (Fig. 8B). Decreases in the proportions of kaolinite and Fe-oxyhydroxides in underlying saprolite correlate with increases in primary silicates and decreases in base cation exchange.

Base cation saturation (BS), defined as the ratio of sum of exchangeable base cations to the sum of total cations, including Al (Driscoll et al., 1985), ranges from between 85% in the shallow soil to 2% in horizons with very high extractable Al concentrations at depths of 0.8, 2.8, and 4.2 m (Fig. 8D). Unlike for base cations, these Al "spikes" do not correspond to kaolinite and goethite increases in the regolith (Fig. 8C). The discrepancy between calculated and measured CECs (Fig. 8D) implies that only a small portion of extractable Al reflects true cation exchange, with the remainder derived from dissolution of a readily soluble aluminum phase (Shanley, 1992). The highest extractable Al occurs near the saprolite/bedrock interface, suggesting that readily soluble Al-containing phases, such as amorphous Al hydroxides or allophane, are forming as the result of extensive K-feldspar weathering at this interface.

4. DISCUSSION

The mineralogical changes and elemental mobility observed in the Panola regolith have resulted principally from long-term fluid flow through the regolith. Solute transport must supply reactants, principally carbonic acid, and organic acids in the shallower soil horizons, that promote weathering reactions. In turn, solute transport is responsible for removing reaction products from the weathering environment. It is this latter transport which forms the basis for evaluating current weathering rates in the Panola regolith, and permits comparison with past long-term rates based on elemental and mineralogical changes.

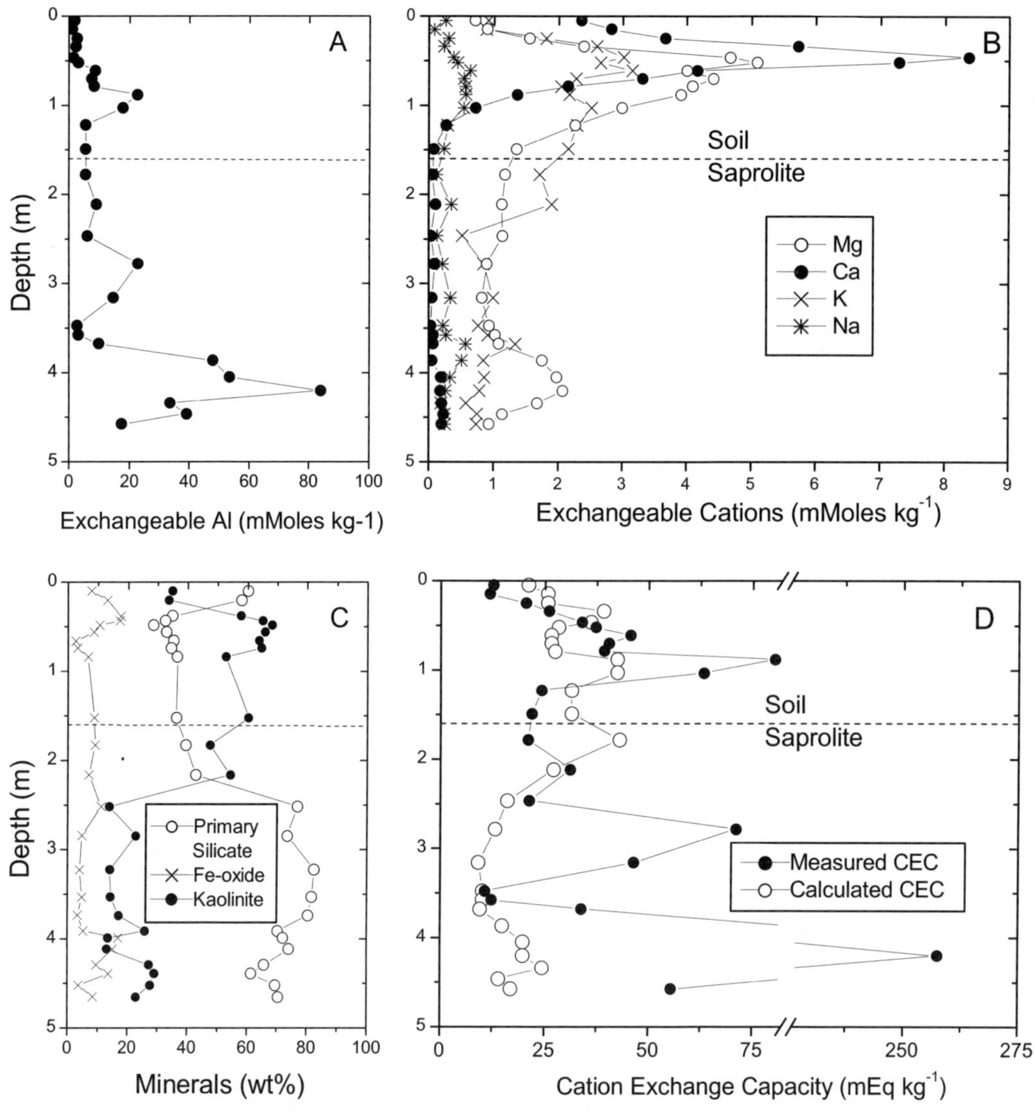

Fig. 8. Cation exchange versus regolith depth. (A) Exchangeable Al. (B) Exchangeable base cations. (C) Wt % of primary silicates and secondary minerals. (D) Comparison of measured cation exchange capacities (solid circles) and calculated CEC based on percentage of mineral phases present.

4.1. Characterization of Solid State Mobility

The data presented in the preceding sections are sufficient to construct mass balances describing element mobility in the Panola saprolite (White et al., 2001). Briefly, quantitative gains or losses of an element or mineral *j*, due to weathering, can be obtained from the regolith concentration $C_{j,w}$ ratioed against the concentration $C_{j,p}$ in the underlying unweathered protolith. This ratio $C_{j,w} / C_{j,p}$ is dependent on three parameters as defined in Eqn. 1 (Brimhall and Dietrich, 1987)

$$\frac{C_{j,w}}{C_{j,p}} = \frac{\rho_p}{\rho_w} \frac{1}{(\varepsilon_i + 1)} (1 + \tau_j) \tag{1}$$

The ratio of the density of the protolith to regolith, ρ_p / ρ_w, is associated with the dissolution and precipitation of mobile elements. The volume ratio V_w/V_p, resulting from soil compaction or expansion, determined for a conservative chemical species *i*, is described by the strain factor $1/(\varepsilon_i+1)$, where $\varepsilon_i = V_w/V_p - 1$. Together, density and volume changes describe closed-system contributions, which occur

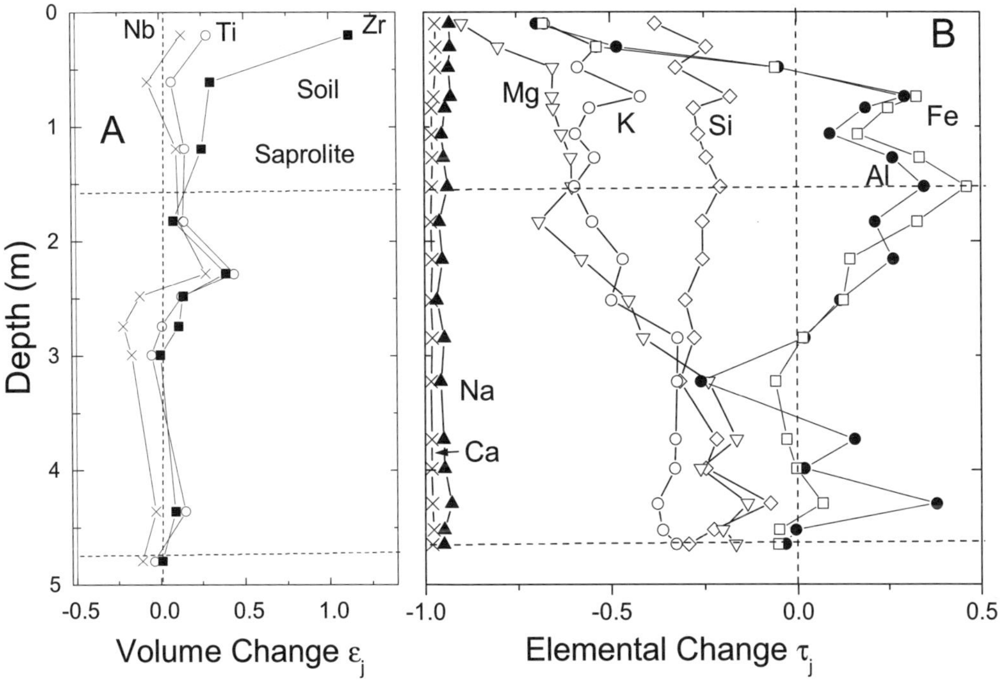

Fig. 9. Volume changes and elemental mass transfers within the Panola regolith. (A) Volumetric compaction and dilation are denoted by values of $\varepsilon_j < 0$ and $\varepsilon_j > 0$, respectively. A value of $\varepsilon_j = 0$ denotes isovolumetric weathering. (B) Elemental changes with regolith depth. A value $\tau_j = 0$ denotes no elemental loss, $\tau_j = -1$ indicates complete elemental loss, and $\tau_j > 1$ denotes elemental additions.

without movement of the component under consideration. The mass transfer coefficient τ_j is an open-system contribution that describes mobility of component j across a unit volume of regolith.

The density ratios ρ_p/ρ_w in Eqn. 1 are calculated directly from the bulk density measurements in the weathered regolith and an assumed granodiorite density of 2.6 g cm^{-3}. The volumetric strain can than be determined from the ratios of densities and concentrations of an inert element i in the soil and protolith (Eqn. 1) to give

$$\varepsilon_i = \frac{\rho_p C_{i,p}}{\rho_w C_{i,w}} - 1 \quad (2)$$

Positive values of ε_i indicate expansion of the regolith and negative values indicate collapse. A value of $\varepsilon_i \approx 0$ is indicative of isovolumetric weathering.

The extent of volumetric changes in the regolith is best established by considering a suite of inert elements. Except for the shallowest soils, which may have undergone volume increases due to bioturbation and the introduction of organic matter, ε_i values calculated from Zr, Ti, and Nb soil concentrations center close to zero (Fig. 9A), indicating that weathering is essentially iso-volumetric. Isovolumetric weathering is commonly observed for saprolites developed on granites and gneisses in the Piedmont of southeastern United States (Cleaves, 1993; Markewich and Pavich, 1991).

Elemental mobility during regolith weathering is characterized by the mass transfer coefficient τ_j (Eqn. 1), which is computed from density and chemical composition data in combination with volume change derived from the strain calculations.

$$\tau_j = \frac{\rho_w C_{j,w}}{\rho_p C_{j,p}} (\varepsilon_i + 1) - 1 \quad (3)$$

The following calculations assume that Ti is conservative. When $\tau_j = -1$, element j is completely removed during weathering. If $\tau_j = 0$, the element is immobile during weathering with respect to the volume of regolith considered.

Ca and Na are almost completely mobilized in the saprolite and soil relative to the fresh bedrock (Fig. 9B, $\tau_{Na,Ca} \approx -1$). This is attributed to near-complete weathering of plagioclase to kaolinite in the upper Panola bedrock (White et al., 2001). The significant losses of K ($\tau_K = -0.33$) and Si ($\tau_{Si} = -0.29$) at the bedrock/saprolite interface correspond to extensive K-feldspar weathering (Fig. 9B). Mg loss, attributed to biotite weathering at this interface, is less ($\tau_{Mg} =$

–0.16). Mg decreases more rapidly than K upwards in the saprolite and is strongly depleted in the soil A horizon (τ_{Mg} = -0.89). Fe and Al are also strongly depleted in the upper A soil horizon (τ_{Fe} = -0.67 and τ_{Al} = -0.70 at 0.1 m), but are significantly enriched relative to fresh bedrock concentrations in the underlying B-horizon and in the upper saprolite (τ_{Fe} = +0.45 and τ_{Al} = +0.30 at 1.6 m). This distribution is indicative of downward Fe and Al mobilization and subsequent re-precipitation as goethite and kaolinite.

The elemental loss or gain ΔM_j (moles) in the saprolite and soil, normalized to unit regolith surface area (m^2), are calculated from the expression (Chadwick et al., 1990; White et al., 2001):

$$\Delta M_{j,l} = \left(\rho_p \frac{C_{j,p}}{w_j} x10^4\right) \int_{Z=d_1}^{Z=d_2} -(\tau_{j,s})dZ \qquad (4)$$

in which an interval of regolith thickness is defined as ΔZ (m), ρ_p is density (cm^{-3}), and w_j is the atomic weight of element j. Table 6 indicates that significantly more K than Mg is lost during the conversion of bedrock to saprolite, but that on a percentage basis the losses of K and Mg are comparable. These base cations are effectively removed from the regolith through solute transport. Si is partly removed from the regolith, with the remainder re-precipitated as kaolinite. Although Fe and Al are mobilized in the upper soil zone (Table 6), their total percentages in the total thickness of regolith increase slightly relative to the protolith due to re-precipitation of Fe and Al as Fe-oxyhydroxides in the lower soil and saprolite (Fig. 9).

4.2. *Characterization of Solute Fluxes*

The mass loss Δm_j (moles) of an element j from the regolith is related to the rate of pore water movement through a unit surface area of regolith (m^2) by the expression (White et al., 1996):

$$\Delta m_j = 1000\, \Delta c_j\, q_h\, t \qquad (5)$$

where Δc_j is the change in the molar aqueous concentration and t is time (s). The fluid flux density q_h ($m\ s^{-1}$), which is the rate of fluid transported over unit area in Eqn. 5, is given by

$$q_h = -K\nabla H \qquad (6)$$

where K is the hydraulic conductivity and ∇H is the hydraulic gradient. For unsaturated flow, which dominates in the Panola regolith, ∇H is also dependent on variations in regolith saturation. The change in head with depth can be described such that (Hillel, 1982)

$$\frac{dH}{dz} = \frac{dh_g}{dz} + \frac{dh_p}{dz} \qquad (7)$$

where h_g is the gravitational head at any point and h_p is the head related to the matric potential of the soil mineral surfaces. Determinations of fluid, and conversely chemical fluxes, based on Eqns. 6 and 7, are difficult for most unsaturated conditions due to the non-linear dependence of conductivity and head potentials on moisture content. Special cases for Eqns. 6 and 7 are discussed by Stonestrom et al. (1998) in which constant matric potentials or tensions exist within a vertical section of a weathering regolith (i.e., $dh_p/dz = 0$).

The duripan at the soil/saprolite interface in the Panola regolith produces a unique hydrology conducive to estimating unsaturated flow in the underlying saprolite. Although LiBr tracer studies have demonstrated significant ET and lateral flow off the ridge top above the duripan (Huntington et al., 1994), negligible lateral flow is assumed to occur in the saprolite beneath the duripan. If evapotranspiration is minor at depth, the mass of water can be assumed to be conservative and the unsaturated flux density q_h, due to drainage anywhere in the saprolite, will approximate the saturated flux density in the overlying duripan. Under saturated conditions h_p = 0 and dh_g/dz = -1 (Eqn. 7). The flux density is approximated as the inverse of the saturated hydraulic conductivity

$$q_h = -K_{sat} \qquad (8)$$

The solute transported in the unsaturated saprolite can be directly calculated if the saturated hydraulic conductivity of the duripan is known.

Table 6. Elemental mobilities (moles m^{-3}) and % total loss in the Panola regolith.

	Saprolite	Soil	% loss
Depth	1.2 - 4.7 m	0.0 - 1.6 m	
Si	-5520	-160	22.7
K	3360	120	43.1
Mg	630	-20	43.8
Fe	-120	540	-8.9
Al	-2030	1780	-10.7
Na	670	-90	94.6
Ca	40	-14	97.9

4.3. *Estimates of Fluid Fluxes and Residence Times*

Saturated hydraulic conductivities K_{sat} ($m\ s^{-1}$) measured on core material, inclusive of the duripan (0.78 to 2.35 m), vary between 1.3×10^{-9} to 5.0×10^{-6} $m\ s^{-1}$ (Table 7 and Fig 10). These conductances are comparable to those measured in soil/saprolites elsewhere in the southeastern United States (Vapraskas and Williams, 1995; Schoenberger et al., 1995). A minimum conductivity of $K_{sat} = 1.3\times10^{-9}\ m\ s^{-1}$ corresponds to an infiltration rate of $4.1\times10^{-2}\ m\ yr^{-1}$ in the

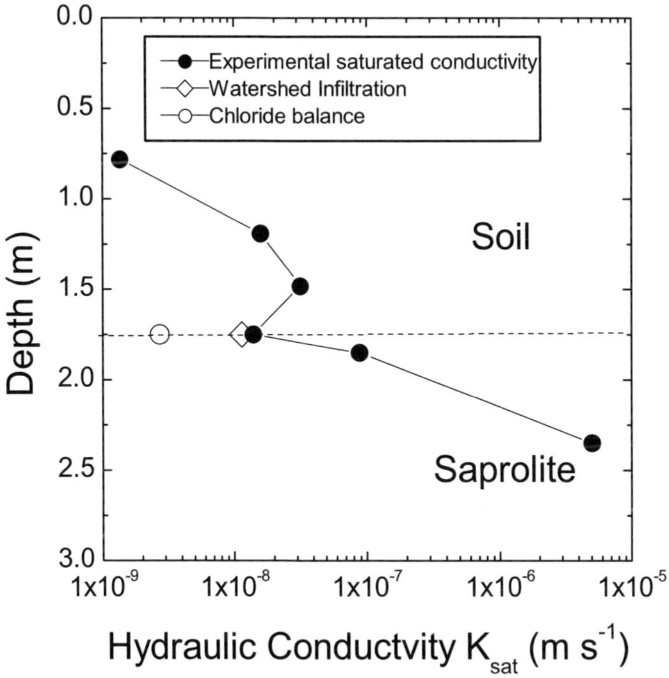

Fig. 10. Laboratory-determined saturated hydraulic conductivities for the soil and saprolite. Also included are estimated infiltration rates based on Cl mass and watershed water balances.

shallow soil (Fig. 10). This rate is clearly contradicted by rapid soil moisture and tension responses to precipitation events (Figs. 4 and 5). Water movement in the shallow soil at Panola occurs as macropore flow, principally along root channels and worm borrows (Flury and Fluhler, 1994). This macropore flow short-circuits the much slower matrix flow measured in the small-scale hydraulic conductivity experiments presented in the present study.

Deeper in the regolith, isovolumetric conditions (Fig. 9A) denote a lack of extensive bioturbation and macropore flow. The measured K_{sat} in the duripan is 1.4×10^{-8} m s^{-1} or 0.44 m yr^{-1} (Fig. 10). This low permeability is responsible for near-saturated to saturated conditions which prevail at the soil/saprolite interface. Due to increased porosity, higher saturated conductivities occur in the saprolite directly beneath the duripan. A maximum value of $K_{sat} = 5.0\times10^{-6}$ m s^{-1} at 2.35 m (Table 7) corresponds to an infiltration rate of >150 m yr^{-1}, thus explaining the lack of saturated conditions in the saprolite (Fig. 4A). Unsaturated conductivities K_{unsat} decrease exponentially with decreases in regolith water content (Stonestrom et al., 1998). Therefore, K_{unsat} can be several orders of magnitude lower than $K_{sat.}$ In the Panola regolith, the assumption of water conservation beneath the soil horizon requires that unsaturated flow in the saprolite must equal saturated infiltration through the overlying duripan.

Experimental conductivities (Table 7) can be compared against independent estimates for field infiltration. Average precipitation P in the Panola watershed is 1.15 m yr^{-1} and average runoff, normalized to watershed surface area, is 0.34 m yr^{-1} (Huntington et al., 1993). The resultant water loss due to ET is 71% . Assuming that the remaining runoff infiltrates through the saprolite, the resulting conductivity for the watershed would be 1.2×10^{-8} m s^{-1}, which is essentially identical to the experimental hydraulic conductivity for the duripan (1.4×10^{-8} m s^{-1}). This is a maximum estimate for the fluid flux because negligible lateral flow is assumed to occur above the saprolite.

An additional method for estimating field flux density is based on ET-corrected precipitation using Cl as a conservative tracer (Erikkson and Kunakasen, 1969)

$$q_h = P\,(Cl_p/Cl_s) \qquad (9)$$

where Cl_p and Cl_s are the respective Cl concentrations in precipitation and in the pore water. Average

Table 7. Saturated hydraulic conductivities in Panola and comparable regoliths.

	Depth (cm)	K_{sat} ($m\ s^{-1}$)
Granodiorite, Panola, Georgia, USA[1]		
B horizon	0.78	1.34×10^{-9}
B horizon	1.19	1.58×10^{-8}
B horizon	1.48	3.16×10^{-8}
B horizon	1.75	1.39×10^{-8}
B/Saprolite	1.85	8.84×10^{-8}
Saprolite	2.35	5.08×10^{-6}
Quartz Diorite, North Carolina, USA[2]		
Saprolite	2.0	1.03×10^{-7}
Felsic Gneiss, North Carolina, USA[3]		
B horizon	0.28	1.62×10^{-6}
B/Saprolite	0.6	7.00×10^{-8}
Saprolite	1.87	8.10×10^{-7}
Quartz Diorite, Luquillo Mtns., PR[4]		
Saprolite	0.68	5.5×10^{-6}

[1] This study, [2]Vapraskas & Williams (1995),
[3] Schoenberger et al., 1995, [4]White et al. (1998).

Cl in precipitation is 14 μM (Table 3) and average pore water Cl in the saprolite is 200 μM (Fig. 6). The resultant flux density is 2.6×10^{-9} m s^{-1}, or about a factor of 6 less than the saturated conductivity K_{sat} of the duripan (1.4×10^{-8} m s^{-1}).

The residence time of pore water in the saprolite can be calculated from the relationship

$$t_r = \frac{\phi s d}{q_h} \tag{10}$$

where ϕ is the average saprolite porosity (0.45), s is the average fractional saturation (0.7), and d is the saprolite thickness (3.0 m). The average pore water residence time is 3.6×10^{8} s (12 years), based on the minimum Cl-derived flux density (2.6×10^{-9} m s^{-1}). These residence times are long when compared to hill slope waters which have been age-dated in the watershed (Burns et al., 1998). Unlike the hillslope flow, which responds rapidly to individual storm events through macropore flow, the deeper ridge-top saprolites are much less responsive to individual or even longer-term season patterns in precipitation and ET (Fig. 5).

4.4. Comparison of Short- and Long-Term Rates of Silica Weathering

Of the major elements associated with silicate weathering, only Si exhibits consistent pore water increases with depth in the saprolite (Fig. 6). Assuming that the present short-term weathering flux r_{Si} (moles m^{-2} s^{-1}) is comparable to past long-term weathering flux R_{Si} (moles m^{-2} s^{-1}), the relationship can be developed

$$r_{Si} = \frac{\Delta m_{Si}}{t} = \frac{\Delta M_{Si}}{t^*} = R_{Si} \tag{11}$$

where Δm_{Si} (Eqn. 5) is the Si loss over the fluid residence time t (s) and ΔM_{Si} (Eqn. 4) is the solid-state Si loss over the duration of saprolite weathering (t*). The maximum pore water residence time is 12.6 yrs (Eqn. 10) and the average increase in pore water Si in the saprolite is Δc_{Si} = 255 µM (Fig. 6). Therefore the present day Si flux r_{Si} (Eqn. 11) is 2.0×10^{-2} moles $m^{-2}yr^{-1}$.

Bierman et al. (1995) measured ^{36}Cl concentrations in surface outcrops of granodiorite within the Panola watershed and calculated exposure ages of between 250–500 kyrs. These distributions produced a mean model erosion rate of 7 m my^{-1}, which is in agreement with steady-state chemical weathering rates calculated from solute fluxes. The time span required to produce 3 m of saprolite is estimated to be 4.3x 10^5 yrs. The long term Si loss is ΔM_{Si} =5.5 x 10^3 moles (Eqn. 4 and Table 6). The resulting long-term Si flux is R_{Si} = 1.3×10^{-2} moles m^{-2} yr^{-1} (Eqn. 11), which is similar to the present day rates based on Cl balances (r_{Si} = 2.0×10^{-2} moles m^{-2} yr^{-1}).

The effect of infiltration rates on estimates for solute weathering is emphasized by plotting weathering trajectories (White et al., 1998). The mass loss ΔM_{Si} resulting from long-term regolith weathering is plotted in Fig. 11. The upward trending lines are weathering trajectories calculated by the extrapolation of solute loss Δm_{Si} to long times based on Eqn. 5 and various rates of infiltration. Sustained saturated flow in the saprolite replicates total Si loss in less than 2×10^3 years, an unreasonable time span.

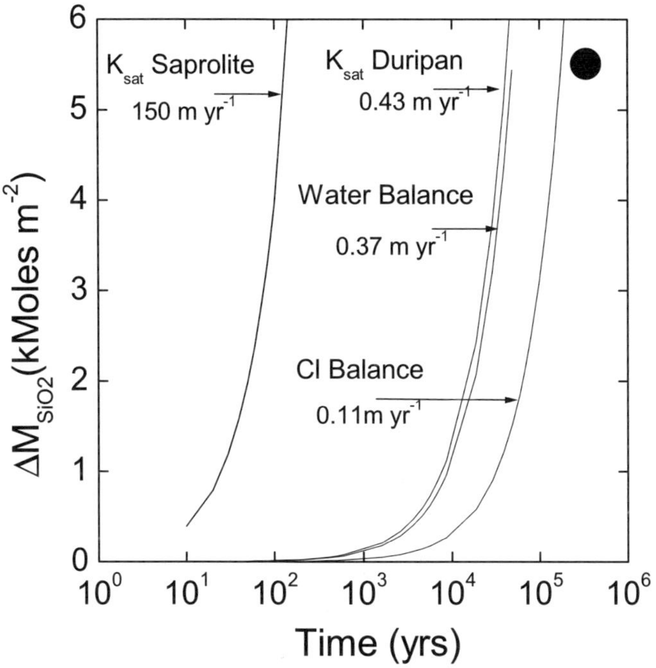

Fig. 11. Comparison of short- and long-term Si weathering rates for the Panola saprolite. The solid symbol corresponds to regolith mass loss ΔM_{Si} calculated from changes in chemistry and profile age (Eqn. 4). Lines correspond to extrapolated weathering trajectories calculated from solute mass loss Δm_{Si} (Eqn. 5) and the indicated infiltration rates.

Slower unsaturated flow in the saprolite, estimated from watershed balances and from the saturated conductivity of the overlying less-permeable duripan produce trajectories that are more realistic, overestimating the Si loss from the saprolite by less than a factor of 10 (Fig. 11). Estimated Si fluxes based on Cl balances most clearly agree with long-term Si mobility (Fig. 11). An infiltration rate of 4×10^{-7} m s^{-1} over 3×10^5 years produces a calculated total Si loss of $\Delta m_{Si} = 7.6\times10^3$ moles m^{-2}, which is comparable to the measured long-term loss of 5.5×10^3 moles m^{-2}.

This study concludes that present day Si fluxes in pore waters are indistinguishable from long-term average Si fluxes derived from saprolite weathering over the last several hundred thousand years. This implies that parameters influencing Si weathering rates, such as precipitation, temperature, and vegetative cover, while not necessarily constant, have not varied enough to significantly impact Si leaching rates during this time. Similar close agreements between short- and long-term weathering fluxes have been reported for other saprolitic regoliths (Velbel, 1985; Pavich et al., 1995; White et al., 1998).

4.5. Rates of Biotite Weathering

Dissolved Si in the pore waters of the Panola regolith is derived principally from silicate weathering reactions. The extent to which primary silicates are weathered can be calculated from mass balance considerations (White et al., 2001). The mass of weathered plagioclase is first calculated based on plagioclase stoichiometry, fresh bedrock Na concentrations (Table 2), and τ_{Na} values for the regolith (Fig. 9B). Next, all of the Mg is similarly partitioned into the mica phases. It is not possible to distinguish computationally between muscovite and biotite because each mineral contains the same set of elements, i.e., K, Mg, Fe, Ti, Si, and Al, although at different concentrations. Based on petrographic and chemical evidence (Figs. 2 and 3), muscovite is assumed to be unreactive in the regolith and equal in concentration to that in the fresh bedrock. The change in Mg in the regolith is thereby apportioned solely to biotite (Velbel, 1985). The final step of the calculation is to distribute the remaining K into the K-feldspar phase.

Computations show that the amount of plagioclase lost in the saprolite is essentially constant with depth and equal to the initial amount of plagioclase present in the unweathered Panola Granite (Fig. 12A, vertical line). Plagioclase is nearly completely converted to kaolinite at the weathering front in the underlying bedrock and is unavailable as a reactant in the overlying soil and saprolite (White et al., 2001). Approximately half of the K-feldspar weathers at the bedrock/saprolite interface (Fig. 12B), thus accounting for most of the K loss tabulated in Table 6. Losses of K-feldspar vary significantly in the overlying saprolite and soil. This may, in part, be an artifact attributable to compounding errors in the mass balance calculations, which first must distribute K into the muscovite and biotite phases before the mass of residual K-feldspar can be calculated. The overall trend in the data suggests that residual K-feldspar continues to weather, albeit at a slower rates in the overlying saprolite and soil.

Biotite losses can be calculated based on Mg distributions (Table 2 and Fig. 9B), which decrease consistently with decreasing depth in the regolith (Fig. 12C). Based on the biotite stoichiometry determined by microprobe analysis (Table 1), the epitaxial reaction of biotite to kaolinite can be written as follows

$$K_2(Mg_{1.78}Fe(II)_{3.05}Al_{0.51}Ti_{0.30})(Al_{2.40}Si_{5.59})O_{20}(OH)_4 + 5.56\,H^+ + 0.76\,O_2 + 4.98\,H_2O \rightarrow$$

$$2K^+ + 1.78\,Mg^{2+} + 2.67H_4SiO_4 + 0.73\,Al_4Si_4O_{10}(OH)_8 + 3.05\,FeOOH + 0.30\,TiO_2 \quad (12)$$

This reaction assumes that all of the Fe in the biotite is in the ferrous state, which is in accord with charge-balance constraints. For every mole of Mg lost, 0.56 moles of biotite have reacted. The diagonal line in Fig. 12 C corresponds to a linear regression fit for depth versus biotite loss (slope $b = 1.74\times10^{-2}$ m^4 $moles^{-1}$ and $r^2 = 0.90$). Under isovolumetric conditions, the mineral weathering rate is inversely proportional to the slope of the mineral distribution with depth; the shallower the slope, the more rapid the weathering reaction (White et al., 2001). The long-term dissolution rate constant for biotite, $k_{biotite}$ (moles m^{-2} s^{-1}), can be derived from this relationship, such that

$$k^*_{Biotite} = \frac{w_{reg}}{b}\left(\frac{1}{\rho_{reg} f_{biotite} S_{biotite}}\right) \quad (13)$$

where w_{reg} (m s^{-1}) is the rate of regolith formation, estimated to be 7 m/10^6 yrs based on ^{36}Cl age dating (Bierman et al., 1995). The terms in parentheses define the volumetric surface area of biotite (m^2m^{-3}). The average saprolite density $\rho_{sap} = 1.85\times10^6$ g m^{-3} (Table 2). The average volume fraction of biotite $f_{biotite}$ present during regolith weathering is assumed to be 0.05, or one-half the fraction of biotite found in the fresh bedrock (Table 1). The specific surface area of the biotite is taken as the average of the BET analyses reported in Table 4 (5.0 m^2 g^{-1}). The resulting long-term biotite dissolution rate constant is $k_{biotite} = 2.8\times10^{-17}$ moles $m^{-2}s^{-1}$.

The dissolution rate constant for long-term biotite weathering in the Panola saprolite $k_{biotite}$ is compared to both natural and experimental dissolution rates reported elsewhere in the literature (Table 8). Reported pH ranges, methods of surface area determination, and elemental basis for the biotite rate constants are also tabulated. The Panola rate is the slowest yet recorded for biotite weathering and is almost 5 orders

Table 8. Weathering rates of biotite measured in field and laboratory studies

log rate R moles m^{-2} s^{-1}	pH	Location or experiment	Surface area	Basis of rate determination	
Field					
-16.4	**5.0 - 5.5**	**Panola, GA USA**	**BET**	**Mg regolith loss**	**This study**
-15.0	4.5	Luquillo Mtn., PR	BET	Mg solute flux	Murphy et. al., 1998
-14.0	4.5	Bear Brook, MA USA	Geometric	Si solute flux	Swoboda-Colberg & Drever (1993)
-13.0	6.0	Coweeta, NC USA	Geometric	Mg release	Velbel (1985)
Laboratory					
-11.9	4.5	experimental	BET	Cation release	Kalinowski & Schweda (1996)
-11.5	4.0 - 4.5	experimental	BET	K & Mg release	Sverdrup and Warfvinge (1995)
-11.4	4.5	experimental	Geometric	K release	Swoboda-Colberg & Drever (1993)
-10.7	5.0	experimental	BET	Mg release	Acker and Bricker (1992)

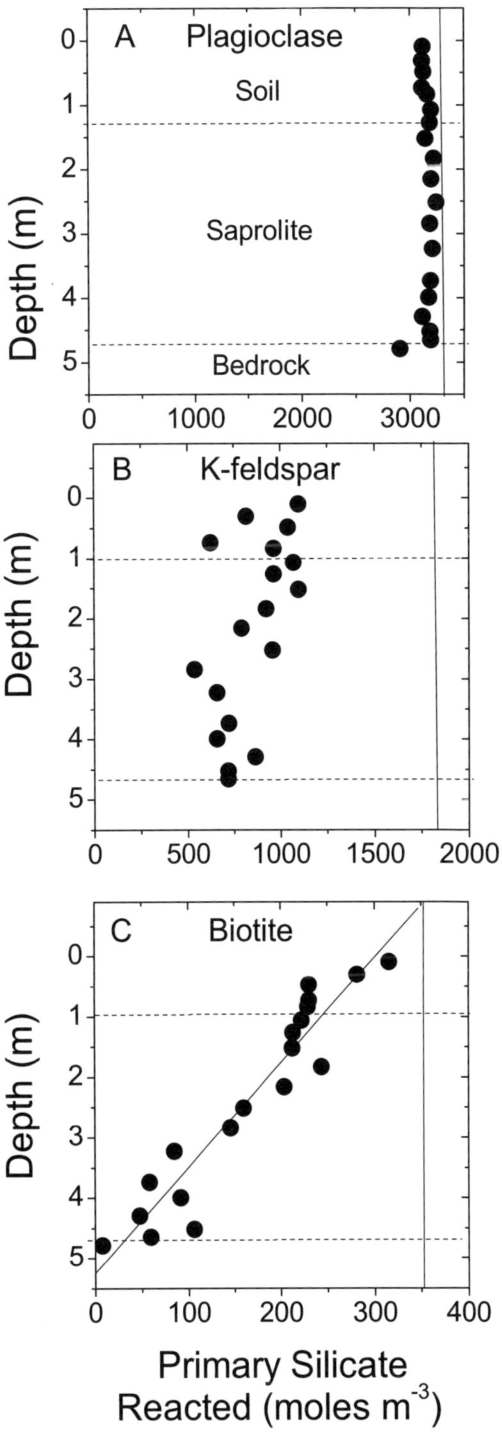

Fig. 12. Masses of primary silicates reacted in the Panola regolith. Vertical lines correspond to the initial protolith compositions. Diagonal line is the linear regression fit to the biotite weathering data

of magnitude slower than the fastest experimental rates, recorded by Acker and Bricker (1992).

In general, experimental rates of biotite weathering are faster than the natural weathering rates, as is commonly observed for most silicate minerals (White and Brantley, 1995; and papers therein). Much of the variation in reported rates are inevitably due to differences in methodologies used in the weathering studies. One of the most sensitive variables is specific surface area. Field studies by Velbel (1985) and Swodoba-Colberg and Drever (1993) used geometric surface areas, which in the case of naturally weathered silicates, have been shown to be 1 to 2 orders of magnitude less than BET surface areas (White and Peterson, 1990; Brantley et al., 1999). The only other natural biotite-weathering rate using BET surface areas was that for the Luquillo watershed in Puerto Rico (Murphy et al., 1998). This biotite rate, based on present day solute transport rates of Mg and K, is about 25 times faster than that at Panola. At least a part of this difference can be attributed to accelerated silicate weathering under significantly wetter and warmer climatic conditions in Puerto Rico (White et al., 1998).

4.6. Solute Fluxes Related to Cation Exchange and Biotic Cycling

Biotite and K-feldspar weathering mobilizes Mg and K in the Panola regolith as indicated by long-term losses from the saprolite (Table 6). However, the solute concentrations of these elements decrease as functions of depth and residence times in the saprolite (Fig. 6), implying that significant sinks for these elements exist in the regolith. The roles of ion exchange and biological cycling are often neglected in attempts to explain such apparent discrepancies in weathering processes. The assumption most often made is that ion exchange is at equilibrium and no net gain or loss of cations occurs in the solute phase (White and Blum, 1995). Over long-term weathering, this assumption is clearly incorrect because the exchange capacity of the regolith increases as the percentage of clays and secondary oxides increase (Fig. 8). On a short-term basis, cation exchange disequilibrium can occur due to external perturbations, such as changes in climate, deforestation, and acid precipitation, all of which can potentially impact the pore water solutes in the Panola regolith.

At shallow soil depths, exchangeable Ca dominates over lesser amounts of Mg and K, and very low concentrations of Na (Fig. 8B). The regolith contains very little Ca due to the near-complete weathering of plagioclase at depth in the bedrock. High concentrations of exchangeable Ca in the shallow soils reflect the preferential cycling and retention of Ca during biological cycling. Wet precipitation, which is relatively enriched in Ca (Table 3), is the ultimate source of much of this recycled Ca (Huntington et al., 2000). In the deeper soils and saprolite (>0.5 m), the proportions of Mg and K are much higher than Ca, a trend reflecting the diminished role of biological cycling and the increased role of chemical weathering. Exchangeable Mg is derived principally from biotite and K is derived from both biotite and K-feldspar (Fig. 12).

Total exchangeable Mg and K in the Panola regolith are 1.9 moles and 1.4 moles per m^2 of land surface, respectively (Fig. 8B). Based on the biotite weathering rate ($k^*_{biotite} = 2.8\times10^{-17}$ moles m^{-2} s^{-1}), the time intervals required to replace this exchangeable Mg and K are 270 and 180 years, respectively. The actual replacement time for K is expected to be less due to an additional contribution from K-feldspar weathering. In contrast, pore water Mg and K concentrations produced by biotite weathering are limited to the fluid residence time in the regolith, which at a maximum is calculated to be 12 years (Eqn. 10). Resulting total solute Mg and K in pore waters, 0.083 and 0.093 moles per m^2 of land surface respectively, are correspondingly much less than the exchangeable Mg and K. The magnitude of these differences implies that solute concentrations are significantly impacted by even small perturbations in the cation exchange reservoir. Historically, the Panola watershed has been subjected both to past agricultural development (Huntington et al., 2000), and more recently, acidic precipitation (Shanley, 1992). Long-term cation exchange re-equilibration in response to these impacts could serve as a significant sink by which solute Mg and K, produced by silicate weathering, are effectively removed.

In many mass-balance studies in pristine watersheds, the forest ecosystem is assumed to be steady state, i.e., there is neither degradation nor aggradation. As discussed by Taylor and Velbel (1991) and Velbel (1995), the neglect of botanical factors in geochemical estimates of weathering stems from the misunderstanding of steady-state conditions in terms of temporal and spatial scales. On a watershed scale, biological cycling generally removes inorganic nutrients such as K, Mg, and Ca from the system, resulting in an underestimation of weathering rates by a factor 2 to 5 for these elements. Part of this net biologic loss can be attributed to net erosional loss of particulate organic matter that is not considered in solute mass-balance calculations. In addition, as discussed by Huntington (1995), the Panola watershed is considered an aggrading ecosystem in which the forest cover is recovering from logging and agricultural practices employed up to the earlier part of the twentieth century.

On the smaller regolith scale, the concept of closed- and open-system interaction between weathering and biologic uptake is also important (Burnham, 1989). In a closed system, biologic uptake of solutes resulting from mineral weathering is negligible. Forest fertility is maintained by low level atmospheric inputs and by efficient retention and recycling of these nutrients in near-surface soil horizons. This is the case for Ca cycling in the Panola soil (Huntington et al., 2000), and is demonstrated for Mg and K cycling in a parallel study of the tropical regolith in the Luquillo Mountains in Puerto Rico (White et al., 1998; Murphy et al., 1998). As at Panola, solute Mg and K levels are high in shallow Luquillo soil and low in the underlying shallow saprolite. However with increasing depth in the Luquillo saprolite, solute K and Mg increased linearly, denoting contributions from their stoichiometric release from biotite over progressively longer residence times. The deeper Luquillo regolith is closed to biological uptake of K and Mg produced by weathering due to the shallow rooting depth of the rainforest vegetation. This closed system cycling is consistent with other tropical environments in which soil mineral nutrient balances for vegetative covers are shown to be spatially consistent over differing bedrock types (Hamdan and Burnham, 1996).

In contrast to Luquillo, the Panola regolith is an example of an open system in which base cations, principally K and Mg released from biotite weathering in the saprolite, are removed due to biological uptake. In the Panola regolith, tree roots predominate in the soil horizon, but a lower density of roots is also observed to penetrate to the bedrock. The uptake by this deeper rooting system, coupled with the shallower thickness of saprolite, is sufficient to quantitatively remove mineral nutrients. Based on the biotite weathering

Table 9. Annual fluxes in the Panola Watershed (mmoles m^{-2} yr^{-1})

	Si	K	Mg	Ca	Na
Wet Precipitation					
	0.0	0.1	0.18	0.08	6
Weathering of biotite in saprolite					
	4.8	2.5	2.3	0.0	0.0
Vegetative Fluxes					
Through fall[1]	na	36.5	7.2	12.9	6.6
Stem flow[1]	na	2.1	0.6	0.9	0.5
Foliar runoff[1]	na	149	83.3	112	2.6
Stream Discharge	67.6	1.8	8.2	12.1	32

[1]Cappellato et al. (1993) & Cappellato and Peters (1995).

rate ($k^*_{biotite}$ = 3.9×10^{-17} moles m^{-2} s^{-1}), the annual solute flux of Mg and K from biotite weathering is 2.0 and 2.3 mmoles m^{-2} yr^{-1}, respectively (Table 9). The annual export from the forest canopy to the soil surface in the forms of through fall, stem flow, and foliar deposition are also summarized in Table 9. The sum of the Mg and K fluxes are respectively 36 and 75 greater than are the corresponding fluxes from biotite weathering. At most, on an annual basis, biotite weathering can contribute 1 to 3% of the total K and Mg cycled through the biosphere. Soluble K and Mg, made available during chemical weathering of silicates, are immediately included in biologic uptake, which explains the lack of Mg and K in the saprolite pore waters.

5. CONCLUSIONS

The Panola regolith possesses unique features that optimize the comparison of past long-term weathering rates, based on solid-state compositions, with present short-term weathering rates based on solute fluxes. Unlike soils developed on bedrocks exposed to periods of relatively recent glaciation, the long duration of *in situ* isovolumetric weathering at Panola has produced a thick regolith, which has undergone significant mineral weathering and elemental mobility. These conditions are ideal for applying mass-balance constraints using conservative elements such as Zr, Ti, and Nb. The use of cosmogenic ^{36}Cl to estimate the steady rate of regolith propagation permits the determination of long-term elemental and mineral weathering rates. As demonstrated, the weathering rate constant for biotite can be directly calculated from the rate of mineral loss with regolith depth. This rate is up to 5 orders of magnitude slower than rates for experimental biotite dissolution, implying significantly different reaction mechanisms and/or conditions.

Long-term weathering has led to the development of significant regolith structure that influences present day weathering, in particular, the duripan at the soil/saprolite boundary which strongly attenuates both fluid flow and mass transfer into the underlying saprolite. Such steady-state conditions permit characterizing solute fluxes based on pore water concentrations and regolith permeability determined from experimental conductivities and water and Cl mass balances. Fluid residence times on the order of a decade are contrasted to time spans of several hundred thousands of years for regolith weathering. However, when the respective Si losses in the regolith and pore waters are normalized with respect to these times, the Si weathering rates differ by less than a factor of two. This similarity in Si rates implies that past changes in the watershed, including climate, vegetation, and anthropogenic influences have not significantly influenced the rates of chemical weathering. The calculation of mineral dissolution rates based on increases in solute cation concentrations is not possible because, unlike the solid phases, these species decrease with depth and residence times in the regolith. When the magnitude of the weathering rates are compared to cation exchange capacities and biologic cycling, it is apparent that small variations in these parameters can effectively eliminate the solute cation weathering signature.

Acknowledgements—The Water Energy Biogeochemical Budgets (WEBB) program of the Water Resources Division of the U. S. Geological Survey funded this work. The authors express their gratitude to Drs. Michael Velbel, James Drever, Susan Anderson, and Roland Hellmann for helpful reviews and suggestions pertaining to this paper.

Editorial handling: Roland Hellmann

REFERENCES

Acker J. G. and Bricker O. P. (1992) The influence of pH on biotite dissolution and alteration kinetics at low temperature. *Geochim. Cosmochim. Acta* **56**, 3073-3092.

Al-Kanani T. and MacKenzie A. F. (1990) Phosphate hydrolysis in mineral fractions of soils, goethite, kaolinite and montmorillionite. *Soil Science* **149**, 239-247.

April R., Newton R., and Coles L. T. (1986) Chemical weathering in two Adirondack watersheds: Past and present-day rates. *Geol. Soc. Amer. Bull.* **97**, 1232-1238.

Bierman P., Gillespie A., Caffee M., and Elmore D. (1995) Estimating erosion rates and exposure ages with ^{36}Cl produced by neutron activation. *Geochim. Cosmochim. Acta* **59**, 3779-3798.

Brantley S. L., White A. F., and Hodson M. E. (1999) Surface area of primary silicate minerals. In *Growth, Dissolution and Pattern Formation in Geosystems* (eds. B. Jamtveit and P. Meakin). Kluwer Academic Publishers, Dordrecht. pp 291-326.

Brimhall G. H. and Dietrich W. E. (1987) Constitutive mass balance relations between chemical composition, volume, density, porosity, and strain in metasomatic hydrochemical systems: Results on weathering and pedogenesis. *Geochim. Cosmochim. Acta* **51**, 567-587.

Buffle J. (1988) *Complexation Reactions in Aquatic Systems: An Analytical Approach.* John Wiley and Sons, New York.

Buol S. W. and Weed S. B. (1991) Saprolite-soil transformations in the Piedmont and Mountains of North Carolina. *Geoderma* **51**, 15-28.

Burnham C. P. (1989) Pedological processes and nutrient supply from parent material in tropical soils. In *Mineral Nutrients in Tropical Forest and Savanna Ecosystems* (ed. J. Proctor). Blackwell Scientific Publications, Oxford. pp. 27-40.

Burns D. A. (1998) The hydrochemical evolution of stormflow in a forested Piedmont catchment, Ph.D. thesis, State University of New York.

Burns D. A., Hooper R. P., McDonnell J. J., Freer J. E., Kendall C., and Beven K. (1998) Base cation concentrations in subsurface flow from a forested hillslope-The role of flushing frequency. *Water Resources Res.* **34,** 3535-3544.

Cappellato R., Peters N. E., and Ragsdale H. L. (1993) Acidic atmospheric deposition and canopy interactions of adjacent deciduous and coniferous forests in the Georgia Piedmont. *Canadian J. Forest Res.* **23**, 1114-1124.

Cappellato R. and Peters N. P. (1995) Dry deposition and canopy leaching in deciduous and coniferous forests of the Georgia Piedmont: an assessment of a regression model. *J. Hydrology* **169**, 131-150.

Chadwick O. A., Brimhall G. H., and Hendricks D. M. (1990) From black box to a grey box: A mass balance interpretation of pedogensis. *Geomorphology* **3**, 369-390.

Cleaves E. T. (1993) Climatic impact on isovolumetric weathering of a coarse-grained schist in the northern Piedmont Province of the central Atlantic states. *Geomorphology* **8**, 191-198.

Dixon J. B. and Weed S. B. (1977) *Minerals in Soil Environments.* Soil Science Society of America, Madison.

Dong H., Peacor D. R., and Murphy S. F. (1998) TEM study of progressive alteration of igneous biotite to kaolinite throughout a weathered soil profile. *Geochim. Cosmochim. Acta* **62**, 1881-1887.

Driscoll C. T., van Breemen N., and Mulder J. (1985) Aluminum chemistry in a forested spodosol. *Soil Sci. Soc. Am. J.* **49**, 437-444.

Erikkson E. and Kunakasen V. (1969) Chloride concentrations in groundwater, recharge rate and rate of deposition of chloride in the Israel Coastal Plain. *J. Hydrology* **7**, 178-197.

Flury M. and Fluhler H. (1994) Susceptibility of soils to preferential flow of water: A field study. *Water Resources Res.* **30**, 1945-1954.

Grant W. H. (1975) Chemical Weathering of Panola Adamellite with special reference to apatite. *Southeastern Geology* **17**, 15-25.

Hamdan J. and Burnham C. P. (1996) The contribution of nutrients from parent material in three deeply weathered soils of Peninsular Malaysia. *Geoderma* **74**, 219-233.

Higgins M. W., Atkins R. L., Crawford T. J., Crawford R. F., III, Brooks R., and Cook R. (1988) The structure, stratigraphy, tectonostratigraphy and evolution of the southernmost part of the Appalachian orogen. *U. S. Geological Survey Prof. Paper* **1475**, 173p.

Hillel D. (1982) *Introduction to Soil Physics*. Academic Press, San Francisco.

Huntington T. G. (1995) Carbon sequestration in an aggrading forest ecosystem in the Southeastern USA. *Soil Sci. Soc. Am. J.* **59**,1459-1467.

Huntington T. G., Johnson A. H., and Schwartzman T. N. (1990) Mechanical vacuum extraction vs. batch equilibrium for estimation of exchangeable cations. *Soil Sci. Soc. Am. J.* **54**, 381-385.

Huntington T. G., Hooper R. P., Peters N. E., Bullen T. D., and Kendall C. (1993) Water, energy and biogeochemical budgets investigation at Panola Mountain Research Watershed, Stockbridge, Georgia - A research plan. *U. S. Geological Survey Open File Report.* **93-55**, 49.

Huntington T. G., Blum A. E., and White A. F. (1994) Migration of a bromide tracer in a forest soil in the Georgia Piedmont. *EOS* **75**(16), 150.

Huntington T. G., Hooper R. P., Johnson C. E., Aulenbach B. T., Cappellato R., and Blum A. E. (2000) Calcium depletion in forest ecosystems of the southeastern United States. *Soil Sci. Soc. Amer. J.* **64** 1845-1858.

Jardine P. M., Parker J. C., and Zelazny L. W. (1985) Kinetics and mechanisms of aluminum adsorption on kaolinite using two-site non-equilibrium transport model. *Soil. Sci. Soc. Am. J.* **49**, 867-872.

Jeong G. Y. (1998) Vermicular kaolinite epitactic on primary phyllosilicates in the weathering profiles of anorthosite. *Clays & Clay Minerals* **46**, 509-520.

Kalinowski B. E. and Schweda P. (1996) Kinetics of muscovite, phlogopite and biotite dissolution and alteration at pH 1-4, room temperature. *Geochim. Cosmochim. Acta* **60**, 367-385.

Kenoyer G. J. and Bowser C. J. (1992) Groundwater chemcial evolution in a sandy silicate aquifer in northern Wisconsin 2. Reaction modeling. *Water Resources Res.* **28**, 591-600.

Kirkwood S. E. and Nesbitt H. W. (1991) Formation and evolution of soils from an acidified watershed: Plastic Lake, Ontario, Canada. *Geochim. Cosmochim. Acta* **55**, 1295-1308.

Klute A. and Dirksen C. (1986) Hydraulic conductivity and diffusivity: Laboratory methods. In *Methods of Soil Analysis* (ed. A. Klute). Soil Sci. Soc. America, Madison. pp. 687-699.

Kretzschmar R., Robarge W. P., Amoozegar A., and Vepraskas M. J. (1997) Biotite alteration to halloysite and kaolinite in soil-saprolite profiles developed from mica schist and granite gneiss. *Geoderma* **75**, 155-170.

Land M., Ingri J., and Ohlander B. (1999) Past and present weathering rates in northern Sweden. *Applied Geochem.* **14**, 761-774.

Lee M. R. and Parsons, P. I.. (1999) Biomechanical and biochemical weathering of lichen-encrusted granite: Textural controls on organic-mineral interactions and deposition of silica-rich layers. *Chem. Geology* **161**, 385-397.

Levy G. J., van der Watt H. H., Shainberg I., and Plessis H. M. (1988) Potassium-calcium exchange and sodium-calcium exchange on kaolinite and kaolinitic soils. *Soil Sci. Soc. Am. J.* **52**, 1259-1264.

Markewich H. W. and Pavich M. J. (1991) Soil chronosequence studies in temperate to subtropical, low-latitude, low-relief terrain with data from the eastern United States. *Geoderma* **51**, 213-239.

Mehra O. P. and Jackson M. L. (1960) Iron oxide removal from soils and clays by a dithionite-citrate system buffered with sodium bicarbonate. In *Proceedings 7th National Conference on Clays and Clay Minerals* (ed. A. Swineford). Pergamon Press, New York. pp. 317-327.

Murphy S. F., Brantley S. L., Blum A. E., White A. F., and Dong H. (1998) Chemical weathering in a tropical watershed, Luquillo Mountains, Puerto Rico. II Rate and mechanism of bioitite weathering. *Geochem. Cosmochim. Acta* **62,** 227-243.

Nixon R. A., III. (1981) Rates and mechanisms of chemical weathering in an organic environment at Panola Mountain, Georgia. Ph.D. thesis, Emory University, GA (USA).

O'Brien E. L. and Buol S. W. (1984) Physical transformations in a vertical soil-saprolite sequence. *Soil Sci. Amer. J.* **48**, 354-357.

Pačes T. (1986) Rates of weathering and erosion derived from mass balance in small drainage basins. In *Rates of Chemical Weathering of Rocks and Minerals* (eds. S. M. Colman & D. P. Dethier). Academic Press, San Francisco. pp. 531-550.

Pavich M. J., Brown L., and Valette-Silver H. N. (1995) ^{10}Be analysis of a Quaternary weathering profile in the Virginia Piedmont. *Geology* **13**, 39-41.

Schoenberger P. J., Amoozegar A., and Buol S. W. (1995) Physical properties of a soil and saprolite continuum at three geomorphic positions. *Soil Sci. Soc. Amer. J.* **59**, 1389-1397.

Schroeder P. A. and Melear N. D. (1999) Stable carbon isotope signatures preserved in authigenic gibbsite from a forested granitic regolith: Panola Mt. Georgia, USA. *Geoderma* **91**, 261-279.

Schulz M. S. and White A. F. (1999) Chemical weathering in a tropical watershed, Luquillo Mountains, Puerto Rico III: Quartz dissolution rates. *Geochim. Cosmochim. Acta* **63**, 337-350.

Shanley J. B. (1992) Sulfate retention and release in soils at Panola Mountain, Georgia. *Soil Science* **153**, 499-508.

Simpson G. G. (1986) Hydraulic characteristics of soil-saprolite profiles from North Carolina Piedmont. *Proc. 29th Meeting of the Soil Sci. Soc., North Carolina*, 147-200.

St. Pierre T. G., Singh B., Webb J., and Gilkes B. (1992) Mössbauer spectra of kaolins from south-western Australia. *Clay and Clay Minerals* **40**, 341-346.

Stonestrom D. A., White A. F., and Akstin K. C. (1998) Determining rates of chemical weathering in soils-solute transport versus profile evolution. *J. Hydrology* **209**, 331-345.

Sverdrup K. and Warfvinge P. (1995) Estimating field weathering rates using laboratory kinetics. In *Chemical Weathering Rates of Silicate Minerals* (eds. A. F. White and S. L. Brantley). Reviews in Mineralogy **31**, Mineralogical Society of America, Washington, D.C. pp. 485-541.

Swodoba-Colberg N. G. and Drever J. D. (1993) Mineral dissolution rates in plot-scale field and laboratory experiments. *Chem. Geology* **105**, 51-69.

Taylor A. B. and Velbel M. A. (1991) Geochemical mass balances and weathering rates in forested watersheds of the southern Blue Ridge II. Effects of botanical uptake terms. *Geoderma* **51**, 29-50.

Vapraskas M. J. and Williams J. P. (1995) Hydraulic conductivity of saprolite as a function of sample dimensions and measurement technique. *J. Soil Sci. Amer. J.* **59**, 975-981.

Velbel M. A. (1985) Geochemical mass balances and weathering rates in forested watersheds of the southern Blue Ridge. *Amer. J. Sci.* **285**, 904-930.

Velbel M. A. (1995) Interaction of ecosystem processes and weathering processes. In *Solute Modelling in Catchment Systems* (ed. S. T. Trudgill). John Wiley and Sons, New York. pp. 193-209.

White A. F. (1995) Chemical weathering rates in soils. In *Chemical Weathering Rates of Silicate Minerals* (eds. A.F. White and S.L. Brantley). Reviews in Mineralogy **31**, Mineralogical Society of America, Washington, D.C. pp. 407-458.

White A. F. and Blum A. E. (1995) Effects of climate on chemical weathering rates in watersheds. *Geochim. Cosmochim. Acta* **59**, 1729-1747.

White A. F. and Brantley S. L. (1995) Chemical weathering rates of silicate minerals: An overview. In *Chemical Weathering Rates of Silicate Minerals* (eds. A.F. White and S.L. Brantley). Reviews in Mineralogy **31**, Mineralogical Society of America, Washington, D.C. pp. 1-22.

White A. F. and Claassen H. C. (1978) Dissolution kinetics of silicate rocks, application to solute modeling. In *Speciation, Sorption, Solubility and Kinetics in Aqueous Systems* (ed. E. Jenne). American Chemical Society Symposium Series **93**, American Chemical Society, Washington, D.C. pp. 771-793.

White A. F. and Peterson M. L. (1990) Role of reactive surface area characterization in geochemical models. In *Chemical modeling of aqueous systems II* (eds. D. C. Melchior and R. L. Bassett), American Chemical Society Symposium Series **416**, American Chemical Society, Washington, D.C. pp. 771-793.

White A. F., Blum A. E., Schulz M. S., Bullen T. D., Harden J. W., and Peterson M. L. (1996) Chemical weathering of a soil chronosequence on granitic alluvium 1. Reaction rates based on changes in soil mineralogy. *Geochim. Cosmochim. Acta* **60**, 2533-2550.

White A. F., Blum A. E., Schulz M. S., Vivit D. V., Larsen M., and Murphy S. F. (1998) Chemical weathering in a tropical watershed, Luquillo Mountains, Puerto Rico: I. Long-term versus short-term chemical fluxes. *Geochim. Cosmochim. Acta* **62**, 209-226.

White A. F., Blum A. E., Bullen T. D., Vivit D. V., Schulz M., and Fitzpatrick J. (1999a) The effect of temperature on experimental and natural weathering rates of granitoid rocks. *Geochim. Cosmochim. Acta* **63**, 3277-3291.

White A. F., Bullen T. D., Vivit D. V., and Schulz M. S. (1999b) The role of disseminated calcite in the chemical weathering of granitoid rocks. *Geochim. Cosmochim. Acta*, **63**, 1939-1953.

White A. F., Bullen, T. D, Schulz M. S., Blum A. E., Huntington T. G., and Peters N. E. (2001) Differential rates of feldspar weathering in granitic regoliths. *Geochim. Cosmochim. Acta* **65**, 847-869.

Water-Rock Interactions, Ore Deposits, and Environmental Geochemistry: A Tribute to David A. Crerar

Editors: Roland Hellmann and Scott A. Wood

Nonsteady-state dissolution of goethite and hematite in response to pH jumps: the role of adsorbed Fe (III)

SHERRY D. SAMSON* AND CARRICK M. EGGLESTON

Department of Geology and Geophysics, University of Wyoming, Laramie, WY 82071 USA

* Author to whom correspondence should be addressed (samson@colorado.edu). Present address: Department of Geological Sciences, University of Colorado, Boulder CO 80309 USA

Abstract—Dissolution transients following downward pH jumps to pH 1 from a variety of higher pH values during dissolution in a mixed-flow reactor contain information about dissolution processes and mechanisms. Despite more than an order of magnitude difference in steady-state dissolution rates, the transients for goethite (α-FeOOH) (this study) and hematite (α-Fe_2O_3) (Samson and Eggleston, 1998) are similar in their pH dependence and relaxation times. After a pH jump, the time required to reach a new steady state is 40 to 50 hours. The amount of excess Fe released in the transients (defined as the amount of Fe released in excess of that released in an equivalent length of time during steady-state dissolution at pH 1) increases with increasing initial (i.e., pre-jump) pH, and is dependent on initial pH in a manner similar to the pH dependence of Fe^{3+} adsorption to other oxides. We suggest that the excess Fe released in the transients is derived from partial dissolution or depolymerization of the iron (hydr)oxide at pH $\geq$ 1 and the transition of such Fe into the adsorbed state on the mineral surface. Following a pH jump to pH 1, this adsorbed Fe is desorbed by stepwise depolymerization.

1. INTRODUCTION

Mineral weathering has been a subject of geochemical study for at least 150 years because weathering is a key component of elemental cycling at and near Earth's surface, and because there are also a number of industrial and technological reasons for understanding such processes. Among these are mitigating the deleterious effects of scale deposits, through either inhibited growth or engineered dissolution, in settings ranging from production wells to power plants; corrosion control (e.g., iron oxides in particular play an important role in the passivation of metal surfaces); and understanding how pollutants participate in mineral dissolution and growth, and may, through solid solution, become sequestered in secondary products of the weathering of primary minerals.

Great progress has been made in understanding the environmental factors that affect mineral dissolution rates (e.g., pH, saturation state, secondary precipitates, and the presence of organic and other ligands; see Hochella and White, 1990, and White and Brantley, 1995, for reviews). A virtually universal observation made in experimental studies is that dissolution rates often start high and decay toward a steady state. Such transients have been reported for many oxide and silicate minerals: e.g., quartz (Holt and King, 1955), corundum and kaolinite (Carroll-Webb and Walther, 1988), goethite and δ-Al_2O_3 (Kraemer and Hering, 1997), hematite (Maurice et al., 1995), albite (Holdren and Berner, 1979; Chou and Wollast, 1985; Hellmann, 1995), biotite (Malmström and Banwart, 1997), diopside (Knauss et al., 1993), and enstatite (Schott et al., 1981). In a few cases, dissolution rates accelerate to a steady state, e.g., quartz at pH > 10 (Knauss and Wolery, 1988).

Transients have often been thought to be limited to the onset of dissolution, and, depending on the solid or the experiment, have been variously attributed to one or more of the following: artifacts of sample preparation, e.g., grinding (Schott et al., 1981; Eggleston et al., 1989), ultra-fine particles (Holdren and Berner, 1979), development of a leached layer (Casey and Bunker, 1990), and nonstoichiometry of the unreacted surface (Burch et al., 1993). Information on rates obtained during the nonsteady-state interval was often discarded as unrepresentative of the reaction of interest, e.g., the long-term dissolution rate of the mineral, and pretreatments such as etching with HF or preconditioning (e.g., 3-4 days of pre-experiment dissolution) have been employed specifically to avoid initial transients (Zinder et al., 1986).

However, observations of recurrent transients in response to cycles in pH (Chou and Wollast, 1984; Spokes and Jickells, 1996), increases in ligand concentration (Mast and Drever, 1987; Wieland and Stumm, 1992; Kraemer and Hering, 1997), and changes in electrolyte composition (Sjöberg, 1989) suggest that in some cases transients contain useful information about dissolution processes. For example, Chou and Wollast (1984) induced recurring dissolution transients for albite by instantaneously modifying the pH of the input solution in a fluidized-bed reactor, and demonstrated that transients need not be limited to the onset of dissolution. They interpreted the transients as indications of readjustment of altered or leached layers on the feldspar surface in response to the pH jumps. Furthermore, the transition from one steady state to another takes time; the time required for albite to reach a new steady state in response to pH jumps can be 200 hours or more at 25 °C (Chou and Wollast, 1984).

Here, we report the results of pH-jump experiments with goethite (α-FeOOH), compare them to previous results for hematite (Samson and Eggleston, 1998), and use this comparison to explore the role of adsorbed Fe(III) as a reaction intermediate in the dissolution of goethite and hematite. Because the solutions reach a steady state and closely approach equilibrium during the intervals at pH > 1, we show how the idea of adsorbed Fe(III) on hematite surfaces can be fit within a relatively simple surface complexation model.

2. EXPERIMENTAL APPROACH

2.1. Goethite Preparation

Goethite was prepared according to a hydrothermal method using 1 M $Fe(NO_3)_3$ (prepared from $Fe(NO_3)_3 \cdot 9H_2O$; Spectrum reagent A.C.S.) and 5 M KOH (Spectrum reagent, A.C.S.) (Schwertmann and Cornell, 1991). The goethite was suspended in deionized water, centrifuged and decanted several times to remove the finest particles, and aged in deionized water for approximately 30 days. It was again washed and centrifuged twice and dried overnight in a 70°C oven prior to XRD and SEM. Powder XRD (Scintag XDS 2000 Diffractometer equipped with a Cu anode tube) confirmed the goethite structure and SEM (JEOL 35CF) revealed uniform acicular crystals of approximately 1.5 µm in length with no visible amorphous material. The specific surface area, as measured with a Micromeritics TriStar 3000 multipoint N_2 BET surface area analyzer, was 36 m^2 g^{-1}.

2.2. Analytical Techniques and Experimental Conditions

Total dissolved Fe effluent concentrations were measured by flame atomic absorption (Perkin Elmer Model 2380 AA spectrophotometer), with a detection limit of 10 ppb Fe. Separate standards were prepared for each experiment from the inlet solutions used in that experiment.

The pH-jump experiments were conducted in a polycarbonate, continuous-flow overhead-stirred tank reactor (CSTR) at ambient laboratory temperature (22.7 ± 1° C, with the exception of 1 day where the temperature reached 25.8° C) with a Mettler Toledo titration system stirrer motor and stir paddle (input power 4.5V, ~3100 rpm). A discussion of the reactor properties as they relate to mineral dissolution may be found in Samson et al. (2000). CO_2 was not excluded. The reactor was pH-statted (i.e., pH maintained at constant values) using HNO_3 (TraceMetal grade, Fisher Scientific) and NaOH (reagent special grade, A.C.S., Spectrum), with software, burettes, and a voltmeter from McIntosh Analytical Systems. The Ag/AgCl combination electrode (Ross Orion semi-micro model 8305BN) used was calibrated at pH 2.0, 4.0, and 7.0.

A flow rate of 0.6 mL min^{-1} was maintained with variable flow peristaltic pumps. Cell volume was approximately 100 mL, with a goethite concentration of 10 g L^{-1}. Effluent was filtered through a 0.2-µm cellulose acetate membrane (Micro Filtration Systems) mounted in an in-line 47 mm diameter Gelman filter holder at the base of the reactor cell, and collected continuously in acid-washed polypropylene test tubes, 10 minutes per tube for the first two hours following the pH jump and 20 minutes per tube thereafter, with an ISCO 500 Retriever fraction collector.

All solutions were prepared with high-purity, 18.2 MΩ-cm water with total organic carbon ≤ 10 ppb (Milli-Q Plus, Millipore). pH-adjusted input solutions were stored in VWRbrand™ TraceClean amber glass jugs. The pH in the reactor and the input reservoir was changed nearly simultaneously. The pump was attached to a new inlet solution reservoir pre-adjusted to the new pH and the pH change in the reactor was made by pipettor with concentrated acid (16 N HNO_3) or base (8 N NaOH) in order to accomplish the change as quickly as possible (within seconds to minutes) with a minimum increase in volume.

There were two pH-jump experiments in the following sequence of pH: 4.5, 1, 2, 1 for the first experiment and 3, 1 for the second. In each experiment, the first pH jump to pH 1 was preceded by six days of flow at pH 4.5 or pH 3, respectively. Subsequent pH jumps in the first experiment were at 72-hour intervals. The second experiment was unintentionally terminated on the seventh day, six hours after the jump from pH 3 to pH 1.

2.3. Modeling

The kinetic model used to fit our transient data has been justified and described previously (Samson et al., 2000). The overall dissolution rate, $Rate_{diss}$ [mol m^{-2} s^{-1}], is the sum of the rates for parallel dissolution reactions (assumed to be first-order) taking place at various types of surface sites. In the model, these var-

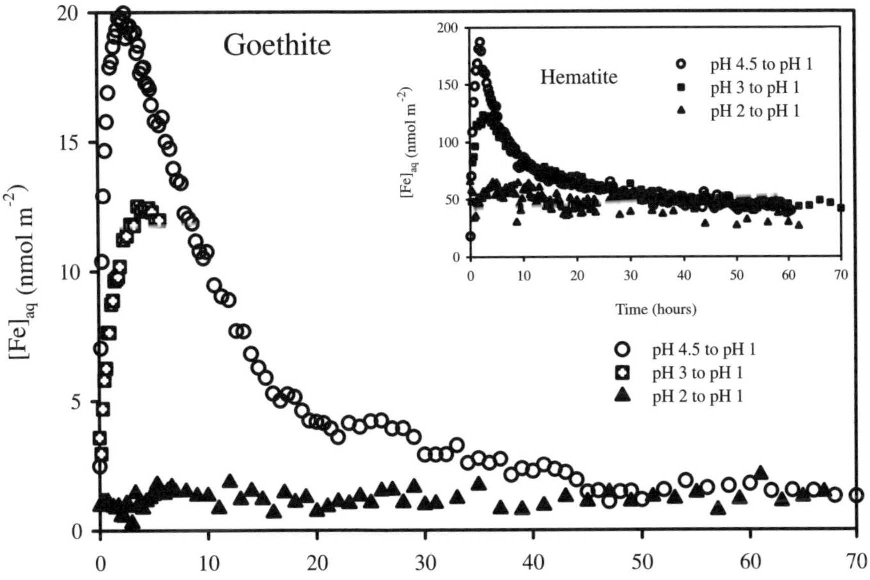

Fig. 1. Results of α-FeOOH pH-jump experiments (this study) with results of α-Fe_2O_3 pH-jump experiments (Samson and Eggleston, 1998) included for comparison. Each data series represents the Fe outlet concentrations following a pH jump at time 0 from the initial pH indicated to pH 1.

ious surface sites are represented by four groupings of surface sites with similar rate coefficients:

$$Rate_{diss} = k_{SS1}[>Fe_{SS1}] + k_{PI}[>Fe_{PI}] + k_{EXF}[>Fe_{EXF}] + k_{EXS}[>Fe_{EXS}] \quad (1)$$

where $[>Fe_{SS1}]$ [mol m^{-2}] is the constant concentration of Fe surface sites active for dissolution at steady state at pH 1 (> denotes bonds to the surface), $[>Fe_{PI}]$, $[>Fe_{EXF}]$, and $[>Fe_{EXS}]$ [mol m^{-2}] are the concentrations of dissolution-active Fe sites in excess of those present at steady state at pH 1 that are consumed in a pH jump from higher pH to pH 1 (pulse input, fast-dissolving excess, and slow-dissolving excess, respectively), and k_{SS1}, k_{PI}, k_{EXF}, and k_{EXS} [s^{-1}] are the corresponding rate coefficients.

The four groups of sites represent conceptual groupings of surface sites with similar rate coefficients. We are not suggesting that there are only four structural types of surface sites; our macroscopic data provide us with kinetic, not structural, information. The pulse input parameter represents sites whose dissolution is indistinguishable from a pulse input, i.e., the dissolution rate is faster than can be mathematically resolved given the reactor residence time and the sampling interval. The remaining excess dissolution-active sites are consumed at rates which are also indicative of their relative reactivity, one type dissolving rather quickly ($[>Fe_{EXF}]$) and the other more slowly ($[>Fe_{EXS}]$).

Although we found in previous work (Samson et al., 2000) that at least four groups of surface sites were needed to simultaneously model data from three different solids (α-Fe_2O_3, α-Al_2O_3, and In_2O_3), not all the parameters are needed to describe the data set presented here. Specifically, it was not necessary to use the pulse input parameter, as all rate coefficients were resolvable. We nevertheless include all parameters in the derivation for completeness and applicability to transients other than those presented here.

The total aqueous Fe concentration in the reactor effluent, $[Fe]_{TOT}$ [mol L^{-1}], is the sum of contributions from the four dissolution reactions:

$$[Fe]_{TOT} = \tau A Rate_{SS1}\left(1-e^{-t/\tau}\right) + A[>Fe_{PI}]e^{-t/\tau} + \tau_{EXF} A k_{EXF}[>Fe_{EXF}]_o e^{-k_{EXF}t}\left(1-e^{-t/\tau_{EXF}}\right) + \tau_{EXS} A k_{EXS}[>Fe_{EXS}]_o e^{-k_{EXS}t}\left(1-e^{-t/\tau_{EXS}}\right) \quad (2)$$

where $Rate_{SS1}$ [mol m^{-2} s^{-1}] is the steady-state dissolution rate at pH 1, A [m^2 L^{-1}] is the goethite surface area per unit of solution volume, and τ is a time constant defining the rate of accumulation of Fe within the reactor. Equation 2 is derived from the model (Eqn. 1) and standard chemical engineering approaches to nonsteady-state processes in mixed-flow reactors (see Samson et al., 2000). For both steady-state dissolution and a pulse input, the time constant equals reactor volume divided by flow rate ($\tau = V\, q^{-1}$), while the time constant for dissolution of the fast- and slow-dissolving excess sites is a function of not only volume and flow rate, but also of the dissolution reaction rate coefficient: $\tau_{EX(F\ or\ S)} = V/(q - Vk_{EX(F\ or\ S)})$.

3. RESULTS AND MODEL FITS

In spite of dissolution rates more than an order of magnitude slower than those of hematite, the goethite dissolution transients following downward pH jumps to pH 1 were remarkably similar to those of hematite in their pH dependence and relaxation times (Fig. 1). We emphasize that the data presented here for both goethite and hematite are for nonsteady-state dissolution *at pH 1*; the periods of time spent at the higher pH conditions serve to allow the surface to generate or regenerate the active sites involved in dissolution at pH 1.

Following the reactor's mixing interval of ~5τ (based on CSTR theory), the reactor outlet concentration directly reflects the dissolution rate. Furthermore, any solute derived from sites whose dissolution is perceived as a pulse input, $[>Fe_{PI}]$, will have been washed out of the reactor, and the outlet concentration attributable to steady-state dissolution at pH 1 will have reached a constant value. Therefore, by subtracting the steady-state outlet concentration from the observed outlet concentrations, the amount of solute derived from the dissolution of $[>Fe_{EXF}]$ and $[>Fe_{EXS}]$ sites can be determined. Assuming the goethite dissolution rate in the interval 5τ to 7.5τ (hours 15 to 22.5) is dominated by dissolution of the slowly dissolving sites, a value for the parameter k_{EXS} can be obtained from the slope of a plot of the natural log of the excess Fe concentration (the measured outlet concentration minus the steady-state concentration) vs. time. The value for k_{EXS} for the pH jump from pH 4.5 to pH 1 is 0.11 h^{-1}. An independent value for k_{EXS} for the pH 3 to pH 1 transient cannot be determined due to the lack of data in the interval 5τ to 7.5τ. However, previously the hematite results had been successfully modeled with a single value of k_{EXS} for all data sets (see Table 1). Therefore, a value of 0.11 h^{-1} for k_{EXS} was assumed for both goethite transients (pH 4.5 to pH 1, and pH 3 to pH 1), and the remaining model parameters were obtained by nonlinear least-squares fitting of each data set to Eqn. 2.

Figure 2 shows model fits for the transients following pH jumps from pH 4.5 and pH 3 to pH 1 (there was no transient following the pH jump from pH 2) including the relative contributions from two of the three available types of excess dissolution-active sites considered in the model; the use of the pulse input parameter was unnecessary as indicated by zeros in Table 1. The hematite model fits are shown in Fig. 3. Model parameters for the goethite transients are provided in Table 1 along with hematite parameters (Samson et al., 2000) for comparison. Clearly, different amounts of Fe surface sites active for dissolution at pH 1 are generated by aging at different pH values greater than pH 1. Below, we explore how adsorbed Fe might relate to these amounts.

4. DISCUSSION

4.1. Dissolution-active Sites

As in this study of goethite, previous experiments designed to investigate the response of hematite (α-Fe_2O_3), corundum (α-Al_2O_3), and cubic In_2O_3 dissolution rates to abrupt changes in pH (Samson et al., 2000) revealed that following pH jumps to pH 1 from higher pH solutions, a reproducible and regenerable nonsteady-state period of elevated dissolution rate occurs. The results are consistent with the dissolution (at low pH) and rapid regeneration (at higher pH) of a

Table 1. Model parameters (units of $Rate_{SS1}$ are in terms of moles of Fe) with hematite parameters (Samson et al., 2000) for comparison. A value of zero indicates use of the parameter was not required in order to fit the data.

Mineral	α-FeOOH	α-FeOOH	α-Fe_2O_3	α-Fe_2O_3	α-Fe_2O_3	α-Fe_2O_3	α-Fe_2O_3
Initial pH	3	4.5	2	2.5	3	4.5	6
k_{EXF} [h^{-1}]	1.29	1.29	0	0.78	0.78	0.78	0.78
k_{EXS} [h^{-1}]	0.11	0.11	0.07	0.07	0.07	0.07	0.07
$[>Fe_{PI}]_o$ [nmol m^{-2}]	0	0	43	0	0	0	0
$[>Fe_{EXF}]_o$ [nmol m^{-2}]	4	19	0	100	96	249	316
$[>Fe_{EXS}]_o$ [nmol m^{-2}]	51	57	217	363	717	772	804
$Rate_{SS1}$ [pmol m^{-2} s^{-1}]	0.17	0.17	5.6	7.2	7.8	8.6	6.9
Total Excess [nmol m^{-2}]	55	76	260	463	813	1021	1120

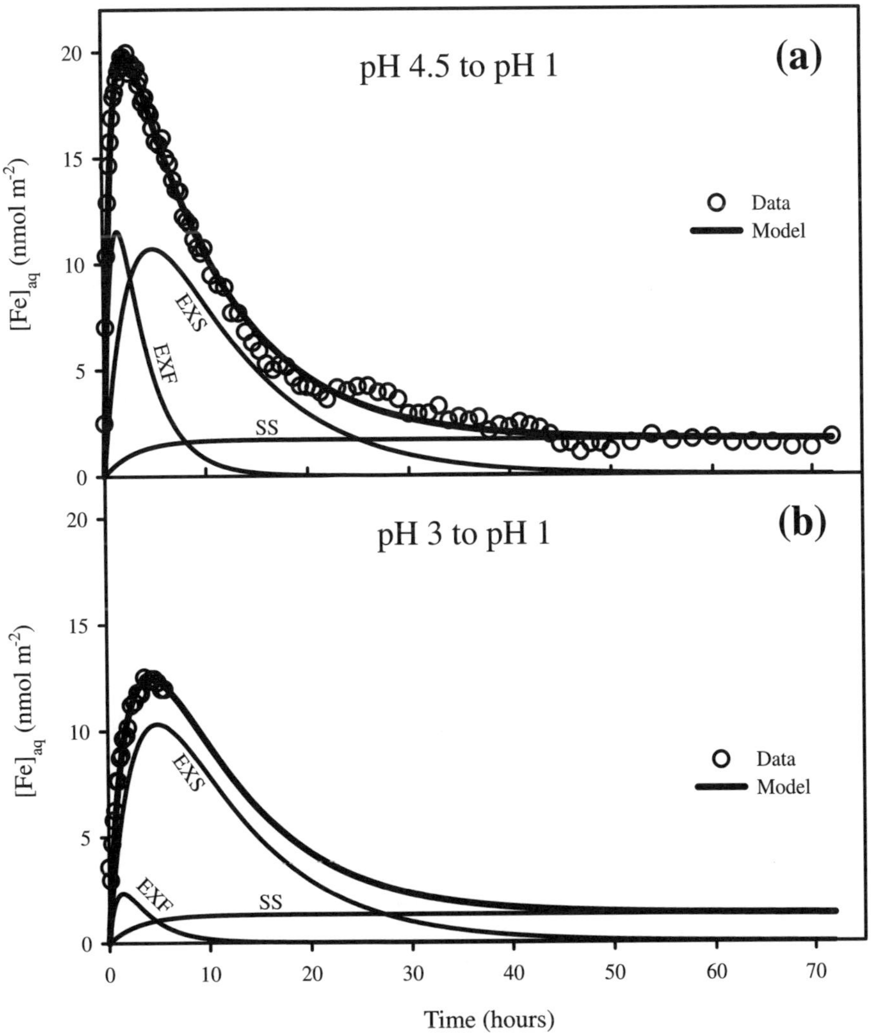

Fig. 2. Model fits for α-FeOOH transients following pH jumps to pH 1 from (a) pH 4.5 and (b) pH 3. EXF, EXS, and SS represent contributions to the total modeled dissolution from fast- and slow-dissolving excess sites and steady-state dissolution, respectively.

form of dissolution-labile metal ion at the oxide surface. We equate such dissolution-labile sites with active sites for dissolution.

Surface complexation models for steady-state dissolution assume a constant ratio of active sites to total (active and less active) sites, with the understanding that dissolution is favored at a relatively small fraction of active sites with relatively low activation energies (Furrer and Stumm, 1986; Wieland et al., 1988). A departure from a steady-state rate of dissolution in response to a change in conditions implies a corresponding change (i.e., increase or decrease) in the concentration of dissolution-active sites. Where do such sites originate? A number of possibilities have been suggested, including the formation of altered or leached layers on the feldspar surface (Chou and Wollast, 1984), aspects of surface topography such as surface roughness and crystal defects (Helgeson et al., 1984; Lasaga and Blum, 1986), and the presence of secondary phases.

Another possibility is the presence of adsorbed "nutrient," e.g., Fe^{3+} on hematite, as postulated by Burton-Cabrera-Frank (BCF)-type theories of crystal growth and dissolution (Burton et al., 1951). Here, the term "adsorbed" refers to surface Fe^{3+} that has no more than three bonds to the surface (i.e., no greater

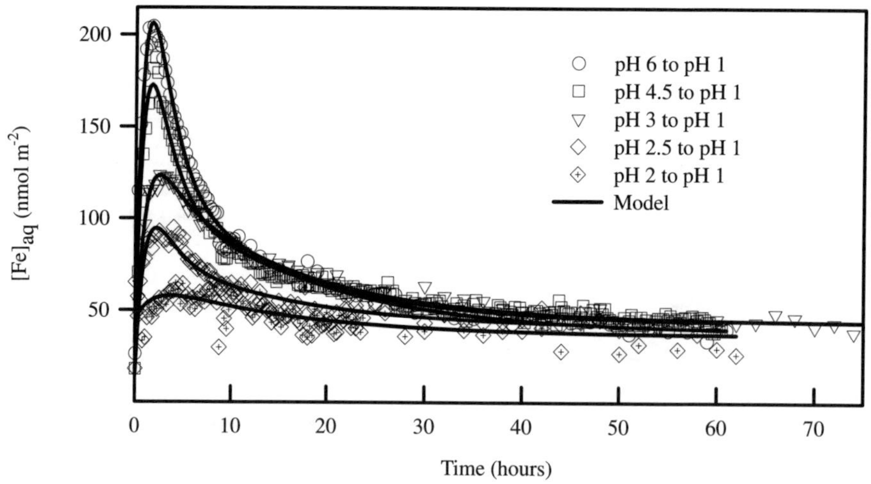

Fig. 3. Model fits for α-Fe_2O_3 transients following pH jumps to pH 1 at time 0 from the initial pH indicated.

polymerization than Fe^{3+} at kink sites) and that may occupy surface sites that are distinct from the terminated bulk hematite structure. Although by definition this includes Fe^{3+} that may be adsorbed or precipitated from solution during an upward jump in pH, we must emphasize that this source of Fe alone cannot account for the amount of Fe that is released in the transients following subsequent downward pH jumps (Samson and Eggleston, 2000; Samson et al., 2000). Iron released in the transients must also include Fe extracted from the terminated bulk hematite structure by depolymerization into the more labile adsorbed state (e.g., surface complexes such as kink sites, ledges, and adatoms). To support these ideas, and to introduce the basic issues involved, we briefly review earlier investigations that attributed dissolution transients to adsorbed "nutrient".

4.2. Lattice Metal Ions as Adsorbates

Holt and King (1955) noted that quartz and amorphous silica dissolve with an initial rapid dissolution transient. Artifactual disturbed layers due to grinding and other theories for the high-rate transient were proposed. However, surface electron diffraction showed no disordered surface layer (Holt and King, 1955). They dissolved a "disturbed" layer on a pretreated quartz powder in alkaline (pH ~13) solution and found that immersion of the resulting powder in an *undersaturated* silica solution at pH 8 produced a gradual *decrease* in the silica concentration over a period of 16 to 20 hours, which they attributed to silica adsorption. Redissolution of this powder at pH 13 gave another initial transient. Therefore, the initial transient was restored by adsorption of silicic acid. They also found, using ^{31}Si, that the original powder would exchange with silica in solution, i.e., the ^{31}Si was diluted by adsorption-exchange with the surface, suggesting that an adsorbed-silica layer forms.

Bergman and Patterson (1961) also found that silica would adsorb to etched quartz surfaces. The model for dissolution and precipitation of amorphous silica by Fleming (1986) explicitly includes adsorption of silicic acid, and Thornton and Radke (1988) also invoked a silica adsorption step in their kinetic model for silica dissolution and condensation. Gratz and Bird (1993) gave a detailed steady-state model for quartz dissolution based on BCF theory (Burton et al., 1951) and consideration of different types of adsorbed silica groups at the quartz surface.

These observations imply that lattice ions of oxides can adsorb to the oxide surface; further, they can be potential-determining ions. The idea that Fe(III) can be a potential-determining ion on Fe(III) oxides, for example, is not new. This possibility was one reason for Parks' (1965) examination of the relationship between the pH of point of zero charge for Fe(III) oxide minerals and the isoelectric points, or equivalence points, of aqueous Fe(III) species. Parks and de Bruyn (1962) and Parks (1965) suggested that one way to view surface charge on hematite is to consider slight dissolution of hematite, followed by readsorption of hydrolyzed Fe ions. Fuerstenau (1970) pointed out that surface charge on oxides could be accounted for by "Adsorption-dissociation of H^+ from surface hydroxyls…[or] partial dissolution of the oxide and

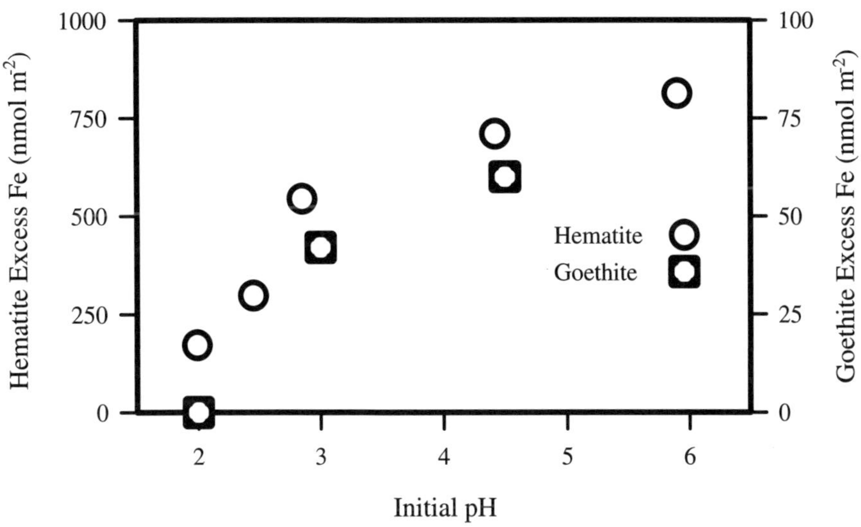

Fig. 4. The amount of excess Fe released in 15 h following downward pH jumps to pH 1, from the initial pH indicated, for goethite (this study) and hematite (Samson and Eggleston, 1998). The amount of excess Fe released following the goethite pH jump from pH 3 is an estimate based on the available 6 hours of data and the model projections (see Fig. 2). Analytical error is within the size of the data points; the determination of excess Fe, however, is also sensitive to the choice of steady state.

formation of hydroxyl complexes in solution, followed by adsorption of these complexes." These earlier ideas have not been greatly explored in the past two decades, but the nonsteady-state dissolution observations have forced us to re-examine them.

4.3. Adsorption of Fe(III) on Hematite

A Mössbauer study of Fe^{3+} adsorption on hematite (Ambe and Ambe, 1990) reported 60% adsorption of Fe^{3+} at pH 2.5. Furthermore, adsorption of Fe(III) on several other solids, including α-SiO_2 and α-Al_2O_3, has been studied (Parks, 1990), and the adsorption edge (i.e., the pH at which 50% of the total amount of a solute cation is adsorbed) is between pH 2 and pH 4, independent of the solid. This is consistent with the general observation that metal ions adsorb roughly at the pH where they start to hydrolyze (James and Healy, 1972; Hachiya et al., 1984a;b); the onset of hydrolysis appears to be more important than the properties of the surface (Fuerstenau, 1970).

Consistent with the latter observation, despite the differences between goethite and hematite crystal structure, dissolution rate, and the amounts of Fe released in the transients following downward pH jumps, the transients exhibit the same pH dependence (Fig. 4). Moreover, the amount of excess Fe released in the transients for both minerals is pH-dependent in a way similar to the pH dependence of Fe^{3+} adsorption to other oxides (Fig. 4). A similar relation to the pH dependence of adsorption was observed for the release of excess Al in pH jump experiments with corundum (Samson et al., 2000).

Here we consider, within the context of existing surface complexation ideas, the formation of adsorbed Fe(III) on iron oxide, and its subsequent dissolution. Because (1) the solubilities of hematite and goethite are extremely low at pH > 3, (2) pH jumps back to pH > 1 from pH 1 involve transient supersaturation, (3) it is extremely difficult to make solutions with less than ppb Fe impurities, and (4) we have observed very little difference in the dissolution transients from non-flowing versus flowing reactors at pH > 3, we assume that equilibrium is closely approached during the times spent at pH > 1 in order to show how adsorbed Fe(III) can be understood within an equilibrium surface complexation model.

Consider an aqueous solution in equilibrium with hematite at pH 1. The predominant aqueous species of Fe(III) in equilibrium with hematite under these conditions is Fe^{3+}(aq). Small amounts of strong base are added to adjust the pH upward and, as pH rises, the amount of Fe^{3+} that can exist in solution in equilibrium with hematite falls at a rate of 3 orders of magnitude per pH unit. Thus, raising the pH from 1 to 2 will cause Fe(III) to attach to the hematite surface. This at-

Table 2. Reactions (> denotes a surface site) and equilibrium constants used for fitting in MICROQL (diffuse double layer model, capacitance 1 F m-2, site density 9.4 g L-1, surface area 4.8 m2 g-1, ionic strength 0.1).

Reaction	Log K	Ref.
$0.5\alpha\text{-}Fe_2O_3 + 3H^+ = Fe^{3+} + 1.5H_2O$	–2.057	(a)
$0.5\alpha\text{-}Fe_2O_3 + 2H^+ = FeOH^{2+} + 0.5H_2O$	–4.247	(b)
$0.5\alpha\text{-}Fe_2O_3 + H^+ + 0.5H_2O = Fe(OH)^{2+}$	–7.727	(b)
$0.5\alpha\text{-}Fe_2O_3 + 1.5H_2O = Fe(OH)_3^0$	–15.79	(c)
$0.5\alpha\text{-}Fe_2O_3 + 2.5H_2O = Fe(OH)_4^- + H^+$	-23.657	(b)
$>FeOH = >FeO^- + H+$	-11.3	(d)
$>FeOH + H^+ = >FeOH_2^+$	5.7	(d)
$0.5\alpha\text{-}Fe_2O_3 + >FeOH + H^+ + H_2O = (>FeO)FeOH^+ + 1.5H_2O$	3.4	(e)
$0.5\alpha\text{-}Fe_2O_3 + >FeOH + 2H^+ = (>FeO)Fe^{2+} + 1.5H_2O$	7.5	(e)
$0.5\alpha\text{-}Fe_2O_3 + 2>FeOH = (>FeO)_2FeOH^0 + 0.5H_2O$	2.67	(e)
$0.5\alpha\text{-}Fe_2O_3 + 3>FeOH + H^+ = (>FeO)_3FeH^+ + 1.5H_2O$	12.9	(e)
$0.5\alpha\text{-}Fe_2O_3 + 3>FeOH = (>FeO)_3Fe^0 + 1.5H_2O$	6.8	(e)

(a) Calculated from the Gibbs free energy value for hematite from Hemingway (1990) and free energy values for Fe^{3+} and H_2O from Wagman et al. (1982).
(b) Calculated using hydrolysis constants from Baes and Mesmer (1976).
(c) Calculated using the hydrolysis constant from the MINTEQA2 database (Allison et al., 1991).
(d) Sahai and Sverjensky (1997).
(e) Obtained by trial-and-error fitting (this study).

tachment must in all likelihood proceed by the stepwise attachment of single bonds. Adopting a traditional 2-pK model for surface sites (i.e., the surface is assumed to consist of $>FeOH_2^+$, $>FeOH$ and $>FeO^-$ sites), the first adsorbed Fe complex to form might be written stoichiometrically as:

$$> FeOH + Fe^{3+} = (> FeO)Fe^{2+} + H^+ \quad (3)$$

This adsorption process forms a positively charged surface complex despite releasing a proton. Of course, >FeOH sites are scarce at low pH, but the adsorption of a trivalent cation to other, positively charged surface sites is much more energetically unfavorable. After the formation of one bond to the solid, more bonds to the surface can form:

$$(> FeO)Fe^{2+} + > FeOH = (> FeO)_2 Fe^+ + H^+ \quad (4)$$

that repolymerize the adsorbed Fe(III) to the solid. Taken one step further, we can form a neutral surface Fe(III) site according to:

$$(> FeO)_2 Fe^+ + > FeOH = (> FeO)_3 Fe^0 + H^+ \quad (5)$$

Equations 3-5 release three protons as part of polymerizing aqueous Fe^{3+} onto a hypothetical hematite surface. If we add Eqn. 6 for the dissolution of hematite (in order to create aqueous Fe^{3+}) to the sum of reactions 3-5 (Eqn. 7), the resulting equation (Eqn. 8) describes the formation of neutral adsorbed-Fe surface sites via the consumption of hematite (we emphasize that all equations here are equilibrium, not kinetic, expressions):

$$0.5\alpha\text{-}Fe_2O_3 + 3H^+ = Fe^{3+} + 1.5H_2O \quad (6)$$

$$Fe^{3+} + 3 > FeOH = (> FeO)_3 Fe^0 + 3H^+ \quad (7)$$

$$0.5\alpha\text{-}Fe_2O_3 + 3 > FeOH = (> FeO)_3 Fe^0 + 1.5H_2O \quad (8)$$

This reaction does not involve H^+ directly, although the number of available >FeOH sites is pH-dependent. Equations 3-8 provide an equilibrium framework for the adsorption of Fe(III) to hematite surfaces that is consistent with the overall stoichiometry of hematite dissolution and with solubility equilibrium between hematite and solution. In essence, they allow for a form of Fe that is neither hematite bulk solid nor aqueous solute. It should be remembered that prior to pH jumps from higher pH conditions to pH 1, after which dissolution output is monitored, the samples have been aged in the higher pH solutions for periods of up to a week. Experiments in which the samples were aged at the higher initial pH for various amounts of time show that the dissolution transients can be regenerated in a very short time, after which aging time does not have much effect on the amount of iron involved in a subsequent transient induced by a jump to pH 1 (Samson and Eggleston, 2000). As we have stated above, there is thus a very defensible basis for assuming that equilibrium was closely approached during aging at the initial pH conditions, and thus the use of equilibrium adsorption modeling to better understand the role of adsorbed Fe in the dissolution transients is reasonable.

In addition to the reactions above that depict changes in the polymerization of surface or adsorbed Fe(III) on hematite, there are protonation-deprotonation reactions that change the charge of adsorbed-Fe(III) sites in a manner analogous to hydrolysis of aqueous species in solution, e.g.:

$$(>FeO)_2Fe^+ + H_2O = (>FeO)_2FeOH^0 + H^+ \quad (9)$$

and

$$(>FeO)Fe^{2+} + H_2O = (>FeO)FeOH^+ + H^+ \quad (10)$$

The product of Eqn. 9 (resulting from hydrolysis on the surface) is the same as the product of Eqn. 11 below (adsorption of hydrolyzed species):

$$2>FeOH + FeOH^{2+} = (>FeO)_2FeOH^0 + 2H^+ \quad (11)$$

By simply reversing the sequence of polymerization above (i.e., by adding a proton) we can force surface Fe closer to the dissolved state, e.g.:

$$(>FeO)_2FeOH^0 + H^+ = >FeOH + (>FeO)FeOH^+ \quad (12)$$

or we may create charged surface sites, e.g.:

$$(>FeO)_2FeOH^0 + H^+ = (>FeO)_2Fe^+ + H_2O \quad (13)$$

Thus protonation can have different effects depending upon whether a bridging oxygen or terminal hydroxyl is protonated, and these differences are very likely to have significant kinetic consequences. Finally, and in analogy to the structure of goethite, adsorbed Fe may be bonded to the surface via one or more bridging hydroxyls according to:

$$(>FeO)_2Fe^+ + (>FeOH) = (>FeO)_2(>FeOH)Fe^+ \quad (14)$$

The surface species $(>FeO)_2(>FeOH)Fe^+$ formed in Eqn. 14 is indistinguishable from that formed in the forward progress of Eqn. 13 on the basis of any aqueous species, but the presence or absence of a bridging hydroxyl can have important kinetic implications. Therefore we include such species in the surface complexation modeling below.

Hachiya et al. (1984a,b) have shown that the kinetics of metal ion adsorption to alumina are best described by the following sequence of reactions:

$$>MOH + M'(H_2O)_6{}^{z+} \xleftrightarrow{K_{OS}} >MOH...M'(H_2O)_6{}^{z+} \quad (15)$$

$$>MOH...M'(H_2O)_6{}^{z+} \xleftrightarrow{k_{ads}} >MOHM'(H_2O)_5{}^{z+} + H_2O \quad (16)$$

$$>MOHM'(H_2O)_5{}^{z+} \xleftrightarrow{fast} >MOM'(H_2O)_5{}^{(z-1)+} + H^+ \quad (17)$$

where M is the metal ion of the solid and M' is the sorbing metal ion. The formation of an outer-sphere complex (denoted by OS in Eqn. 15) is followed by the loss of a water molecule from the inner sphere of the adsorbing metal ion (in general, even if a metal is hydrolyzed, water loss from the inner hydration sphere is faster than hydroxyl loss; Margerum et al., 1978). This leads to the formation of an inner-sphere hydroxyl-bridged complex that may in turn deprotonate to form an oxygen-bridged inner-sphere complex. Apparently bridging OH^- is an important intermediate in the overall adsorption-desorption process and there is no *a priori* reason that the same process should not also apply to iron oxides.

One way to view active sites in the nonsteady-state dissolution of hematite, then, is as adsorbed Fe that desorbs at pH 1 in response to a downward pH jump from a higher initial pH. The source of such adsorbed Fe would be reactions in which Fe(III) adsorbs to hematite surface sites at slightly acidic to neutral pH (whether by adsorption from solution or by depolymerization of the solid, i.e., Eqn. 8). By aging the hematite and goethite suspensions at various pH values before a pH jump, we may have approached equilibrium with respect to a surface configuration of Fe(III) sites of the types proposed in Eqns. 3-5, 9, and 10. The rapidity with which dissolution-active sites on hematite are regenerated upon return to higher pH from pH 1 (within minutes or less; Samson and Eggleston, 2000) suggests that the intervals at higher pH in these studies (from 3 to 6 days) should be more than adequate for the distribution of surface sites to equilibrate. Therefore, we can inquire experimentally whether the amount of Fe released in a nonsteady-state transient bears any relation to calculated amounts of adsorbed Fe.

Figure 5 shows the result of trial-and-error fitting of calculated abundances of various Fe surface sites to the amount of Fe released after a pH jump that, by our model, corresponds to pulse input (*PI*), fast-dis-

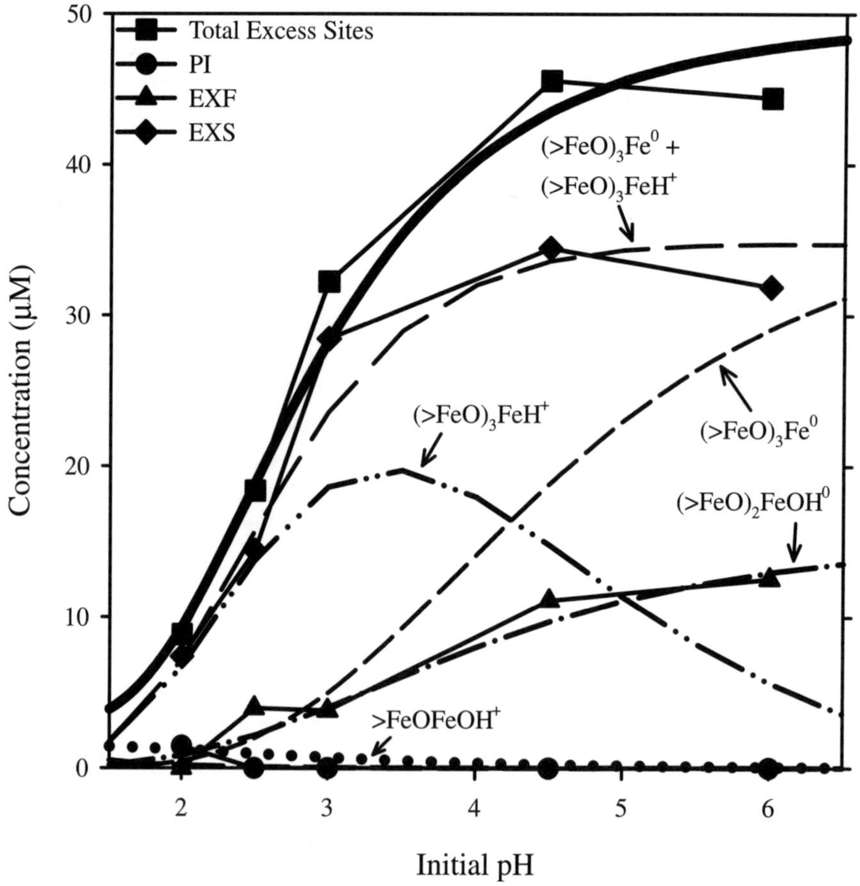

Fig. 5. MICROQL (Westall, 1979) calculations of hypothetical adsorbed Fe surface sites on hematite, as a function of initial pH, and fitted model parameters from hematite pH-jump experiments (Samson and Eggleston, 1998): pulse input (PI), fast-dissolving (EXF) and slow-dissolving (EXS) excess sites, and total excess sites. The heavy solid line represents the calculated total of the Fe surface sites illustrated. Reactions, equilibrium constants, and model assumptions are provided in Table 2.

solving (*EXF*), and slow-dissolving (*EXS*) excess sites (using MICROQL; Westall, 1979). Note that the data points in Fig. 5 are the initial surface concentrations of the different types of excess surface sites derived from our experimental data; another way to view this is that the integrated amount of Fe released in each dissolution transient represents an amount of what we have described as adsorbed Fe created by virtue of the time spent at the initial, higher pH. There is a correspondence between the behavior of *PI* sites and the calculated $>FeOFeOH^+$ sites, between *EXF* sites and the calculated $(>FeO)_2FeOH^0$ sites, and between *EXS* sites and the calculated sum of the triply-bonded sites $(>FeO)_3Fe^0$ and $(>FeO)_3FeH^+$. The latter site represents Fe adsorbed to two $>FeO^-$ sites and one >FeOH site, i.e., $(>FeO)_2(>FeOH)Fe^+$ with one bridging hydroxyl. The rate coefficients for the surface sites proposed in the model correspond with the degree of polymerization of the hypothetical adsorbed-Fe surface sites: the greater the degree of polymerization, the smaller the rate coefficient (i.e., the slower the dissolution).

Use of a greater variety of surface sites did not improve the fit substantially. In addition, fitting of the slow-dissolving excess sites (*EXS*) with less-polymerized species was much less satisfactory. While it is clearly hazardous to make too much of the structural nature of stoichiometric combinations of Fe, surface sites and protons, we did find that any stoichiometric combination involving only two surface sites (i.e., doubly polymerized to the surface) could not adequately fit the concentration versus pH dependence of the total or *EXS* sites active for dissolution at pH 1.

4.4. *Effect of Structural OH^- in Goethite on Surface Ligand Exchange*

The idea that surface Fe sites attached to the solid via bridging OH^- may be subject to particularly slow dissolution rates is intriguing. First, goethite, with structural OH^-, dissolves more slowly at steady state on a total-surface-area-normalized basis than does hematite. In fact, in mixtures of goethite and hematite, all the hematite dissolves before the goethite, even when the hematite is in excess (Cornell and Giovanoli, 1993). Goethite also dissolves more slowly than does hematite during nonsteady-state dissolution transients on a per site basis. However, rate constants for nonsteady-state dissolution vary by less than a factor of two between hematite and goethite. The major difference appears to lie simply in the lower number of dissolution-active sites for goethite.

Dissolution of hematite does not appear to take place preferentially at specific crystal faces. Cornell and Giovanoli (1993), using seven different samples of hematite prepared by different methods and with different morphologies, found a linear relationship between rates of acid dissolution and surface area, independent of surface morphology. However, it has been shown that dissolution of acicular synthetic goethite crystals similar to those used in this study is anisotropic and takes place preferentially on the (001) and (010) faces compared to the (100) face (Cornell et al., 1974). Thus, one explanation for the smaller number of active sites on goethite is simply that less of the surface is participating in active dissolution. This, however, requires an explanation of why some goethite surfaces are apparently inactive. Cornell et al. (1974) point out that the (001) face contains only singly coordinated hydroxyls (i.e., coordinated to one iron atom). The (010) face contains singly and doubly coordinated hydroxyls, whereas the (100) face includes hydroxyls coordinated to 1, 2, and 3 Fe atoms. This suggests the possibility that dissolution-active sites are associated with the singly coordinated hydroxyls, and that slow dissolution is associated with bridging hydroxyls. Whether or not this is due to effects upon local ligand exchange rates is unknown.

We cannot measure directly the ligand exchange rates of surface metal centers, and the coordinative possibilities are more numerous and complex than for aqueous ions. Nevertheless, the reactivity trends of aqueous ions can provide clues to trends in oxide mineral dissolution kinetics. For example, the steady-state proton-promoted dissolution rates of some simple oxides, nesosilicates, and trivalent sesquioxides correlate with the rates of water exchange at the corresponding aqueous ions (Casey, 1991; Casey and Westrich, 1992; Samson et al., 2000). Westrich et al. (1993) also found that the dissolution rates at pH 2 of mixed-cation nesosilicates scale in the same manner as the stoichiometrically-weighted average water exchange rates of the corresponding aqueous cations. Since mineral dissolution is fundamentally a process of ligand exchange, the inference is that the rates of surface ligand exchange for these minerals trend with the rates of water exchange for the aqueous ions.

Hydrolysis, in general, has a labilizing effect on the rates of water exchange at aqueous cations (Margerum et al., 1978), resulting in a trend of increasing rates of water exchange with increasing solution pH, and observations from previous studies (Samson and Eggleston, 1998; Samson et al., 2000) suggest that hydroxylation of the mineral surface with increasing pH has a similar labilizing effect on the rates of ligand exchange of surface metal centers. Consistent with this idea, a study exploring the kinetics of the regeneration of dissolution-active sites on hematite at pH 3 and pH 6 (Samson and Eggleston, 2000) demonstrated that dissolution-active sites are regenerated within minutes or less upon a return to higher pH from pH 1. This labilization, however, may not lead to dissolution unless there is a sufficient supply of protons to aid net depolymerization rather than merely the rapid exchange of surface Fe sites between different polymerized configurations. Protonation, therefore, can have opposing effects. It can participate in the depolymerization of structural or adsorbed Fe, facilitating dissolution, but it may also decrease the ligand exchange rate of the surface metal center, inhibiting dissolution. Here, the response of hematite and goethite dissolution rates to pH jumps allows us to compare, to a first approximation, a case in which the surface ligands that coordinate adsorbed Fe(III) surface sites are likely to be protonated (i.e., bridging hydroxyl in goethite) with a case in which they likely are not (bridging oxygen in hematite). We speculate that the slower dissolution rates for goethite as compared to hematite, under both steady- and nonsteady-state conditions, may be due to inhibition of surface ligand exchange rates by structural OH^- in goethite.

5. SUMMARY AND CONCLUSIONS

Although the dissolution rates of goethite are more than an order of magnitude less than those of hematite, under both steady- and nonsteady-state conditions, the relaxation times of dissolution transients following downward pH jumps to pH 1 from various higher pH values are nearly identical. Furthermore, the amount of excess Fe released in the transients for both minerals is pH-dependent in a way similar to the pH dependence of Fe^{3+} adsorption to other oxides.

We suggest that the excess Fe released in the transients is derived from partial dissolution or depolymerization of the iron (hydr)oxide at pH ≥ 1 followed by readsorption of Fe to the mineral surface. Following a pH jump to pH 1, this adsorbed Fe is desorbed by stepwise depolymerization. The slower dissolution rates of goethite relative to hematite appear to be related to bridging hydroxyl; our speciation of adsorbed Fe on hematite also associates slowly dissolving sites with bridging OH^-. The exact reason for this association is not clear, but is likely related to effects upon local acidity constants, ligand exchange rate, and rates of back-reaction.

Acknowledgments—We thank NSF for funding under grant #EAR-9527031 to CME. We acknowledge reviews by Roland Hellmann, Gerhard Furrer, Christian Ludwig, Laurent Charlet, and two anonymous reviewers.

Editorial handling: R. Hellmann

REFERENCES

Allison J. D., Brown D. S., and Novo-Gradac K. J. (1991) MINTEQA2/PRODEFA2, A geochemical assessment model for environmental systems. U.S. Environmental Protection Agency.

Ambe S. and Ambe F. (1990) Mössbauer study of ferric ions adsorbed at α-ferric oxide/aqueous solution interface. *Langmuir* **6**, 644-649.

Baes C. F. Jr. and Mesmer R. E. (1976) *The Hydrolysis of Cations*. John Wiley & Sons, Inc., New York.

Bergman I. and Paterson M. S. (1961) Silica powders of respirable size. I. Preliminary studies of dissolution rates in dilute sodium hydroxide. *J. Appl. Chem.* **11**, 369-376.

Burch T. E., Nagy K. L., and Lasaga A. C. (1993) Free energy dependence of albite dissolution kinetics at 80 °C and pH 8.8. *Chem. Geol.* **105**, 137-162.

Burton W. K., Cabrera N., and Frank F. C. (1951) The growth of crystals and the equilibrium structure of their surfaces. *Phil. Trans. Roy. Soc. London Ser. A.* **243** (A.866), 299-358.

Carroll-Webb S. A. and Walther J. V. (1988) A surface complex reaction model for the pH-dependence of corundum and kaolinite dissolution rates. *Geochim. Cosmochim. Acta* **52**, 2609-2623.

Casey W. H. (1991) On the relative dissolution rates of some oxide and orthosilicate minerals. *J. Colloid Interface Sci.* **146**, 586-589.

Casey W. H. and Bunker B. (1990) Leaching of mineral and glass surfaces during dissolution. In *Mineral-Water Interface Geochemistry* (eds. M. F. Hochella Jr. and A. F. White). Reviews in Mineralogy Vol. 23, Mineralogical Society of America, Washington, D.C. pp. 397-426.

Casey W. H. and Westrich H. R. (1992) Control of dissolution rates of orthosilicate minerals by divalent metal-oxygen bonds. *Nature* **355**, 157-159.

Chou L. and Wollast R. (1984) Study of the weathering of albite at room temperature and pressure with a fluidized bed reactor. *Geochim. Cosmochim. Acta* **48**, 2205-2217.

Chou L. and Wollast R. (1985) Steady-state kinetics and dissolution mechanisms of albite. *Amer. J. Sci.* **285**, 963-993.

Cornell R. M. and Giovanoli R. (1993) Acid dissolution of hematites of different morphologies. *Clay Minerals* **28**, 223-232.

Cornell R. M., Posner A. M., and Quirk J. P. (1974) Crystal morphology and the dissolution of goethite. *J.Inorg. Nucl. Chem.* **36**, 1937-1946.

Eggleston C. M., Hochella M. F., Jr., and Parks G. A. (1989) Sample preparation and aging effects on the dissolution rate and surface composition of diopside. *Geochim. Cosmochim. Acta* **53**, 797-804.

Fleming B. A. (1986) Kinetics of reaction between silicic acid and amorphous silica surfaces in NaCl solutions. *J. Colloid Interface Sci.* **110**, 40-64.

Fuerstenau D. W. (1970) Interfacial processes in mineral/water systems. *Pure Appl. Chem.* **24**, 135-164.

Furrer G. and Stumm W. (1986) The coordination chemistry of weathering: I. Dissolution kinetics of δ–Al_2O_3 and BeO. *Geochim. Cosmochim. Acta* **50**, 1847-1860.

Gratz A. J. and Bird P. (1993) Quartz dissolution: Theory of rough and smooth surfaces. *Geochim. Cosmochim. Acta* **57**, 977-989.

Hachiya K., Sasaki M., Ikeda T., Mikami N., and Yasunaga T. (1984a) Static and kinetic studies of adsorption-desorption of metal ions on a γ-Al_2O_3 surface. 2. Kinetic study by means of pressure-jump technique. *J. Phys. Chem.* **88**, 27-31.

Hachiya K., Sasaki M., Saruta Y., Mikami N., and Yasunaga T. (1984b) Static and kinetic studies of adsorption-desorption of metal ions on a γ-Al_2O_3 surface. 1. Static study of adsorption-desorption. *J. Phys. Chem.* **88**, 23-27.

Helgeson H. C., Murphy W. M., and Aagaard P. (1984) Thermodynamic and kinetic constraints on reaction rates among minerals and aqueous solutions. II. Rate constants, effective surface area, and the hydrolysis of feldspar. *Geochim. Cosmochim. Acta* **48**, 2405-2432.

Hellmann R. (1995) The albite-water system: Part II. The time-evolution of the stoichiometry of dissolution as a function of pH at 100, 200, and 300 °C. *Geochim. Cosmochim. Acta* **59**, 1669-1697.

Hemingway B. S. (1990) Thermodynamic properties for bunsenite, NiO, magnetite, Fe_3O_4, and hematite, Fe_2O_3, with comments on selected oxygen buffer reactions. *Amer. Mineral.* **75**, 781-790.

Hochella M. F., Jr. and White A. F. (1990) *Mineral-Water Interface Geochemistry*, *Reviews in Mineralogy*, Vol. 23 (ed. P. H. Ribbe). Mineralogical Society of America, Washington, D.C.

Holdren G. R., Jr. and Berner R. A. (1979) Mechanism of feldspar weathering-I. Experimental studies. *Geochim. Cosmochim. Acta* **43**, 1161-1171.

Holt P. F. and King D. T. (1955) The chemistry of silica surfaces. *J. Chem. Soc. London,* 773-779.

James R. O. and Healy T. W. (1972) Adsorption of hydrolyzable metal ions at the oxide-water interface: I, II and III. *J. Colloid Interface Sci.* **40**, 42-81.

Knauss K. G., Nguyen S. N., and Weed H. C. (1993) Diopside dissolution kinetics as a function of pH, CO_2, temperature, and time. *Geochim. Cosmochim. Acta* **57**, 285-294.

Knauss K. G. and Wolery T. J. (1988) The dissolution kinetics of quartz as a function of pH and time at 70^oC. *Geochim. Cosmochim. Acta* **52**, 43-53.

Kraemer S. M. and Hering J. G. (1997) Influence of solution saturation state on the kinetics of ligand-controlled dissolution of oxide phases. *Geochim. Cosmochim. Acta* **61**, 2855-2866.

Lasaga A. C. and Blum A. E. (1986) Surface chemistry, etch pits and mineral-water reactions. *Geochim. Cosmochim. Acta* **50**, 2363-2379.

Malmström M. and Banwart S. (1997) Biotite dissolution at 25°C: The pH dependence of dissolution rate and stoichoimetry. *Geochim. Cosmochim. Acta* **61**, 2779-2799.

Margerum D. W., Cayley G. R., Weatherburn D. C., and Pagenkopf G. K. (1978) Kinetics and mechanisms of complex formation and ligand exchange. In *Coordination Chemistry, Vol. 2*, ACS Monograph 174 (ed. A. E. Martell). American Chemical Society, Washington, D.C. pp. 1-220.

Mast M. A. and Drever J. I. (1987) The effect of oxalate on the dissolution rates of oligoclase and tremolite. *Geochim. Cosmochim. Acta* **51**, 2559-2568.

Maurice P. A., Hochella M. F. Jr., Parks G. A., Sposito G., and Schwertmann U. (1995) Evolution of hematite surface microtopography upon dissolution by simple organic acids. *Clays Clay Mineral.* **43**, 29-38.

Parks G. A. (1965) The isoelectric points of solid oxides, solid hydroxides, and aqueous hydroxo complex systems. *Chem. Rev.* **65**, 177-198.

Parks G. A. (1990) Surface energy and adsorption at mineral-water interfaces: An introduction. In *Mineral-Water Interface Geochemistry* (eds. M. F. Hochella Jr. and A. F. White). Reviews in Mineralogy Vol. 23, Mineralogical Society of America, Washington, D.C. pp. 133-175.

Parks G. A. and de Bruyn P. L. (1962) The zero point of charge of oxides. *J. Phys. Chem.* **66**, 967-973.

Sahai N. and Sverjensky D. A. (1997) Evaluation of internally consistent parameters for the triple-layer model by the systematic analysis of oxide surface titration data. *Geochim. Cosmochim. Acta* **61**, 2801-2826.

Samson S. D. and Eggleston C. M. (1998) Active sites and the non-steady-state dissolution of hematite. *Environ. Sci. Technol.* **32**, 2871-2875.

Samson S. D. and Eggleston C. M. (2000) The depletion and regeneration of dissolution-active sites at the mineral-water interface: II. Regeneration of active sites on α-Fe_2O_3 at pH 3 and pH 6. *Geochim. Cosmochim. Acta* **64**, 3675-3683.

Samson S. D., Stillings L. L., and Eggleston C. M. (2000) The depletion and regeneration of dissolution-active sites at the mineral-water interface: I. Fe, Al, and In sesquioxides. *Geochim. Cosmochim. Acta* **64**, 3471-3484.

Schott J., Berner R. A., and Sjöberg E. L. (1981) Mechanism of pyroxene and amphibole weathering-I. Experimental studies of iron-free minerals. *Geochim. Cosmochim. Acta* **45**, 2123-2135.

Schwertmann U. and Cornell R. M. (1991) *Iron Oxides in the Laboratory: Preparation and Characterization.* VCH Publishers, Inc., New York.

Sjöberg L. (1989) Kinetics and non-stoichiometry of labradorite dissolution. In *Water-Rock Interaction WRI-6* (ed. D. L. Miles). A.A. Balkema, Rotterdam. pp. 639-642.

Spokes L. J. and Jickells T. D. (1996) Factors controlling the solubility of aerosol trace metals in the atmosphere and on mixing into seawater. *Aquatic Geochem.* **1**, 355-374.

Thornton S. D. and Radke C. J. (1988) Dissolution and condensation kinetics of silica in alkaline solution. *SPE Res. Eng.* **May**, 743-752.

Wagman D. D., Evans W. H., Parker V. B., Schumm R. H., Halow I., Bailey S. M., Churney K. L., and Nuttall R. L. (1982) The NBS tables of chemical thermodynamic properties: Selected values for inorganic and C_1 and C_2 organic substances in SI units. *J. Phys. Chem. Ref. Data* **11**(Supplement No. 2), 1-392.

Westall J. C. (1979) MICROQL II. Calculation of adsorption equilibria in BASIC. Swiss Federal Institute of Technology, EAWAG.

Westrich H. R., Cygan R. T., Casey W. H., Zemitis C., and Arnold G. W. (1993) The dissolution kinetics of mixed-cation orthosilicate minerals. *Amer. J. Sci.* **293**, 869-893.

White A. F. and Brantley S. L. (1995) *Chemical Weathering Rates of Silicate Minerals, Reviews in Mineralogy*, Vol. 31 (ed. P. H. Ribbe). Mineralogical Society of America, Washington, D.C.

Wieland E. and Stumm W. (1992) Dissolution kinetics of kaolinite in acidic aqueous solutions at 25°C. *Geochim. Cosmochim. Acta* **56**, 3339-3357.

Wieland E., Wehrli B., and Stumm W. (1988) The coordination chemistry of weathering: III. A generalization on the dissolution rates of minerals. *Geochim. Cosmochim. Acta* **52**, 1969-1981.

Zinder B., Furrer G., and Stumm W. (1986) The coordination chemistry of weathering: II. dissolution of Fe(III) oxides. *Geochim. Cosmochim. Acta* **50**, 1861-1869.

Water-Rock Interactions, Ore Deposits, and Environmental Geochemistry: A Tribute to David A. Crerar

Editors: Roland Hellmann and Scott A. Wood

Corrosion forms and twinning in zeolite ZSM-5 crystals

CLARENCE D. CHANG*#, CYNTHIA T.-W. CHU**, AND JOHN L. SCHLENKER†

Mobil Technology Company, Strategic Research Center, Paulsboro, NJ 08066, USA
*Present address: 11 Murray Place, Princeton, NJ 08540 USA
***Present address*: AI Technology, Inc., Princeton Jct., NJ 08550 USA
† deceased

corresponding author (e-mail: cdchang@nji.com)

Abstract—The dissolution (corrosion) of twinned and untwinned zeolite ZSM-5 crystals by carbonate solutions often produces distinctive hollow crystals with the appearance of "open sarcophagi". This phenomenon is examined in light of the distribution of framework aluminum coupled with symmetry constraints imposed by the point-group symmetry of the crystals. The ZSM-5 morphology and possible macroscopic twin laws are also briefly discussed. Dissolution under highly alkaline conditions (1N Na_2CO_3, 90 °C) is incongruent and directional. Scanning electron micrographs show that the carbonate solution selectively attacks the pinacoid form {010}, which comprises (010) and $(0\bar{1}0)$ crystal faces. Incongruent behavior is attributed to the relative corrosion resistance of the {100} and {102} forms, which arises from Al-enrichment at the associated faces. Selective Al-enrichment in this zeolite occurs during crystallization under hydrothermal conditions. The observations provide a basis for some speculations on possible mechanisms of Al incorporation during crystal growth. It is hypothesized that the rate of dissolution is accelerated at regions of high strain energy and retarded at regions of relatively higher Al concentration. Thus, selective dissolution can be seen to initiate at twin boundaries. In contrast, dissolution of untwinned crystals is observed to be more uniform.

1. INTRODUCTION

The synthetic zeolite ZSM-5 (Argauer and Landolt, 1972; Kokotailo et al., 1978) and its recently discovered natural analogue, mutinaite (Galli et al., 1997), are highly siliceous, microporous tectoaluminosilicates. The ZSM-5 structure is a negatively-charged network of four-connected tetrahedra bridged by corner-sharing oxygen atoms charge-balanced by cations situated within the zeolite pore structure. Macroscopic twinning is frequent in synthetic ZSM-5, but is not yet reported for the mineral analog mutinaite. The synthetic material is known to be ferroelastic. Zeolite ZSM-5 has found widespread industrial application due to a unique suite of catalytic and sorptive properties combined with high thermal and hydrothermal stability (Olson et al., 1981).

The dissolution of ZSM-5 in alkaline solutions is incongruent (Čižmek et al., 1994) and directionally non-uniform, often producing hollow crystals with the characteristic appearance of "open sarcophagi", as exemplified in Fig. 1b. This distinctive phenomenon, along with incongruent dissolution, has been linked to "aluminum zoning" (von Ballmoos and Meier, 1981; Derouane et al., 1981; Althoff et al., 1993) in the zeolite crystal. The preferential removal of framework Si was attributed to kinetically-favored base-catalyzed hydrolysis of siloxy bridges, whereas negatively-charged Al centers were assumed to be more resistant to hydroxyl attack (Dessau et al., 1993; Čižmek et al., 1995).

This can be compared to the dissolution behavior of feldspars, a group of commonly-occurring aluminosilicate minerals. Taking albite feldspar as an example, the initial stages of dissolution at 25 °C and at basic pH are characterized by essentially congruent dissolution (Holdren and Speyer, 1985; Chou and Wollast, 1984; Casey et al., 1989). On the other hand, at elevated temperatures (100-300 °C), there is evidence of incongruent dissolution at moderately basic conditions ($r_{Al} > r_{Si}$); this trend is amplified ($r_{Al} >> r_{Si}$) at extremely basic pH conditions and temperatures >200 °C (r = rate of release; see details in Hellmann et al., 1990; Hellmann, 1995).

The present study provides further elaboration of the selective dissolution phenomenon in ZSM-5 zeolite in a crystallographic-morphological context. Several possible reasons for the observed differences in the dissolution behavior of twinned and untwinned ZSM-5 crystals are discussed.

2. COMPOSITION AND STRUCTURE OF ZSM-5 ZEOLITE: EXPERIMENTAL METHODS

The ZSM-5 (MFI framework) structure is characterized by a 10-ring system of interpenetrating channels. A sinusoidal channel parallel to the a-axis has

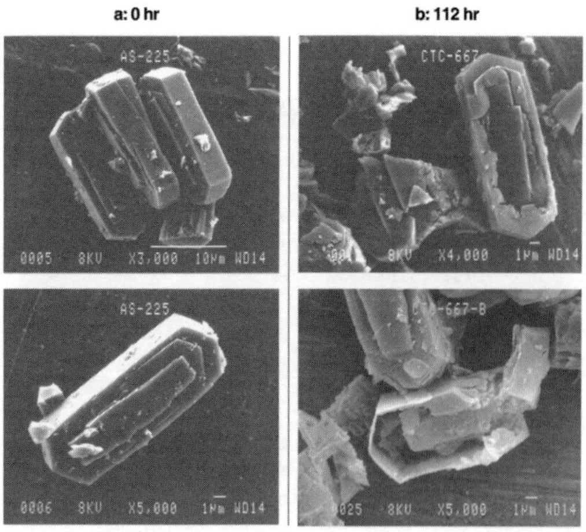

Fig. 1: Scanning electron micrographs of macroscopically twinned ZSM-5 crystals (Si/Al = 110) before and after alkaline treatment

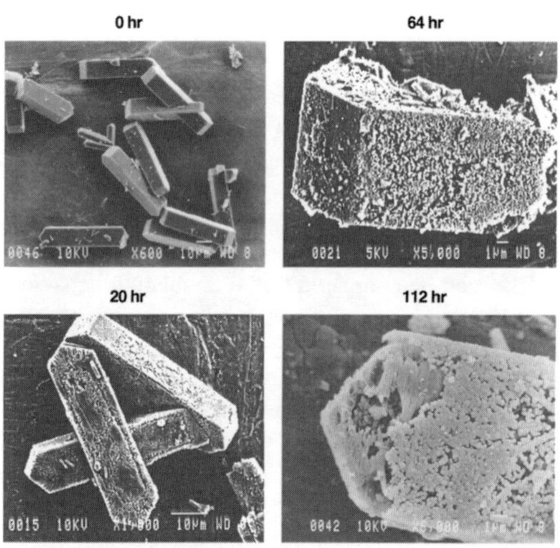

Fig. 2: Scanning electron micrographs of untwinned ZSM-5 crystals (Si/Al = 1500) during alkaline treatment

dimensions 5.1 × 5.5 Å and a second linear channel, parallel to the b-axis, is 5.3 × 5.6 Å (Olson et al., 1981). The crystal structure (framework topology) of ZSM-5 is characterized by a large number of five-membered rings, but there is also a small fraction of four-membered rings. The ideal unit cell formula is represented by $M_{x/n}(AlO_2)_x(SiO_2)_{96-x}\cdot zH_2O$, where M is a cation of valence *n*. Occluded water may be replaced by various organic substances. The composition is highly variable, but ZSM-5 with end-member composition (x = 0) $Si_{96}O_{192}$ is readily synthesized.

Crystals of zeolite ZSM-5 were synthesized with Si/Al 110 and 1500 following published procedures (Argauer and Landolt, 1972). The as-synthesized zeolite crystals were air-calcined to remove the tetrapropylammonium directing agent, and digested (in stainless steel Parr® autoclaves under static conditions) with excess aqueous 1N Na_2CO_3 at 90 °C for varying lengths of time up to 112 hr. Samples of the crystals were periodically removed, washed with deionized water, dried, and examined with a scanning electron microscope (SEM).

3. RESULTS

Scanning electron micrographs of zeolite crystals before and after alkali treatment (up to 112 hr) are shown in Figs. 1 and 2. Figure 1 shows results for twinned crystals, whereas Fig. 2 shows results for untwinned crystals. Selective etching-dissolution is most prominent in the macroscopically twinned crystals (Fig. 1b). These samples also show evidence of accelerated, possibly discontinuous, dissolution along the twin boundary. In contrast, the results of alkaline treatment of the untwinned crystals, displayed in Fig. 2, show their progressive dissolution at 0, 20, 64, and 112 hr. Here the process appears somewhat slower and more uniform, though incipient selective etching-dissolution is visible at 20 hr. There is also evidence of secondary nucleation on the crystal surface at 64-112 hr. The morphology of these exterior crystals, coupled with their generation presumably by a seeding mechanism (Ostwald ripening) or epitaxy, indicate that they are ZSM-5 zeolite, though their compositions were not directly analyzed.

4. DISCUSSION

4.1. Mechanism of Alkaline Dissolution

On the basis of kinetic studies, Čižmek et al. (1994) concluded that the alkaline dissolution of ZSM-5 involves at least three main steps: (i) breaking of Si-O-Si and Si-O-Al bonds, with production of silicate and aluminate monomers in solution; (ii) formation of an amorphous SiO_2 layer on the crystal surface by reaction of silicate anions; (iii) dissolution of the amorphous SiO_2. We assume that the reaction initiates at the silanols on the crystal surface. These can be terminal (non-acidic) I, or bridging (acidic) II:

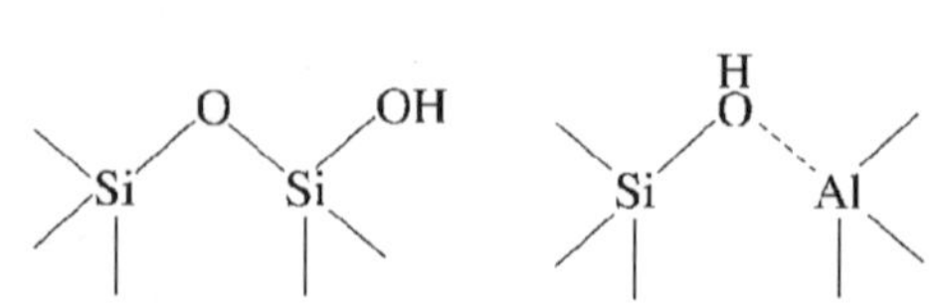

I II

For a working model of the initiation reaction, the mechanism of Dove and Crerar (1990) for quartz dissolution appears reasonable (see also Xiao and Lasaga, 1996, for a quantum mechanical treatment). Dove and Crerar observed dramatic enhancement of the dissolution rate upon addition of small concentrations of alkali cations, even at near-neutral pH conditions. A transition state interpretation (Lasaga, 1981) was invoked, where the activation entropy is increased by surface-coordinated cations, which in turn increases the frequency of species crossing the activated barrier associated with Si-O bond breakage. Given the elevated pH, 1N Na_2CO_3 solutions used in the present study, zeolitic protons (those associated with the bridging siloxyl groups II) will be essentially completely exchanged for Na^+ at equilibrium.

The influence of Al centers in the selective dissolution of ZSM-5 may be given two functionally equivalent interpretations: (*i*) Al centers may be regarded as strongly adsorbed "impurities" (a particularly apt model for the highly siliceous zeolites of this study), which may serve as rate inhibitors (Lasaga, 1997) by virtue of their low reactivity; (*ii*) Al centers (T-sites, II) have higher surface energy due to steric and electronic differences with respect to their neighboring Si T-sites, and serve therefore as "active sites" for initiating silica dissolution. The selective dissolution of minerals at active sites has been extensively investigated (for examples in the geochemistry literature, see e.g., Blum and Lasaga, 1987; Schott et al., 1989).

The marked difference in dissolution behavior between the twinned and untwinned crystals may be attributed to a number of factors, including surface area and Al content. A further consideration is based on differences in surface strain energy. According to Schott et al. (1989), the net surface reaction-limited rate of dissolution *r* of a flat crystal may be expressed as follows:

$$r = \prod_i a_i^{n_i} \left[\nu \left\{ \sum_{perfect\ surface\ sites} \exp\left(\frac{-\Delta G_p^{+}}{kT}\right) + \sum_{dislocation\ surface\ sites} \exp\left(\frac{-\Delta G_d^{+}}{kT}\right) \right\} \right] \quad (1)$$

Here a_i and n_i are the activity and reaction order of the *i*th species, v is a frequency factor, and k is the Boltzmann constant. The difference between ΔG_p^{+}, the activation energy for dissolution of the perfect crystal surface, and that of a surface with dislocations, ΔG_d^{+}, is the strain energy u associated dislocations. Thus, for a single dislocation, the ratio of the dissolution rate constants for a perfect crystal k_p and a crystal with dislocations k_d can be written:

$$k_p / k_d = exp\,(-u / \mathrm{k}T) \quad (2)$$

Referring to the selective dissolution of twinned crystals shown in Fig. 1b, it appears that twinning makes the twin plane surfaces more susceptible to dissolution. Theoretically, a perfectly coherent boundary associated with twinning should not be the locus of any additional strain. However, in reality, it is commonly observed that twin boundaries are areas that are preferentially attacked in solutions; in fact, preferential etching of twin boundaries is a well known method for delimiting twin boundaries (see examples in Heiman et al., 1975; and references therein). Thus, it remains an open question as to whether twinning itself results in elevated strain at the twin boundaries in ZSM-5 crystals.

If twinning is not directly responsible, it is possible that elevated impurity concentrations at twin boundaries are a possible mechanism for increasing the strain energy. Studies on the incorporation of trace elements on mineral surfaces have revealed the complex nature of how surface structure and the kinetic behavior of surface processes strongly affect elemental partitioning (e.g., Paquette and Reeder, 1995; Rakovan and Reeder, 1994; Reeder, 1996). In the case of ZSM-5, Al may play the role of an impurity which is preferentially incorporated at the twin boundary, thereby increasing the strain energy.

Even though Al is known for its role as a dissolution-inhibitor, this effect may be offset by the subsequent increase in the strain energy due to preferential Al incorporation. The net effect would be a higher rate of dissolution along the twin boundaries. In view of Eqns. 1 and 2, the more "perfect" untwinned crystals would be expected to dissolve at a relatively slower rate.

Because the exact nature of the twin boundary in ZSM-5 is not known, another possible reason (E. Dowty, private communication) for the increased rates of dissolution of twinned ZSM-5 crystals may be due to an irregular contact surface associated with interpenetrative twins (in contrast to contact twins, such as albite feldspar twins). Thus, the twin contact surface may be composed of multiple reentrants (alternating {100} and {010} surfaces), which are not present on single crystals. Multiple reentrants have been implicated in influencing crystal growth rates, and therefore, it is possible that they also influence dissolution rates.

4.2. Symmetry and Morphology of ZSM-5 Crystals

ZSM-5 is ferroelastic with a high temperature orthorhombic, o-form, and a low temperature monoclinic, m-form. The ferroelastic transition temperature

Table 1. Symmetry distinct morphological forms of the orthorhombic bipyramidal class (2/m 2/m 2/m)

Form Designation	Designation	Number of Faces	Present on ZSM-5	ZSM-5 form
bipyramid	{hkl}	8	no	-------
a-dome	{h0l}	4	yes	{102}
b-dome	{0kl}	4	yes	{012}
prism	{hk0}	4	no	-------
front-pinacoid	{100}	2	yes	{100}
side-pinacoid	{010}	2	yes	{010}
basal pinacoid	{001}	2	no	-------

is near room temperature and is influenced by both the Si/Al ratio and by applied mechanical stress.

The space group symmetry of o-ZSM-5 is P 2_1/n 2_1/m 2_1/a. The real symmetry of the high temperature or o-form of ZSM-5 coincides with the maximum topological symmetry (MTS) of the aluminosilicate framework. Space group P 2_1/n 2_1/m 2_1/a is isogonal with the centrosymmetric point group 2/m 2/m 2/m . This is a point group of the orthorhombic crystal system traditionally designated the orthorhombic bipyramidal class. A tabulation of forms for this point group symmetry is given in Table 1.

The general {hkl} form (bipyramid) of the orthorhombic bipyramidal crystal class has never been observed with ZSM-5 crystals. For ZSM-5, the dominant morphological forms are the front- and side-pinacoids: {100} and {010}, and the dome {102}. The typical ZSM-5 crystal habit is illustrated in Fig. 3. The ZSM-5 crystal morphology is a result of the appearance of three distinct (symmetrically non-equivalent) forms. These forms appear in combination, because no one form has sufficient faces to enclose space. Symmetrically equivalent faces belonging to each of the three forms are stippled in Fig. 3. The habit of

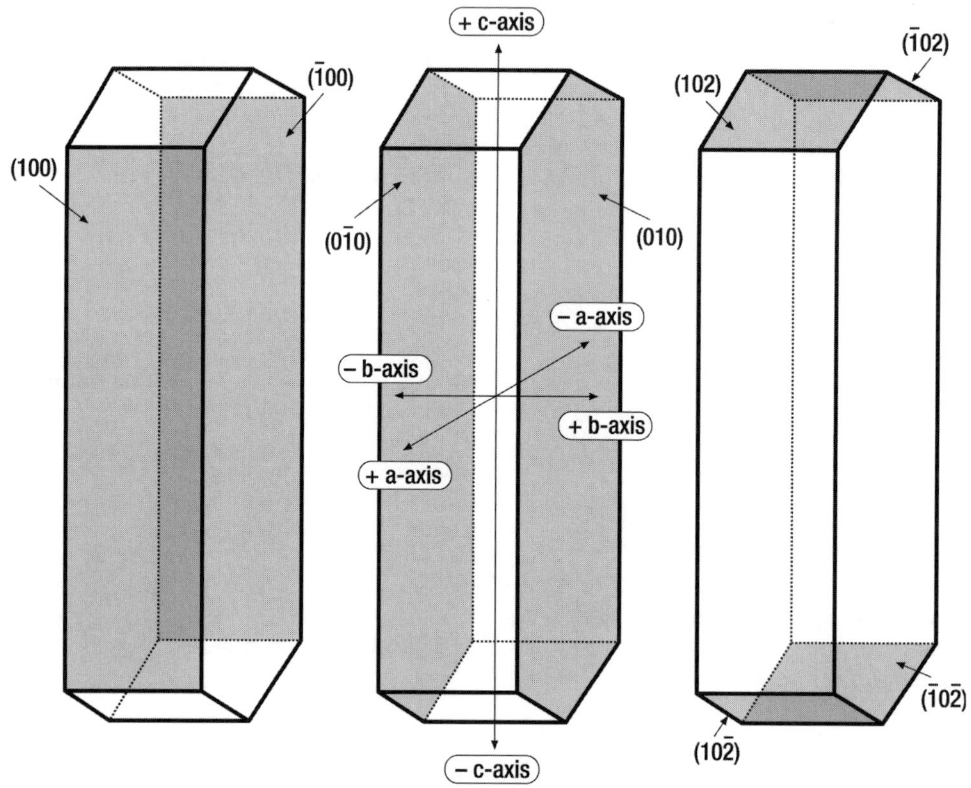

Fig. 3: Crystal habit of ZSM-5 typically exhibits three morphological forms; these are, from left to right, the front-pinacoid, the side-pinacoid, and the a-dome.

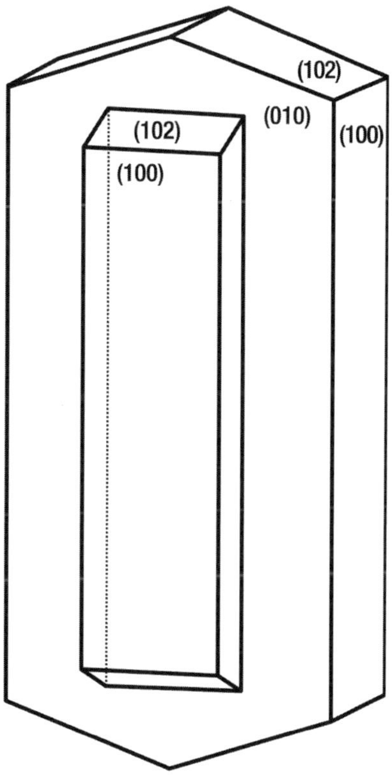

Fig. 4: Typical appearance of a macroscopically-twinned crystal of ZSM-5

ZSM-5 is not always the same since the relative development of the symmetrically distinct morphological forms is variable within limits and can be influenced by the conditions prevailing during crystal growth.

4.3. Macroscopic Twinning and the Twin Law in ZSM-5

In macroscopic twinning, the individuals constituting the twin are related by symmetry transformation operators not formally present in the point group symmetry of the crystal. This form of twinning is favored by the presence of pseudo-symmetry axes and pseudo-mirror planes, and the presence of such pseudo-symmetry axes and pseudo-mirror planes is often indicated by the near equality of two or more symmetry-independent cell parameters of a material. In ZSM-5: a = 20.022 Å and b = 19.899 Å.

In ZSM-5, individual macroscopic twins appear to be related by a 180° rotation about [110]. The usual appearance of such twins is illustrated in Fig. 4. The precise nature of the twin boundary is unclear. Scanning electron micrographs demonstrate that macroscopic twinning is ubiquitous in many, but by no means all, ZSM-5 preparations. Some preparations contain relatively few macroscopically twinned crystals, as untwinned ZSM-5 crystals are not a rarity. Although the ferroelastic transition is immediately detectable using x-ray powder diffraction, there is no obvious difference between powder diffraction patterns of ZSM-5 preparations consisting primarily of macroscopically twinned crystallites and batches where this macroscopic twinning is relatively infrequent. Macroscopic twinning, of course, has presented difficulties for single crystal x-ray diffraction studies.

4.4. Symmetry Information from Dissolution Forms of ZSM-5

The mode of dissolution of ZSM-5 allows us to make certain inferences regarding its habit. The dissolution forms observed here definitely establish that the faces of the principle zone of ZSM-5 cannot belong to a single prismatic {110} form, for, if such were the case, all four faces in this zone should be symmetrically equivalent and react at an equivalent rate. The implication is that the four faces of this prism-like zone are equivalent in pairs and belong to two distinct morphological (pinacoidal) forms, the front-pinacoid {100} and the side pinacoid {010}. It is the faces of the {010} form, that is faces (010) and $(0\bar{1}0)$, that are rapidly attacked during dissolution in the Na_2CO_3 solution. These are the faces opening on the straight zeolite channel. Symmetry theory tells us that because faces (010) and $(0\bar{1}0)$ belong to the same form, they must interact with the solvent in an equivalent way. The two faces of the {100} form, (100) and $(\bar{1}00)$, are observed to be corrosion-resistant. The {102} form is also resistant to corrosion. Because three distinct forms commonly appear on ZSM-5 crystals, three different dissolution rates must be involved; however symmetry theory makes no statement regarding the extent to which these rates should differ. An idealized depiction of the effect of selective dissolution appears in Fig. 5.

The observed habit of ZSM-5 is frequently slightly platy with predominant development of the {010} pinacoid. This habit suggests that after a certain point in their crystallization history ZSM-5 crystals grow primarily by accretion of material to the {100} side-pinacoid and {102} dome faces. Addition of material to the faces of the {010} pinacoid seems to slow or cease. This development may be connected to Si depletion of the surrounding solution and could constitute a clue as to the way in which Al is incorporated into ZSM-5 during growth. As a result of this mode of growth, the {100} and {102} faces should be uniformly Al-enriched (presumably to different extents,

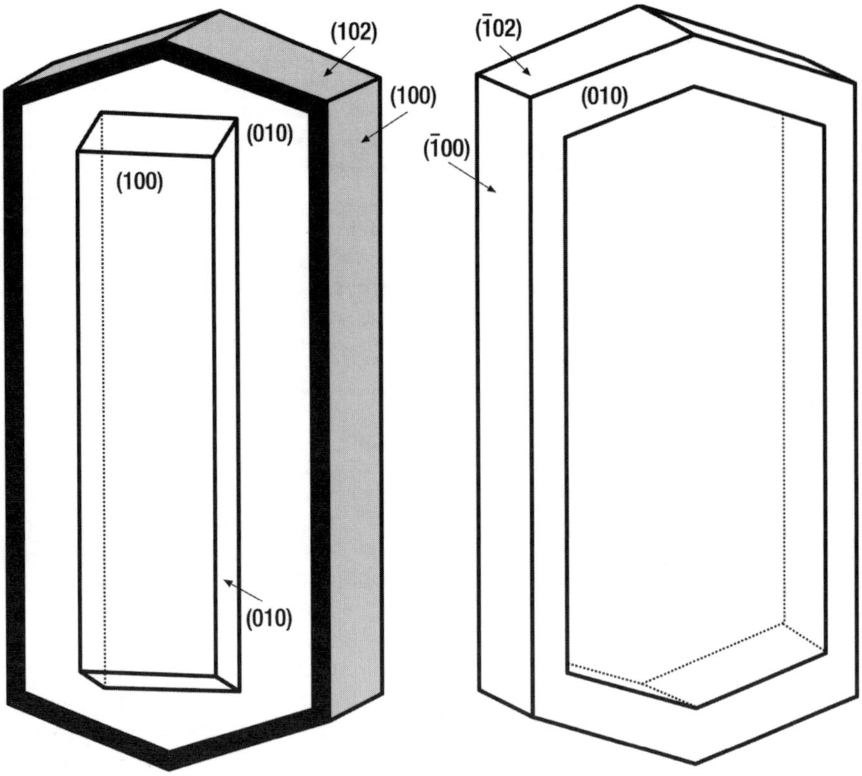

Fig. 5: Idealized depiction of a macroscopically-twinned crystal of ZSM-5 before (on right) and after dissolution (on left) in Na_2CO_3 solution. The aluminum enriched rim is stippled in the left portion of the figure.

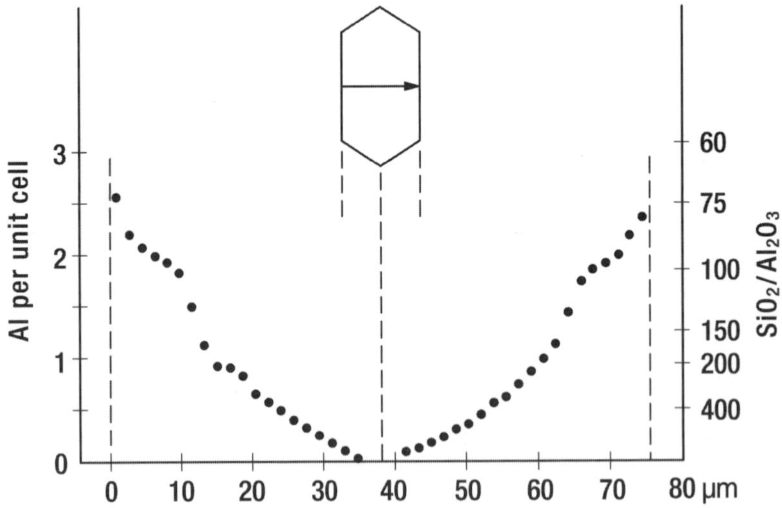

Fig. 6: Al-zoning in a ZSM-5 crystal as revealed through electron microprobe scan (von Ballmoos and Meier, 1981). Reprinted by permission from *Nature* 1998, Macmillan Magazines, Ltd.

since they are symmetrically nonequivalent) and evidently corrosion-resistant. Corrosion resistance is clearly associated with elevated Al concentration. The aluminum incorporation on the crystal faces of the {100} and {102} forms appears compositionally uniform. Corrosion rates must differ for the {100} and {102} forms, but this difference is evidently small.

As demonstrated in Fig. 6 (von Ballmoos and Meier, 1981), the mode of crystal growth described above causes faces of the slowly accreting {010} pinacoidal form to exhibit Al zoning, where the {010} pinacoid faces have a Si-enriched and reactive inner core and an Al-enriched and corrosion resistant outer rim. Curiously, in certain instances the composition change at the core/rim composition boundary appears rather sharp and possibly discontinuous.

5. CONCLUSIONS

The dissolution of macroscopically twinned and untwinned ZSM-5 crystals in Na_2CO_3 solutions shows a directional dependence. In addition, dissolution proceeds more rapidly in macroscopically twinned crystals. A consequence of this behavior is the production of distinctive hollow crystals shaped like "open sarcophagi". This directional solubility is tied to symmetry constraints imposed by the point-group symmetry of the crystals. Scanning electron micrographs demonstrate that the carbonate solution selectively attacks the pinacoid form {010}, which comprises (010) and $(0\bar{1}0)$ faces, of the ZSM-5 crystals. The relative corrosion resistance of the {100} and {102} forms is associated with Al-enrichment of these faces during their hydrothermal synthesis. In some crystals, dissolution reveals a sharp core/rim boundary suggesting that the difference in solubility rates is discontinuous. It is hypothesized that dissolution is governed by the interplay of opposing factors, namely surface strain energy, local Al concentration, and dissolution rate retardation by Al. The rate of dissolution is accelerated at regions of high strain energy at twin boundaries, and retarded at regions of relatively higher Al concentration. In contrast, dissolution of untwinned crystals is seen to be more uniform.

Acknowledgments—The authors express their appreciation to R. Hellmann, D. Bosbach, E. Dowty, and H. Teng for constructive comments. John Schlenker, co-author of this paper, passed away on September 20, 2000. Dr. Schlenker was an eminent crystallographer, and was writing a book on the theory of x-ray powder diffraction before his untimely death. David Crerar was a treasured friend of CDC's, who still misses his intellectual fervor and inspiration, as well as the many gastronomical adventures shared through the years.

Editorial handling: R. Hellmann

REFERENCES

Althoff R., Schultz-Dobrick B., Schüth F., and Unger K. (1993) Controlling the spatial distribution of aluminum in ZSM-5 crystals. *Microporous Mater.* **1**, 207-218.

Argauer R. J. and Landolt G. R. (1972) Crystalline zeolite ZSM-5 and method of preparing the same. *U. S. Patent* 3702886.

Blum A. E. and Lasaga A. C. (1987) Monte Carlo simulations of surface reaction rate laws. In *Aquatic Surface Chemistry: Chemical Processes at the Particle-Water Interface* (ed. W. Stumm). John Wiley & Sons, Inc., New York. pp. 255-292.

Casey W. H., Westrich H. R., Arnold G. W., and Banfield J. F. (1989) The surface chemistry of dissolving labradorite feldspar. *Geochim. Cosmochim. Acta* **53**, 821-832.

Chou L. and Wollast R. (1984) Study of the weathering of albite at room temperature and pressure with a fluidized bed reactor. *Geochim. Cosmochim. Acta* **48**, 2205-2217.

Čižmek A., Komunjer L., Subotic B., Aiello R., Crea F., and Nastro A. (1994) Kinetics of zeolite dissolution: Part 4. Influence of the concentration of silicon in the liquid phase on the kinetics of ZSM-5 dissolution. *Zeolites* **14**, 182-189.

Čižmek A., Komunjer L., Subotic B., Aiello R., Crea F., Nastro A., and Tuoto C. (1995) Dissolution of high-silica zeolites in alkaline solutions I. Dissolution of silicalite-1 and ZSM-5 with different aluminum content. *Microporous Mater.* **4**, 159-168.

Derouane E. G., Detremerrie S., Gabelica Z., and Blom N. (1981) Synthesis and characterization of ZSM-5 type zeolites I. Physico-chemical properties of precursors and intermediates. *Appl. Catal.* **I**, 201-224.

Dessau R. M., Valyocsik E. W., and Goeke N. H. (1993) Aluminum zoning in ZSM-5 as revealed by selective silica removal, *Zeolites* **12**, 776 – 779.

Dove P. M. and Crerar D. A. (1990) Kinetics of quartz dissolution in electrolyte solutions using a hydrothermal mixed flow reactor. *Geochim. Cosmochim. Acta* **54**, 955-969.

Galli E., Vezzalini G., and Quartieri S. (1997) Mutinaite, a new zeolite from Antarctica: the natural counterpart of ZSM-5. *Zeolites* **19**, 318–322.

Heiman R. B. (1975) *Auflösung von Kristallen.* Springer-Verlag, Wien.

Hellmann R. (1995) The albite-water system Part II. The time-evolution of the stoichiometry of dissolution as a function of pH at 100, 200 and 300 °C. *Geochim. Cosmochim. Acta* **59**, 1669-1697.

Hellmann R., Eggleston C. M., Hochella M. F., and Crerar D. A. (1990) The formation of leached layers on albite surfaces during dissolution under hydrothermal conditions. *Geochim. Cosmochim. Acta* **54**, 1267-1281.

Holdren G. R. Jr. and Speyer P. M. (1985) Reaction rate-surface area relationships during the early stages of weathering- **I**. Initial observations. *Geochim. Cosmochim. Acta* **49**, 675-681.

Kokotailo G. T., Lawton S. L., Olson D. H., and Meier W. M. (1978) Structure of synthetic ZSM-5. *Nature* **272**, 437-438.

Lasaga A. C. (1981) Transition state theory. In *Kinetics of Geochemical Processes* (eds. A. C. Lasaga and R.J. Kirkpatrick), Reviews in Mineralogy **8**, Mineralogical Society of America, Washington, D.C. pp. 135-169.

Lasaga A. C. (1997) *Kinetic Theory in the Earth Sciences*, Princeton University Press, Princeton.

Olson D. H., Kokotailo G. T., Lawton S. L., and Meier W. M.

(1981) Crystal structure and structure-related properties of ZSM-5. *J. Phys. Chem.* **85**, 2238-2243.

Paquette J. and Reeder R. J. (1995) Relationship between surface structure, growth mechanism, and trace element incorporation in calcite. *Geochim. Cosmochim. Acta* **59**, 735-749.

Rakovan J. and Reeder R. J. (1994) Differential incorporation of trace elements and dissymmetrization in apatite: The role of surface structure during growth. *Amer. Mineral.* **79**, 892-903.

Reeder R. J. (1996) Interaction of divalent cobalt, zinc, cadmium, and barium with the calcite surface during layer growth. *Geochim. Cosmochim. Acta* **60**, 1543-1552.

Schott J., Brantley S., Crerar D., Guy C., Borcsik M., and Willaime C. (1989) Dissolution kinetics of strained calcite, *Geochim. Cosmochim. Acta* **53**, 373-382.

von Ballmoos R. and Meier W. M. (1981) Zoned aluminum distribution in synthetic zeolite ZSM-5, *Nature* **289**, 782 – 783.

Xiao Y. and Lasaga A. C. (1996) Ab initio quantum mechanical studies of the kinetics and mechanisms of quartz dissolution. *Geochim. Cosmochim. Acta* **60**, 2283-2295.

Water-Rock Interactions, Ore Deposits, and Environmental Geochemistry: A Tribute to David A. Crerar
© The Geochemical Society, Special Publication No. 7, 2002
Editors: Roland Hellmann and Scott A. Wood

Topography of polished plates of albite crystal and glass during dissolution

NATHAN P. MELLOTT [1, 3], SUSAN L. BRANTLEY [1,*], AND CARLO G. PANTANO [2]

[1] Department of Geosciences, The Pennsylvania State University, University Park, Pennsylvania 16802 USA
[2] Department of Material Science and Engineering, The Pennsylvania State University, University Park, Pennsylvania 16802 USA
[3] Present address: Department of Materials Science and Engineering, The Pennsylvania State University, University Park, Pennsylvania, 16802

* corresponding author: brantley@geosc.psu.edu

Abstract—Polished surfaces of glass and crystal of albite composition were analyzed using atomic force microscopy (AFM) before and after dissolution at pH 2.0 and 9.0, and changes in topography were compared to dissolution rates of powders of the same phase. The roughness of the crystal plates was greater when dissolved at pH 2.0 than at pH 9.0, reflecting greater extent of etching at intrinsic crystalline and polishing defects at lower pH. In contrast, the RMS roughness of the glass plate was identical within error when dissolved at pH 2.0 and pH 9.0. The similarity in roughness of the polished glass plate as a function of pH of dissolution may be related to lack of reproducibility in polishing, different controls on dissolution for glass vs. crystalline substrates, or the presence of a thick, porous leached layer developed on the glass during dissolution at pH 2.0. The AFM only documented an increase in RMS roughness by a factor of 7 on the glass dissolved at pH 2.0 after 5000 h, while BET surface area increased by a factor of 30 on glass powder dissolved at pH 2.0 after 4000 h of dissolution. However, total roughness of the surfaces could not be quantified by AFM through analysis of only RMS roughness; spatial distribution (fractal dimension *D*) also had to be measured. *D* was calculated from the roughness exponent and from the power spectral density on several samples. These two methods of calculating *D* did not always yield comparable values for all samples. The surface area calculated from the RMS roughness and fractal dimension measured for several samples increased only slightly over 5000 h of dissolution at pH 2.0 for both phases, demonstrating that the AFM documented only insignificant topographic effects related to nonstoichiometric dissolution (leaching). Feldspar dissolution rates estimated from AFM measurements were within a factor of 10 of dissolution rates based upon solution results, even though rate of retreat of entire surface layers could not be quantified. Further research should continue to emphasize using AFM to document changes in roughness for comparison to dissolution measured from solution chemistry.

1. INTRODUCTION

The development of surface topography related to crystallographic features during dissolution of minerals has been noted by many workers (e.g. Helgeson et al. 1984; Brantley et al., 1986; Lasaga and Blum, 1986; Casey et al., 1988; Blum et al., 1990; White and Peterson, 1990; Anbeek, 1992; Lee and Parsons, 1995; Lee et al., 1998; Fenter et al., 2000), but quantitative measurement of the evolution of topography during dissolution is generally lacking. Instead, dissolution rates are usually related simply to the surface area as determined through gas adsorption using the Brunauer, Emmet, and Teller (BET) isotherm (Brunauer et al., 1938). The BET measurement provides total surface area accessible to the adsorbate molecule (for a review, see Brantley et al., 1999; Brantley and Mellott, 2000). New surface techniques such as scanning tunneling microscopy (STM) and atomic force microscopy (AFM) now make documentation of microscopic (μm to nm scale) to atomic (~1-5 Å) roughness possible on both conductive and nonconductive surfaces. Few comparisons have been made between changes in BET surface area of minerals during dissolution and changes in surface topography as measured by AFM.

In order to investigate the evolution of feldspar surfaces with dissolution, atomic force microscopy was used to analyze the surface topography of a unique suite of glass and crystal samples prepared by Hamilton et al. (2000). In these preparations, polished plates of albite glass and crystal (compositions summarized in Hamilton et al., 2000) were dissolved at pH 2.0 or 9.0 in batch reactors under ambient conditions far from equilibrium. Based upon comparisons of dissolved solute chemistry and thermodynamic data, Hamilton et al. (2000) concluded that secondary phases did not precipitate during dissolution. Polished plates were used so as to provide flat surfaces amenable for elemental analyses (x-ray photoelectron spectroscopy (XPS) and secondary ion mass spectrometry (SIMS)) and topographic analyses. Significant effort was expended to polish the surface with little chemical alteration.

Hamilton et al. (2000) also dissolved powders (grain size = 75-150 μm) in flow experiments at room temperature and reported that the albite crystal and glass powders dissolved at the same rate (± 40%) at pH 2.0 (HCl-H_2O) when rates were normalized by initial surface area. Error was determined by propagating uncertainties in the measurement of rates (Hamilton et al., 2000). Although the crystal and glass powders had similar initial BET surface areas (Table 1), after dissolution for 4000 h at pH 2.0, the surface areas of the glass and crystal powders increased by a factor of 30 and 2, respectively. Structures of altered, leached layers on albite crystal and glass are presumed to be more open and porous than the unaltered surface (Casey and Bunker, 1990; Hellmann, 1995, 1997; Hamilton et al., 2000), allowing proton-bearing species access deep into the near surface. However, as per the observations of Casey et al. (1989) for a different feldspar composition, Hamilton and coworkers saw no increase in dissolution rate with time for the glass dissolved at pH 2.0. They therefore interpreted the increase in specific surface area of the glass dissolved at pH 2.0 as the result of the formation of a porous leached near-surface layer. This increase in specific surface area was not thought to represent an increase in *reactive* area, and rates were thus normalized by initial surface area. Although most other researchers investigating feldspar dissolution have similarly normalized rates by initial surface area, some have normalized by final surface area (see Stillings and Brantley, 1995).

Hamilton et al. (2000) also observed that these albite crystal and glass powders dissolved in LiOH-H_2O at the same rate within error (± 40%) at pH 8.4 (= outlet pH, inlet pH = 9.3). For these powder samples, BET surface areas differed minimally after dissolution for 5400 h (reacted/unreacted surface area = 1.5 for glass and 1.1 for crystal).

AFM is used here to analyze surface topography of the plates prepared by Hamilton in order to address the following questions: How does crystallinity affect RMS roughness and power spectral density of roughness during dissolution of albite? How do AFM measurements of bearing volume correlate with observed dissolution based upon solution chemistry? Can AFM-measured surface area correlate with BET surface area?

2. ANALYTICAL METHODS

2.1. Sample Preparation

Samples analyzed in this study are some of the identical plates described by Hamilton et al. (2000). Albite crystal was obtained from Wards Scientific and albite glass was prepared by melting crystalline Amelia albite and annealing it at 700 °C for ~12 hours (Zellmer, 1986; Hamilton et al., 2000). Compositions of albite crystal and glass are equal to within better than 1 wt. % for all major oxides (Hamilton et al., 2000).

Albite crystal and glass plates were cut with a diamond saw to 2 cm x 1 cm x 0.2 cm. The plates were polished to 0.1 μm using a LECO AP-60 automatic polisher in an oil-based diamond spray, followed by a final polishing to 0.05 μm with a cerium oxide/chrome oxide spray. No attempt to orient the polycrystalline samples prior to polishing was made. Based on limited observations under an optical microscope, grains of Amelia albite varied from at least a few mm to much larger in diameter. The sample preparation procedure was chosen from among many others because it provided glass and crystals with surface chemistries, as measured by x-ray photoelectron spectroscopy (XPS), that were as close as possible to the bulk stoichiometry for all elements (Hamilton et al., 2000).

Crystal plates used for this study came from batch experiments run for either 1000 or 5000 hrs (note that plates run for 5000 h were separate plates from those run for 1000 h). Glass plates were similarly reacted, but only unleached plates and plates dissolved for 5000 h were available for this study. Solutions used for dissolution experiments included: 1) trace-metal grade HCl and reverse-osmosis (RO) water at pH 2.0 2) LiOH and RO water at pH 9.0. All vessels were held at 25±1 °C in an oven and pH was constant to within ±0.3 pH units. Each reaction vessel had an initial surface area to volume ratio of 5.0×10^{-3} cm^{-1}. Following dissolution experiments, plates were rinsed with RO water, ultrasonically cleaned in acetone, blown dry with nitrogen, and stored in a vacuum dessicator. Scanning electron microscopy (SEM) revealed no evidence for the cracking of the leached layers.

2.2. Microscopy

Plates were imaged using differential interference contrast microscopy (DIC) and a Digital Instruments Nanoscope IIIa Dimension 3100 Atomic Force Microscope before and after dissolution. Both height and amplitude images were collected in Tapping Mode®, with a sharpened tip, at a scan rate of 1.0–1.5 Hz. Multiple scans (~20) were completed on each sample at different locations on the surface. From these, three to ten representative 10 μm x 10 μm areas of each sample were then chosen for further analysis. In choosing scanning sites, cracks, grain boundaries, rip-up clasts, etc. were avoided where possible. While scanning, the cantilever records a fixed number (256) of height measurements throughout each scan line. Therefore, for a 10 μm x 10 μm scan, the lateral resolution is about 40 nm per pixel. The AFM tip also impedes resolution below approximately 10-30 nm,

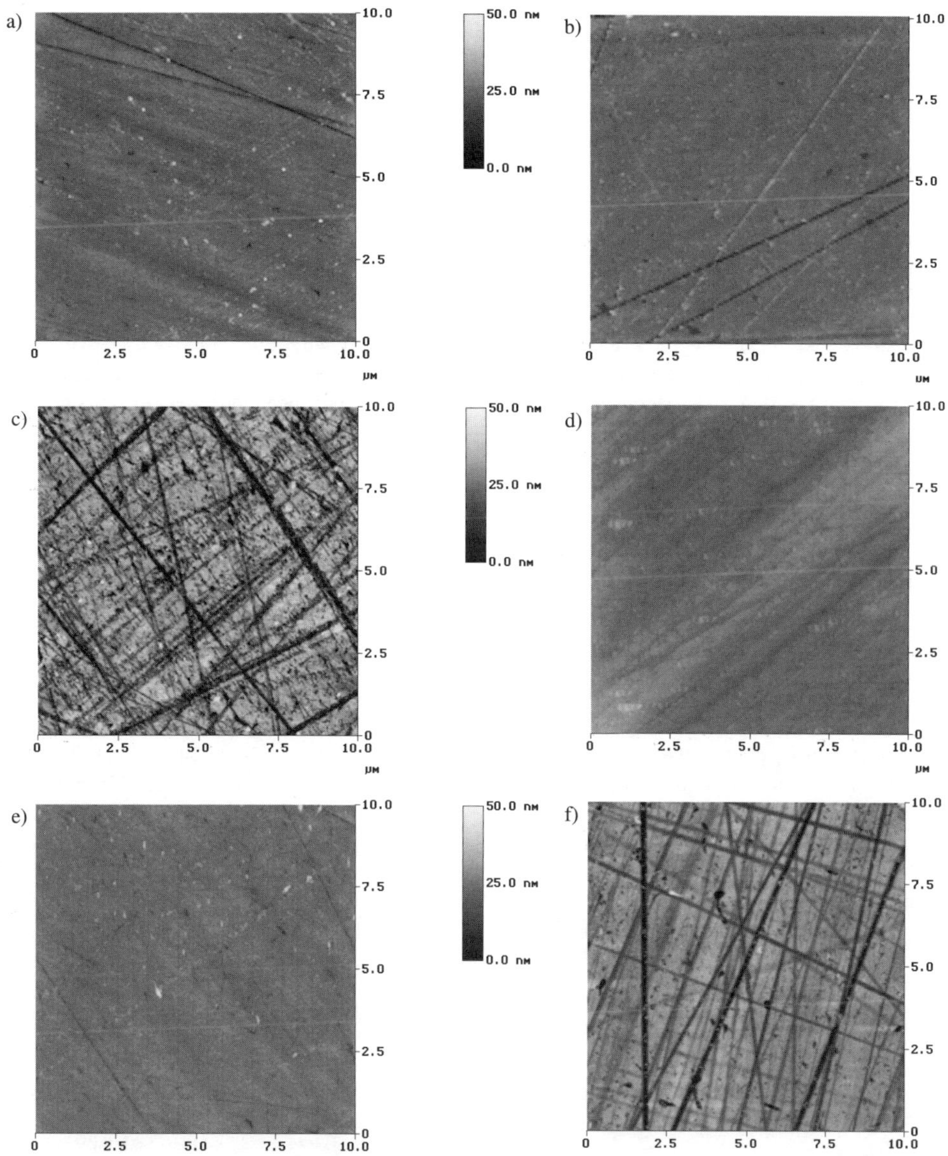

Fig. 1. Atomic force microscope images of polished albite crystal and glass samples showing distribution of surface features: (a) albite crystal unleached, (b) albite glass unleached, (c) albite crystal leached at pH 2 for 5000 h, (d) albite glass leached at pH 2 for 5000 h, (e) albite crystal leached at pH 9 for 5000 h, and (f) albite glass leached at pH 9 for 5000 h. Note the prevalence of polishing lines on most surfaces. All images are 10 µm x 10 µm.

depending upon whether a new or duller tip is used. Except where noted, all roughness values referred to in the text were derived from images of 10 µm x 10 µm areas of the surface.

RMS roughness calculations were performed with Digital Instruments software (Digital Instruments, 1997). Third-order plane fitting was performed on each image prior to analysis to remove any image distortions caused by the curvature of the piezoelectric material, thermal drift, or lateral forces from sample images (Ruppe and Duparee, 1996). Third-order plane fitting calculates a third-order polynomial fit from the image, removing tilt and S-shaped bow (Digital Instruments, 1997).

Root mean squared roughness (RMS) is defined as the root mean square average of height deviations from the mean elevation plane (standard deviation of the height measurements), calculated from the relative heights of each pixel in the image:

$$RMS = \sqrt{\left(\left(z_1^2 + z_2^2 + ... + z_N^2 \right) \Big/ N \right)} \quad (1)$$

where z_i equals the difference in height from the mean plane for each point *i*, and *N* equals the number of points measured.

Surface area, *A*, was calculated on three to seven 10 μm x 10 μm scans of each albite glass and crystal plate using a tiling method, wherein the areas of triangles drawn between measured points are summed (Digital Instruments, 1997). Roughness ratio (λ), was calculated for one representative scan so as to be similar to the roughness as defined by Helgeson (1984):

$$\lambda = \left(A \Big/ A_{geo} \right) \quad (2)$$

where A_{geo} is the surface area calculated from the product of the length and width of the scanned field.

Random surface topography can be modeled as a summation of sine waves through two-dimensional Fast Fourier Transform (FFT) analysis; the power spectral density (PSD) function is defined as the square of the FFT (Bennet and Mattson, 1989). Power spectral density (PSD) plots describe the spatial distribution of roughening by plotting the roughness amplitude squared as a function of frequency or wavelength (Digital Instruments, 1997). Power spectra were determined on leached (pH 2.0) and unleached samples as an average over all directions in the data. In contrast to the images used for RMS roughness, the images used for PSD plots were made of different scan areas (1 μm x 1 μm, 5 μm x 5 μm, etc.) representing different resolutions chosen from different places on each sample. Every attempt was made to use images for power spectral determinations which were void of identifiable tip artifacts, noise, and "dust".

3. RESULTS

Unleached albite crystal plates are polycrystalline, and even after polishing, twin planes, cleavage planes, scars from rip-up clasts, and grain boundaries are imaged with DIC microscopy. Albite crystal plates dissolved in all solutions show similar features. In contrast, polished albite glass exhibits no such features under DIC before or after dissolution (DIC images not shown).

Under the AFM, polished, unleached mineral and glass plates show smooth surfaces with few polishing scratches (Figs. 1a, b). The number, width, and depth of pits or scratches all increased with dissolution (Figs. 1c, d, e, and f). In general, crystal plates dissolved at pH 9.0 showed fewer pits or scratches than crystal plates dissolved at pH 2.0 for comparable times. In contrast, glass plates dissolved at pH 9.0 showed more pits or scratches than glass plates dissolved at pH 2.0 for comparable times. Twin boundaries were observed under SEM (Hamilton et al., 2000) and AFM (this study) on crystalline samples leached in pH 9.0 solution for 1000 hours, but not on other samples.

As described previously, topographic data were obtained on 10 μm x 10 μm areas of the surface in order

Table 1. Surface measurements on albite crystal and glass

Sample	pH (± 0.3)	Duration of leaching (h)	RMS roughness[1] (nm)	λ^2	λ^3	H	D^4	D^5	A_{BET} (m^2/g)[6]
Crystal	--	0	1.23±0.14	1.02	1.006	0.13±0.11	2.87±0.11	2.80±0.08	0.06±0.02
	2.0	1000	7.37±0.24						
	2.0	5000	8.10±0.37	1.03	1.189	0.02±0.07	2.92±0.07	2.29±0.07	0.12±0.04
	9.0	1000	4.47±0.33						
	9.0	5000	1.94±0.04						0.06±0.02
Glass	--	0	2.02±0.19	1.03	1.003	0.18±0.01	2.82±0.01	2.67±0.05	0.04±0.01
	2.0	5000	5.04±0.24	1.02	1.005	0.05±0.06	2.95±0.11	2.74±0.06	1.2±0.4
	9.0	5000	6.31±0.50						0.06±0.02

[1]Calculated as the average for 3-20, 5 μm x 5 μm sub-areas, as described in text.
[2]Calculated using a tiling algorithm (Digital Instruments, 1997).
[3]Calculated on one 10 μm x 10 μm area using the algorithm of Lai and Irene (1999a) and Eqn. 2 in the text.
[4]Calculated using the equation: $D = 3\text{-}H$ (Eqn. 4 in text).
[5]Calculated from the slope of the power spectral density plot, as described in the text.
[6]As measured using BET Kr adsorption by Hamilton et al. (2000) on powders dissolved in flow experiments for 3000-5000 h. Final pH of solutions was either pH 2.0 or 8.4 (inlet pH was 2.0 or 9.3, respectively).

to maintain constant resolution. However, sub-areas of these scans were typically analyzed in order to avoid dust and tip artifacts in the larger scans. Even with constant resolution, measured RMS roughness was observed to increase with the size of the sub-area analyzed (Fig. 2a). Above 5 μm x 5 μm, the increase in RMS roughness with sub-area dimension was minimal for both glass and crystal; for this reason, roughness was compared for different samples as measured on 5 μm x 5 μm scans.

In general, RMS roughness was larger on samples after dissolution, but varied depending upon where it was measured on the sample. To compare RMS roughness among samples, a best estimate for RMS roughness for each sample was determined as the mean of RMS roughness calculations from 3 – 20, 5 μm x 5 μm sub-areas, from at least 3 representative 10 μm x 10 μm images of each sample (Table 1). Implicitly, we have assumed that the measured values of RMS roughness for a sample form a normal distribution and that the mean of these measurements is the best estimate of the true RMS roughness. Uncertainties in the determinations of the means for samples (Table 1) represent the standard deviations of the 3 – 20 RMS measurements on all sub-areas divided by the square root of the number of sub-areas.

RMS roughness of unleached glass and crystal plates was 1 – 2 nm (Table 1) and changes during dissolution were on the order of 1 to 8 nm for all samples (Figs. 3a, b). In acidic solutions, the RMS roughness of the crystal plates increased to a greater extent (by a factor of 6.6) than the glass (factor of 2.5) after 5000 hours of dissolution (Fig. 3a, b). In contrast, during dissolution in basic solutions for 5000 h, the glass plate roughened to a greater extent (factor of 3.1) than the crystal (factor of 1.6) after 5000 hours of dissolution (Fig. 3a, b). As a result, RMS roughness is greater on crystalline samples dissolved for 5000 hours at pH 2.0 than pH 9.0, while RMS roughness is greater on glass plates dissolved for 5000 hours at pH 9.0 than 2.0.

For the crystal plates, the RMS roughness increased with dissolution at pH 2.0 up to 1000 h, and then increased or stayed roughly constant to 5000 h. In contrast, the RMS roughness of the crystal plate was lower after dissolution for 5000 h at pH 9.0 than after 1000 h. The crystal plate dissolved for 1000 h at pH 9.0 was also the only sample to show etching of twin boundaries.

Little variation in power spectral density for different images of the same sample at the same scan size was observed (Fig. 4a). Thus, the power spectral density plots showed stationarity at this scale (see Dumas et al., 1993).

To compare 4 samples (crystal and glass, unleached and leached), 5 representative plots, with

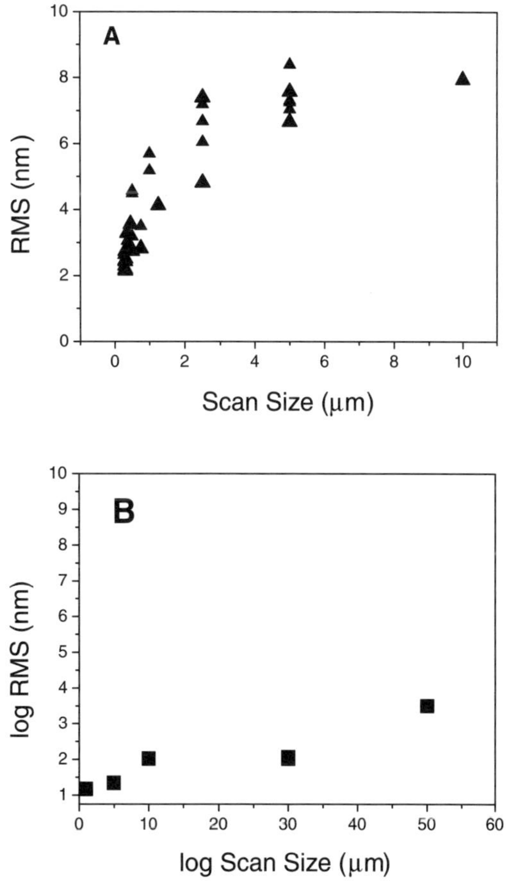

Fig. 2. (a) RMS roughness of albite crystal leached at pH 2 for 1000 h as a function of scan size measured at the same resolution. Each symbol represents a different scan on a different area of the sample. (b) Dependence of RMS roughness on scan size measured at different resolutions on unleached albite glass. A linear relationship implies a self-affine surface

scan dimensions of 1 μm x 1 μm, 5 μm x 5 μm, 10 μm x 10 μm, 30 μm x 30 μm, and 50 μm x 50 μm were analyzed and their power spectra plotted (Figs. 4b, c). For each plot, data for frequencies greater than or equal to the Nyquist frequency for that resolution (0.5 x number of pixels/linear scan dimension) were not used since aliasing can bias the measurement of such signals (Russ, 1994). Unleached glass plates exhibited slightly higher or equal power compared to the unleached crystal plates at all frequencies. For both the crystal and glass plates leached at pH 2.0, increased power with dissolution was observed at all frequencies (Figs. 4b, c); however, the roughness increase at frequencies <10 μm^{-1} was greater for the

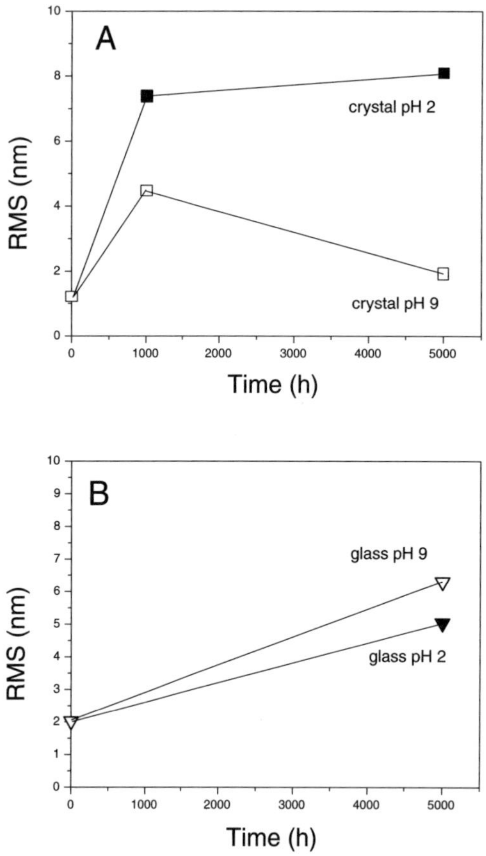

Fig. 3. RMS roughness as a function of dissolution time: (a) albite crystal dissolved at pH 2 (closed squares) and pH 9 (open squares) (b) albite glass at pH 2 (closed triangles) and pH 9 (open triangles).

crystal plates than for the glass plates. RMS roughness determined for each sample (see Fig. 2b for albite glass unleached) for images collected at these five resolutions (1 μm x 1 μm, 5 μm x 5 μm, 10 μm x 10 μm, 30 μm x 30 μm, and 50 μm x 50 μm) increased nonlinearly with scan size.

4. DISCUSSION

4.1. Roughness

Only one sample (albite crystal dissolved at pH 9.0) showed higher roughness after 1000 h than after 5000 h (Fig. 3a). This sample was also the only sample where etched twin boundaries were imaged under AFM. Roughness as measured by AFM for the crystal dissolved at pH 9.0 may therefore have been affected by variability in crystallographic orientation or non-reproducibilities in surface preparation. If it is assumed that the crystal RMS values at a given pH are the same (within error) after 1000 and 5000 h of dissolution (i.e., assuming that steady-state conditions of dissolution were achieved after 1000 h, see pH 2 results, Fig. 3a), then it is possible that the difference in RMS measured at pH 9 for the crystal (RMS = 4.47 nm after 1000 h, RMS = 1.94 nm after 5000 h, see Table 1) may also simply be due to the variability inherent to crystal orientation and surface preparation. Thus, based on this difference in RMS values, we assume an error of approximately ±2.5 nm with respect to measured values of roughness for all crystal samples. Therefore, based on this estimated error, the difference in RMS of unleached glass and crystal plates (Table 1) is insignificant.

Increased roughness values (compared to unleached samples) for the samples dissolved at pH 2.0 for 5000 h were documented in both RMS roughness values (Table 1) and in increased power at all frequencies in the power spectra (Fig. 4). Simple geometric arguments might suggest that roughness develops on a dissolving surface as a balance between roughening due to pitting, and smoothing due to annihilation of pits through overlapping growth. For example, Affatigato et al. (1996) argued that RMS roughness on polished glass developed due to etching of flaws caused by diamond abrasive particles, but that annihilation of flaws due to lateral growth of these features smoothed the surface. A third process that could smooth surfaces is surface retreat; for a pitted surface in which pits are not deepening as fast as the overall surface is retreating, a smoother surface should develop.

On crystalline surfaces, pits can develop at point, line, and planar defects, as well as on perfect surfaces. Etching of point defects creates pits which disappear with continued etching (e.g., MacInnis and Brantley, 1992). In contrast, etching of line defects continues until the line defect terminates on another linear or planar defect; however, pits develop at dislocations only if the solution concentration is less than the critical concentration (e.g., Brantley et al., 1986). All solutions at pH 2.0 and 9.0 were far from equilibrium (Hamilton et al., 2000), and development of pits at intrinsic and polishing-induced defects on the crystal plates is therefore expected to have been spontaneous. The difference in RMS roughness on crystals dissolved at pH 2.0 and 9.0 therefore probably results from the difference in the *extent* of etching at crystallographic defects rather than differences in the *initiation* of etch pits on the surface. The greater extent of etching is consistent with the faster observed rate of dissolution at pH 2.0 as compared to pH 9.0 (Hamilton et al.,

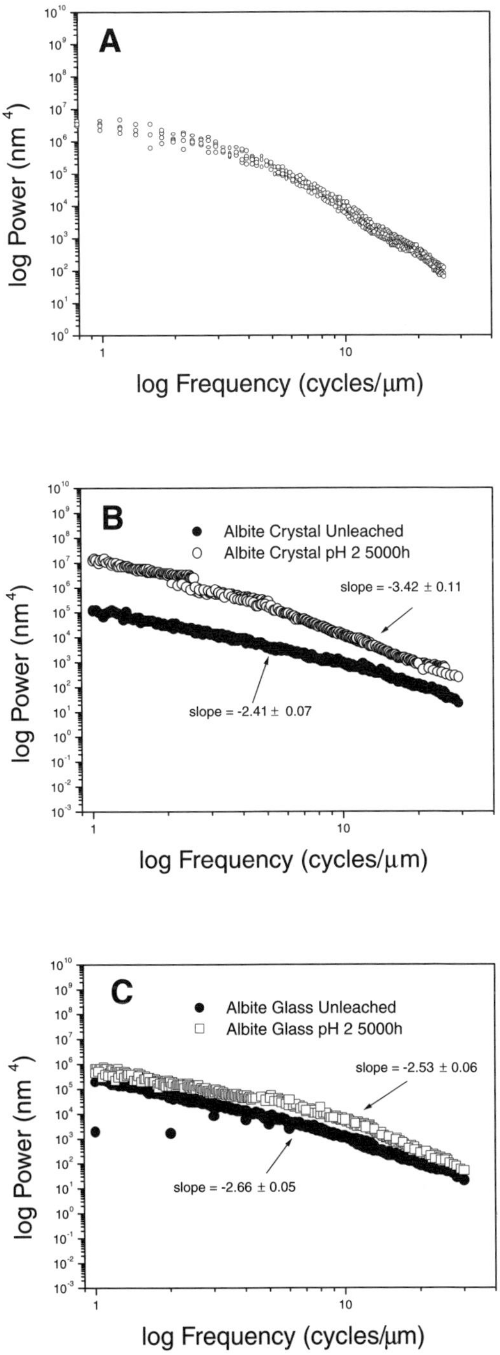

Fig. 4. Characteristic log power vs. log frequency plots of five 10 μm x 10 μm scans from 3 separate spots on (a) an albite crystal leached in a pH 2 solution for 5000 hours. Log power vs. log frequency plots of (b) albite crystal unleached and leached in a pH 2 solution for 5000 h (note: a different spot than shown in part (a)), and (c) glass samples unleached and leached for 5000 h in a pH 2 solution.

2000). However, in relatively strain-free crystals such as the Amelia albite used here, the density of line defects is ~10^6 cm^{-2} (e.g. Casey et al., 1988) and only about 1 defect outcrop per 10 μm x 10 μm image would be expected. Roughness of the surface measured by AFM was therefore most likely related to pitting on perfect surfaces, or at defects induced by polishing. However, variability in crystallographic orientation from one spot to another on the Amelia albite samples makes unambiguous interpretation of roughness impossible for these polycrystalline samples.

The greater roughness on the crystal plates dissolved at pH 2.0 as compared to pH 9.0 might also be hypothesized to be related to nonstoichiometric leaching of the surface (and near surface) and structural changes in that leached (near) surface. Hamilton et al. (2000) reported that at pH 2.0, both Na and Al were significantly depleted in the upper 10 nm of the crystal plate surface after dissolution for a thousand hours. The extent of Na and Al depletion was much smaller at pH 9.0 than at pH 2.0. If RMS roughness correlated with surface leaching, however, then we would have expected a much larger roughness on the acid-dissolved glass plates than was observed, since the upper 10 nm of the albite glass was almost 100% depleted after dissolution at pH 2.0 compared to significantly less depletion after dissolution at pH 9.0 (Hamilton et al., 2000). Despite this enhanced surface leaching at low pH, the glass plates dissolved at pH 9.0 showed the same RMS roughness as the glass plates dissolved at pH 2.0 (Table 1), assuming an uncertainty of ±2.5 nm. Preferential leaching at the surface and near surface apparently does not introduce significant RMS roughness on albite glass or crystal at the scale of measurement of the AFM.

In contrast, adsorption of Kr on the glass powder (Table 1) after 4000 h of dissolution at pH 2.0 documented the development of a leached layer which is porous to gas adsorbate atoms. If the leached layer on the dissolved plate was also porous, the pores that outcropped on the surface were only minimally accessible to the AFM tip or were not present in great abundance. The cross-sectional area of Kr adsorbate gas is 15.2 $Å^2$ (at 78 K, Gregg and Sing, 1982), while the diameter of the AFM tip is on the order of 10-20 nm.

While the RMS roughness of the crystal plates was larger when dissolved at pH 2.0 than at pH 9.0, the roughness of the glass plate was the same within error (±2.5 nm) after dissolution at these two values of pH. For both crystal and glass powders, the difference in the dissolution rates at pH 8.4 and 2.0 in the flow experiments was approximately an order of magnitude (Hamilton et al., 2000). Affatigato et al. (1996) ob-

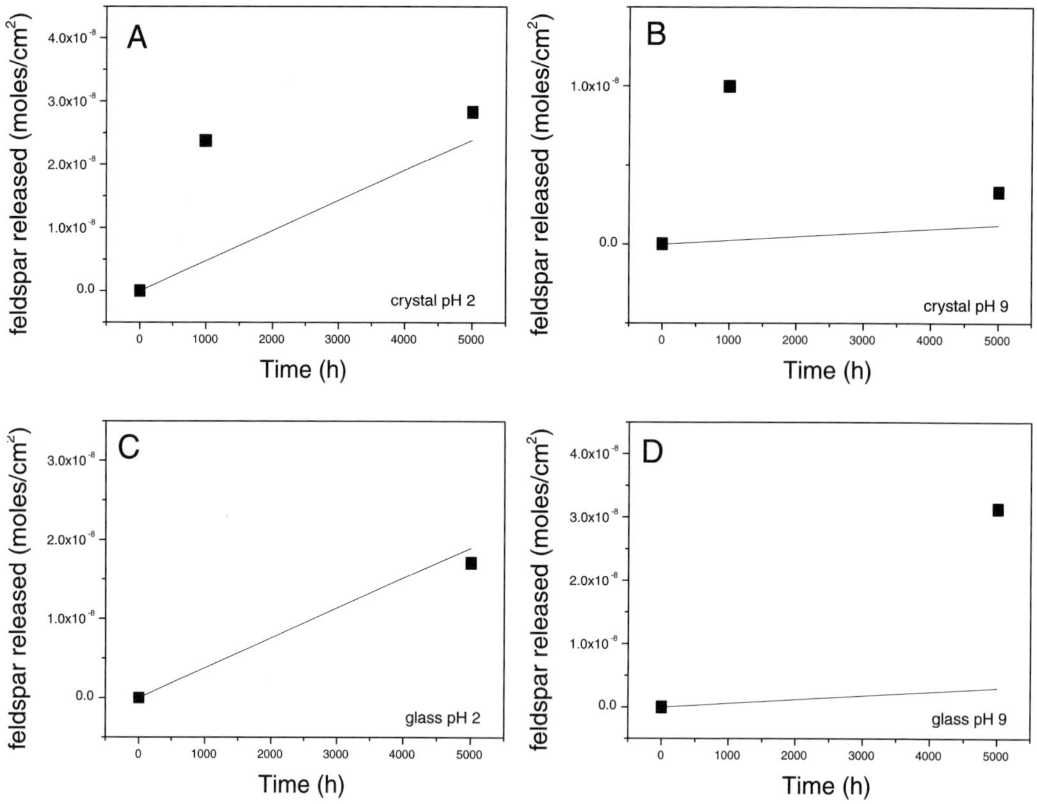

Fig. 5. Moles of released feldspar ($NaAlSi_3O_8$) calculated as a function of time (closed squares) as described in text for A) albite crystal dissolved at pH 2, B) albite crystal dissolved at pH 9, C) albite glass dissolved at pH 2, D) albite glass dissolved at pH 9. Solid line represents expected steady-state release of feldspar as a function of time as calculated from powder dissolution rates from Hamilton et al. (2000) for crystal or glass in flow experiments with outlet values of pH of 2.0 or 8.4 (inlet pH of 2.0 and 9.0, respectively). Moles of feldspar calculated from AFM measurements have been corrected so that the release at time zero equals zero (i.e., calculations were corrected for starting roughness).

served that deepening of initial flaws on silica glass occurred quickly during dissolution, while lateral growth (smoothing) occurred more slowly and became more important in the latter stages of dissolution. It is possible that such smoothing occurred to a greater extent on the acid-dissolved than on the base-dissolved sample, explaining the similarity in roughness of the two samples.

It is also possible that the presence of a leached, porous layer, if it developed on the glass plates, encouraged smoothing instead of roughening. Jordan et al. (1999) observed dissolution of anorthite and labradorite at pH 2.0 with AFM at above ambient temperatures and concluded that dissolution occurred mainly beneath an altered surface layer. They concluded that the topography of the outer surface mimicked that of the interface beneath the layer. Similar results were also found in a study of in situ albite dissolution at 25 °C with AFM by Hellmann (1992). If such a dissolving interface exists below a porous surface layer on the albite glass studied here, and if the dissolved concentration of solutes in the pore solutions within the porous layer are higher than in bulk solution, then the surface topography beneath the porous layer might be expected to be smoother than if no layer were present. In other words, within the porous layer, the driving force for dissolution would be lower because of the higher activity of dissolved solutes than at the outer surface–aqueous solution interface, where the activity of dissolved solutes might be relatively lower. With a lower driving force for dissolution, the dissolution would not open up pits on the surface, but would instead occur equivalently at all surface sites, smoothing the interface. This effect, if important, would be expected to be most important on the glass sample leached at pH 2.0, possibly explaining the relatively smooth surface of that sample.

Removal of complete layers of material (i.e., without roughening) will contribute to dissolution. However, such removal is difficult to quantify by the AFM without a reference for the initial surface height. For our samples, the apparent rate of dissolution was estimated using AFM by calculating the bearing volume. The bearing volume is an estimate of the minimum volume of material removed from the plates (Digital Instruments, 1997); this volume is defined by the difference between a reference plane and the geometric surface area of the sample. The bearing volume was determined for ten to twenty 5 μm x 5 μm scans on at least 4 different images of each crystal and glass sample. Of these measurements, one representative determination of bearing volume was chosen. For this image, the reference plane for the calculation was set as the highest point on the image. The bearing, or "removed" volume, was converted into moles of released feldspar per unit area for each batch experiment using the molar volume of albite crystal or glass. These removed volumes were normalized by the scan area for calculation of dissolution rates. However, Si concentrations in solution as measured by Hamilton et al. (2000) for the plates dissolved here were too low for detection by ICP-AES (<0.1 ppm Si) and solutions were not stored for other types of analysis. For comparison, however, the extent of dissolution of the samples is compared to the measured rates of dissolution on the powders as reported by Hamilton et al. (2000) at steady state for similar values of pH in flow-through reactors (e.g., Fig. 5).

Calculated feldspar released per unit area based on AFM is observed to be always within about an order of magnitude of feldspar released as calculated from solution chemistry. In most cases, the number of moles of feldspar released based on AFM is larger than that based on the powder dissolution experiments (Hamilton et al., 2000). The surprisingly close agreement between rates based on solution chemistry and surface topography may suggest that AFM could be used to measure dissolution rates using this methodology for these phases. Possibly, with these slowly dissolving phases, no overall surface retreat had occurred, even over 5000 h. However, the relative agreement documented by Fig. 5 may also be fortuitous given that the powder dissolution rates were normalized by BET surface area but were extrapolated to this system using the geometric surface area of the AFM scanned image (25 μm^2). The discrepancy between the BET and AFM surface areas for these plates is unknown. The relative agreement exhibited in Fig. 5 suggests, however, that future investigations might eventually yield meaningful dissolution rates based on such AFM measurements, especially if the initial surface height is measured.

4.2. Fractal Dimension

Several authors have pointed out (Varnier et al., 1989; Dumas et al., 1993; Lai and Irene, 1999a) that RMS roughness is only a measure of the vertical component of roughness and does not describe the distribution of topography at the surface. To understand one such measure of the distribution, the fractal dimension, it is helpful to note that RMS roughness of images of the same resolution increased non-linearly with image size (Fig. 2a). Other workers have similarly reported a critical scan dimension above which the change in RMS roughness of a polished surface became minimal with image dimension (e.g. Dumas et al., 1993; Affatigato et al., 1996). By definition, if a surface shows self-affine fractal behavior, the RMS roughness values for the surface should scale as a power law of the image size (Dumas et al., 1993; Gollion and Grenet, 1996):

$$RMS \;\alpha\; L^H \tag{3}$$

where L is the linear dimension of the scan, and H is a constant known as the scaling factor or roughness index. Gollion and Grenet (1996) point out that L in Eqn. 3 can be varied using AFM by changes in imaged area at one resolution (Fig. 2a) or by a variation in the resolution of images (Fig. 2b). They also point out that the latter method is preferable for calculation of H. The second method, when applied to the unleached albite glass surface (Fig. 2b), yielded H = 0.18±0.01. When applied to a limited data set collected for the three other samples, the linear model could not be rejected, suggesting these four surfaces (surfaces of glass and crystal, unleached and leached for 5000 h) are self-affine (Table 1).

Dumas et al. (1993) showed that this scaling factor H is related to the fractal dimension D:

$$D = 3 - H \tag{4}$$

Thus, the fractal dimension defines the scaling law relating roughness and sampled area (Fig. 2). Fractal dimensions for rough surfaces are non-integral numbers between 2 and 3, where a value of 2 represents the dimension of a Euclidean plane and 3 an infinitely rough surface (e.g., Dumas et al., 1993; Spanos and Irene, 1994; Lai and Irene, 1999a). Surfaces with low fractal dimension are simpler, with more surface features of low aspect ratio, while surfaces with fractal dimension near 3 approach volume-filling surfaces, dominated by surface features of high aspect ratio. Spanos and Irene (1994) point out that both the irregularity and the volume-filling nature of a surface increases with increasing D. They also give examples of surfaces with different total roughness that have identical RMS roughness but differing D.

Values of D calculated from Eqn. 4 are identical within error for the unleached glass and crystal plates (Table 1). D increases slightly for the glass plates after dissolution at pH 2.0 (Table 1), while the crystal shows only a slight increase or no change in D with dissolution.

Numerous other methods of quantifying the fractal dimension of a surface are available, and their respective advantages and disadvantages have been investigated (Pfeifer, 1984; Dubuc et al., 1989; Mitchell and Bonnell, 1990). In this study, the 2-D Fast Fourier method is applied (Russ, 1994; Dumas et al., 1993). According to Dumas et al. (1993) and Vatel et al. (1993), when log power is plotted against log frequency for scans of different areas at different resolution, the slope (S_{PSD}) of the line defines D:

$$S_{PSD} = -8 + 2D \tag{5}$$

To estimate D using Eqn. 5, the slope of the power spectral data (Fig. 4) was fit over the entire frequency range, discarding data at < 1 cycle per micron frequency. Furthermore, no data above 30 cycles per micron, where tip dimension would impair resolution, were used. However, because of curvature in the PSDs, several fits were also made, including smaller subsets of the data, to estimate the error associated with truncation of the frequency range. Error estimates for values of D summarized in Table 1 therefore represent variability in the slope as calculated over different ranges of data.

In three of four samples, the value of D calculated from H using Eqn. 4 was close in magnitude to the value calculated from Eqn. 5 (Table 1). Only the crystal plate dissolved at pH 2.0 yielded values of D that differed markedly when calculated with the two equations.

The values of D calculated from Eqn. 5 show that the 2 starting materials have identical D within error and that D may have increased slightly for the glass after dissolution (Table 1). However, D calculated from Eqn. 5 decreased significantly for the crystal plate after dissolution at pH 2.0. Given the curvature in the PSDs, especially for the PSD for the crystal plate dissolved at pH 2.0, estimation of a slope is difficult and this decrease in fractal dimension may not be accurate. Furthermore, this value of D is calculated from data for five different images on the surface. Given that the surface was polycrystalline, calculation of D may be suspect or meaningless for this sample.

Both calculations of D show that the fractal dimension of the glass plates remains unchanged or increases after dissolution at pH 2.0. This observation can be seen qualitatively in Fig. 4C, based on the similarity in slopes for both glass samples. Henderson et al. (1997) also observed no change in D for glasses after dissolution. Again, assuming that the fractal dimension of a porous surface layer should be significantly closer to 3 than D for the polished surface, we infer that the AFM is not measuring the internal surface area of such a layer on the acid-dissolved glass sample, as expected based on the geometry of measurement by an AFM tip. Hamilton et al. (2000) concluded from XPS and SIMS data that the thickness of the leached layer was in the range 9 - 50 nm on glass plates leached at pH 2.0.

4.3. Increases in Surface Area

Although RMS roughness increased with dissolution for all samples, λ, the ratio of measured surface area to geometric surface area (see Eqn. 2), calculated using a tiling algorithm (Digital Instruments, 1997; see also, discussion in Brantley et al., 1999), increased for the crystal plate but decreased for the glass plate after dissolution (Table 1). This observation is puzzling since measured BET surface areas increased for all samples after dissolution (Table 1). However, Lai and Irene (1999a) have suggested that the tiling method may underestimate surface area for surfaces with RMS roughness values such as those reported here (Table 1), where the magnitude of deviations in the z direction (vertical) are much less than the horizontal resolution. The algorithm for calculation of surface area by tiling is neither accurate nor sensitive due to round-off error (Lai and Irene, 1999a). These workers suggest an alternative algorithm to calculate surface area, $A_{(rms,D_f)}$, using AFM data:

$$A_{\left(rms,D_f\right)} = \left(a'\,RMS\;e^{bD}\right)^2 \tag{6}$$

where a' and b are weighting factors dependent upon scan dimensions, D is the fractal dimension, and RMS is RMS roughness. This equation should yield a more accurate estimation of surface area for samples with microscopic roughness such as those in this study (Lai and Irene, 1999a), as compared to the tiling calculation (Digital Instruments, 1997). Using Eqn. 6, Lai and Irene (1999a) have shown that two surfaces with identical RMS roughness can differ in surface area and total roughness, due to the spatial distribution of roughness (fractal dimension). The authors show that for surfaces with identical RMS roughness, the area increases as fractal dimension increases (Lai and Irene, 1999b).

Using the Lai and Irene algorithm (1999a), and calculating values for a' and b according to the empirical methods outlined in Lai and Irene (1999a), the surface area was calculated based upon D and RMS roughness measured on one 10 μm x 10 μm area on both glass and crystal plates. For both cases surface area increased with dissolution, albeit by a small fraction (Table 1). In contrast to the BET data for similarly dissolved powder, however, the surface area of the crystal increased more drastically than the surface area of the glass (Table 1). Again, this re-emphasizes that BET measures the outer surface and some part of the porous leached layer internal near surface, whereas the AFM only measures some part of the outer surface.

Mineral dissolution rates are almost always cited as moles of phase "*x*" released per unit time per unit surface area (as measured by BET), even though many workers have suggested that the *reactive* surface area might differ from the BET measured area (see Brantley and Mellott, 2000 for a review). However, Hamilton et al. (2000) point out that Si release rates of albite glass did not increase during dissolution even while surface area increased by a factor of 30x. Casey et al. (1989) made similar conclusions based on observations of increased surface area without an increase in dissolution rate for labradorite feldspar. The main arguments for normalization of most mineral dissolution rates by BET surface area of the mineral instead of reactive surface area are: i) there is no agreement about how to assess reactive surface area, and ii) reactive surface area is assumed to be proportional to BET surface area. Apparently, the ratio of reactive/total (BET) surface area for the albite glass changed during dissolution at low pH. Other authors have observed similar phenomena for naturally weathered phases (Lee and Parsons, 1995). One positive attribute of the AFM measurement of surface area is that this surface area remained relatively constant during dissolution of the albite glass plate at pH 2.0, instead of increasing drastically, as was observed for the BET surface area of the albite glass powder. Brantley et al. (1999) discussed the difference between surface area measured using BET and using AFM on naturally etched quartz grains and pointed out that little is known about the topographic features that are measured by both tools. Further assessment of surface area algorithms and measurements of roughness (e.g., RMS, power spectral density, correlation lengths) based on AFM measurements may be useful in clarifying the notion of *reactive* surface area.

5. CONCLUSIONS

Atomic force microscopy (AFM) of polished surfaces of glass and crystal of albite composition were analyzed before and after dissolution at pH 2.0 and 9.0. The roughness of the crystal plates dissolved at pH 2.0 was larger than the same phase dissolved at pH 9.0. Enhanced RMS roughness on the dissolved crystal plates at pH 2.0 is interpreted as evidence of the greater extent of etching at defects at the lower pH, consistent with faster dissolution at lower pH. In contrast, the RMS roughness of glass plates dissolved at pH 2.0 was identical within error to that phase dissolved at pH 9.0, despite faster dissolution at low pH. The difference in roughness developed on crystal and glass plates cannot be explained unambiguously given the polycrystallinity of the crystal plates and the possibility of variations in polishing effects. However, AFM did not document large increases in RMS roughness due to nonstoichiometric dissolution in either crystal or glass.

Total roughness of the surfaces could not be measured without measuring the spatial distribution (fractal dimension D) as well as the vertical component of roughness (RMS). RMS roughness and fractal dimension were used to calculate surface area of plates dissolved at pH 2.0. The surface area measured by the AFM at this scale increased only slightly over 5000 h of dissolution at pH 2.0 for both phases. In contrast, the BET surface area of glass powder dissolved at pH 2.0 showed a 30-fold increase in BET surface area after 4000 h dissolution, despite the observation that the Si release rate of the glass showed no significant increase with extent of dissolution. Feldspar dissolution rates calculated from AFM measurements were within a factor of 10 of inferred dissolution rates, even though no attempt was made to measure the initial surface height nor the rate of surface retreat by removal of entire atomic layers. If results from polished samples can be extrapolated to powders, then the AFM-measured surface area may be a better proxy for reactive surface area on albite samples than BET surface area. Significant problems remain, however, in calculating surface area from AFM measurements for polished samples or powders. Additional research should focus on comparing solution measurements with AFM measurements of topography including RMS roughness, correlation length, and power spectral density.

Acknowledgments—J. Hamilton, L. Lai, G. Irene, D. Voigt, V. Bojan, and members of the Mineral Surface Science Laboratory at V.P.I., especially M. Hochella and B. Bickmore, are acknowledged for access to data and/or advice. Funding from the Dept of Energy Office of Basic Energy Sciences Grant

DE-FG02-95ER14547 (to SLB and CGP) is also acknowledged. Extremely helpful reviews were received from R. Hellmann, R. Wogelius, G. Jordan, and one anonymous reviewer. Finally, SLB acknowledges her incalculable debt to David Crerar, who introduced her to geochemistry and convinced her to become an aqueous geochemist.

Editorial handling: R. Hellmann

REFERENCES

Affatigato M., Osborne D.H., and Haglund R.F. (1996) Changes in the surface morphology of glass due to abrasion and the deposition of sol-gel thin films. *J. Non-Crys. Solids* **181**, 27-38.

Anbeek C. (1992) The dependence of dissolution rates on grain size for some fresh and weathered feldspars. *Geochim. Cosmochim. Acta* **56**, 3957-3970.

Bennet J.M. and Mattson L. (1989) *Surface Roughness and Scattering*. Optical Society of America Publication.

Blum A.E., Yund R.A., and Lasaga A.C. (1990) The effect of dislocation density on the dissolution rate of quartz. *Geochim. Cosmochim. Acta* **54**, 283-297.

Brantley S.L., Crane S.R., Crerar D.A., Hellmann R., and Stallard R. (1986) Dissolution at dislocation etch pits in quartz. *Geochim. Cosmochim. Acta* **50**, 2349-2361.

Brantley S.L., White A.F., and Hodson M.E. (1999) Surface area of primary silicate minerals. In *Growth and Dissolution in Geosystems* (eds. B. Jamtveit and P. Meakin). Kluwer Academic Publishers, Dordrecht. pp. 291-326.

Brantley S.L. and Mellott N.P. (2000) Surface area and porosity of primary silicate minerals. *Amer. Mineral.* **85**, 1767-1783.

Brunauer S., Emmett P.H., and Teller E. (1938) Adsorption of gases in multimolecular layers. *J. Am. Chem. Soc.* **60**, 309-319.

Casey W. H. and Bunker B. (1990) Leaching of mineral and glass surfaces during dissolution. In *Mineral-Water Interface Geochemistry* (eds. M. F. Hochella Jr. and A. F. White). Reviews in Mineralogy, Vol. 23, Mineralogical Society of America, Washington, D.C. pp. 397-425.

Casey W.H., Carr M.J., and Graham R.A. (1988) Crystal defects and dissolution kinetics of rutile. *Geochim. Cosmochim. Acta* **52**, 1545-1556.

Casey W.H., Westrich H.R., Massis T., Banfield J.F., and Arnold, G.W. (1989) The surface of labradorite after acid hydrolysis. *Chem. Geology* **78**, 205-218.

Digital Instruments (1997) *Nanoscope Command Reference Manual*, V. 4.31, Rev. B.

Dubuc B., Zucker S.W., Tricot C., Quiniou J.F., and Wehbi D. (1989) Evaluating the fractal dimension of surfaces. *Proc. R. Soc. London, Ser A* **370**, 113-127.

Dumas P.H., Bouffakhreddine B., Amra C., Vatel O., Andre E., Galindo R., and Salvan F. (1993) Quantitative microroughness analysis down to the nanometer scale. *Europhys. Lett.* **22**, 717-722.

Fenter P., Teng H., Geissbühler P., Hanchar J.M., Nagy K.L., and Sturchio N.C. (2000) Atomic-scale structure of the orthoclase (001)-water interface measured with high-resolution x-ray reflectivity. *Geochim. Cosmochim. Acta* **64**, 3663-3673.

Gollion P. and Grenet G. (1996) Determination of fractured steel surface roughness by atomic force microscopy using fractal-based approaches. *Surface Interface Anal.* **24**, 282-285.

Gregg S.J. and Sing K.S.W. (1982) *Adsorption, Surface Area and Porosity*. Academic Press, London.

Hamilton J.P., Pantano C.G., and Brantley S.L. (2000) A comparison of the dissolution and leaching behavior of albite crystal and glass in acidic, neutral, and basic conditions. *Geochim. Cosmochim. Acta*. **64**, 2603-2616.

Helgeson H.C., Murphy W.M., and Aagaard, P. (1984) Thermodynamic and kinetic constraints on reaction rates among minerals and aqueous solutions II. Rate constants, effective surface area, and the hydrolysis of feldspar. *Geochim. Cosmochim. Acta* **48**, 2405-2432.

Hellmann R. (1995) The albite-water system: Part II. The time-evolution of the stoichiometry of dissolution as a function of pH at 100, 200, and 300 °C. *Geochim. Cosmochim. Acta* **59**,1669-1697.

Hellmann R. (1997) The albite-water system: Part IV. Diffusion modeling of leached and hydrogen-enriched layers. *Geochim. Cosmochim. Acta* **61**, 1595-1611.

Hellmann R., Drake B., and Kjoller K. (1992) Using atomic force microscopy to study the structure, topography and dissolution of albite surfaces. In *Water-Rock Interaction 7*, Vol. I (eds. Y. K. Kharaka and A. S. Maest). A.A. Balkema, Rotterdam. pp. 149-152.

Henderson S., Castle J., and Zhdan P. (1997) The SPM and XPS characterization of the corrosion resistance of glasses. In *Advances in the Characterization of Ceramics* (eds. R. Freer, B. Derby, and B.H. Lewis). British Ceramics Proc., pp. 54-66.

Jordan G., Higgins S.R., Eggleston C.M., Swapp S.M, Janney D.E., and Knauss K.G. (1999) Acidic dissolution of plagioclase: In-situ observations by hydrothermal atomic force microscopy. *Geochim. Cosmochim. Acta* **63**, 3183-3193.

Lai L. and Irene E.A. (1999a) Area evaluation of microscopically rough surfaces. *J. Vac. Soc. Tech. B* **17**, 33-39.

Lai L. and Irene E.A. (1999b) Limiting Si/SiO_2 interface roughness resulting from thermal oxidation. *J. App. Phys.* **86**, 1729-1739.

Lasaga A.C. and Blum A.E. (1986) Surface chemistry, etch pits and mineral-water reactions. *Geochim. Cosmochim. Acta* **50**, 2363-2379.

Lee M.R. and Parsons I. (1995) Microtextural controls of weathering of perthitic alkali feldspars. *Geochim. Cosmochim. Acta* **59**, 4465 - 4488.

Lee M.R., Hodson M.E., and Parsons I. (1998) The role of intragranular microtextures and microstructures in chemical and mechanical weathering: direct comparisons of experimentally and naturally weathered alkali feldspars. *Geochim. Cosmochim. Acta* **62**, 2771-2788.

MacInnis I.N. and Brantley S.L. (1992) The role of dislocations and surface morphology in calcite dissolution. *Geochim. Cosmochim. Acta* **56**, 1113-1126.

Mitchell M.W. and Bonnell D.A. (1990) Quantitative topographic analysis of fractal surfaces by scanning tunneling microscopy. *J. Mat. Res.* **5, 10**, 2244-2254.

Pfeifer P. (1984) Fractal dimension as a working tool for surface-roughness. *Applic. Surface Sci.* **18**, 146-164.

Ruppe C. and Duparee A. (1996) Roughness analysis of optical films and substrates by atomic force microscopy. *Thin Solid Films* **288**, 8-13.

Russ J. (1994) *Fractal Surfaces*. Plenum Press, New York.

Spanos L. and Irene E.A. (1994) Investigation of roughened silicon surfaces using fractal analysis I. Two-dimensional variation method. *J. Vac. Soc. Tech. A* **12**, 2646-2652.

Stillings L.L. and Brantley S.L. (1995) Feldspar dissolution at 25 °C and pH 3: Reaction stoichiometry and the effect of cations. *Geochim. Cosmochim. Acta* **59**, 1483-1496.

Varnier F., Mayani N., and Rasigni G. (1989) Surface roughness and the fractal nature of thin films of MgF_2 and Ag/MgF_2. *J. Vac. Sci. Technol. A* **4**, 1289-1293

Vatel O., Dumas P., Chollet F., Salvan F., and Andre E. (1993) Roughness assessment of polysilicaon using power spectral density. *J. App. Phys.* **32**, 4921-4922.

White A.F. and Peterson M.L. (1990) Role of reactive-surface-area characterization in geochemical kinetic models. In *Chemical Modeling of Aqueous Systems II.* Amer. Chem. Soc. Symp. Ser. 416, American Chemical Society, Washington, D.C. pp. 461-475.

Zellmer L.A. (1986) Dissolution kinetics of crystalline and amorphous albite. M. Sc. thesis, The Pennsylvania State University, University Park, PA (USA).

Water-Rock Interactions, Ore Deposits, and Environmental Geochemistry: A Tribute to David A. Crerar

Editors: Roland Hellmann and Scott A. Wood

Linking molecular-scale barite precipitation mechanisms with macroscopic crystal growth rates

DIRK BOSBACH[1]

Institut für Mineralogie, Westfälische Wilhelms-Universität Münster, Corrensstr. 24, 48149 Münster, Germany
[1]Current address: Forschungszentrum Karlsruhe, Institut für Nukleare Entsorgung, Postfach 3640
76021 Karlsruhe, Germany
email: bosbach@ine.fzk.de

Abstract—Barite growth in aqueous solution at room temperature was observed in-situ using Atomic Force Microscopy. Various molecular scale growth mechanisms were identified on {001} and {210} cleavage surfaces. The advancement of monolayer steps depends on the crystallographic orientation and increases linearly with the degree of supersaturation. The formation rate of two-dimensional nuclei increases exponentially with increasing supersaturation. The interfacial energy was determined from two-dimensional nucleation data to be 105 mJ/m^2 for the {001} surface and 76 mJ/m^2 for the {210} surface. The critical radius for two-dimensional nuclei in a supersaturation range of $10 < S < 70$ varies between 3–6 Å ($S = [Ba^{2+}][SO_4^{2-}]/K_{sp}$; $[Ba^{2+}]$, $[SO_4^{2-}]$: Ba^{2+} and SO_4^{2-} activities in solution; K_{sp}: solubility product). Up to 10^{13} growth spirals/m^2 have been observed. Their lateral spreading rate and the interstep spacing were determined at various degrees of supersaturation. An overall growth rate was determined at various degrees of supersaturation (up to $S = 50$) by combining the intrinsic $BaSO_4$ attachment rate for the observed microscopic growth mechanisms and the associated reactive surface area. The reactive surface area for these growth mechanisms is not a constant parameter but increases nonlinearly with increasing supersaturation. The macroscopic growth rate as determined from mixed-flow reactor experiments follows a second order rate law, $R = k\times(S\text{-}1)^2$ up to a supersaturation of $S = 50$, with a reaction rate constant of $k = 1.5\times10^{-11} \pm 0.2\times10^{-11}$ mol/m^2 s. An activation energy of 35±8 kJ/mol indicates a surface reaction-controlled growth process. Comparing the macroscopically-determined growth rates with microscopic AFM observations allowed an estimation of the contribution of individual growth mechanisms. Above a supersaturation of $S > 40\text{-}50$ two-dimensional nucleation contributes more than 50% to the macroscopic growth rate.

1. INTRODUCTION

Understanding and modeling geochemical processes in low-temperature aqueous environments and hydrothermal systems of the earth's crust requires a detailed knowledge of mineral/water interfacial reactions and the reactivity of mineral surfaces (Hochella, 1990). Sorption processes at mineral surfaces control the distribution and mobility of elements via various uptake and release mechanisms (e.g., Brown et al., 1995). Among these mechanisms, mineral precipitation from aqueous fluids is an important sink for a number of major elements as well as toxic heavy metals. The impact of such a complex process on natural systems can be modeled quantitatively with respect to, for example, the mobility of trace elements using reactive transport computer codes (Lichtner et al., 1996). Ignoring chemical reaction rates can lead to incorrect predictions of chemical mass transfer in sedimentary and metamorphic systems (Cathles, 1983; Steefel & Lasaga, 1992; Lefèvre et al., 1993). Since such codes rely heavily on kinetic data for heterogeneous reactions, reliable data on mineral precipitation rates are needed (Johnson et al., 1998) in order to describe the mobility of aqueous fluids and dissolved species in chemically reactive porous rocks.

Such kinetic data as a function of the chemical affinity can be obtained from macroscopic growth experiments in batch reactor or mixed-flow reactor systems (Zhang & Nancollas, 1990). Rate laws obtained from such experiments have been used to identify rate controlling mineral precipitation mechanisms, such as spiral growth or two-dimensional (heterogeneous) nucleation in solutions (Nielsen, 1984; Brady & House, 1996). However, mineral growth is a very complex process with several independent molecular scale mechanisms on various crystallographic surfaces, with each surface having a unique atomic bonding environment contributing to the macroscopic growth rate (Lasaga, 1995). With a changing degree of supersaturation the contribution of individual mechanisms on certain surfaces to the macroscopic growth rate can change (Sunagawa, 1987). The morphology and texture of the precipitated mineral reflects this behavior macroscopically (Rimstidt, 1997).

The aim of this paper is the identification of molecular-scale precipitation mechanisms under various degrees of supersaturation using Atomic Force Microscopy (AFM). In addition, the reactive surface area with respect to a particular mechanism under a certain degree of supersaturation has been determined. Combining macroscopic mineral growth experiments and AFM observations allows an estimation of the contribution of various microscopic growth mechanisms to the total reaction rate. Precipitation of barite, $BaSO_4$, and isostructural celestite, $SrSO_4$, are used as a model system. Relatively pure barite precipitates are frequently found as a gangue mineral formed under hydrothermal conditions (Rimstidt, 1997); barite precipitates are also the major Ba sink in marine environments (Paytan et al., 1996). Furthermore, barite scale formation during secondary oil recovery is a costly problem (Benton et al., 1993). Barite precipitates over a broad supersaturation range due to temperature or pressure drops, or due to mixing of two fluid reservoirs. For example, during secondary oil recovery, ocean water with 28.1 mmol/L SO_4^{2-} (Stumm & Morgan, 1995) mixes with formation waters containing Ba^{2+} at concentrations up to 18.2 mmol/L (Hardy & Simm, 1996), thereby producing high degrees of supersaturation (dependent on mixing proprotions) with respect to the barite equilibrium solubility concentration of 0.01 mmol/L .

2. EXPERIMENTAL METHODS

Crystal growth was observed in-situ on a molecular level using a Digital Instruments NanoScope III MultiMode AFM operating in contact mode at room temperature with a commercially available fluid cell. Si_3N_4 cantilevers with a length of 100 µm with integrated pyramidal TwinTip™ tips were used (spring constant: 0.6 N/m). The maximum scan rate used in this study was about 15 scan line/s. Therefore, it took about 17 s to aquire an AFM image consisting of 256 x 256 data points. The observable reaction rate is in the range of 10^{-6} - 10^{-10} mol/m^2 s (Dove & Platt, 1996). Freshly cleaved, optically clear, natural barite crystals from various locations were mounted in a cup-like Teflon sample holder (as part of the AFM fluid cell) with a carbon-based adhesive (Leit-C, Plano). The crystallographic orientation of the sample crystal with respect to the AFM scan direction was determined after the growth experiment by X-ray defraction using an Enraf-Nonius precession camera with the Mo Kα line (40 kV, 30 mA). Before running the actual growth experiment, the sample was initially exposed to deionized water in order to dissolve and clean the topmost mineral surface layer, thus reducing the risk of surface contamination which could affect the growth experiments.

Due to strong interactions between the AFM tip and the sample surface, with a loading force in the nanonewton range, it has been shown that under certain circumstances the dissolution rate of molecular steps on a calcite cleavage surface can be mechanically enhanced (Park et al., 1996). Although contact mode AFM potentially affects the observed precipitation processes, it allows fast scan rates, which is crucial for imaging rapid precipitation mechanisms such as two-dimensional nucleation. In addition, the faster the inherent surface reaction, the smaller the relative effect the AFM tip potentially has on the observed kinetics. Comparing the microtopography of surface areas which were imaged over a period of time with surface areas that were imaged only once at the end of an experiment allows a determination of whether the observed crystal growth processes were obstructed by the AFM tip.

Step velocities of advancing monolayer steps were determined from flow-through experiments with a constant flow rate of up to 10 µL/s. Measured step velocities were independent of the flow rate, indicating that the growth rate was not transport controlled. In contrast, two-dimensional nucleation experiments in the AFM fluid cell were done in free drift experiments where a supersaturated solution was injected into the AFM fluid cell. At high degrees of supersaturation the nucleation rate was too high to be observable with AFM. Steady-state nucleation rates could not be obtained with a flow-through setup. Therefore, during the actual growth experiments the fluid flow was stopped after injection of the fluid into the AFM fluid cell. As a consequence, the fluid composition changed with time.

Supersaturated $BaSO_4$ and $SrSO_4$ were prepared by mixing equimolar Na_2SO_4 and $Ba(NO_3)_2$/$Sr(NO_3)_2$ solutions made of analytical grade chemicals and deionized water immediately before the experiment. The degree of supersaturation was determined using the aqueous speciation code in PHREEQC (Parkhurst, 1995). In this study, the supersaturation is expressed as $S = IAP/K_{sp}$, where IAP is the ion activity product of the ionic species in solution and K_{sp} is the solubility product (log K_{sp}(barite) = -9.96 (Blount, 1977) and log K_{sp}(celestite) = -6.62: Reardon & Armstrong, 1987). Macroscopic barite growth experiments were done in a mixed-flow reactor. Steady-state growth rates were obtained after initially elevated reaction rates.

Barite has an orthorhombic structure (space group: Pmna, a = 8.87 Å, b = 5.45 Å, c = 7.14 Å, Colville, 1967). Each unit cell consists of two $BaSO_4$ layers parallel to {001}, which are symmetrically related to each other via a 2_1 diad axis perpendicular to {001} (Fig. 1a). Good cleavage occurs parallel to {001} and {210} (Deer et al., 1992) (Fig. 1b). A natural hydrothermally-formed barite sample from Bad Lauterberg (Germany) was ground and a grain size fraction between 65 – 200 µm was separated in various stages. The smallest particles were removed by washing with water. Impurities were removed with an isodynamic magnetic separator and heavy liquid separation. ICP

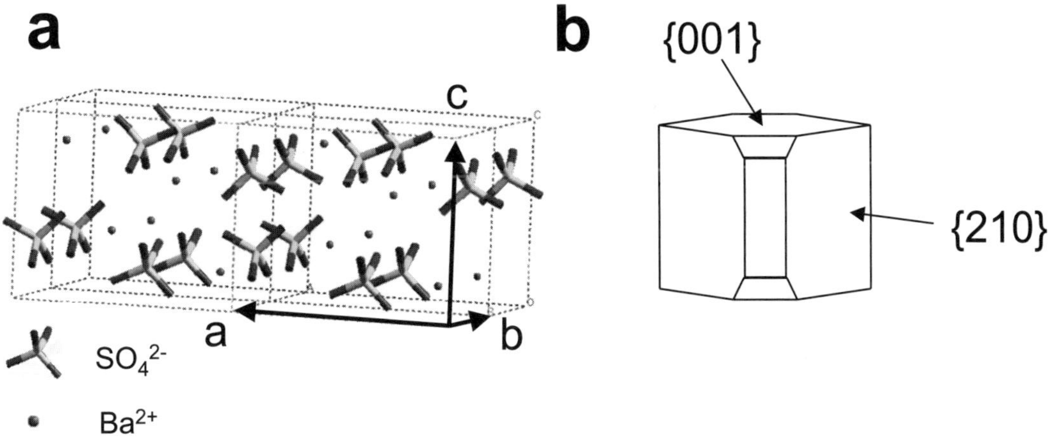

Figure 1: a) Barite has an orthorhombic crystal structure (space group: Pnma). b) The dominant crystal faces of naturally occurring barite crystals are parallel to {001} and {210}. They are energetically the most stable barite crystal faces, based on periodic bond chain theory (Hartman & Strom, 1989).

analysis showed no significant contamination by Sr or other trace elements. Only a very small amount of intergrown quartz could not be removed.

In order to derive a surface area normalized-growth rate, the initial specific surface area of barite was determined to be 0.5 m^2/g, using N_2 and the BET method (Haul & Dümbgen, 1960). The sample was heated to 300 °C prior to N_2 adsorption. However, scanning electron microscope images showed that the surfaces of the barite particles were irregular. Therefore, crystal growth rates were initially elevated with respect to euhedral crystals. The fluid composition during the experiment was monitored via solution conductivity measurements, using a 4-electrode open-geometry conductivity cell (WTW TetraCon 96). The experiments were done in a Teflon mixed flow reactor with a suspended stirring bar in order to exclude grinding effects during the experiment. The observed barite growth rate was independent of the stirring rate, thus excluding a diffusion-controlled reaction. A typical reactor experiment ran for about 24 hours with a fluid volume of 150 mL and 200 mg of barite. Depending on the degree of supersaturation, steady state was reached after 1-5 hours.

3. MOLECULAR-SCALE PRECIPITATION MECHANISMS

Mineral growth from aqueous solution is a complex process which involves several independent microscopic mechanisms. Growth units (ions or molecules) are transported by diffusion or convection to the mineral/fluid interface, where they adsorb to the mineral surface and partially dehydrate. Adsorbed species might move along the surface via surface diffusion. Eventually, they become integrated into the crystal lattice upon complete dehydration. The most likely surface site for the integration step is a kink-site along a monolayer step. As a consequence, monolayer steps advance parallel to the mineral surface. They represent the most reactive parts of a mineral surface exposed to a supersaturated solution.

3.1. Two-dimensional Island Morphology and Molecular Step Growth

In a supersaturated solution growth units tend to form islands on a mineral surface, which grow laterally with a distinct step velocity that depends on both the supersaturation in the aqueous fluid and the crystallographic orientation. As a consequence of this the islands develop a distinct morphology. In supersaturated solutions, islands with a height of one $BaSO_4$ layer form on barite {001} and {210} surfaces (Fig. 2). On the {001} surface, islands develop a circular sector shape, which is defined by molecular step edges parallel to <120> and a curved step edge (Fig. 3a) (Bosbach et al., 1998; Pina et al., 1998a, b). The curvature can be described in terms of the island diameter parallel to <120> and to [100]. For a perfectly circular

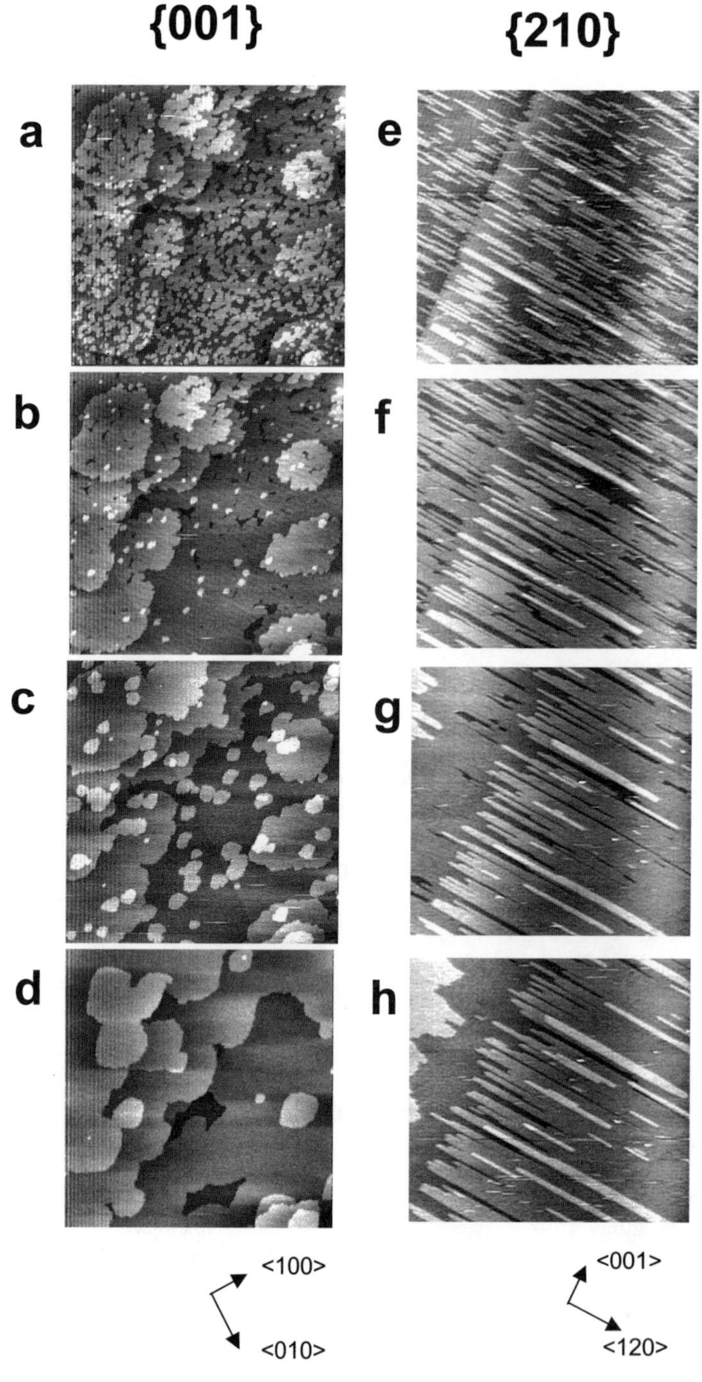

Figure 2: Growth sequences on the barite {001} (a-d) and {210} (e-h) surface. All images were taken in contact mode, in situ, under free-drift conditions. a) The initial supersaturation was $S = 70$ (scan size 1.8 x 1.8 μm, scan rate: 15 Hz, 256 x 256 pixel). The island height of 3.5 Å represents one $BaSO_4$ layer. b)-d) After the formation of two-dimensional islands monolayer step growth is the dominant growth mechanism. e) The initial supersaturation was $S = 41$ (scan size 6.0 x 6.0 μm, scan rate: 10.2 Hz, 256 x 256 pixel). f)-h) The lateral spreading due to monolayer step growth is highly anisotropic, resulting in an elongated, needle-like morphology. The island height also represents one $BaSO_4$ layer.

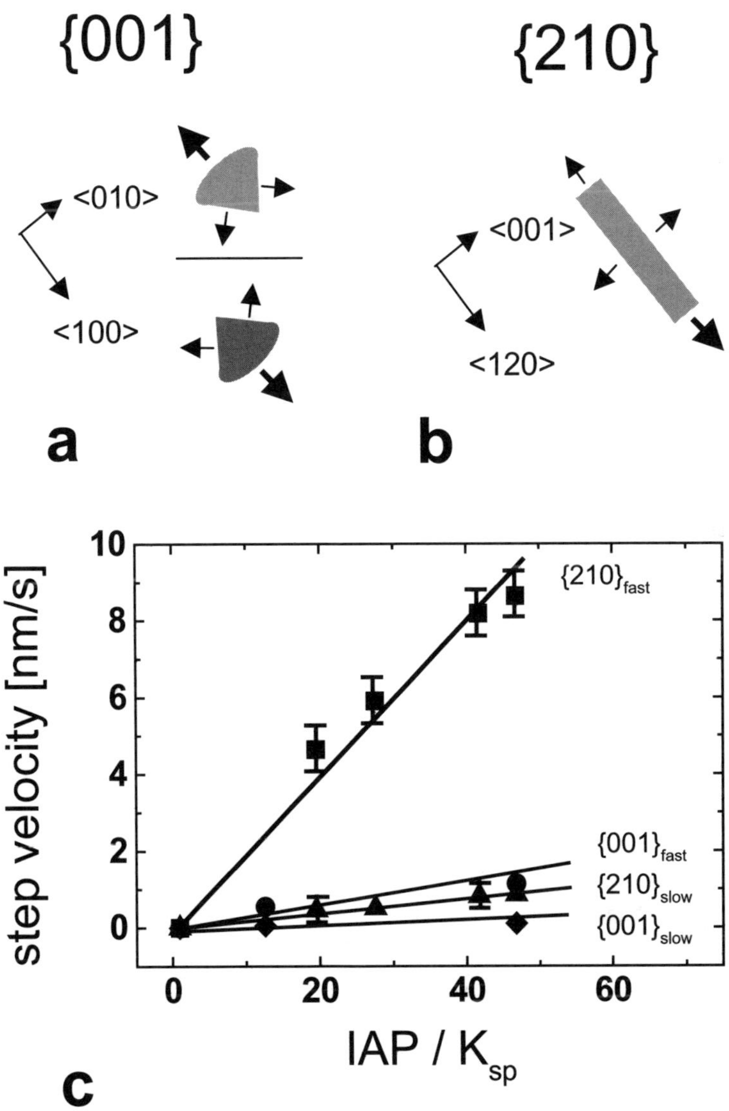

Figure 3: a) Islands on the {001} surface with a height of one $BaSO_4$ layer have a sector shape morphology, with step edges parallel to <120> and a curved step edge. In alternating layers at c = 0 and c = 0.5 islands point in the opposite direction as a consequence of a 2_1 diad axis perpendicular to the {001} surface. The curved step edges grow significantly faster than the step edges parallel to <120>, as indicated by arrows. b) On the {210} surface molecular islands have an elongated needle-like morphology, with step edges parallel to <001> and <120>. One of the short step edges parallel to <001> advances significantly faster compared to the monolayer step parallel to <001> growing in the opposite direction. Molecular steps parallel to <120> grow with a similar rate. c) The growth rate of molecular steps increases linearly with supersaturation. Anisotropic monolayer step growth can be observed on the {001} as well as on the {210} surface.

curved step edge the ratio would be 1. However, on the barite {001} surface the ratio of <120> diameter/[100] diameter is consistently about 1.1 for islands with an average diameter between 100 and 1500 nm. The stability of steps parallel to <120> can be explained by periodic bond chain (PBC) theory (Hartman, 1987). Strong bonds parallel to <120> relative to bonds perpendicular to these directions result in a low kink density along these steps. Monte-Carlo calculations on the basis of attachment energies for various sites on a barite {001} surface and the resulting double kink site formation probabilities can explain the curved step edge (Pina et al., 1998a). The deviation from the perfectly circular sector morphology is also a consequence of this selective attachment of growth units. Steps grow anisotropically, with the curved step edges advancing about 10 times faster than the steps parallel to <120> (Fig 3a). In alternating $BaSO_4$ layers the orientation of the sector is reversed due to the 2_1 diad axis perpendicular to the {001} surface (Bosbach et al., 1998; Pina et al., 1998a). If a fast growing step edge joins a slow growing step in the underlying $BaSO_4$ layer, the step velocity of the slow step controls the advancement of the resulting double step. Thus, a complex interaction between step growth kinetics and microtopography occurs.

The growth of islands on the {210} surface is different compared to the {001} surface (Pina et al., 1998b). Islands grow as long needles with a height of 3.4 Å, this representing one $BaSO_4$ layer (Fig. 2e-h). The elongated morphology is again a consequence of an anisotropic monolayer step growth (Fig. 3c). Along the long axis of the needles parallel to <001>, step growth is significantly faster in one direction relative to the opposite direction (Fig. 3b). The fastest growth direction is not reversed in alternating $BaSO_4$ layers on the {210} surface. Step growth of steps parallel to <120> is comparable to the slow growing steps parallel to <001>.

Monolayer step growth increases linearly with supersaturation for monolayer steps observed on the {001} surface, as well as on the {210} surface (Fig. 3c). In general, step edges are more reactive with respect to molecular step growth on the {210} surface relative to the {001} surface. In order to link molecular step growth with the macroscopic growth rate it is critical to estimate the rate at which new steps are generated.

3.2. Two-dimensional Nucleation Kinetics

The formation rate and the dimensions of two-dimensional nuclei strongly depend on the interfacial energy, σ, between the mineral surface and the adjacent fluid, as well as on the degree of supersaturation. Energetically, this process can be discussed in terms of the change in Gibbs free energy of formation of two-dimensional nuclei, ΔG:

$$\Delta G = n \cdot k \cdot T \cdot \ln(S) + \sigma \cdot 2 \cdot \pi \cdot r \cdot h \qquad (1)$$

In the first term of Eqn. 1, n represents the number of growth units ($BaSO_4$) in the island, k is the Boltzmann constant, and T the temperature in K. It will be assumed that one growth unit for barite growth consists of ($BaSO_4$), which represents a quarter of the unit cell. In the second term of Eqn. (1), r is the radius and h the height of a pancake-like island. Clearly, the formation of nuclei is energetically not favored at low degrees of supersaturation. Only above a certain size are nuclei stable and able to grow laterally. Above this critical size islands are called nuclei, whereas below this they are called embryos (Chernov, 1984). Due to the energy barrier for forming stable nuclei, their formation is thought to be an efficient growth mechanism only at high degrees of supersaturation. The critical size of nuclei is supersaturation-dependent. Although a theoretical framework has been developed for describing the thermodynamics of two-dimensional nucleation in aqueous solution (Nielsen, 1964; Walton, 1965; Hillig, 1966), it is still not well understood. In many cases, the interfacial energy and the critical dimensions of such nuclei are not known. Unfortunately, these processes are not accessible directly with light microscopic techniques, such as differential interference contrast microscopy (DICM). Despite their excellent vertical resolution, which allows for the imaging of molecular scale steps, the lateral resolution is limited by the wavelength of the light source. Nevertheless, these methods have provided detailed in-situ information on spiral growth and step growth in aqueous fluids (Sunagawa, 1987).

Figure 2 shows growth sequences on barite {001} and {210} cleavage surfaces. At the beginning of a growth experiment in the AFM fluid cell, 2 mL of a supersaturated solution was injected in less than one second, thus ensuring that the entire volume of the fluid cell of about 100 µl had been replaced by the supersaturated solution. After a short induction time, two-dimensional islands formed on atomically flat terraces (Fig 2). On the {001} surface (Fig. 2a-d), upon formation of the first layer, the islands spread laterally and subsequently only a few new islands form. Since the fluid flow was stopped after the initial injection of the supersaturated solution, the fluid composition became depleted in Ba^{2+} and SO_4^{2-}, which resulted in a reduced degree of supersaturation and consequently in a reduced growth rate. Due to the fast reaction rate relative to the scan speed of the AFM, the formation of islands could not be observed when constantly feeding the fluid cell with supersaturated solution. Although this would ensure constant composition conditions, the experiments were performed under free drift conditions. Therefore, steady-state nucleation rates could not be extracted.

During the period of initially high degree of supersaturation, the lateral spreading of the islands is fast relative to the scan speed of the AFM, and it is therefore difficult to follow microtopographic changes (Figs. 2 a and e). After injection of the supersaturated solution, the first image (Fig. 2a) was generated (from bottom to top), requiring about 17 sec for the complete image. At the bottom of the image (Fig. 2a), individual islands can be clearly distinguished. Upon formation, these islands grow laterally via step growth. Therefore, in the top region of the image (Fig. 2a), individual islands can no longer be distinguished because they have spread laterally and cover almost the entire crystal surface. Furthermore, a few islands have formed on top of the first islands (Fig. 2b), but at a lower density, as a consequence of depletion of the $BaSO_4$ solution. Repeating such experiments with various initial supersaturations reveals that the initial island density increases exponentially with increasing supersaturation (Fig. 4a), as one would expect according to the theoretical work by Nielsen (1964), Walton (1965) and Hillig (1966), which predicts an exponential increase of the steady state nucleation rate, J, with increasing supersaturation:

$$J = J_0 \times \exp\left(-\frac{\pi \times h \times \omega \times \sigma^2}{(R \times T)^2 \times \ln S} \right) \qquad (2)$$

where ω represents the molecular volume. Eqn. (2) also assumes that the two-dimensional nuclei have pancake-like morphology with height h. The pre-exponential factor J_0 is related to the efficiency of ion collisions with the mineral surface.

The island density, which was estimated after injecting a supersaturated solution into the AFM fluid cell (Figs. 2a and 2e) can be referred to as an initial surface nuclei density, N_i. In the case of three-dimensional nucleation it has been shown that such an initial nucleation rate is directly proportional to the nucleation rate under steady state conditions (Nielsen, 1964; Lasaga 1998). A linear correlation between $\ln N_i$ vs. $(\ln S)^{-2}$ for the case of three-dimensional nucleation has been interpreted as an indication that the above given thermodynamic approach applies to the precipitation of barite and isostructural crystals (Nielsen, 1964). Similarly, in the case of the two-dimensional nucleation data presented here, $\ln N_i$ for barite and celestite can be plotted versus $(\ln S)^{-1}$, yielding a good linear correlation (Fig. 4b). The interfacial energy for the mineral/fluid interface can be determined from the slope of the linear fits in Fig. 4b based on equation (2). In case of barite one obtains 105 mJ/m^2 from the observations on the {001} and 76 mJ/m^2 on the {210} surfaces. However, these values do not represent the interfacial enrergies of the {001} and {210} surface, but rather they are related to the edge surfaces of the nuclei. Recently, He et al. (1995) found a value of 93.4 mJ/m^2

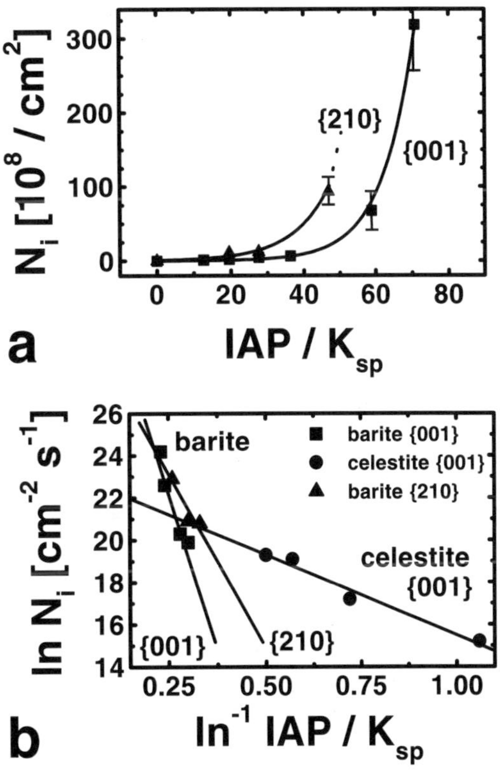

Figure 4: a) The observed initial density of islands, N_i, after injecting a supersaturated solution into the AFM fluid cell increases exponentially on the {001} and {210} barite surfaces. b) A linear correlation between $\ln N_i$ versus $(\ln IAP/K_{sp})^{-1}$ for the barite {001} and {210} surfaces, as well as for the celestite {001} surface, suggest that the observed island density can be related to a steady-state nucleation rate.

for barite from macroscopic three-dimensional (homogeneous) nucleation experiments at low ionic strength at room temperature. However, the macroscopically determined value integrates over several crystal surfaces with various crystallographic orientations and interfacial energies. Based on PBC theory (Hartman, 1987), the crystal morphology for a barite crystal grown at low to moderately high degrees of supersaturation has been predicted to be defined by a combination of {001} and {210} crystal faces. Furthermore, a statistical evaluation of the crystal morphology of naturally occuring barite crystals showed that {001} and {210} are the most dominant crystal faces (Hartman & Strom, 1989). Therefore, the interfacial energy derived from macroscopic three-dimensional nucleation experiments should be equal to the weighted sum of the interfacial energies obtained on the {001} and {210} surfaces. For celestite one ob-

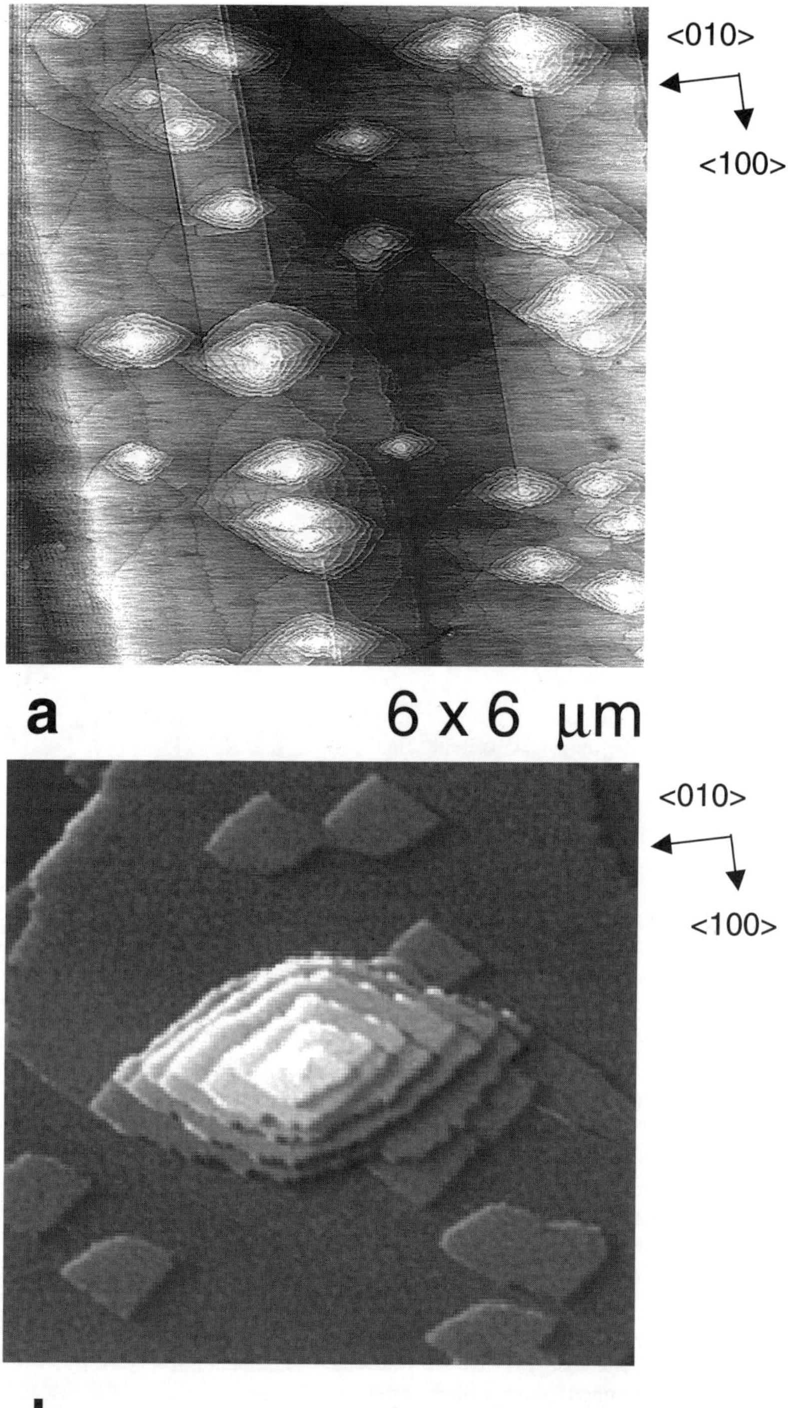

Figure 5: On the barite {001} surface growth spirals develop in the vicinity of dislocation cores in supersaturated solutions. a) The image was taken 30 min after exposure of a freshly cleaved sample to a supersaturated solution with $S = 25$. The spiral density varies between (0 to 10^{13} m^{-2} on the same crystal. b) Due to the growth anisotropy in alternating $BaSO_4$ layers, the shape and the lateral spreading of the growth spirals are controlled by the slowest growing molecular steps. Each spiral consists of monolayer and double layer steps. (after Pina et al., 1998a)

tains 38 mJ/m^2 for the interfacial energy from AFM observations on the {001} surface. This value is significantly smaller than the 105 mJ/m^2 obtained for the barite {001} surface, as one would expect following the empirical relation by Nielsen & Söhnel (1979), which relates the solubility to the interfacial energy.

In addition to the kinetic aspects of nuclei formation, the critical size of nuclei is another important parameter for understanding two-dimensional nucleation. Assuming a pancake-like morphology, the number of growth units, n_{crit}, in a nucleus with critical size is given by (Lasaga, 1998):

$$n_{crit} = \frac{\omega \cdot \sigma^2 \cdot h \cdot \pi}{(k \cdot T \cdot \ln S)^2} \qquad (3)$$

In the supersaturation range of $10 < S < 70$ the number of growth units in a nucleus with critical size varies between 7-5 on the barite {001} surface and between 6–4 on the {210} surface. Assuming a pancake-like morphology with a height of one $BaSO_4$ layer, the corresponding critical radius varies between 6 and 3 Å on the {001} surface and 5–2 Å on the {210} surface. These values agree well with the predicted critical size of nuclei by Nielsen (1964), who estimated that two dimensional nuclei forming at a mineral/fluid interface consist of 10 or fewer ions.

3.3. Spiral Growth

Due to the interfacial energy that is involved in the formation of two-dimensional nuclei, this mechanism is the predominant growth mechanism only at high degrees of supersaturation. Following Eqn. (2), at near-equilibrium conditions the critical radius for nuclei would be large (e.g., for barite {001} 150 Å at S = 5) and the probability for their formation is small. However, screw dislocations intersecting a mineral surface provide a continous source of monolayer steps. The subsequent attachment of growth units to these steps result in the development of growth spirals (BCF model, Burton et al., 1951).

The development of growth spirals on the barite {001} surface was monitored over several hours (Fig. 5). As a consequence of the monolayer step growth anisotropy, fast growing monolayer step edges parallel to <120> catch up with underlying slow growing monolayer step edges and form a two-layer step. The lateral spreading of growth spirals is controlled by such two-layer step edges parallel to <120>. The lateral spreading of the growth spirals agrees within the limits of error with the slowest step velocity observed for monolayer steps parallel to <120>. The morphology of growth spirals should be defined by steps parallel to these directions. However, this is only the case for small growth spirals and the 3–4 topmost $BaSO_4$ double layers of larger spirals (Fig. 5b). At the bottom of a spiral, double steps are curved and not strictly aligned parallel to <120>. The curvature can be explained by the preferential attachment kinetics of growth units and the resulting probability for the formation of double kink sites at certain sites along these steps (Pina et al., 1998a). The curvature of the <120> double layer steps can only be found within growth spirals. On the contrary, for large islands with a height of a single $BaSO_4$ layer, steps with the slowest growth rate have crystallographically aligned steps parallel to <120> , independent of the size.

While a growth spiral develops on a freshly cleaved barite surface, its growth rate (in mol/L) at a given degree of supersaturation is not a constant but increases as it becomes larger because it generates new molecular steps. A rough estimation based on the dimensions of single spirals shows that a typical growth spiral that has developed within one hour at a supersaturation of S=10 has a growth rate of about 10^{-21} mol/s. In order to determine a macroscopic surface area-normalized steady-state growth rate from microscopic experiments under non-steady-state conditions, one has to determine the dislocation density. Unfortunately, the defect density is not always homogeneously distributed over the entire barite surface (Bosbach et al., 1998). On {001} cleavage surfaces the observed dislocation density was up to 10^{13} m^{-2}. Assuming this defect density, one can calculate a macroscopic growth rate of about 10^{-8} mol/m^2 s, based on the observed growth rate of individual spirals within one hour under non-steady state conditions. However, under steady-state conditions, where the entire barite surface is covered by monolayer or two-layer steps, the macroscopic growth rate is controlled by the step spacing or step density and the step growth.

In contrast to the {001} surface, the formation of growth spirals was never observed on the {210} surface, even after several hours of exposure to a supersaturated solution. However, dislocations with a Burgers vector equivalent to one $BaSO_4$ layer parallel to {210} could be identified. Spiral growth on the {210} surface seems to be a structurally self-inhibited mechanism as a consequence of a strong anisotropy for the growth of molecular steps. Consequently, the contribution of spiral growth on the {210} surface is negligible for growth at room temperature and for supersaturations of $S \leq 20$.

Based on in-situ observations with AFM, molecular scale barite growth mechanisms can be identified at various degrees of supersaturation. However, in order to evaluate the importance of these observed growth mechanisms with respect to the macroscopic crystal growth rate, surface area-normalized reaction rates have to be determined at various degrees of supersaturation as a reference.

4. MACROSCOPIC BARITE GROWTH RATES

At a macroscopic scale, barite growth in dilute aqueous solutions at an ionic strength of 10^{-4} M follows a second order growth rate law according to:

$$R = k \cdot (S-1)^2 \qquad (4)$$

within the investigated supersaturation range (up to S=50) with a rate constant of $1.5\times10^{-11} \pm 0.2\times10^{-11}$ mol/m^2 s (Fig. 6). The growth rate varies between about 10^{-11} mol/m^2 s under near equilibrium conditions to 1.2×10^{-8} mol/m^2 s at S = 50. The activation energy for barite growth in a supersaturation range of $8 < S < 26$ was determined from simple batch reactor experiments at various temperatures from 21-50 °C to be 35±8 kJ/mol, following an Arrhenius approach.

A 2nd order rate law has been interpreted as an indication that the direct integration of growth units into the crystal lattice at molecular steps associated with growth spirals is the rate controlling step (Zhang & Nancollas, 1990; Brady & House, 1996). Below a threshold supersaturation where two-dimensional nucleation becomes the rate-controlling mechanism, spiral growth according to the BCF model should be the dominant precipitation mechanism under near equilibrium conditions (Burton et al., 1951; Nielsen, 1984). A rate law with a reaction order greater than 2 has been interpreted as an indication that two-dimensional nucleation is the dominant mechanism, which is known as the birth & spread model (O'Hara & Reid, 1973). The 2nd order rate law observed for barite suggests, therefore, that spiral growth is the dominant growth mechanism within the investigated supersaturation range, up to S = 50. Furthermore, similar macroscopic experiments by Doremus (1958) and by Liu et al. (1976) show a 2nd-order rate law up to S = 230.

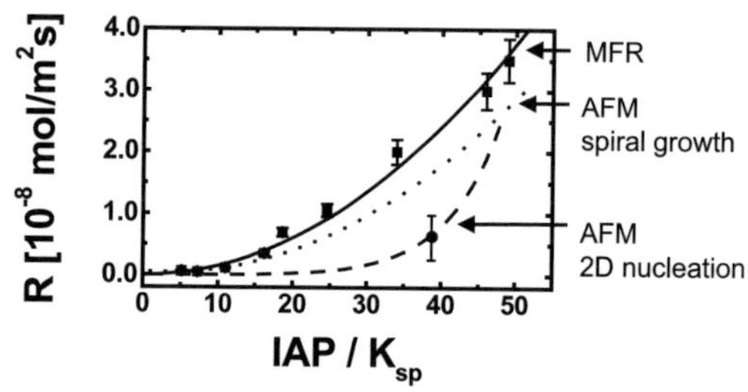

Figure 6: The macroscopic barite steady-state growth rate as determined from mixed flow reactor (MFR) experiments follows a second order rate law (solid line). A macroscopic growth was calculated on the basis of the island density (two dimensional nucleation) and the monolayer step growth data obtained from AFM observations (dashed line). An error bar at a selected degree of supersaturation indicates the uncertainty based on the errors for the individual measurements for step growth and two-dimensional nucleation. The contribution of spiral growth (dotted line) has been predicted based on AFM observations at low degrees of supersaturation up to S = 20. Above a supersaturation range of $40 < S < 50$ growth associated with two-dimensional nucleation becomes the predominant precipitation mechanism.

5. LINKING MICROSCOPIC MECHANISMS AND MACROSCOPIC PRECIPITATION RATES

Relating the observed molecular-scale barite growth mechanism to the macroscopic rate law can be achieved by estimating the contribution of: (i) step growth of cleavage steps, (ii) spiral growth, and (iii) two-dimensional nucleation on every crystal face. Linking molecular-scale growth mechanisms and macroscopic reaction rates ultimately requires that the reactive surface area be determinable. This has recently been demonstrated for the case of dissolution of structurally highly anisotropic sheetsilicates (Rufe & Hochella, 1999; Bosbach et al., 2000). Reactive surface area is not a constant parameter but depends on the monolayer step density in the case of step growth, the dislocation density in the case of spiral growth under non-steady state conditions, on the nucleation rate in the case of two dimensional nucleation, and on the lateral step velocity. In addition, the step density associated with growth spirals growing under steady state conditions depends on the degree of supersaturation.

5.1. Monolayer Step Growth

In the case of monolayer step growth only a small fraction of the entire mineral surface is reactive with respect to this mechanism. If one considers only the mineral surface adjacent to monolayer steps as the reactive part of the mineral surface for the attachment of one growth unit, one can estimate a macroscopic growth rate on the basis of measured step velocities (Fig. 3c) and observed step densities. Freshly cleaved barite {001} surfaces have a reactive surface area that represents up to 0.1% of the total surface area. Since the cleavage process is not well defined, the values scatter significantly and the value of 0.1% represents the highest observed monolayer step density. Under near equilibrium conditions at S = 2 the reactive surface area-normalized growth rate can be estimated to be about 10^{-9} mol/m^2 s. Therefore, the growth rate normalized to the total surface area should be about 10^{-12} mol/m^2 s. The macroscopically-determined total (BET) surface area normalized-growth rate under this condition is about 10^{-11} mol/m^2 s. The barite particles used for the macroscopic growth experiments were not as flat as the cleavage surfaces used for the AFM experiments. Although the intrinsic $BaSO_4$ attachment rate was comparable in the macroscopic and microscopic (AFM) experiments, the macroscopically-determined growth rate is faster than the growth rate obtained from AFM observations. However, comparing these rates allows an estimation of the reactive surface area of the material used in macroscopic experiments.

5.2. Spiral Growth

The development of growth spirals on freshly cleaved barite surfaces increases the reactive surface area significantly, although the total surface area as determined from BET measurements would only increase slightly. However, under steady-state conditions, growth spirals on a crystal surface cover the entire crystal surface and the microtopography of the crystal surface would establish a constant step density with a constant reactive surface area/total surface area ratio, which only depends on the degree of supersaturation. The BCF model (Burton et al., 1951) predicts, and AFM observations confirm, that the step density associated with growth spirals increases with increasing supersaturation. Therefore, the number of reactive sites along step edges depends on the degree of saturation and cannot be considered as a constant.

The growth rate of spirals normal to the surface can be used to estimate the steady-state normal growth of a crystal. At a supersaturation of $S = 10$ the height of growth spirals on the barite {001} surface increases at about 5×10^{-4} nm/s. A macroscopic growth rate can be estimated to be 5×10^{-9} mol/m^2 s, by assuming an average diameter of 100 μm for cubic particles. The macroscopically-determined growth rate at $S = 10$ is 1.0×10^{-9} mol/m^2 s. Considering the fact that AFM observations suggest that spiral growth does not contribute significantly to the growth of the {210} surface, an estimation for a macroscopic growth rate based on AFM observations on the barite {001} surface agrees very well with the macroscopically observed rate (Fig. 6). Spiral growth normal to the {001} surface was only determined over a narrow supersaturation range, up to $S < 20$.

5.3. Two-dimensional Nucleation

The formation of two-dimensional nuclei and lateral spreading of islands was imaged on the {001} and the {210} surface. The entire barite substrate is covered with islands, which implies that 100% of the crystal surface can be considered to be reactive with respect to nucleation events.

Combining the observed island density (Fig. 4a) and the monolayer step growth rates on the {001} and {210} surfaces (Fig. 3c), one can calculate a macroscopic rate with respect to two-dimensional nucleation (Fig. 6). Depending on the relative size of the {001} and {210} crystal faces, growth associated with two-dimensional nucleation contributes more than 50% to the total growth rate above a threshold supersaturation range of $40 < S < 50$. Assuming that the microscopic in-situ AFM experiments were performed under conditions comparable to the macroscopic mixed-flow reactor experiments, the difference between the macroscopically-obtained growth rate and the predicted growth rate based on AFM observations (of two-dimensional nucleation) is due to spiral growth.

In the case of an ideal system, where the precipitating crystal and the substrate are identical materials, the macroscopic crystal growth rate at high degrees of supersaturation can be predicted on the basis of microscopically-characterized two-dimensional nucleation processes. However, the composition of aqueous fluids in nature can be very complex and the heterogeneous precipitation on a foreign substrate is very common. As a simple example, the precipitation of a solid solution system such as barite growth on a celestite ($SrSO_4$) {001} substrate was imaged in-situ with AFM (Fig. 7). The lattice constants of isostructural celestite (a = 8.38 Å, b = 5.37 Å, c = 6.85 Å) and barite (a = 8.87 Å, b = 5.45 Å, c = 7.14 Å) differ by up to 6%. AFM observations showed no indication for a structural surface reconstruction on the barite {001} surface (Bosbach et al., 1998). Although there is no experimental evidence available, this will also be assumed for celestite. There is a lattice mismatch of 6% parallel to <100>, 2% parallel to <010>, and 3% parallel to <120>. Therefore, epitaxial growth of pure barite or a $(Ba,Sr)SO_4$ solid solution on the celestite substrate are affected by these differences.

Heterogeneous precipitation in the $(Ba,Sr)SO_4$ solid solution aqueous system was observed in situ at various degrees of supersaturation. In these AFM growth experiments, monolayer step edges were generated by etching the celestite {001} surface in deionized water. Shallow triangular etch pits formed within 5 minutes. The morphology of the etch pits is defined by molecular steps parallel to <010> and <120> (Fig. 7), which has also been observed on barite {001} (Putnis et al., 1995; Bosbach et al., 1998; Higgins et al., 1998). The orientation of the etch pits is reversed in alternating $SrSO_4$ layers as a consequence of the 2_1 diad axis perpendicular to

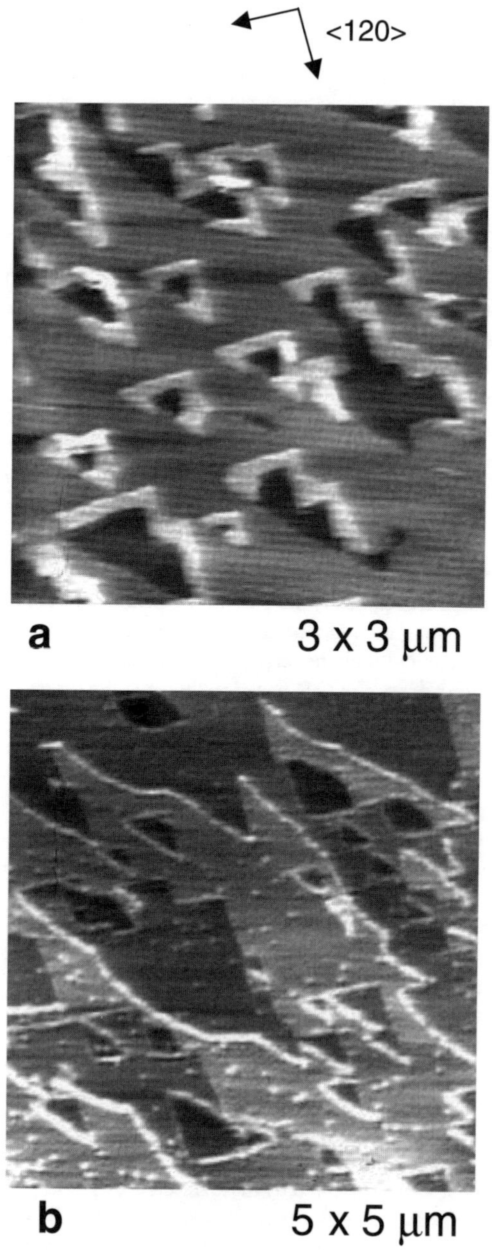

Figure 7: a) Precipitation of $BaSO_4$ (barite) onto an isostructural $SrSO_4$ (celestite) substrate at a supersaturation of $S = 170$ only occurs along monolayer step edges. The celestite (001) substrate was etched in pure water before the supersaturated solution was injected. As a consequence of the etching process, shallow triangular etch pits with a depth of one $SrSO_4$ layer have developed. The triangular shape of the etch pits is defined by molecular step edges parallel to [010] and <120>. $BaSO_4$-rich precipitates nucleate preferentially on step edges parallel to <120>. Flat celestite terraces are not covered with islands. b) At a higher supersaturation of $S = 380$, two-dimensional nucleation also occurs on flat terraces. However, the nucleation process is still enhanced along monolayer step edges.

{001}. After injecting a $(Ba,Sr)SO_4$ solution with 30 µM $BaSO_4$ and 550 µM $SrSO_4$ into the AFM fluid cell, islands with a height of about 3.5 Å form (Fig. 7a). These islands do not cover the entire crystal surface, but rather their formation is restricted to a narrow area close to the rim along monolayer step edges parallel to <120>. Unfortunately, AFM does not provide information about the chemical composition of the precipitates. However, following Lippman's approach to describe solid solution aqueous systems (Lippmann, 1980; Glynn & Reardon, 1990), one can calculate a supersaturation function with respect to the $(Ba,Sr)SO_4$ solid solution. The given solution has the highest supersaturation (S = 171) with respect to an almost pure $BaSO_4$ endmember. It has been shown that two-dimensional nucleation in the $(Ba,Sr)SO_4$ system is consistent with Lippmann's thermodynamic approach (Pina et al., 1999). Therefore, it will be assumed that the islands shown in Fig 7a consist of a $BaSO_4$-rich phase. The fact that two-dimensional nucleation under the given conditions is restricted to monolayer step edges suggests that these sites are energetically favorable with respect to the formation of critical nuclei. The lattice mismatch between barite and celestite can be overcome by elastic deformation of the precipitating phase as well as the substrate (Walton, 1965). Considering the lattice parameters, the formation of two-dimensional barite nuclei on a celestite {001} substrate implies that the barite lattice is compressed, whereas the celestite lattice is expanded. Quantum mechanical calculations of stepped mineral surfaces (Vaughan et al., 1997) indicate that the atomic structure is relaxed in the vicinity of molecular steps. Therefore, small lattice mismatches between the precipitating phase and the substrate can be accommodated most easily near step edges. Similar nucleation behavior has been observed for various heterogeneous precipitation processes, such as the epitaxial formation of $NaNO_3$ on a $CaCO_3$ substrate (Chernov, 1984), or the precipitation of MnOOH onto an albite substrate (Junta-Rosso & Hochella, 1997). The reactive surface area for these heterogeneous reactions is only a small fraction of the entire crystal surface. In contrast, in the case of barite precipitation onto a barite {001} substrate, there are no preferred nucleation sites.

At a higher supersaturation of $S = 380$, two-dimensional nucleation also occurs on atomically flat terraces, however at a smaller rate compared to monolayer step edges (Fig. 7b). In addition to elastic deformation the lattice mismatch between precipitating barite and celestite substrate can also be overcome by the formation of dislocations at the interface. Apparently, this requires a higher degree of supersaturation and consequently a faster precipitation rate.

The supersaturation-dependent limits of the reactive surface area for two-dimensional nucleation in solid solution aqueous systems are an excellent example of crystal growth complexity in natural systems. At a supersaturation of S = 171 the reactive surface area depends on the step density. Under these conditions only about 2% of the total surface area was reactive with respect to the formation of two-dimensional nuclei. The reactive surface area increases with the degree of supersaturation. At S = 380 the entire crystal surface becomes reactive. Due to the variations in reactive surface area, macroscopically-derived growth rate laws under such conditions cannot easily be interpreted with respect to the dominant precipitation mechanism. Deriving surface area-normalized growth rates based on BET surface areas can vary significantly due to the nonlinear variation of the reactive surface area with changing degrees of supersaturation.

6. CONCLUSIONS

Macroscopically-derived crystal growth rates cannot be understood without microscopic information about the actual molecular scale growth mechanisms. However, fluid cell AFM observations open the possibility of obtaining details on monolayer step growth, spiral growth, and two-dimensional nucleation under controlled conditions and determining their intrinsic growth rate. In the case of barite, the contribution of these growth mechanisms to the overall growth rate changes with increasing degree of supersaturation. Monolayer step growth associated with cleavage steps and growth spirals is the predominant mechanism below a supersaturation of $40 < S < 50$. Above a supersaturation of $S > 40$-50, two-dimensional nucleation and the subsequent lateral spreading of islands contributes more than 50% to the total crystal growth rate. The intrinsic rate of $BaSO_4$ attachment to molecular steps changes with increasing supersaturation. In addition, the attachment of growth units does not involve the entire mineral surface since only a small fraction is reactive with respect to growth. The reactive surface area changes nonlinearly with the dominant growth mechanism and the degree of supersaturation. Fluid cell AFM experiments simultaneously provide quantitative information about the reaction rate at molecular scales and the reactive surface area. The macroscopic mineral growth rate is a function of both the intrinsic molecular scale attachment of growth units and the reactive surface area. Their complex relation can be unraveled by combining macroscopic growth experiments with microscopic AFM observations.

Acknowledgements—Many thanks are due to the Münster Mineral Surface group, which includes U. Becker, F. Brandt, C.M. Pina, A. Putnis, and P. Risthaus. The help of H. Pieper and V. Rapelius in the chemistry lab is also greatly appreciated. Financial support by the Deutsche Forschungsgemeinschaft is gratefully acknowledged. Constructive reviews by M. F. Hochella, Jr., G. Jordan, H. H. Teng, and an anonymous reviewer helped to improve the manuscript.

Editorial handling: Roland Hellmann

REFERENCES

Benton W. J., Collins I. R., Grimsey I. M., Parkinson G. M., and Rodger S. A. (1993) Nucleation, growth and inhibition of barium sulfate controlled modification with organic and inorganic additives. *Faraday Disc.* **95**, 281-297.

Blount C. W. (1977) Barite solubilities and thermodynamic quantities up to 300 °C and 1400 bars. *Am. Mineral.* **62**, 942-957.

Bosbach D., Hall C., and Putnis A. (1998) Mineral precipitation and dissolution in aqueous solution: in-situ microscopic observations on barite (001) with atomic force microscopy. *Chem. Geol.* **151**, 143-160.

Bosbach D., Charlet L., Bickmore B. R., and Hochella M. F. Jr. (2000) The dissolution of hectorite: An in-situ, real-time study using Atomic Force Microscopy. *Am. Mineral.* **85**, 1209-1216.

Brady P. V. and House W. A. (1996) Surface controlled dissolution and growth of minerals. In *Physics and Chemistry of Mineral Surfaces* (ed. P.V. Brady). CRC Press, New York. pp. 225-305.

Brown G. E., Parks G. A., and O'Day P. A. (1995) Sorption at mineral – water interfaces: macroscopic and microscopic perspectives. In *Mineral Surfaces* (eds. D.J. Vaughan & R.A.D. Pattrick). Chapman & Hall, London. pp. 129-184.

Burton W. K., Cabrera N., and Frank F. C. (1951) The growth of crystals and the equilibrium structure of their surfaces. *Philos. Trans. Royal Soc.* **A243**, 299-358.

Cathles L. M. (1983) An analysis of the hydrothermal system responsible for massive sulphide deposition in the Hokuroku Basin of Japan. *Econ. Geol. Monogr* **5**, 439-487.

Chernov A. A. (1984) *Modern Crystallography III. Crystal Growth*. Springer. Heidelberg.

Colville A. A. (1967) A refinement of the structure of barite. *Am. Mineral.* **52**, 1877-1880.

Deer W. A., Howie R. A., and Zussman J. (1992) *An Introduction to the Rock Forming Minerals*. Longman. Essex.

Doremus R. H. (1958) Precipitation kinetics of ionic salts from solution. *J. Phys. Chem.* **62**, 1068-1073.

Dove P. M. and Platt F. M. (1996) Compatible real-time rates of mineral dissolution by Atomic Force Microscopy (AFM). *Chem. Geol.* **127**, 331-338.

Glynn P. D. and Reardon E. J. (1990) Solid solution aqueous solution equilibria: thermodynamic theory and representation. *Am. J. Sci.* **290**, 164-201.

Hardy J. A. and Simm I. (1996) Low sulfate seawater mitigates barite scale. *Oil & Gas Journal* **94**, 64-67.

Hartman P. (1987) Modern PBC theory. In *Morphology of Crystals* (ed. I. Sunagawa). Terra Scientific Publishing Company. Tokyo. pp 209-253.

Hartman P. and Strom C. S. (1989) Structural morphology of

crystals with barite ($BaSO_4$) structure: A revision and extension. *J. Crystal Growth* **97**, 502-512.

Haul R. and Dümbgen G. (1960) Vereinfachte Methode zur Messung von Oberflächengrößen durch Gasadsorption. *Chem. Ing. Tech.* **32**, 349-354

He S., Oddo J. E., and Tomson M. B. (1995) The nucleation kinetics of barium sulfate in NaCl solutions up to 6 m and 90 °C. *J. Colloid Interface Sci.* **174**, 319-326.

Higgins S. R., Jordan G., Eggleston C. M., and Knauss K. G. (1998) Dissolution kinetics of the barium sulfate (001) surface by hydrothermal atomic force microscopy. *Langmuir* **14**, 4967-4971.

Hillig W. B. (1966) A derivation of classical two-dimensional nucleation kinetics and the associated crystal growth laws. *Acta Metallurgica* **14**, 1868-1869.

Hochella M. F. Jr. (1990) Atomic structure , microtopography, composition and reactivity of mineral surfaces. In *Mineral-Water Interface Geochemistry* (eds. M.F. Hochella and A.F. White). Reviews in Mineralogy **23**. Mineralogical Society of America, Washington D.C. pp. 87-132.

Johnson J. W., Knaus K. G., Glassley W. E., DeLoach L. D., and Tompson A. F. B. (1998) Reactive transport modeling of plug-flow reactor experiments: quartz and tuff dissolution at 240 °C. *Journal of Hydrology* **209**, 81-111.

Junta-Rosso J. L. and Hochella M. F. Jr. (1997) Linking microscopic and macroscopic data for heterogeneous reactions illustrated by the oxidation of manganese (II) at mineral surfaces. *Geochim. Cosmochim. Acta* **61**, 149-159.

Lasaga A. C. (1995) Fundamental approaches in describing mineral dissolution and precipitation rates. In *Chemical Weathering Rates of Silicate Minerals* (eds. A.F. White and S.L Brantley). Reviews in Mineralogy **31**. Mineralogical Society of America. Washington D.C. pp 23-86.

Lasaga A. C. (1998) *Kinetic Theory in the Earth Sciences*. Princeton University Press. Princeton.

Lefèvre F., Sardin M., and Schweich D. (1993) Migration of strontium in clayey and calcareous sandy soil: precipitation and ion exchange. *J. Contaminant Hydrology* **13** (1993) 215-229.

Lichtner P. C., Steefel C. I., and Oelkers E. H. (1996) *Reactive Transport in Porous Media*. Reviews in Mineralogy, Vol. 34, Mineralogical Society of America, Washington D.C.

Lippmann F. (1980) Phase diagrams depicting aqueous solubility of binary carbonate systems. *N. Jb. Mineral. Abh.* **139**, 1-25.

Liu S. T., Nancollas G. H., and Gasiecki E. A. (1976) Scanning electron microscopic and kinetic studies of the crystallization and dissolution of barium sulfate crystals. *J. Crystal Growth* **33**, 11-20.

Nielsen A. E. (1964) *Kinetics of Precipitation*. Pergamon Press, Oxford.

Nielsen A. E. (1984) Electrolyte crystal growth mechanisms. *J. Crystal Growth* **67**, 289-310.

Nielsen A. E. and Söhnel O. (1979) Interfacial tensions of electrolyte crystal – aqueous solutions from nucleation data. *J. Crystal Growth* **11**, 233-242.

O'Hara M. and Reid R. C. (1973) *Modeling Crystal Growth Rates from Solution*. Prentice-Hall, Englewood-Cliffs.

Park N. S., Kim M.-W., Langford S. C., and Dickinson J. T. (1996) Atomic layer wear of single crystal calcite in aqueous solution using scanning force microscopy. *J. Appl. Phys.* **80**, 2680-2686.

Parkhurst D. L. (1995) *User's guide to PHREEQC—a computer program for speciation, reaction-path, advective-transport, and inverse geochemical calculations*. U.S. Geological Survey Water-Resources Investigations Report 95-4227.

Paytan A., Moore W. S., and Kastner M. (1996) Sedimentation rate as determined by ^{226}Ra activity in marine barite. *Geochim. Cosmochim. Acta.* **60**, 4113-4319.

Pina C. M., Becker U., Risthaus P., Bosbach D., and Putnis A. (1998a) Molecular-scale mechanisms of crystal growth in barite. *Nature* **395**, 483-486.

Pina C. M., Bosbach D., Prieto M., and Putnis A. (1998b) Microtopography of the barite (001) face during growth: AFM observations and PBC theory. *J. Crystal Growth* **187**, 119-125.

Pina C. M., Enders M., and Putnis A. (1999) The composition of solid solutions crystallizing from aqueous solutions: the influence of supersaturation and growth mechanisms. *Chem. Geol.* **168**, 195-210.

Putnis A., Junta-Rosso J. L., and Hochella M. F. Jr. (1995) Dissolution of barite by a chelating agent: an atomic force microscopy study. *Geochim. Cosmochim Acta* **59**, 4623-4632.

Reardon E. J. and Armstrong D. K. (1987) Celestite ($SrSO_4$(s)) solubility in water, seawater and NaCl solution. *Geochim. Cosmochim. Acta* **51**, 942-957.

Rimstidt J. D. (1997) Gangue Mineral Transport and Deposition. In *Geochemistry of Hydrothermal Ore Deposits* (ed. H.L. Barnes). Wiley, New York..

Rufe E. and Hochella M. F. Jr. (1999) A quantitative assesment of reactive surface area in silicate dissolution. *Science* **285**, 874-876.

Steefel C. I. and Lasaga A. C. (1992) Putting transport in water-rock interaction models. *Geology* **20**, 680-684.

Stumm W. and Morgan J. J. (1995) *Aquatic Chemistry*. John Wiley, New York.

Sunagawa I. (1987) Surface microtopography of crystal faces. In *Morphology of Crystals* (ed. I. Sunagawa). Terra Scientific Publishing, Tokyo. pp 509-553.

Vaughan D. J., Becker U., and Wright K. (1997) Sulphide mineral surfaces: theory and experiment. *Int. J. Miner. Process.* **51**, 1-14.

Walton A. G. (1965) Nucleation of crystals from solution. *Science* **148**,601-607.

Zhang J.-W. and Nancollas G. H. (1990) Mechanisms of growth and dissolution of sparingly soluble salts. In *Mineral-Water Interface Geochemistry* (eds. M.F. Hochella, Jr. and A.F. White). Reviews in Mineralogy **23**, Mineralogical Society of America, Washington D.C. pp 365-396.

Water-Rock Interactions, Ore Deposits, and Environmental Geochemistry: A Tribute to David A. Crerar
© The Geochemical Society, Special Publication No. 7, 2002
Editors: Roland Hellmann and Scott A. Wood

Probing molecular-scale adsorption and dissolution-growth processes using nonlinear optical and scanning probe methods suitable for hydrothermal applications

STEVEN R. HIGGINS[1*], ANDREW G. STACK[1], KEVIN G. KNAUSS[2], CARRICK M. EGGLESTON[1], GUNTRAM JORDAN[3]

[1]Department of Geology and Geophysics, University of Wyoming, Laramie, Wyoming 82071-3006 USA
[2]L-202, Geosciences and Environmental Technology, Lawrence Livermore National Laboratory
Livermore, California 94550 USA
[3]Institut für Mineralogie, Ruhr-Universität Bochum, 44780 Bochum, Germany

* corresponding author: shiggins@uwyo.edu

Abstract—Experimental hydrothermal geochemistry has made great strides on the basis of stoichiometric equilibrium studies, measured kinetic rates, and bulk-sensitive spectroscopic data, but there are current problems in hydrothermal geochemistry ranging from deep aquifer chemistry to nuclear waste tank chemistry that can benefit from new techniques able to provide direct spectroscopic and microscopic information about processes operative at mineral surfaces. We present and discuss applications of two inherently interface-sensitive techniques that are amenable to application in hydrothermal experimentation: optical second harmonic generation (SHG) and hydrothermal atomic force microscopy (HAFM). Results of experimental SHG from a corundum-water interface are presented, along with a model relating SHG signal in part to the structure of water in the electric double layer. Although the SHG signal is closely related to surface potential, direct extraction of surface potential over a wide range of pH and ionic strength will depend upon better understanding of water structure near oxide-water interfaces. We present applications of HAFM to dissolution and growth of barite, and in particular show that the velocity and spacing of kink sites along [010] and [120] steps can be extracted from the HAFM imagery using a simple step model. The model enables us to conclude that although steps may appear straight in AFM images, the atomic scale roughness can be large, giving rise to kinks at a large percentage of step sites. Both techniques are promising ways to extend our ability to probe and better understand the nature of mineral-water interface chemistry.

1. INTRODUCTION

Chemical processes occurring at mineral-water interfaces are central to understanding how geologic materials regulate and modulate the chemical environment in which we all live. Many of these interactions occur at the Earth's surface where temperatures and pressures do not differ greatly from those with which we have direct experience, but many also occur in deep aquifers, sediments, hydrocarbon reservoirs, buried radioactive waste tanks, and geothermal systems that are considerably more experimentally challenging.

A primary tool of hydrothermal geochemistry appropriately continues to be the analysis of aqueous solutions from "quenched" field samples or from either static or flow-through laboratory experiments in order to gain information about chemical equilibrium or about reaction rates in particular systems.

Increasingly, geochemists must confront questions related to the kinetics and mechanisms of mineral-water interactions. Indirect wet-chemical methods remain a primary tool, but we need additional information to reliably interpret such data. For example, given a set of macroscopic concentration and stoichiometry data, there are often several models (e.g., diffuse layer or triple layer; single site or two site models) that apply more or less equally well (Westall and Hohl, 1980) but that differ in the surface structures implied. Interface-sensitive spectroscopic and microscopic techniques, judiciously applied, can allow us to test hypotheses pertaining to reaction mechanisms. Further, such direct information often shows that no existing model captures the complexity of the processes found to be in operation.

Many spectroscopic techniques have been applied to better understanding the nature of mineral-water interaction in aqueous systems (e.g., electronic and vibrational spectroscopies (Buckley and Woods, 1994; Hug and Sulzberger, 1994; Ainsworth et al., 1998; Eggleston et al., 1998), as well as X-ray diffraction and absorption techniques (Bargar et al., 1997; Cheng et al., 1997), white light interferometry (Lüttge et al., 1999), and nuclear magnetic resonance (NMR) (Casey et al., 2000)). However, there is a lack of techniques that are both 1) inherently sensitive to the solid-water interface rather than to the bulk of the material, and 2) amenable to direct in-situ application, particularly within hydrothermal experiments. In this paper, we describe two techniques that are relatively new to geochemistry and that satisfy both of these requirements. These are the laser-based technique of optical second harmonic generation (SHG) (for a review, see Corn and Higgins, 1994), and hydrothermal atomic force microscopy (HAFM) (Higgins et al., 1998b). We describe how each technique works, and give examples of the kinds of information that have been, and can be, obtained from them.

SHG can be used to study the adsorption of both organic and inorganic species to mineral surfaces; we choose to concentrate here on the pH dependence of SHG from an alumina-water interface. Prior work by Ong et al. (1992) demonstrated that SHG is a viable and direct probe of interface potential, particularly for the silica/water interface. In this work, we apply this same idea to a corundum/water interface to determine if the potential-sensitive SHG mechanism is appropriate for other interfaces. The results show that relatively direct measures of surface potential may be made without the need to "filter" stoichiometric and charge-density data through models that relate charge to potential. These results however, are not free of uncertainties in that the location of a true point of zero charge requires a rather simple SHG mechanism, whereas our results suggest the mechanism may be more complicated.

In the HAFM section, we show how direct in-situ imaging of step motion on barite ($BaSO_4$) surfaces allows us to extract information pertaining to the rates and mechanisms of such microscopic processes as the motion of kink sites along steps during dissolution or growth. In general, the results point the way to improving upon long-standing equilibrium- and diffusion-based models of crystal growth (e.g., Burton et al., 1951; Chernov, 1963b).

The two experimental methods above can be used to obtain complementary information that could ultimately lead to more consistent descriptions of specific mineral-water interfaces. Resonant SHG can provide adsorbate coverage information regardless of the dynamics of the system, and thus, presents itself as a useful tool for testing the equilibrium assumption associated with modeling a net reaction rate using equilibrium surface complexation models. Whereas SHG provides important information on the adsorption of aqueous constituents, HAFM, as we will discuss below, provides dynamic information down to the molecular level (within the confines of a particular step kinetic model). As both methods are suitable for investigations under identical experimental conditions, it should be possible to make direct correlation between adsorption data from SHG and dynamic data from HAFM.

2. SECOND HARMONIC GENERATION (SHG)

2.1. SHG Theory

The electric field portion of light at a fixed location varies sinusoidally with time, creating an AC polarization of the electron cloud of a molecule exposed to it. If the magnitude of the electric field is small, the electron cloud behaves harmonically. That is, as the magnitude of the electric field changes, there is a proportional change in the position of the electrons. If the incident light's electric field is large, such as that created by a high power laser, the resulting polarization can be anharmonic and include a significant component with twice the incident light's frequency. This component is known as the second-order polarization, $P^{(2)}(2\omega)$. As an oscillating electric charge, $P^{(2)}(2\omega)$ leads to generation of light with twice the incident light's frequency. $P^{(2)}(2\omega)$ can be described as:

$$E(2\omega) \propto P^{(2)}(2\omega) = \chi^{(2)}\ E(\omega)E(\omega) \tag{1}$$

where $E(2\omega)$ is the second harmonic field strength at frequency 2ω, $E(\omega)$ is the incident light field (at frequency ω), and $\chi^{(2)}$ is the non-linear susceptibility:

$$\chi^{(2)} = n_s\left\langle \alpha^{(2)} \right\rangle \tag{2}$$

where n_s is the surface concentration of molecules and $\alpha^{(2)}$ is the second-order polarizability (or hyperpolarizability) of the molecules; $\alpha^{(2)}$ is a third-rank tensor and is averaged over the molecular orientation distribution denoted by the brackets.

Within the electric dipole approximation, SHG is forbidden in a centrosymmetric phase because induced polarizations will be equal in opposite directions and destructively interfere. Thus, all elements of Eqn. 2 above would be zero if the molecules were randomly oriented in the bulk phase in which the driving fields (fundamental light at frequency ω) exist. At an interface between two centrosymmetric phases, however, the centrosymmetry is broken and SHG is allowed. This imparts to SHG its inherent interface sensitivity in otherwise centrosymmetric systems.

The polarizabilities, and thus the $\chi^{(2)}$ of a given collection of molecules, is a function of the energy of the light interacting with them. Near energies at which a given molecule has an optical resonance due to transitions between electronic states, the electron cloud (or part of it) may be thought of as more deformable than at energies far from resonance. The electronic polarizabilities that give rise to SHG thus also tend to be relatively large near optical absorption energies. It is then possible to enhance the SHG signal by tuning the input laser light such that the SH light generated occurs at a wavelength where the subject molecule has an absorption band. This sets up resonance between the electronic transition at a particular energy and the polarization of electrons at the *second harmonic* (rather than primary) frequency (Corn and Higgins, 1994).

Often, however, resonant SHG is not possible. In the case of the oxide-water interfaces discussed below, there is negligible optical absorption at frequencies used in the experiments. This type of experimental arrangement is usually referred to as non-resonant SHG and commonly a total internal reflection (TIR) experimental geometry is employed. In this case, the evanescent electric field associated with light undergoing TIR from an interface can give rise to enhanced SHG (Ong et al., 1992; Conboy et al., 1994). The TIR geometry restricts the range of solid-water interfaces that can be probed because the input laser light must approach the interface from the side that is optically more dense. This is most often from the solid side, requiring an optically transparent single crystal (or amorphous material) of up to centimeter dimensions. The source of SHG from these non-resonant systems is often ascribed to the static electric field at the interface (Ong et al., 1992; Lantz et al., 1993; Lantz and Corn, 1994). However, it is important to clarify that it is not the static field that gives rise to the SHG, but that there is some material property that changes with the static field. Thus, it may be more correct to state that the non-linear susceptibility of the material, immersed in the field, changes with the field strength. For the case of aqueous interfaces, it is believed that the orientation of the water molecules in the static field gives rise to a significant contribution to the non-linear susceptibility of the interface (Ong et al., 1992; Zhao et al., 1993a).

Resonant SHG can be used to probe the relative amounts of, as well as the average orientation of, organic molecules adsorbed at interfaces given appropriate molecular characteristics (i.e., large non-linear molecular polarizability). In recent work, we have used SHG to study the adsorption of rhodamine dye molecules to single-crystal corundum surfaces (Higgins et al., 1998d; Higgins and Eggleston, submitted). Other studies of dye molecule adsorption to various fused silica interfaces have appeared in the literature (e.g., Heinz et al., 1982; 1983; Higgins et al., 1991; Kikteva et al., 1999). These studies demonstrate how Eqns. 1 and 2 may be utilized for determining the average molecular orientation of the adsorbates. A recent paper by Simpson and Rowlen (1999), however, point out some potential flaws in the assumptions associated with molecular orientation calculations. Aside from dye molecules, whose large non-linear polarizabilities make them attractive for SHG studies, many adsorption studies of para-substituted benzene have been successful due to large changes in the molecular dipole moment associated with electronic transitions (e.g., Bhattacharyya et al., 1987; Castro et al., 1991; Higgins et al., 1992; Zhao et al., 1993b).

Recent SHG developments have shown that SHG may also probe the surfaces of particle suspensions, opening up the potential applications of SHG to a large variety of mineral-water interfaces (Wang et al., 1996; Yan and Eisenthal, 2000). Here, however, we concentrate instead on the pH-dependence of SHG from amorphous silica-water and corundum-water interfaces. As mentioned above, this type of study is best performed in the TIR geometry owing to the non-resonant relationship between the interface and the fundamental and second harmonic fields. The TIR geometry may only be applied through bulk solids of optical quality and minimal absorption at the frequencies employed.

2.2. *pH Dependence of SHG from the Oxide-Water Interface*

Figure 1 shows the experimental TIR configuration we used to obtain SHG from oxide-water interfaces. Laser light impinges on the oxide-water interface from the solid side, through a prism comprising the mineral of interest (e.g., silica or corundum). Further experimental detail is presented elsewhere (Stack et al., 2001).

Figure 2 shows the SHG amplitude from a fused silica-water interface as a function of pH. The SHG intensity rises as pH rises, but unevenly; relatively abrupt rises in signal occur at pH < 2, pH ≅ 4.5, and pH ≅ 8. Our measurements were intended to, and indeed do, reproduce the pH-dependent results of Ong et al. (1992) for an amorphous silica/water interface. These authors proposed a model in which the second-order polarization of the interface, $P^{(2)}(2\omega)$, is the sum of a pH-independent baseline and a pH- and ionic strength-dependent component:

$$E(2\omega) \propto P^{(2)}(2\omega) = \chi^{(2)} E(\omega)E(\omega) + \chi^{(3)} E(\omega)E(\omega)\Psi_0 \qquad (3)$$

where $E(2\omega)$ is the second harmonic field, $E(\omega)$ is the field of the fundamental light and contains the Fresnel coefficients relevant to the particular experimental geometry employed (Mizrahi and Sipe, 1988), Ψ_0 is the potential at the interface relative to the bulk solution, and $\chi^{(2)}$ and $\chi^{(3)}$ are second- and third-order non-linear susceptibilities, respectively. The $\chi^{(2)}$ term is considered by Ong et al. (1992) and Zhao et al. (1993a) to be independent of pH and ionic strength. The $\chi^{(3)}$ term accounts for a static electric field con-

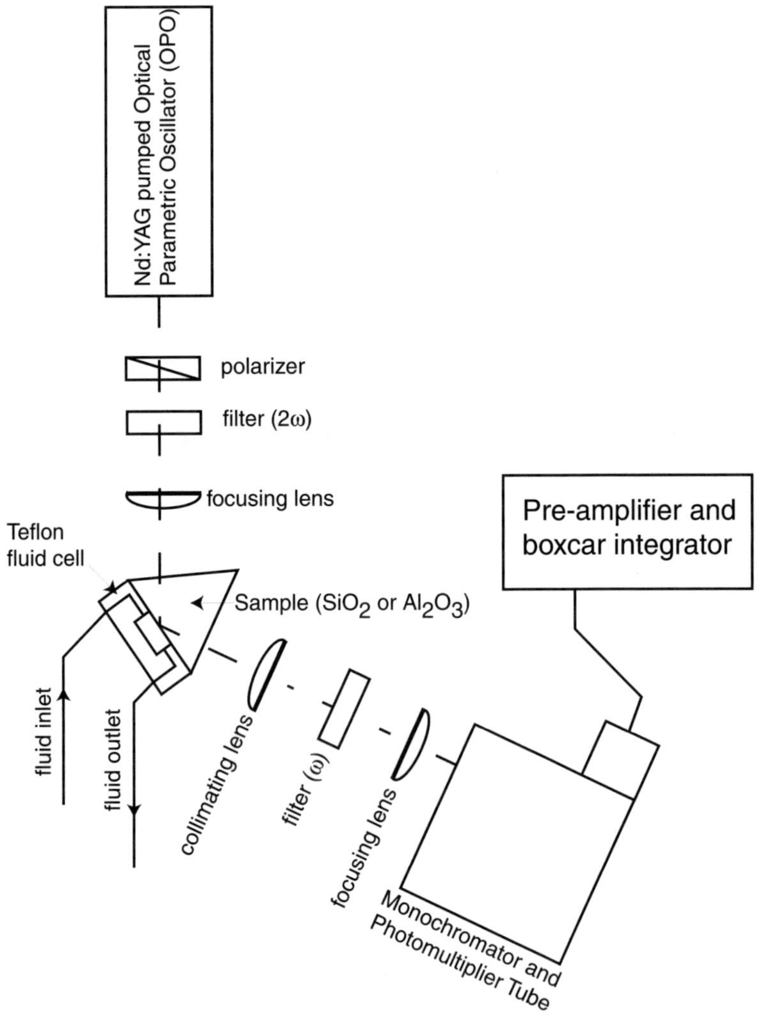

Fig. 1. Experimental configuration for total internal reflection (TIR) SHG investigation of oxide water interfaces.

tributing to the SHG due to Ψ_0. $\chi^{(3)}$ is attributed to water molecules in the diffuse layer that are oriented by the static interfacial electric field.

Equation 3, with the assumptions that $\chi^{(2)}$ and $\chi^{(3)}$ are constant, provides a way of directly probing oxide surface potentials without the need to interpret stoichiometric (surface charge density) data via various possible physical models of the electric double layer. Extracting a surface potential, however, requires a method for determining $\chi^{(3)}$ through variation of Ψ_0. Thus, in order to determine $\chi^{(3)}$ initially, a specific potential-charge density relationship must be applied as a calibration. It has been shown that a determination of $\chi^{(3)}$ by this approach yields model-independent potential measurements for the silica-water interface (Ong et al., 1992). Equation 3 can be solved for Ψ_0 if $\chi^{(2)}$ and $\chi^{(3)}$ are known and are pH-independent, but the assumptions may be questionable. The $\chi^{(3)}$ term, for example, depends upon the overall structure of water in the interfacial region. If the average water orientation does not vary linearly with surface potential, then extraction of Ψ_0 from pH-dependent SHG data will depend upon models of interfacial water structure—an area of current and sometimes controversial research (Toney et al., 1994; Danielwicz-Ferchmin and Ferchmin, 1996).

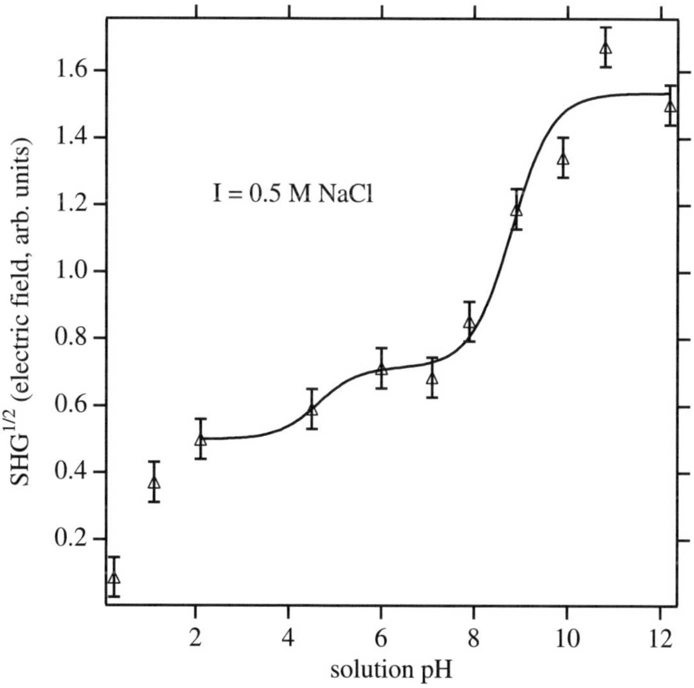

Fig. 2. SH field strength, $E(2\omega)$, vs. pH for the amorphous silica/water interface. The fit was calculated using an acid-base species distribution function.

Whether linear or not, however, it is clear that the surface electric field created by the development of surface charge at high pH also leads to increased SHG at high pH (Fig. 2). Given that $\Psi_0 = 0$ at the pH of point of zero charge (PZC) of silica, then all of the SHG signal at the PZC can be attributed to the $\chi^{(2)}$ term of Eqn. 3. The PZC of amorphous silica is in the pH range 2-3.5 (Parks, 1965; 1990; Sahai and Sverjensky, 1997). Assuming that $\chi^{(2)}$ is indeed pH independent (Ong et al., 1992), then increases in SHG with increased pH are directly proportional to the surface potential Ψ_0 developed as the silica develops negative charge by deprotonation of silanol groups. The relatively abrupt rises in SHG field strength occur at pH values corresponding to the pK_a of surface silanol groups. Ong et al. (1992) determined pK_a values of 4.5 and 8.5, and from our data we obtain comparable values of 4.6 and 8.8 on the basis of fitting of the SHG data to a typical monoprotic acid-base equilibrium. However, it must be emphasized that these surface pK_a values are obtained on the basis of assumptions that remain to be validated.

The lowest pH value explored by Ong et al. (1992) was 2. This is roughly coincident with the amorphous silica PZC. Our data (Fig. 2) show that the SHG signal continues to decrease at pH values below 2, suggesting a protonation reaction with a pK_1 less than 2. This is consistent with reported values for amorphous silica of $pK_1 < 0$ (e.g., Sverjensky and Sahai, 1996). Despite the fact that $|\Psi_0|$ increases as pH increases or decreases away from the PZC, there is no minimum in the SHG signal at the PZC. This can be understood by consideration of the phase relations between different sources of SHG. As the sign of the electric field at the interface changes about the PZC, the phase of SH light due to this static electric field undergoes a 180 degree phase shift. If the static field generated the only source of SH light, then the PZC should correspond to the minimum in SH field strength. However, in Fig. 2, the SH field is not zero at the PZC for silica, so generally, we must regard the phase relationship between multiple sources of SH light.

We turn to our results for SHG from a corundum-water interface to examine these phase relations. Using the same experimental configuration shown in Fig. 1, but replacing the fused silica prism with a corundum prism, we measured the SHG field strength as a function of bulk pH and ionic strength. We assume that the corundum-water interface is hydroxylated (Laiti et al., 1998).

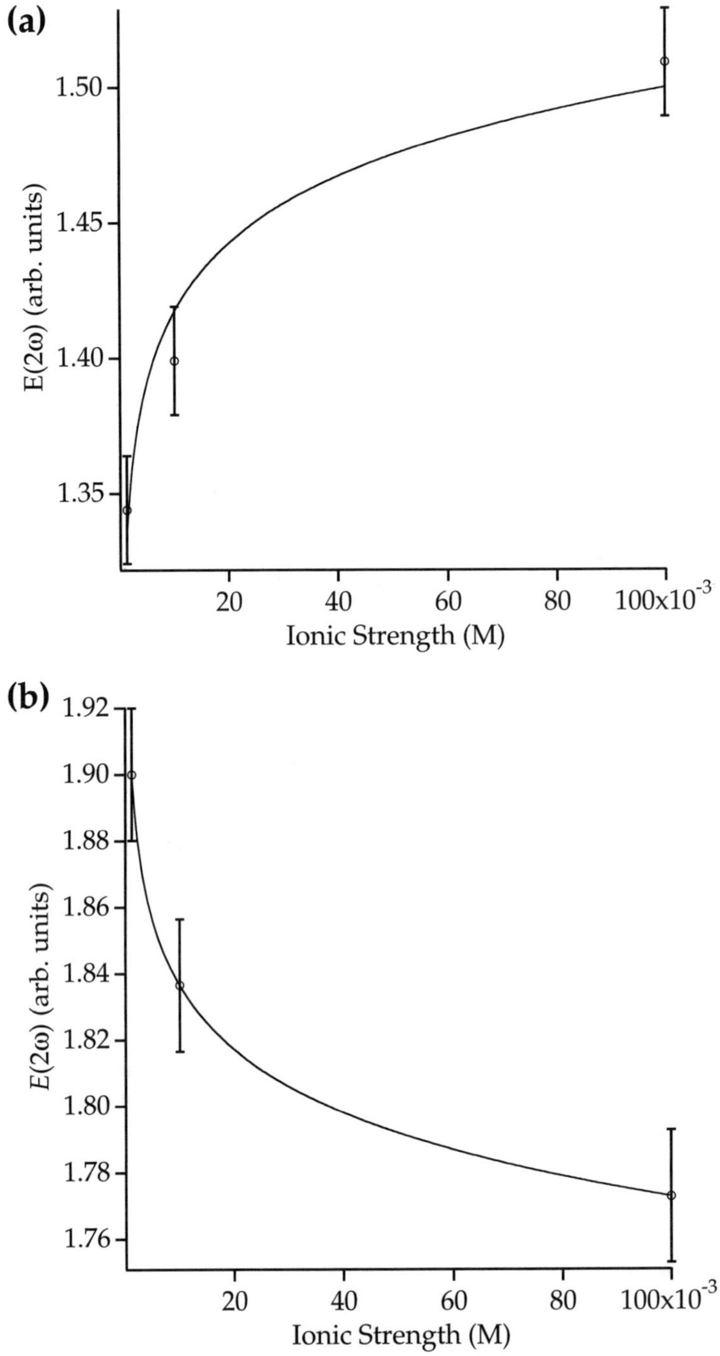

Fig. 3. SH field strength, $E(2\omega)$, as a function of ionic strength for the corundum-water interface. (a) NaCl supporting electrolyte, pH = 3.5. (b) $NaNO_3$ supporting electrolyte, pH = 10.5. The solid lines are two parameter fits to Eqn. 4 using the diffuse layer model for the surface potential (Zhao et al., 1993b).

Figure 3 shows the ionic strength dependence of $E(2\omega)$ from the corundum-water interface at pH 10.3 for $NaNO_3$ electrolyte and pH 3.5 for NaCl electrolyte. We fit the data in Fig. 3 to Eqn. 4, which is functionally equivalent to Eqn. 3 (Zhao et al., 1993b), using the Gouy-Chapman diffuse-layer model for Ψ_0.

$$E(2\omega) = A + B\Psi_0 \tag{4}$$

By choosing extreme pH values, relative to the observed point of zero salt effect (PZSE), for fitting $E(2\omega)$ vs. ionic strength, we assume that the corundum surface charge is near or at the maximum surface charge density, $\sigma \cong \sigma_{max}$. Figure 4 shows some of our results for both the pH and ionic strength dependence of SHG from the corundum-water interface. Using the parameters A, B, and σ_{max} from fitting the Fig. 3 data, Ψ_0 was found as a function of pH as shown on the right-hand axis of Fig. 4. For the NaCl electrolyte, A = 1.9, B = -0.072 ± 0.014, and σ_{max} = 1.4 ± 1.8 C m^{-2}, whereas for the $NaNO_3$ electrolyte, A = 1.5, B = 0.055 ± 0.006, and σ_{max} = 0.7 ± 0.8 C m^{-2}. On both plots in Fig. 4, we have superimposed a line with slope equal to 59.2 mV per pH unit. This slope is expected for the data if the surface protonation reaction can be regarded as potential-determining. The data in both plots have similar slope to the superimposed lines, indicating consistency between the observations and the expected Nernstian behavior of the oxide/water interface.

Aside from the Nernstian behavior, the general trends in the data offer additional points for discussion. As for silica, $E(2\omega)$ increases with pH. In $NaNO_3$ solution, at high pH, $E(2\omega)$ decreases with increasing ionic strength, while at low pH, $E(2\omega)$ increases with increasing ionic strength, though not as markedly. In NaCl solution, at low pH, $E(2\omega)$ increases with increasing ionic strength, but shows no apparent ionic strength dependence for pH > 6. A crossing point of the $NaNO_3$ data allowed us to estimate a PZSE in the region 4 < pH < 6. However, there is no clear crossing point for the NaCl data.

Zhao et al. (1993a) measured the ionic strength dependence of SHG from two similar air-water interfaces to which positively or negatively charged organic molecules had been adsorbed, and found that the direction of the ionic strength dependence reverses with the sign of the charge of the interface. This behavior is attributed to constructive or destructive interference between the $\chi^{(2)}$ and $\chi^{(3)}$ terms of Eqn. 3, an idea supported by Du et al. (1994) who showed that the phase of the sum-frequency generation (SFG) signal from a silica-water interface shifts by 180° between high and low pH. This is as expected from the "flip" of near-surface water molecules from oxygen-down at low pH (positive surface charge) to oxygen-up at high pH (negative surface charge). Such light can interfere constructively or destructively with the fixed-phase light of the $\chi^{(2)}$ term in Eqn. 3.

Thus, the data suggest that at high pH (pH > PZSE), constructive interference between the $\chi^{(2)}$ and $\chi^{(3)}$ terms of Eqn. 3 leads to a relatively large SHG signal. As the ionic strength increases and the depth of the electric double layer decreases, fewer water molecules contribute to SHG via the $\chi^{(3)}$ term and the signal decreases. At pH < PZSE, destructive interference between the $\chi^{(2)}$ and $\chi^{(3)}$ terms leads to relatively small SHG signal. As the ionic strength increases and the depth of the electric double layer decreases, fewer water molecules contribute to SHG via the $\chi^{(3)}$ term and the signal increases because there is *less* destructive interference.

2.3. SHG Model Discussion

One of the main problems with the data in Fig. 4 is the discrepancy between PZSE values for the two electrolytes. Whereas in the NaCl electrolyte a PZSE cannot be discerned above pH ≅ 6, in the $NaNO_3$ electrolyte the PZSE appears at pH ≅ 5. In the absence of specific electrolyte complications to the surface chemistry, or to the surface non-linear susceptibility, the PZSE should correspond to the PZC and is generally accepted to be at pH ≅ 9.3 for α-alumina (Davis and Kent, 1990; Sverjensky, 1994). The NaCl data therefore, are not inconsistent with this literature value. However, Stack et al. (2001) found that AFM force-distance measurements are inconsistent with a PZC of this corundum prism surface located at pH ≅ 9.3. If we accept that the literature value of the PZC for corundum is applicable to this particular interface, the $NaNO_3$ data require further discussion to attempt to examine the model deficiencies. The assumption that the electrolyte concentration only affects the ionic strength may not be valid for $NaNO_3$ if the non-linear polarizability of NO_3^- is non-negligible. If this were true, we would expect the SHG model to deviate at low pH where the positive surface charge will cause a higher concentration of NO_3^- ions at the interface. The result of this additional SHG mechanism will be to shift the observed PZSE from the true PZC. The direction of the shift will depend upon the phase of the SH light generated by the ions. Regardless of the phase, if the NO_3^- ions are contributing to the surface non-linear susceptibility, then the assumption that the $\chi^{(2)}$ term is independent of pH and ionic strength is invalid because the NO_3^- polarizability will contribute to this term. The observed PZSE at pH ≅ 5 suggests that if the NO_3^- contribution dominates

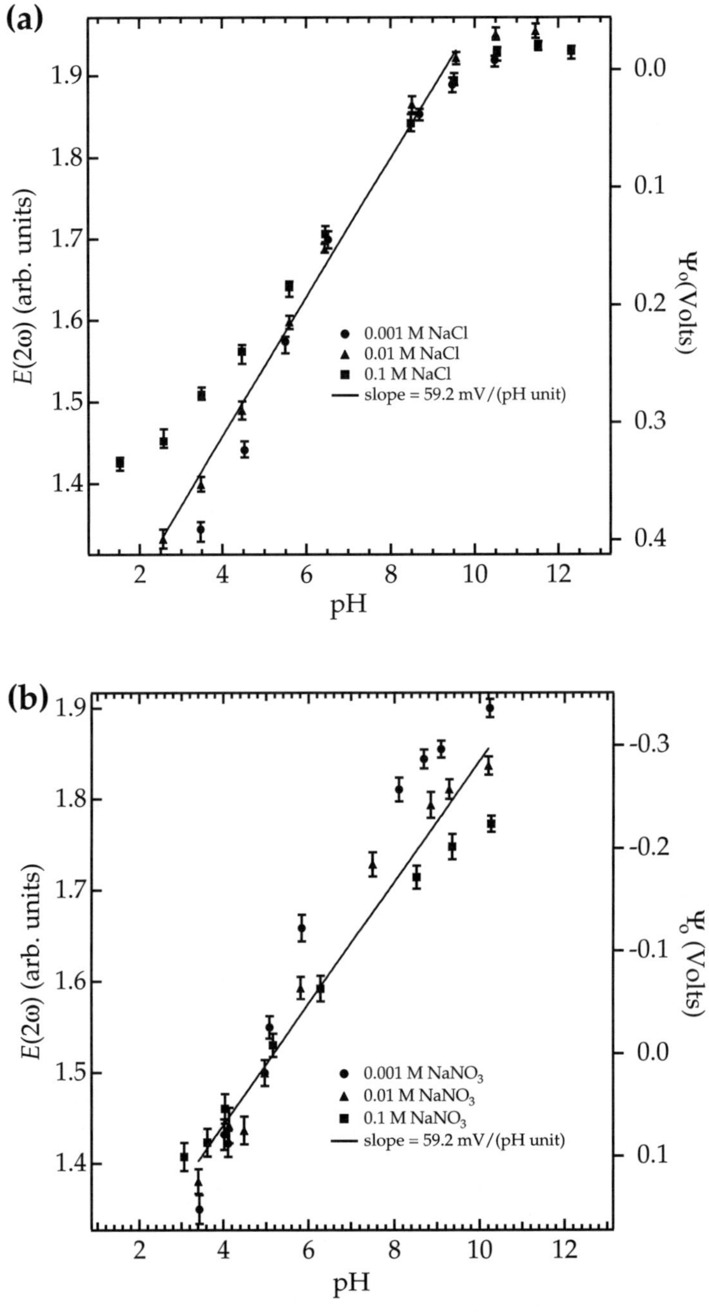

Fig. 4. SH field strength, $E(2\omega)$, as a function of pH for 3 different ionic strengths of (a) NaCl and (b) $NaNO_3$. The right hand axes in both figures correspond to the surface potential determined by the fitting parameters from Fig. 3.

$\chi^{(2)}$ at low pH, it would be 180° out of phase with respect to the $\chi^{(3)}$ term.

The $\chi^{(3)}$ term will be a function of the number of water molecules oriented, their average orientation, and a static electronic polarization stemming from the surface potential. Following Gragson et al. (1997), we propose that $\chi^{(3)}$ of Eqn. 3 cannot be assumed a constant at high surface potentials. This amounts to asserting that near-surface water molecules, once aligned (dipole oriented) by a surface electric field, cannot become *more* aligned with higher potentials. To address these possible model shortcomings, information regarding the interfacial water structure must be provided by more direct methods such as SFG (Du et al., 1994), or far-IR absorption studies (Russell et al., 1993a,b). Until a more quantitative description of water structure at interfaces is developed, we must accept the limitations of the constant $\chi^{(3)}$ assumption.

In summary, according to available models, the PZSE of the SHG data should occur at the PZC if the model for the SHG mechanism by Ong et al. (1992) is correct. The data using NaCl as a background electrolyte was consistent with the SHG mechanism, giving an approximate change in surface potential with pH of 59 mV/pH unit. While this $\Delta\Psi/\Delta pH$ value was similar for the $NaNO_3$ background electrolyte, the PZSE was found to be much lower than expected from literature data on the PZC for corundum. Thus, the $NaNO_3$ results suggest that the SHG mechanism may be more complicated than that described by the model. Clearly a better understanding of SHG mechanisms are vital to further applications of this optical method to mineral/water interfaces. With this knowledge, SHG could be used to obtain complementary information from surfaces undergoing net dissolution or precipitation reactions. Thus, it may be possible to relate surface coverage of ligands to an observed ligand-promoted dissolution rate, all under the same conditions, whereas in current experimental methods, the surface coverage information must be obtained under equilibrated conditions.

3. HYDROTHERMAL ATOMIC FORCE MICROSCOPY (HAFM)

3.1. Introduction to HAFM

Many studies of mineral dissolution and growth have been made using the capabilities provided by atomic force microscopy (AFM), also known as scanning force microscopy (SFM). Calcite has been a particularly popular target of AFM-based dissolution and growth studies primarily because of its geologic importance but also, in part, because of its relative kinetic lability. In other words, calcite dissolves and grows quickly enough to be amenable to ambient temperature AFM imaging of step motion. Such studies have provided information of unprecedented detail about calcite surface and step structure, step kinetic rate constants, activation energies for step and kink dynamics, and even thermodynamic quantities such as step energies (e.g., Gratz et al., 1993; Liang et al., 1996; Britt and Hlady, 1997; Jordan and Rammensee, 1998; Teng et al., 1998; 1999; Shiraki et al., 2000).

In practice, there is a fairly narrow range of dissolution or growth rates that can be studied using AFM. On the one hand, surface features must move far enough to be clearly distinguished from microscope drift over time or from tip-induced effects on the surface (Park et al., 1996). On the other hand, if growth or dissolution rates are sufficiently large, it is difficult to interpret surface kinetic observations without considering the probable limits of mass transport. It has been noted that for practical AFM imaging, the range of amenable dissolution or growth rates lies between 10^{-6} and 10^{-10} mol m^{-2} s^{-1} (Dove and Platt, 1996). Depending upon particular microscopes and upon the particular goals of an experiment, however, this range may be generous as the upper limit may result in mass transport limitations in unpredictable fluid flow conditions and tip influences will likely disturb measurements at the lower limit.

One consequence of this limited range of AFM accessibility is that, despite the near-molecular scale resolution capabilities of AFM, slowly-dissolving or growing minerals, including most oxides and aluminosilicates, cannot be investigated in a practical manner using commercially available AFMs that are limited to near room temperature. The key problem in approaching hydrothermal conditions with AFM is the need to pressurize a fluid cell to elevate temperatures up to and beyond the ambient-pressure boiling point of the solution. Here, we describe the design of an HAFM (Higgins et al., 1998b) capable of imaging at up to about 12 bars and 150 °C, and present results demonstrating its utility for mineral dissolution and growth experiments at elevated temperatures for barite.

Our specific hypothesis for barite is based on our previous observations of differences in the barite (001) monolayer and bilayer step velocities (Higgins et al., 1998c). In these observations, we found that when bilayer steps are formed, there remains on either end of the bilayer step, a monolayer step segment. If the formation of the bilayer step results in a new step velocity (i.e., $v_{st}(monolayer) \neq v_{st}(bilayer)$), then the monolayer segments on either end of the bilayer step should adopt a new step direction. In particular, we present results pertinent to a Kink Donation model that allows us to extract net rates of kink attachment/detachment from HAFM data for dissolving and growing barite ($BaSO_4$). With the typical room temperature growth and dissolution rates of barite falling toward the lower end of observable dissolution by AFM, we have chosen to accelerate these rates, to avoid large tip influences, through the use of elevated temperatures in the HAFM.

3.2. HAFM Design

A principal component of virtually any SFM is the piezoelectric ceramic scanner required for molecular scale, three-dimensional positioning and rastering of the sample beneath the AFM tip. Some AFMs use tracking lenses in association with a scanned tip design and keep the sample stationary, but the HAFM uses the scanned-sample design. We use a 7-electrode lateral-vertical decoupled PZT-5A tube scanner. This type of scanner significantly reduces coupling between horizontal and vertical scanner motions, and has nearly constant sensitivity between 60 and 150 °C. Piezoelectric tube scanners cannot function effectively under pressure differentials of the > 6 bars magnitude needed for imaging in aqueous solution at 150 °C, and they and their associated wiring must be physically separated from the corrosive, hot aqueous solutions of an HAFM fluid cell. These constraints require, first, that both the scanner and the fluid cell be pressurized (i.e., that the piezoelectric scanner and coarse-approach stepper motor be placed within a gas-pressurizable chamber). Second, the material used to separate the fluid cell (containing the sample) from the chamber housing the piezoelectric scanner must be flexible, have low water and gas permeability, and be inert toward hot, corrosive aqueous solutions. We have chosen a material known by its trade name as Kalrez^{TM} (a perfluoro elastomer) to form a flexible, chemically-inert membrane separating the fluid cell from the base chamber. The remainder of this section is a brief description of the HAFM fluid cell, the fluid delivery system, and discussion of recently implemented and planned improvements to the original HAFM design (Higgins et al., 1998b).

The fluid cell, detailed in Fig. 5, consists of a main cell body, a cell cover, and a sample mounting disk, all fabricated from grade 2 titanium. Fluid inlet and outlet ports are provided for flow-through experiments and a third port is available for a temperature sensor. Optical access for laser light (optical detection schemes for cantilever deflection measurements are used in most commercial systems) is provided by sealing a sapphire window to the cell cover. The cell cover also serves as the mount for the cantilever/substrate assembly. The bottom of the cell consists of the sample mounting disk and the Kalrez^{TM} membrane which additionally effects a seal of the fluid cell. We have found this fluid cell design useful for experiments ranging from a few hours to several days duration. Other than periodic calibration checks and routine cleaning, there is no additional maintenance over and above what one would expect from a commercial system.

The fluid delivery system for the HAFM permits controlled, ratioed mass flow from two reservoirs whose contents may be changed without disrupting an experiment in progress. A key to making the HAFM function consistently and nearly free of thermal drift is the control of fluid flow with a controlled leak downstream of the HAFM fluid cell using a liquid mass flow controller. Other means of controlling fluid flow with pumps are not only more expensive, but require two pressure sources that must be carefully controlled in order to keep the pressure differential across the Kalrez^{TM} membrane < 0.05 bars at all times. The HAFM design may be compared to other recent developments that allow sample and fluid cell heating (Butt et al., 1999; Workman and Manne, 2000).

3.3. HAFM Applications

We have made use of the HAFM in dissolution and/or growth studies of such mineral phases as barite (Higgins et al., 1998c; 2000a), magnesite (Jordan et al., in press), anorthite/labradorite (Jordan et al., 1999), and hematite (Higgins et al., 1998a). Some very enlightening barite dissolution and growth studies (e.g., Bosbach et al., 1998; Pina et al., 1998; Bosbach, 2002; Wang et al., 2000) using ambient temperature AFM have been reported. For example, Bosbach et al. (1998) reported that chelating agents such as ethylenediaminetetraacetate (EDTA) increased the rate of pit formation in the dissolution of barite while growth inhibitors such as nitrilotrimethylenephosphonic acid (NTMP) retarded the formation rate of monolayer islands and the growth velocity of steps. Pina et al. (1998) noted a dramatic difference in the growth velocity of straight monolayer steps along <120> on barite (001) in contrast to the approximately ten-fold higher velocity of curved steps opposite these slower, straight steps. These results indicate a marked difference in the molecular kinetics at crystallographically different steps.

In the present work, we focus on some of the subtleties associated with barite dissolution and growth that, at present, are most amenable to study using the HAFM. In particular, we are interested in gaining kinetic information on the rates of attachment and detachment at step edges, thus providing a means to discuss crystal growth and dissolution mechanisms at the molecular scale.

The experimental methods for study of barite dissolution and growth have been reported elsewhere (Higgins et al., 1998c; 2000a). Figure 6 shows a sequence of four HAFM images of dissolving barite (001) taken at 90°C with a constant flow rate of 20 g/hr of de-ionized H_2O. Using the crystallographic axes shown in Fig. 6a, it is clear that there are two principal step di-

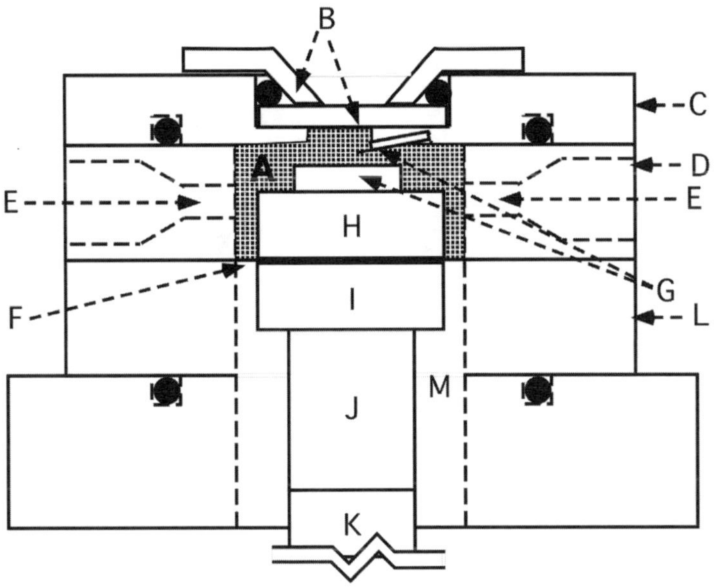

Fig. 5. Schematic of the HAFM fluid cell. Key: A. fluid environment (stippled pattern) B. sapphire window and cover C. Ti cell cover D. Ti cell body E. fluid in/out ports F. Kalrez™ membrane G. sample and cantilever/tip H. Ti sample mount I. sample mount back disk J. Macor™ spacer K. Piezoelectric ceramic scanner tube L. Kel-F™ spacer M. inert gas environment.

rections observed during dissolution under these conditions: the <010> and <120> step directions. Due to the 2_1 screw axis in barite, monolayer (3.5 Å deep) triangular-shaped pits, bounded by <120> steps on two sides and a [010] step on the third side, are oriented in opposite directions in adjacent monolayers along [001]. The <120> monolayer steps retreat at a velocity approximately 2 times greater than that of the [010] monolayer steps and as a result, the pit vertices eventually reach the monolayer [010] step of the next highest layer exposed at the surface. For example, in Fig. 6c, the pit vertex defined by intersecting <120> steps reaches the neighboring layer's [010] step, resulting in the formation of a [010] bilayer (7.1 Å height) step marked as [010]-2 in the figure. A similar process occurs in the formation of <120> bilayer steps (Higgins et al., 1998c). Both types of bilayer steps retreat at a higher velocity than the corresponding monolayer steps, with the <120> bilayer steps retreating 2.5 times faster than the [010] bilayer steps. We turn our discussion toward the fate of the remaining monolayer step segments (denoted [010]-1, in Fig. 6c for example) on either side of the bilayer steps.

Step velocity measurements on images similar to those in Fig. 6 revealed that the [010]-1 and [120]-1 step segments assume the velocity of the respective bilayer segments (Higgins et al., 1998c). From image sequences similar to that in Fig. 6, it can be shown that the monolayer step segments assume the velocity of the bilayer segment by a slight change in the monolayer step orientation.

We describe the above increase in the monolayer step velocity, or change in the monolayer step orientation by means of a Kink Donation (KD) model. In the KD model, kinks reaching the end of the bilayer segment cross the point at which the bilayer step bifurcates (labeled "A" in Fig. 7b) as donated kinks to the monolayer step segment. Thus, the rate at which kinks reach point *A* must equal the frequency of row removal (i.e., retreat rate) from the faster of the two kink sources (i.e., steps). The retreat rate of the monolayer step then is no longer governed by local processes on that segment, but by the processes governing the rate of bilayer step retreat. The increase in the monolayer step velocity then results from the misorientation of the step as shown by the dashed lines in Fig. 7. The KD model is simply a restatement of the Wulff construction in terms of anisotropy in step kinetics (Chernov, 1963a), as opposed to anisotropy in step free energy (Zangwill, 1988). The KD model is particularly useful in making quantitative estimates of the kink dynamics on barite (001). Below is a quantitative description of the KD model.

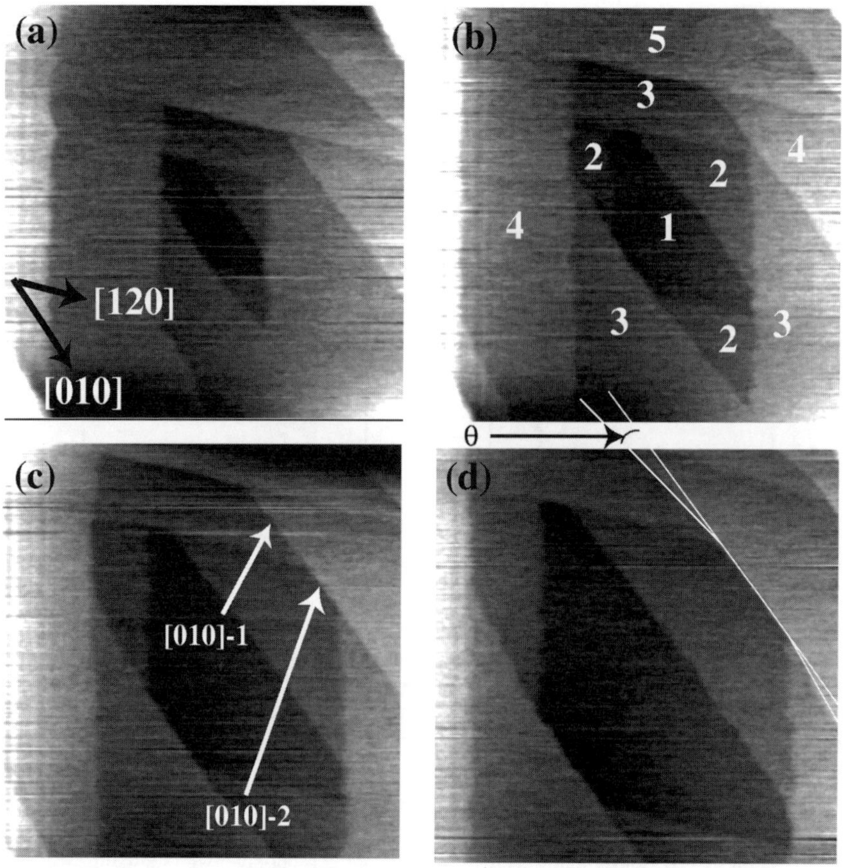

Fig. 6. Sequence of four 1.4 μm x 1.4 μm HAFM images of barite (001) taken at 90 °C with a constant flow of de-ionized water at 20 g/hr. The time between frames was 100 sec. The predominant step orientations were [010] and [120]. The crystallographic orientation is shown in (a). In (b), the terraces are labeled in order of increasing height on the surface. Also in (b), on the right hand side of the image, a pit vertex is about to reach the adjacent layer's [010] step, resulting in the formation of a [010] bilayer step separating terraces 2 and 4 in (c). The acceleration of the newly formed bilayer step segment causes a misorientation of the monolayer step segments to either side of the bilayer segment in (d).

The kinematics of step retreat, pertinent to the KD model, are described by a simple one-dimensional nucleation and growth model. In this model, described by Lauritzen (1973) and Frank (1974), a nucleation event along a step gives rise to two dissolution/growth sites (e.g., kinks) on either side of the nucleus. The primary supposition in this mathematical treatment is that the nucleus has infinitesimally small width compared to the distance to the next kink. The subsequent growth at the kinks then takes place until either (1) the kink reaches the end of the step or (2) the kink annihilates with another kink traveling in the opposite direction. The velocity of steps, v_s, can be expressed exactly if the length density (e.g., μm^{-1}) of kinks, ρ_k, and their velocity, R_k, are known:

$$v_S = b\rho_k R_k \tag{5}$$

where b is the length, normal to the step, of one dissolution/growth unit. However, ρ_k depends upon R_k, prompting an approximate solution involving independent variables. The solution to this problem is approximated by two limiting growth regions (Frank, 1974):

$$v_s \approx \begin{cases} R_{kk}bl, & \text{region I, where } l\left(\dfrac{R_{kk}}{2R_k}\right)^{1/2} < 0.1 \\ b(2R_{kk}R_k)^{1/2}, & \text{region II, where } l\left(\dfrac{R_{kk}}{2R_k}\right)^{1/2} > 10 \end{cases} \tag{6}$$

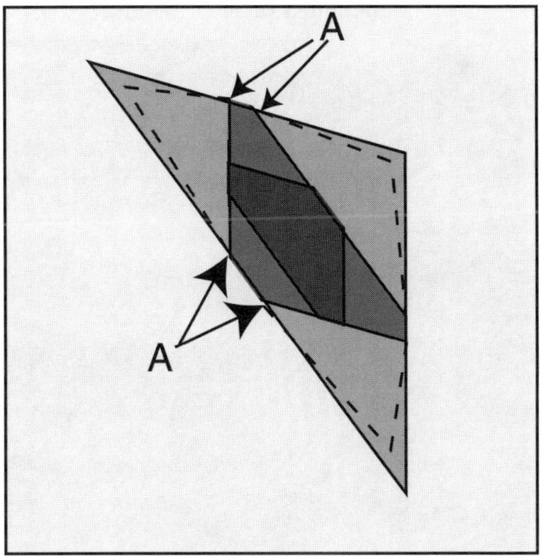

Fig. 7. Observed etch pit shape under the conditions of this study. Monolayer pit vertices are truncated due to bilayer step formation. The actual position of the monolayer step segments is depicted by the dashed lines.

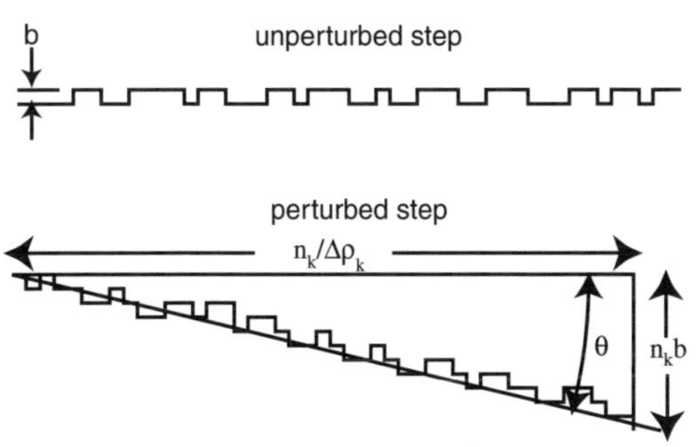

Fig. 8. Illustration of the concept of kink density perturbation.

where l is the length of the step, and R_{kk} is the nucleation rate per unit step length. The equation describing the growth region lying in between these limiting cases is not easily derived. In the step length-dependent growth region, the rate is determined by the nucleation rate. When a step becomes sufficiently long, or if $R_{kk} \approx R_k$, step velocity is nearly independent of step length, and the velocity is approximated by the geometric mean of the kink nucleation and growth rates. In this case, the growth rate is neither limited by nor determined by the nucleation rate. However, the balanced kinetics involving nucleation, growth, and annihilation, is maintained in part by nucleation events.

To understand the perturbation in the monolayer step orientation (Fig. 8), we cast our observations in terms of the KD model. We start by considering two separate growing steps, the only difference being that one step orientation is along a close-packed direction (e.g., the <120> or <010> monolayer steps) whereas the latter step orientation is vicinal relative to the former. That is, the latter step contains a certain number of donated, or excess, unidirectional kinks. Figure 8 illustrates these ideas. Assuming now that the particular kink dynamics associated with the unperturbed step results in a large average number of kinks existing along the step (region II, Eqn. 6), then combining Eqns. 5 and 6-region II, we can express the unperturbed step velocity, v_s as:

$$v_s \approx b\left(\frac{2R_{kk}}{R_k}\right)^{1/2} R_k, \quad \text{where} \left(\frac{2R_{kk}}{R_k}\right)^{1/2} = \rho_k \tag{7}$$

In the KD model, a small perturbation in the kink density, $\Delta\rho_k$, leads to a new step velocity, normal to the original step direction, v_s':

$$v_s' = v_s + \Delta\rho_k b R_k \tag{8}$$

In Eqn. 8, we have assumed that the increase in step vicinality, giving rise to a higher density of unidirectional kinks, does not reduce the density of opposing kinks. In a recent step model directed at determining the interpolation function between a straight, close-packed step velocity and a highly vicinal step velocity (Chernov, 1998), a somewhat different function from that of Eqn. 8 was found. From Eqn. 8, the angle θ that the perturbed step forms with the unperturbed direction (see Fig. 8) is defined by:

$$\theta = \tan^{-1}\frac{n_k b}{n_k / \Delta\rho_k} = \tan^{-1}\Delta\rho_k b \tag{9}$$

where n_k is the average number of excess unidirectional kinks present along the step. The KD model can be used to indirectly determine the net kink velocity by combining Eqns. 7–9 above:

$$R_k = v_s'\left(1 - \frac{v_s}{v_s'}\right)\cot\theta \tag{10}$$

In contrast, the model by Chernov (1998) leads to the calculation of R_k via Eqn. 11:

$$R_k = v_s'\left(1 - \frac{v_s}{v_s'}\right)^{1/2}\cot\theta \tag{11}$$

Equation 10 will always give a value of R_k that is less than that given by Eqn. 11, but for small v_s/v_s', the error is small. This follows from the continuum model treatment of sites parallel to the step edge for Eqn. 10. In arriving at Eqn. 10, it was assumed that the length, parallel to the step, of a double kink is infinitesimally small. Due to the actual size of atoms and molecules along a step, this assumption is only reasonable for small v_s/v'_s. Equation 11 accounts for this finite size and as a result, the forced kinks on a vicinal step reduce the number of opposing kinks due to the limited density of available sites for kinks to exist. It is clear then that the Chernov (1998) model is more realistic by accounting for the discrete size of the step components. Therefore, we will use Eqn. 11 in the following examination of HAFM results, but Eqn. 10 and the simplified model outlined above provide the necessary visualization of the KD problem.

Once R_k is determined, R_{kk} may be found from Eqn. 6. As an example of the application of this KD model, the terrace labeled "3" in Fig. 6b (90 °C in water) is bounded on the right hand side by a [010] monolayer segment perturbed by a bilayer segment during dissolution in Figs. 6c and 6d. The bilayer step velocity perpendicular to the [010] direction is 0.61 ± 0.06 nm s^{-1} = v'_s. The monolayer step velocity in the same direction is $v_s = 0.41 \pm 0.04$ nm s^{-1}, and the angle $\theta = 7.5 \pm 1.8°$. Using Eqn. 11, the kink velocity is therefore 2.7 ± 0.8 nm s^{-1}. Further, using Eqn. 6-(region II) and $b = 0.888$ nm for barite along [100], the double-kink nucleation rate is $R_{kk} = 0.04 \pm 0.01$ nm^{-1} s^{-1}. Equation 7 then gives a kink density of $\rho_k = 0.17 \pm 0.04$ nm^{-1} or a kink-kink spacing of ρ_k^{-1} ~ 6 ± 1 nm. Although the [010] steps appear straight based on the HAFM images in Fig. 6, the KD model predicts that these steps are rather rough on a scale of 1-10 nm. This roughness scale is at or below the true contact-mode resolution limit of an AFM tip assuming a typical tip radius of 5-10 nm. This example demonstrates the importance of suitable models for step motion due to the lack of direct information provided by AFM images. In contrast, for the [120] monolayer steps, $v'_s = 1.1 \pm 0.1$ nm s^{-1}, $v_s = 0.51 \pm 0.05$ nm s^{-1}, $b = 0.7$ nm and $\theta = 2.1 \pm 1.4°$ where we find the kink-kink spacing to be ~ 30 ± 15 nm at 90 °C. This large spacing implies that the rate of detachment at kink sites, relative to kink nucleation, is much higher on the [120] steps when compared to the [010] steps.

A second application of the KD model is demonstrated from a barite growth example. Figure 9a shows one of several hillocks from Fig. 9b growing at 108 °C. When steps of one growth hillock annihilate with those of the neighboring hillock, an *unbounded* step occurs. Specifically defined for growth conditions, an *unbounded* step is a step that is terminated on one or both ends by internal corners which serve as continuous sites for kink attachment. A *bounded* step is then defined as a step that is terminated at both ends by external corners, as in the steps comprising the

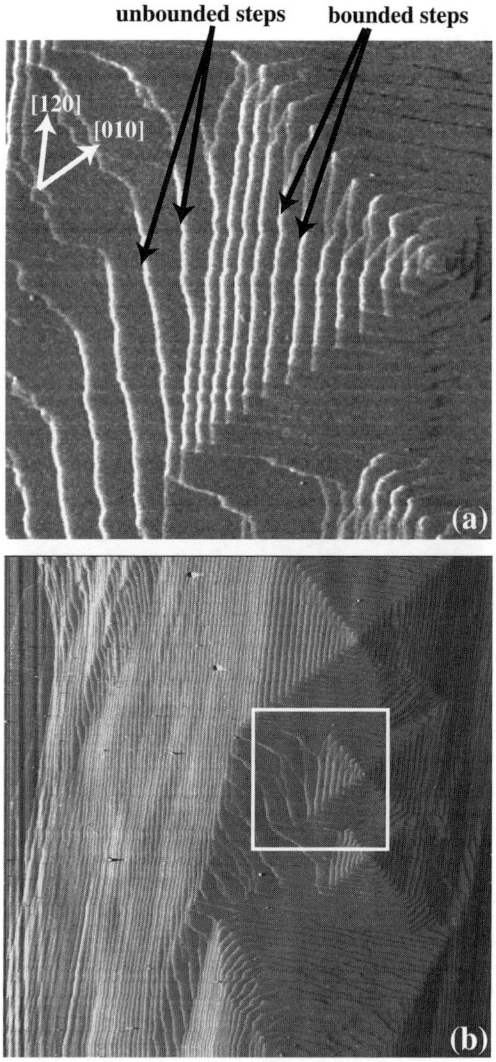

Fig. 9. (a) 5.2 µm x 5.2 µm HAFM image of barite (001) taken at 108 °C with a 20 g/hr flow of supersaturated solution ($\log[(a_{Ba^{2+}} \cdot a_{SO_4^{2-}})/K_{sp}] = 0.155$).

(b) 19 µm x 19 µm HAFM image showing the multiple intersecting growth hillocks in the region surrounding the hillock shown in (a), outlined by the square in (b).

upper tiers of the growth hillock in Fig. 9a. The formation of an unbounded step, as illustrated in Fig. 9a, can remove the nucleation barrier to step advance. As a result, the step advances rapidly and at a new angle determined by the density of kinks produced by the step annihilation event. In the KD model, the unbounded step moves at a velocity (normal to the original step direction) that is directly related to the kink velocity. Applying the KD model to measurements of step angles and growth velocities gives $R_k = 4.2 \pm 0.8$ nm s^{-1} and $R_{kk} = 0.10 \pm 0.03$ nm^{-1}s^{-1} by taking $b = 0.7$ nm for the kink depth on the [120] bilayer steps. The kink spacing is then found to be approximately 5 nm on the bounded [120] bilayer steps. As in the dissolution example for the [010] monolayer step, the KD model suggests an atomically rough [120] bilayer step under the above conditions.

AFM observations allow us to better understand kink dynamics along elementary steps on crystal surfaces, and how these dynamics ultimately lead to predictions of kink formation energies and microscopic equilibrium structures. The estimation of rates of kink attachment and/or detachment is useful for evaluating rate laws for elementary reactions at crystal surfaces. For example, kink attachment may be assumed to follow a first-order rate law whereas kink detachment generally follows a zeroth-order rate law:

$$R_k = b(w_+ - w_-) = bw_-\left(\frac{a}{a_{eq}} - 1\right) \tag{12}$$

In Eqn. 12, R_k is the net kink velocity and is only obtained via models such as the KD model. w_+ and w_- are attachment and detachment frequencies, respectively, a is the experimental activity, and a_{eq} is the equilibrium activity of dissolved growth constituents. To determine w_+ and w_-, it is necessary to obtain data over a range of solute activity or supersaturation. Under local equilibrium conditions, $R_k = 0$, which is impractical to measure with reasonable certainty. However, by determining w_+ and w_- and their temperature dependence, then by applying Eqn. 7 to find the kink density, the kink formation energy, ε, can be determined:

$$\varepsilon = -kT\ln\left(\frac{b\rho_k}{2(a/a_{eq})^{1/2}}\right) \tag{13}$$

The kinetic barrier to exchange at kink sites as represented by the activation energy for w_+ and w_- when compared to ε, allows us to then discuss the appropriate treatment of step dynamics (e.g., Higgins et al., 2000a) from either a kinetically-determined process, or a process based on equilibrium thermodynamics.

The above examples that we have presented show that it is possible to measure net kink motion within the implicit assumptions of the KD model. The key variable that remains uncontrolled in most dynamic AFM experiments, with few exceptions (e.g., Hong et al., 1997; Coles et al., 1998; Suárez and Compton, 1998), is mass transport near the crystal surface. While very sluggish surface reactions generally are accepted as surface-controlled reactions, there is no easy means of distinguishing between reactions that are slightly too fast to be surface-controlled and reactions that are slow enough to be unaffected by variable or unpredictable fluid flow. We have tested a new hydrodynamic inlet system for the HAFM based on the design of Coles et al. (1997) that allows us to assess the validity of proposed reaction mechanisms obtained by HAFM under known hydrodynamic conditions. A report of our initial results is forthcoming (Higgins et al., submitted).

The above application of a simple approach to determining kink dynamics is an important step toward understanding the *chemical kinetics* of mineral dissolution and growth at the ionic/molecular level. By treating the problem initially through the generic *kink* nomenclature, we may begin to assign relative reactivities to various molecular surface structures. In doing so, we may begin looking toward understanding the reasons for a particular ordering of kink reactivity based on molecular structure, steric effects, nearest neighbor interactions, and edge diffusion. The data presented are truly new data in the sense that they cannot be immediately compared to any existing data. Effective rates of kink motion are only amenable to measurement by high-resolution, in situ observations. Therefore, we must regard these data with objectivity when discussing them in light of the model assumptions that were used to obtain the data. The observations reported here for barite add a new branch of information regarding the kinetics of reactions at this surface in that without the use of elevated temperatures, the molecular kinetics on barite surfaces are far too slow to make AFM measurements under a wide range of solution conditions. In addition to further experiments to test various hypotheses regarding reactivity, this level of detail will require computational tools such as *ab-initio* molecular modeling, molecular dynamics, and Monte Carlo simulations. Computational models may be used to evaluate kink energetics obtained by this experimental approach, such as those results recently reported by Pina et al. (1998).

4. SUMMARY

Both techniques we have presented enable the direct (or, strictly speaking, *more* direct) investigation of interfacial processes and properties that have previously been inferred on the basis of macroscopic stoichiometric equilibrium or rate data. These abilities, used in conjunction with stoichiometric and compositional data, can provide key constraints upon a range of possible mechanistic interpretations and models. Further, these techniques provide data that can be more easily and directly related to macroscopic rates on the one hand, and to molecular-modeling calculations on the other. For example, it is increasingly possible to use computational approaches to inquire about local surface structures and reactivities (e.g., pertaining to kink sites or to the orientation and structure of water molecules near a surface) and then to relate such calculations directly to HAFM or SHG measurements that must be interpreted on the basis of the same structures and reactivities. These, in turn, can provide mechanistic insight for reactions at mineral surfaces. New techniques are thus allowing us to integrate our understanding of rock-water interaction in a wide variety of geochemically relevant settings across molecular-to-macroscopic scales.

Acknowledgments-We thank Roland Hellmann, Andreas Lüttge, Don Baer, and an anonymous reviewer for helpful suggestions for improving the manuscript. We gratefully acknowledge the support of grants from the National Science Foundation, #EAR-9634143 and #EAR-9725993 to CME, #CHE-9708451 to CME and SRH, the DOE Office of Basic Energy Sciences, #KC040302 to KGK, #DE-FG03-96SF14623/A000 to CME, and the German Academic Exchange Service (DAAD) to GJ.

Editorial Handling: R. Hellmann

REFERENCES

Ainsworth C. C., Friedrich D. M., Gassman P. L., Wang Z., and Joly A. G. (1998) Characterization of salicylate-alumina surface complexes by polarized fluorescence spectroscopy. *Geochim. Cosmochim. Acta* **62**, 595-612.

Bargar J. R., Towle S. N., Brown G. E., and Parks G. A. (1997) XAFS and bond-valence determination of the structures and compositions of surface functional groups and Pb(II) and Co(II) sorption products on single-crystal α-Al_2O_3. *J. Colloid Interface Sci.* **185**, 473-492.

Bhattacharyya K., Sitzmann E. V., and Eisenthal K. B. (1987) Study of chemical reactions by surface second harmonic generation: p-nitrophenol at the air-water interface. *J. Chem. Phys.* **87**, 1442-1443.

Bosbach D. (2002) Linking molecular scale barite precipitation mechanisms with macroscopic crystal growth rates (this volume).

Bosbach D., Hall C., and Putnis A. (1998) Mineral precipitation and dissolution in aqueous solution: in-situ microscopic observations on barite (001) with atomic force microscopy. *Chem. Geol.* **151**, 143-160.

Britt D. W. and Hlady V. (1997) In-situ atomic force microscope imaging of calcite etch pit morphology changes in undersaturated and 1-hydroxyethylidene-1,1-diphosphonic acid poisoned solutions. *Langmuir* **13**, 1873-1876.

Buckley A. N. and Woods R. (1994) On the characterization of sulfur species on sulfide mineral surfaces by X-ray photoelectron-spectroscopy and Raman-spectroscopy. *J. Electroanal. Chem.* **370**(1-2), 295-296.

Burton W. K., Cabrera N., and Frank F. C. (1951) The growth of crystals and the equilibrium structure of their surfaces. *Phil. Trans. Roy. Soc. London* **A243**, 299-358.

Butt H.-J., Kappl M., Mueller H., Raiteri R., Meyer W., and Rühe J. (1999) Steric forces measured with the atomic force microscope at various temperatures. *Langmuir* **15**, 2559-2565.

Casey W. H., Phillips B. L., Karlsson M., Nordin S., Nordin J. P., Sullivan D. J., and Neugebauer-Crawford S. (2000) Rates and mechanisms of oxygen exchanges between sites in the $AlO_4Al_{12}(OH)_{24}(H_2O)_{12}{}^{7+}$ (aq) complex and water: Implications for mineral surface chemistry. *Geochim. Cosmochim. Acta*, **64**, 2951-2964.

Castro A., Bhattacharyya K., and Eisenthal K. B. (1991) Energetics of adsorption of neutral and charged molecules at the air/water interface by second harmonic generation: Hydrophobic and solvation effects. *J. Chem. Phys.* **95**, 1310-1315.

Cheng L., Lyman P. F., Sturchio N. C., and Bedzyk M. J. (1997) X-ray standing wave investigation of the surface structure of selenite anions adsorbed on calcite. *Surf. Sci.* **382**, L690-L695.

Chernov A. A. (1963a) Crystal growth forms and their kinetic stability. *Sov. Phys. Crystallogr.* **8**, 63-67.

Chernov A. A. (1963b) The kinetics of the growth forms of crystals. *Sov. Phys. Crystallogr.* **7**, 728-730.

Chernov A. A. (1998) Step kinetics by one-dimensional "nucleation". *Materials Sci. Forum* **276-277**, 71-78.

Coles B. A., Compton R. G., Booth J., Hong Q., and Sanders G. H. W. (1997) A hydrodynamic AFM flow cell for the quantitative measurement of interfacial kinetics. *J. Chem. Soc., Chem. Commun.*, 619-620.

Coles B. A., Compton R. G., Suárez M., Booth J., Hong Q., and Sanders G. H. W. (1998) A hydrodynamic atomic force microscopy flow cell for the quantitative measurement of interfacial kinetics: The aqueous dissolution of salicylic acid and calcium carbonate. *Langmuir* **14**, 218-225.

Conboy J. C., Daschbach J. L., and Richmond G. L. (1994) Studies of alkane/water interfaces by total internal reflection second harmonic generation. *J. Phys. Chem.* **98**, 9688-9692.

Corn R. M. and Higgins D. A. (1994) Optical second harmonic generation as a probe of surface chemistry. *Chem. Rev.* **94**, 107-125.

Danielwicz-Ferchmin I. and Ferchmin A. R. (1996) Water density in the double layer. *J. Phys. Chem.* **100**, 17281-17286.

Davis J. A. and Kent D. B. (1990) Surface complexation modeling in aqueous geochemistry. In *Mineral-Water Interface Geochemistry* (eds. M. F. Hochella Jr. and A. F. White). Reviews in Mineralogy vol. 23, Mineralogical Society of America, Washington, D.C. pp. 177-259.

Dove P. M. and Platt F. M. (1996) Compatible real-time rates of mineral dissolution by atomic force microscopy (AFM). *Chem. Geol.* **127**, 331-338.

Du Q., Freysz E., and Shen Y. R. (1994) Vibrational spectra of water molecules at quartz/water interfaces. *Phys. Rev. Lett.* **72**, 238-241.

Eggleston C. M., Hug S., Stumm W., Sulzberger B., and dos Santos Afonso M. (1998) Surface complexation of sulfate by hematite surfaces: FTIR and STM observations. *Geochim. Cosmochim. Acta* **62**, 585-593.

Frank F. C. (1974) Nucleation-controlled growth on a one-dimensional growth of finite length. *J. Cryst. Growth* **22**, 233-236.

Gragson D. E., McCarty B. M., and Richmond G. L. (1997) Ordering of interfacial water molecules at the charged air/water interface observed by vibrational sum frequency generation. *J. Am. Chem. Soc.* **119**, 6144-6152.

Gratz A. J., Hillner P. E., and Hansma P. K. (1993) Step dynamics and spiral growth on calcite. *Geochim. Cosmochim. Acta* **57**, 491-495.

Heinz T. F., Chen C. K., Ricard D., and Shen Y. R. (1982)

Spectroscopy of molecular monolayers by resonant second-harmonic generation. *Phys. Rev. Lett.* **48**, 478-481.

Heinz T. F., Tom H. W. K., and Shen Y. R. (1983) Determination of molecular orientation of monolayer adsorbates by optical second-harmonic generation. *Phys. Rev. A* **28**, 1883-1885.

Higgins D. A., Byerly S. K., Abrams M. B., and Corn R. M. (1991) Second harmonic generation studies of methylene blue orientation at silica surfaces. *J. Phys. Chem.* **95**, 6984-6990.

Higgins D. A., Abrams M. B., Byerly S. B., and Corn R. M. (1992) Resonant second harmonic generation studies of p-nitrophenol adsorption at condensed phase interfaces. *Langmuir* **8**, 1994-2000.

Higgins S. R. and Eggleston C. M. An optical second harmonic generation study of dye molecule sorption at Al_2O_3-water interfaces. *Geochim. Cosmochim. Acta* (submitted).

Higgins S. R., Eggleston C. M., Jordan G., Knauss K. G., and Boro C. O. (1998a) In situ observation of oxide and silicate mineral dissolution by hydrothermal scanning force microscopy: Initial results for hematite and albite. *Mineral. Mag.* **62**, 618-619.

Higgins S. R., Eggleston C. M., Knauss K. G., and Boro C. O. (1998b) A hydrothermal atomic force microscope for imaging in aqueous solution up to 150 °C. *Rev. Sci. Instrum.* **69**, 2994-2998.

Higgins S. R., Jordan G., Eggleston C. M., and Knauss K. G. (1998c) Dissolution kinetics of the barium sulfate (001) surface by hydrothermal atomic force microscopy. *Langmuir* **14**, 4967-4971.

Higgins S. R., Stack A., Eggleston C. M., and Dos Santos Afonso M. (1998d) Proton and ligand adsorption at silica- and alumina-water interfaces studied by optical second harmonic generation (SHG). *Mineral. Mag.* **62**, 616-617.

Higgins S. R., Bosbach D., Eggleston C. M., and Knauss K. G. (2000a) Kink dynamics and step growth on barium sulfate(001): A hydrothermal scanning probe microscopy study. *J. Phys. Chem. B*, **104**, 6978-6982.

Higgins S. R., Boram L., Coles B. A., Compton R. G., Knauss K. G., and Eggleston C. M. Dissolution kinetics, step and surface morphology of magnesite (104) surfaces in acidic aqueous solution at 60 °C by atomic force microscopy under defined hydrodynamic conditions. *J. Phys. Chem. B* (submitted).

Hong Q., Suárez M. F., Coles B. A., and Compton R. G. (1997) Mechanism of solid/liquid interfacial reactions. The maleic acid driven dissolution of calcite: an atomic force microscopy study under defined hydrodynamic conditions. *J. Phys. Chem. B* **101**, 5557-5564.

Hug S. J. and Sulzberger B. (1994) In situ Fourier transform infrared spectroscopic evidence for the formation of several different surface complexes of oxalate on TiO_2 in the aqueous phase. *Langmuir* **10**, 3587-3597.

Jordan G. and Rammensee W. (1998) Dissolution rates of calcite (1014) obtained by scanning force microscopy: Microtopography-based dissolution kinetics on surfaces with anisotropic step velocities. *Geochim. Cosmochim. Acta* **62**, 941-947.

Jordan G., Higgins S. R., Eggleston C. M., Swapp S. M., Janney D. E., and Knauss K. G. (1999) Acidic dissolution of plagioclase. In-situ observations by hydrothermal atomic force microscopy. *Geochim. Cosmochim. Acta* **63**, 3183-3191.

Jordan G., Higgins S. R., Eggleston C. M., Knauss K. G., and Schmahl W. W. Dissolution kinetics of magnesite in acidic aqueous solution. A hydrothermal atomic force microscopy (HAFM) study: Step orientation and kink dynamics. *Geochim. Cosmochim. Acta* (in press).

Kikteva T., Star D., Zhao Z., Baisley T. L., and Leach G. W. (1999) Molecular orientation, aggregation, and order in rhodamine films at the fused silica/air interface. *J. Phys. Chem. B* **103**, 1124-1133.

Laiti E., Persson P., and Öhman L.-O. (1998) Balance between surface complexation and surface phase transformation at the alumina/water interface. *Langmuir* **14**, 825-831.

Lantz J. M., Baba R., and Corn R. M. (1993) Optical second harmonic generation as a probe of electrostatic fields and flatband potential at single-crystal TiO_2 electrodes. *J. Phys. Chem.* **97**, 7392-7395.

Lantz J. M. and Corn R. M. (1994) Electrostatic field measurements and band flattening during electron-transfer processes at single-crystal TiO_2 electrodes by electric field-induced optical second harmonic generation. *J. Phys. Chem.* **98**, 4899-4905.

Lauritzen J. I. (1973) Effect of a finite substrate length upon polymer crystal lamellar growth rate. *J. Appl. Phys.* **44**, 4353-4359.

Liang Y., Baer D. R., McCoy J. M., and LaFemina J. P. (1996) Interplay between step velocity and morphology during the dissolution of $CaCO_3$ surface. *J. Vac. Sci. Tech. A* **14**, 1368-1375.

Lüttge A., Bolton E. W., and Lasaga A. C. (1999) An interferometric study of the dissolution kinetics of anorthite: The role of reactive surface area. *Amer. J. Sci.* **299**, 652-678.

Mizrahi V. and Sipe J. E. (1988) Phenomenological treatment of surface second-harmonic generation. *J. Opt. Soc. Am. B* **5**, 660-667.

Ong S., Zhao X., and Eisenthal K. B. (1992) Polarization of water molecules at a charged interface: second harmonic studies of the silica/water interface. *Chem. Phys. Lett.* **191**, 327-335.

Park N.-S., Kim M.-W., Langford S. C., and Dickinson J. T. (1996) Atomic layer wear of single-crystal calcite in aqueous solution using scanning force microscopy. *J. Appl. Phys.* **80**, 2680-2686.

Parks G. A. (1965) The isoelectric points of solids oxides, solid hydroxides, and aqueous hydroxo complex systems. *Chem. Rev.* **65**, 177-198.

Parks G. A. (1990) Surface energy and adsorption at mineral/water interfaces: An introduction. In *Mineral-Water Interface Geochemistry* (eds. M. F. Hochella Jr. and A. F. White). Reviews in Mineralogy vol. 23, Mineralogical Society of America, Washington, D.C. pp. 133-175.

Pina C. M., Becker U., Risthaus P., Bosbach D., and Putnis A. (1998) Molecular-scale mechanisms of crystal growth in barite. *Nature* **395**, 483-486.

Russell A. E., Lin A. S., and O'Grady W. E. (1993a) In situ far-infrared evidence for a potential dependence of silver-water interactions. *J. Chem. Soc. Faraday Trans.* **89**, 195-198.

Russell A. E., Williams G. P., Lin A. S., and O'Grady W. E. (1993b) Far-infrared synchrotron radiation for in situ studies of the electrochemical interface. *J. Electroanal. Chem.* **356**, 309-315.

Sahai N. and Sverjensky D. A. (1997) Evaluation of internally consistent parameters for the triple-layer model by the systematic analysis of oxide surface titration data. *Geochim. Cosmochim. Acta* **61**, 2801-2826.

Shiraki R., Rock P. A., and Casey W. H. (2000) Dissolution kinetics of calcite in 0.1 M NaCl solution at room temperature: An atomic force microscopic (AFM) study. *Aquatic*

Geochem. **6**, 87-108.

Simpson G. J. and Rowlen K. L. (1999) An SHG magic angle: Dependence of second harmonic generation orientation measurements on the width of the orientation distribution. *J. Am. Chem. Soc.* **121**, 2635-2636.

Stack A. G., Higgins S. R., and Eggleston C. M. (2001) Point of zero charge of a corundum-water interface probed with optical second harmonic generation (SHG) and atomic force microscopy (AFM): Non-stoichiometric approaches to oxide surface charge. *Geochim. Cosmochim. Acta* **65**, 3055-3063.

Suárez M. F. and Compton R. G. (1998) Dissolution of magnesium oxide in aqueous acid: An atomic force microscopy study. *J. Phys. Chem. B* **102**, 7156-7162.

Sverjensky D. A. (1994) Zero-point-of-charge prediction from crystal-chemistry and solvation theory. *Geochim. Cosmochim. Acta* **58**, 3123-3129.

Sverjensky D. A. and Sahai N. (1996) Theoretical prediction of single-site surface-protonation equilibrium constants for oxides and silicates in water. *Geochim. Cosmochim. Acta* **60**, 3773-3798.

Teng H. H., Dove P. M., and DeYoreo J. J. (1999) Reversed calcite morphologies induced by microscopic growth kinetics: Insight into biomineralization. *Geochim. Cosmochim. Acta* **63**, 2507-2512.

Teng H. H., Dove P. M., Orme C. A., and DeYoreo J. J. (1998) Thermodynamics of calcite growth: Baseline for understanding biomineral formation. *Science* **282**, 724-727.

Toney M. F., Howard J. N., Richer J., Borges G. L., Gordon J. G., Melroy O. R., Wiesler D. G., Yee D., and Sorenson L. B. (1994) Voltage-dependent ordering of water molecules at an electrode-electrolyte interface. *Nature* **368**, 444-446.

Wang H., Yan E. C. Y., Borguet E., and Eisenthal K. B. (1996) Second harmonic generation from the surface of centrosymmetric particles in bulk solution. *Chem. Phys. Lett.* **259**, 15-20.

Wang K.-S., Resch R., Dunn K., Shuler P., Tang Y., Koel B. E., and Yen T. F. (2000) Scanning force microscopy study of etch pits formed during dissolution of a barite (001) surface in CDTA and EDTA solutions. *Langmuir* **16**, 649-655.

Westall J. and Hohl H. (1980) A comparison of electrostatic models for the oxide/solution interface. *Adv. Coll. Interface Sci.* **12**, 265-294.

Workman R. K. and Manne S. (2000) Variable temperature fluid stage for atomic force microscopy. *Rev. Sci. Instrum.* **71**, 431-436.

Yan E. C. Y. and Eisenthal K. B. (2000) Rotational dynamics of anisotropic microscopic particles studied by second harmonic generation. *J. Phys. Chem. B* **104**, 6686-6689.

Zangwill A. (1988) *Physics at Surfaces*. Cambridge University Press, Cambridge.

Zhao X., Ong S., and Eisenthal K. B. (1993a) Polarization of water molecules at a charged interface. Second harmonic studies of charged monolayers at the air/water interface. *Chem. Phys. Lett.* **202**, 513-520.

Zhao X., Ong S., Wang H., and Eisenthal K. B. (1993b) New method for determination of surface pKa using second harmonic generation. *Chem. Phys. Lett.* **214**, 203-207.

Water-Rock Interactions, Ore Deposits, and Environmental Geochemistry: A Tribute to David A. Crerar
© The Geochemical Society, Special Publication No. 7, 2002
Editors: Roland Hellmann and Scott A. Wood

Experimental pressure solution compaction of chalk in aqueous solutions Part 1. Deformation behavior and chemistry

ROLAND HELLMANN[1*], PETER J.N. RENDERS[1**], JEAN-PIERRE GRATIER[1], AND ROBERT GUIGUET[1]

[1] Crustal Fluids and Dynamics Group, LGIT, CNRS UMR C5559, Observatoire de Grenoble, Université J. Fourier, BP 53X, 38041 Grenoble Cedex 9, FRANCE
**present address: 211 Beekman St., Saratoga Springs, N.Y., 12866, U.S.A.

* corresponding author (hellmann@obs.ujf-grenoble.fr)

Abstract—To investigate pressure solution creep, chalk from the Paris basin (France) was deformed in a triaxial press with fluids. Experiments were conducted either under chemically closed (no fluid flow) or chemically open (fluid flow) conditions. In all experiments, the vertical stress (σ_1) and the lateral stress (σ_3) were independently controlled, with $4.0 < \sigma_1 \leq 8.0$ MPa and $\sigma_3 = 4.0$ MPa; the pore fluid pressure was ≤ 0.3 MPa. In each experiment, axial strain (ε) was recorded as a function of time; experiments ranged from 30 to almost 700 days.

The major goal of this investigation was to study the physico-chemical reactions that occur when chalk is deformed under different differential stresses, in the presence of various fluids, and at temperatures ranging up to 80 °C. The results show that chalk deformation is characterized by viscous behavior, based on plots of log strain rate ($\dot{\varepsilon}$) vs. log (σ_1) with slopes of 1-2. The chemical nature of the pore fluid also plays a critical role in determining deformation rates; the strain rates decreased in the following order: saline solution > water > propanol. Thus, strain rate correlates with calcite solubility. The dependence of $\dot{\varepsilon}$ on grain diameter could not be evaluated. The temperature dependence of strain rates was found to be very low, with an activation energy estimated to be ≈ 1-2 kJ mol^{-1}. When all rate-influencing parameters were kept constant, the material strain hardens, according to the relation $\dot{\varepsilon} \propto \varepsilon^{-2-5}$. The exact rate-limiting step (intergranular diffusion, dissolution, or precipitation) could not be determined based on the available data.

1. INTRODUCTION

1.1. Importance of Pressure Solution

In the crust, a very important percentage of material transfer occurs by way of water-rock interactions, which can be characterized in terms of both physical erosion processes and chemical weathering reactions. Chemical weathering reactions that occur at the surface and at shallow depths (≤ 1 km) occur predominantly by dissolution reactions, where dissolution takes place at essentially hydrostatically-stressed fluid/solid interfaces. This implies that dissolution reactions occur under isotropic conditions, and therefore, these reactions can be considered to be equally effective in all directions with respect to fluid attack of mineral grains comprising rocks.

Water-rock interactions within the crust, however, become increasingly anisotropic with increasing depth, this being a function of the divergence of the lithologic and hydrostatic gradients with depth. A commonly occurring and important water-rock interaction process at shallow to moderate depths is called pressure solution, or intergranular pressure solution (*ips*) creep. It occurs in geological environments in the upper crust that are characterized both by moderate levels of differential stress (i.e., at stress levels below those leading to brittle or plastic deformation) and the presence of fluids. Due to the fact that rocks at depth are exposed to differential stresses (i.e., vertical stress ≠ lateral stress ≠ pore fluid pressure), chemical instabilities develop on the scale of individual rock grains and the associated fluids in intergranular regions and pore spaces. As a consequence, there is a net transfer of material from regions of high stress along grain contacts to regions of lower stress in pore spaces, as schematically shown in Fig. 1.

The effective result of pressure solution is the deformation of individual mineral grains and precipitation of material in pore spaces, leading to a global compaction of the rock, and in general, a concomitant decrease in permeability. Pressure solution plays an important role with respect to chemical reactions, mass redistribution, and porosity changes associated with diagenesis of sediments and sedimentary rocks. In fact, pressure solution may be one of the primary processes responsible for the local production of dissolved material that precipitates as cements during diagenesis and burial cementation (e.g., Hudson, 1975; Wong and Oldershaw, 1981). Closely related to this is the evolution of permeability and porosity, as well as the channelization of fluids in sedimentary

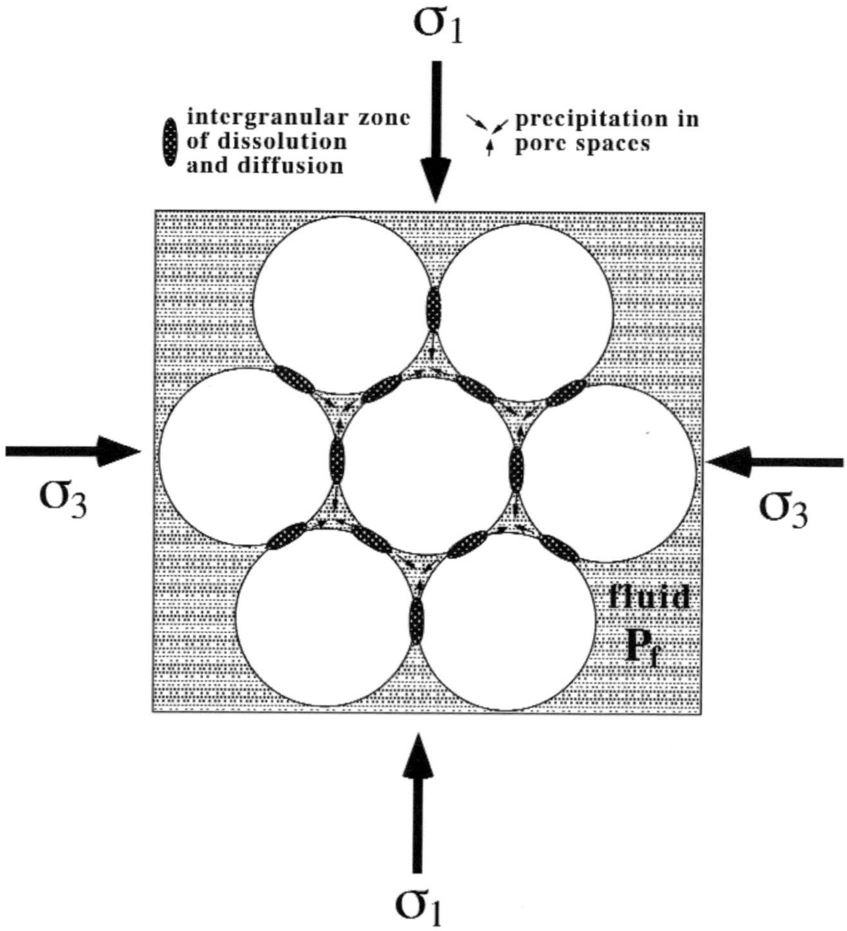

Fig. 1. Rock grains subject to a differential stress, where the vertical stress (σ_1) > lateral stress (σ_3); the interconnected pore spaces contain a fluid at pressure P_f, with $P_f < \sigma_1, \sigma_3$. Grain surfaces in contact with these pore fluids (free faces) are assumed to be in chemical equilibrium, such that no net dissolution occurs. Dissolution and diffusion occur at grain-to-grain boundaries subject to elevated stresses (dark stippled oval areas), precipitation occurs in pore spaces (advection of solutes is not considered). The net result is compaction of the grains and porosity reduction, coupled with a potential decrease in permeability, as well.

basins and other shallow geological environments (e.g., Ortoleva et al., 1987a, b).

Pressure solution creep is considered to be an important mechanism of rock deformation in the upper crust, along with mechanical compaction due to grain sliding, cataclasis (i.e., brittle failure of rocks due to crack propagation or frictional slip), and plastic deformation (i.e., a function of defect motion in crystals). What primarily differentiates pressure solution from these other deformation mechanisms is that it is a chemical process associated with intergranular dissolution and diffusion of material in a fluid phase. Pressure solution deformation is generally accepted to be limited to depths less than 15-20 km, with pressure and temperature (P,T) conditions ranging up to greenschist facies (Tada and Siever, 1989). At these P,T conditions, pressure solution and cataclastic deformation may act in a coupled fashion, such that the crust behaves in both a viscous and brittle manner (see Gratier and Gamond, 1990; Gratier et al., 1999). At higher P,T conditions, plastic deformation becomes the dominant creep mechanism (see overview in Chen and Molnar, 1983; and references therein).

There are at least 11 processes that can lead to deformation and compaction of granular minerals, one of which is pressure solution (described in Fig. 15.1, Evans et al., 1999). Some of the key factors in determining the importance of each type of deformation process are temperature, pressure, applied stress, and chemical fugacities of the phases. With this in mind, it is important to note that both in nature and in the laboratory, deformation and compaction of granular aggregates probably are a function of many simultaneously operating processes, with each process having a different degree of overall importance. Even a single type of process, such as pressure solution, may depend on several distinct mechanisms, as described further on.

Over the past two to three decades, theoretists have developed and refined models treating non-hydrostatically-stressed solids and pressure solution reactions (e.g., Rutter, 1976; Lehner and Bataille, 1985; Paterson, 1973, 1995; Tada et al., 1987; Bayly, 1987, Dewers and Ortoleva, 1989; Gratz, 1991; Lehner, 1995; Renard et al., 1999). Numerical experiments serve as a basis for modeling a variety of geological phenomena at diverse scales, ranging from compaction of sedimentary basins (e.g., Lemée and Guéguen, 1996; Schneider et al., 1996), the segregation of minerals in metamorphic rocks (e.g., Robin, 1979; Dewers and Ortoleva, 1990b), densification and compaction of aggregates and unconsolidated materials in the laboratory (e.g., Elias and Hajash, 1992; Spiers and Brzesowsky, 1993), to the evolution of pore space geometries under differential stress (e.g., Reuschlé et al., 1988; Heidug and Leroy, 1994; Leroy and Heidug, 1994).

It has long been recognized that fluids play an active role in determining the dynamics and strength of faults, mainly by affecting physical mechanisms (e.g., friction, fluid pressure). It has become increasingly evident, however, that the role of fluids in faults is not just limited to mechanical effects, but that the chemical role of fluids is also of great importance. The chemistry of fluids and chemical water-rock interactions may play a role in determining, for example, fault creep behavior, brittle fracture behavior, constitutive response, frictional stability, and long-term fault strength (Hickman et al, 1995; and references therein). More specifically, pressure solution reactions have been postulated to play an important role in the rheology and strength of faults within the framework of the aseismic/seismic cycle of faults (e.g., Rutter and Mainprice, 1979, Angevine et al., 1982, Sleep and Blanpied, 1992; Gratier et al., 1994; Hickman et al., 1995; and references therein). Studies of fault zones occurring in limestones and chalks have revealed that compaction, attributed to pressure solution and cementation, increases to a significant extent close to fault planes (e.g., Carrio-Schaffhauser and Gaviglio, 1990; Gaviglio et al., 1993, 1997).

Experimental approaches to pressure solution have generally been based on the deformation of unconsolidated materials, such as quartz and feldspathic sand (de Boer et al., 1977; Gratier and Guiguet, 1986; Schutjens, 1991; Hajash and Bloom, 1991; Elias and Hajash, 1992; Dewers and Hajash, 1995), aggregates (salt: Raj, 1982; Urai et al., 1986; Spiers et al., 1990; Spiers and Brzesowsky, 1993; Hellmann et al., 1998; gypsum: de Meer and Spiers, 1995, 1997; quartz: Cox and Patterson, 1991), or porous, natural rocks (chalk: Hellmann et al., 1996; quartzite: e.g., den Brok and Spiers, 1991; den Brok, 1996). Another approach that is occasionally used is the indenter technique (Tada and Siever, 1986; Tada et al., 1987; Gratier, 1993; Gratier et al., 1999). Other, more uncommon approaches also have been used to study pressure solution; for example, a study by Hickman and Evans (1995) where deformation of halite-silica lenses was monitored by reflected light interferometry and transmitted light photomicrography.

Despite that many experimental approaches have been used to study pressure solution, there remains a fundamental discrepancy between natural rates of pressure solution, which are extremely slow, and those measured in the laboratory, which by practical necessity, are much faster (de Boer, 1977). Elevated rates of pressure solution deformation are generally realized in the laboratory by applying much higher loading rates than occur naturally. As an example, typical loading rates associated with North Sea oil reservoirs are on the order of 10^{-8} MPa s^{-1}, whereas laboratory loading rates are usually on the order of 10^{-2} -10^{-4} MPa s^{-1} (Monjoie et al., 1991). In addition to the use of high loading rates, experimental studies also rely on elevated temperatures and/or chemically aggressive fluids to accelerate the rates of deformation in the laboratory. This approach of accelerated deformation carries with it the potential danger that measured deformation rates result from mechanisms or reactions that may be different from those occurring naturally.

1.2. Chalk Deformation: Mechanical and Chemical Processes

The present study is based on the experimental deformation of natural chalk in the presence of aqueous and non-aqueous fluids. Chalk is a material which is of broad interest in the fields of geology, petroleum and civil engineering, and radwaste storage. Especially because of its importance with respect to oil reserves (e.g., North Sea), many studies over the past few decades have been dedicated to understanding the mechanical properties and deformation behavior of chalk, both dry and in the presence of fluids (e.g., Botter, 1985; Da Silva et al., 1985; Jones and Leddra, 1989; Monjoie et al., 1990; Shao et al., 1994; Piau and Maury, 1995; Schroeder and Shao, 1996; Homand et al., 1998; Risnes and Flaageng, 1999; see also references in GBMR, 1995).

Taking a recent study by Homand et al. (1998) as a representative example, the deformation of chalk in the presence of fluids is modeled using an elasto-plastic constitutive law, where the total strain tensor is partitioned into elastic ($d\varepsilon_{ij}^{e}$), plastic collapse ($d\varepsilon_{ij}^{c}$), and plastic deviatoric ($d\varepsilon_{ij}^{d}$) components, such that: $d\varepsilon_{ij} = d\varepsilon_{ij}^{e} + d\varepsilon_{ij}^{c} + d\varepsilon_{ij}^{d}$. The elastic strain is more correctly called pseudo-elastic, since not all of the strain is recoverable, and in addition, the strain is time-dependent. The second strain component is due to pore collapse, which is induced by the physico-chemical breaking of grain-to-grain bonds or disruption of attractive forces, this permitting grain sliding to take place (Monjoie et al., 1991). Pore collapse leads to a porosity decrease on the order of just $\approx 1\%$ (Schroeder, 1995). The effective axial stress threshold that leads to pore collapse (also called the yield stress) strongly depends on external parameters, such as effective pressure, deformation rate, and temperature, as well as on internal parameters, such as the porosity, the amount and type of impurities (e.g., silica content, Da Silva et al., 1985), and especially on the nature of the pore fluids (e.g., Schroeder, 1995; Schroeder and Shao, 1996). The yield strength can be as low as 5-7 MPa for a high porosity (>45%), water-saturated chalk (Schroeder and Shao, 1996). The plastic deviatoric strain component is assumed to be due to matrix distortion due to shearing. For strains greater than 10%, a 2nd plastic deformation occurs with a stronger degree of hardening than in the first stage of plastic deviatoric deformation (Monjoie et al., 1991). Note that the use above of the term plastic deformation with respect to mechanical deformation does not imply a specific mechanism (i.e., dislocation flow).

The main reason for undertaking the present study was to examine the wet deformation of chalk at modest stress levels, where deformation can be primarily attributed to pressure solution. Until now, the deformation of chalk has almost exclusively been investigated and interpreted from the standpoint of mechanical deformation processes, whereas pressure solution has received at most only scant attention as a possible deformation mechanism. The stress conditions associated with deformation in the present study overlap with the lower effective stress fields in most of the chalk deformation studies mentioned immediately above. As will be discussed further on in more detail, much of the deformation behavior (at moderate stress levels) that has been traditionally attributed to mechanical processes may actually be more accurately understood in terms of pressure solution, where chemical reactions at stressed grain-to-grain boundaries cause deformation.

In this study, natural chalk samples were deformed in the presence of fluids over long time periods, up to almost 700 days. The axial compaction of the samples was recorded as a function of time. The primary aim of these experiments was to study the physico-chemical reactions that occur when chalk is deformed under different differential stresses in the presence of various types of fluids at temperatures ranging from 25 to 80 °C. This enabled us to investigate the various deformation rate-influencing parameters within the framework of published constitutive pressure solution relations, with the ultimate goal of better understanding the mechanisms of pressure solution. This constitutes Part 1 of this communication.

The pressure solution deformation of chalk was also investigated by examining the redistribution of mass within the matrix, as evidenced by measured changes in porosity and density before and post-deformation. This investigation is primarily based on the results from two experiments (L9-11 and L16), and is detailed in Part 2.

2. MECHANISMS AND MODELS OF PRESSURE SOLUTION: A BRIEF REVIEW

2.1. Thermodynamics of Pressure Solution

Among the first attempts to describe the thermodynamics of solids in contact with fluids at varying pressures were pioneering studies by Sorby (1863) and Gibbs (1878). More than a century later, uncertainties still remain in our understanding of non-hydrostatically stressed solids and the mechanisms of pressure solution (Evans and Kohlstedt, 1995). Even though a detailed analysis of the thermodynamic and kinetic driving forces associated with pressure solution is beyond the scope of this article (see e.g., Paterson, 1973; Bayly, 1985, 1987; Lehner and Bataille, 1985; Lehner, 1990 for detailed discussions; see also transport models in Table 4, Evans and Kohlstedt, 1995), a short review is given below, with the aim of helping readers unfamiliar with intergranular pressure solution creep understand the analysis of the chalk deformation data that is given later on.

As mentioned earlier in the *Introduction*, rocks at depth in the presence of fluids develop chemical instabilities on the scale of individual rock grains. This gives rise to differences in the chemical potentials of the individual solid constituents (e.g., μ_{Si}) along various grain surfaces. The following example (see Fig. 2), based on two interfaces of the same grain that are subject to different stresses, is used to illustrate how these differences in chemical potential arise. Both equilibrium and non-equilibrium thermodynamic ap-

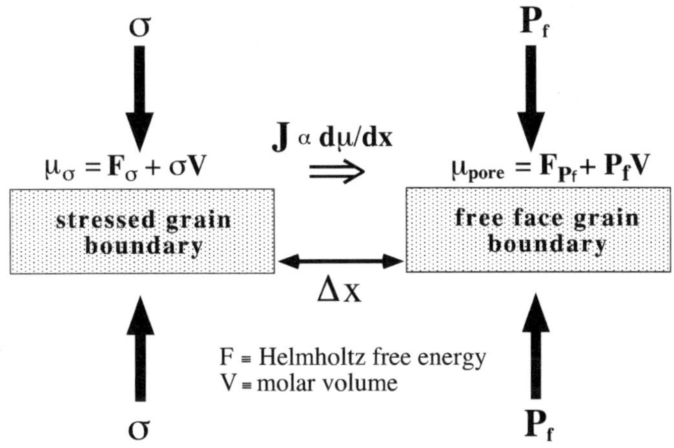

Fig. 2. The main thermodynamic driving force of pressure solution is the difference in chemical potentials ($\Delta\mu = \mu_\sigma - \mu_{pore}$) between a fluid-solid interface, subject to normal stress σ, and a pore fluid-solid interface, subject to hydrostatic stress P_f (P_f = pore pressure). The chemical potentials of the solid interfaces and the respective fluids in contact with them are assumed to be equal, based on a 'local equilibrium' approximation. The flux of dissolved material diffusing between the stressed interface and the pore fluid interface is a function of $\Delta\mu$ and the intergranular diffusion distance Δx.

proaches (e.g., Kamb, 1959, 1961; Paterson, 1973; Lehner and Bataille, 1985) lead to the following expressions (see original derivation by Gibbs, 1878) for the respective chemical potentials of an interface subject to a compressive normal stress σ and an interface subject to a hydrostatic stress P_f:

$$\mu_\sigma = F_\sigma + \sigma V_\sigma \quad ; \quad \mu_{pore} = F_{Pf} + P_f V_{Pf} \quad (1a)$$

In the above equations F is the Helmholtz free energy of the solid (i.e., energy associated with stored elastic and strain energies, which are a function of applied stresses), and V_σ and V_{Pf} are the molar volumes of the σ-stressed and pore fluid pressure-stressed solids, respectively (the small difference in these terms is usually neglected). Equation 1a neglects radius of curvature effects, which are assumed in this case not to be important. The differences in chemical potentials of these two interfaces are also shown graphically in Fig. 2. In Eqn. 1a, $\mu_\sigma > \mu_{pore}$, due to the fact that $F_\sigma > F_{Pf}$ and $\sigma V_\sigma > P_f V_{Pf}$. Note that an assumed condition of "local equilibrium" between a stressed interface and the fluid in contact with it (Kamb, 1961; Paterson, 1973) allows the chemical potential of any given component in the solid to be equated to its chemical potential in the fluid (e.g., $\mu_{Si\ solid\ \sigma} = \mu_{Si\ aq\ \sigma}$).This approximation, however, should not be construed to mean that a *true* chemical equilibrium exists at the solid/fluid phase interface (see Lehner, 1995).

The difference in chemical potentials:

$$\Delta\mu = (\sigma - P_f)V + (F_\sigma - F_{P_f}) \quad (1b)$$

is the essential thermodynamic driving force for the overall process of pressure solution which leads to the enhanced solubility of stressed interfaces and the diffusional transfer of dissolved material to regions of lower stress, such as pore spaces (see Figs. 1, 2).

Since $\Delta\mu$ is the essential energy driving pressure solution, then according to Eqn. 1b, the higher the normal compressive stress is, the greater $\Delta\mu$ will be, due to increases in the terms $(\sigma - P_f)V$ and $(F_\sigma - F_{Pf})$. Thus, referring to Fig. 1, the grain-to-grain contacts perpendicular to σ_1 are theoretically predicted to converge at a faster rate than the contacts perpendicular to σ_3 (i.e., pressure solution deformation will be fastest parallel to σ_1). In general, the overall difference in the chemical potentials is due primarily to $\sigma V > P_f V$, while the difference in the Helmholtz energy terms is significantly less (Paterson, 1973). However, some pressure solution models are based explicitly only on differences in the Helmholtz energy terms. Given that pressure solution is based on the serial processes of dissolution, diffusion, and precipitation, all three processes will consume the total available free energy ($\Delta\mu$). The partitioning of $\Delta\mu$ among the three processes is thought to be a function of which process is rate-determining; the slowest (rate-limiting) step will consume the vast majority of this energy (see Lehner, 1995).

From the above analysis of pressure solution, one could be led to believe that only the thermodynamic force $\Delta\mu$ will determine the rate of grain-to-grain convergence. However, observations based on nature and laboratory results clearly show that for the same differential stress $(\sigma - P_f)$, different materials deform at different rates. This is because the actual rate of grain-to-grain convergence is a function of a force-flux relationship, and not just the force. Thus, considering the dissolution process at a stressed interface, a simple linear force-flux relation of the following form can be postulated (adapted from Lehner, 1995):

$$\dot{m} \propto K\,[\mu_n^s(r) - \mu(r)] \tag{1c}$$

where $\dot{m}$ is the flux, K is a rate constant, and the quantity in brackets is the chemical potential driving force (with $\mu_n^s(r)$ equal to the normal component of the solid phase chemical potential and $\mu(r)$ equal to the fluid phase potential in the grain boundary region). A similar relation can be expressed for precipitation, with the appropriate changes made to the chemical potential driving force. The diffusion process can also be represented in terms of a force-flux relationship, based on Fick's Law. Therefore, for the diffusion flux one can write (adapted from Lehner, 1995):

$$\dot{m} \propto C_0\,D^{gb}\delta / a^2\,[\tilde{\mu} - \mu(a)] \tag{1d}$$

where C_0 is a reference concentration equal to the solubility of the solid in the pore fluid, D^{gb} is the effective grain boundary diffusivity, δ is the mean grain boundary thickness, a is the grain boundary radius, and $[\tilde{\mu} - \mu(a)]$ is the average chemical potential driving force. Equations 1c and 1d form the basis of most constitutive relations for pressure solution deformation. Often just one of these relations is given preference based on an *a priori* choice as to which process is thought to be rate-limiting.

Thus, Eqns. 1c and 1d are the key to understanding why different materials deform at different rates, when all other rate-determining parameters are equal. Taking halite and quartz for example, one can predict that halite grain-to-grain convergence will be faster than quartz convergence, based on the fact that $K_{diss}^{halite} >> K_{diss}^{qtz}$, or $C_0^{halite} >> C_0^{qtz}$. These kinetic and solubility factors far outweigh any potential differences in the constitutive properties of the solid related to the Helmholtz free energy (Eqns. 1a and b).

2.2. Pressure Solution Models

The first model, often called free-face pressure solution, is based on the premise that grain-to-grain contacts are dry, and that dissolution only occurs at grain surfaces in contact with pore fluids (these surfaces are called 'free faces'). Dissolution along free faces at the perimeters of the grain contacts, which are accelerated by elevated surface strain and plastic energies (e.g., Dewers and Ortoleva, 1989), results in undercutting and a reduction in the intergranular contact area, this eventually leading to deformation of intergranular boundaries, either by plastic dislocation flow (e.g., Pharr and Ashby, 1983; Tada and Siever, 1986; Tada et al. 1987) or by brittle failure (Bathurst, 1958).

The second and third models are both based on grain-to-grain contacts that contain a trapped fluid which cannot be squeezed out by the normal stresses imposed on the grains. In addition, both of these models are based on 3 material transfer processes occurring in series:dissolution at stressed intergranular regions, diffusion of dissolved material along stressed grain-to-grain boundaries, and precipitation in pore spaces. It is primarily the physical nature of the intergranular regions that differentiates these two models, which often are referred to as the water-film diffusion and island-channel models, respectively.

The water-film diffusion model is based on the presence of a continuous thin film of fluid at grain-to-grain contacts that is able to support a shear stress (e.g., Weyl, 1959; Elliot, 1973; Rutter, 1976, 1983; Hickman and Evans, 1995; Renard and Ortoleva, 1997). This thin fluid film (≈ 1-30 nm) is postulated to have very special properties with respect to a bulk fluid, both in terms of surface forces and the structured nature of the fluid molecules (see discussion in Heidug, 1995; and references therein), as well as with respect to the thermodynamic and transport properties (e.g., Weyl, 1959; Robin, 1978; Rutter, 1976, 1983; Tada et al., 1987). Experimental evidence for such thin films is quite circumstantial at present, however.

As the name suggests, the island-channel model posits an intergranular region consisting of solid-solid contacts at islands (with μm to atomic scale roughness) that support the imposed normal stresses; these islands are surrounded by a network of interconnected channels containing fluid (e.g., Raj, 1982; Spiers et al., 1990; Cox and Patterson, 1991; Schutjens and Spiers, 1999). If the interconnected channels are directly connected to the pore fluid, then the fluid pressures should be equal. This is supported by intergranular diffusion rates that are estimated to be equivalent to those in bulk fluids or pore spaces (e.g., Raj and Chyung, 1981; Spiers et al., 1990; Lehner, 1990). Given that the normal compressive stresses acting along wetted intergranular surfaces are considered to be equal to the fluid pressure in adjacent pore spaces, the gradient in chemical potentials between the wetted intergranular regions and the pore spaces should logically *not* be due to $(\sigma - P_f)V$, but rather to differences in $(F_\sigma - F_{P_f})$, the Helmholtz energy (refer to Eqn. 1b). Stress concentrations at the islands within the intergranular zone, due in part to undercutting, also lead to deformation, either through brittle failure or plastic dislocation flow. Nonetheless, most studies that model pressure solution using the island-channel model still consider that the primary driving force is $\Delta\mu \propto (\sigma - P_f)V$ (e.g., Raj and Chyung, 1981; Spiers et al., 1990). This contradiction can only be alleviated by assuming that the fluid pressure in the intergranular regions is higher than that in pore spaces and that the intergranular fluid is not directly connected to the pore fluid. Despite similarities with the free-face model, the island-channel model stresses that preferential dissolution and the outward diffusion of dissolved material occur exclusively within the intergranular zone, and not in pore spaces.

The dynamic stability of interface roughness is a subject of debate. There is evidence that surface roughness between wetted grains decreases over time, but that a certain degree of micro-roughness prevails, due to the effects of heterogeneous dissolution (i.e., influence of defects, impurities, etc.) (Spiers and Schutjens, 1990; Lehner, 1990). As an example, the time evolution of a halite-halite interface has been observed to start with μm-scale roughness and progress to <500 nm roughness by Schutjens and Spiers (1999). Based on entropy considerations, Raj (1982) postulated that islands (surface roughness) are thermodynamically stable.

It should be noted that the three pressure solution models presented above are not necessarily mutually exclusive. As an example, Gratz (1991) proposed a hybrid water-film/island-channel model, where load-supporting islands are capped by a thin fluid film. In addition, in laboratory studies (e.g., Gratier, 1993) and in natural cases (e.g., Engelder, 1982), both intergranular and free-face pressure solution mechanisms have been inferred to be operative. In addition to the above factors, studies have shown that pressure solution deformation rates are usually significantly faster when solids of different composition are in contact, in comparison to solids having the same composition (e.g., Dewers and Ortoleva, 1991; Hickman and Evans, 1995). This has important ramifications with respect to rates of pressure solution in nature where, for example, it is often observed that clays occupy grain-to-grain contacts in sandstones.

Both Figs. 1 and 2 show the three principal material transfer processes that operate in series during pressure solution: dissolution, diffusion, and precipitation. It is assumed that advective transport of material between pore spaces does not occur. Since in a serial process the slowest step is rate-controlling, one of the 3 material transfer steps will determine the overall rate of pressure solution deformation of an aggregate. In free-face pressure solution, the interfacial kinetics (i.e., dissolution or precipitation in pore spaces) or the constitutive properties of the solid are rate-limiting (Tada et al., 1987). On the other hand, with respect to intergranular pressure solution (water-film and island-channel models), most studies in the literature postulate diffusion as the rate-limiting step (e.g., Weyl, 1959; Elliot, 1973; Rutter, 1976; Gratier and Guiguet, 1986; Hickman and Evans, 1995). However, in the case of sparingly soluble minerals (e.g. quartz, gypsum), surface processes, such as dissolution or precipitation, may instead be the rate-limiting process (e.g., Raj and Chyung, 1981; Mullis, 1991, 1993; de Meer and Spiers, 1995).

The rate-limiting step in pressure solution is either assumed *a priori* in a model (e.g., NaCl, Spiers et al., 1990), or alternatively, no *a priori* choice of rate-limiting process is imposed (e.g., Merino et al., 1983; Lehner, 1995). The criterion of rate control (i.e., the rate-limiting step), as well as the mechanism of deformation, may also change during deformation, depending on the exact conditions and rates of deformation, burial depth, grain diameters, and the chemical and physical characteristics of the fluids and solids (e.g., Lehner and Bataille, 1985; Tada et al. 1987; Spiers and Brzesowsky, 1993; Mullis, 1993; Renard et al., 1999).

The transfer of material from intergranular regions subject to elevated stress to pore spaces occurs by a grain boundary diffusional process, within a fluid phase (see dark gray oval zones in Fig. 1). Fick's first law can be used as an approximation to link the diffusive flux (J) from stressed intergranular regions to pore spaces as a function of the difference in chemical potentials in the fluid and the distance of diffusion (Δx) separating both regions: $J \propto \partial\mu/\partial x$. This simple diffusional flux relationship (see Fig. 2) is an integral part of most intergranular pressure models. The diffusional distance (Δx), which is often taken to be equal to the radius of grains, is not always easily constrained, however. One reason for this is experimental evidence for stress-induced micro-cracking on intergranular surfaces (e.g., Gratz, 1991; den Brok, 1998; Gratier et al., 1999). Micro-cracking may considerably diminish the distance of diffusion within the hypothesized thin fluid film. This phenomenon can also be considered to be a mechanism of strain.

3. EXPERIMENTAL MATERIALS AND PROCEDURES

3.1. Characteristics of the Chalk

The chalk used in this study was quarried from the Guerville formation (Carrière de Mantes, Guerville, France), a Campanian age chalk that crops out in the Mantes region within the Paris basin (for additional details, see Megnien, 1980; and references therein). There was no visible stratigraphic layering within the 1/3 m^3 block from which the cores were extracted. Examination by optical microscopy and SEM reveals that the chalk is composed largely of coccolith skeletal debris and other shell fragments (see SEM image, Fig. 3a). The SEM image reveals numerous voids with diameters ranging from ≈1 to several μm. The magnification of the images was not sufficient to characterize the nature of the intergranular cement. X-ray diffraction results of 4 post-run samples taken from the same core (L16) show that the chalk is composed essentially of calcium carbonate: 3 samples show ≈100% calcite, whereas one sample shows the presence of a minor amount of dolomite (≈1%). The occurrence of dolomite may be related to Mg-rich fluids injected during the experiment. The representative surface area of the undeformed chalk, as measured by the BET method, is ≈ 1.4 m^2 g^{-1} (±15%).

The chalk is heterogeneous (visible even to the naked eye), and contains mm- to cm-sized chert fragments dispersed randomly throughout the chalk. Apparently non-oriented thin calcite veins cross-cut the sample over a cm-scale interval. These veins are composed of the same material as the chalk matrix (as confirmed by SEM/EDX analyses), but have more calcite cement, and are more compact and less porous than the matrix. Figure 3b is an SEM image of a typical vein cross cutting the chalk matrix; x-ray computerized tomography images in Fig. 15 (Part 2) show a macroscopic view of the vein network.

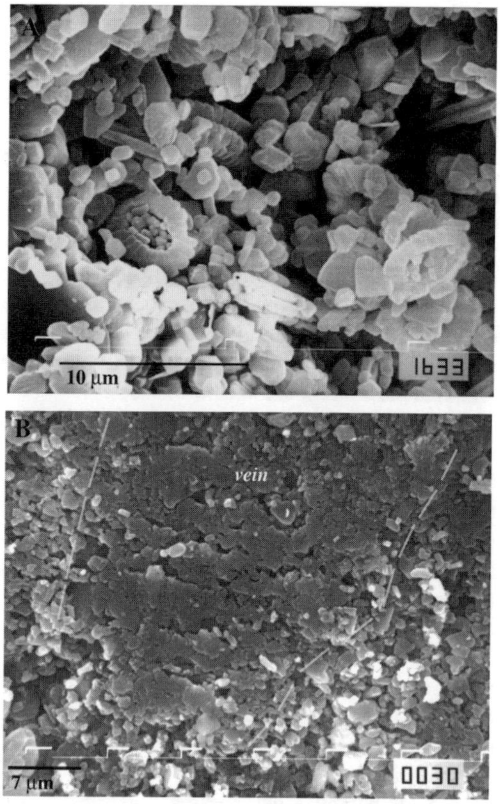

Fig. 3. SEM images of Guerville chalk used in study. (A) The chalk is composed primarily of coccolith skeletal debris and other shell fragments; note the high porosity inherent to this chalk. (B) Vein crosscutting the matrix; the porosity is considerably diminished in comparison to surrounding matrix.

3.2. Experimental Apparatus

Two core sizes were used for the deformation experiments: $\phi = 25$ mm, $l = 50$ mm and $\phi = 100$ mm, $l = 200$ mm. Figure 4 shows a schematic design of the triaxial deformation apparatus used in the large core experiments; the apparatus used for short cores is very similar. The vertical stress (σ_1) is imposed by gas pressure on a piston in contact with the core; the lateral stress (σ_3) is imposed directly by gas pressure acting on the entire vertical surface of the sheathed core. Note that the imposed vertical and lateral stresses are independent of each other. The indicated values of σ_1 have not been corrected for the effect of o-ring friction, as it is estimated to be ≈ 0.035 MPa.

A pump (HPLC or syringe design) serves to move fluids from bottom to top through the sample at a precise and constant rate. Injection rates (v) were variable, but in general, they did not exceed 0.1- 0.2 ml min^{-1}; the measured inlet pressures were in general ≤ 0.3 MPa. Based on the aforementioned flow rates, the volumetric flow rates per unit area of connected pore space were 0.0032 - 0.0064 cm min^{-1} (i.e., effective Darcy velocities based on large core dimensions, and an average chalk porosity of 40%). As an example, a Darcy velocity of 0.0032 cm min^{-1} corresponds to a residence time of 6250 mins (104 hours) for fluids traversing the large cores. The fluid outflow is at ambient pressure; however, if a higher fluid pressure (P_f) is desired, the outflow tube is easily connected to a backpressure regulator which controls the fluid pressure. The top and bottom faces of the chalk cylinder are fitted with stainless steel cap plates incorporating a rectilinear grid of holes (mm-spacing between holes) to optimize the homogeneous infiltration of fluid into the sample. The entire pressure vessel, which incorporates 4 horizontal thermocouple temperature sensors, is heated by an external furnace (thermal stability ± 1 °C).

The vertical displacement of the piston is measured continuously using a mechanical micrometer with a resolution of 0.0002 mm. Radial strain could not be

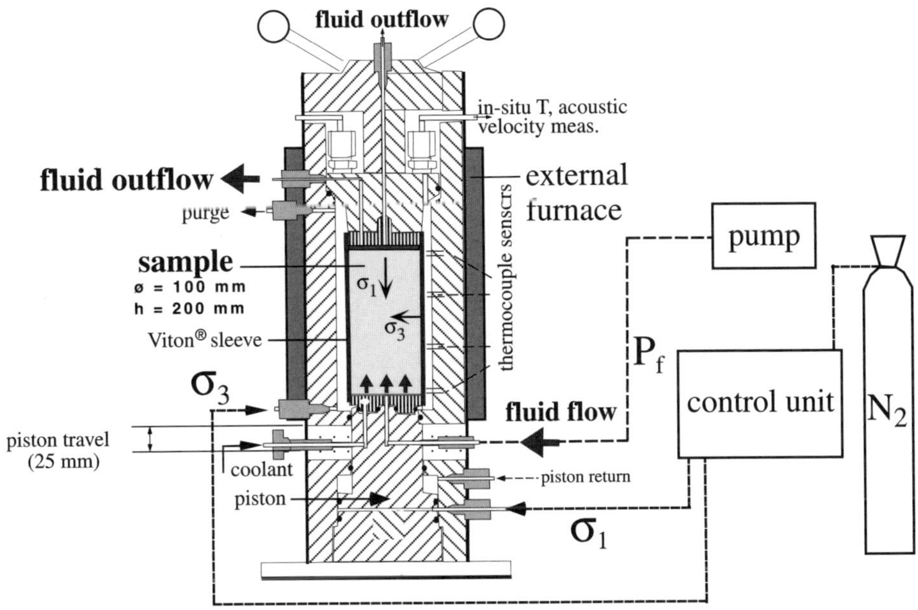

Fig. 4. Triaxial flow apparatus used in deformation experiments of large samples (200 mm length). The vertical and lateral stresses are independently controlled. Fluids are injected into the sample with a syringe or HPLC pump, fluid flow is from bottom to top. The fluid pressure can be controlled by connecting the outflow tubing to a back pressure regulator.

measured continuously during the experiment, but measurements with calipers at the end of experiments did not indicate any measurable radial strain. Measurement of the vertical displacement allows for determination of the axial strain (ε) at any specified time using the relation:

$$\varepsilon = \frac{l_0 - l}{l_0} \qquad (2)$$

where l_0 is the original length of the sample and l is the length at a specified time. The vertical strain rate ($\dot{\varepsilon}$) at any specified time is equal to the time derivative of ε, normalized to the instantaneous length of the sample:

$$\dot{\varepsilon} = \frac{\partial l / \partial t}{l} \qquad (3)$$

3.3 Experimental Procedures

Because the chalk is very fragile, it was not possible to core a single 200-mm long cylinder; instead, two smaller cores, approximately the same length, were placed end-to-end. The top and bottom faces of each core were cut and sanded perpendicular to the core axis. Before the cores were deformed, they were dried in an oven at ≈70-110 °C over periods ranging from several hours up to 2 days. This drying step greatly decreased the probability of core collapse during pressure loading. Before insertion into the deformation apparatus, the cylindrical body of chalk (composed of 2 cores) was sheathed with either a rubber or a Viton® sleeve (rubber used for T < 50 °C). This sleeve serves a double purpose: first, it provides an impermeable surface for transmitting the lateral stress to the sample; second, it prevents the escape of fluids from within the chalk core via the cylindrical walls of the core.

Initial pressure loading of the sample was performed in the absence of fluids at room temperature over the course of approximately 1 day. Loading was nominally non-hydrostatic ($\sigma_1 \geq \sigma_3$), with loading rates less than 0.1 MPa min^{-1}. Once the desired value of σ_3 was reached, the axial stress σ_1 was slowly increased to the final value, such that $\sigma_1 > \sigma_3$. After stabilization of the samples ($\dot{\varepsilon} \approx 0$) during the initial phase of dry deformation (over periods of up to 10 days), pressure solution deformation was initiated by the injection and infiltration of fluids into the samples. Aqueous fluids were degassed by boiling ≈ 1

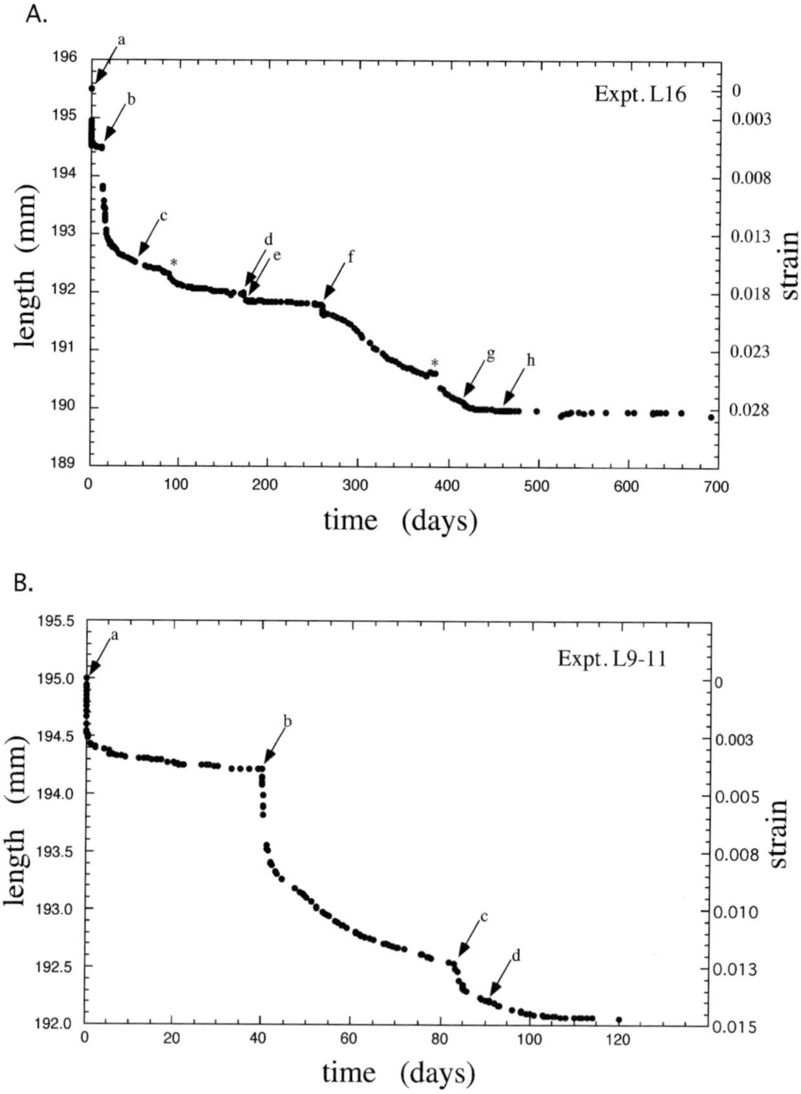

Fig. 5. Deformation curves for experiments L16 and L9-11 (note that strain (ε) increases in negative direction along ordinate axis). Changes in various strain rate-influencing parameters during the course of each experiment served to perturb local "steady-state" deformation behavior. (A) L16 parameter changes a: start of dry deformation, $\sigma_1 = 6.0$, $\sigma_3 = 4.0$ MPa, 25 °C b: injection H_2O, $v = 0.5 \rightarrow 0.1$ ml min^{-1}, T = 25 → 50 °C c: $v = 0$ over t = 50-172 days; *instability, gas bottle change d: injection saline solution over t = 179-189 days, v = variable e: $v = 0$ over t = 189-259 days f: T = 50 → 77 °C, $v = 0$ over t = 259-417 days; * instability, gas bottle change, $\sigma_1 = 7.0$, $\sigma_3 = 4.0$ MPa g: $v = 0.1$ ml min^{-1} over t = 417-461 days h: injection propanol, $v = 0.1 \rightarrow 0$ ml min^{-1}. (B) L9-11 parameter changes a: dry deformation: 0 - 0.08 days, followed by start of fluid-assisted deformation, injection H_2O, $v = 0.5 \rightarrow 0.2$ ml min^{-1}, $\sigma_1 = 4.9$, $\sigma_3 = 4.0$ MPa b: $\sigma_1 = 7.0$, $\sigma_3 = 4.0$ MPa, $v = 0.2 \rightarrow 0.1$ ml min^{-1} c: T = 25 → 50 °C d: injection of saline solution, $v = 0.1$ ml min^{-1}.

hour prior to injection into the sample; non-aqueous fluids (propanol) were not degassed and were assumed to be in equilibrium with $P_{CO_2 atm}$. Fluids were periodically sampled at the exit and analyzed for Ca and Mg by standard analytical techniques.

Experimental run times ranged from 30 to almost 700 days. The general experimental procedure consisted of allowing the system to attain a "steady-state" rate of deformation, over periods up to several weeks. Deliberate changes in any one of the deformation rate-influencing parameters (σ_1, temperature, chemistry of input solution, injection rate, fluid pressure) served to perturb the steady-state behavior of the system, producing in most cases very abrupt changes in the deformation rate. These abrupt changes gave way to progressive stabilization and the establishment of new steady-state rates of deformation. Thus, the overall deformation behavior of samples can be described by a series of transitions from one local "steady state" to another, separated by abrupt changes in the deformation rate (see Figs. 5a, b). The exact meaning of "steady-state" deformation behavior refers to instantaneous deformation rates that are very slowly varying- this is discussed further on in more detail.

3.4. Experimental Conditions

In total, 5 experiments were run with small samples, 4 with large samples. Two experiments, L16 and L9-11, are described in detail below. The post-deformation cores of these two experiments were also analyzed in detail in terms of their porosity and density by various analytical techniques, as reported in Part 2.

Experiment L16 was run with the following applied stresses: $\sigma_3 = 4.0\pm0.2$ MPa, $\sigma_1 = 6.0\pm0.2$ MPa initially; towards the end of the experiment, σ_1 was increased to 7.0±0.2 MPa. The instabilities associated with σ_1 and σ_3 reflect the effects of slow pressure changes (i.e., non-optimal performance of pressure regulation system), as well as more sudden pressure fluctuations resulting from the occasional exchange of gas bottles that were used as the pressure medium. Four experimental parameters that influence the rate of axial deformation were varied separately during the course of the experiment: axial stress (σ_1), temperature (25→50→77 °C), fluid composition, and injection rate (Fig. 5a). The wet deformation of the sample was initiated with the injection of pure water, followed by saline fluid (0.5 M NaCl + 0.05 M $MgSO_4 \cdot 7H_2O$), and terminated with propanol. After each period of fluid injection (injection at variable rates; median rate 0.1 ml/min), fluid flow was stopped over extended periods of time to allow for the establishment of steady-state deformation rates. The total duration of experiment L16 was 691 days.

The experimental parameters for expt. L9-11 were slightly different (Fig. 5b). The initial pressure loading of the sample under dry conditions (0-0.08 days) was followed by deformation in the presence of pure water ($\sigma_1 = 4.9\pm0.1$ MPa, $\sigma_3 = 4.0\pm0.1$ MPa, injection rate: 0.5, 0.2 ml min^{-1}). After the rate of deformation stabilized (over $t \approx$ 30-40 days), σ_1 was increased to 6.9±0.1 MPa (water injection rate: 0.2, 0.1 ml min^{-1}). Before the attainment of steady-state deformation, the temperature was increased from 25 to 50 °C, and the injected fluid was changed from water to saline solution (t ≈ 82 days and 90 days, respectively; fluid injection rate: 0.1 ml min^{-1}). The experiment was terminated during a regime with a "steady-state" rate of deformation at t = 120 days. The chemistry and pH of the exit fluids were semi-continuously measured. As opposed to experiment L16, fluids were continuously injected into the sample, without intermittent periods of zero flow.

4. RESULTS AND DISCUSSION

4.1. General Observations

4.1.1. Dry vs. fluid-assisted deformation; "steady-state" deformation

The pressure loading of all samples resulted in an initial period of dry deformation. As an example, Fig. 6 shows the preliminary deformation of sample L16 in the absence of fluids (note: Fig. 6 is a partial enlargement of Fig. 5a, which shows the entire deformation history of L16). The strain vs. time curve shows that the initial deformation rate of the sample rapidly decreases, to the point where $\dot{\varepsilon} \approx 0$. The total axial strain (ε) associated with dry deformation for experiment L16 is on the order of 0.005. This compaction is purely mechanical and can be attributed to crushing of grains at the end of the sample when the piston seats itself during loading. There may also be a small contribution due to partial pore collapse within the sample; however, pore collapse in dry chalks is generally attributed to much higher stresses, on the order of at least 10-20 MPa (the minimum amount of stress depends on the nature of the chalk, its porosity, amount and chemistry of pore fluids, etc.; e.g., Schroeder 1995; Pecqueur et al., 1995). The results in this study show no evidence for measurable steady-state deformation under dry conditions after the initial phase of mechanical compaction. On the other hand, the injection of fluids always triggered rapid deformation. Deformation associated with fluids is not transitory, but is a long term, time-dependent phenomenon (e.g., sample L16, t ≥ 12 days, Fig. 5a), suggesting a fluid-enhanced creep mechanism.

The fluids that were injected into the chalk cores were not pre-saturated with respect to calcite. On the other hand, exit fluids were saturated with respect to calcite, as determined by measurements of [Ca], the exit pH (9.1-9.8), and experimental data in the literature concerning calcite solubility (Plummer and Busenberg, 1982). Several lines of evidence show

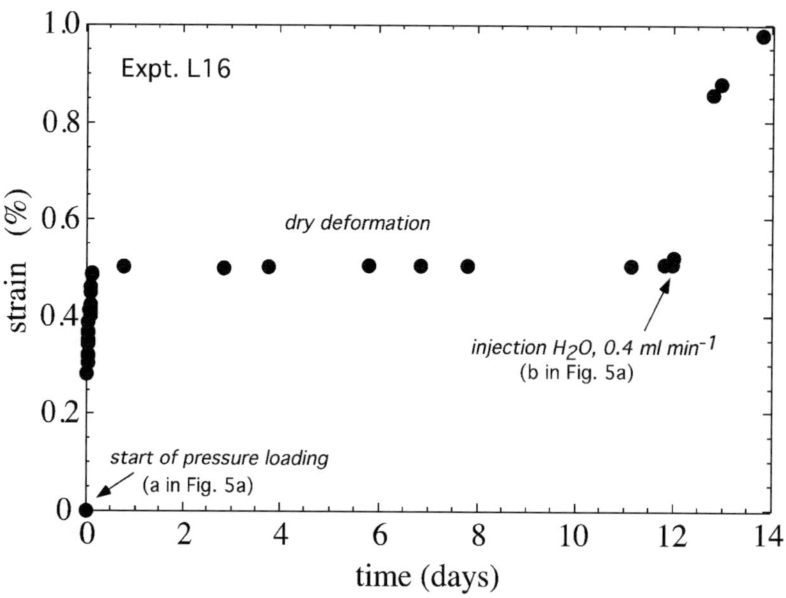

Fig. 6. Enlargement of Fig. 5a (note change in ordinate axis with respect to ε), showing rapid stabilization of initial dry deformation of L16 during stress loading. Long term, fluid-assisted deformation, which is attributed to pressure solution creep, is initiated with the injection of water into the sample.

that injected solutions become rapidly saturated with respect to calcite:

a. Exit calcium concentrations were invariant with respect to injection rate (e.g., change in injection rate from 0.025 to 0.007 ml min^{-1}).
b. Kinetic experiments of chalk in degassed H_2O and saline solution were performed and show that saturation with respect to calcite is achieved within ≤ 2 days at 20-25 °C. This is significantly less than the mean residence time of fluids in the chalk cores (> 4 days).
c. Studies in the literature confirm that the dissolution rate of calcite in water is very rapid at far from equilibrium conditions (log k = -6.63 mmol cm^{-2} s^{-1} at 25 °C; Busenberg and Plummer, 1986).

In referring to the general style of deformation behavior recorded in this study, initially rapid strain rates progressively stabilized with time, leading to local "steady-state" strain rates. The term local "steady state" refers to regimes of deformation where ε vs. t was approximately linear over periods of several tens of days (e.g., Fig. 9). True steady-state conditions ($\partial\varepsilon / \partial t = 0$) were never attained, even though at the end of some experiments the change in ε was too small to be measured (e.g., $t >$ 500 days, Fig. 5a). The stabilization of rates occurred when the experiments were configured both as open or closed systems (i.e., with and without flow- e.g., expt. L9-11, Fig. 5b).

4.1.2. Fluid chemistry behavior

When an aqueous fluid is first injected, or when the fluid injected is changed to one in which calcite is more soluble (e.g., water replaced by saline solution), there is always an initial rapid and transitory rate of deformation, accompanied by an elevated pulse of Ca in the exit solution (e.g., see 2 [Ca] spikes in Fig. 7; compare to deformation curves in Fig. 5b; both figures are based on expt. L9-11). In either case, the injection of fluid is predicted to lead to dissolution reactions at both free-face interfaces (i.e., grain surfaces in contact with pore space fluids) and stressed grain-to-grain interfaces. Both types of reactions are probably responsible (to varying degrees) for grain compaction and deformation of the matrix. Fluid access to pore spaces containing surface fines created by grain crushing during the initial phase of dry deformation may be the main reason for the behavior of Ca shown in Fig. 7 (i.e., [Ca] spike during first 20 days; [Ca] spike associated with injection of saline solution, $t > 90$ days). In the same figure, the elevated, steady-state [Ca] ($t > 115$ days) is a function of the increased solubility of calcite in the saline solution. In contrast to the behavior shown in Fig. 7, a change in fluid chemistry from high calcite solubility to lower solubility does not lead to a [Ca] spike.

The experimental results show that the exit solution chemistry is only nominally affected by changes in the

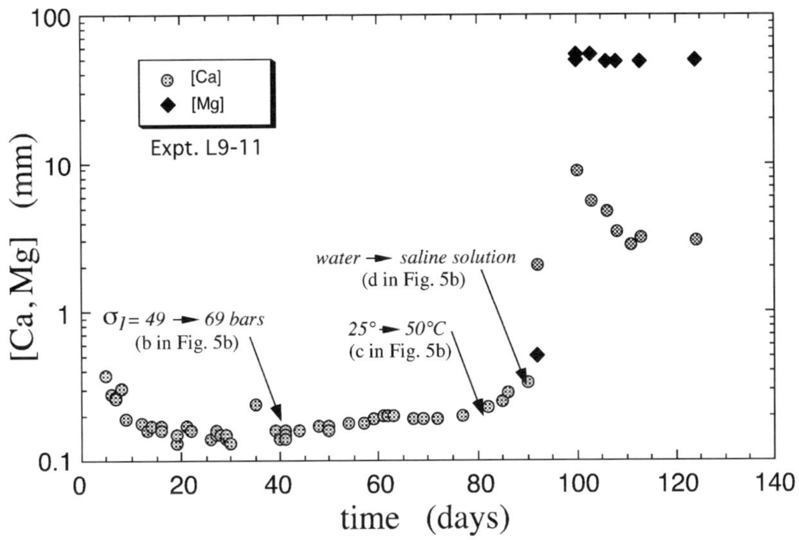

Fig. 7. Evolution of Ca and Mg content in output fluid, L9-11. Response of exit [Ca] to change in vertical stress is small, whereas changes in temperature and fluid chemistry significantly influence [Ca]. The magnitude of these changes is a function of calcite solubility. The response of $[Mg]_{exit}$ after the injection of Mg saline solution reveals that dispersion of fluids occurs within the matrix.

applied stress. The very small change in calcite solubility as a function of applied stress can be calculated using Eqn. 1b. A significant change in the vertical stress (several MPa) had no measurable effect on the exit solution [Ca] in most experiments; however, in some cases the effect was small, but nonetheless measurable. One such example is shown in Fig. 7, where [Ca] increases slightly due to a change in axial stress ($\sigma_1 = 4.9 \rightarrow \sigma_1 = 6.9$ MPa, $t > 40$ days). On the other hand, the aforementioned change in the axial stress induces a large increase in the strain and the strain rate, as can be seen in Fig. 5b. The various examples discussed above reveal an important point: the behavior of [Ca] in the exit solution does not correlate in a simple manner with changes in the deformation rate.

4.1.3. Fluid flow effects

Even though the influence of fluid flow on deformation rates was not explicitly studied in the present study, the effect was not overly significant. In general, when the experimental configuration was changed from open to closed conditions, there was no measurable effect that could be differentiated from the evolution of the 'background' compaction rate (i.e., stabilization). In one case in particular, however, there was a measurable effect when the system was changed from closed to open conditions: the steady-state rate of deformation (at t = 400 days, expt. S17) jumped from 2.4 x 10^{-10} to 4.8 x 10^{-10} s^{-1} when fluid flow was induced at υ = 0.5 ml $mins^{-1}$; the latter rate stabilized to 1.5 x 10^{-10} s^{-1} after a period of approximately 100 days. The sudden increase in the deformation rate may have been the direct result of an increase in the importance of free-face dissolution (due to a higher degree of pore fluid undersaturation), this resulting in undercutting of grains and matrix deformation.

The effect of fluid flow on deformation rates may vary, depending on the process that is rate-limiting. In the current experiments, the dependence was in most cases difficult to quantify and not significant. On the other hand, a study of the deformation of wet gypsum aggregates showed increases in the rate of compaction creep by a factor of 10-30x with fluid flow (de Meer and Spiers, 1997); this behavior was attributed to precipitation in pore spaces being the inferred rate-limiting step. Based on the above example, it is evident that fluid flow may change the rates of pressure solution deformation. For this reason, when examining strain vs. time data in experimental studies such as these (e.g., Figs. 5a, b), it is important to exclusively use values of $\dot{\varepsilon}$ that are derived from steady-state ε vs. time curves at conditions of zero fluid flow; this increases the probability that the fluids in the interconnected pore spaces are saturated (with respect to calcite in present study), such that measured deformation rates are a function only of pressure solution reactions (intergranular and/or free-face).

4.2. Overview of Rate-Determining Parameters

4.2.1. Constitutive relations

As was previously discussed, the current literature contains numerous models describing pressure solution deformation of geological materials. A brief discussion concerning an intergranular pressure solution model is presented in order to qualitatively analyze the various deformation rate-influencing parameters based on the present results. The constitutive relations developed by Rutter (1976) and Dewers and Ortoleva (1990a, 1990b) can be used as a representative example. Their models are based on diffusion in the trapped intergranular fluid being the rate-controlling step; a condensed version that was applied to quartz aggregates (Eqn. 7, Dewers and Hajash, 1995) shows that the strain rate ($\dot{\varepsilon}$) can be expressed in terms of the following parameters:

$$\dot{\varepsilon} = \frac{k' \, C^{eq}}{d^n \, (1-\varepsilon) A_D} \{ \exp(\alpha' P_e f) - 1 \} \tag{4}$$

where k' is a thermally-activated diffusion coefficient, C^{eq} is the solid solubililty at fluid P and T, d is the initial grain diameter (with $n = 1$), ε is the strain, A_D is the area of contact (equivalent to square of island diameter), α' is a temperature and molar volume-dependent coefficient, P_e is the effective stress, and f is a stress magnification term for relating the far field stress to mineral grains. The stress magnification term f is not a constant since it decreases as a function of increasing strain, such that the effective stress decreases as well, thereby causing the rate of strain $\dot{\varepsilon}$ to decrease with ongoing compaction. Considering d and A_D as constants is an approximation since they will increase during compaction, which in turn decreases the local stresses and the rate of strain $\dot{\varepsilon}$. The use of ε in constitutive equations is avoided by some workers in the field, who instead use porosity which is considered a state variable. However, the use of ε has the advantage that it is a directly measurable quantity during deformation, whereas porosity is not.

Equation 4, which contains most of the fundamental rate-determining parameters that are associated with pressure solution, can be used as a guide for examining the experimental results from the present study, based on data from a total of 9 experiments, including L9-11 and L16. A much more detailed analysis of these data, in the context of constitutive relations as well as the mechanisms of pressure solution, is the subject of a forthcoming communication (Hellmann et al., in prep.).

4.2.2. Effective stress

In the present study, it was assumed that the measured vertical strain is accommodated by mass transfer from the grain contacts perpendicular to σ_1 to free faces associated with pore spaces containing pore fluid at pressure P_f. For this reason, the effective stress has been defined as: $\sigma_{eff} = \sigma_1 - P_f \approx \sigma_1$; the effect of the pore pressure (in connected pore spaces) was neglected since it generally did not exceed ≈ 0.3 MPa at the inflow end of the core. In one experiment (S8-9), however, the effect of the pore pressure was investigated. A change in the pore pressure from 2.0 to 0.5 MPa (measured and controlled at the outflow end by the backpressure regulator, constant injection rate of 0.007 ml min^{-1}, σ_1 = 8.8 MPa) resulted in a sudden and transitory increase in the rate of deformation, followed by rapid stabilization. The steady-state strain rates before and after the pressure decrease were approximately the same.

The relation between the strain rate and the effective stress was determined by plotting $\dot{\varepsilon}$ vs. σ_1 in log-log space ($\dot{\varepsilon}$ at constant T, ε, chemistry of solutions). The instantaneous strain rates were based on tangents to strain-time curves (i.e., $\partial l / \partial t$ in Eqn. 3). The resulting slopes, in the majority of cases, ranged from 1-2. This range in values appears to be normal for pressure solution experiments at moderate effective stresses. As a comparison, Spiers et al. (1990) determined slopes of ≈ 0.8-1.8, based on the densification of salt at $\sigma_{eff} \leq 2.1$ MPa; at higher σ_{eff} (> 4 MPa), the dependence of $\dot{\varepsilon}$ on σ_{eff} was much higher, with slopes ranging from 2-4. Creep experiments using wet gypsum aggregates (de Meer and Spiers, 1995, 1997) produced slopes of ≈ 1.5-3.5. An investigation of the pressure solution deformation of quartz sand at 200 °C (Dewers and Hajash, 1995) also showed a large variation in slopes, ranging from 0.2-2.65 (based on linear slope segments in log $\dot{\varepsilon}$ - log σ_{eff} space, $7 < \sigma_{eff} < 48$ MPa). Based on all of the above data (except that of gypsum), it appears that there is quasi-linear relation between $\dot{\varepsilon}$ on σ_{eff} at low effective stresses; this has been pointed out by Rutter (1976), for example, who developed a constitutive relation based on $\dot{\varepsilon} \propto \sigma_{eff}$ at low effective stress values ($\sigma_{eff} < 30$ MPa). Nonetheless, beyond a certain critical limit in effective stress (which is dependent on the material, ε, T , etc.), the linear relation between $\dot{\varepsilon}$ vs. σ_{eff} gives way to a power law or exponential relationship, which appears to be compatible with the $\dot{\varepsilon}$ vs. σ_{eff} relationship given in Eqn. 4.

The relation $\dot{\varepsilon} \propto A\sigma_{eff}^m$, where A and m are constitutive parameters, is useful in elucidating the mechanisms of deformation. Solids displaying Newtonian rheological (viscous) behavior, by definition, are characterized by $m = 1$. Low values of m are typically related to diffusional transfer processes (Evans and Kohlstedt, 1995), such as Nabarro-Herring creep (matrix diffusion), Cobble creep (grain boundary diffusion), or pressure solution creep (dissolution and fluid diffusion in grain boundaries). However, values of $m > 2$ are commonly reported for pressure solution at elevated stress values. Whether this change in m is strictly linked to the influence of stress on solubility (as in Eqn. 4) or to evolution towards another deformation mechanism is not yet clear. For example, m = 3 or 5 is attributed to crystalline plasticity or creep via dislocation movement (see original references in section 4, Evans and Kohlstedt, 1995).

4.2.3. Grain size

According to most models in the literature where diffusion in the intergranular regions is the rate-limiting step, the strain rate varies inversely with the cube of the radius of the grains: $\dot{\varepsilon} \propto 1/d^{n=3}$. (Raj, 1982; Rutter, 1983; Lehner, 1990; Spiers and Brzesowsky, 1993). On the other hand, if interfacial kinetics are rate-controlling, the strain rate simply varies inversely with radius, and $n = 1$ (Raj, 1982). In Eqn. 4 the former relation between strain rate and grain size is implied (i.e., $n = 3$), but not explicitly formulated, due to the separate d and A_d terms. Halite densification data from Spiers and Brzesowsky (1993), for example, confirm the relation $\dot{\varepsilon} \propto 1/d^3$, indicating pressure solution creep rates limited by diffusion. In contrast, pressure solution experiments on wet gypsum aggregates yielded variable results, with $n \approx 1$ (de Meer and Spiers, 1995) and $n \approx 2$-4 (de Meer and Spiers, 1997). A quartz sand study by Dewers and Hajash (1995) showed that $\dot{\varepsilon} \propto 1/d^{2.3}$. Since there was no way to systematically vary the grain size in the chalk cores in the present study, the dependence of the strain rate on grain diameter could not be determined. In Eqn. 4, the diffusion parameter k (which includes the diffusion coefficient D and the width of the intergranular fluid film w), which in general cannot be measured directly, was not evaluated in the present study.

4.2.4. Strain and strain hardening

It is important to note that periods of deformation termed "steady-state" are not truly invariant with time. Taking an example from Fig. 5a, based on a regime of "steady-state" deformation in the presence of water (σ_1 = 6.0 and σ_3 = 4.0 MPa, T = 50 °C, zero fluid flow), the strain rate ($\dot{\varepsilon}$) decreases almost an order of magnitude with increasing time and compaction (ε), from 2.71 x 10^{-10} s^{-1} (ε = 0.0160, t = 78.20 days) to 3.62 x 10^{-11} s^{-1} (ε = 0.0178, t = 144.55 days). Based on an extrapolation of this strain vs. time data to 700 days, ε = 0.0185 and $\dot{\varepsilon}$ = 7.24 x 10^{-12} s^{-1} are predicted. This demonstrates that even over very long time periods (at least with respect to laboratory time scales), apparent "steady-state" rates of wet chalk deformation continue to evolve with time and therefore, $\dot{\varepsilon}$ is never constant. This strain hardening behavior demonstrates that the influence of strain on strain rates over long periods of time is not negligible, and hence, the term "steady-state" strain rate is actually only a valid approximation over small (local) time intervals, extending not more than a few tens of days.

In the constitutive relation given by Eqn. 4, the effect of the strain on strain rates is given by $\dot{\varepsilon} \propto 1/(1-\varepsilon)$. The strain magnification term (f in Eqn. 4) also influences the stress as a function of the strain. A constitutive relation developed by Spiers and Brzesowsky (1993) for salt aggregates is based on an inverse relation between the volumetric strain rate ($\dot{\beta}$) and strain (ε), such that $\dot{\beta} \propto 1/\varepsilon^c$ (note that $\dot{\beta} = 3\dot{\varepsilon}$). According to their experimental data, $c \approx 2$ for strains up to 10%, and $c \approx 4$-5 for strains of 15-20% (Spiers et al., 1990, Spiers and Brzesowsky, 1993). Thus, the inverse dependence of the strain rate on strain is much greater in their model than that proposed in Eqn. 4. In a similar manner, two gypsum aggregate studies yielded highly variable values, with c ≈ 4.3-7.5 (from Fig. 3, de Meer and Spiers, 1995), but were much lower in a latter study using gypsum, with c ≈ 2.4-4.2 (from Fig. 3, de Meer and Spiers, 1997). Quartz sand compaction rates measured by Dewers and Hajash (1995) showed a much smaller dependence on strain, with $\dot{\varepsilon} \propto 1/(1-\varepsilon)$, which is equivalent to the relation in Eqn. 4. In the present study, the dependence of the axial strain rate on strain (constant T, σ_1, chemistry of solutions) yields values of $c \approx 2$-5. The range in c reflects both experimental scatter and the heterogeneous deformation behavior of the chalk; nonetheless, this result shows the strong inverse dependence of the rate of deformation on strain. At first glance, comparing the above results for halite, quartz, and chalk compaction suggests that the solubility of the material may affect the dependence of the strain rate on strain and the strain hardening behavior. The different dependencies of $\dot{\varepsilon}$ on σ_{eff} may also result from different assumptions concerning the fundamental mechanisms of pressure solution (see discussion in Dewers and Hajash, 1995).

4.2.5. Chemistry of solutions

The effect of solution chemistry, which controls the solubility of chalk, is given by the C^{eq} term in Eqn. 4. In fact, most intergranular pressure solution models in the literature generally incorporate a linear dependence of the strain rate on the non-stressed solubility of the solid (see Eqn. 1d). This follows from the fact that most *ips* models equate the strain rate with the outward flux of dissolved material, which is a function of the concentration gradient.

A few studies have examined the effect of solution chemistry on strain rates. A study by Rutter (1976) showed that 3 salts were found to deform at rates qualitatively in accord with their solubilities, as predicted by Eqn. 4. A gypsum study by de Meer and Spiers (1997) showed that aggregates flooded with oil, which is a non-polar solvent, showed no measurable creep. Studies concerning the mechanical behavior of chalks also show that the chemical nature of the fluid strongly affects the yield strength (see section *4.3.1* below). As an example, triaxial tests using saturated chalk samples demonstrated that the salt content of aqueous fluids is directly related to the degree of weakening of the chalk (Da Silva et al., 1985; see Figs. 4-6 therein).

In the present study, the effect of solubility was investigated by changing the chemical nature of the fluid (water vs. Mg-saline solution, polar vs. non-polar solvent). The results show that chalk deformation rates are a strong function of the fluid chemistry. In comparison to calcite solubility in pure water, calcite is more soluble in the saline solution used, and less soluble in propanol (the changes in solubility are approximately one order of magnitude with respect to pure water). The influence of solution chemistry on the deformation behavior of chalk for expts. L16 and L9-11 was investigated by changing the chemical composition of the injected fluids several times (water $\rightarrow$ saline solution $\rightarrow$ propanol).

The first change in chemistry involved the replacement of water by the injection of a saline solution ("d" in Figs. 5a and b). The non-steady-state deformation behavior is characterized by an abrupt increase in the strain and strain rate, followed by a stabilization of the strain rate (Figs. 5a, b, and 8a). The change in deformation behavior is due to the higher solubility of calcite in saline solution vs. water. To quantify the effect of chemistry on steady-state strain rates, we examine the switch from water to saline solution in L16 (Fig. 5a, ε vs. time data on both sides of point "d"). The steady-state $\dot{\varepsilon}$ in water is $\approx 3.62 \times 10^{-11}$ s^{-1} ($\varepsilon = 0.0178$, $t = 144$ days, zero flow); after the switch to saline solution, $\dot{\varepsilon} \approx 6.03 \times 10^{-11}$ s^{-1} ($\varepsilon = 0.0189$, $t = 260$ days, zero flow). However, a meaningful comparison of $\dot{\varepsilon}_{water}$ *vs.* $\dot{\varepsilon}_{saline}$ must be made for the same strain (see discussion concerning strain hardening). One way this can be determined is by extrapolation of the $\dot{\varepsilon}_{water}$ vs. t curve forward in time, up to the point where $\varepsilon = 0.0189$; this results in $\dot{\varepsilon} \approx 2.78 \times 10^{-12}$ s^{-1}. Thus, comparing the two strain rates at the same strains reveals the dramatic effect of the solution chemistry: $\dot{\varepsilon}_{\text{saline}} \approx 20\dot{\varepsilon}_{\text{water}}$.

The second change in chemistry that is examined concerns L16 and a switch from saline solution to propanol. The effect that injection of a non-aqueous (non-polar) solvent, propanol, has on the rate of deformation is shown in Figs. 5a (point "h") and 8b (an enlargement around point "h"). In this case, the effect on the deformation rate is the opposite, the axial strain rate diminishes to the point where $\dot{\varepsilon}$ is no longer measurable (making a quantitative comparison of strain rates impossible). The drastic reduction in the strain rate is due to the very low solubility of calcite in propanol, a non-polar solvent.

The dependence of $\dot{\varepsilon}$ on the chemical nature of the fluid (solubility of solid) measured in the present study supports the relation $\dot{\varepsilon} \propto C^{eq}$ that is expressed in Eqn. 4. However, whether this relation is perfectly linear remains to be determined. Perhaps the most important implication of these results is that they offer additional evidence that pressure solution creep of chalk (and other materials) is not solely a mechanical process, but also depends on chemical reactions and transport processes occurring in intergranular interface regions.

4.2.6. Temperature

Temperature was the last parameter investigated. An increase in temperature has two main consequences with respect to the rate of pressure solution and chalk deformation. Since calcite solubility decreases with temperature (due to retrograde solubility), the solubility term acts to decrease the strain rate with increasing temperature. On the other hand, the rate of diffusion, D, increases with temperature. The temperature dependence of D can be expressed in terms of the Arrhenius relationship, $D \propto \exp(-E_a / RT)$, where E_a is the activation energy and R is the universal gas constant.

Diffusional processes generally have a weak dependence on temperature as a consequence of low activation energies. In the present study, an increase in temperature by several tens of °C always resulted in an abrupt and significant increase in $\dot{\varepsilon}$, followed by the establishment of a steady-state rate of deformation that was just slightly higher than the original, lower temperature rate (see Fig. 5a- point "f", Fig. 5b-point "c", and Fig. 9). Applying the Arrhenius relationship to deformation rates (i.e., plotting log $\dot{\varepsilon}$ vs. $T(\mathrm{K})^{-1}$) results in a preliminary estimate of $\approx$ 1-2 kJ mol^{-1} for E_a, using data from one experiment (S17) where the temperature was raised sequentially from 25 to 60 to 80 °C (Fig. 9). This estimate is based on rates of deformation that were normalized to the same strain; however, the rates were not normalized with respect to solubility as a function of temperature.

This estimate can be compared with activation energies in the literature. A study of grain boundary diffusion under wet conditions determined an activation energy of 15 kJ mol^{-1} (Nakashima, 1995). Two studies based on salt pressure solution yielded higher activation energies: 19 kJ mol^{-1} (KCl, Pharr and Ashby, 1983) and 24 kJ mol^{-1} (NaCl, Spiers et al., 1990). On the other hand, a study by Hickman and Evans (1995-see description in *Introduction*) found no temperature dependence. Nonetheless, as Hickman and Evans (1995) point out, great care must be taken when comparing pressure solution activation energies in the literature. In addition, many estimates of E_a are based on strain rate-temperature data which include not only the effects due to the thermal activation of diffusion, but also the effects of temperature on solubility; this is the case for the estimate from this study, for example.

The weak dependence of strain rates on temperature reflects the low E_a associated with diffusional processes in a fluid (E_a < 20-30 kJ mol^{-1} for diffusion-controlled kinetic processes; e.g., Lasaga, 1981). This implies diffusional transport in intergranular fluids, and not just in pore fluids. Not all studies have found low activation energies, however; Dewers and Hajash (1995) determined an activation energy of 73 kJ mol^{-1} for quartz sand compaction. This value is the same as that measured for quartz dissolution. This high activation energy indicates that for the case of quartz sand pressure solution, dissolution rather than diffusion should be the rate-determining step, at least to moderately elevated temperatures.

4.3. Fluid Flow and Interactions with Chalk Matrix

4.3.1. Fluid percolation behavior

The penetration of fluids into the matrix of samples is important for understanding the dynamics of fluid flow, fluid communication between pore spaces, as well as intergranular chemical interactions. Results from experiment L9-11 show (see Fig. 7), for example, how the output chemistry changes as a function of time when water (calcite-saturated) is replaced by Mg-bearing saline fluid (constant injection rate of 0.1 ml min^{-1}, theoretical Darcy velocity ≈ 0.0032 cm min^{-1}, total pore space volume ≈ 628 cm^3, avg. porosity ≈ 40%, 20 cm. core length). After switching from water to Mg-saline solution, the first Mg signal in the exit fluid is recorded after ≈ 2.1 days. The time necessary for [Mg] in the exit solution to match the [Mg] of the input solution is ≈ 9.8 days. Plotting $[Mg]_{exit}$ / $[Mg]_{input}$ vs. time results in what is termed a cumulative residence-time distribution curve *F(t)*; analysis of this *F(t)* curve (see details in Denbigh and Turner, 1984) yields a mean residence time of ≈ 5 days. On the other hand, based on an average porosity of 40%, and assuming that all pore spaces are accessible to the injected fluid and that pore spaces fill at an equal rate from bottom to top in the core, an injection rate of 0.1 ml min^{-1} corresponds to a theoretical residence time of ≈ 4.3 days. This situation corresponds to a chalk core that behaves as a plug flow reactor with no internal dispersion, such that a step change in the input chemistry remains a step change at the exit of the core. The behavior of the measured Mg *F(t)* curve (i.e., spread of *F(t)* about the mean) indicates that dispersion occurs within the chalk, such that some fluids bypass a significant portion of the matrix, whereas others percolate much more slowly. This indicates that a portion of the fluid flow is localized in channels and zones of enhanced porosity (e.g., Fig. 2 - Part 2), whereas the remaining fluid flow infiltrates the micropore space structure over much longer time scales (t > 10 days for the above conditions) than the mean residence time.

It is interesting to note that changes in the chemistry of the injected fluid do not induce an immediate change in $\dot{\varepsilon}$, but rather $\dot{\varepsilon}$ changes after a time lag. Based on several experiments where the fluid chemistry was changed, the time lag is variable, depending on the injection rate, the nature of fluid replacement (polar vs. non-polar), as well as such factors as the strain; two examples illustrate the influence of the two latter factors. The first example concerns the replacement of water by Mg-saline solution (two polar fluids) at a constant injection rate of 0.1 ml min^{-1} (point "d" - Fig. 5b, expt. L9-11). The strain vs. time curve (Fig. 8a) shows a modest increase (perturbation) in the strain rate ≈ 2.5 days after the injection of saline fluid, followed by a gradual stabilization of the rate. The second example that is discussed differs from the first in that Mg-saline fluid, which has a polar nature, is replaced by propanol, which is much less polar (point "h"- Fig. 5a, expt. L16; replacement at constant injection rate of 0.1 ml min^{-1}). As can be seen more clearly in Fig. 8b, the time lag associated with a measurable change in the strain rate is much greater, on the order of weeks. The difference in the response rates in these two examples is most probably a function of two variables: first, the difference in the polar nature of the injected fluid vs. the in-situ pore fluid, and second, the differences in the strain at the time of injection (0.0142 in expt. L9-11 and 0.0282 in expt. L16). The influence of the first factor can be attributed to the fluid's wettability characteristics; in addition, the miscibility of both fluids may also play a role. The second factor, the degree of strain, influences the porosity and permeability, which in turn controls the rate of access of injected fluids into pore spaces.

When a change in the fluid chemistry is effected, the rate of infiltration of new fluids into the micro-pore space matrix occurs over very long time scales, on the order of days to weeks (with respect to the two examples discussed earlier). The first measurable changes in strain rates most probably occur when only a fraction

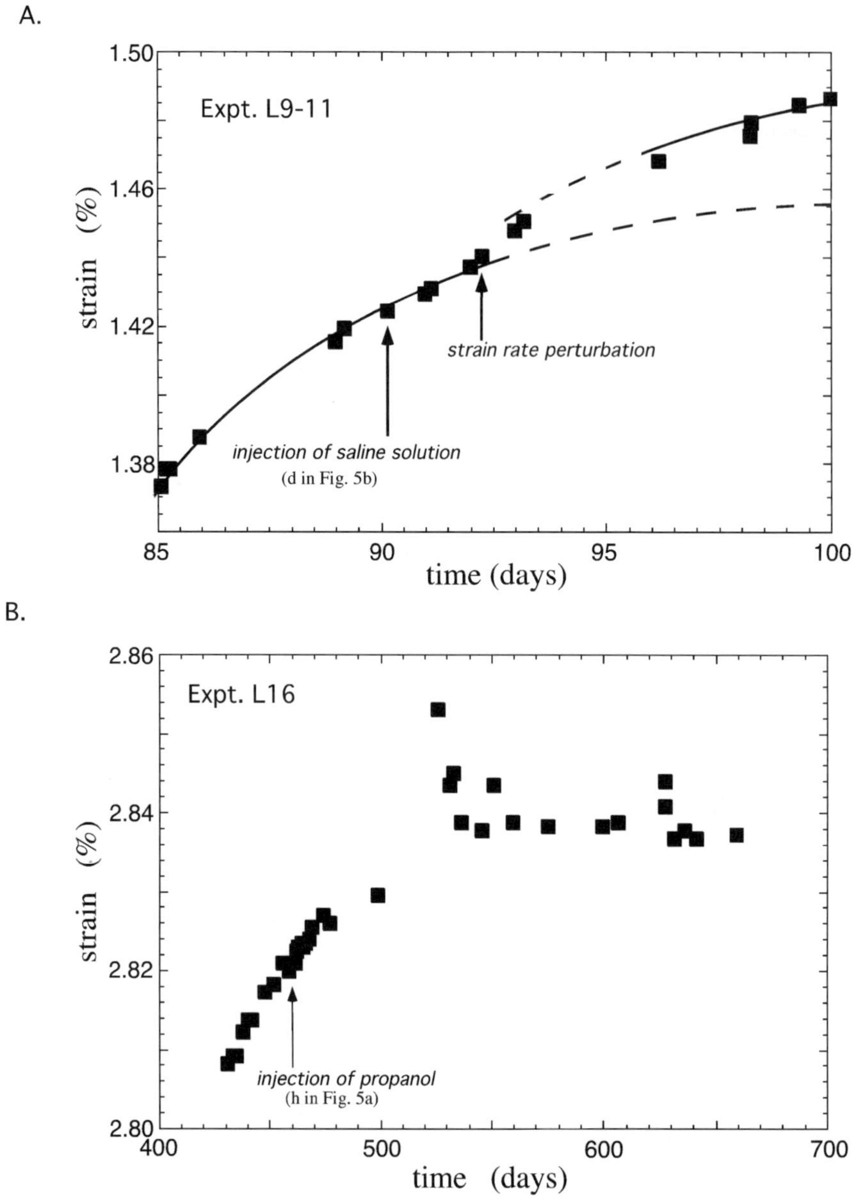

Fig. 8. Evolution of strain as function of chemistry of input solution. (A) Replacement of water by saline fluid results in an increase in strain rate, due to an increase in the solubility of chalk in saline solution. (B) Replacement of saline solution by propanol significantly decreases the rate of deformation, due to the low solubility of chalk in propanol. Excursions of data at $t > 500$ days due to unknown reason.

of the pore spaces have had their fluids exchanged, as estimated from cumulative residence time curves. As time progresses, strain rates continuously adjust to the saturation of all of the pore spaces with the injected fluid. It is, nonetheless, difficult to estimate what percentage of pore spaces is initially affected and the rate at which complete pore saturation occurs. Because of the continual compaction of the chalk, it is also difficult to deconvolute the effects of fluid chemistry changes with those due to increasing strain and compaction, especially since they operate in an inverse manner with respect to the strain rate (see Eqn. 4).

For a change in solution chemistry to affect the overall pressure solution process, not only must fluids infiltrate all of the pore spaces, but in addition, a diffusional process must occur from pore spaces to the

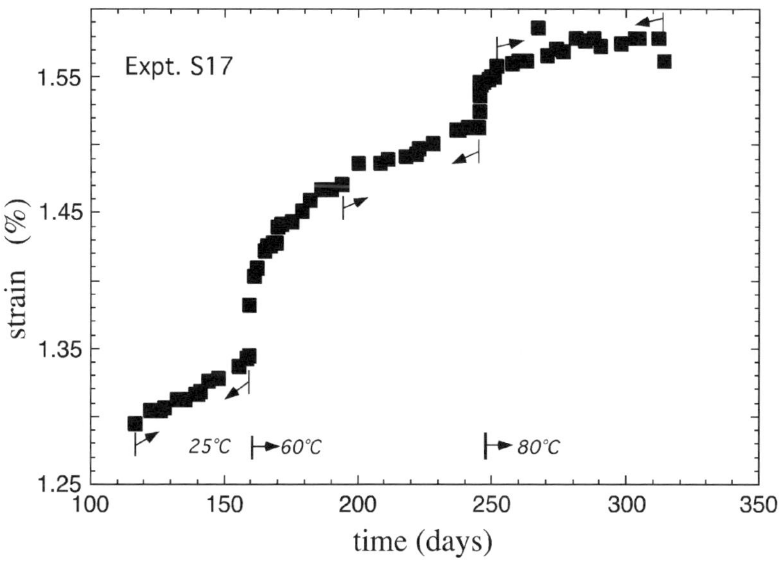

Fig. 9. Evolution of strain with increases in temperature; note the rapid changes in strain rate, followed by stabilization (local "steady states" delimited by arrows). Based on this data and the Arrhenius equation, the estimated activation energy is very low: ≈1-2 kJ mol^{-1}.

intergranular regions at stressed grain-to-grain contacts. However, the rate at which fluids penetrate and wet stressed grain-to-grain boundaries is uncertain, especially since the rates of diffusion in thin, trapped fluid films remain one of the biggest uncertainties with respect to the pressure solution mechanism. An additional complication arises from the fact that the progressive replacement of one fluid by another at stressed contacts requires the diffusion of aqueous species in the opposite direction with respect to the diffusion of dissolved material from intergranular regions to pore spaces (refer to Fig. 2); this raises the question whether the inward and outward diffusion of species are possibly coupled (note: coupled diffusion processes occur in a variety of geochemical processes, for example, the formation of leached layers in certain minerals and glasses- see e.g., Hellmann, 1997; and references therein). Fluid miscibility may also be an additional factor of importance with respect to the rate of exchange of fluids in pore spaces and intergranular regions.

To summarize the fluid percolation behavior in the present study, the slow response of strain rates (days to weeks) to changes in fluid chemistry is evidence that the deformation behavior of chalks is controlled primarily by intergranular reactions, and not rapid dissolution of free-face interfaces that are in contact with pore space fluids (refer to discussion on mechanisms of pressure solution). Free-face dissolution reactions in pore spaces should lead to changes in deformation rates with response times not longer than the mean residence times of fluids in the chalk cores; this estimate is based in part on the kinetics of chalk dissolution under isotropic conditions.

4.3.2. Chemical nature of fluids: mechanical vs. chemical effects

Because of the importance of oil reserves in chalk reservoirs (e.g., North Sea), many recent investigations have aimed to understand the role of different types of fluids in chalks, especially in terms of their ability to penetrate pore spaces and the effect they have on their mechanical strength. Among the key properties studied have been the wettability of interfaces and capillary forces (e.g., Andersen et al. 1992; Piau and Maury, 1994, 1995; Brignoli et al., 1995; Schroeder, 1995; Delage et al., 1996; Homand et al., 1998; Schroeder et al., 1998; Risnes and Flaageng, 1999; and references therein); these two properties are a function of the dielectric constant, the polarity, and the surface tension of the fluid.

Delage et al. (1996) and Schroeder and Shao (1996), for example, determined that the polar nature of the fluid affects the mechanical resistance of chalk; their results show that the upper limit of elastic deformation behavior (yield strength) is dependent on the nature of the pore fluid, and increases in the following order: seawater < water < methanol < oil, dodecane < absence of fluid. In quantitative terms, it was determined that the pore collapse threshold (i.e., this corresponds to the upper limit of elastic behavior) occurs at 5-7 MPa in water-saturated chalk samples (Schroeder, 1995) and 10-15 MPa in paraffin or kerosene-saturated samples (Delage et al., 1996). Delage et al. (1996) and Schroeder and Shao (1996) reason that chalks containing oil or dodecane are stronger than those containing seawater or water because of capillary effects, where the attractive forces between grains are due to menisci formed by connate water bridges between grains (Schroeder et al., 1998; Homand et al., 1998). Fluids that have a high wettability potential (e.g., water, seawater, and similar polar fluids) tend to decrease local capillary attraction between grains by destroying grain-to-grain menisci, which destabilizes the grain structure and results in pore collapse. Apparently when a non-polar fluid (oil) is replaced by a polar fluid (water), the transition from oil-saturated behavior (high mechanical resistance) to water-saturated behavior (lower mechanical resistance) is abrupt; 2% water saturation is sufficient for chalk to behave as if it were completely water-saturated (Schroeder et al., 1998).

Another series of tests concerning the mechanical strength of chalks was carried out with varying methanol-water mixtures (Risnes and Flaageng, 1999). Their study measured a monotonic increase in the yield strength of chalk (11 to 18 MPa, based on compression tests) as the mole % of methanol was varied from 0-100%. Risnes and Flaageng (1999) propose two mechanisms to explain their results: 1. water weakening of silica cement bonds within the chalk matrix 2. development of repulsive forces of electrostatic origin between grains (measured zeta potential values of chalk powder: water = -20mV, methanol = +10 mV, oil ≈ 0 mV).

It is interesting to note that the order of fluids with respect to increasing mechanical resistance of chalk (see above, Delage et al., 1996 and Schroeder and Shao, 1996) follows qualitatively the solubility of chalk in these fluids, in decreasing order; in other words, the mechanical strength varies inversely with solubility. Based on the results in the present study and many other pressure solution models, the rate of deformation is proportional to the solubility (see Eqn. 4), such that chalk deforms at a much higher rate in the presence of seawater than in the presence of an alcohol, for example (see evidence for this in Figs. 8a, b). Thus, the overall effect of the chemistry of fluids on deformation behavior is similar between models based on capillary forces and pressure solution (i.e., as in the present study), the main difference being that the former is based on *physico-mechanical interactions* between grains, whereas pressure solution is based on *chemical dissolution reactions* in stressed intergranular regions.

4.3.3. Application of numerical model results: dissolution and precipitation in porous media

The dissolution and precipitation behavior of solutes in porous media and fracture systems has recently been examined in a series of numerical modeling studies (e.g., Sallès et al., 1993; Békri et al., 1995, 1997; Mourzenko et al., 1996). The results from these studies show that the geometry of pore and fracture spaces, as well as two associated dimensionless parameters, the Péclet and Péclet-Damköhler numbers, are important in the evolution of the hydrodynamic and chemical behavior of the system. The Péclet number (Pe) is a comparative measure of the convective vs. the diffusion characteristics of the system; Pe < 1 indicates diffusion is predominant, whereas convection predominates when Pe > 1. The Damköhler number (Da) is a comparative measure of kinetics (rates of dissolution or precipitation) vs. convection; when Da < 1 convection predominates, and for Da > 1, kinetics predominates. Finally, the Péclet-Damköhler number is a measure of kinetics vs. diffusion, such that if PeDa < 1, diffusion is more rapid and therefore the process is reaction rate-limited, and conversely, when PeDa > 1, kinetics are more rapid, such that diffusion is rate-limiting. The respective definitions of Pe, Da, and PeDa, as applied to porous media, are defined as follows:

$$\mathrm{Pe} = \frac{\bar{v}_0^* b_m}{D}; \qquad \mathrm{Da} = \frac{K}{\bar{v}_0^*}; \qquad \mathrm{PeDa} = \frac{K b_m}{D} \tag{5}$$

where $\bar{v}_0^*$ is the average interstitial velocity in the porous medium (or fracture), b_m is the initial aperture opening of the pore space or fracture, K is the kinetic rate constant, and D is the coefficient of diffusion (Mourzenko et al., 1996). The fluxes associated with either dissolution or precipitation at a connected pore space or fracture wall have been related to the interfacial solute concentration by PeDa (Békri et al., 1995, 1997). According to the results of Sallès (1993) and Békri et al. (1995), when PeDa < 1, precipitation is reaction-limited and occurs uniformly throughout the medium; on the other hand, if PeDa > 1, precipitation is predicted to occur either in the widest parts of pore spaces (Pe < 1) or along the main flow paths (Pe > 1).

In the present study, Pe was very small, on the order of 10^{-4}, based on $b_m \approx 1$ μm, $\bar{v}_0^* \approx 10^{-7}$ m s^{-1}, and $D \approx 10^{-9}$ m^2 s^{-1}. The determination of PeDa is more difficult, since the rate of precipitation (K) is not known. However, K is with certainty < 10^{-3} m s^{-1}, such that even if an exact value cannot be determined, PeDa << 1. According to the numerical results discussed immediately above, this corresponds to a situation of uniform precipitation throughout the matrix. This predicted result has relevance to the present study with respect to the high density/low porosity zones observed in the interfacial regions of L9-11 and L16. The CT image in Fig. 16-Part 2 (and Fig. 18-Part 2 to a lesser degree) does show a uniform zone of high density which can be attributed to precipitation. The localized nature of these zones, as well as their positions, cannot however be predicted with the models discussed above. Despite the mathematical complexity of these numerical approaches, precipitation processes are dependent on many critical factors that have not been incorporated, such as kinetic and nucleation parameters.

5. CONCLUSIONS

Chalk subjected to moderate levels of differential stress in the presence of fluids undergoes time-dependent deformation and compaction that can be primarily attributed to pressure solution creep. The strain and the derived strain rates were measured as a function of several variables: effective differential stress, strain, solubility of the solid, and temperature. The linear relationship between the strain rate and vertical stress (in log-log space, with slopes $m \approx$ 1-2) and the direct influence of the chemistry of the fluid on the rate of deformation strongly support a pressure solution mechanism. The inverse relation between the strain rate and the strain points out that time-dependent strain hardening is an important experimental parameter during the compaction of porous materials such as chalk. The dependence of the deformation rate on temperatures is very weak (measured up to 80 °C in the present study). The chemical nature of chalk deformation, based on pressure solution creep, is a concept which has been largely ignored in most mechanical models of chalk deformation. Perhaps one of the most important contributions of this study is to enlarge the debate concerning the nature of chalk deformation mechanisms.

While the long-term creep behavior of chalk points to a pressure solution mechanism, it is possible that this mechanism is assisted by others when the chalk-fluid 'system' is perturbed (i.e., non-steady-state behavior). As shown by the strain vs. time curves in the present study, the strain rates nearly always changed in an abrupt manner after a change in temperature, solution chemistry, or effective stress. These periods of rapid deformation were always transitory, since they gave way to very slowly varying, time-dependent creep (often referred to as "steady-state"). The rapid and very transitory nature of these perturbations may be related to grain-grain sliding, due to the partial dissolution and subsequent breakage of cement bridges connecting the grains. The loss of cohesion of cement bridges can be envisioned to be a physico-chemical process, where rapid free-face dissolution of bridges leads to mechanical failure and concomitant grain-grain sliding. In natural systems, rate-influencing parameters should not be subject to rapid changes, such that pressure solution creep should be the dominant deformation process at moderate levels of stress in wet chalks. However, rapidly applied external perturbations may be pertinent under some conditions, such as associated with oil exploitation operations in chalks (e.g., North Sea) or also with certain tectonic conditions (e.g., seismic activity, pulsed fluid flow in fault planes). In such cases, transient strain rate jumps are potentially a function of multiple deformation mechanisms, such as pressure solution creep, grain-grain sliding, brittle failure, and perhaps other modes of deformation, as well. The partitioning of strain among these modes of deformation merits additional study.

Acknowledgements—This research was supported by generous grants from Elf Aquitaine (Pau, France) and the GdR-Géomécanique program (CNRS). Comments from M. Jessell, C. Schroeder, F. Renard, and A.-M. Boullier aided in clarifying certain points in the original manuscript. We thank Brian Evans, Enrico Merino, and Andrew Hajash for their constructive reviews of the manuscript (note: the original reviews were based on a single manuscript, before it was divided into Parts 1 and 2; however, the scientific content and interpretations remain unchanged). R.H. also thanks Brian Evans for assuming the task of editorial handling.

Finally, I would like to acknowledge David Crerar's role as teacher, advisor, and friend during the 6 years of my graduate studies in his geochemistry group in the Department of Geology and Geophysics, Princeton University. -Roland Hellmann

Editorial handling: Brian Evans

REFERENCES

Andersen M. A., Foged N., and Pedersen H. F. (1992) The link between waterflooding-induced compaction and rate sensitive behavior in a weak North Sea Chalk. *4th North Sea Chalk Symposium*. Deaville, France.

Angevine C. L., Turcotte D. L., and Furnish M. D. (1982) Pressure solution lithification as a mechanism for the stick-slip behavior of faults. *Tectonics* **1**, 151-160.

Bathurst R. C. G. (1958) Diagenetic fabrics in some British Dinantian limestones. *Liv. and Manch. Geol. J.* **2**, 11-36.

Bayly B. (1985) Deformation with simultaneous chemical change: The thermodynamic basis. In *Metamorphic Reactions Kinetics, Textures, and Deformation* (eds. A. B. Thompson and D. C. Rubie). Advances in Physical Geochemistry 4, Springer Verlag, New York. pp. 269-277.

Bayly B. (1987) Nonhydrostatic thermodynamics in deforming rocks. *Can. J. Earth Sci.* **24**, 572-579.

Békri S., Thovert J. F., and Adler P. M. (1995) Dissolution of porous media. *Chem. Eng. Sci.* **50**, 2765-2791.

Békri S., Thovert J.-F., and Adler P. M. (1997) Dissolution and deposition in fractures. *Eng. Geol.* **48**, 283-308.

Botter B. J. (1985) Pore collapse measurements on chalk cores. *2nd North Sea Chalk Symposium*. Deauville, France.

Brignoli M., Papamichos E., and Santarelli F. J. (1995) Capillary effects in sedimentary rocks: application to reservoir water-flooding. *35th U.S. Symposium on Rock Mechanics*. Lake Tahoe, California.

Busenberg E. and Plummer L. N. (1986) A comparative study of the dissolution and crystal growth kinetics of calcite and aragonite. In *Studies in Diagenesis* (ed. F. A. Mumpton). U.S.G.S. Bulletin 1578, U.S. Geological Survey. pp. 139-168.

Carrio-Schaffhauser E. and Gaviglio P. (1990) Pressure solution and cementation stimulated by faulting in limestones. *J. Struct. Geol.* **12**, 987-994.

Chen W.-P. and Molnar P. (1983) Focal depths of intracontinental and intraplate earthquakes and their implications for the thermal and mechanical properties of the lithosphere. *J. Geophys. Res.* **88, B5**, 4183-4214.

Cox S. F. and Patterson M. S. (1991) Experimental dissolution-precipitation creep in quartz aggregates at high temperatures. *Geophys. Res. Letters* **18**, 1401-1404.

Da Silva F., Monjoie A., Debande G., Schroeder C., Poot B., Detiege C., and Halleux L. (1985) Mechanical behaviour of chalks. *2nd North Sea Chalk Symposium*. Stavanger, Norway, **2**, 1-10.

de Boer R. B. (1977) Pressure solution: Theory and experiments. *Tectonophysics* **39**, 287-301.

de Boer R. B., Nagtegaal P. J. C., and Duyvis E. M. (1977) Pressure solution experiments on quartz sands. *Geochim. Cosmochim. Acta* **41**, 257-264.

Delage P., Cui Y. J., and Schroeder C. (1996) Subsidence and capillary effects in chalks. *Eurock '96, ISRM International Symposium*. Torino, Italy, 1291-1298.

de Meer S. and Spiers C. J. (1995) Creep of wet gypsum aggregates under hydrostatic loading conditions. *Tectonophysics* **245**, 171-183.

de Meer S. and Spiers C. J. (1997) Uniaxial compaction creep of wet gypsum aggregates. *J. Geophys. Res.* **102, no. B1**, 875-891.

den Brok S. W. J. (1996) The effect of crystallographic orientation on pressure solution in quartzite. *J. Struct. Geol.* **18**, 859-860.

den Brok S. W. J. (1998) Effect of microcracking on pressure-solution strain rate: The Gratz grain-boundary model. *Geology* **26**, 915-918.

den Brok S. W. J. and Spiers C. J. (1991) Experimental evidence for water weakening of quartzite by microcracking and solution creep. *Geol. Soc. J.* **148**, 541-548.

Denbigh K. G. and Turner J. C. R. (1984) *Chemical Reactor Theory*. Cambridge University Press, Cambridge.

Dewers T. and Hajash A. (1995) Rate laws for water-assisted compaction and stress-induced water-rock interaction in sandstones. *J. Geophys.Res.* **100, no. B7**, 13093-13112.

Dewers T. and Ortoleva P. J. (1989) Mechano-chemical coupling in stressed rocks. *Geochim. Cosmochim. Acta* **53**, 1243-1258.

Dewers T. and Ortoleva P. (1990a) A coupled reaction/transport/mechanical model for intergranular pressure solution, stylolites, and differential compaction and cementation in clean sandstones. *Geochim. Cosmochim. Acta* **54**, 1609-1625.

Dewers T. and Ortoleva P. (1990b) Geochemical self-organization. III. A mechano-chemical model of metamorphic differentiation. *Am. J. Sci.* **290**, 473-521.

Dewers T. and Ortoleva P. (1991) Influences of clay minerals on sandstone cementation and pressure solution. *Geology* **19**, 1045-1048.

Elias B. P. and Hajash A. J. (1992) Changes in quartz solubility and porosity due to effective stress: An experimental investigation of pressure solution. *Geology* **20**, 451-454.

Elliot D. (1973) Diffusion flow law in metamorphic rocks. *Geol. Soc. Am. Bull.* **84**, 2645-2664.

Engelder T. (1982) A natural example of the simultaneous operation of free-face dissolution and pressure solution. *Geochim. Cosmochim. Acta* **46**, 69-74.

Evans B. and Kohlstedt D. L. (1995) Rheology of rocks. In *Rock Physics and Phase Relations* (ed. T. J. Ahrens). AGU Reference Shelf 3, American Geophysical Union, Washington, D.C. pp. 148-165.

Evans B., Bernabé Y., and Zhu W. (1999) Evolution of pore structure and permeability of rocks in laboratory experiments. In *Growth, Dissolution and Pattern Formation in Geosystems* (eds. B. Jamtveit and P. Meakin). Kluwer Academic Publishers, Dordrecht. pp. 327-344.

Gaviglio P., d'Albissin M. C., Bergerat F., and Vandycke S. (1993) Modifications de texture dans la craie au contact de failles normales: un exemple de graben dans le basin de Mons (Belgique). *Bull. Soc. géol. France* **164**, 565-575.

Gaviglio P., Adler P., Thovert J.-F., Vandycke S., Bergerat F., Békri S., and Lestideau R. (1997) Grain-scale microstructure and physical properties of faulted chalk. *Bull. Soc. géol. France* **168**, 727-739.

GBMR (1995) Craies et Schistes Krijt en Leisten Kreide und Schiefer Chalk and Shales. Colloquium Mundanum Proceedings, Brussels, Belgium.

Gibbs J. W. (1878) On the equilibrium of heterogeneous substances. Trans. Conn. Academy, III. In *The Scientific Papers of J. Willard Gibbs, Vol. 1*. Longman, Green, and Co., Toronto. pp. 343-524.

Gratier J.-P. (1993) Experimental pressure solution of halite by an indenter technique. *Geophys. Res. Letters* **20**, 1647-1650.

Gratier J.-P. and Gamond J. F. (1990) Transition between seismic and aseismic deformation in the upper crust. In *Deformation mechanisms, rheology and tectonics* (eds. R. J. Knipe and E. H. Rutter). Geological Society of London Special Publication 54, The Geological Society, London. pp. 461-473.

Gratier J.-P. and Guiguet R. (1986) Experimental pressure solution-deposition on quartz grains: the crucial effect of the nature of the fluid. *J. Struct. Geol.* **8**, 845-856.

Gratier J.-P., Chen T., and Hellmann R. (1994) Pressure solution as a mechanism for crack sealing around faults: Natural and experimental evidence. In *Proceedings of Workshop LXIII The Mechanical Involvement of Fluids in Faulting* (eds. S. Hickman, R. Sibson, and R. Bruhn). Open-file Report 94-228, U.S.G.S., Menlo Park. pp. 279-300.

Gratier J.-P., Renard F., and Labaume P. (1999) How pressure solution creep and fracturing processes interact in the upper crust to make it behave in both a brittle and viscous manner. *J. Struct. Geol.* **21**, 1189-1197.

Gratz A. J. (1991) Solution-transfer compaction of quartzites: Progress toward a rate law. *Geology* **19**, 901-904.

Hajash A. J. and Bloom M. (1991) Marine diagenesis of feldspathic sand: A flow-through experimental study at 200 °C, 1 kbar. *Chem. Geol.* **89**, 359-377.

Heidug W. K. (1995) Intergranular solid-fluid phase transformations under stress: The effect of surface forces. *J. Geophys. Res.* **100**, 5931-5940.

Heidug W. K. and Leroy Y. M. (1994) Geometrical evolution of stressed and curved solid-fluid phase boundaries 1. Transformation kinetics. *J. Geophys. Res.* **99**, 505-515.

Hellmann R. (1997) The albite-water system Part IV. Diffusion modeling of leached and hydrogen-enriched layers. *Geochim. Cosmochim. Acta* **61**, 1595-1611.

Hellmann R., Gratier J.-P., and Renders P. (1996) Deformation of chalk by pressure solution. *V.M. Goldschmidt Conference*. Heidelberg, Germany, **1**, 248.

Hellmann R., Gratier J.-P., and Chen T. (1998) Mineral-water interactions and stress: Pressure solution of halite aggregates. In *Water-Rock Interaction WRI-9* (eds. G. B. Arehart and J. R. Hulston). A.A. Balkema, Rotterdam. pp. 777-780.

Hickman S. and Evans B. (1995) Kinetics of pressure solution at halite-silica interfaces and intergranular films. *J. Geophys. Res.* **100**, 13113-13132.

Hickman S., Sibson R., and Bruhn R. (1995) Introduction to special section: Mechanical involvement of fluids in faulting. *J. Geophys. Res.* **100**, no. B7, 12831-12840.

Homand S., Shao J. F., and Schroeder C. (1998) Plastic modelling of compressible porous chalk and effect of water injection. *Eurock '98*. Trondheim, Norway, **2**, 495-504.

Hudson J. D. (1975) Carbon isotopes and limestone cement. *Geology* **3**, 19-22.

Jones M. E. and Leddra M. J. (1989) Compaction and flow characteristics of porous chalks. *Journée Craie*. Université de Lille, Lille, France.

Kamb W. B. (1959) Theory of preferred crystal orientation developed by crystallization under stress. *J. Geol.* **67**, 153-170.

Kamb W. B. (1961) The thermodynamic theory of nonhydrostatically stressed solids. *J. Geophys. Res.* **66**, 259-271.

Lasaga A. C. (1981) Rate laws of chemical reactions. In *Kinetics of Geochemical Processes* (eds. A. C. Lasaga and R. J. Kirkpatrick). Reviews in Mineralogy 8, Mineralogical Society of America, Washington, D.C. pp. 1-68.

Lehner F. K. (1990) Thermodynamics of rock deformation by pressure solution. In *Deformation Process in Minerals, Ceramics, and Rocks* (eds. D. J. Barber and P. G. Meredith). Unwin-Hyman, London. pp. 296-333.

Lehner F. K. (1995) A model for intergranular pressure solution in open systems. *Tectonophysics* **245**, 153-170.

Lehner F. K. and Bataille J. (1985) Nonequilibrium thermodynamics of pressure solution. *Pure Appl. Geophys.* **122**, 53-85.

Lemée C. and Guéguen Y. (1996) Modeling of porosity loss during compaction and cementation of sandstones. *Geology* **24**, 875-878.

Leroy Y. M. and Heidug W. K. (1994) Geometrical evolution of stressed and curved solid-fluid phase boundaries 2. Stability of cylindrical pores. *J. Geophys. Res.* **99**, 517-530.

Megnien C. (1980) Synthèse géologique du Bassin de Paris, *Mém. BRGM*, 101.

Merino E., Ortoleva P., and Strickholm P. (1983) Generation of evenly-spaced pressure-solution seams during (late) diagenesis: A kinetic theory. *Contrib. Mineral. Petrol.* **82**, 360-370.

Monjoie A., Schroeder C., Prignon P., Yernaux C., Silva F. D., and Debande G. (1990) Establishment of constitutive law of chalk and long term test. *3rd North Sea Chalk Symposium*. Copenhagen, Denmark, 1-17.

Monjoie A., Schroeder C., and Da Silva F. (1991) Testing procedure for time-dependent behaviour of chalk. *7th Intl. Congr. International Congress on Rock Mechanics*. Aachen, Germany, **7**, 565-567.

Mourzenko V. V., Békri S., Thovert J.-F., and Adler P. M. (1996) Deposition in fractures. *Chem. Eng. Communications* **148-150**, 431-464.

Mullis A. J. (1991) The role of silica preciptation kinetics in determining the rate of quartz pressure solution. *J. Geophys. Res.* **96**, 10007-10013.

Mullis A. M. (1993) Determination of the rate-limiting mechanism for quartz pressure solution. *Geochim. Cosmochim. Acta* **57**, 1499-1503.

Nakashima S. (1995) Diffusivity of ions in pore water as a quantitative basis for rock deformation rate estimates. *Tectonophys.* **245**, 185-203.

Ortoleva P., Chadam J., Merino E., and Sen A. (1987a) Geochemical self-organization II: The reactive-infiltration instability. *Am. J. Sci.* **287**, 1008-1040.

Ortoleva P., Merino E., Moore C., and Chadam J. (1987b) Geochemical self-organization I: Reaction-transport feedbacks and modeling approach. *Am. J. Sci.* **287**, 979-1007.

Paterson M. S. (1973) Nonhydrostatic thermodynamics and its geologic applications. *Rev. Geophys. Space Phys.* **11**, 355-389.

Paterson M. S. (1995) A theory for granular flow accomodated by material transfer via an intergranular fluid. *Tectonophys.* **245**, 135-151.

Pecqueur G., Mikolajczak A., and Siwak J. M. (1995) Etude expérimentale de deux craies en torsion. In *Craies et Schistes Krijt en Leisteen Kreide und Schiefer Chalk and Shales* (ed. GBMR). Colloquium Mundanum Proceedings, Brussels, Belgium. pp. 1.1.44-1.1.52.

Pharr G. M. and Ashby M. F. (1983) On creep enhanced by a liquid phase. *Acta Mettallica* **31**, 129-138.

Piau J. M. and Maury V. (1994) Mechanical effects of water injection in chalk reservoirs. *Proceedings Eurock '94* SPE/ISRM Intl. Conference. pp. 819-828.

Piau J. M. and Maury V. (1995) Basic mechanical modelisation of chalk/water interaction. In *Unsaturated Soils* (eds. Alonso and Delage). pp. 775-783

Plummer L. N. and Busenberg E. (1982) The solubilities of calcite, aragonite and vaterite in CO_2-H_2O solutions between 0 and 90 °C, and an evaluation of the aqueous model for the system $CaCO_3$ -CO_2 - H_2O. *Geochim. Cosmochim. Acta* **46**, 1011-1040.

Raj R. (1982) Creep in polycrystalline aggregates by matter transport through a liquid phase. *J. Geophys. Res.* **87 no. B6**, 4731-4739.

Raj R. and Chyung C. K. (1981) Solution-precipitation creep in glass ceramics. *Acta Metall.* **29**, 159-166.

Renard F. and Ortoleva P. (1997) Water films at grain-grain contacts: Debye-Hückel, osmotic model of stress, salinity, and mineralogy dependence. *Geochim. Cosmochim. Acta* **61**, 1963-1970.

Renard F., Park A., Ortoleva P., and Gratier J.-P. (1999) An integrated model for transitional pressure solution in sandstones. *Tectonophys.* **312**, 97-115.

Reuschlé T., Trotignon L., and Gueguen Y. (1988) Pore shape evolution by solution transfer: thermodynamics and mechanics. *Geophys. J.* **95**, 535-547.

Risnes R. and Flaageng O. (1999) Mechanical properties of chalk with emphasis on chalk-fluid interactions and micromechanical aspects. *Oil & Gas Science and Technology - Rev. IFP* **54**, 751-758.

Robin P.-Y. F. (1978) Pressure solution at grain-to-grain contacts. *Geochim. Cosmochim. Acta* **42**, 1383-1389.

Robin P.-Y. F. (1979) Theory of metamorphic segregation and related processes. *Geochim. Cosmochim. Acta* **43**, 1587-1600.

Rutter E. H. (1976) The kinetics of rock deformation by pressure solution. *Phil. Trans. R. Soc. Lond.* **A, 283**, 203-219.

Rutter E. H. (1983) Pressure solution in nature, theory and experiment. *J. Geol. Soc. London* **140**, 725-740

Rutter E. H. and Mainprice D. H. (1979) On the possibility of slow fault slip controlled by diffusive mass transfer processes. *Gerlands Beitr. Geophys.* **88**, 154-162.

Sallès J., Thovert J. F., and Adler P. M. (1993) Deposition in porous media and clogging. *Chem. Eng. Sci.* **48**, 2839-2858.

Schneider F., Potdevin J. L., Wolf S., and Faille I. (1996) Mechanical and chemical compaction model for sedimentary basin simulators. *Tectonophys.* **263**, 307-317.

Schroeder C. (1995) Le "pore collapse": Aspect particulier de l'interaction fluide-squelette dans les craies? In *Craies et Schistes Krijt en Leisteen Kreide und Schiefer Chalk and Shales*. (ed. GBMR). Colloquium Mundanum Proceedings, Brussels, Belgium. pp. 1.1.53-1.1.60.

Schroeder C., Bois A.-P., Maury V., and Hallé G. (1998) Water/chalk (or collapsible soil) interaction: Part II. Results of tests in laboratory on Lixhe chalk to calibrate water/chalk models. *Eurock '98*. Trondheim, Norway, **2**, 505-514 (SPE 47587).

Schroeder C. and Shao J. (1996) Plastic deformation and capillary effects in chalks. *5th North Sea Chalk Symposium*. Reims, France, 1-14.

Schutjens P. M. T. M. (1991) Experimental compaction of quartz sand at low effective stress and temperature conditions. *J. Geol. Soc. London* **148**, 527-539.

Schutjens P. M. T. M. and Spiers C. J. (1999) Intergranular pressure solution in NaCl: Grain-to-grain contact experiments under the optical microscope. *Oil & Gas Science and Technology - Rev. IFP* **54**, 729-750.

Shao J. F., Bederiat M., and Schroeder C. (1994) Elasto-viscoplastic modelling of a porous chalk. *Mechanics Res. Comm.* **21**, 63-75.

Sleep N. H. and Blanpied M. L. (1992) Creep, compaction and the weak rheology of major faults. *Nature 359*, 687-692.

Sorby H. C. (1863) On the direct correlation of mechanical and chemical forces. *Proc. Roy. Soc.* **12**, 538.

Spiers C. J. and Brzesowsky R. H. (1993) Densification behaviour of wet granular salt: Theory versus experiment. In *Seventh Symposium on Salt*, 1. Elsevier Science Publishers, Amsterdam. pp.82-92.

Spiers C. J. and Schutjens P. M. T. M. (1990) Densification of crystalline aggregates by fluid-phase diffusional creep. In *Deformation Processes in Minerals, Ceramics and Rocks* (eds. D. J. Barber and P. G. Meredith). Unwin Hyman, London. Chap. 12.

Spiers C. J., Schutjens P. M. T. M., Brzesowsky R., Peach C. J., Liezenberg J. L., and Zwart H. J. (1990) Experimental determination of constitutive parameters governing creep of rocksalt by pressure solution. In *Deformation Mechanisms, Rheology and Tectonics* (eds. R. J. Knipe and E. H. Rutter). Geological Society Special Publication No. 54, The Geological Society, London. pp. 215-227.

Tada R. and Siever R. (1986) Experimental knife-edge pressure solution of halite. *Geochim. Cosmochim. Acta* **50**, 29-36.

Tada R. and Siever R. (1989) Pressure solution during diagenesis: A review. *Ann. Rev. Earth Planet. Sci.* **17**, 89-118.

Tada R., Maliva R., and Siever R. (1987) A new mechanism for pressure solution in porous quartzose sandstone. *Geochim. Cosmochim. Acta* **51**, 2295-2301.

Urai J. L., Spiers C. J., Zwart H. J., and Lister G. S. (1986) Water weakening effects in rock salt during long term creep. *Nature* **324**, 554-557.

Weyl P. K. (1959) Pressure solution and the force of crystallization - a phenomenological theory. *J. Geophys. Res.* **64**, 2001-2025.

Wong P. K. and Oldershaw A. (1981) Burial cementation in the Devonian, Kaybob reef complex, Alberta, Canada. *J. Sediment. Petrol.* **51**, 507-520.

Water-Rock Interactions, Ore Deposits, and Environmental Geochemistry: A Tribute to David A. Crerar
© The Geochemical Society, Special Publication No. 7, 2002
Editors: Roland Hellmann and Scott A. Wood

Experimental pressure solution compaction of chalk in aqueous solutions. Part 2. Deformation examined by SEM, porosimetry, synthetic permeability, and X-ray computerized tomography

ROLAND HELLMANN[1*], PATRICK GAVIGLIO[2], PETER J.N. RENDERS[1**], JEAN-PIERRE GRATIER[1], SAMIR BÉKRI[3 ***], PIERRE ADLER[3]

[1]Crustal Fluids and Dynamics Group, LGIT, CNRS UMR C5559, Observatoire de Grenoble, Université J. Fourier, BP 53X, 38041 Grenoble Cedex 9, FRANCE
[2]Département de Géosciences, Université de Franche-Comté, 16, route de Gray, 25030 Besançon Cedex, FRANCE
[3]Institut de Physique du Globe de Paris, 4, place Jussieu, 75252 Paris Cedex 05, FRANCE
** present address: 211 Beekman St., Saratoga Springs, N.Y., 12866, U.S.A.
*** present address: Institut Français du Petrole, Rueil Malmaison, France

* corresponding author (hellmann@obs.ujf-grenoble.fr)

Abstract—To investigate pressure solution creep, chalk from the Paris basin (France) was deformed in a triaxial press with fluids. Experiments were conducted either under chemically closed (no fluid flow) or chemically open (fluid flow) conditions. In all experiments, the vertical stress (σ_1) and the lateral stress (σ_3) were independently controlled, with $4.0 < \sigma_1 \leq 8.0$ MPa and $\sigma_3 = 4.0$ MPa; the pore fluid pressure was ≤ 0.3 MPa. In each experiment, axial strain (ε) was recorded as a function of time; experiments ranged from 30 to almost 700 days.

Two experiments were examined in detail in order to study porosity reduction associated with pressure solution deformation and mass redistribution. Porosities and pore diameter distributions were measured analytically; porosities were also simulated using binarized SEM images. Porosity is reduced in the direction of fluid flow; however, the changes in porosity (up to $\approx$10 porosity %) are heterogeneously distributed. Computerized x-ray tomography (CT) of the same two samples reveals the presence of high-density/low-porosity regions adjacent and upstream of the interface between the two cores making up each of the samples. These low-porosity regions are interpreted to be the result of precipitation in pore spaces, but this interpretation is not certain due to the initial heterogeneous nature of the chalk. The binarized SEM images were also used to calculate the transport properties of the samples, based on the method of reconstructed porous media (Adler et al., 1990). The simulated permeabilities are the same order of magnitude as many chalk permeabilities published in the literature.

1. INTRODUCTION

Pressure solution is a commonly occurring water-rock interaction process at shallow to moderate depths. Perhaps more importantly, pressure solution (which is also referred to as intergranular pressure solution (*ips*) creep) is considered to be an important mechanism of rock deformation in the upper crust, along with mechanical compaction, cataclasis, and plastic deformation. The effective result of pressure solution is the deformation of individual mineral grains due to intergranular dissolution, and precipitation of material in pore spaces, leading to a global compaction of the rock matrix, and in general, a concomitant decrease in permeability. A general review of pressure solution can be found in Part 1.

Pressure solution creep has been extensively documented in the field. Stylolites are perhaps the most obvious feature and are especially noticeable in limestones, sandstones, and evaporites (see extensive discussions in Bathurst, 1995; Merino, 1992; for theoretical stylolite models, see Merino et al., 1983, Dewers and Ortoleva, 1990). Other pressure solution features include pressure solution cleavage (Onasch, 1983; Gratier, 1987), differentiated crenulation cleavage (Gray and Durney, 1979), and secondary mineral growth in pressure shadows and along fault planes (Bayly, 1987; Gratier, 1987; Gratier and Gamond, 1990). On a smaller scale, grain indentation (e.g., Gratier et al., 1999), fitted intergranular rock fabrics, and secondary overgrowths on grains (e.g., Dewers and Hajash, 1995) are features typically associated with pressure solution deformation. Taken together, all of these geological features document the preferential dissolution of minerals in zones of elevated stress and precipitation in zones of minimum stress. An extensive review of low-temperature deformation features (from outcrop to thin section scale) and their mechanisms of formation, including many due to pressure solution, can be found in Groshong (1988).

Pressure solution has also been shown to apply to weathering profiles and pseudomorphic replacement reactions. Based on authigenic fabrics of mineral grains, it has been suggested that when one mineral grain grows at the expense of another (i.e., dissolution via pressure solution), a kinetic-rheological feedback establishes itself such that the rate of pressure solution and the rate of crystallization eventually become equal, such that the overall volume of rock is preserved (Merino et al., 1993, 1994; Nahon and Merino, 1997; Merino and Dewers, 1998).

Numerous experimental studies of pressure solution deformation on a variety of geological materials have been carried out (see *Introduction*, Part 1), mainly with the aim of establishing constitutive relationships. In a few of these studies, detailed microscopic examinations have been carried out on the deformed aggregates. These aggregates often show grains with indentation, truncation, and overgrowth features (e.g., halite: Spiers and Brzesowsky, 1993; gypsum: de Meer and Spiers, 1995, 1997; sandstone: Dewers and Hajash, 1995). These features are interpreted to be due to pressure solution: truncation and indendation from preferential intergranular dissolution, and grain overgrowths due to precipitation in pore spaces (for details, see section 2, Part 1).

Truncation, indentation, and overgrowths are markers of mass redistribution that occur during aggregate compaction. Mass redistribution induced by pressure solution has been the subject of both theoretical and experimental studies. Modeling the evolution of curved solid-fluid phase boundaries associated with fluid-filled pores embedded in a stressed solid matrix (Reuschlé et al., 1988; Heidug and Leroy, 1994; Leroy and Heidug, 1994) is an example of a theoretical approach to understanding mass redistribution. Experimental studies have dealt with closely related subjects, such as changes in fluid inclusion geometry (Gratier and Jenatton, 1984) and crack healing (Smith and Evans, 1984; Brantley et al., 1990; Gratier et al., 1994).

An important reason for studying deformation and mass redistribution is understanding how these result in changes of pore geometry and total porosity. Pore structure and porosity are important since they affect the physical and fluid transport properties of rocks (e.g.,Walsh, 1965; Heard and Page, 1982; Evans et al., 1999). Fluid transport in rock is pertinent not only with respect to promoting chemical reactions between fluids and rock grains (e.g., pressure solution), but also because local changes in connected porosity and permeability directly affect the fluid pore pressure, and therefore the effective state of stress (Brace, 1972).

Dissolution and precipitation reactions in porous media have been addressed in many numerical modeling studies (e.g., Sallès et al., 1993; Békri et al., 1995, 1997; Mourzenko et al., 1996). Such numerical models allow for the prediction of how pore geometries and total porosity evolve under various conditions, and how this affects the fluid transport properties of the medium. Other studies have modeled porosity reduction in porous media/sediments as a function of depth, and implicitly include pressure solution creep (e.g., Lemée and Guéguen, 1996; Renard et al., 1999). Experimental pressure solution creep experiments that have measured total porosity reduction as a function of strain are quite numerous. A few of these studies have also measured reductions in aggregate porosity at the thin section scale (e.g., Dewers and Hajash, 1995). In general, however, attempts to quantify the temporal and spatial evolution of porosity during deformation are not numerous, and for this reason, detailed studies are needed.

In this study (Part 2), we examine the spatial changes in porosity in deformed chalk samples. The specific questions we were aiming to resolve concern both the degree of total porosity reduction as a function of strain and the spatial distribution of porosity reduction. This has been effected by measuring in detail the porosity changes of two experiments, L9-11 and L16, based on three commonly used porosity measurement techniques. Another important aim of this research was to apply the powerful computerized x-ray tomography technique to study post-deformation porosity and compaction, and to tie this in with the porosity data obtained by classical techniques. This technique also allowed us to make inferences concerning fluid percolation. Finally, we also applied a theoretical approach that allowed for fluid transport properties to be estimated. This approach is based on image analyses of SEM photomicrographs and the application of the "reconstructed porous media" method (Adler et al., 1990). This multi-technique approach has provided semi-quantitative information concerning the coupled processes of compaction, deformation, and mass redistribution that characterize the experimental deformation of chalk by pressure solution.

2. METHODS AND TECHNIQUES

This communication (Part 2) is based on SEM, porosity, and X-ray tomography measurements of deformed chalk, and specifically, on two experiments that used 200-mm long cores (expts. L9-11 and L16). Because the chalk is very fragile, it was not possible to core a single 200-mm long cylinder; instead, two smaller cores, approximately the same length, were placed end-to-end. The top and bottom faces of each core were cut and sanded perpendicular to the core axis. The deformation history of these two experiments is detailed in Part 1. Detailed information concerning the experimental materials and procedures is also given in Part 1.

2.1. Porosimetry

The porosity of the chalk was measured by 3 separate techniques. The connected and occluded porosity of the chalk were measured both before and after deformation using the Hg injection technique on sample volumes of ≈ 0.5-1 cm^3. The average connected and occluded porosities of undeformed chalk are 41.5 and 10.7%, respectively. Another technique (using the same Hg injection apparatus) is based on a determination of the density of samples, assuming that samples were only composed of calcite. A third technique, based on saturation of samples with water, used much larger samples, with volumes on the order of 6-8 cm^3. The porosity of a reference, undeformed sample ranged from 39.9-41.5% (based on these 3 techniques). The measurement of post-deformation porosities was effected by cutting the chalk cores in half along the core axis and extracting samples from the inside surface of each half-core.

The Hg injection technique was also used for the determination of pore space diameter distributions. The accessibility of pore spaces with a given radius (r) is a function of the Hg injection pressure (P), the surface tension of mercury (γ), and the solid-Hg interfacial angle (θ); a commonly used equation is $r = 2\gamma \cos \theta / P$ (Rootare, 1970; Gregg and Sing, 1982). In this study, the maximum Hg injection pressure was 150 MPa, which permitted Hg access to pores with $\phi \approx 0.01$ μm. After injection of Hg at high pressures, the exterior pressure relative to the sample is lowered to ambient pressure, forcing most of the Hg in the sample pore spaces to be evacuated. The fraction of Hg that remains trapped in very small pore spaces is the basis for the calculation of the occluded porosity. However, the exact physical meaning of occluded porosity measurements is not straightforward. Occluded porosity can be interpreted to be related to disconnected porosity (i.e., the porosity which is not effective in transmitting fluids), but verification of this is not possible without an appropriate and independent measurement technique. In addition, it should be noted that occluded and disconnected porosity measurements depend on the physical characteristics (e.g., viscosity, surface tension) of the injected fluid; thus, results based on Hg measurements may differ quantitatively from those based on aqueous fluids, for instance.

2.2. SEM Image Analysis

SEM photomicrographs from an undeformed reference sample, as well as from post-deformation samples, were used to investigate compaction; these are discussed further on in the *Results and Discussion* section. A large series of SEM images realized of experiments L16 and L9-11 (post deformation) were also analyzed for the derivation of simulated porosities and permeabilities. This series of images was based on a sampling technique that minimizes the danger of mechanically disturbing the matrix. Each plug that was extracted from a sample was cut into two parts, one for SEM images, the other for porosity measurements. Each SEM plug was impregnated with a low-viscosity resin that imparted enough rigidity to the matrix so that it could be dimensioned for SEM use. A second impregnation step filled all remaining pores. In both impregnation steps, only capillary forces were responsible for the penetration of pore spaces (i.e., no vacuum techniques were used, this ensured that the matrix remained undisturbed). The impregnation steps were followed by gentle polishing of the surfaces that were viewed by SEM. The color difference between the resin and the solid matrix was sufficient for the discrimination of pore spaces and grains.

Four corner areas on each sample plug were chosen for SEM imaging. Using the code Visilog, each SEM image was binarized, with white representing pore spaces and black representing matrix material. The process of binarization was interactive (i.e., grayscale threshold, erosion and dilatation cycles) to ensure that the binarized image resembled as much as possible the spatial porosity characteristics of the original SEM image (i.e., by visual comparison, comparison of measured and 2-D binarized porosity). Even though this step introduces a certain subjectivity to the procedure, experience has shown this to be the most precise manner for faithfully reproducing the real pore space structure (Adler, 1994; see also discussion in Gaviglio et al., 1997 on the effects of interactive processing).

SEM images were taken at 50x and 2000x magnification. Porosities based on the 50x images are termed "mesoporosity", while those based on the 2000x images are termed "microporosity". The mesoporosity is on the order of ≤ 1%, while the microporosity is ≈ 34%. This extreme difference in values shows the importance of the pixel to pore size ratio in determining porosities. It should be noted, however, that porosity and magnification are not correlated beyond a certain level of image magnification (as shown in Section *3.3*). Since the microporosity values were closest to the measured values of bulk porosity, only microporosities were considered to be useful for this study.

2.3. Simulated Porosity and Permeability Calculations

Each binary image was analyzed using the "reconstructed porous media" method developed by Adler and coworkers (Adler et al., 1990, 1992; Adler, 1994; for the development of this technique, see Joshi, 1974; Quiblier, 1984); a brief overview of this method follows. For any point x within a binary image, a phase function $Z(x)$ can be used to differentiate between solids and pore spaces:

$$Z(x) = 1 \quad x \in \text{pore space}$$
$$0 \quad x \in \text{solid phase} \tag{1a}$$

The porosity can then be defined as

$$\phi = \overline{Z(x)} \tag{1b}$$

where the overbar denotes a statistical average. Another important quantity is the correlation function $R_z(u)$, which is defined by the second moment of the phase function:

$$R_z(u) = \overline{[Z(x) - \phi] \cdot [Z(x+u) - \phi]} \,/\, (\phi - \phi^2) \tag{2}$$

where u is the displacement vector. The correlation function quantifies the statistical organization of the porous medium; it also can be defined by the probability that two points belong to the same phase.

The subsequent generation of a three-dimensional medium with porosity ϕ and correlation function $R_z(u)$ is based on the generation of a random function of space $Z(x)$, where $Z(x) = 0$ in a solid phase and $Z(x) = 1$ in a liquid phase. In addition, $Z(x)$ must satisfy the conditions given by Eqns. 1b and 2. The resulting reconstructed porous medium is portrayed in terms of small cubes that are either liquid or solid-filled. Transport behavior associated with reconstructed media can be calculated from the usual Stokes equations; this permits the determination of the flow field (assuming Newtonian fluids at low Reynold's numbers). Once the flow field is determined, the calculation of various transport properties follows (e.g., permeability, conductivity, etc.; for details on this method, see Adler et al., 1990).

2.4. X-Ray Computerized Tomography

In order to investigate the 3-D spatial density of the chalk samples, x-ray computerized tomography (CT) images were obtained of an undeformed chalk sample, as well as samples L16 and L9-11 after deformation. The imaging was carried out on a 3rd generation medical x-ray tomography unit located in a hospital (Siemens Somatom Plus S, Radiologie GHB K.U. Leuven, Belgium; instrument settings: 137 kV, 75 mA for L9-11 and non-deformed reference chalk; 120 kV, 240 mA for L16). The X-ray CT method is based on the attenuation of x-rays by the sample and the subsequent reconstruction of a section through the sample, which is portrayed in terms of the spatial variation of the radiological density (see e.g., Hounsfield, 1973; Herman, 1980; Wellington and Vinegar, 1987; and references therein; for recent geological applications, see e.g., Raynaud et al., 1989; Verhelst et al., 1995; Géraud et al., 1998; Brown et al., 1999). For monochromatic x-rays, the linear attenuation coefficient is a material constant, such that x-ray density variations are due to differences in the bulk density. The main advantage of this method, which is completely non-destructive, is that it can provide a series of sequential density images (i.e., density slices) of a sample.

The radiological density variations of the bulk chalk matrix can be directly related to porosity variations, based on the fact that the matrix is composed exclusively of calcite (as confirmed by x-ray diffraction and SEM/EDX results; with the exception of chert fragments). The application of this technique in the present study is qualitative, however, for the following reasons: first, the CT scanner reconstruction algorithms (ultra-high resolution AH_07541) were fixed for applications related to medical reconstructions of the human head; no modifications were attempted for this study; and second, a quantitative approach requires the calibration of the gray-scale densities to an exact porosity scale. Nevertheless, this technique proved to be very powerful in that it yielded important qualitative information on the initial heterogeneous structures of the chalk, as well as the changes in porosity that resulted from the combined effects of pressure solution deformation, fluid percolation, and mass redistribution by precipitation in pore spaces.

3. RESULTS AND DISCUSSION

3.1. SEM Images

SEM images were taken with several objectives in mind: (a) recognition of matrix compaction and porosity reduction, (b) identification of possible fluid pathways, and (c) the numerical 3-D reconstruction of porous media after deformation, based on binarized SEM images (discussed further on). The interpretation of SEM images, based on both undeformed and deformed samples, proved to be difficult with respect to identifying fluid flow pathways and the effects of compaction, as discussed below.

Both undeformed and deformed chalk samples show the presence of pore spaces, with typical diameters ranging from one to several μm, as seen in Fig. 1a. Figure 1b shows a thin vein that cross-cuts an undeformed chalk matrix. These veins are composed of the same material as the chalk matrix (as confirmed by SEM/EDX analyses), but have more calcite cement, and are more compact and less porous than the matrix.

Fig. 1. SEM images of Guerville chalk used in study. (a) The chalk is composed primarily of coccolith skeletal debris and other shell fragments; note the high porosity inherent to this chalk (≈ 40%). (b) Vein crosscutting the matrix; the porosity is considerably diminished in comparison to surrounding matrix.

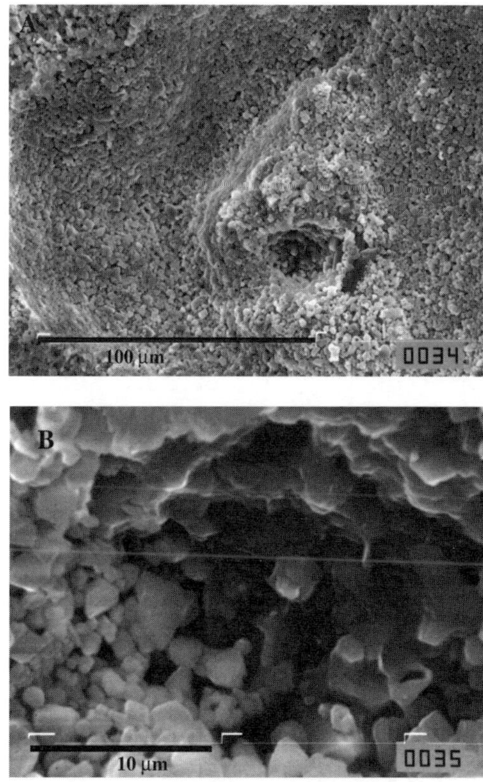

Fig. 2. (a) An example of a large channel present in deformed chalk (expt. S8-9); (b) an enlarged view. Large channels such as these are present both in undeformed and deformed samples. Their axial continuity is not known, but they may play a significant role in preferential fluid flow.

Occurring less frequently are large channel-like structures with diameters often exceeding 10 μm; one such channel is shown in Fig. 2. These structures were observed by SEM both in undeformed and deformed chalk samples. The origin of such large channels is not known; they may represent a primary feature, having formed during sedimentation and/or diagenesis; alternatively, they may be secondary dissolution features, having formed in situ due to groundwater percolation, for example. These channels apparently were not destroyed during deformation in the laboratory. In fact, it is possible that they may have been enlarged by fluid circulation during the experiments, this having occurred concomitantly with compaction, reduction in pore space volumes, and decreasing permeability of the matrix. The exact relationship between these channels, pressure solution, and fluid flow during experimental deformation remains an open question, however.

According to studies by Gaviglio et al. (1993, 1997) on the Mons basin chalks, large channels (ϕ = 1- 10 μm, and in some cases $\phi > 10$ μm) most commonly occur immediately adjacent to fault planes; these authors postulate that channels form (or are enlarged) as a consequence of the initial stages of deformation in the presence of fluids, and are kept open by fluid circulation even during compaction of the surrounding matrix. Their observations imply that large channels characteristically form in chalks that have undergone deformation in the presence of fluids. Nonetheless, the two dimensional nature of SEM images in this study makes the interpretation of these channels and pore spaces tenuous, especially since it is not possible to ascertain their connectivity. For this reason, it is difficult to determine the exact pathways and fluxes associated with fluids percolating through the samples during experimental deformation runs.

It is interesting to note that SEM images of the interface between the two cores comprising sample L9-

Fig. 3. (a) SEM image of the irregular interface (dashed line) between the 2 cores of L9-11; the irregular nature suggests stylolite formation. The large object that straddles the interface has acted as a resistant indenter and has penetrated both cores (i.e., by pressure solution). (b) Examples from L9-11 of grain-to-grain indentation and truncation due to pressure solution, as indicated by arrows. (c) SEM image of interface between the two cores in L9-11 (dashed line). The upstream core, shown on the right, shows the presence of large, cemented areas; these are interpreted to be due to precipitation as a consequence of pressure solution deformation and mass redistribution.

11 reveal it to be highly irregular (see Figs. 3a, c) after deformation, perhaps indicating the onset of stylolite formation. The irregularity is most probably due to a combination of mechanical and chemical processes (i.e., pressure solution reactions) acting upon the interfacial boundary. Figure 3a shows the presence of a large, resistant object situated at the interface that has acted as an indenter with respect to both cores during the experimental deformation. The penetration of this object into both cores can be explained in terms pressure solution leading to preferential dissolution of the matrices perpendicular to the direction of major stress (this can be compared to laboratory and natural examples - e.g., Gratier et al., 1999). Evidence for pressure solution at the scale of individual grains, as evidenced by grain indentation and truncation, is shown in Fig. 3b. It should be noted, however, that even though this image was taken of grains from the interfacial region of L9-11 after deformation, there is no proof that these pressure solution features do not owe their origin to the some kind of paleo-pressure solution deformation (i.e., pressure solution features possibly present in original samples).

Perhaps the most predictable consequence of deformation associated with fluid flow and pressure solution is compaction of the granular chalk matrix. However, a comparison of SEM images of undeformed and deformed chalk samples does not reveal any discernable differences in the degree of matrix compaction. A qualitative comparison of the undeformed chalk matrix in Fig. 1 with the deformed chalk matrix of L9-11 (Fig. 3) underscores this point. This is most probably due to the very modest degree of compaction experienced by the sample (1.5% strain, see Fig. 5b, Part 1).

Figure 3c also clearly shows the exact nature of the interface between the two cores of L9-11. The void space that defines the interface in the image has a thickness < 1 μm, which is roughly the same dimension as many of the pore spaces found throughout the matrix. The chalk matrices on both sides of the interface do not reveal any obvious, qualitative differences in compaction or porosity. Nonetheless, a few of the areas adjacent to the upstream side of the interface do show grain "welding" and coalescence due to cement overgrowths (see Fig. 3c). These low-porosity regions, which always end abruptly at the interface, resemble very closely the veins shown in Fig. 1b. In addition, the cemented grains shown in Fig. 3c resemble very closely coalesced grains found in chalks that have been deformed and compacted adjacent to fault planes (Fig. 4, Gaviglio et al., 1993). This close resemblance suggests a transformation of the matrix, due most probably to precipitation of calcium carbonate in pore spaces. The interfacial region of L9-11 proved to be of

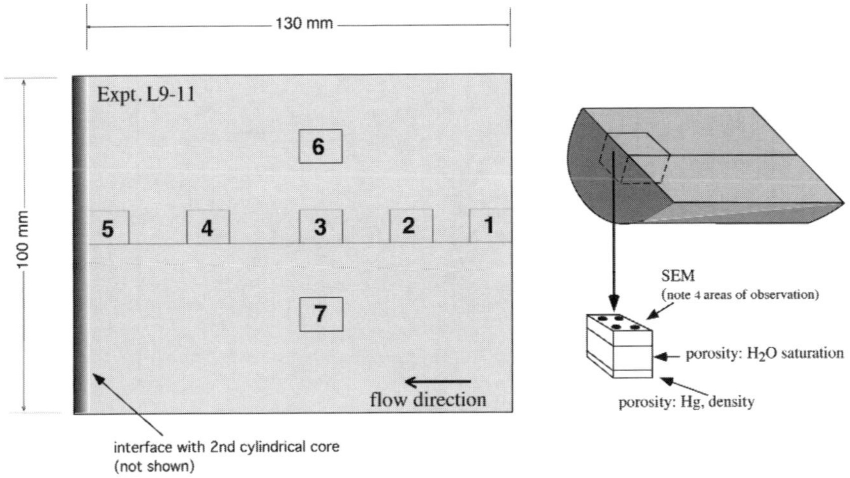

Fig. 4. The figure on the right shows how all cores were cut along the axis and sampled after deformation, as well as the partitioning of samples for SEM imaging and porosity measurements. The left-hand figure shows the sample localities within the upstream core of L9-11.

special interest since this same area corresponds to a large contrast in x-ray tomographic density (see Fig. 16). As discussed further on, this density contrast corresponds to a difference in matrix porosities on adjacent, but opposite sides of the interface.

3.2. Porosimetry

3.2.1. Reference and experiment L9-11 porosities

The average porosities of undeformed chalk samples, based on the Hg injection, density, and water saturation methods, are 41.5, 41.3% (avg. of 3 samples, $1\sigma = 2.1$ porosity %), and 39.9% (avg. of 7 samples, $1\sigma = 0.9$ porosity %), respectively. Multiple Hg injection measurements were not performed on a reference sample, but the accuracy and precision of this technique is roughly the same as the density method, based on measurements of other chalks. In the description of porosities below, simulated porosities, based on binarized images and the method of reconstructed porous media, are also presented for comparison with measured values (see section *3.3* for more details). The average simulated porosity of an undeformed chalk sample is 33.6% (avg. from 4 separate images, $1\sigma = 4.5$ porosity %); note that this is considerably lower than the average of the measured values. The uncertainties of porosities associated with deformed samples (measured and simulated) are given directly in the pertinent figures.

Figure 4 shows schematically the manner in which extracted plugs were divided up for SEM and porosity measurements. In addition, the locations of the plugs, the direction of fluid flow, and the interface between the 2 cores for experiment L9-11 are shown.

Figure 5 represents the cumulative distribution curves of the pore diameters (Hg-injection method) associated with the 7 locations given in Fig. 4, as well as for a non-deformed sample (reference). Even though there is some variation in the pore distribution curves of samples 1, 2, 3, 6, and 7 with respect to the reference, they are statistically equivalent, based on the analytical uncertainties. On the other hand, samples 4 and 5 show diminished porosities and a noticeable and significant shift in their pore distributions towards smaller diameters (i.e., increase in pore space population: $0.1 < \phi < 1$ μm).

Figure 6a shows the evolution of the porosities as a function of axial distance and flow direction (i.e., samples 1-5 in Fig. 4). The porosities are based on Hg-injection and density measurements, as well as simulated values. The simulated porosities have uncertainties ranging from 0.9 to 6.1 porosity %, with an average uncertainty of 3.6 porosity % (based on sample nos. 1-7, 1σ uncertainties). For any given sample location, the maximum difference between the various porosities (analytical vs. simulated) varies from 2 to 8 porosity %. However, the trends of all of the profiles reveal a decrease of roughly 5 porosity % between the upstream and downstream ends of the core. The occluded porosity, also shown in Fig. 6a, reveals a modest increase along the core axis and direction of flow, which is in accord with the shift towards smaller

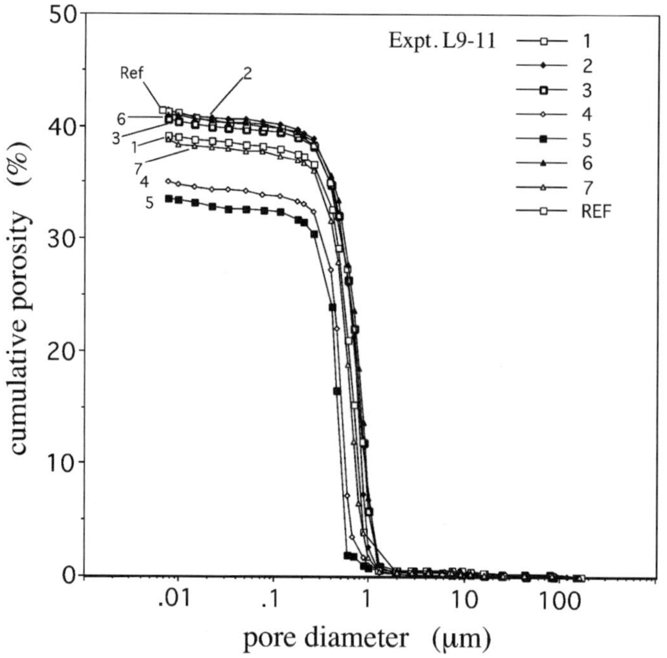

Fig. 5. Cumulative pore diameter distribution curves for samples 1-7 of L9-11 and an undeformed reference. Samples 4 and 5 show diminished porosities and a shift in pore distributions towards smaller pore diameters ($0.1 < \phi < 1$ µm).

pore diameters shown by the pore distribution curves in Fig. 5 (i.e., distributions 1, 2, 3, ref. vs. 4, 5). A graph of the ratio of occluded porosity to total porosity as a function of axial position (Fig. 6b) shows more concisely the significant diminution of the total connected pore space volume towards the downstream end of the sample L9-11, close to the interface with the other core (see Fig. 4).

3.2.2. Experiment L16 porosities

The porosities of L16 were more extensively sampled and measured in comparison to expt. L9-11. Figure 7 portrays the 55 sampling locations where the porosities where measured by the water saturation method; in addition, the porosities at 8 of these locations were also measured by the Hg injection and density methods, and simulated by the reconstructed porous media method. Also shown is the location of the interface between the two cores of expt. L16, which are designated as the 'upstream' and 'downstream' cores. The 8 simulated porosities have uncertainties (1σ) ranging from 2.1 to 8.7 porosity %, with an average of 4.9 porosity % (based on sample nos. 3, 18, 28, 33, 43, 53).

The pore distribution curves associated with the aforementioned 8 localities, shown in Fig. 8a, show a general shift towards reduced porosities and smaller pore diameters with respect to the reference. The most pronounced differences were measured for the samples adjacent to the interface between the two cores (nos. 33 and 35-see Fig. 7). The reduced porosity and the shift towards smaller pore diameters associated with sample 33 vs. a reference sample is indicated more clearly in Fig. 8b. This example shows that the median pore size diameter decreases from 0.8 to 0.3 µm. The pore diameter shifts (with respect to the reference) of all of the distribution curves in Fig. 8a do not follow any obvious generalized trend, as evidenced, for example, by a strong shift associated with sample no. 3 (located at the inflow end of the upstream core) and no shift for no. 28 (adjacent to the upstream side of the interface).

The 55 water saturation-based porosities are shown graphically in Fig. 9 in terms of a gray scale, with black representing the maximum porosity (40.4%) and white the minimum porosity (29.5%) measured. The highest variation in porosities occurs in the upstream core, whereas the downstream core generally shows a higher total porosity. The overall gray scale pattern does not indicate any obvious zones of connected high porosity which might be associated with preferential fluid flow (i.e., on a cm-scale). Nonetheless, two patterns stand out: a. the left-hand side of both cores shows a continu-

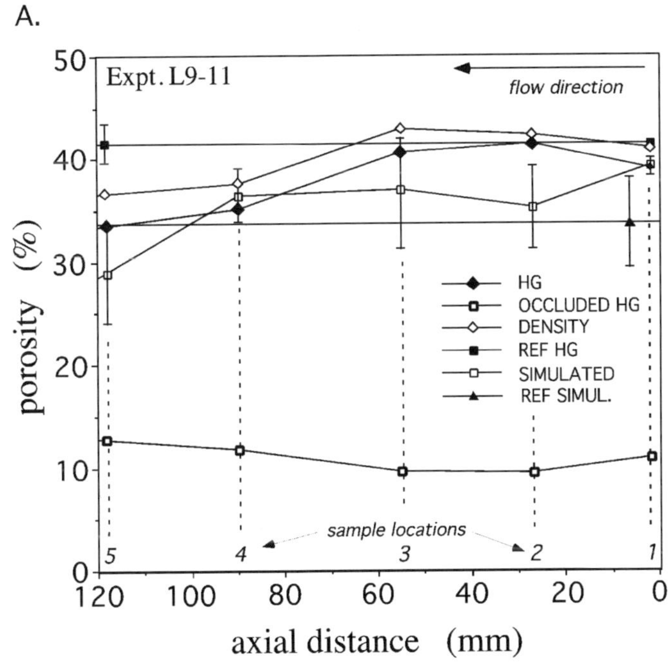

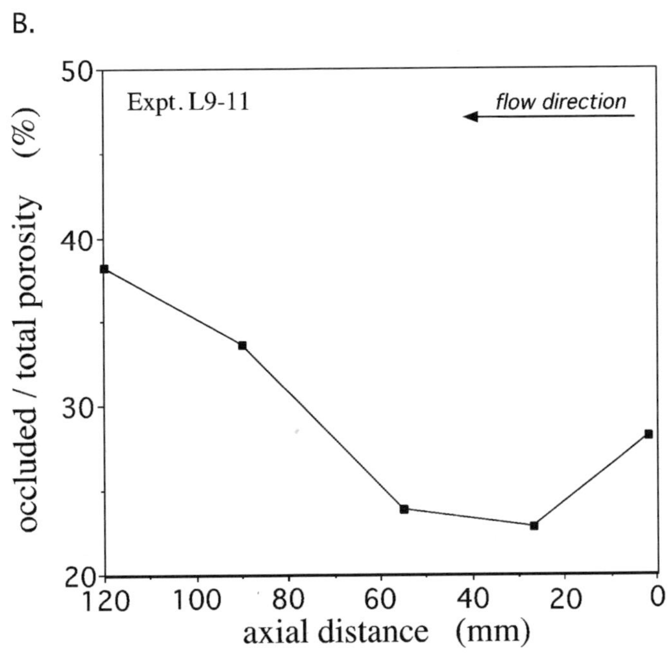

Fig. 6. (a) Axial porosity trends for L9-11, based on samples 1-5. Porosity measurements are based on Hg injection, density, and water saturation methods; in addition, simulated porosities are shown for comparison. All of the trends show a general decrease in porosity towards the downstream direction, with the exception of the occluded porosity, which increases in the direction of flow. Error bars represent $\pm 1\sigma$. (b) The marked increase in the ratio of occluded to total porosity towards the downstream end of the core reveals the combined effect of decreasing porosity and increasing unconnected pore space volume.

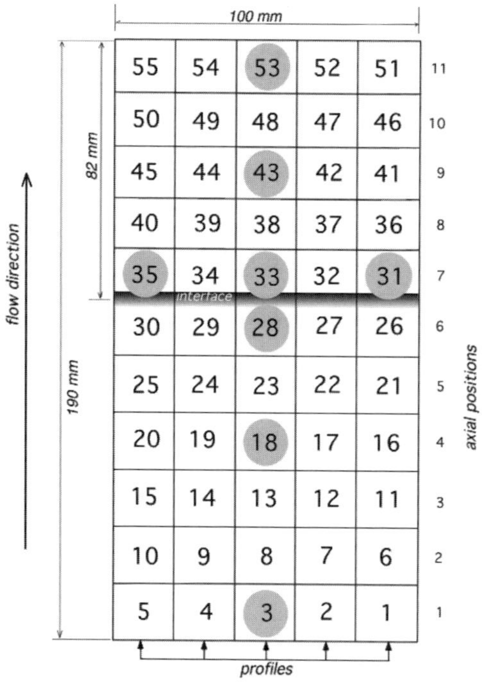

Fig. 7. Expt. L16 sample locations (cut-away view along axis - as in Fig. 4): porosities 1-55 were measured by the water saturation method, whereas samples 3, 18, 28, 31, 33, 35, 43, and 53 were also measured by the Hg injection and density methods; the porosities of these samples were also simulated. The indicated axial positions and profiles refer to Figs. 10 and 11a.

ous zone of intermediate to low porosity b. the upstream core immediately adjacent to the interface between the two cores is marked by a continuous band of very low porosity, ranging from 30.1 to 31.1%.

Five longitudinal profiles corresponding to the gray-scale porosities in Fig. 9 are shown in Fig. 10. The upstream profiles show a general decrease in porosities in the direction of flow, up to the interface, followed by a sudden increase on the downstream side of the interface. Figure 11a also portrays longitudinal profiles, based on measured and simulated porosities corresponding to the central axis (profile 3, Fig. 7). As opposed to the measured and simulated porosities of L9-11 shown in Fig. 6a, the correlation between the measured and simulated porosities is poor, and in addition, the measured porosities do not agree with each other (i.e., water saturation vs. Hg injection and density measurements). Nevertheless, the Hg injection and density porosity profiles do show a marked minimum adjacent to the interface, but on the downstream side (as opposed to the upstream side, as with the water saturation results). The poor correlation of the measured porosities may be due to real differences (i.e., due to the different volumes analyzed - see Fig. 4). On the other hand, there may have been a slight inaccuracy in the exact location of sampling for each method with respect to the interface, such that the minima in porosities actually correspond to the same location (i.e., upstream and immediately adjacent to the interface); if this is the case, then the trends of the measured porosities are in better accord.

Figure 11a shows that the simulated porosities range from ≈ 20-50% over the entire length of sample L16; this variation is significantly higher than the variations in the measured porosities. It is interesting to note that for sample L9-11, the variations in the measured and simulated porosities over the entire length of the upstream core are approximately equivalent (Fig. 6a). This difference between the two experiments may reflect a greater amount of heterogeneous porosity reduction in L16 vs. L9-11.

The large variation in simulated porosities that was determined over the entire length of sample L16 also occurs at a single sampling locality. Figure 11b represents simulated porosities of sample 53, based on a tightly-spaced (0.33 mm-step size) lateral SEM traverse (11 mm total distance, traverse from left to right, refer to Fig. 7). Figure 11b shows that the porosity varies from ≈ 30-57%, which represents a variation similar to that determined over the entire length of L16 (i.e., simulated porosity profile in Fig. 11a). There is, however, an important difference between the simulated porosities calculated in Figs. 11a and b; they are based on SEM images with significantly different magnifications. In Fig. 11a, sample 53 has a simulated porosity of ≈ 19 % (based on 2000x images), whereas in Fig. 11b the average porosity of sample 53 is ≈ 40-45 % (based on 250x images). This example shows that the SEM image scale has a significant effect on the simulated porosities (this is discussed in more detail in the next section).

3.2.3. Comparison of techniques

Based on the results discussed above for expts. L9-11 and L16, the various analytical methods used in this study are not all equally suitable for porosity measurements of deformed porous materials, such as chalk. The Hg injection technique is invaluable due to its ability to measure the distribution of pore sizes in a sample. An additional positive point is that is possible to analyze relatively small samples using this technique (this applies to the density method, as well). On the other hand, the greater sampling volume associ-

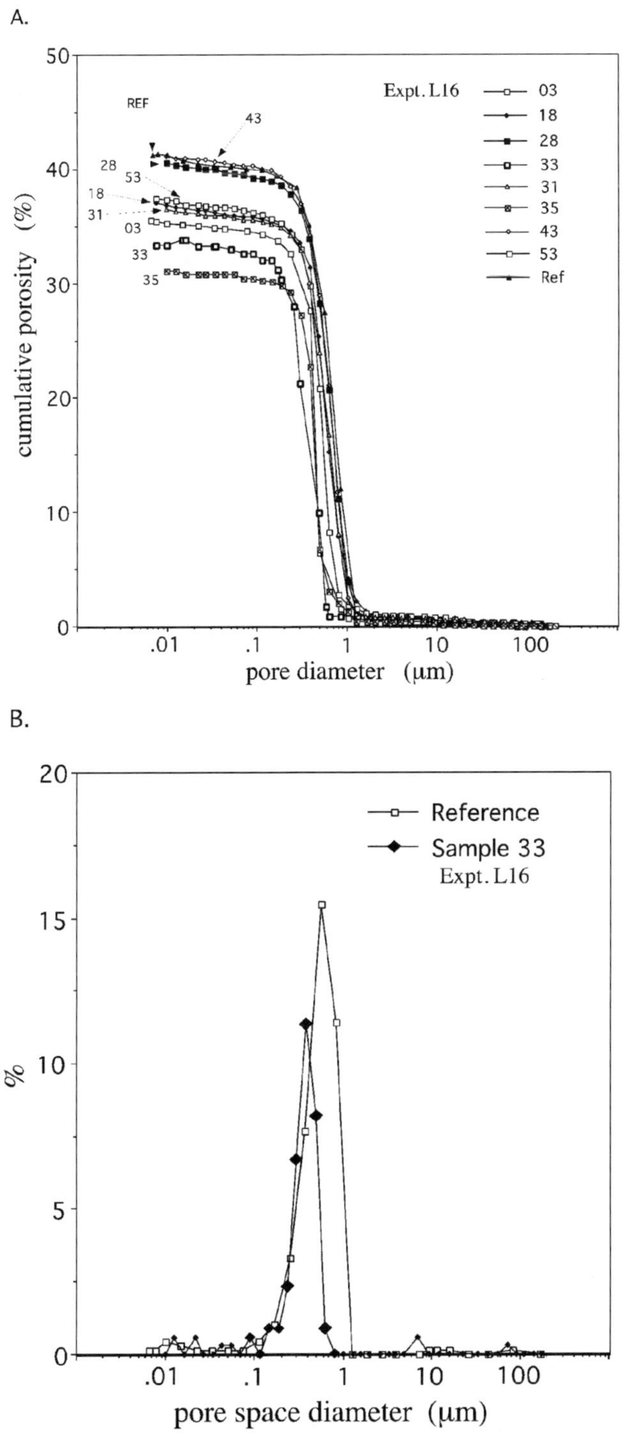

Fig. 8. (a) Cumulative pore diameter distribution curves measured by Hg injection for samples from L16 and an undeformed reference sample. The two samples adjacent to the interface (samples 33, 35) show the largest decrease in porosity and the most significant pore diameter shifts with respect to the reference. (b) Individual normalized porosity % vs. pore space diameters for sample 33 and a reference. These curves show more clearly the overall decrease in porosity of sample 33, as well as a shift in the pore diameter populations, from a median of 0.8 μm for the reference to 0.3 μm for sample 33.

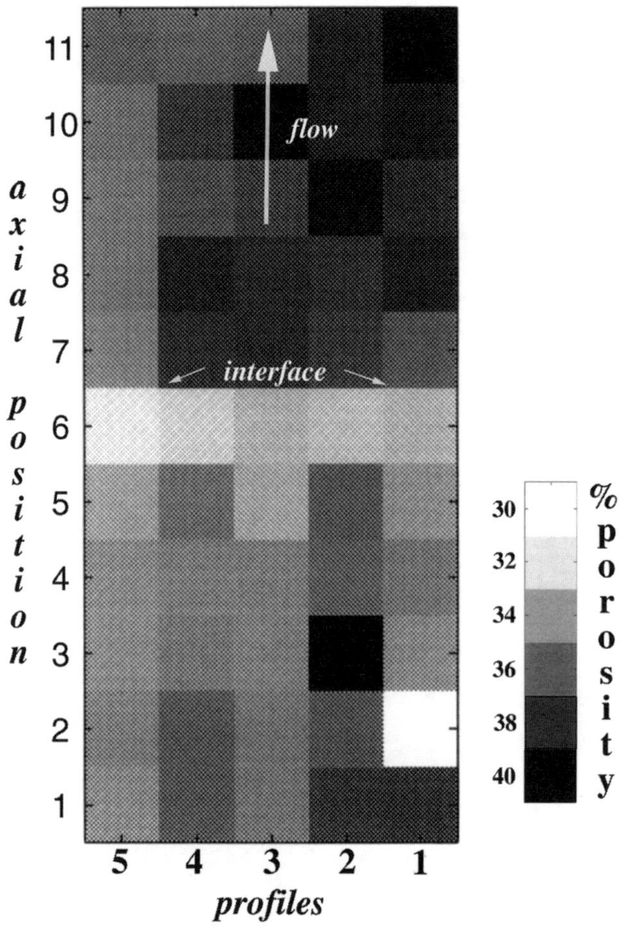

Fig. 9. Gray scale representation of water saturation porosities measured for samples 1-55, L16 (cut-away view along axis- compare to Fig. 7). The maximum measured porosity (40.4%) is indicated in black, the lowest (29.5%) is indicated in white. The overall pattern does not indicate any obvious zones of connected high porosity at the cm-scale. The most obvious feature is a zone of very low porosity (ranging from 30.1-31.1%) immediately adjacent to the interface of the two cores, on the upstream side. The indicated profiles and axial positions refer to Figs. 10 and 11a.

ated with the water saturation method presents an advantage in terms of a reduced sensitivity to local porosity heterogeneities. This translates to measurements that are more representative for heterogeneous materials. The poor accord in the measured porosity results for expt. L16 (i.e., Fig. 11a) potentially reflects an elevated ratio of local heterogeneities to volume analyzed by each technique. It is not certain why the measured porosities for L9-11 are in much better accord than those for L16, but the greater degree of deformation of L16 may have played a role in accentuating the heterogeneities already present in the chalk (L9-11: 0.015 strain vs. L16: 0.028 strain).

As shown in Figs. 6a and 11a, the accord between the measured and the simulated porosities differs with respect to expts. L9-11 and L16; in the former the accord is relatively good, whereas in the latter, the accord is poor. Local porosity heterogeneities and the scale of measurement exert a large influence on simulated porosities (e.g., Fig. 13a), probably even more so than with analytical porosity measurements. Since the simulated porosities are based on binarized SEM images, the process of binarization is the limiting factor with respect to the accuracy and the representative nature of the simulated porosities. The scale of the binarized images is critical since it influences the over-

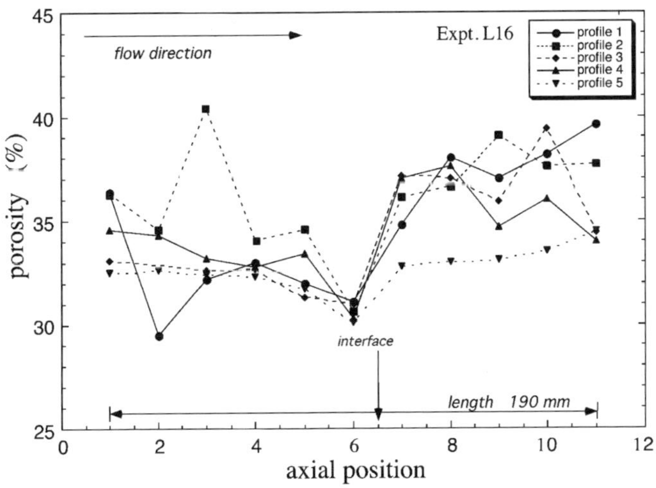

Fig. 10. Longitudinal profiles of water saturation porosities for expt. L16, as shown graphically in Fig. 9. The profiles show the heterogeneous nature of the porosities, a general decreasing porosity trend with flow direction in the upstream core, and a local porosity minima adjacent to the interface.

all values of porosity (i.e., in the extreme case, micro vs. mesoporosities, see Fig. 12), as well as their sensitivity to local heterogeneities. A comparison of the simulated porosities for sample 53 (expt. L16) shows quite well the trade-offs associated with the different scales of the binarized images. Thus, increasing the magnification of SEM images increases the resolution of the binarized images, which is crucial for accurately reproducing the subtleties of the microporous matrix structure. However, the higher the magnification, the greater the sensitivity of the simulated porosities to local heterogeneities (e.g., Fig. 13a), which in turn renders them less representative for the whole of the sample. Conversely, binarized images that are based on low magnifications cannot systematically resolve the pore spaces, resulting in simulated porosities that are perhaps less sensitive to heterogeneities, but are nonetheless, too low compared to those measured analytically.

The complex nature of deformed and heterogeneous porous materials, such as the chalk in the present study, are perhaps too complex for accurate, and at the same time, representative binarization. This is not a problem with more simple materials, such as well-sorted sandstones, where simulated porosities are very close to those measured analytically (e.g., Fontainebleau sandstone, Adler et al., 1990). Perhaps the only viable solution to increasing the accuracy of simulated porosities and transport properties of complex and heterogeneous materials is significantly increasing the number of binarized images per sample.

3.2.4. Comparison with porosities in naturally-deformed chalks

It is worthwhile at this point to compare the porosity results in this study with those from naturally deformed chalks. Extensive work in the Mons basin, Belgium (geographically close to Guerville chalk quarry-the provenance of chalk in Parts 1, 2) by Gaviglio and coworkers (Gaviglio et al., 1993, 1997), for example, has documented the physical and chemical changes induced in chalks that have undergone modest deformation due to tectonic movements along faults. A study of chalk within a graben structure (Gaviglio et al., 1993) shows that with increasing proximity to a normal fault (over a lateral distance of ≈ 4 m), chalk porosities systematically decrease by approximately 5-10 porosity %. In addition, chalk samples immediately adjacent to the fault display a much higher degree of variation in porosities as compared to non-deformed chalk, ranging from 30.3 to 37 porosity %. The effect of deformation on porosities was not found to be homogeneous, however, since weakly deformed chalks actually show slightly higher porosities (on the order of 0.7-5.5 porosity %), due apparently to grain dilatancy. Pore distribution curves (see Fig. 6, Gaviglio et al., 1993) also reflect the heterogeneous nature of deformation. Weakly deformed chalks show a modest increase in pore diameters (ϕ 1-10 μm) with respect to non-deformed chalk; strongly deformed samples (adjacent to fault) show that a large percentage of large pores (ϕ 1-10 μm) remain, even

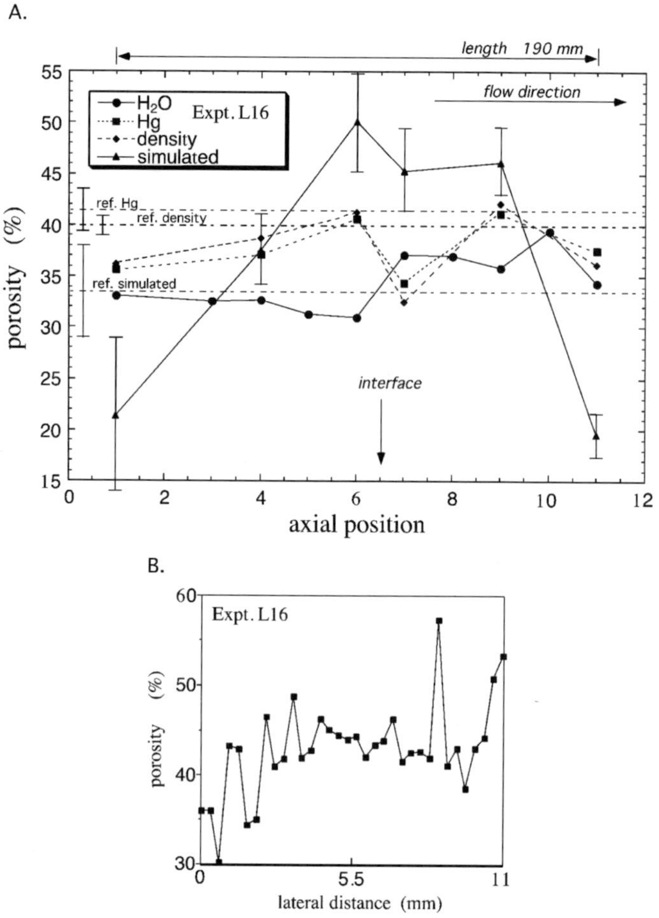

Fig. 11. (a) Comparison of various porosity profiles along the central axis of L16, based on analytical and simulated data. There is a strong discordance between the simulated data and analytical measurements; this is most probably a function of the high sensitivity of the simulated porosity determinations to local porosity heterogeneities. Error bars represent $\pm 1\sigma$. (b) Lateral simulated porosities for sample 53, L16 (refer to Fig. 7). The scattered nature of the data shows the elevated sensitivity of simulated porosities to local heterogeneities. Note that the average porosity (≈43%) is much higher than for sample 53 in (a) (data point lower right corner); this is a function of different SEM magnifications- see text for details.

though the median pore distribution is shifted to smaller pore diameters (in the range of ϕ 1- 0.1 μm) with respect to non-deformed chalk (see also Fig. 10, Gaviglio et al., 1997). Gaviglio et al. (1993) attribute the overall porosity changes being due to initial dilatation of grains and enlargement of flow channels, followed by progressive porosity reduction due to pressure solution compaction and cementation. These results are complementary to those of the present study in that they show that the effects of deformation and pressure solution on chalk in both laboratory studies and natural environments can be quite heterogeneous.

3.3. Simulated Porosities and Reconstructed Porous Media

Two representative SEM images of sample 4 (expt. L9-11, see Fig. 4) are shown in Figs. 12a, b. These images represent the same area at two different magnifications, 50x and 2000x, respectively. The respective binarized images show the dramatic difference in porosities associated with these different image scales. The porosity associated with the 50x binarized image is 1%; this "mesoporosity" essentially represents large void spaces due to macroscopic fossils (e.g., upper right hand corner, Fig.

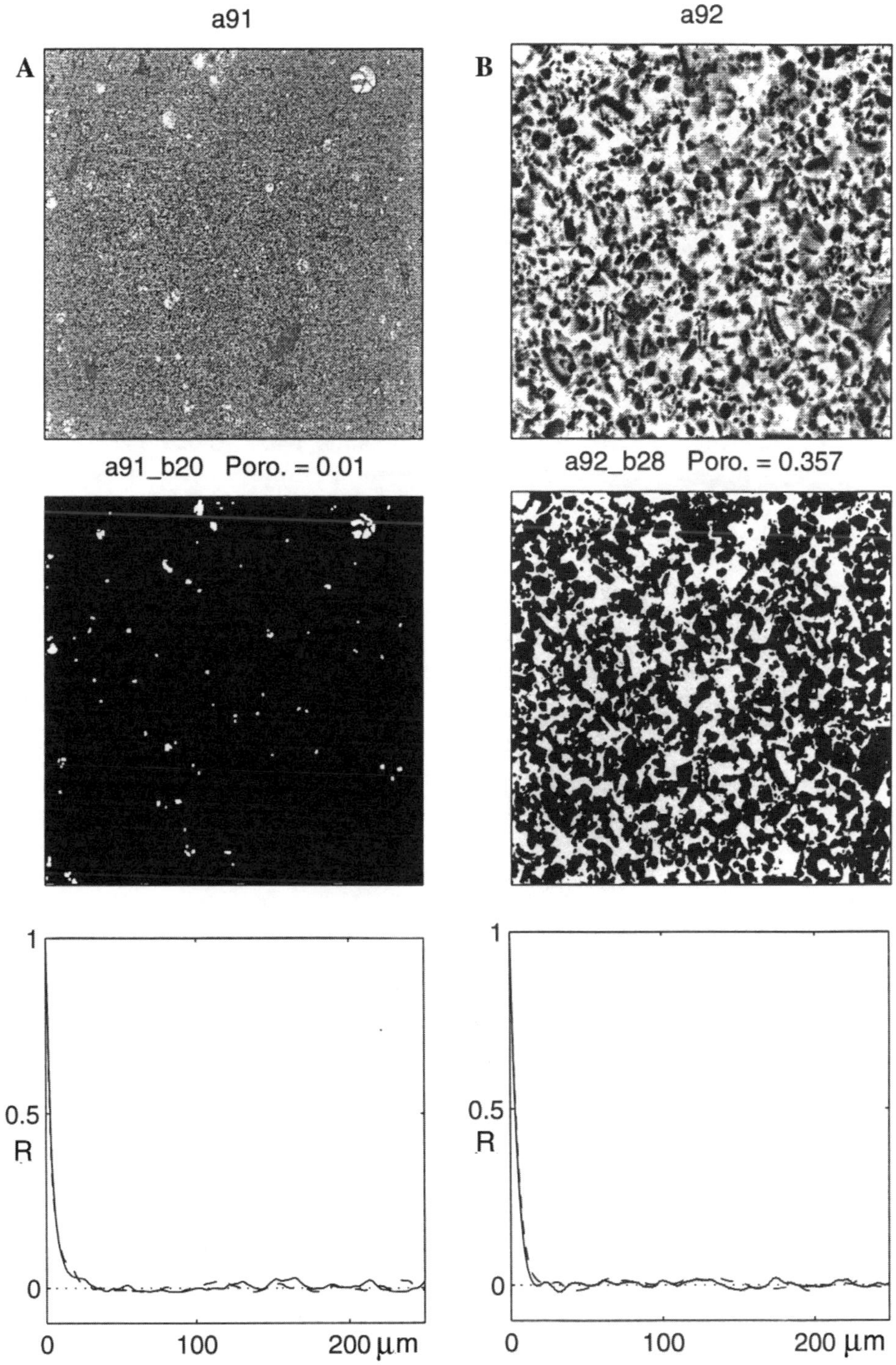

Fig. 12. SEM images (top), binarized equivalents (middle), and autocorrelation functions (bottom), based on one corner area, sample 4, L9-11. Pore spaces indicated in white, solid matrix in black. (a) Image areas 1600 x 1600 µm, top and middle. The simulated "mesoporosity" is 1%, this representing only large fossil voids (e.g., upper right hand corner) (b) Image areas 40 x 40 µm, top and middle. The simulated "microporosity" is 35.7%, in close agreement with analytical values. In both (a) and (b), the autocorrelation functions R_x (—) and R_y (---) are similar, indicating insignificant pore space anisotropy.

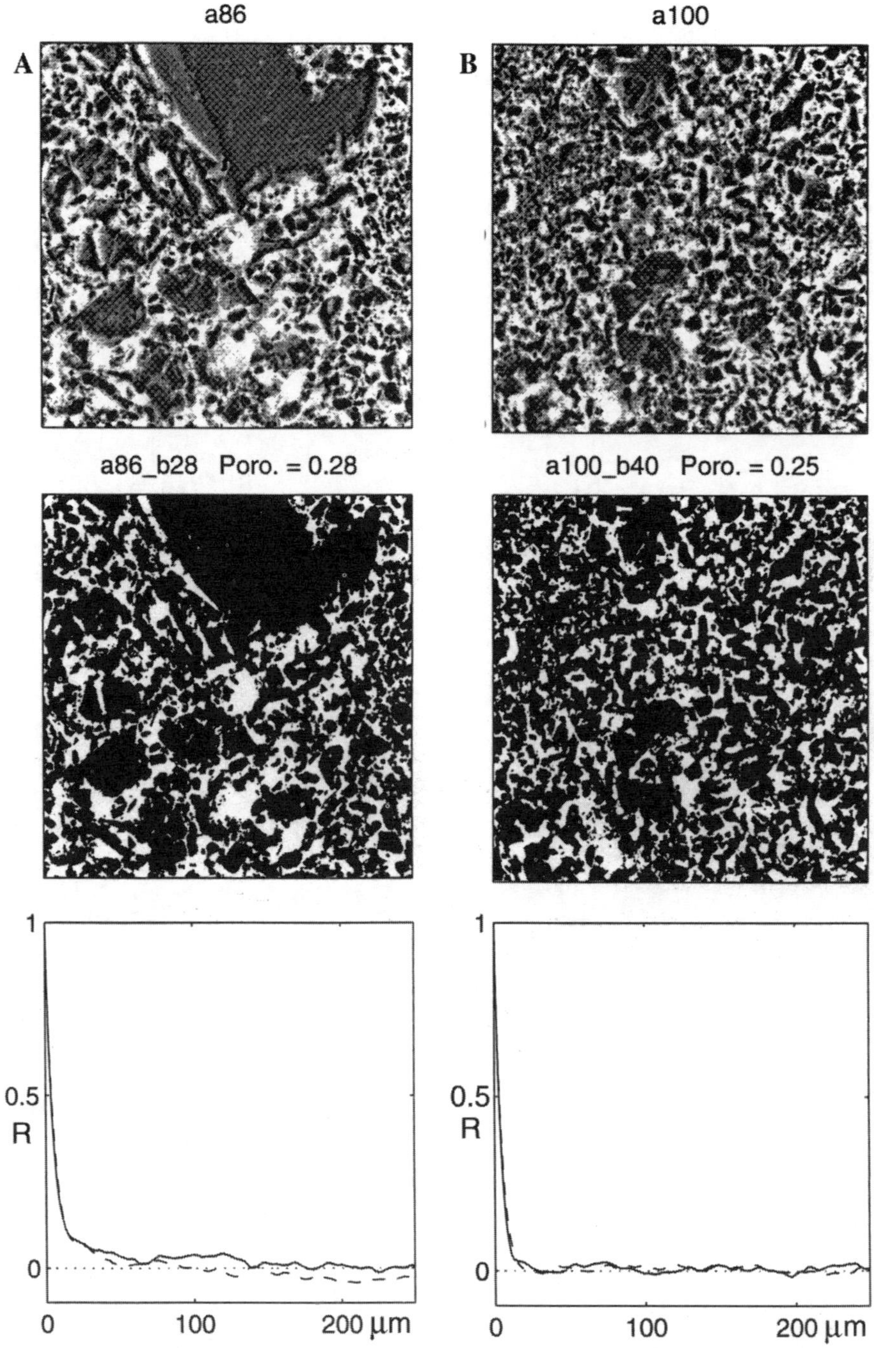

Fig. 13. SEM images (top), binarized equivalents (middle), and autocorrelation functions (bottom), based on one corner area, samples 3 and 6 in L9-11, (a) and (b) respectively. Pore spaces indicated in white, solid matrix in black. Image areas top and middle: 40 x 40 µm. (a) This image shows the sensitivity of simulated porosities to heterogeneities; in this case the simulated porosity is reduced to 28% by the large object visible in the image. The autocorrelation functions R_x (—) and R_y (---) show pore space anisotropy. (b) This example reveals a simulated porosity of only 25%, despite a qualitative similarity with the images in Fig. 12 b, with a simulated porosity of 35.7%. To compare, the analytical porosities for samples 3 (a) and 6 (b) were ≈ 12-15 porosity % higher. The autocorrelation functions R_x (—) and R_y (---) reveal low pore space anisotropy

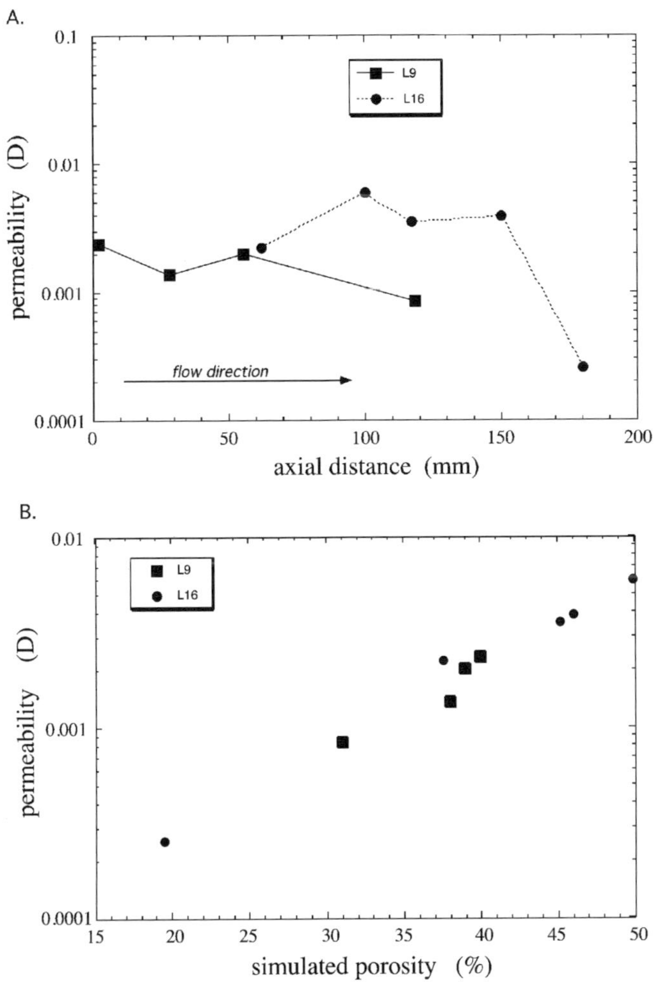

Fig. 14. (a) Simulated permeabilities along the central axes of L9-11 and L16 were derived using a reconstructed porous media method (Adler et al., 1990). Axial distance increases with direction of flow. The non-linear trend of L16 follows as a consequence of the irregular nature of the simulated porosities. (b) Simulated permeabilities as a function of simulated porosities for both L9-11 and L16 follow the same linear trend. Note that the permeabilities of L16 in particular are subject to error, given the uncertainties in the simulated porosities.

12a). The porosity of the 2000x binarized image is 35.7%. This simulated "microporosity" is in very good agreement with the measured porosities for sample no. 4 (35.1% by Hg injection and 37.6% by density technique). The autocorrelation functions R_x and R_y in both Figs. 12a and b are almost equivalent, indicating that the pore spaces show no preferred orientation.

Figure 13a shows a SEM image and its binarized equivalent for sample 3, L9-11 (see Fig. 4). This pair of images shows the sensitivity of simulated porosities to the presence of heterogeneities, in this case, a large object. The effect is dramatic, the porosity is reduced to 28%, which is much lower than the measured values of 40.6% and 43.0% (Hg injection and density methods, respectively). In the same figure the autocorrelation functions reflect a degree of anisotropy with respect to the pore spaces; this is most probably due to the influence of the large object at the top of the image. Figure 13b shows a set of SEM/binarized images (sample 6, L9-11, see Fig. 4) with a very low simulated porosity of only 25%. In comparison, the measured porosity values are significantly higher for sample 6: 41.0% and 44.0% by the Hg injection and density methods, respectively. Despite the presence of some medium-sized grains in

Fig. 13b, it is surprising that the porosity is significantly lower than that shown in Fig. 12b. The reason for the lower simulated porosity is most probably due to small pore spaces remaining unresolved during binarization of the SEM image.

Using the method of reconstructed porous media (Adler et al., 1990), the permeabilities of deformed chalk were calculated, based on the simulated porosities discussed above (using only 2000x images). The permeabilities for experiments L9-11 (samples 1, 2, 3, 5) and L16 (samples 18, 28, 33, 43, 53) as a function of axial distance and simulated porosity are shown in Figs. 14a, b, respectively. In Figure 14a, the (log) permeabilities show a general decrease as a function of axial distance; the non-linearity of the trends follows as a consequence of the irregular nature of the simulated porosity vs. distance profiles (especially expt. L16, see Fig. 11a). Figure 14b shows that the log permeability vs. porosity data for L9-11 and L16 are linear.

In analyzing the permeability relationships illustrated in Fig. 14, it is important to bear in mind that these simulated permeability estimates are subject to large errors, given the discrepancies between the simulated and the measured porosities (i.e., refer to Fig. 11a) and the fact that the permeability varies as the square of the correlation length, which is a function of the pore space diameters. Therefore, a small change in porosity translates to a large change in permeability. This has been observed experimentally with North Sea chalks, where measured permeabilities decrease sharply with increasing confining pressure (P > 8 MPa; Da Silva et al., 1985).

It is interesting to compare these simulated permeabilities with those for natural samples determined analytically. Permeabilities measured in natural chalk samples are approximately of the same order of magnitude. As an example, the permeabilities of Lixhe chalk (> 40% porosity, quarried close to Guerville chalk) are on the order of ≈ 1 mD (Schroeder et al., 1998). Similarly, chalk cores from the North Sea oil fields have permeabilities ranging from ≈ 1-7 mD (Monjoie et al., 1990).

3.4. X-Ray Computerized Tomography

In the present study, the x-ray density in the CT scans has been represented in terms of a gray scale tonality, such that high-density objects appear white and low-density objects appear black. The white rims associated with the CT images are an artifact of the technique. In order to better evaluate the post-deformation tomography images of experiments L9-11 and L16, the images obtained for an undeformed (reference) chalk sample are described first. The

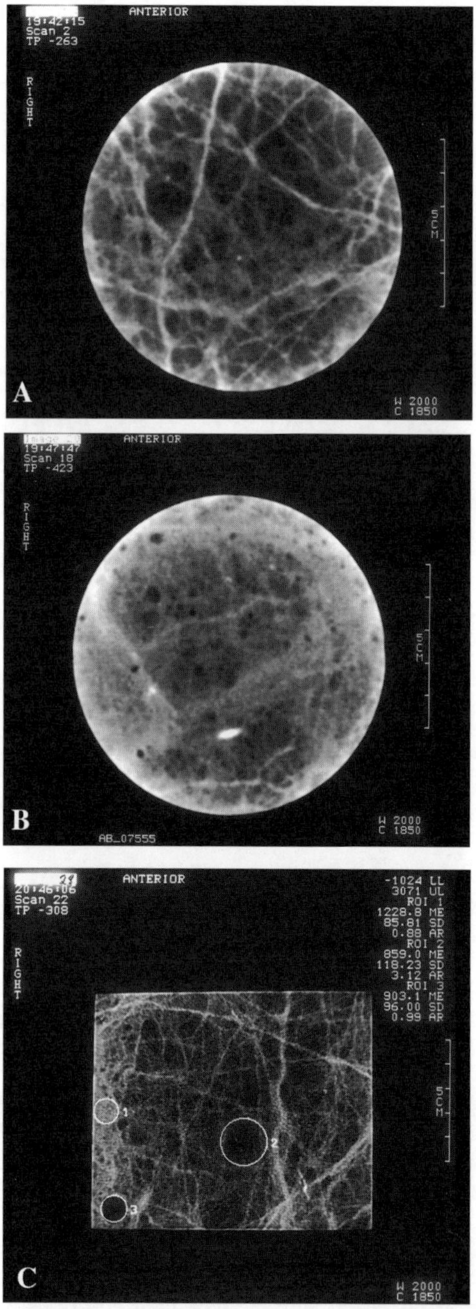

Fig. 15. X-ray computerized tomography (CT) images of an undeformed, reference sample. High and low radiological density regions are portrayed in white and black, respectively. The top and middle figures show radial images representing the far right-hand (a) and far left-hand side (b) of the axial image shown in (c). Because the samples are almost entirely composed of calcite, the radiological density can be qualitatively interpreted in terms of porosity. All images show the heterogeneous nature of the chalk, especially notable are the system of cross-cutting veins and the black "worm hole channels."

structures discussed with respect to the reference are either primary sedimentary and/or diagenetic features, or secondary in situ dissolution/precipitation features.

3.4.1. Reference sample

Sequential radial images of the reference core show that the original spatial density of the chalk is very heterogeneous, which supports the SEM observations discussed previously. Two representative radial images and an axial image are shown in Figs. 15a-c, respectively. The image in Fig. 15a corresponds to a radial slice on the far right-hand side of the axial image (Fig. 15c), whereas Fig. 15b is from the far left-hand side. All of the images display internal structures with large radiological density variations. In Fig. 15c, for example, 3 selected areas reveal large differences in density (expressed in terms of mean Hounsfield units: ME), ranging from 1229 ME (area no. 1, which shows an area of high vein density) to 859 ME (area no. 2), which corresponds to a dark region of low density. Note that the two radial images show a difference in overall density of the matrix, as evidenced by the slightly lighter color of the matrix in Fig. 15b; the lighter color translates qualitatively to a reduced porosity and more compact structure than that in Fig. 15a.

Both radial images contain numerous dark, circular to oblong-shaped regions, which are perhaps more visible in Fig. 15b, due to the lighter matrix tone. When closely comparing Figs. 15a and b, it appears that the spatial density and distribution of these regions is more or less the same throughout the reference. These dark regions have variable dimensions (up to ≈ 1 cm in diameter), and appear to be randomly distributed in all of the images; the smallest of these may correspond to the very large channels seen in SEM images- see Fig. 2. These zones appear to be spatially isolated entities since they are not sequentially continuous over more than 1 or 2 radial slices (slice separation = 7 mm). Note that in Fig. 15c, there is no evidence for single, long tubes of low porosity. Nonetheless, some of the channels may be oblique to the slices, thereby masking their connectivity. These dark, low density regions or channels resemble very closely dissolution features (often called "worm holes") created by acid percolation in limestones (e.g., see CT images of worm holes in Bazin et al., 1996). This suggests that these zones are due to dissolution, which results in the formation of short, circular to tubular zones of elevated porosity. In the study by Bazin et al. (1996), the continuity of the worm holes was a function of the rate of fluid flow; high flow rates favored single, continuous worm holes, whereas slow flow rates favored multiple-branching worm holes. Thus, the lack of single, continuous worm holes in the present study (reference and deformed samples) suggests that flow rates were slow enough to promote branching. If channel branching did occur, then some form of connectivity between some of the vertical channels cannot be excluded.

Equally ubiquitous is the system of veins, which cross cut all of the radial and axial images of the reference. As opposed to the dark round zones, the veins show a certain degree of continuity from one radial image to the next. Note how the system of veins is more prominent in Fig. 15a vs. b; this is simply a function of a higher density contrast due to a less dense matrix (i.e., higher porosity matrix) in the former image.

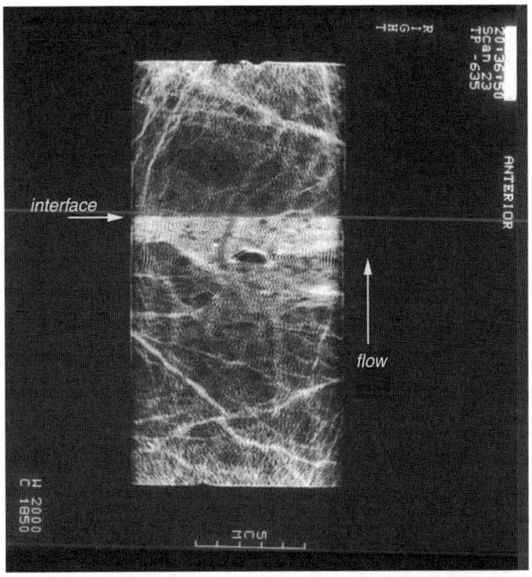

Fig. 16. Axial CT image of L9-11 after deformation. The most notable feature is a region of high density/low porosity adjacent and upstream of the interface between the two cores. The exact location of the interface is defined by the sharp and linear density contrast. The low porosity, tabular region may be the result of precipitation that occurred during deformation- see text for details.

3.4.2. Deformed samples

CT images of deformed chalks are shown in Figs. 16-19. The comments about channels (worm holes) discussed immediately above also apply to all of the deformed samples. A *possible* redistribution of mass due to pressure solution deformation and precipitation can be noted in an axial CT image of sample L9-11, shown in Fig. 16. In this image, fluids percolated from bottom to top in the vertically-oriented core. The most

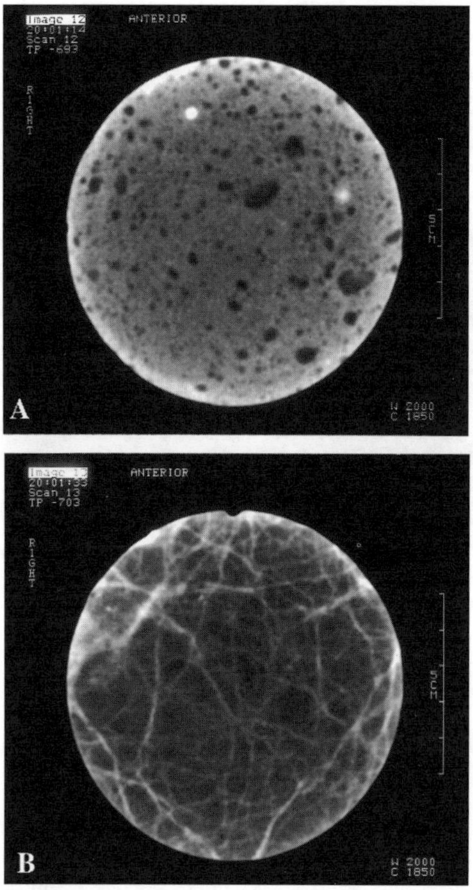

Fig. 17. Radial CT images of L9-11 after deformation, representing the interfacial area between the 2 cores. (a) CT image directly adjacent to the interface, on the upstream side (see Fig. 16). The high porosity, dark circular objects probably represent dissolution features; their connectivity is difficult to determine, however. (b) CT image directly adjacent to the interface, on the downstream side (see Fig. 16). The majority of radial CT slices, in both cores of L9-11, bear a strong resemblance to this image.

notable feature is a tabular, ≈1-1.5 cm-thick low porosity zone that occurs exclusively in the upstream core, and is perpendicular to the core axis and fluid flow direction. The sharp, linear contrast in density delimits the interface between the two cores.

Sequential radial CT images of the entire L9-11 core were also realized, with a 1-cm interval. This sequence of radial images displays a sudden change in density (and porosity) that correlates exactly to that shown in the axial image. Figures 17a, b show adjacent (1-cm separation) core slices on opposite sides of the interface, the upstream slice (Fig. 17a) is within the high-density, low-porosity zone; Fig. 17b represents the adjacent downstream slice with lower density and higher porosity. A comparison of Figs. 17a and b reveals an apparently higher density of worm holes in the region of low porosity, shown in the former image. The regions of higher density evident in Figs. 16 and 17a may correspond to low porosity cemented regions that were evidenced by SEM, as shown in Fig. 3c. The matrix shown in Fig. 17b (downstream core) is very much like that found in the majority of the upstream core, to the right (upstream) of the high-density zone shown in Fig. 16.

CT images of L16 show that this core displays roughly the same heterogeneous density variations associated with cross-cutting veins and dark, circular to oblong-shaped, low density regions as are present in L9-11. Figure 18 shows two axial images; fluid flow was from bottom to top. The interface between the two parts of the core is easily discernable in terms of a density difference; the downstream core is interpreted to have a higher porosity adjacent to the interface of the two cores. On the other side of the interface (upstream side) is a lenticular region characterized by matrix of higher density and lower porosity (Fig. 18a). The same region in Fig. 18b has a far more regular, tabular geometry. The density contrast of this region with respect to the surrounding matrix is simi-

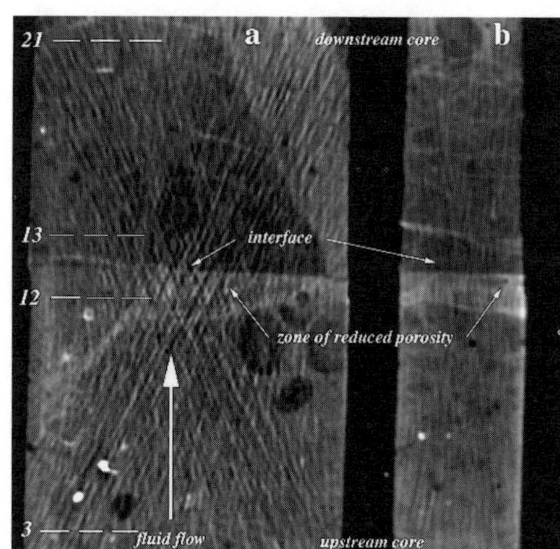

Fig. 18. Axial CT images of L16 after deformation; the interface between the cores is indicated. Image (a) is close to the axis center, image (b) is far off-center (hence the reduced thickness). Both (a) and (b) are characterized by a high density/low porosity region in the upstream core, adjacent to the interface (compare to Fig. 16). The thick diagonal band in the downstream core (image a) is interpreted to be a primary sedimentary feature; the faint diagonal bands in both images are an artifact. Nos. 3, 12, 13, and 21 (image a) refer to the locations of the radial slices shown in Fig. 19.

A.

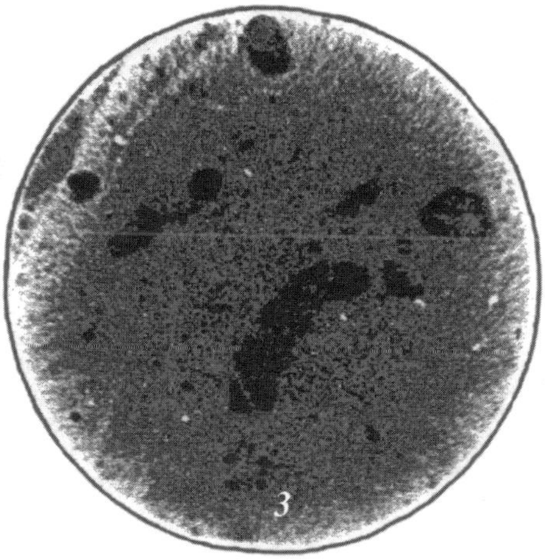

D.

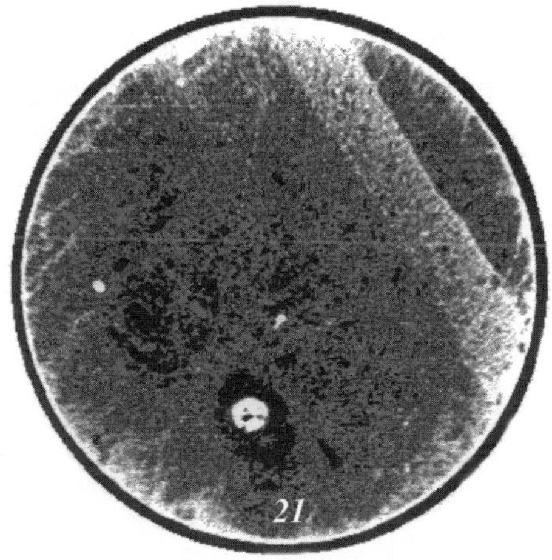

Fig. 19. Sequential radial CT images (ϕ = 10 cm) of L16 after image treatment (positions of slices 3, 12, 13, and 21 (a, b, c, d, respectively) shown in Fig. 18a). The relative amount of "intermediate porosity" (shown by medium gray tone) decreases sharply immediately adjacent and upstream of the interface (slice 12, image b), which is in accord with the axial image shown in Fig. 18. This is interpreted to be a result of precipitation in pore spaces during deformation.

B.

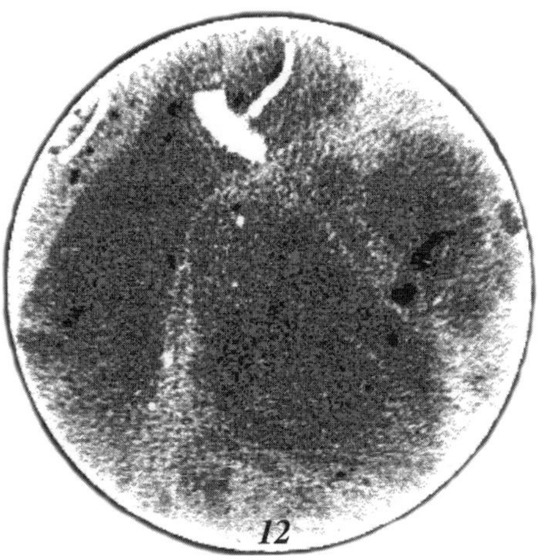

C.

lar to the density contrast described for L9-11; in addition, the general geometry with respect to the interface, the dimensions, and the irregular nature of this region are similar to that in L9-11 (compare with Fig. 16). The downstream core also contains a high-density region (but much larger), in the form of a continuous diagonal band (Fig. 18a). This diagonal band appears in nearly all radial slices of the downstream core. The band's diagonal orientation to the fluid flow direction, its large size, as well as its sharp boundaries (i.e., on both edges-see radial slice 21 in Fig. 19) strongly suggest that this feature is a syn-sedimentary feature, and is most probably not attributable to fluid percolation, deformation, and mass redistribution within the chalk core.

In order to further investigate the spatial density variations associated with L16, 21 sequential radial tomographic images (1 cm-interval along core axis) were analyzed with image treatment software (NIH Image, release 1.61). Each radial tomography image was converted into a selective density slice, where each slice was defined by the same gray-scale density interval. The grayscale interval defines a constant interval or range of densities for any given image; the interval (which is user-adjustable) was chosen to be representative of an 'intermediate' chalk matrix density, such that high-density objects (chert fragments, veins, and bands unrelated to deformation) and low-density objects (dark, circular to oblong regions)

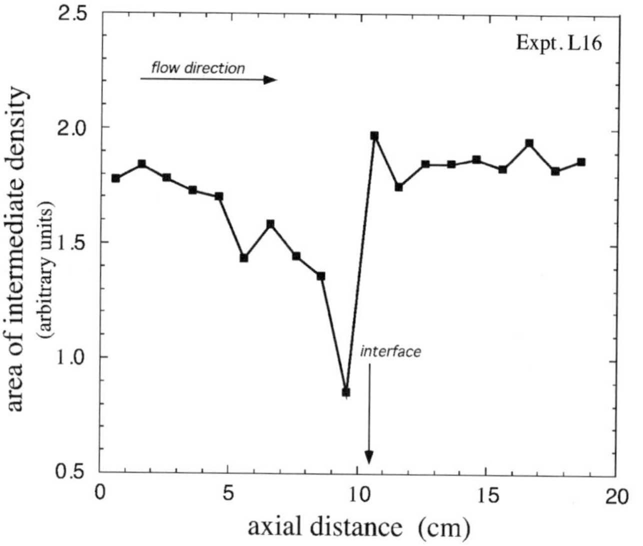

Fig. 20. Axial profile of 'intermediate' porosity of L16 (see caption Fig. 19). The decrease in 'intermediate' porosity as a function of flow direction in the upstream core is similar to overall porosity trends shown for L9-11 (all measurements) and L16 (water saturation method).

were excluded from the selective density area determinations. Figure 19 shows a sequential series of 4 selective density slices, starting from the inflow end of the upstream core (no. 3) and ending at the outflow end (no. 21) of the downstream core (i.e., bottom to top on axial image, Fig. 18a); the gray-scale density interval used for image analysis corresponds to the areas of intermediate gray tone visible in each slice. The amount of intermediate density matrix decreases significantly in slice 12; this slice corresponds to the region of high density, low porosity adjacent to the interface between the two cores (see axial image in Fig. 18). A profile representing the amount of the intermediate density (i.e., intermediate gray tone pixels of each slice) as a function of axial length is shown in Fig. 20. This profile, which should be interpreted only in a qualitative sense, shows a sharp minimum at the axial position corresponding to the upstream side of the interface between the two cores. This minimum corresponds to a sharp decrease in the area associated with 'intermediate' matrix porosity; this is interpreted to be due to a decrease of the overall porosity of the matrix by precipitation of material in pore spaces close to the interface.

Since it was not possible to obtain CT scans of each core before deformation, it is difficult, if not impossible, to positively ascribe each of the high-density/low-porosity features observed in the CT scans of L9-11 and L16 to processes of mass redistribution and precipitation that occurred as a consequence of pressure solution deformation. One can, for example, take the position that the high-density/low-porosity regions in L9-11 and L16 were already present before deformation. In support of this argument is the qualitative similarity between these regions and the high-density/low-porosity region of the reference (left-hand side of axial image, Fig. 15c). To counter this argument, it is rather improbable (but not impossible) that pre-existing high-density/low-porosity regions in L9-11 and L16 should always occur adjacent and parallel to the artificially-created boundary of the two cores, and always perpendicular and upstream with respect to the fluid flow direction. The high-density/low-porosity region of L9-11, in particular, appears to be more uniform than that in the reference, this supporting the idea of syn-deformational mass redistribution and uniform precipitation in pore spaces. It can also be noted that the upstream boundaries of the high-density regions in L9-11 and L16 are sharp, whereas in the reference, this region is diffuse and vein-like. This also supports the idea that the high-density regions in the reference and those in L9-11 and L16 were formed by different processes. Of course, it is also possible that the high-density/low-porosity regions of L9-11 and L16 started out as porosity regions with an elevated density/low porosity (e.g., left-hand side of axial image, Fig. 15c), and that precipitation was localized in these areas, such that during deformation the already present heterogeneities were accentuated. The various scenarios proposed above, unfortunately, cannot easily be resolved based solely on the data in the present study.

4. CONCLUSIONS

An important goal of this study was an investigation into how a porous material such as chalk compacts as a function of pressure solution deformation. The porosities of several reference samples (non-deformed), as well as post-deformation samples from experiments L9-11 and L16, were extensively measured in order to document the changes in porosity due to compaction. Radiological density measurements (CT scans) provided complementary information concerning density and porosity variations in a reference sample, as well as L9-11 and L16. As was discussed in detail in section *3.4.*, the interpretation of the post-deformation data is difficult because of the initial heterogeneous nature of the chalk. The ambiguities associated with the interpretation of the data make it difficult to resolve one of the key aspects of the pressure solution process - the redistribution of mass associated with compaction and precipitation in pore spaces. It is relatively certain that the primary deformation mechanism observed in the present experiments is pressure solution creep (as discussed in detail in Part 1). An important question that must be asked is what happened to the dissolved material that was removed from stressed intergranular regions?

It is evident that the process of mass redistribution, from stressed intergranular boundaries to pore spaces, is not simple. The precipitation of dissolved material can theoretically take place in the adjacent pore spaces. Many studies have documented grain overgrowths that have been interpreted to result from precipitation within the context of the pressure solution process (e.g., Dewers and Hajash, 1995). The provenance of grain overgrowth precipitates is not always clear, unless however, the pore spaces are completely unconnected, such that dissolved material can only precipitate in adjacent pore spaces. The SEM images in the present study do not show any obvious individual grain overgrowths, even though it is possible that precipitation occurred at a sub-micron scale and was therefore not observed. On the other hand, many regions containing cemented grains on the order of 10 μm in size were observed by SEM (see Fig. 3c). While such regions also were observed in undeformed chalk samples (see vein in Fig. 1b), the frequency of occurrence was higher in deformed samples, and was most notably evident in the upstream area adjacent to the two-core interface of L9-11. This area corresponds exactly to the region of high-density and low-porosity that was observed in the CT scans (Fig. 16) and measured analytically by various porosity techniques. Thus, it is very likely that the redistribution of mass in the chalk samples occurred by a deposition process that resulted in a precipitation front, producing large-scale cementation of grains. Taking chalks in the Mons basin as a comparative example, the localization of calcium carbonate precipitation, in the form of overgrowths and cemented grains, immediately adjacent to fault zones is evidence that mass redistribution associated with pressure solution deformation can be very localized in natural systems (Gaviglio et al., 1993, 1997). On a more general scale, the presence of precipitation fronts is a commonly observed geochemical phenomenon; this type of behavior has been shown to be associated with many types of ore deposits, for example.

If the high-density/low-porosity zones in L9-11 and L16 are interpreted to be due to precipitation-mass redistribution associated with pressure solution creep of chalk, then it is possible that the presence of a continuous void space, such as was present at the interfaces of the two cores in L9-11 and L16, controlled in some manner the localization of preferential precipitation. A possible change in the hydrodynamic behavior and/or a decrease in the pore pressure of the saturated solution upon entering such void regions may have had an effect on the precipitation kinetics of calcite. The onset of precipitation may in turn have favored continuous precipitation upstream from the initial regions of precipitation, thereby resulting in thick and tabular low-porosity regions. The occurrence of tabular low-porosity regions, due to uniform precipitation within the matrix, is also in accord with results from numerical models (e.g., Sallès et al., 1993; Békri et al., 1995, 1997; Mourzenko et al., 1996), based on the hydrodynamics of fluid flow, diffusivity, and kinetic rate parameters (see section *4.3.3*, Part 1 for details).

There are naturally occurring cases that may be analogous to what was observed in the present study; two examples are given. In the first example, Dunnington (1967) noted that stylolites tend to start in regions that originally have a higher degree of porosity. The formation of stylolites and the concomitant transfer and precipitation of dissolved material in turn results in regions with decreased porosity, adjacent to the higher porosity zones. Using the Mons chalk study by Gaviglio et al. (1993) as the second example, porosity reduction and cementation of grains occurred predominantly and immediately adjacent to the normal fault plane (i.e., on both sides of fault plane). It is hypothesized that fluid flow occurred from the surrounding chalk matrix to the fault planes, given that the fluid pressure in the matrix was probably higher than within the fault plane. In both of these examples, it appears that pressure solution is favored in areas or along planes that originally had a higher degree of porosity than the surrounding rock. The higher porosity zones were a locus for pressure solution, and the surrounding areas of lower porosity were the locus of local precipitation of the dissolved material.

The type of scenario described above correlates with theoretical self-organizing systems that operate on a positive feedback system (e.g., Merino et al., 1983; Merino, 1992; for a general overview, see Ortoleva, 1994). The central idea is that the inherent stress-strain energy of grain aggregates is a function of texture or porosity. Areas of slightly higher porosity (such as bedding planes, fault planes, or in the present case, the interface between the two cores) have a higher average stress-strain energy, and this causes preferential dissolution to become localized there at an accelerating rate. In a similar manner, areas of slightly lower porosity have less stress-strain energy, and thus are more likely to be a locus of precipitation. The overall result is that local porosity fluctuations tend to be amplified with time because of a positive feedback relationship that exists between porosity and preferential dissolution and precipitation. An additional factor that can be considered is the possibility that regions or zones of local precipitation behave in a more brittle manner during deformation as compared to the surrounding matrix (i.e., core interfacial regions in L9-11 and L16). The development of fractures, as a result of brittle behavior, is thought to be responsible for increasing diffusional transfer of material during deformation, thus amplifying the efficacy of pressure solution (see Gratier et al., 1999). A positive feedback process, which includes all of the factors described above, is a plausible explanation for the porosity profiles that developed during the deformation chalk, as exemplified by experiments L9-11 and L16.

Acknowledgements—This research was supported by generous grants from Elf Aquitaine (Pau, France) and the GdR-Géomécanique program (CNRS). The assistance of F. Verhelst, A. Timmerman, and K. Vandersteen (all from K.U. Leuven, Belgium) in obtaining the CT images is gratefully acknowledged. Comments from M. Jessell, C. Schroeder, F. Renard, and A.-M. Boullier aided in clarifying certain points in the original manuscript. P. Gaviglio acknowledges assistance from M. Boué in the preparation of samples. We thank Brian Evans, Enrico Merino, and Andrew Hajash for their constructive reviews of the manuscript (note: the original reviews were based on a single manuscript, before it was divided into Parts 1 and 2; however, the scientific content and interpretations remain unchanged). R.H. also thanks Brian Evans for assuming the task of editorial handling.

Finally, I would like to acknowledge David Crerar's role as teacher, advisor, and friend during the 6 years of my graduate studies in his geochemistry group in the Department of Geology and Geophysics, Princeton University. -Roland Hellmann

Editorial handling: Brian Evans

REFERENCES

Adler P. M. (1994) The method of reconstructed porous media. *Current Topics Phys. Fluids* **1**, 277-306.

Adler P. M., Jacquin C. G., and Quiblier J. A. (1990) Flow in simulated porous media. *Int. J. Multiphase Flow* **16**, 691-712.

Adler P. M., Jacquin C. G., and Thovert J.-F. (1992) The formation factor of reconstructed porous media. *Water Resources Res.* **28**, 1571-1576.

Bathurst R. G. C. (1995) Burial diagenesis of limestones under simple overburden. Stylolites, cementation, and feedback. *Bull. Soc. géol. France* **166**, 181-192.

Bayly B. (1987) Nonhydrostatic thermodynamics in deforming rocks. *Can. J. Earth Sci.* **24**, 572-579.

Bazin B., Bieber M. T., Roque C., and Boutéca M. (1996) Improvement in the characterization of the acid wormholing by "in-situ" X-ray CT visualizations, SPE Paper 31073. *International Symposium on Formation Damage Control*, Lafayette, Louisiana, U.S.A.

Békri S., Thovert J. F., and Adler P. M. (1995) Dissolution of porous media. *Chem. Eng. Sci.* **50**, 2765-2791.

Békri S., Thovert J.-F., and Adler P. M. (1997) Dissolution and deposition in fractures. *Eng. Geol.* **48**, 283-308.

Brace W. F. (1972) Pore pressure in geophysics. In *Flow and Fracture of Rocks, Geophys. Monogr. Ser.* Vol. 16 (eds. H. C. Heard, I. Y. Borg, N. L. Carter, and C. B. Rayleigh). AGU, Washington, D.C., pp. 265-273.

Brantley S. L., Evans B., Hickman S. H., and Crerar D. A. (1990) Healing of microcracks in quartz: implication for fluid flow. *Geology* **18**, 136-139.

Brown M. A., Brown M., Carlson W. D., and Denison C. (1999) Topology of syntectonic melt-flow networks in the deep crust: Inferences from three-dimensional images of leucosome geometry in migmatites. *Amer. Mineral.* **84**, 1793-1818.

Da Silva F., Monjoie A., Debande G., Schroeder C., Poot B., Detiege C., and Halleux L. (1985) Mechanical behaviour of chalks. *2nd North Sea Chalk Symposium*. Stavanger, Norway, **2**, 1-10.

de Meer S. and Spiers C. J. (1995) Creep of wet gypsum aggregates under hydrostatic loading conditions. *Tectonophysics* **245**, 171-183.

de Meer S. and Spiers C. J. (1997) Uniaxial compaction creep of wet gypsum aggregates. *J. Geophys. Res.* **102, no. B1**, 875-891.

Dewers T. and Hajash A. (1995) Rate laws for water-assisted compaction and stress-induced water-rock interaction in sandstones. *J. Geophys.Res.* **100, no. B7**, 13093-13112.

Dewers T. and Ortoleva P. (1990) A coupled reaction/transport/mechanical model for intergranular pressure solution, stylolites, and differential compaction and cementation in clean sandstones. *Geochim. Cosmochim. Acta* **54**, 1609-1625.

Dunnington H. V. (1967) Aspects of diagenesis and shape change in stylolitic limestone reservoirs. *Seventh World Petroleum Congress*, 339-352.

Evans B., Bernabé Y., and Zhu W. (1999) Evolution of pore structure and permeability of rocks in laboratory experiments. In *Growth, Dissolution and Pattern Formation in Geosystems* (eds. B. Jamtveit and P. Meakin). Kluwer Academic Publishers, Dordrecht. pp. 327-344.

Gaviglio P., d'Albissin M. C., Bergerat F., and Vandycke S. (1993) Modifications de texture dans la craie au contact de failles normales: un exemple de graben dans le basin de Mons (Belgique). *Bull. Soc. géol. France* **164**, 565-575.

Gaviglio P., Adler P., Thovert J.-F., Vandycke S., Bergerat F., Bekri S., and Lestideau R. (1997) Grain-scale microstructure and physical properties of faulted chalk. *Bull. Soc. géol. France* **168**, 727-739.

Géraud Y., Mazerolle F., Raynaud S., and Lebon P. (1998) Crack location in granitic samples submitted to heating, low confining pressure and axial loading. *Geophys. J. Int.* **133**, 553-567.

Gratier J.-P. (1987) Pressure solution-deposition creep and associated tectonic differentiation in sedimentary rocks. In *Deformation of Sediments and Sedimentary Rocks No. 29* (eds. M. E. Jones and R. M. F. Preston). Geological Society, London. pp. 25-38.

Gratier J.-P. and Gamond J. F. (1990) Transition between seismic and aseismic deformation in the upper crust. In *Deformation mechanisms, rheology and tectonics* (eds. R. J. Knipe and E. H. Rutter). Geological Society of London Special Publication 54, The Geological Society, London. pp. 461-473.

Gratier J.-P. and Jenatton L. (1984) Deformation by solution-deposition, and re-equilibration of fluid inclusions in crystals depending on temperature, internal pressure and stress. *J. Struct. Geol.* **6**, 189-200.

Gratier J.-P., Chen T., and Hellmann R. (1994) Pressure solution as a mechanism for crack sealing around faults: Natural and experimental evidence. In *Proceedings of Workshop LXIII The Mechanical Involvement of Fluids in Faulting* (eds. S. Hickman, R. Sibson, and R. Bruhn). Open-file Report 94-228, U.S.G.S., Menlo Park. pp. 279-300.

Gratier J.-P., Renard F., and Labaume P. (1999) How pressure solution creep and fracturing processes interact in the upper crust to make it behave in both a brittle and viscous manner. *J. Struct. Geol.* **21**, 1189-1197.

Gray D. R. and Durney D. H. (1979) Crenulation cleavage differentiation: implication of solution-deposition processes. *J. Struct. Geol.* **1**, 73-80.

Gregg S. J. and Sing K. S. W. (1982) *Adsorption, Surface Area and Porosity*. Academic Press, San Francisco.

Groshong R. H. Jr. (1988) Low-temperture deformation mechanisms and their interpretation. *Geol. Soc. Amer. Bull.* **100**, 1329-1360.

Heard H. C. and Page L. (1982) Elastic moduli, thermal expansion and inferred permeability of two granites to 350 °C and 55 MPa. *J. Geophys. Res.* **87**, 9340-9348.

Heidug W. K. and Leroy Y. M. (1994) Geometrical evolution of stressed and curved solid-fluid phase boundaries 1. Transformation kinetics. *J. Geophys. Res.* **99**, 505-515.

Herman G. T. (1980) *Image Reconstruction from Projections: The Fundamentals of Computerized Tomography*. Academic Press, San Francisco.

Hounsfield G. N. (1973) Computerized transverse axial scanning tomography. *Brit. J. Radiol.* **46**, 1016-1022.

Joshi M. Y. (1974) A class of stochastic models for porous media. Ph.D. thesis, University of Kansas, Kansas, U.S.A.

Lemée C. and Guéguen Y. (1996) Modeling of porosity loss during compaction and cementation of sandstones. *Geology* **24**, 875-878.

Leroy Y. M. and Heidug W. K. (1994) Geometrical evolution of stressed and curved solid-fluid phase boundaries 2. Stability of cylindrical pores. *J. Geophys. Res.* **99**, 517-530.

Merino E. (1992) Self-organization in stylolites. *Amer. Scientist* **80**, 466-473.

Merino E. and Dewers T. (1998) Implications of replacement for reaction-transport modeling. *J. Hydrology* **209**, 137-146.

Merino E., Ortoleva P., and Strickholm P. (1983) Generation of evenly-spaced pressure-solution seams during (late) diagenesis: A kinetic theory. *Contrib. Mineral. Petrol.* **82**, 360-370.

Merino E., Nahon D., and Wang Y. (1993) Kinetics and mass transfer of pseudomorphic replacement: application to replacement of parent minerals and kaolinite by Al, Fe and Mn oxides during weathering. *Am. J. Sci.* **293**, 135-155.

Merino E., Wang Y., Wang Y., and Nahon D. (1994) Implications of pseudomorphic replacement for reaction-transport modeling of rocks. *Mineral. Mag.* **58A**, 599-600.

Monjoie A., Schroeder C., Prignon P., Yernaux C., Silva F. D., and Debande G. (1990) Establishment of constitutive law of chalk and long term test. *3rd North Sea Chalk Symposium.* Copenhagen, Denmark, 1-17.

Mourzenko V. V., Békri S., Thovert J.-F., and Adler P. M. (1996) Deposition in fractures. *Chem. Eng. Communications* **148-150**, 431-464.

Nahon D. and Merino E. (1997) Pseudomorphic replacement in tropical weathering: Evidence, geochemical consequences, and kinetic-rheological origin. *Am. J. Sci.* **297**, 393-417.

NIH Image, 1996, release 1.61.

Onasch C. M. (1983) Origin and significance of microstructures in sandstones of the Martinsburg formation, Maryland. *Am. J. Sci.* **283**, 936-966.

Ortoleva P. J. (1994) *Geochemical Self-Organization*. Oxford University Press, New York.

Quiblier J. A. (1984) A new three-dimensional modeling technique for studying porous media. *J. Colloid Interface Sci.* **98**, 84-102.

Raynaud S., Fabre D., Mazerolle F., Geraud Y., and Latière H. J. (1989) Analysis of the internal structure of rocks and characterization of mechanical deformation by a non-destructive method: X-ray tomodensitometry. *Tectonophys.* **159**, 149-159.

Renard F., Park A., Ortoleva P., and Gratier J.-P. (1999) An integrated model for transitional pressure solution in sandstones. *Tectonophys.* **312**, 97-115.

Reuschlé T., Trotignon L., and Gueguen Y. (1988) Pore shape evolution by solution transfer: thermodynamics and mechanics. *Geophys. J.* **95**, 535-547.

Rootare H. M. (1970) A review of mercury porosimetry. In *Advanced Experimental Techniques in Powder Metallurgy* (eds. J. S. Hirschhorn and K. H. Roll). Plenum Press, New York. pp. 225-252.

Sallès J., Thovert J. F., and Adler P. M. (1993) Deposition in porous media and clogging. *Chem. Eng. Sci.* **48**, 2839-2858.

Schroeder C., Bois A.-P., Maury V., and Hallé G. (1998) Water/chalk (or collapsible soil) interaction: Part II. Results of tests in laboratory on Lixhe chalk to calibrate water/chalk models. *Eurock '98*. Trondheim, Norway, **2**, pp. 505-514 (SPE 47587).

Smith D. L. and Evans B. (1984) Diffusional crack healing in quartz. *J. Geophys. Res.* **89**, 4125-4135.

Spiers C. J. and Brzesowsky R. H. (1993) Densification behaviour of wet granular salt: Theory versus experiment. In *Seventh Symposium on Salt* **1**. Elsevier Science Publishers, Amsterdam. pp.82-92.

Verhelst F., Vervoort A., De Bosscher P., and Marchal G. (1995) X-ray computerized tomography: determination of heterogeneities in rock samples. *8th International Congress on Rock Mechanics*. Tokyo, Japan.

Walsh J. B. (1965) The effcts of cracks on the compressibility of rock. *J. Geophys. Res.* **70**, 381-389.
Wellington S. L. and Vinegar H. J. (1987) X-ray computerized tomography. *J. Petrol. Eng.* **39**, 885-898.

Part B.

Thermodynamics, Hydrothermal Geochemistry, and Ore Deposits:

Water-Rock Interactions, Ore Deposits, and Environmental Geochemistry: A Tribute to David A. Crerar

Editors: Roland Hellmann and Scott A. Wood

Stable and metastable equilibrium: The third constraint

GREG M. ANDERSON

Department of Geology, University of Toronto, 22 Russell Street, Toronto, Ontario M5S 3B1, Canada
e-mail: greg@galena.geology.utoronto.ca

...fundamental equations do not govern objects in reality; they govern only objects in models

—*Cartwright (1983).*

1. INTRODUCTION

Most people who have tried it would agree that teaching thermodynamics is a difficult task. Yet exactly why this is so is hard to explain. Certainly the mathematics involved is not the problem—it includes little more than relatively simple differential and integral calculus. The nature of the experiments involved is not the problem—simple piston-cylinder arrangements, calorimeters and chemical reactions are not usually mystifying to students. The problem seems to lie somewhere in the union of these two aspects—the physical meaning of the thermodynamic variables and processes, and how they are related to the real processes we observe around us every day. The questions that seem to bother us all when learning thermodynamics boil down to wondering how on Earth this collection of hypothetical systems, unattainable states, unknowable parameters and impossible processes can be of any use in the real world. Yet we know it can, and the learning process for most of us probably involves "getting used to it" more than getting to understand it. As Dickerson (1969) said, "It is quite possible to know thermodynamics without understanding it".

I describe here an approach to understanding the subject which has helped me. This approach emphasizes the role of models in science, and hence thermodynamics as a mathematical model of energy relationships, rather than as any description of real states or processes. In this, the role of *constraints* plays a central role, in particular, the *third constraint*, and the emphasis is on the meaning of metastable states.

2. SYSTEM

Virtually every textbook on thermodynamics begins by defining a system as some arbitrary part of the universe that interests us; a crystal, or a beaker containing some solution, or a gas, and so on.[1] Then it goes on to differentiate between closed, open and isolated systems, depending on the type of walls the system has, or how energy is able to be transferred to and from the system. Later, we have discussions of piston-cylinder arrangements, falling weights, and other very simple devices, all presumably real objects.

What's wrong with that? The problem is that thermodynamics, in common with virtually all physical theories, does not deal with any part of the real universe, but with highly simplified models, abstract objects characterized or described by systems of equations, which attempt to represent the behavior of selected parts of the universe. That it is reasonably successful at doing so makes it useful. The definition of system as some part of the universe is fine, if it is made clear that this is the system we are interested in. However, it is not really the system that thermodynamics deals with.

Every scientist is familiar with the concept of models, and normally will get impatient at any suggestion that he/she is deficient in this respect. Most scientists spend their lives devising models of physical or biological behavior, and the most successful ones, such as Newton's Laws of motion, Einstein's relativity, and quantum theory are justly famous. But consider the following:

> Strictly speaking, Newtonian physics must rank as the biggest confidence trick in the history of human learning: it makes all kinds of unrealistic assumptions about the existence of perfect vacuums, ideal gases, and frictionless processes, none of which ever occurs in nature.
>
> —*Dunbar (1995), p.98.*

This is a bit of hyperbole, used to grab the attention of the lay reader, but it illustrates a point well known to scientists, but not, generally, to others. Obviously, "Newtonian physics" does not apply to the real world, except to the extent that the simplified models that it does apply to mimic behavior in the real world. It works wonderfully well in guiding satellites in outer space, but not so well if you drop a feather and a rock together in a windstorm. All scientists know this.

Then why is it so common to pretend that "Gibbsian physics" applies to real geological systems?[2]

Well, you might say, true enough, but why make such a fuss about it? Can it be that important? Let's look a little further.

3. EQUILIBRIUM

Following the assertion that thermodynamics deals with real systems is generally the definition of equilibrium as a state of rest, of no perceptible change, in real systems (see variations of this in the Quotations section). However, some definitions are a little different, in that they mention the equality of temperature, pressure and chemical potentials throughout the system. Both definitions may occur in the same text.

A state of rest, or of returning to an initial state after a disturbance, can be used in observing real systems, and follows nicely from the definition of a system as a real crystal or solution, or whatever. A state of equal chemical potentials, however, can only be *inferred* from other observations, and clearly refers to models of systems, not real systems. In other words, the definition of equilibrium is quite different in real and model systems. As a matter of fact, it might be argued that no real system ever reaches equilibrium by the criteria applied to model systems. That is, there are always impurities, inhomogeneities and gradients of various sorts, however small, in all real systems.

3.1. Stable vs. Metastable

The distinction between stable and metastable equilibrium is generally that the stable equilibrium state is "truly unchanging", or unchanging given indefinite time, whereas the metastable state may be changing, but too slowly to be observed (see Quotations). This distinction clearly refers to real systems, and really misses the essential point.

In working with real systems it is quite common to be not at all sure whether equilibrium has been achieved, or if it has, whether it is stable or metastable. We even have systems which we know are not at equilibrium, but we apply thermodynamics to them anyway, using the local equilibrium concept (Knapp, 1989).[3] On the other hand, our thermodynamic models of these systems are invariably simple, and they are always at (perfect) equilibrium. As many authors have remarked, some thermodynamic quantities are not even defined in unstable states.[4] Furthermore, the distinction between stable and metastable is different, and completely under our control (because we make the model, we control every aspect of it).

The distinction between stable and metastable states in real systems is often difficult, and the usual definitions don't help much. We are quite sure that diamond is a metastable form of carbon at Earth surface conditions, but whether or not it will recrystallize to graphite given "indefinite time" is quite uncertain. On the other hand, aragonite is a metastable form of $CaCO_3$, but although common in recent rocks, it becomes progressively less common in older rocks, so maybe it is changing very slowly. Who knows? Obviously a messy topic, involving kinetics (Fyfe and Bischoff, 1965; Bischoff and Fyfe, 1968). But in both cases we know with certainty that there is a crystal form for that composition with lower energy. How to best express this in thermodynamic terms?

The best way is to discuss thermodynamic models of these systems, in which a metastable equilibrium state is a state of true or unchanging equilibrium, having a higher energy content than another state of the same composition and under the same conditions. It is prevented from spontaneously changing to this lower energy state by a *constraint*. In other words, the real system may be truly unchanging and metastable, or it may be unstable and changing very, very slowly. We don't know. But our model of this system has no such uncertainty. The system is modeled as metastable and unchanging. If the real system is really unstable, our model is in this respect incorrect, but probably useful nonetheless.

The subject of constraints deserves more discussion than is usually devoted to it. I originally realized the importance of constraints from reading Callen (1960) and Reiss (1965), but I now use the term slightly differently. This requires a bit of explanation.

3.2. Constraints

Constraint, like several other terms, has a slightly different meaning in real and model systems. Real systems, such as aragonite, are said to be constrained from reacting to a more stable state (calcite) by an activation energy barrier. The usage is rather imprecise, because we don't actually know if the aragonite is changing on some very long time scale or not. In model systems, the meaning is much more exact.

Thermodynamics deals with systems in various energy states. So we must begin (after defining our system) by defining how energy changes in these systems can occur. This in turn boils down to exchanges of heat and work, which we now know, after a great deal of confusion in the 19th century, are two forms of the same thing—energy. This results in the First Law,

$$\Delta U = q + w \quad (1)$$

where ΔU is the "internal" energy of a system, q is the energy transferred as heat, and w is the energy transferred as work.

There are many forms of work, but if we choose only the most common one, pressure-volume work, and after introducing entropy, equation (1) becomes

$$dU = T\,dS - P\,dV \quad (2)$$

There are two terms on the right side, because we chose two ways of changing energy (this is the origin of the '2' in the Phase Rule). If we instead chose two forms of work capable of changing our system energy, say pressure-volume work and work done by a magnetic field, equation (2) would have another term on the right side. We choose these terms (out of many possible choices), so this is already a model equation, not one about real systems.

To change the energy of a system at equilibrium, you must change a state variable. Therefore, each way of changing the energy of a system is associated with a state variable. A corollary is that, in model systems, equilibrium states can be characterized by their number of constraints, defined as state variables associated with the ways a closed system can change its energy content. The minimum number of constraints required to define an equilibrium state is two [e.g., S and V in equation (2)]. Equilibrium states having only two constraints have the lowest energy for given conditions and are stable equilibrium states. Equilibrium states having more than two constraints are *metastable* equilibrium states. Calcite, for example, has stable equilibrium states at chosen values of T and P within certain ranges. These are the two constraints. At these T and P values, aragonite has a third constraint, an energy barrier, which prevents it from recrystallizing to calcite. A system consisting of a beaker of water and a gram of solid salt also has three constraints—T, P, and the separation of water and salt. When the third constraint is released by adding the salt to the water, the system reaches stable equilibrium as a salt solution with only two constraints. At the risk of laboring the point, we are talking about model calcite and model aragonite. Real calcite and real aragonite have many other factors to consider, such as grain size, defects, impurities, strain energy, inhomogeneities, and so on. Normally, the simple model works well enough.

It may not be obvious what state variable could be associated with a "third constraint" consisting of an energy barrier or a separation of two parts of a system, but there is one. Whatever the nature of the third (or higher) constraint, the state variable used to describe it is the progress variable. If there are in fact three constraints, then there are three terms on the right side of equations dealing with energy change.

Switching variables from S and V to T and P, equation (2) becomes

$$dG = -S\,dT + V\,dP \quad (3)$$

for stable equilibrium states, and

$$dG = \left(\frac{\partial G}{\partial T}\right)_{P,\xi} dT + \left(\frac{\partial G}{\partial P}\right)_{T,\xi} dP + \left(\frac{\partial G}{\partial \xi}\right)_{T,P} d\xi \quad (4)$$

or

$$dG = -S\,dT + V\,dP - \mathbf{A}\,d\xi \quad (5)$$

for systems with a third constraint. T and P are the state variables associated with the two constraints defining the stable equilibrium state. S, V, and **A** define the slopes of the plane defined by equation (5) in the T, P, and ξ directions, and the progress variable ξ is the state variable controlling departure from the most stable state. It is the "third constraint". The term $\mathbf{A}d\xi$ thus determines how far removed the state is from the stable equilibrium state, in terms of joules. We have complete control over ξ in our model (e.g., in reaction path programs), though often not in real systems. Thus, although the distinction between stable and metastable states is often difficult in real systems, it is perfectly clear cut in model systems.

This kind of definition then leads to a very simple distinction and rationale for reversible *vs.* irreversible processes. In real systems, there are innumerable examples of irreversible processes, providing the motive for introducing entropy. They often involve quite complex situations. In model systems, an irreversible process is what happens when you release a third constraint, and the system is usually very simple.

Of course, in a general sense, everyone knows this. In fact, many topics in thermodynamics can hardly be discussed at all without admitting in some way or other that we are using idealized models. These subjects include, in addition to reversible processes,

- isolated systems
- infinite dilution
- frictionless pistons
- ideal gases
- ideal one-molal standard states
- absolute zero temperature, and
- ideal (Henryan and Raoultian) solutions.

So the problem is perhaps not that we don't use idealized models in teaching and discussing, but that we don't admit that the whole thing is an idealized model, in common with all physical theory.

4. REACTION PATH MODELS

Although there are many ramifications of the "model vs. reality" theme throughout thermodynamics, the one most closely related to the subject of sta-

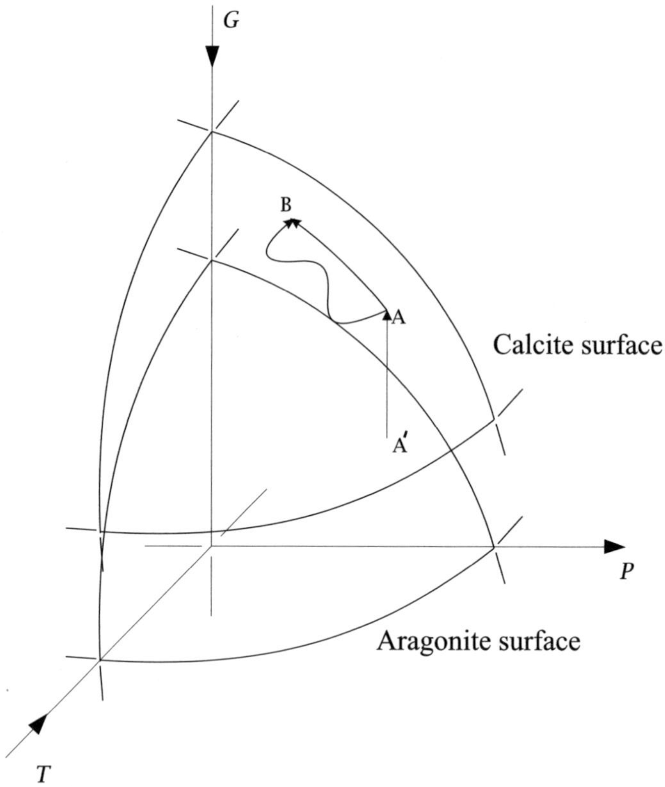

Figure 1. G–T–P surfaces for stable calcite and metastable aragonite. Note that for ease of representation, the G axis increases downwards. The two surfaces intersect somewhere off the diagram. Paths on a surface such as A → B represent reversible processes. Path A′ → A represents an irreversible process.

ble *vs.* metastable equilibrium is a proper understanding of the progress variable and reaction path models. Introduced by De Donder (1920), used extensively by Prigogine and Defay (1965) and made widely useful in the geochemical literature by Helgeson (1968, 1970), the progress variable has by now been discussed by many authors. Nevertheless, perhaps some confusion remains about how an irreversible reaction can be represented or calculated usefully by using strictly equilibrium thermodynamics.

It is here that a clear understanding of metastable equilibria is most useful. Figure 1 shows the G–*T*–*P* surface for a stable and a metastable form of the same compound, $CaCO_3$. At a given value of the two constraints *T* and *P*, aragonite has a more positive G than calcite, and is prevented from recrystallizing to calcite by a third constraint. The nature of the third constraint need not be made explicit, but in this case it is an activation energy barrier. Any line on either surface, such as the two shown (A → B), represents a reversible process, a continuous succession of equilibrium states. Within the calcite stability field, any change of aragonite to calcite (e.g., along A′ → A) is irreversible.

In real life, controlling this irreversible process would be extremely difficult, and more complex examples would certainly be impossible to control. In the model, we simply consider any number of intermediate metastable states, as shown in Figure 2, consisting in this case of various mass ratios of the two minerals.[5]

The classic case used by Helgeson (1979) is K-feldspar dissolving in water. The point of view expressed here is illustrated in Figure 3. The system as a whole is metastable, the third constraint being the separation of the feldspar and water. This separation is altered in small steps, represented by small grains dropping into the water. After each grain is added we have a new metastable state at complete equilibrium.

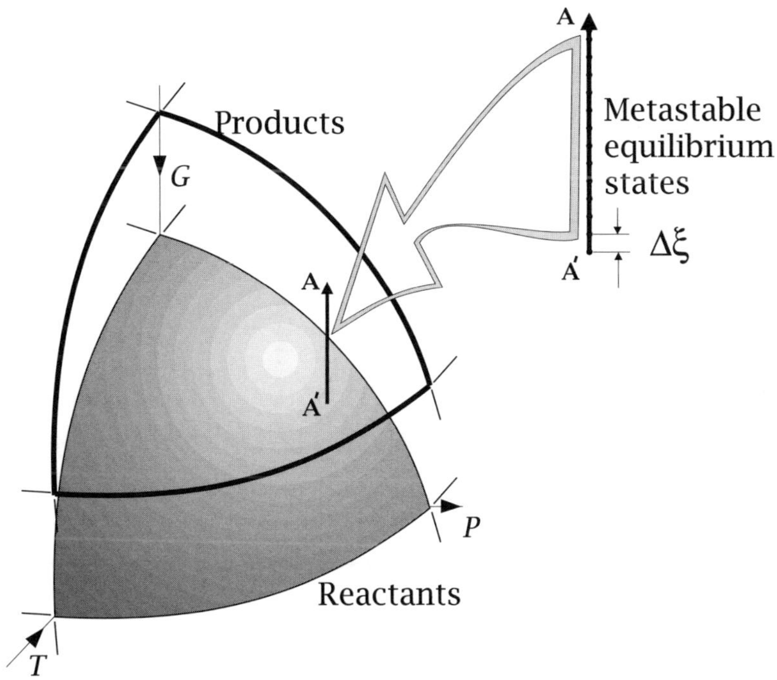

Figure 2. A sequence of metastable equilibrium states for the reaction A′ → A at constant T, P. The progress variable is ξ. Note that for ease of representation, the G axis increases downwards.

The common alternative to this explanation is a discussion of the dissolution of real feldspar in water, during which the feldspar dissolves slowly, and the system achieves what is called "partial equilibrium", because all solutes in the solution are at equilibrium, but the solution is not at equilibrium with the feldspar. This is simply another confusion of real and model systems. There is no such thing as partial equilibrium in model systems except in special cases, such as osmotic or semi-permeable membrane systems. All parts of model systems are in complete (stable or metastable) equilibrium.

5. PROGRESS PATH MODELS

Actually, equation (5) is quite general as to the nature of the third constraint, and the progress variable can be used to represent any kind of third constraint.

For example, if the system consists of an electrochemical cell, it will have an equilibrium voltage, which will gradually decrease to zero as the cell reaction takes place. In the model, this cell voltage can be controlled, and is the third constraint. It matters not how this is done in the model, but in real life it might be done with a potentiometer. This is a case where the metastable → stable transition is easily controlled, in contrast to many other situations.

If the system consists of a crystal having an equilibrium number of defects, we can consider cases with a greater or fewer number of defects, which will be metastable states. The third constraint is whatever we imagine as the control we have over the number of defects, which we can change at will using the progress variable. You might imagine this control being an ability to warm the crystal until the desired number of defects has formed, then quenching it so that they are "frozen in", but any discussion of this sort also misses the point, which, to repeat, is that we are dealing with model defects, not real ones. There is no mention of how one actually controls the number of defects in the model. In fact, one can usually detect "reality *vs.* model" confusion by any mention of time ('slow dissolution'; 'move the piston very slowly'; 'release the constraint for a very brief period'; 'quenching', 'freezing', 'thawing'; and so on). Time is important in real systems, but it has no place in equilibrium thermodynamics, McGlashan to the contrary notwithstanding.[6]

Many seemingly more bizarre cases (in the sense that they do not conform to normal lab practices) are actually quite common in the modeling world. In geochemical modeling, you can use practically any state variable as a third (or higher) constraint, and vary it at will. For example you can specify a *p*H or a particular gas fugacity. Controlling these state vari-

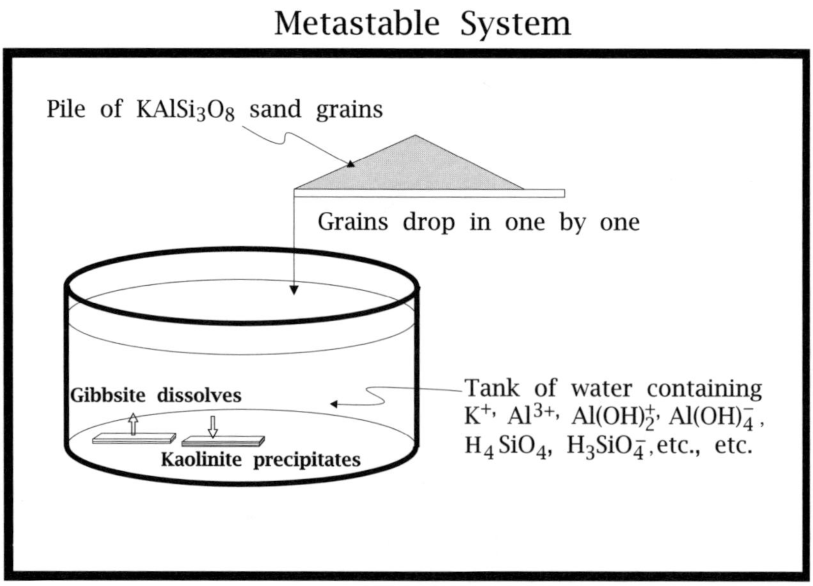

Figure 3. A possible conceptualization of the reaction path model for the dissolution of K-feldspar in water.

ables independently, rather than allowing them to be controlled by, say, T, P, and bulk composition, means they are additional constraints. This is discussed fully by Bethke (1996).

Of course, there may be more than one extra constraint. For example a system having four constraints (e.g., a battery in which the electrodes have a variable amount of strain energy), there would be two affinity variables and two progress variables, e.g.,

$$dG = -S\,dT + V\,dP - \mathbf{A}_1\,d\xi_1 + \mathbf{A}_2\,d\xi_2 \qquad (6)$$

where $\mathbf{A}_1$ represents the work energy that could be released by the battery running down to equilibrium, $d\xi_1$ represents increments of this process, $\mathbf{A}_2$ represents the amount of strain energy in the electrodes, and $d\xi_2$ represents increments of this strain energy.

6. CONSTRAINTS AND DEGREES OF FREEDOM

Is there any difference between these terms? Yes. First, it must be noted that the derivation of the Phase Rule, in which the concept of Degrees of Freedom normally appears, is based on systems at complete chemical, thermal, and mechanical equilibrium, in which chemical potentials are equal throughout.

Imposing a third constraint (imposing a voltage; changing the degree of order or of stress, recrystallizing to a different polymorphic form, etc.) on a closed system in a stable equilibrium state (one at equilibrium and having only two constraints, such as T and P) always requires doing work on the system, and releasing that constraint always allows the system to do work, whatever the nature of the constraint (Reiss, 1965). The result of this work being done sometimes results in a system in which the chemical potentials are not equal throughout, in which case the Phase Rule does not apply, and the constraint cannot be considered to be a Degree of Freedom, quite apart from the fact that Degrees of Freedom are defined as intensive variables, while **A** is extensive.

For example in the system $CaCO_3$, the chemical potential of $CaCO_3$ in calcite and aragonite are only equal on the two-phase equilibrium boundary. If we change some calcite to aragonite (or vice-versa) within the calcite stability field, the chemical potentials are clearly not equal in the two minerals, so the Phase Rule does not apply, and the constraint (preventing aragonite from changing spontaneously to calcite) is not a Degree of Freedom.

However, in some cases the resulting system does have equality of chemical potentials (as well as of T and P) throughout, so that the system does obey the Phase Rule, but with no added degrees of Freedom. For example aragonite, without calcite, in a phase assemblage, could be at equilibrium with equality of chemical potentials in all phases. Similarly, a granite in which all the crystals of SiO_2 happened to be cristobalite instead of quartz, could have μ_{SiO_2} (as well as all other potentials) equal throughout, and the Phase Rule would apply. The assemblage is metastable, a constraint is applied, but there is no added Degree of Freedom in the Phase Rule sense.

Constraints, as they are defined here, are certainly parameters which must be specified to completely define the system, but they do not qualify as Degrees of Freedom in the Phase Rule sense. Thus constraints are independent variables, but not every independent variable is a Degree of Freedom.

7. SUMMARY

It has long seemed to me that the distinction between metastable and stable systems is very central to an understanding of thermodynamics, and that the distinction is made easier by clearly distinguishing between real systems and model systems. In my view, making this distinction is one of the secrets to understanding thermodynamics, because the hypothetical systems, impossible processes, and unknowable parameters are then seen to be parts of a mathematical framework, a model, related to real systems only by virtue of using measured parameters as variables. We make simple models using thermodynamics, and we compare them to much more complex natural systems. In the process, we use many of the same terms in both kinds of systems. As a result, some simple thermodynamic concepts can get unnecessarily confused.

Acknowledgments—This article seems appropriate for a volume dedicated to the memory of David Crerar. David was my first graduate student. He was intensely interested in thermodynamics. We discussed the books by Reiss (1965) and by Callen (1960) many times, the ideas in which eventually led to those expressed here. I thank Tom Giordano and three anonymous reviewers for their comments, which led to improvements in the manuscript.

Editorial handling: Scott A. Wood

REFERENCES

Anderson G. M. (1995) Thermodynamics of Natural Systems. John Wiley & Sons, New York. 382 pp.

Anderson G. M. (1998) The Thermodynamics of Hydrothermal Systems. Chapter 1 in Reviews in Economic Geology, Vol. 10, *Techniques in Hydrothermal Ore Deposits Geology.* J.P. Richards and P. Larson, Eds.; Published by the Society of Economic Geologists, pp. 1–32.

Berlinski D. (1995) A Tour of the Calculus. Vintage Books, Random House Inc., New York, 331 pp.

Bethke C. M. (1996) Geochemical Reaction Modeling. Oxford University Press, New York, 397 pp.

Bischoff J.L. and Fyfe W.S. (1968) Catalysis, inhibition, and the calcite-aragonite problem. *Amer. Jour. Sci.* **266**, 65-79.

Callen H. B. (1960) Thermodynamics. New York, Wiley and Sons, 376 pp.

Cartwright N. (1983) How the Laws of Physics Lie. Clarendon Press, Oxford.

De Donder Th. (with F. H. van den Dungen and G. van Lerberghe) (1920) Leçons de Thermodynamique et de Chimie-Physique. Gauthiers-Villars, Paris.

Dickerson R. E. (1969) Molecular Thermodynamics. New York, W.A. Benjamin, 452 pp.

Dunbar R. (1995) The Trouble With Science. Faber and Faber Ltd., London, 213 pp.

Fletcher P. (1993) Chemical Thermodynamics for Earth Scientists. Longman Group UK Ltd., Essex.

Fyfe W.S. and Bischoff J.L. (1965) The calcite - aragonite problem. *Soc. Econ. Paleontologists and Mineralogists, Spec. Pub.* **13**, 3-13.

Giere R. N. (1999) Science Without Laws. The University of Chicago Press, Chicago and London, 285 pp.

Greenwood H. J. (1989) On models and modeling. *Can. Mineral.* **27**, 1-14.

Guggenheim E. A. (1959) Thermodynamics. 4th Edition. Amsterdam, North-Holland Publishing Co., 476 pp.

Helgeson H. C. (1968) Evaluation of irreversible reactions in geochemical processes involving minerals and aqueous solutions. I. thermodynamic relations. *Geochim. Cosmochim. Acta* **32**, 853–877.

Helgeson H. C., Brown T. H., Nigrini A., and Jones T. A. (1970) Calculation of mass transfer in geochemical processes involving aqueous solutions. *Geochim. Cosmochim. Acta* **34**, 569–592.

Helgeson H. C. (1979) Mass transfer among minerals and hydrothermal solutions. In *Geochemistry of Hydrothermal Ore Deposits*, 2nd ed., (ed. H.L. Barnes). New York, Wiley-Interscience, pp. 568–610.

Herzfeld C. M. (1962) The thermodynamic temperature scale, its definition and realization. Chapter 6, pp. 41–50 In *Temperature, Its Measurement and Control in Science and Industry*, C.M. Herzfeld, editor. Vol. 1, Basic Standards and methods, F.G. Brickwedde, editor. New York, Rheinhold Publishing Co., 848 pp.

Hillert M. (1998) Phase equilibria, Phase Diagrams and Phase Transformations. Their Thermodynamic Basis. Cambridge University Press, Cambridge, 538 pp.

Klotz I. M. (1964) Chemical Thermodynamics. Basic Theory and Methods, Revised Edition. W.A. Benjamin,Inc., New York, 468 pp.

Knapp R. A. (1989) Spatial and temporal scales of local equilibrium in dynamic fluid-rock systems. *Geochim. Cosmochim. Acta* **53**, 1955–1964.

McGlashan M. L. (1979) Chemical Thermodynamics. Academic Press, London, 345 pp.

Moore W. J. (1962) Physical Chemistry, 3rd Edition. Prentice-Hall, Inc., Englewood Cliffs, N.J.,844 pp.

Nordstrom D. K. and Munoz J. L. (1958) Geochemical Thermodynamics. The Benjamin/Cummings Publishing Co., Inc. California, 477 pp.

Pippard A .B. (1966) Elements of Classical Thermodynamics. Cambridge University Press, 165 pp.

Prigogine I. and Defay R. (1954) Chemical Thermodynamics (translated by D.H. Everett). London, Longmans Green, 543 pp.

Reiss H. (1965) Methods of Thermodynamics. New York, Blaisdell, 217 pp.

APPENDIX

Some Illustrative Quotations

Models

What is the status of claims that are typically cited as "laws of nature"—Newton's Laws of Motion, the Law of Universal Gravitation, Snell's Law, Ohm's law, the Second law of Thermodynamics, the Law of Natural Selection? Close inspection, I think, reveals that they are neither universal nor necessary—they are not even true.

—Giere (1999), p. 90.

What one learns about the world is not general truths about the relationship between mass, force, and acceleration, but that the motions of a vast variety of real-world systems can be successfully represented by models constructed according to Newton's principles of motion.

—Giere (1999), p. 95.

What one learns about the world is not general truths about the relationship between systems and energy, but that a vast variety of real-world systems can be successfully represented by models constructed according to equilibrium thermodynamics.

—A paraphrase of Giere (1999), p. 95.

The calculus is a mathematical theory, a set of connected concepts, but it emerges in human history as the expression of a fantastic and unprecedented ambition, which is nothing less than to represent or re-create the real world in terms of the real numbers.

...The world that the calculus reflects cannot be discerned by the senses, and the world that can be discerned by the senses this mirror does not reflect.

—Berlinski (1995), p. 288.

A model is a simplified version of reality that is useful as a tool.

—Bethke (1996), p. 9.

Let us not grace loose thinking with the term `model'.

—Greenwood (1989), p. 12.

Constraints

The basic problem of thermodynamics is the determination of the equilibrium state that eventually results after the removal of internal constraints in a closed composite system.

—Callen (1960), p. 24.

Every state of equilibrium is subject to certain constraints which are imposed upon the system.

...Constraints are always associated with variables. The addition of a constraint implies the addition of an independent variable.

...Just as constraints and variables are in one-to-one correspondence with one another, they are each in one-to-one correspondence with the kinds of work which a system may perform on its environment or vice versa.

—Reiss (1965), p. 11–13.

System

...a system, which is any region of the universe, large or small, that is being considered in our analysis.

—Klotz (1964), p. 35.

A thermodynamic system is any part of the universe we choose to study.

—Fletcher (1993), p. 3.

A natural system is any part of the universe we choose to consider, such as the contents of a beaker, a crystal of quartz, the solar system, or a bacterium. Thermodynamic systems, on the other hand, are not real but conceptual and mathematical, and are of three types. The three types are used to distinguish between the ways that changes in composition and energy content can be effected, and therefore they are defined basically by the nature of their boundaries.

—Anderson (1998)

Equilibrium

...none of its thermodynamic properties are changing at a measurable rate.

—Reiss (1965), p. 3.

If a system is in complete equilibrium, any conceivable infinitesimal change in it must be reversible.

—Guggenheim (1959), p. 15.

A system is said to have attained a state of equilibrium when it shows no further tendency to change its properties with time.

—Moore (1962), p. 6.

...a state in which no further change is perceptible, no matter how long one waits. The system is then said to be in equilibrium.

—Pippard (1966), p. 6.

For any given closed or isolated system, each substance and group of substances can be in its most stable state, which would be the state of stable equilibrium. If any substance or assemblage of substances had not reached that state but showed no apparent change with time, then it would be in a state of metastable equilibrium.

—Nordstrom and Munoz (1985), p. 12.

We are thus led to consider a system not in chemical equilibrium in which however the chemical reactions leading towards its attainment have been virtually suppressed. The system is then in a special kind of metastable equilibrium sometimes called frozen equilibrium.

—Guggenheim (1959), p. 42.

Experimental conditions under which an internal variable ξ does not change will be called freezing-in conditions and an internal variable that does not change due to such conditions will be regarded as being frozen-in.

—Hillert (1998), p. 11.

The concept of metastability seems to have been employed mainly to indicate the degree of control the experimenter has over the constraint.... However, from the thermodynamic point of view, differences in degree of control are meaningless.

—Reiss (1965), p. 17.

But no real measurement is reversible, and no real system is completely in equilibrium.

—Herzfeld, (1962), p. 41.

ENDNOTES

1 A few examples are quoted at the end of this article. The quotations are illustrative; they do not necessarily reflect the present viewpoint.

2 Confusion about the use of models in science is also at the heart of much controversy in the philosophy of science, usefully discussed by Giere (1999). See the quotations at the end of this article.

3 Local equilibrium is a feature that real systems must have, to be modeled successfully by thermodynamics. It is not a part of thermodynamics itself.

4 By this I mean that no equations in classical thermodynamics refer to unstable states. There are inequality relations, but these are always between metastable equilibrium states, or metastable and stable equilibrium states. Kinetics, on the other hand, always deals with unstable states.

5 A more detailed mathematical treatment of this simple example is found in Anderson (1995), Chapter 12.

6 "There are those who say that time has no place in thermodynamics. They are wrong." —McGlashan (1979), p.102.

Water-Rock Interactions, Ore Deposits, and Environmental Geochemistry: A Tribute to David A. Crerar

Editors: Roland Hellmann and Scott A. Wood

Complexation of the rare earth elements with aqueous chloride at 200 °C and 300 °C and saturated water vapor pressure

CHRISTOPHER H. GAMMONS[1*], SCOTT A. WOOD[2], AND YOUNING LI[3]

[1]Dept. of Geological Engineering, Montana Tech of the University of Montana, Butte, MT 59701
[2]Dept. of Geological Sciences, University of Idaho, Box 443022, Moscow, Idaho 83844-3022
[3]Dept. of Chemistry and Geochemistry, Montana Tech of the University of Montana, Butte, MT 59701

[*]Author to whom correspondence should be addressed (e-mail: cgammons@mtech.edu)

Abstract—The stabilities of aqueous chloride complexes of the rare earth elements (REE) at elevated temperature have been evaluated using the indirect solubility method. The solubility of silver chloride was measured at 200 °C and 300 °C, saturated vapor pressure, 0.1 or 1.0m NaCl + HCl, and 0 to 0.3m $LnCl_3$, where Ln stands for any trivalent lanthanide. The change in AgCl solubility in the presence of an incremental increase in $LnCl_3$ was used to quantify the extent of complexation between chloride ion and each REE cation. Combined with our previous results for neodymium (Gammons et al., 1996), we now have experimentally-derived estimates for the formation constants of Ln-chloride complexes at elevated temperature for all of the rare earth elements, with the exception of promethium and dysprosium.

This study shows that all of the rare earth elements have a similar affinity for aqueous chloride in high-temperature, acidic solutions. The average ligand number for REE-chloride complexes at 300 °C in 0.1 and 1.0m chloride solutions is 2.0±0.2, indicating that $LnCl_2^+$ is the dominant species under these conditions, in agreement with our previous results for neodymium. At 200 °C, 1.0m chloride, the average ligand number is 1.14±0.24, indicating that $LnCl^{2+}$ is dominant. Inter-element deviations in ligand number show no consistent variation across the periodic table at 300 °C, but show a weak trend towards decreased complexation with higher atomic number at 200 °C. The data at 200 °C were used to estimate log K for the reaction $Ln^{3+} + 2Cl^- \leftrightarrow LnCl_2^+$ (β_2). The average value for 11 elements was 2.73±0.47. This compares favorably with an estimate of log β_2 = 2.52 obtained by Gammons et al. (1996) from a more detailed study of neodymium. Attempts to calculate log β_1 were hindered by a lack of data over a sufficient range in chloride concentration.

Our experimental results are in close agreement with the theoretical estimates of Haas et al. (1995) at 200 °C; however, at 300 °C, our study indicates a wider predominance of the $LnCl_2^+$ species. Both our study and that of Haas et al. (1995) show small systematic variations in the stability of chloride complexes of the lanthanide elements across the periodic table. If chloride complexes are the dominant form of REE transport in hydrothermal systems, aqueous complexation by itself should not result in large REE fractionation trends.

1. INTRODUCTION

The hydrothermal geochemistry of the rare earth elements (REE) has relevance to a wide spectrum of geological disciplines, including igneous and metamorphic petrology, isotope geochemistry, economic geology, and environmental geochemistry. Much effort has been put forth in recent years to document the REE concentrations of naturally-occurring hydrothermal systems (see Michard et al., 1983; Michard and Albarède, 1986; Michard, 1989; Klinkhammer et al., 1994; Bau and Dulski, 1999; Douville et al., 1999; and references therein). Most of these studies have focused on deep-sea hydrothermal vents. In a compilation of existing data, Klinkhammer et al. (1994) showed that the REE concentrations of hydrothermal fluids near oceanic spreading centers are remarkably similar, with pronounced enrichments in LREE relative to chondrite compositions, and a very strong positive europium anomaly. Based on a limited database, Michard (1989) showed that continental geothermal waters are much more variable in terms of their REE content. Recent research (e.g., van Middlesworth and Wood, 1998; Shannon et al., 1999) is improving our understanding of REE in continental geothermal systems.

In order to fully understand the behavior of REE in hydrothermal systems, data are needed on the form and stability of aqueous REE complexes at elevated temperature. Using different approaches, Wood (1990) and Haas et al. (1995) derived theoretical estimates for the high-temperature stability constants of a large number of potentially important aqueous REE species. However, theoretical studies of this nature are prone to uncertainty due to the difficulty of extrapolating low-temperature data that are, in many cases, incomplete or of dubious accuracy. Gammons et al. (1996) used an experimental solubility method to derive stability constants for the chloride complexes of neodymium up to 300 °C. This study showed that the Nd-chloride complexes, although quite weak at ambient temperature, become much stronger as temperature increases. The stability constants at T > 200 °C were higher than originally predicted by Wood (1990), but were in closer agreement with the estimates of Haas et al. (1995). Based on these results, Gammons et al. (1996) suggested that chloride complexes could dominate the speciation of neodymium and other REE in high-temperature brines, such as sea-floor vents. A similar conclusion has recently been reached by Douville et al. (1999) and Bau and Dulski (1999).

The purpose of this paper is to expand on the experimental study of Gammons et al. (1996), to include all of the lanthanide elements. The same experimental approach and methods of interpretation are employed. The results are then used to identify inter-element variation in the stability of REE-chloride complexes at elevated temperature, and to predict systematic changes in the behavior of REE in chloride-rich hydrothermal fluids.

2. METHODOLOGY

The approach used in this study is identical to that of Gammons et al. (1996). Briefly, the effect of incremental addition of lanthanide chloride salt on the solubility of AgCl (the mineral chlorargyrite) is used to derive the stoichiometry of the dominant REE(III) chloride complexes. A number of previous experimental studies of hydrothermal complexation have used a similar technique, including Ruaya and Seward (1986), Williams-Jones and Seward (1989), Gammons (1995), and Gammons and Seward (1996).

2.1. Solution Preparation

A stock solution containing 0.9 molal NaCl and 0.1 molal HCl (1.0 molal total chloride) was prepared gravimetrically from analytical grade NaCl and a 1.0M HCl volumetric standard. A second set of stock solutions containing 1.0 molal $LnCl_3{\cdot}xH_2O$ in a 0.9m NaCl + 0.1m HCl matrix was prepared for Ln = La, Ce, Pr, Sm, Eu, Gd, Tb, Dy, Ho, Er, Tm and Yb. The lanthanide salts were purchased from Alfa Aesar, and had a quoted purity of > 99.9% in all cases. Neodymium was excluded from the list, as we have previously studied this element in detail (Gammons et al., 1996). Also, we did not attempt to investigate promethium, the radioactive lanthanide. With the exception of $DyCl_3{\cdot}xH_2O$, all of the solids dissolved completely to yield solutions that were clear and of variable color. The holmium and erbium solutions were deep pink, the praseodymium solutions were deep green, samarium was pale yellow, and the remainder were colorless, or nearly so. Because the $DyCl_3$ solid did not dissolve completely, the concentration of this stock sample was reduced to 0.5 molal.

To confirm the stoichiometry of the lanthanide reagents, small aliquots of each salt were digested in dilute nitric acid and analyzed for their REE concentration by inductively coupled plasma-atomic emission spectroscopy (ICP-AES) and their chloride content by ion chromatography (IC). The results showed that all of the salts had the correct 1:3 REE:Cl stoichiometry, within the analytical precision (roughly 5%). Because of the hygroscopic nature of many of the salts (some were a watery crystal mush), the exact REE concentration of each nominal 1.0-molal stock solution was determined by ICP-AES after preparation of the solutions.

For each rare earth element, a set of 8 experimental solutions was prepared by mixing the 1.0m NaCl+HCl stock with the 1.0m lanthanide chloride stock to create solutions with final REE concentrations ranging from 0 to roughly 0.3 molal $LnCl_3$, all with a matrix of exactly 0.9m NaCl + 0.1m HCl. The acid was added to avoid potential complications associated with hydrolysis. A second set of experimental solutions containing variable amounts of REE in a 0.09m NaCl + 0.01m HCl matrix was prepared for the elements lanthanum, holmium, and lutetium by dilution of the corresponding 1.0m samples. Each solution was then used in separate solubility experiments at 300 °C and 200 °C, as discussed below. This procedure created a (15 x 8 x 2) data set, for a total of 240 individual samples. In addition, replicate samples were run for roughly 10% of the total matrix.

2.2. Indirect Solubility Experiments

For the solubility experiments, approximately 3 mL of each NaCl/HCl/$LnCl_3$ solution were transferred to a silica-glass tube (9mm O.D., 7mm I.D., 15cm length) which contained a pre-weighed pellet of AgCl and a narrow constriction at a point roughly 25-35% from one end (see Seward, 1976). The tubes were sealed under a slight vacuum, and were placed inside small stainless steel autoclaves containing 20 mL of distilled water. The autoclaves were then placed vertically inside an air-thermostated oven having a maximum temperature variation of ±1.0 °C at 300 °C. The experiments were equilibrated at 200 °C or 300 °C for a period of 1 to 2 weeks. This was considered ample time to achieve equilibrium, based on our previous experience and on the earlier study of Seward (1976). At the end of the experiment, the autoclaves were removed from the oven and immersed in water. Before quenching, the autoclaves were inverted and shaken, allowing the solutions to separate from the AgCl pellets (which remained trapped above the constrictions) during cooling. Within minutes of the quench, most of the dissolved silver had precipitated as fine-grained, feather-shaped AgCl crystals in the bottom of the tubes. This precipitate, like the AgCl pellets, was pure white in color.

After opening the tubes, the AgCl pellets were recovered, rinsed, dried, and weighed to ±0.1 mg to determine the high-temperature silver solubility by mass loss. An independent measure of the AgCl solubility was obtained by digesting the fresh precipitate in the bottom of each tube with concentrated NH_4OH solution. The tube walls and precipitate had to be carefully rinsed with water at least twice before this step to rid the solutions of aqueous lanthanide (otherwise, a fine-grained hydrolysis precipitate formed when the basic NH_4OH was added). These solutions were diluted with 1% NH_4OH prior to analysis for silver via flame atomic absorption spectroscopy (detection limit 0.1 mg/kg, precision ±2%), using matrix-matched standards. In most cases, the agreement between the mass loss and AAS results was better than ±5%. Because the AAS results were more precise (especially at 200 °C), they were deemed more reliable.

2.3. Data Interpretation

The methods used to interpret our solubility data are identical to those described in Gammons et al. (1996). The reader is referred to that paper for details that are not covered in the following summary.

Two methods were used to interpret our solubility results. The first was a forward modeling procedure in which the change in AgCl solubility due to addition of $LnCl_3(s)$ was calculated for a given set of initial conditions. Parallel calculations were made assuming integral values of the average ligand number of the Ln(III) chloride complexes (n = 0 to 4). By comparing these results with the measured solubilities, the average n values of the experimental solutions were obtained. In the second set of calculations, the measured solubility data were regressed to derive cumulative formation constants for the Ln(III) chloride complexes. The program employed at this step was a modification of the non-linear least squares algorithm HACK (Aldridge and Seward, 1976). The user first inputs a set of isothermal solubility data, along with a set of initial guesses for the $LnCl_n^{3-n}$ formation constants. For each data entry, a model AgCl solubility is calculated (S_{calc}), which is compared to the measured AgCl solubility (S_{meas}). On the basis of these results, new estimates of the Ln(III) formation constants are obtained, and the process is repeated until the sum of the squares of the residuals $[(S_{calc}-S_{meas})/S_{meas}]^2$ is minimized. Mass-balance relations for ΣH^+, ΣCl and ΣH_2O in each experiment were adjusted for dissolution of the solids.

To perform the modeling calculations, a complete aqueous speciation needed to be calculated for each solution. To be consistent with our previous study of neodymium, the same procedures were used. Individual ion activity coefficients for all charged species were calculated using the following version of the extended Debye Hückel equation (Helgeson, 1969):

$$\log \gamma_n = -\frac{A|Z_n|^2 I^{1/2}}{1 + BaI^{1/2}} + bI \qquad (1)$$

Values for a (distance of closest approach) for H^+ and Cl^- were taken from Kielland (1937); those for dissolved Ag species were taken from Seward (1976); those for all aqueous lanthanide species were set at 9 Ångstroms. Values for b (the deviation function) were taken from Table 26 of Helgeson et al. (1981), using NaCl as the model solute. Thermodynamic data for the Ag(I) chloride complexes were taken from Zotov et al. (1986a,b) and Levin (1991), and are similar to the results of the previous solubility study of Seward (1976). Data for NaCl and HCl ion-pairing were taken from Mesmer et al. (1988) and Frantz and Marshall (1984), respectively. Although considerable controversy remains as to the extent of HCl ion-pairing in hydrothermal solutions (cf. Pokrovskii, 1999), our results were not highly sensitive to uncertainty in this parameter due to the relatively low temperatures involved and the fact that NaCl was the dominant electrolyte. In addition, no attempt was made to adjust equilibrium constants for the slight deviation between the saturated vapor pressure of pure water vs. the NaCl-HCl-$LnCl_3$ solutions used in this study. Our most concentrated solutions were 1.9 molal in combined salts, which corresponds to roughly 10 wt% NaCl equivalent. Using tabulated data from Haas (1971), the saturated vapor pressure at 300 °C for pure water is 85.9 bars, vs. 80.0 bars for a 10 wt% NaCl solution. At 200 °C, the difference is 15.6 bars vs. 14.5 bars. Thus, we feel that any pressure effect in our study due to changes in salt concentration is negligible.

In previous studies employing the indirect solubility approach, effort has been made to perform experiments over as wide a range as possible in temperature and total chloride concentration. The data obtained were then regressed to determine the association constants for the aqueous metal chloride complexes as a function of temperature. A detailed study of this type has been performed for the neodymium(III) chloride system (Gammons et al., 1996). However, the number of experiments necessary to accomplish this task for all 12 of the remaining stable lanthanide elements was deemed prohibitive. Therefore, instead of repeating the entire range of experimental conditions for each element, we decided to select one initial experimental condi-

Table 1. Solubility data for 0.9m NaCl + 0.1m HCl matrix solutions.[1]

$LnCl_3$	AgCl solubility 200° C	AgCl solubility 300° C
lanthanum (La)		
0.0000	0.0179	0.115
0.0250	0.0198	0.116
0.0500	0.0184	0.123
0.1000	0.0197	
0.1500	0.0214	0.125
0.2000	0.0240	0.135
0.2500	0.0280	0.142
0.3000		0.147
cerium (Ce)		
0.0000	0.0191	0.114
0.0218	0.0198	0.117
0.0436	0.0208	0.121
0.0873	0.0215	0.131
0.1309	0.0241	0.133
0.1746		0.135
0.2182	0.0277	0.143
0.2618	0.0299	0.150
praseodymium (Pr)		
0.0000	0.0193	
0.0219	0.0228	0.121
0.0438	0.0197	0.119
0.0876	0.0235	0.122
0.1314	0.0251	0.125
0.1752	0.0276	0.131
0.2190	0.0292	0.139
0.2628	0.0314	0.137
samarium (Sm)		
0.0000	0.0193	0.102
0.0230	0.0183	0.100
0.0460	0.0188	0.110
0.0920	0.0213	0.114
0.1380	0.0196	0.125
0.1840	0.0267	0.138
0.2300	0.0295	0.140
0.2760	0.0296	0.143
europium (Eu)		
0.0000	0.0206	0.111
0.0225	0.0249	0.114
0.0450	0.0218	0.125
0.0900	0.0237	0.122
0.1350	0.0255	0.121
0.1800	0.0277	
0.2250	0.0296	0.133
0.2700	0.0317	0.139
gadolinium (Gd)		
0.0000	0.0164	0.100
0.0250	0.0178	0.114
0.0500	0.0194	0.108
0.1000	0.0189	0.112
0.1500	0.0239	0.120
0.2000	0.0279	0.131
0.2500	0.0269	0.134
0.3000	0.0313	0.144
terbium (Tb)		
0.0000	0.0212	0.105
0.0250	0.0221	0.111
0.0500	0.0221	0.104
0.1000	0.0268	0.116
0.1500	0.0255	0.135
0.2000	0.0273	0.134
0.2500	0.0296	0.143
0.3000	0.0321	0.148
holmium (Ho)		
0.0000	0.0183	0.106
0.0227		0.109
0.0454	0.0211	0.110
0.0907	0.0235	0.115
0.1361	0.0250	0.132
0.1814	0.0272	0.137
0.2268	0.0297	0.143
0.2721	0.0324	0.147
erbium (Er)		
0.0000	0.0199	0.117
0.0216	0.0214	0.119
0.0431	0.0208	0.121
0.0862	0.0240	0.126
0.1293	0.0253	
0.1724	0.0298	
0.2155	0.0297	
0.2586	0.0315	
thulium (Tm)		
0.0000	0.0193	0.125
0.0250	0.0207	0.132
0.0250	0.0203	
0.0500	0.0223	0.127
0.1000	0.0235	0.130
0.1500	0.0257	
0.2000	0.0306	0.145
0.2000	0.0282	
thulium (cont.)		
0.2500	0.0292	0.151
0.3000	0.0354	0.162
0.3000	0.0368	
ytterbium (Yb)		
0.0000	0.0193	0.112
0.0000	0.0193	0.112
0.0218	0.0205	0.114
0.0218	0.0204	
0.0435	0.0216	0.117
0.1740	0.0271	0.130
0.0870	0.0222	0.125
0.0870	0.0238	
0.1305	0.0248	0.124
0.1305	0.0253	
0.1740		0.128
0.1740		0.128
0.2175	0.0297	0.135
0.2610		0.141
0.2610		0.140
lutetium (Lu)		
0.0000	0.0212	0.115
0.0250	0.0207	0.119
0.0500	0.0243	0.130
0.1000	0.0274	0.127
0.1000	0.0274	0.127
0.1500	0.0315	0.137
0.2000	0.0319	0.142
0.2500	0.0322	0.148

[1] All concentrations are in molal (moles/kg H_2O).

tion (1.0 molal NaCl + HCl matrix), and perform a series of identical experiments for all of the REE. Additional experiments in a 0.1-molal NaCl + HCl matrix were performed for lanthanum (a light REE), holmium (a middle REE) and lutetium (a heavy REE). Two temperatures were selected for examination: 300 °C, where complexation is strong, and 200 °C, where interactions are weaker, but still measurable. The data so obtained provide an internally consistent comparison of the extent of interaction of each trivalent lanthanide with aqueous chloride.

3. RESULTS

The experimental data are summarized in Tables 1 and 2. The data have been arranged by element, with

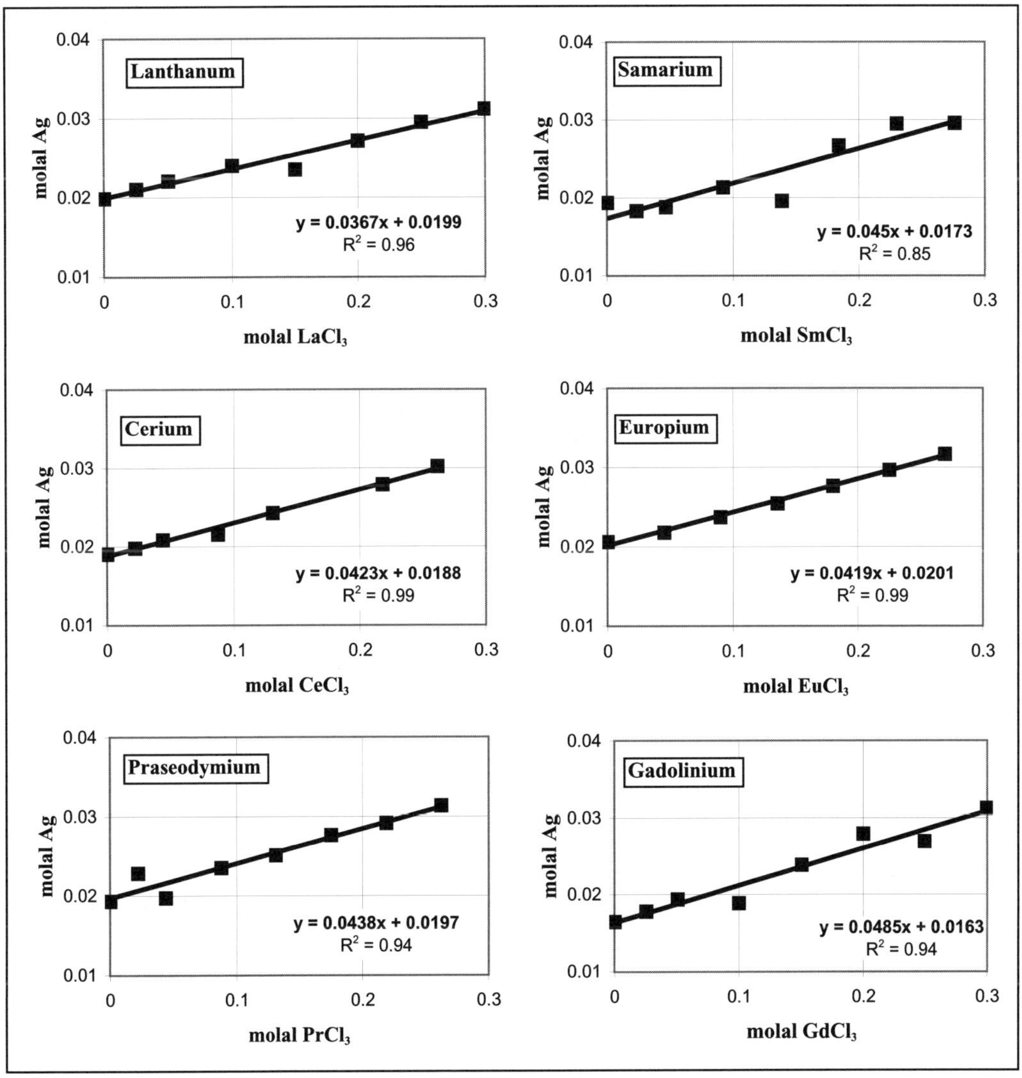

Fig. 1 The solubility of AgCl as a function of $LnCl_3$ concentration (Ln = La, Ce, Pr, Sm, Eu, Gd) at 200 °C, in 0.9m NaCl + 0.1m HCl matrix solutions.

the first column showing the total concentration of the 1:3 $LnCl_3$ salt, and the next two columns giving the measured AgCl solubility at 200 °C and 300 °C. Experiments using $DyCl_3$ gave results that were erroneous when compared to the other elements, and therefore data for dysprosium are not shown. This problem was traced to an error in preparation of the 0.5m $DyCl_3$ stock solution. Within a given series (i.e., same element, temperature, and NaCl/HCl concentration), all of the samples were run simultaneously in the same laboratory oven. The total initial chloride concentration of each sample (before interaction with AgCl) can be computed by the expression:

$$m\Sigma Cl_i = mNaCl_i + mHCl_i + 3mLnCl_{3,i} \qquad (2)$$

where $mLnCl_{3,i}$ is the $LnCl_3$ molality of the starting solutions. Lanthanide was added to the initial solutions as a chloride salt with a REE:Cl ratio of 1:3. This salt dissolves completely into the solutions by the general reaction:

$$LnCl_3(s) \rightarrow LnCl_n^{3-n} + (3-n)Cl^- \qquad (3)$$

where Ln stands for any trivalent rare earth element. The general reaction can be recast as follows, assuming integral values of n, the average ligand number, ranging from 0 to 4:

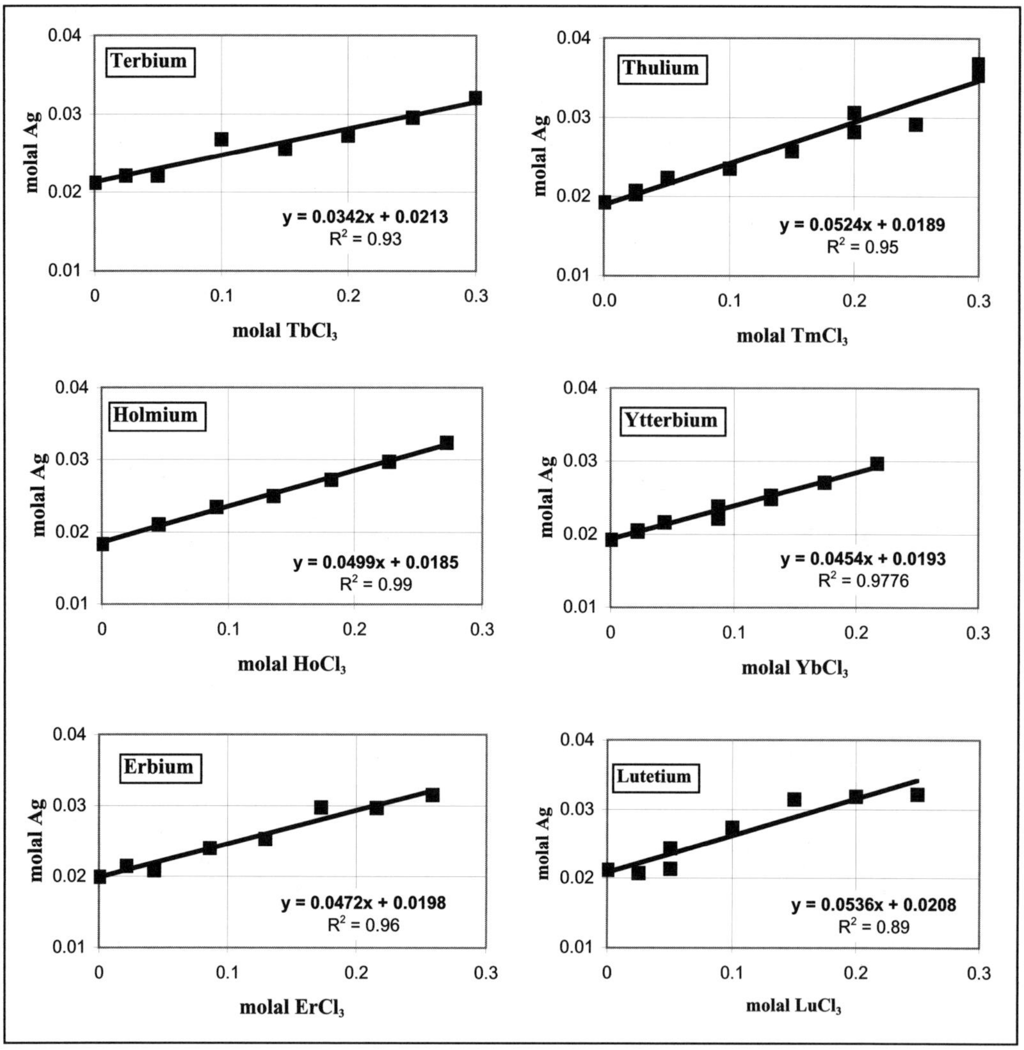

Fig. 2 The solubility of AgCl as a function of $LnCl_3$ concentration (Ln = Tb, Ho, Er, Tm, Yb, Lu) at 200 °C, in 0.9m NaCl + 0.1m HCl matrix solutions.

$$LnCl_3(s) \rightarrow Ln^{3+} + 3Cl^- \quad (4)$$

$$LnCl_3(s) \rightarrow LnCl^{2+} + 2Cl^- \quad (5)$$

$$LnCl_3(s) \rightarrow LnCl_2^{+} + Cl^- \quad (6)$$

$$LnCl_3(s) \rightarrow LnCl_3(aq) \quad (7)$$

$$LnCl_3(s) \rightarrow LnCl_4^{-} - Cl^- \quad (8)$$

Addition of lanthanide changes the total chloride mass balance to an extent that is proportional to the average ligand number of the dominant aqueous Ln species at the temperature of the experiment. Because the solubility of AgCl is sensitive to any such change in m_{Cl^-}, the data in Table 1 may be used to constrain the stoichiometry of the REE complexes.

3.1. Graphical Interpretation

Solubility data for each series of experiments are plotted vs. ΣLn molality in Figures 1 and 2 (1.0 molal NaCl/HCl at 200 °C), Figures 3 and 4 (1.0 molal NaCl/HCl at 300 °C), and Fig. 5 (0.1 molal NaCl/HCl at 200 °C and 300 °C). Linear regression was employed on each data set. The derived slope and intercept values are shown on each diagram, and are summarized in Table 3. All of the experiments at the

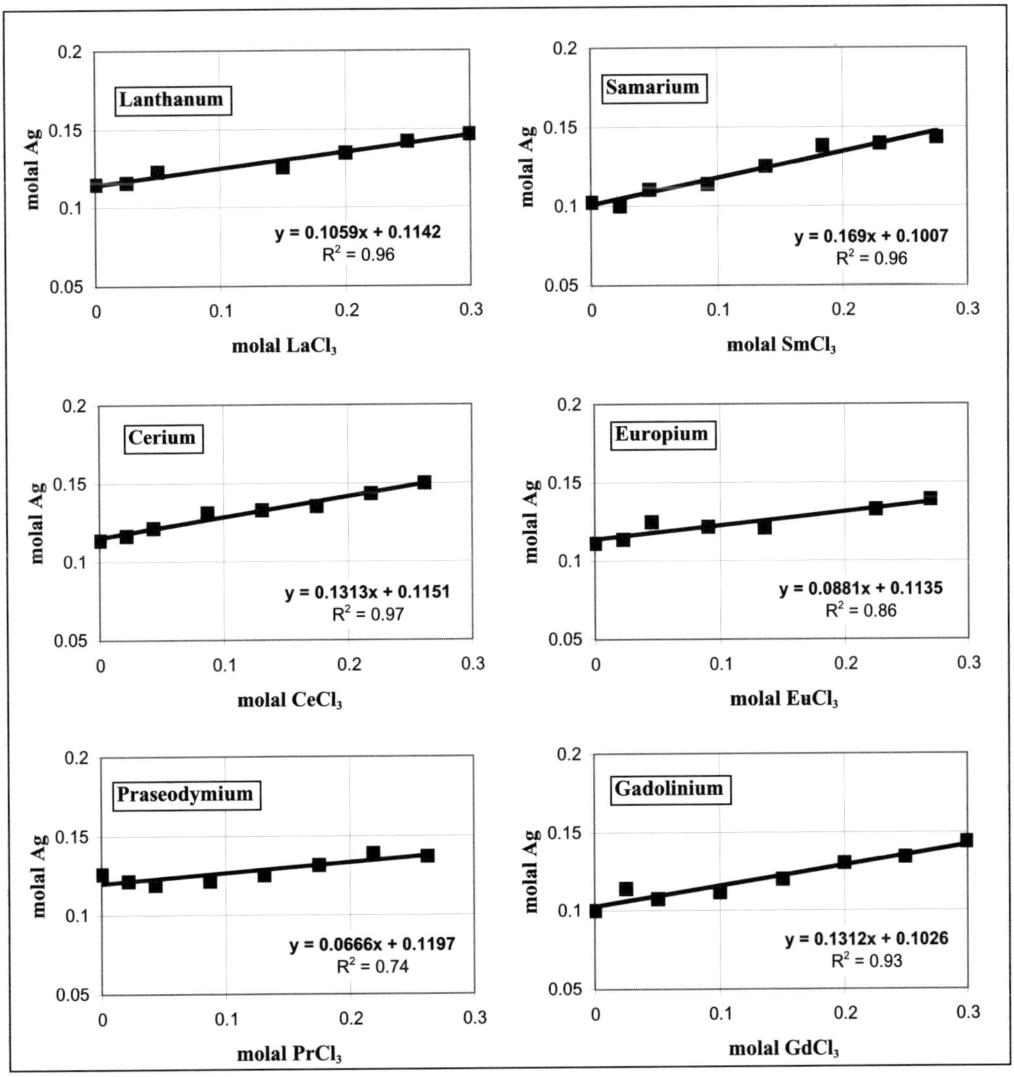

Fig. 3 The solubility of AgCl as a function of $LnCl_3$ concentration (Ln = La, Ce, Pr, Sm, Eu, Gd) at 300 °C, in 0.9m NaCl + 0.1m HCl matrix solutions.

same temperature and total chloride molality should theoretically give the same y-intercept, which is the Ag solubility value at $\Sigma Ln = 0$. Deviation in this case can be used to estimate the experimental error. At 200 °C, 1.0m ΣCl, the geometric mean intercept value was 0.0191±0.0014 molal Ag, with a relative standard deviation (1σ/mean) of 7.3%. At 300 °C, 1.0m ΣCl, the mean value was 0.112±0.0069 molal Ag, with a relative standard deviation of 6.2%. The corresponding relative standard deviations for experiments at 0.1 molal ΣCl were 4.1% and 3.3% at 200° and 300 °C, respectively. The smaller deviations at 0.1 molal are partly due to the smaller number of experiments run at this condition, but also to the fact that all of the 0.1-molal experiments were run simultaneously in the same oven, and therefore equilibrated at precisely the same temperature.

As a check on the accuracy of our experimental methods, our results for AgCl solubility at $\Sigma Ln = 0$ were compared with previous solubility studies of Seward (1976) and Zotov et al. (1986b). Seward (1976) obtained AgCl solubilities of 0.00196 and 0.0199 molal at 200 °C, 0.1m and 1.0m NaCl, respectively. These are both in excellent agreement with our mean intercept values of 0.0020 and 0.0190 molal, respectively. Seward (1976) did not perform

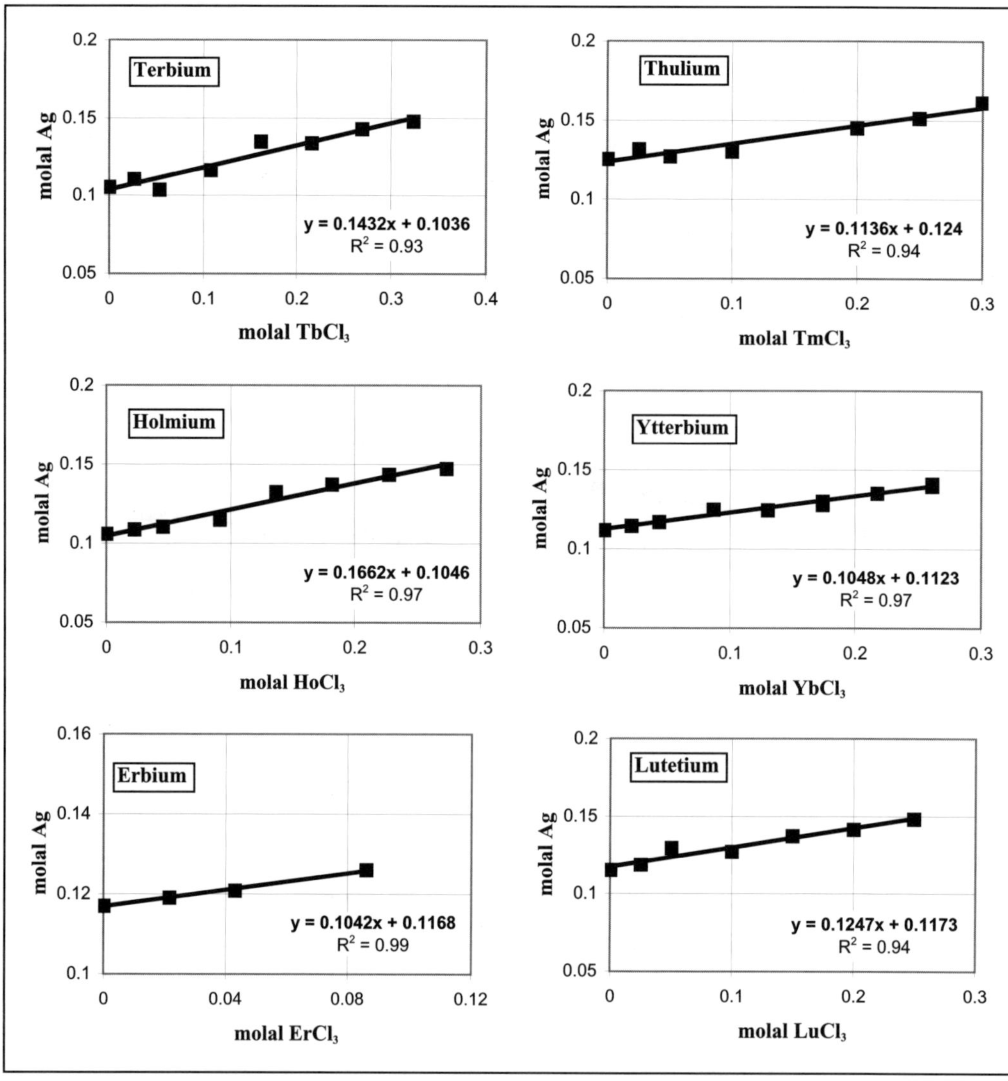

Fig. 4 The solubility of AgCl as a function of $LnCl_3$ concentration (Ln = Tb, Ho, Er, Tm, Yb, Lu) at 300 °C, in 0.9m NaCl + 0.1m HCl matrix solutions.

experiments at 300 °C, so a direct comparison cannot be made at this temperature. However, our results are in very close agreement with the data of Zotov et al. (1986), who obtained AgCl solubilities of 0.0166 and 0.080 molal at 300 °C and 0.1m and 1.0m NaCl, respectively.

The results of forward modeling calculations are summarized in Figures 6 through 9. Each diagram includes a group of curves emanating from the y-intercept (ΣLn = 0.0). These represent hypothetical AgCl solubilities calculated by the forward modeling procedure assuming integral values of n (see Section 2.3). The curves have slopes that are positive (n = 0,1,2), zero (n = 3) or negative (n = 4). These contrasting effects are understood by examination of reactions (4) to (8). For example, dissolution of $LnCl_3$ as Ln^{3+} (n = 0) releases 3 moles of chloride ion for each mole of lanthanide salt added (reaction 4), resulting in a very sharp increase in AgCl solubility. In contrast, dissolution of $LnCl_3$ as $LnCl_4^-$ (n = 4) consumes free chloride (reaction 8), resulting in a proportional decrease in AgCl solubility.

The shaded regions in Figs. 6 to 9 show the total range in solubility for *all* of the REE-NaCl-HCl experiments. This information can be used to quickly assess what the dominant REE species are likely to be

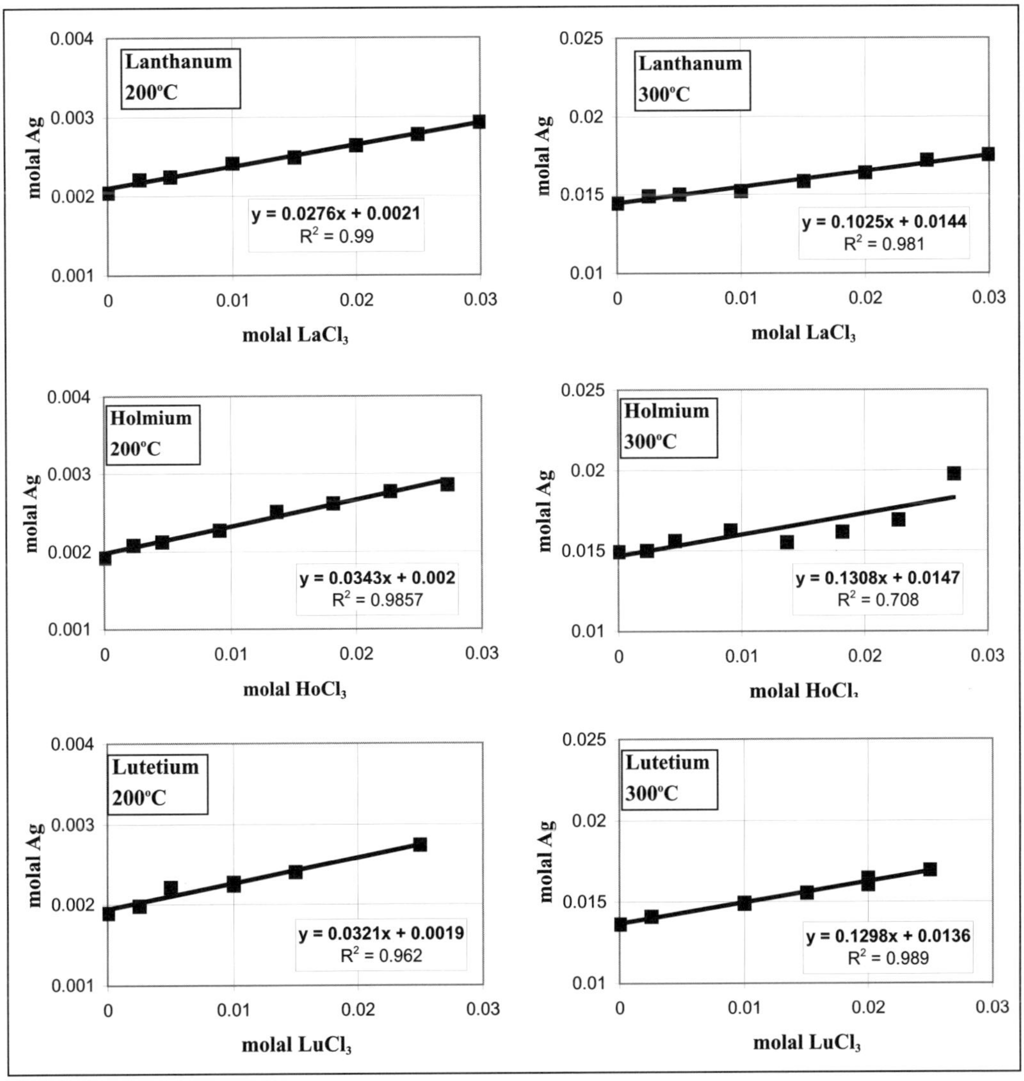

Fig. 5 The solubility of AgCl as a function of $LnCl_3$ concentration (Ln = La, Ho, Lu) in 0.09m NaCl + 0.01m HCl matrix solutions at both 200 °C and 300 °C.

at each temperature and chloride concentration. From Figures 6 and 7, it is evident that the ligand numbers for the REE at 200 °C and 0.1 to 1.0 molal NaCl/HCl matrix solutions tend to cluster around an average value slightly greater than one. This indicates that $LnCl^{2+}$, or the 1:1 species, is the dominant aqueous complex under these conditions, but that $LnCl_2^+$ is also present. In contrast, ligand numbers at 300 °C and 0.1 to 1.0 molal NaCl/HCl matrix (Figs. 8 and 9) are centered near a value of 2, suggesting that $LnCl_2^+$, the 1:2 species, is dominant.

The average ligand numbers (n) for each set of experimental data were quantified by comparing the experimental and forward modeling results using a simple spreadsheet. The derived n values are summarized in the final column of Table 3. At 200 °C, 1.0m ΣCl, the mean n value is 1.14 ± 0.24, with a relative standard deviation of 21%. At 300 °C, 1.0m ΣCl, the mean n value increases to 2.03 ± 0.21, with a relative standard deviation of 10%. Although fewer experiments were performed at 0.1 molal ΣCl, the results suggest that average ligand numbers are not appreciably different. This implies that, at each temperature, a single complex is dominant over a fairly large range in chloride concentration. At 200 °C, this complex must be the 1:1 species; at 300 °C, it is the 1:2 species.

Table 2. Solubility data for 0.09m NaCl + 0.01m HCl matrix solutions. [1]

$LnCl_3$	AgCl solubility 200° C	AgCl solubility 300° C	$LnCl_3$	AgCl solubility 200° C	AgCl solubility 300° C	$LnCl_3$	AgCl solubility 200° C	AgCl solubility 300° C
lanthanum (La)			**holmium (Ho)**			**lutetium (Lu)**		
0.0000	0.00204	0.0145	0.00000	0.00192	0.0150	0.0000	0.00189	0.0136
0.0025	0.00221	0.0149	0.00228	0.00208	0.0150	0.0025	0.00198	0.0141
0.0050	0.00224	0.0150	0.00455	0.00212	0.0156	0.0050	0.00221	
0.0100	0.00241	0.0152	0.00910	0.00227	0.0163	0.0100	0.00224	0.0149
0.0150	0.00249	0.0158	0.01365	0.00251	0.0155	0.0100	0.00228	0.0149
0.0200	0.00264	0.0164	0.01820	0.00261	0.0161	0.0150	0.00240	0.0156
0.0250	0.00278	0.0172	0.02275	0.00277	0.0169	0.0200		0.0164
						0.0200		0.0160

[1] All concentrations are in molal (moles/kg H_2O).

Table 3. Summary of graphical parameters derived from solubility data shown in Figures 1 to 5.

element	slope	intercept	lig.[1]	slope	intercept	lig.[1]
	200 °C, 0.9m NaCl + 0.1m HCl			**300 °C, 0.9m NaCl + 0.1m HCl**		
La	0.0324	0.0175	1.62	0.1059	0.1142	2.16
Ce	0.0423	0.0188	1.23	0.1313	0.1151	1.96
Pr	0.0438	0.0197	1.17	0.0837	0.1165	2.34
Sm	0.0450	0.0173	1.13	0.1690	0.1007	1.66
Eu	0.0419	0.0201	1.25	0.0881	0.1135	2.30
Gd	0.0485	0.0163	1.00	0.1312	0.1026	1.96
Tb	0.0342	0.0213	1.55	0.1432	0.1036	1.86
Ho	0.0499	0.0185	0.94	0.1662	0.1046	1.68
Er	0.0472	0.0198	1.04	0.1042	0.1168	2.17
Tm	0.0524	0.0189	0.85	0.1136	0.1240	2.10
Yb	0.0454	0.0193	1.11	0.1048	0.1123	2.17
Lu	0.0536	0.0208	0.81	0.1247	0.1173	2.01
mean	0.0447	0.0190	1.14	0.1220	0.1118	2.03
1σ	0.010	0.0014	0.24	0.026	0.0069	0.21
r.s.d.	14%	7.3%	21%	22%	6.2%	10%
	200 °C, 0.09m NaCl + 0.01m HCl			**300 °C, 0.09m NaCl + 0.01m HCl**		
La	0.0276	0.0021	1.40	0.1025	0.0144	2.20
Ho	0.0343	0.0020	1.01	0.1308	0.0147	2.00
Lu	0.0321	0.0019	1.14	0.1298	0.0136	2.00
mean	0.0313	0.002	1.18	0.121	0.0142	2.07
1σ	0.0028	8.2E-05	0.16	0.013	4.6E-04	0.09
r.s.d.	8.9%	4.1%	14%	10.8%	3.3%	4.6%

[1] Average ligand number of REE-Cl complexes, derived by graphical interpretation.

Trends in average ligand number across the periodic table are shown in Fig. 10. At 200 °C, 1.0m ΣCl, there is an apparent trend to lower n value with increase in atomic number. Assuming that this trend is real, this could account for some of the scatter (in terms of relative standard deviation) in the "averaged" n values at 200 °C. As discussed below, previous theoretical studies have predicted that REE-Cl complex stability should decrease with increase in atomic number, which is in agreement with the general trend we have obtained. However, at 300 °C, no consistent trend is seen between average ligand number and atomic number. There also appears to be no consistent "odd-even" or "tetrad" effects. This implies that all of the REE are behaving similarly at these conditions (300 °C, 1.0m ΣCl). Experimental uncertainty may be the main reason for the observed deviations.

3.2. Equilibrium Constants

Table 4 summarizes the results of least squares regression of equilibrium constants for the formation of

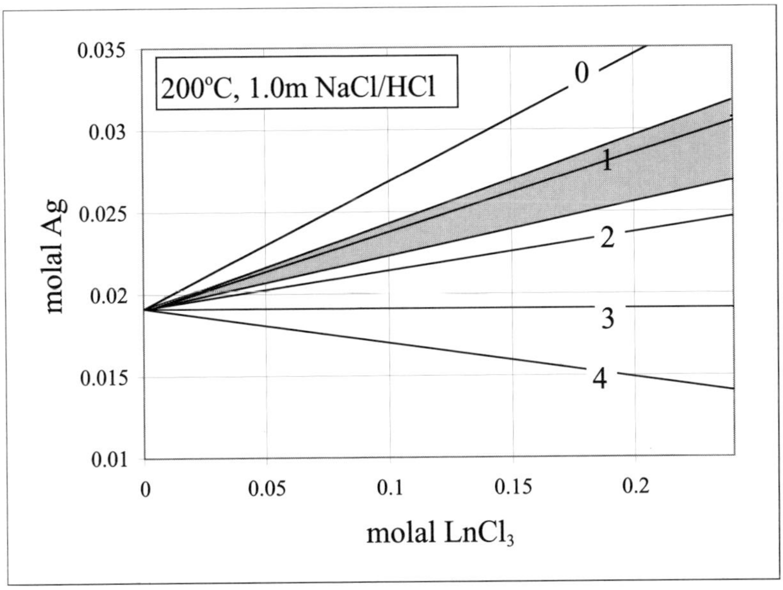

Fig. 6 Predicted vs. measured AgCl solubility, 200 °C, 1.0m NaCl/HCl. The lines labeled 0 to 4 correspond to predicted AgCl solubilities, assuming REE-chloride ligand numbers of 0 to 4. These lines were calculated using the forward modeling procedure (see text). The shaded sector shows the total range in slope and ligand number derived from solubility experiments for all of the rare earth elements investigated in this study.

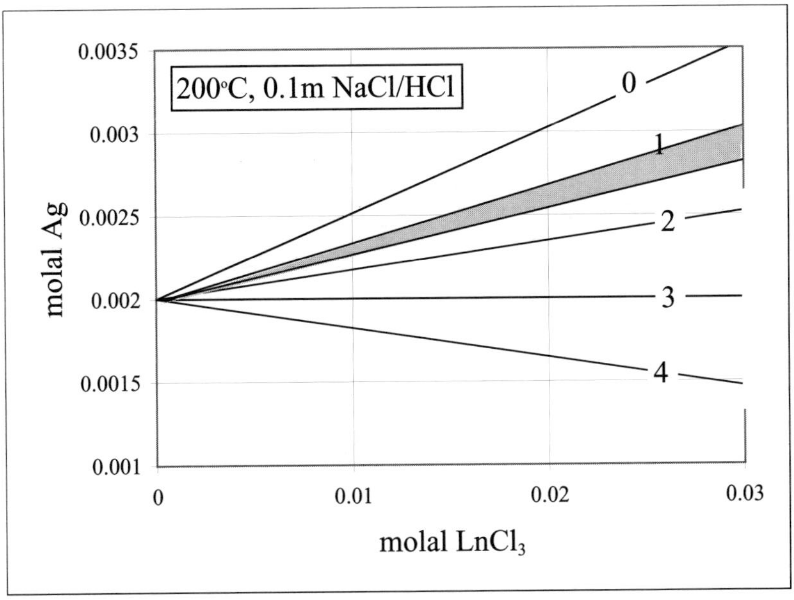

Fig. 7 Predicted vs. measured AgCl solubility, 200 °C, 0.1m NaCl/HCl (see caption to Fig. 4 for more details).

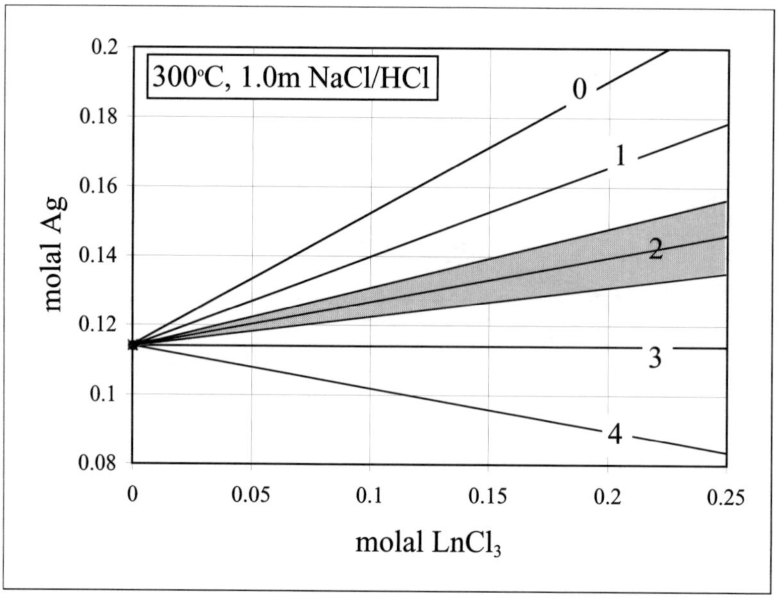

Fig. 8 Predicted vs. measured AgCl solubility, 300 °C, 1.0m NaCl/HCl (see caption to Fig. 4 for more details).

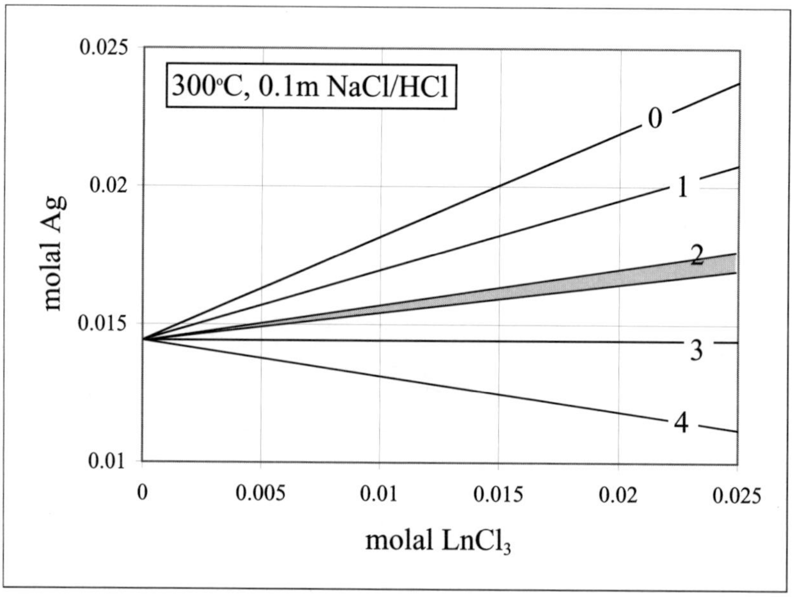

Fig. 9 Predicted vs. measured AgCl solubility, 300 °C, 0.1m NaCl/HCI (see caption to Fig. 4 for more details).

REE-chloride complexes at 200 °C (see section 2.3 for details). The headings β_1, β_2, and K_2 refer to the following reactions, respectively:

$$Ln^{3+} + Cl^- \leftrightarrow LnCl^{2+} \quad (9)$$

$$Ln^{3+} + 2Cl^- \leftrightarrow LnCl_2^+ \quad (10)$$

$$LnCl^{2+} + Cl^- \leftrightarrow LnCl_2^+ \quad (11)$$

For most of the elements of interest, it was not possi-

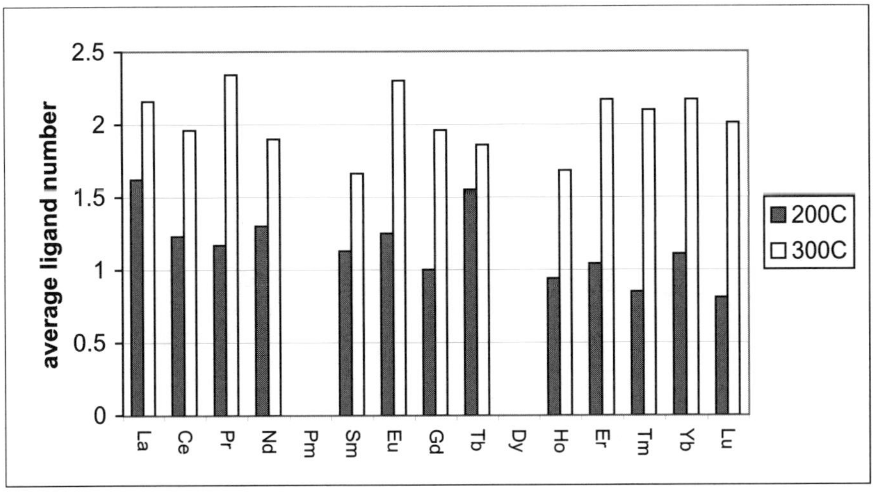

Fig. 10 Variation of experimentally determined chloride ligand numbers across the lanthanide series at 200 °C and 300 °C.

ble to refine β_1, the association constant for reaction (9). This is not surprising, considering the observation that derived ligand numbers were in most cases greater than 1, indicating a very minor contribution to the speciation of each lanthanide from the completely dissociated Ln^{3+} ion. An independent estimate of β_1 was obtained for gadolinium and thulium, but only by assuming $\beta_2 = 0$. Lutetium was the only element for which it was possible to refine *both* β_1 and β_2. Including our previous results for neodymium, the average of the 4 refined log β_1 estimates is 2.16±0.29 (1 standard deviation error). The limited data available preclude any attempt to identify trends in β_1 vs. atomic number. Additional indirect solubility experiments would be needed at lower chloride concentration (< 0.1 molal), but this was deemed beyond the scope of the present study.

Table 4. Cumulative and stepwise formation constants for REE-chloride complexes at 200 °C. [1]

element	log β_1	log β_2	log K_2
La	2.17 (fixed)	3.45	1.28
Ce	2.17 (fixed)	2.83	0.66
Pr	2.17 (fixed)	2.80	0.63
Nd	2.17 [2]	2.52 [2]	0.35
Sm	2.17 (fixed)	2.40	0.23
Eu	2.17 (fixed)	2.94	0.77
Gd	1.81	could not be refined	—
Tb	2.17 (fixed)	3.64	1.47
Ho	2.17 (fixed)	1.96	-0.21
Er	2.17 (fixed)	2.57	0.40
Tm	2.61	could not be refined	—
Yb	2.17 (fixed)	2.57	0.40
Lu	2.06	2.31	0.25
mean	2.16 [3]	2.73	0.57
1σ	0.29	0.47	0.46

[1] P = SWVP, I = 0.0; [2] from Gammons et al. (1996); [3] mean value calculated from refined estimates only.

Estimates of β_2 are given in the 3rd column of Table 4. For most of the elements of interest, it was impossible to derive a β_2 value unless the input file contained a fixed initial guess at β_1. For the latter, we chose the log β_1 value of 2.17 obtained for neodymium by Gammons et al. (1996), which is virtually identical to the average value of log β_1 reported above. The log β_2 values derived by least squares regression for 11 rare earth elements have a mean value of 2.73±0.47, which is in good agreement with our previous estimate of log β_2 = 2.52 for neodymium only (Gammons et al., 1996). Table 4 shows a weak trend to lower β_2 values with increase in atomic number. This is consistent with the previously mentioned trend to lower average ligand number across the periodic table.

The final column in Table 4 lists our estimates for the second stepwise formation constant, K_2, or reaction (11). These constants were found to be much less sensitive to our initial guess for log β_1. The reason for this is subtle, but can be explained as follows. If the average ligand number of a particular group of complexes over a range of chloride concentration is greater than 1 (as was the case in this study for most elements at 200 °C), a mixture of the $LnCl^{2+}$ and $LnCl_2^+$ species is most likely present. Equilibrium

Table 5. Comparison of log β_2 and log K_2 values at 200 °C obtained in this study vs. values from Haas et al. (1995).

REE	log β_2			log K_2		
	HSS [a]	this study	deviation	HSS [a]	this study	deviation
La	2.867	3.45	0.58	0.523	1.28	0.76
Ce	3.117	2.83	–0.29	0.597	0.66	0.06
Pr	2.839	2.80	–0.04	0.511	0.63	0.12
Nd	2.710	2.52 [b]	–0.19	0.43	0.35 [b]	–0.08
Sm	2.611	2.40	–0.21	0.348	0.23	–0.12
Eu	2.868	2.94	0.07	0.519	0.77	0.25
Gd	3.065			0.633		
Tb	2.863	3.64	0.78	0.561	1.47	0.91
Dy	3.034			0.669		
Ho	2.986	1.96	–1.03	0.633	–0.21	–0.84
Er	2.775	2.57	–0.21	0.463	0.40	–0.06
Tm	2.785			0.518		
Yb	2.589	2.57	–0.02	0.381	0.40	0.02
Lu	2.339	2.31	-0.03	0.277	0.25	–0.03

[a] Haas et al. (1995); [b] Gammons et al. (1996).

Table 6. Average ligand numbers obtained in this study vs. those derived from the estimates of Haas et al. (1995).

	200 °C, 1.0m ΣCl			300 °C, 1.0m ΣCl		
	HSS [a]	this study	deviation	HSS [a]	this study	deviation
La	1.23	1.62	0.39	1.50	2.16	0.66
Ce	1.29	1.23	–0.06	1.54	1.96	0.42
Pr	1.21	1.17	–0.04	1.44	2.34	0.90
Nd	1.16	0.99 [b]	-0.17	1.35	1.99 [b]	0.64
Sm	1.12	1.13	0.01	1.30	1.66	0.36
Eu	1.22	1.25	0.03	1.50	2.30	0.80
Gd	1.31	1.00	-0.31	1.61	1.96	0.35
Tb	1.24	1.55	0.31	1.53	1.86	0.33
Dy	1.32			1.67		
Ho	1.29	0.94	–0.35	1.62	1.68	0.06
Er	1.19	1.04	–0.15	1.48	2.17	0.69
Tm	1.20	0.85	–0.35	1.51	2.10	0.59
Yb	1.13	1.11	–0.02	1.41	2.17	0.76
Lu	1.04	0.81	–0.23	1.39	2.01	0.62

[a] Haas et al. (1995); [b] Gammons et al. (1996).

between these two species is governed by K_2, the stepwise formation constant for $LnCl_2^+$. In this case, the solubility data can be interpreted with only a very approximate knowledge of β_1, the first association constant. In contrast, derivation of estimates for β_2 were very strongly dependent on our initial guess of β_1, since β_2 also involves the dissociated Ln^{3+} species. The precision in log K_2 is virtually the same as that for log β_2 (based on the standard deviations). The mean value of log K_2 for 11 rare earth elements is 0.57±0.46. Again, this compares favorably with our previous result of log K_2 = 0.35 in the more detailed study of neodymium (Gammons et al., 1996).

Because of the complexity of the indirect solubility method, it is difficult to attach a precise uncertainty to the equilibrium constants reported in Table 4. A liberal estimate of error in the reported values for most of the elements of concern (excluding neodymium and lutetium) is ± 0.5 log units. Because it was possible to refine both β_1 and β_2 for lutetium, the recommended error in this case is lowered to ± 0.3 log units.

Attempts to derive equilibrium constants at 300 °C from the solubility data obtained in this study were unsuccessful. Again, the main problem was a lack of information over a sufficiently wide range of chloride concentration. In addition, the fact that the derived ligand numbers were close to 2 in many cases suggests that a single complex, the $LnCl_2^+$ species, was dominant in the 300 °C experiments. This being the case, it was not possible to derive either stepwise or cumulative formation constants with an acceptable degree of precision.

4. DISCUSSION

4.1. Comparison with Previous Work

The only high-temperature experimental study with which to compare our results is that of Gammons et al. (1996). Whereas Gammons et al. (1996) investigated the chloride complexation behavior of a single element (neodymium) over a wide range of experimental conditions, in the present study we have looked at all of the rare earth elements for a limited range of conditions. Our results reinforce the earlier conclusions of Gammons et al. (1996), and also show only minor systematic variations in REE-chloride complexation behavior across the periodic table. This is consistent with the lack of strong variation in β_1 of REE-chloride complexes at 25 °C, as reported in the literature. For example, the compilation of Millero (1992) shows only 0.35 log unit variation in log β_1 for all of the rare earth elements, with a trend to lower β_1 values for higher atomic number.

Using theoretical estimation methods, the stability of rare earth element complexes at elevated temperature was predicted by Wood (1990) and Haas et al. (1995). Wood (1990) used a combination of the isocoulombic technique and the electrostatic approach of Helgeson and co-workers to estimate the stability of the $LnCl^{2+}$species to 300 °C. Haas et al. (1995) derived HKF (modified Helgeson-Kirkham-Flowers equation of state) parameters for Ln^{3+}, $LnCl^{2+}$, $LnCl_2^+$, $LnCl_3(aq)$ and $LnCl_4^-$, from which it is possible to estimate stepwise or cumulative formation constants over a wide range of temperature and pressure. Gammons et al. (1996) showed that the agreement between their experiments and the theoretical predictions of Haas et al. (1995) was quite good up to approximately 200 °C. Above this temperature, both Haas et al. (1995) and Wood (1990) underestimated the degree of Nd-chloride complexation.

Table 5 compares our experimentally-derived estimates for log β_2 and log K_2 at 200 °C vs. those calculated using the HKF parameters in Haas et al. (1995). For 7 of the 10 elements in which β_2 and K_2 could be refined (not counting neodymium), the agreement between our work and that of Haas and co-workers falls well within our experimental error (± 0.5 log units). Exceptions include lanthanum, terbium, and holmium. With the data in hand, it is difficult to say whether the disparities for these elements are real, or an artifact of experimental scatter. However, in the absence of a theoretical justification for the anomalous behavior of La, Tb, and Ho, we are inclined to implicate our own experimental uncertainty to explain the discrepant results.

Because we were not able to derive stability constants at 300 °C, a direct comparison cannot be made at this temperature with the study of Haas et al. (1995). However, the results can be indirectly compared by computing average chloride ligand numbers for each rare earth element based on the cumulative formation constants calculated from the data in Haas et al. (1995). The aqueous speciation of each lanthanide was computed for a theoretical solution containing 0.9m NaCl, 0.1m HCl, and 10^{-5} molal ΣLn. The activity of free chloride and activity coefficients for all REE complexes were estimated using the same computer program employed to interpret our own solubility data. The speciation calculations were used to calculate average ligand numbers at 200 °C and 300 °C. The results, summarized in Table 6, show very good agreement at 200 °C, as expected. However, at 300 °C, the theoretically estimated ligand numbers are substantially lower than those obtained from our own experiments, by an average of ~ 0.5 units. In other words, the estimates of Haas et al. (1995) are apparently underestimating the degree of chloride complexation at 300 °C. This reinforces the conclusion of Gammons et al. (1996), who found that their own estimates of log β_2 for neodymium agreed with those of Haas et al. (1995) up to 200 °C, but began to deviate above this temperature. At 300 °C, log β_2 was found to differ by 1.74 log units (6.87 vs. 5.13, respectively, for the two studies). The qualitative results of the present study suggest that a similar comparison would be obtained if it were possible to derive stability constants for the other rare earth elements at 300 °C.

We conclude that Haas et al.'s estimates of the stability of the chloride complexes of the rare earth elements are valid up to 200 °C, but that their formation constants for the 1:2 species are underestimated at T > 200 °C. We suggest that these stability constants should instead be approximated by the experimentally-determined values reported in Gammons et al. (1996) for neodymium. The lack of any systematic differences in average ligand number with atomic number at 300 °C supports the idea that log β_2 for the REE do not deviate greatly at this temperature, as is the case at lower temperature.

4.2. Speciation of Rare Earth Elements in Seafloor Hydrothermal Fluids

Recently, Douville et al. (1999) have reported REE concentrations from deep-sea hydrothermal fluids from the Mid-Atlantic Ridge, the East Pacific Rise, and the Lau and Manus Back-Arc Basins. Chondrite-normalized plots of their data reveal a general light REE-enrichment coupled with a positive Eu anomaly. However, when normalized to the REE content of related plagioclase from altered basalt, the patterns become much flatter (although still slightly LREE-enriched) with much weaker Eu anomalies. Douville et al. (1999) carried out calculations of the distribution of REE species using stability constants derived from the HKF equation parameters in Haas et al. (1995). These calculations suggest that REE-chloride complexes are the most important REE species in solution. Douville et al. (1999) note that previous workers have shown that chloride complex stability decreases across the REE series from light to heavy, and that chloride complex stability increases strongly with temperature. They therefore suggest that the remaining LREE enrichment observed in their plagioclase-normalized REE patterns and the apparent increase in this LREE-enrichment with increasing temperature both reflect the influence of chloride complexation.

The results from our study, combined with calculations in Gammons et al. (1996), confirm that the REE-chloride complexes are likely to be the dominant species in the fluids studied by Douville (1999). There is no question that complexation of the REE by chloride increases in magnitude with temperature. In fact, REE-chloride complexes are probably even more dominant than suggested by the speciation calculations of Douville et al. (1999), because experimental work has shown that the HKF parameters of Haas et al. (1995) underestimate the stability of chloride complexes at T > 200 °C (Gammons et al., 1996; this study), and overestimate the stability of hydroxide complexes (Wood et al., this volume). Our data at 200 °C are consistent with a modest overall decrease in the stability of the REE-chloride complexes from light to heavy, which could partly explain the LREE-enrichment observed by Douville et al. (1999) in their plagioclase-normalized hydrothermal fluids. However, our data suggest that it is unlikely that chloride complexation alone can induce *strong* LREE-enrichment in hydrothermal fluids.

Our results also support the conclusions made by Wood and Williams-Jones (1994), and reiterated by Gammons et al. (1996), with respect to enrichments and depletions of REE in the alteration zones beneath ancient seafloor massive sulfide deposits. The extent of REE mobility in such deposits is difficult to explain without invoking unusually high water/rock ratios (> 1000) or fluid chemistries significantly different than modern deep-sea hydrothermal fluids.

Acknowledgments—Funding for this project was made possible through NSF Grant EAR-9714631 to CHG, and a Montana Tech Faculty Seed Grant to CHG. The manuscript was improved by the helpful comments and suggestions of Tom Giordano, Don Palmer, and an anonymous reviewer.

Editorial Handling: T. H. Giordano

REFERENCES

Aldridge, L.P. and Seward, T.M. (1976) HACK - A Fortran program for calculating equilibrium constants from solubility data at elevated temperatures. *Dept. Scientific Industrial Res.* (New Zealand), *Rept. CD 2227*, 10p.

Bau, M. and Dulski, P. (1999) Comparing yttrium and rare earths in hydrothermal fluids from the Mid-Atlantic Ridge: Implications for Y and REE fractionation during near-vent mixing and for the Y/Ho ratio of Proterozoic seawater. *Chem. Geol.* **155**, 77-90.

Douville, E., Bienvenu, P., Charlou, J. L., Donval, J. P., Fouquet, Y., Appriou, P., and Gamo, T. (1999) Ytrrium and rare earth elements in fluids from various deep-sea hydrothermal systems. *Geochim. Cosmochim. Acta* **63**, 627-644.

Frantz, J.D. and Marshall, W.L. (1984) Electrical conductances and ionization constants of salts, acids, and bases in supercritical aqueous fluids: 1. Hydrochloric acid from 100 ° to 700 °C and at pressures to 4000 bars. *Amer. J. Sci.* **284**, 651-667.

Gammons, C.H. (1995) Experimental investigations of the hydrothermal geochemistry of Pt and Pd. IV. The stoichiometry of Pt(IV) and Pd(II) chloride complexes at 100° to 300 °C. *Geochim. Cosmochim. Acta* **59**, 1655-1668.

Gammons, C.H. and Seward, T.M. (1996) The stability of manganese(II) chloride complexes at 25° to 300 °C. *Geochim. Cosmochim. Acta* **60**, 4295-4311.

Gammons, C.H., Wood, S.A., and Williams-Jones, A.E. (1996) The aqueous geochemistry of the rare earth elements. V. Stability of neodymium-chloride complexes to 300 °C. *Geochim. Cosmochim. Acta* **60**, 4615-4630.

Haas, J.L., Jr. (1971) The effect of salinity on the maximum thermal gradient of a hydrothermal system at hydrostatic pressure. *Econ. Geol.* **66**, 940-946.

Haas, J.R., Shock, E.L., and Sassani, D.C. (1995) Rare earth elements in hydrothermal systems: estimates of standard partial molal thermodynamic properties of aqueous complexes of the rare earth elements at high pressures and temperatures. *Geochim. Cosmochim. Acta* **59**, 4329-4350.

Helgeson, H.C. (1969) Thermodynamics of hydrothermal systems at elevated temperatures and pressures. *Amer. J. Sci.* **63**, 622-635.

Helgeson, H.C., Kirkham, D.H., and Flowers, G.C. (1981) Theoretical prediction of the thermodynamic behaviour of aqueous electrolytes at high pressures and temperatures. IV. Calculation of activity coefficients, osmotic coefficients, and apparent molal and standard and relative partial molal properties to 600 °C and 5 kb. *Amer. J. Sci.* **281**, 1249-1516.

Kielland, J. (1937) Individual activity coefficients of ions in aqueous solutions. *J. Amer. Chem. Soc.* **59**, 1675-1678.

Klinkhammer, B.P., Elderfield, H., Edmond, J.M., and Mitra, A. (1994) Geochemical implications of rare earth element patterns in hydrothermal fluids from mid-ocean ridges. *Geochim. Cosmochim. Acta* **58**, 5105-5113.

Levin, K.A. (1991) An experimental and thermodynamic study of the stabilities of silver chloride complexes in KCl and NaCl solutions up to 7m at 300 °C. *Geokhimiya* **10**, 1463-1468.

Mesmer, R.E., Marshall, W.L., Palmer, D.A., Simonsen, J.M., and Holmes, H.F. (1988) Thermodynamics of aqueous association and ionization reactions at high temperatures and pressures. *J. Sol. Chem.* **17**, 699-718.

Michard, A. (1989) Rare earth element systematics in hydrothermal fluids. *Geochim. Cosmochim. Acta* **53**, 745-750.

Michard, A. and Albarède, F. (1986) The REE content of some hydrothermal fluids. *Chem. Geol.* **55**, 51-60.

Michard, A., Albarède, F., Michard, G., Minster, J.F., and Charlou, J.L. (1983) Rare-earth elements and uranium in high-temperature solutions from East Pacific Rise hydrothermal vent field (13°N). *Nature* **303**, 795-797.

Millero, F.J. (1992) Stability constants for the formation of rare earth inorganic complexes as a function of ionic strength. *Geochim. Cosmochim. Acta* **56**, 3123-3132.

Pokrovskii, V.A. (1999) Calculation of the standard partial molal thermodynamic properties and dissociation constants of aqueous HCl and HBr at temperatures to 1000 °C and pressures to 5 kbar. *Geochim. Cosmochim. Acta* **63**, 1107-1115.

Ruaya, J.R. and Seward, T.M. (1986) The stability of chloro-zinc (II) complexes in hydrothermal solutions up to 350 °C. *Geochim. Cosmochim. Acta* **50**, 651-661.

Seward, T.M. (1976) The stability of chloride complexes of silver in hydrothermal solutions up to 350 °C. *Geochim. Cosmochim. Acta* **40**, 1329-1341.

Shannon, W.M., Wood, S.A., Brown, K., and Arehart, G. (1999) Preliminary measurements of concentrations of lanthanide elements in geothermal fluids from the Taupo Volcanic Zone, New Zealand. Proceedings, Twenty-Fourth Workshop on Geothermal Reservoir Engineering, Stanford University, Stanford, California, January 25-27, 1999. Stanford Geothermal Program Workshop Report SGP-TR-162, p. 227-235.

van Middlesworth, P.E. and Wood, S.A. (1998) The aqueous geochemistry of the rare earth elements and yttrium: Part 7: REE, Th and U contents in thermal springs associated with the Idaho Batholith. *Appl. Geochem.* **13**, 861-884.

Williams-Jones, A.E. and Seward, T.M. (1989) The stability of calcium chloride ion pairs in aqueous solutions at temperatures between 100 and 360 °C. *Geochim. Cosmochim. Acta* **53**, 313-318.

Wood, S.A. (1990) The aqueous geochemistry of the rare-earth elements and yttrium. 2. Theoretical predictions of speciation in hydrothermal solutions to 350 °C at saturation water vapor pressure. *Chem. Geol.* **88**, 99-125.

Wood, S.A. and Williams-Jones, A.E. (1994) The aqueous geochemistry of the rare-earth elements and yttrium. 4. Monazite solubility and REE mobility in exhalitive massive sulfide-depositing environments. *Chem. Geol.* **115**, 47-60.

Wood, S.A., Palmer, D.A., Wesolowski, D.J., and Bénézeth, P. (this volume) The aqueous geochemistry of the rare earth elements and yttrium. Part XI. The solubility of $Nd(OH)_3$ and hydrolysis of Nd^{3+} from 30 to 290 °C at saturated water vapor pressure with in-situ pH measurment.

Zotov, A.V., Levin, K.A., Khodakovskiy, I.L. and Kozlov, V.K. (1986a) Thermodynamic parameters of Ag^+ in aqueous solution at 273-573K. *Geochem. Intl.* **23**(9), 23-33.

Zotov, A.V., Levin, K.A., Khodakovskiy, I.L. and Kozlov, V.K. (1986b) Thermodynamic parameters of Ag(I) chloride complexes in aqueous solution at 273-623°K. *Geochem. Intl.* **23**(9), 103-116.

Water-Rock Interactions, Ore Deposits, and Environmental Geochemistry: A Tribute to David A. Crerar
© The Geochemical Society, Special Publication No. 7, 2002
Editors: Roland Hellmann and Scott A. Wood

The aqueous geochemistry of the rare earth elements and yttrium. Part X. Potentiometric determination of stability constants of acetate complexes of La^{3+}, Nd^{3+}, Gd^{3+} and Yb^{3+} at 25-70 °C and 1 bar

Ru Ding and Scott A. Wood*

Department of Geological Sciences, University of Idaho, Moscow, ID 83844-3022 USA

* corresponding author. E-mail: swood@iron.mines.uidaho.edu

Abstract - The stability constants of acetate (Ac^-) complexes of four trivalent REE (rare earth element) ions (La^{3+}, Nd^{3+}, Gd^{3+} and Yb^{3+}) were determined in aqueous NaCl solutions at temperatures from 25-70 °C and ionic strengths from 0.1 to 2.0 mol kg^{-1} via a potentiometric method. At the relatively low total acetate concentrations employed in this study (≤0.02 m), only the first acetate complex, $LnAc^{2+}$ (where Ln^{3+} represents any trivalent REE ion) was required to fit the data. The data obtained are adequately described by the following equations:

$$\log Q_1(La^{3+}) = 7.081 - 1326/T + 125.2 \cdot I/T - 1.892\times10^{-4} \cdot T \cdot \exp(I) - 6 \cdot A_\gamma \cdot I^{1/2}/(1+I^{1/2})$$

$$\log Q_1(Nd^{3+}) = 6.671 - 1108/T + 141.5 \cdot I/T - 1.856\times10^{-4} \cdot T \cdot \exp(I) - 6 \cdot A_\gamma \cdot I^{1/2}/(1+I^{1/2})$$

$$\log Q_1(Gd^{3+}) = 7.092 - 1258/T + 166.5 \cdot I/T - 2.239\times10^{-4} \cdot T \cdot \exp(I) - 6 \cdot A_\gamma \cdot I^{1/2}/(1+I^{1/2})$$

$$\log Q_1(Yb^{3+}) = 8.196 - 1659/T + 137.4 \cdot I/T - 1.146\times10^{-4} \cdot T \cdot \exp(I) - 6 \cdot A_\gamma \cdot I^{1/2}/(1+I^{1/2})$$

where Q_1 is the conditional stability constant for the association reaction $Ln^{3+} + Ac^- \leftrightarrow LnAc^{2+}$, I is the ionic strength in mol kg^{-1}, T is the temperature in Kelvin and A_γ is the Debye-Hückel limiting slope. Conditional stability constants measured in this study at 25 °C and at ionic strengths ≤ 1.0 mol kg^{-1} were in good agreement (generally less than ±0.2 log unit difference) with most values reported previously in the literature for $NaClO_4$ media, suggesting that competition from chloride complexation was negligible. However, stability constants determined in NaCl solutions of higher concentrations (2.0 mol kg^{-1}) were typically lower (by more than 0.2 log unit for La^{3+}) than those reported in the literature for $NaClO_4$ solutions of the same ionic strength, suggesting that the influence of chloride complexation is significant at this NaCl concentration.

Stability constants obtained in this study were extrapolated to 250 °C using an isocoulombic approximation, and were found to be in fair agreement (within ±0.2 log unit) with the few experimental data available at elevated temperatures. However, our extrapolated values are orders of magnitude greater than previously published theoretical estimates of these constants at 100-250 °C. Our stability constants, combined with additional thermodynamic data from the literature, should permit quantitative modeling of the influence of acetate on REE mass transfer in a variety of geologic environments and fine-tuning of models designed to predict REE-acetate stability constants at elevated temperatures and pressures.

1. INTRODUCTION

Rare earth elements (REE) and Nd isotopes increasingly are being employed as tracers of processes occurring during the formation and diagenesis of sedimentary rocks, and the genesis of petroleum and base-metal deposits (e.g., Mississippi Valley-Type) that these rocks may contain (Scherer and Seitz, 1980; Graf, 1984; Banner et al., 1988a,b; Dorobek and Filby, 1988; McLennan, 1989; Bock et al., 1994; McDaniel et al., 1994; Ohr et al., 1994; Kontak and Jackson, 1995). The REE are usually considered to be relatively immobile, and it is generally held that normalized REE patterns and Nd isotope ratios, which are used widely as petrogenetic tracers, are comparatively immune to changes during weathering, diagenesis, metamorphism and hydrothermal alteration. However, a number of studies have documented disturbance of Nd isotope and REE systematics by fluid-mediated processes in sedimentary rocks (Bock et al., 1994; McDaniel et al., 1994; Ohr et al., 1994; Kontak and Jackson, 1995). Mobility, and particularly fractionation, of the REE in aqueous fluids can greatly complicate the use of these elements as tracers. Assessment of the degree of mass transfer of the REE by aqueous fluids in any given geological environment requires accurate stability constants for relevant REE complexes and solubility products for relevant REE-bearing phases, but such data are often unavailable or of uncertain quality.

The mobility of the REE in hydrothermal solutions also is relevant to the safe disposal and effective remediation of nuclear waste materials in geological environments. The REE are produced during the fission of uranium and plutonium (Rard 1988), and are analogues of trivalent elements of the actinide series (Moeller et al. 1965; Choppin 1983). Moreover, a method which is being evaluated for disposal of Pu from dismantled nuclear warheads involves the production of a waste form containing Pu and the REE gadolinium (Gd) which acts as a neutron moderator to prevent criticality. Thus, data are required to evaluate the solubility of Gd and Pu under the P-T conditions of burial of the waste.

Acetate occurs in high concentrations in some oilfield brines (up to 10,000 ppm; Willey et al., 1975; Carothers and Kharaka, 1978) and in sediments affected by seafloor hydrothermal fluids (Martens, 1990). It is a product of thermal degradation of more complex organic matter (Surdam et al., 1984; Kawamura et al., 1986), and can persist metastably in aqueous solution for geologically long periods of time at temperatures between 100 and 200 °C (Kharaka et al., 1983; Drummond and Palmer, 1986; Palmer and Drummond, 1986; Bell et al., 1993). Acetate also is produced by corn roots and the roots of grasses and herbs, and occurs in soil solutions in concentrations of 2-5 x 10^{-3} mol L^{-1} (Bruckert, 1970; Sposito, 1989). Carboxylate ligands such as acetate may also be present in mixed nuclear-organic wastes such as those occurring at Oak Ridge, Tennessee and Maxey Flats, Kentucky (cf. Means et al., 1978; Cleveland and Rees, 1981). Furthermore, acetate can be used as an analogue of monocarboxylate groups in general, both in individual low-molecular weight ligands and in complex humic and fulvic acids (Wood, 1996; 2000).

The points discussed in the preceding paragraph suggest that acetate is a potentially important ligand for the transport of metals in geological environments, and acetate complexes of the following metals have been studied at elevated temperatures: Al^{3+} (Fein, 1991; Bénézeth et al., 1994; Palmer and Bell, 1994), Zn^{2+} (Hennet et al., 1988; Yang et al., 1989; Giordano and Drummond, 1991), Fe^{2+} (Palmer and Drummond, 1986; 1988; Palmer and Hyde, 1993), Mg^{2+} (Semmler et al., 1990; Fein, 1991), Ca^{2+} (Fein, 1991; Seewald and Seyfried, 1991) and Pb^{2+} (Hennet et al., 1988; Giordano, 1989; Yang et al., 1989). Acetate complexes of the REE have been investigated using potentiometric, spectrophotometric, calorimetric, ion-exchange and solvent extraction techniques (e.g., Sonesson, 1958a,b; 1959; 1960; Grenthe, 1962; 1964; Manning and Monk, 1962; Kolat and Powell, 1962; Kovar and Powell, 1966; Archer and Monk, 1964; 1966; Choppin and Schneider, 1970; Antipenko et al., 1973; Gruzdev and Ermolaev, 1974; Bukietynska et al., 1977; 1981a,b; Daniele et al., 1985; Ilnitskaya and Ternovaya, 1988), and these studies are in reasonably good agreement (cf. review by Wood, 1993). However, the above studies are restricted to low temperatures (20-25 °C) and, although they collectively cover a range of ionic strength from <0.1 to 2.0 mol L^{-1}, individual studies have dealt generally with only a single ionic strength, making problematic the extrapolation of the conditional stability constants to infinite dilution. Because of the lack of experimentally measured values at elevated temperatures and pressures, Shock and Koretsky (1993) estimated the stability constants of REE-acetate complexes using the modified Helgeson-Kirkham-Flowers (HKF) equation. Recently, Deberdt et al. (1998; 2000) measured the stability constants of La^{3+}-, Gd^{3+}- and Yb^{3+}-acetate complexes at temperatures up to 80 °C, and Wood et al. (2000) determined the stability constants of Nd^{3+}-acetate complexes in 0.1 mol kg^{-1} NaCl from 25 to 225 °C. Both Deberdt et al. (1998; 2000) and Wood et al. (2000) used potentiometric methods. Also, Wruck et al. (1997) determined stability constants of Nd^{3+}-acetate complexes from 20 to 70 °C by means of photoacoustic spectroscopy and conventional spectrophotometry. Because the ranges of ionic strengths and/or temperatures covered in these studies are somewhat restricted, additional measurements are warranted.

This paper is the tenth in a series of contributions to improved understanding of the aqueous geochemistry of the REE (Wood, 1990a,b; Wood, 1993; Wood and Williams-Jones, 1994; Wood et al., 1995; Gammons et al., 1996; van Middlesworth and Wood, 1998; Gammons and Wood, 2000; Wood et al., 2000). Here we report the potentiometric determination of the stability constants of the monoacetato ($LnAc^{2+}$) complexes of La^{3+}, Nd^{3+}, Gd^{3+} and Yb^{3+} in 0.1, 0.2, 1.0 and 2.0 mol kg^{-1} NaCl from 25 to 70 °C at 1 bar. Preliminary results of this investigation were reported by Ding et al. (1998). The potentiometric measurements were conducted in a different manner from those of either Deberdt et al. (1998; 2000) or Wood et al. (2000), and should provide an independent check on their results.

2. METHODOLOGY

2.1. Theoretical

In this study, cumulative stability constants (β_n) are defined with respect to the following generalized relations:

$$Ln^{3+} + nAc^{-} \leftrightarrow LnAc_n^{3-n} \quad (1)$$

$$\beta_n = \frac{a_{LnAc_n^{3-n}}}{a_{Ln^{3+}} a_{Ac^-}^n} \quad (2)$$

where Ln^{3+} is any trivalent REE ion, Ac^- is the acetate ion, a_i is the activity of species i, and n is the ligand number and is assumed to range from 1 to 4. The potentiometric method employed is based on the fact that, complexation of free acetate with Ln^{3+} according to reaction (1) above will decrease the free acetate activity, causing the proton association reaction of acetate:

$$H^+ + Ac^- \leftrightarrow HAc^0 \quad (3)$$

to shift to the left, thus liberating protons. Therefore, complex formation can be monitored by measuring the pH of a solution containing REE and acetate, when neutralized by stepwise addition of a strong base (see Baes and Mesmer, 1976 and Martell and Motekaitis, 1992).

In this study, we carried out our experiments at a constant ionic strength maintained by the addition of NaCl. As a result, activity coefficients are maintained approximately constant, and we can define a conditional, cumulative stability constant according to the following expression:

$$Q_n = \frac{m_{LnAc_n^{3-n}}}{m_{Ln^{3+}} m^n_{Ac^-}} \quad (4)$$

These conditional stability constants are truly constant only at the ionic strength at which they were determined. Determination of conditional stability constants at a series of ionic strengths permits the determination of the stability constant at infinite dilution by extrapolation to zero ionic strength. In this study we define pH_m to be the negative logarithm of the stoichiometric hydrogen ion molality, i.e., $pH_m = -\log m_{H^+}$. The convention employed is that H^+ is not complexed by the chloride ion and ion-pairing is treated implicitly by the activity coefficient model used.

We have carried out these experiments in NaCl medium as recommended by Millero (1992), rather than $NaClO_4$ or Na-trifluoromethanesulfonate media, so that the stability quotients determined would be directly applicable to natural hydrothermal fluids, which are dominated by NaCl. Furthermore, REE-chloride complexes are known to be relatively weak at the temperatures employed in this investigation (Wood, 1990a,b; Gammons et al., 1996 and this volume). As will be seen below, our results indicate that chloride complexation was insignificant in our lower-ionic strength experiments but not at the highest NaCl concentration employed.

Measurements by Sonesson (1958b) indicate that small amounts of polynuclear acetate complexes are formed by the heavy REE at high REE concentration. However, they are not expected to be important at the low total REE concentrations employed in our experiments. The potentiometric study of Nd-acetate complexes by Wood et al. (2000) confirmed that polynuclear species are not significant at REE concentrations of 0.005 mol kg^{-1}, the concentration employed in the present study, so such species were not considered in the treatment of our data.

2.2. Materials

All chemicals used in the experiments were reagent grade, or better, and all solutions were prepared with 18-MΩ·cm deionized, degassed water. The following stock solutions were prepared: 1) 0.097 mol kg^{-1} HCl, prepared from concentrated HCl (37 wt %, Mallinckrodt), and standardized by titration against certified 0.1 N NaOH (Fisher); 2) 0.0966 mol kg^{-1} NaOH, which was prepared from 50 wt % NaOH (Fisher) and standardized against the 0.097 mol kg^{-1} HCl by titration; and 3) 4.868 mol kg^{-1} NaCl, which was prepared from A.C.S. grade crystals (Fisher), standardized against silver nitrate by titration using a chloride-ion selective electrode as endpoint detector. The above three stock solutions of HCl, NaOH, and NaCl were sparged with argon and stored under positive argon pressure in 20-L polyethylene containers to exclude CO_2. Solutions were transferred by positive argon pressure to minimize contact with air. Stock solutions of REE chlorides, prepared from hydrated crystalline solids (Alfa Aesar), were standardized by acidimetric titration of the eluent obtained by passing an aliquot of the REE stock solution through a cation-exchange column (Dowex 50WX2-200 ion-exchange resin in H^+ form, Aldrich). Excess acid was required in the REE stock solutions to prevent hydrolysis. In our case, sufficient concentrated HCl was added to the stock solution to keep the pH near 2. A stock solution of 0.196 mol kg^{-1} sodium acetate, which was prepared from A.C.S. grade crystals (Sigma Chemical Co.), was standardized against 0.097 mol kg^{-1} HCl by titration.

2.3. Potentiometric Measurements

The stability constants of La^{3+}, Nd^{3+}, Gd^{3+}, and Yb^{3+} complexes with acetate were determined at ionic strengths of 0.1, 0.2, 1.0 and 2.0 mol kg^{-1} and at temperatures of 25, 40, 50, 60, and 70 °C. The maximum total acetate and REE concentrations employed were 0.02 and 0.005 mol kg^{-1}, respectively. Separate experiments to measure the ionization constants of acetic acid were conducted over the same range of temperature and ionic strength. All titrations were repeated at least twice. The procedure employed closely follows that described by Martell and Motekaitis (1992) and is described briefly below. Additional details of the titration procedure are given by Ding (1999).

The experiments were conducted using two jacketed glass reaction vessels. One of these vessels

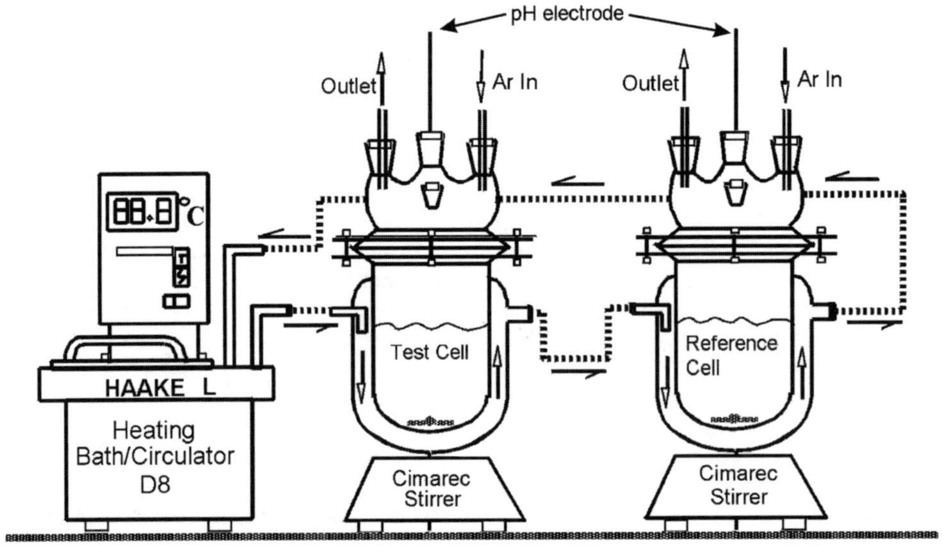

Figure 1: Schematic diagram of the apparatus employed in this study for potentiometric titrations. The reference cell is used for calibration of the pH electrode and contained solutions with precisely known concentrations of NaCl and strong acid or base. The test cell was used for the stability constant measurements and contained solutions with precisely known concentrations of $LnCl_3$, NaCl, HCl and NaAc. After calibration in the reference cell (as described in the text), the electrode was transferred immediately into the test cell and the measurement of stability constants was initiated.

served as the reference cell in which the pH electrode was calibrated, and the other served as the test cell in which the stability constants were determined. The apparatus used to carry out the potentiometric titrations is illustrated schematically in Figure 1. The caps for each reaction vessel have five openings to allow a combination glass pH electrode (Orion, ROSS), PEEK (polyetheretherketone) tubing (the titrant inlet), a mercury thermometer, and gas inlet and outlet tubes to be introduced into the cell. Ultrapure (99.99%) argon was continually bubbled through the cells to exclude CO_2. Before being introduced into the reaction cells, the argon was passed through a polyethylene drying tube containing ASCARITE II (to absorb any traces of CO_2) and bubbled through an aqueous solution at the same ionic strength and a temperature near that of the reaction cell (to inhibit evaporation of water from the cell). A 0.8-mm thick Teflon gasket was used to effect a seal between each cell and its lid. The temperature was controlled by circulation of thermostatted water through the jacket of the cell using a circulating water bath (Haake Model D8). The desired temperature could be maintained throughout the titration experiment to within ±0.05 °C. The solutions were mixed thoroughly during titration by means of a Teflon-coated magnetic stirring bar. When not in use, the glass electrode was stored in a NaCl solution containing 0.001 mol kg^{-1} H^+ with a similar ionic strength to that of the next experiment.

At the start of each experiment, weighed amounts of reference solution (containing a known amount of HCl) and test solution (containing known amounts of $LnCl_3$, HCl, NaCl and acetate) were placed into the calibration and test cells, respectively, using polyethylene transfer pipettes. After sealing the cells, air was removed by sparging with water-saturated Ar and the solutions were allowed to attain thermal equilibrium. The potential of the glass-electrode in the calibration cell was monitored with a pH meter (Radiometer Copenhagen Model PHM84). Once a stable cell potential was attained (< 0.01 mV/min), the potential was recorded. Then, a precise amount of standard base was titrated into the cell to change its pH a known amount, and a second potential reading was taken once a stable value was attained. This procedure was repeated until a calibration curve of mV vs. calculated pH_m, containing 5-10 points, was obtained. Then the electrode was transferred immediately to the test cell and measurement of the stability constant was begun. The measured potential was converted to pH_m of the test solutions using the calibration curve described above. This calibration procedure is based on the Nernst Equation:

$$E = E^o + E_{LJ} + \frac{RT}{2.3025F} pH_m \quad (5)$$

where E is the measured potential, E^0 is a constant depending on the glass and reference electrodes, E_{LJ} is the liquid junction potential, R is the gas constant and F is Faraday's constant. If E^0 and E_{LJ} together are constant in both the calibration cell and the test cell, then a calibration plot of potential vs. pH_m can be used to determine the pH_m of the test cell during stability constant measurements.

Calibration was conducted under the same conditions of ionic strength and temperature as the test cell. The major ion compositions in the calibration and test cells were made as similar as possible to minimize liquid junction and activity coefficient effects. The ionic strength of the solution and the concentration of the titrant-base were also matched to avoid changes in ionic strength during titration. Thus, we believe the E_{LJ} term in equation (5) to be small and constant in our experiments. Therefore, no explicit corrections were made for liquid-junction effects.

During both calibration and stability constant measurements, the titrant solution (NaOH) was transferred by volume using a syringe pump (KD Scientific, Model 210) which was connected via PEEK tubing to the cell. This syringe pump is capable of delivering volumes down to 0.001 mL, and was calibrated by weighing amounts of deionized water delivered. In initial experiments it was found that addition of the base titrant to the test solution caused local supersaturation with respect to solid REE hydroxide, which then precipitated and could not be redissolved. This problem was eliminated by employing PEEK tubing with a very fine inner diameter, relatively slow titrant addition rates, a relatively low concentration of NaOH in the titrant and vigorous mixing in the cell.

During titrations of REE-bearing solutions with NaOH, eventually a pH was reached where precipitation of $Ln(OH)_3$ was observed. This precipitation usually occurred where the rate of change of pH with titrant concentration suddenly increased. Once precipitation occurred, the titration was stopped, and several data points taken prior to precipitation were deleted from the data set to insure that hydrolysis did not affect the results.

2.4. Determination of Conditional Stability Constants

Conditional stability constants of REE-acetate complexes and dissociation constants of acetic acid were obtained from the titration data in this study using the program "BEST" provided by Martell and Motekaitis (1992). This program obtains the best-fit stability constants for a system by minimizing the sum of the squares of the differences between calculated and measured pH for all the points in the titration, i.e., the standard error (see Martell and Motekaitis, 1992 for details). Raw titration data are input into "BEST", and calculated stability constants are output, in volume and molarity (mol L^{-1}) units. All concentrations and conditional stability constants reported in this paper are given in molality units (mol kg^{-1}). Conversions between weight and volume, and molarity and molality were accomplished using the density of aqueous sodium chloride solutions (calculated using the expression of Fabuss et al., 1966), assuming that, because NaCl is the dominant component in solution, the density of the solution is equal to the density of an NaCl solution of the same concentration. Additional data required by "BEST" are the dissociation products of water, which have been reported by Busey and Mesmer (1978).

In order to determine the number and identity of species present in the titration experiments, a number of models containing various combinations of the species $LnAc^{2+}$, $LnAc_2^+$, $LnAc_3^0$ were fit to the titration data. The best fit was deemed to be the one that minimized the overall standard error of the fit using the fewest species. Moreover, for a species to be included in the model, it had to account for a significant percent-

Table 1. Comparison of dissociation quotients of acetic acid in aqueous NaCl media measured in this study with those of Mesmer et al. (1989).

Ionic strength (mol kg^{-1})	T(°C)	log Q[1]	log Q[2]	Diff.
0.1	25	-4.52	-4.54	-0.02
	30	-4.54	-4.54	-0.00
	40	-4.54	-4.55	-0.01
	50	-4.54	-4.56	-0.02
	60	-4.58	-4.58	0.00
	70	-4.58	-4.60	-0.02
0.2	25	-4.50	-4.50	-0.00
	40	-4.49	-4.50	-0.01
	50	-4.52	-4.51	0.01
	60	-4.53	-4.53	0.00
	70	-4.55	-4.54	0.01
1.0	25	-4.49	-4.49	-0.00
	40	-4.47	-4.48	-0.01
	50	-4.48	-4.47	0.01
	60	-4.49	-4.48	0.01
	70	-4.49	-4.48	0.01
2.0	25	-4.61	-4.61	0.00
	40	-4.58	-4.57	0.01
	50	-4.57	-4.56	0.01
	60	-4.56	-4.55	0.01
	70	-4.56	-4.55	0.01

[1]This study – average uncertainty is ±0.03 log units.
[2]Mesmer et al. (1989) – average uncertainty is ±0.02 log units.

a)

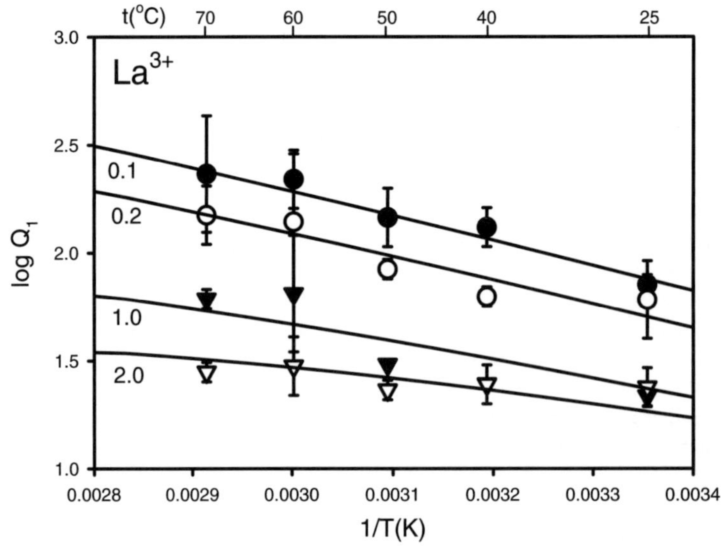

b)

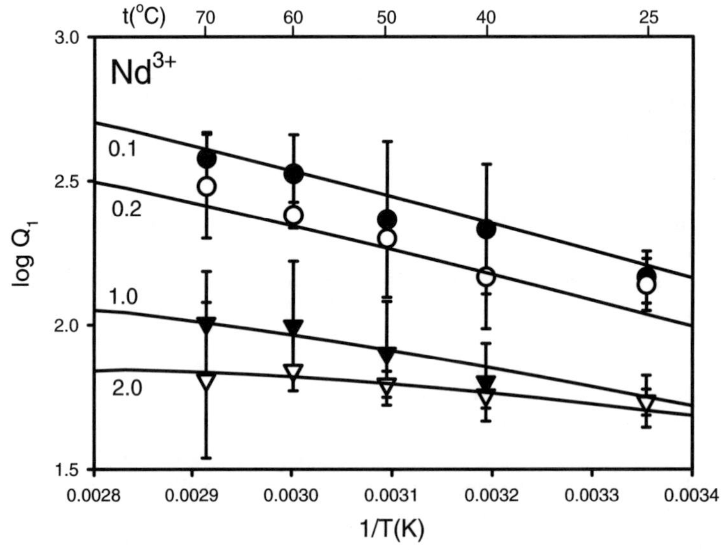

Figure 2: Plots of log Q_1 vs. 1/T(K) for Ln^{3+}-acetate complexes in NaCl solutions at various ionic strengths (in mol kg^{-1}). The error bars represent three standard deviations about the mean of at least two replicates. The solid curves are the predicted values of log Q_1 calculated from equations (7-10) in the text, which represent global fits to all the stability constant data for each REE: a) La^{3+}; b) Nd^{3+};

age of the total REE present over at least some portion of the titration and the standard error of the conditional stability constant for the species had to be relatively small. The details on the mechanism by which the best fit stability constants are calculated by "Best" are given in Martell and Motekaitis (1992).

3. RESULTS

3.1. Conditional Dissociation Constants of Acetic Acid

The conditional dissociation constants of acetic acid in NaCl media were measured in this study as a

c)

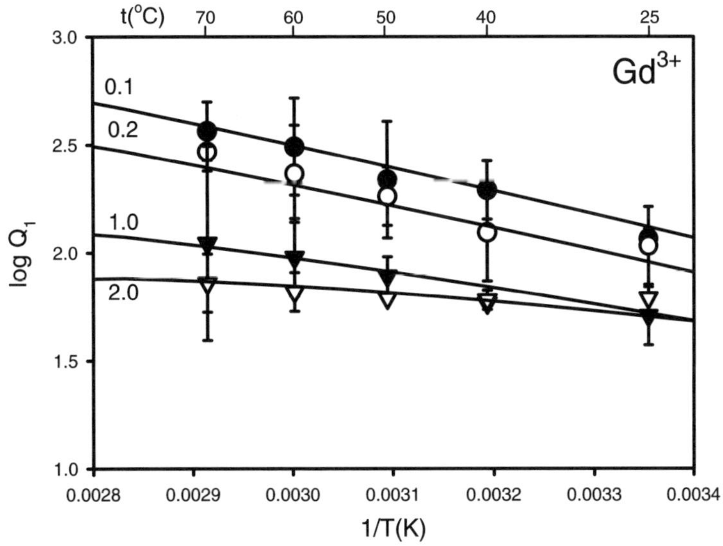

d)

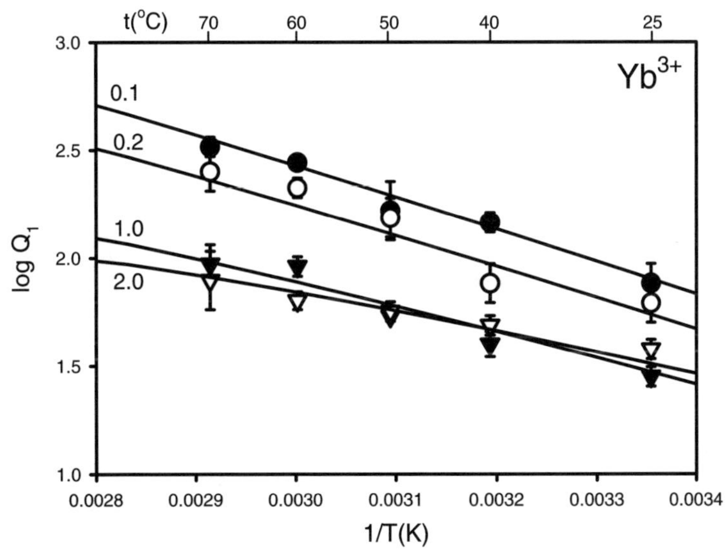

Figure 2 (continued): c) Gd^{3+}; d) Yb^{3+}.

check on our experimental methodology and data reduction procedures. The results of these measurements are given in Table 1, where they are compared to the results of Mesmer et al. (1989). It can be seen that the two sets of measured dissociation quotients are in excellent agreement at all temperatures. This agreement suggests that our experimental apparatus and procedures are capable of yielding high-quality conditional equilibrium constants.

3.2. Conditional Stability Constants of REE-Acetate Complexes

Under all experimental conditions and for all REE studied, only the monoacetato complex, $LnAc^{2+}$, was required to fit the titration data. Other complexes such as $LnAc_2^+$ were too low in concentration to measure accurately at the acetate concentrations studied. This was particularly evident because, on fitting a two-

Table 2. Unsmoothed conditional stability constants for $LnAc^{2+}$ in NaCl solutions obtained in this study at various temperatures and ionic strengths.

t (°C)	I (mol kg^{-1})	La	Nd	Gd	Yb
25	0.1	1.85±0.05	2.17±0.09	2.07±0.05	1.88±0.09
	0.2	1.78±0.18	2.14±0.09	2.03±0.18	1.79±0.09
	1.0	1.34±0.05	1.73±0.09	1.71±0.14	1.45±0.05
	2.0	1.38±0.09	1.73±0.05	1.79±0.02	1.58±0.05
40	0.1	2.12±0.09	2.33±0.22	2.29±0.14	2.16±0.05
	0.2	1.80±0.05	2.17±0.18	2.09±0.22	1.88±0.09
	1.0	1.39±0.02	1.80±0.14	1.77±0.02	1.60±0.14
	2.0	1.39±0.09	1.76±0.05	1.78±0.05	1.69±0.05
50	0.1	2.16±0.14	2.37±0.27	2.34±0.27	2.22±0.14
	0.2	1.92±0.05	2.30±0.02	2.26±0.14	2.19±0.09
	1.0	1.48±0.02	1.90±0.18	1.89±0.09	1.73±0.02
	2.0	1.36±0.05	1.79±0.05	1.79±0.02	1.75±0.05
60	0.1	2.34±0.14	2.52±0.14	2.49±0.22	2.44±0.02
	0.2	2.14±0.32	2.38±0.05	2.37±0.22	2.33±0.05
	1.0	1.81±0.27	2.00±0.22	1.98±0.18	1.96±0.05
	2.0	1.48±0.14	1.84±0.02	1.82±0.09	1.80±0.04
70	0.1	2.37±0.27	2.58±0.09	2.56±0.14	2.52±0.05
	0.2	2.17±0.14	2.48±0.18	2.47±0.09	2.40±0.09
	1.0	1.79±0.05	2.01±0.18	2.04±0.45	1.97±0.09
	2.0	1.45±0.05	1.81±0.27	1.86±0.14	1.90±0.14

Errors given represent three standard deviations from the mean of two or more replicate measurements.

species model ($LnAc^{2+}$ and $LnAc_2^+$) to the data and calculating the distribution of species based on the stability quotients obtained, it was found that the maximum percentage of total REE for the second complex was always very low (≤0.18%) and the standard error for the estimate of Q_2 was often larger than the value of Q_2 itself. Clearly, at the acetate concentrations employed in this study, the diacetato and higher complexes were present in insufficient quantity to obtain their conditional stability constants. Thus, we report conditional stability constants only for the monoacetato $LnAc^{2+}$ complex.

The unsmoothed, conditional stability constants for the $LnAc^{2+}$ complexes (i.e., Q_1) determined in this study are given as a function of temperature and ionic strength in Table 2. In each case the quoted error represents three standard deviations from the mean of replicate measurements. The temperature and ionic strength dependencies of the stability constants are also shown in Figure 2. For the most part, the measured conditional stability constants increase steadily with increasing temperature. However, for La, Nd and Gd, log Q_1 values are nearly independent of temperature at I = 2.0 mol kg^{-1}. It is also evident that, in general, the conditional stability constants increase with decreasing ionic strength. However, the difference in conditional stability constants at I = 1.0 and I = 2.0 mol kg^{-1} is relatively small, especially for Gd and Yb.

Table 3. Values of log β_1 for REE-acetate complexes at zero ionic strength calculated in this study from equations (7-10).

t(°C)	La	Nd	Gd	Yb
25	2.58	2.90	2.81	2.60
40	2.79	3.08	3.00	2.86
50	2.92	3.18	3.13	3.03
60	3.04	3.28	3.24	3.18
70	3.15	3.38	3.35	3.32

Average uncertainty is ±0.1 log units.

For each REE studied, an equation of the following form was fit to the entire set of conditional stability constants:

$$\log Q_1 = A + B/T + C\cdot I/T + D\cdot T\cdot \exp(I) + \Delta Z_1^2\cdot A_\gamma\cdot I^{1/2}/(1+I^{1/2}) \quad (6)$$

where A_γ is the Debye-Hückel limiting slope (taken from Ananthaswamy and Atkinson, 1984), I is the ionic strength in mol kg^{-1}, T is the temperature in Kelvin, ΔZ_1^2 is the change in the charges squared for the complexation reaction (here $\Delta Z_1^2 = -6$), and A, B, C and D are fit parameters. The first two terms in equation (6) are of the van't Hoff type. The last term represents the Debye-Hückel contribution to activity coefficients. The remaining two terms were found, after an exhaustive search, to provide the overall best fit to the stability constant data as judged by the standard error of fit and examination of the residuals. The best fit A, B, C and D parameters were obtained using the Marquardt-Levenberg least squares algorithm in Sigmaplot® (Jandel Scientific). The resulting equations for the conditional stability constants for each of the REE studied are:

$$\log Q_1(La^{3+}) = 7.081 - 1326/T + 125.2\cdot I/T - 1.892\times10^{-4}\cdot T\cdot \exp(I) - 6\cdot A_\gamma\cdot I^{1/2}/(1+I^{1/2}) \quad (7)$$

$$\log Q_1(Nd^{3+}) = 6.671 - 1108/T + 141.5\cdot I/T - 1.856\times10^{-4}\cdot T\cdot \exp(I) - 6\cdot A_\gamma\cdot I^{1/2}/(1+I^{1/2}) \quad (8)$$

$$\log Q_1(Gd^{3+}) = 7.092 - 1258/T + 166.5\cdot I/T - 2.239\times10^{-4}\cdot T\cdot \exp(I) - 6\cdot A_\gamma\cdot I^{1/2}/(1+I^{1/2}) \quad (9)$$

$$\log Q_1(Yb^{3+}) = 8.196 - 1659/T + 137.4\cdot I/T - 1.146\times10^{-4}\cdot T\cdot \exp(I) - 6\cdot A_\gamma\cdot I^{1/2}/(1+I^{1/2}) \quad (10)$$

The solid curves in Figure 2 represent the fit to the conditional stability constant data provided by the above equations. It can be seen that equations (7-10) fit the majority of the data within their respective uncertainties, and almost all the data within ±0.1 log units. Smoothed values of the stability constants at 25-70 °C and infinite dilution, calculated from equations (7-10), are given in Table 3.

Table 4. Thermodynamic data obtained in this study for the formation of $LnAc^{2+}$ complexes at zero ionic strength and 25°C.

Property	La	Nd	Gd	Yb
$\Delta G^0_{r,298.15}$ (kJ mol^{-1})	-14.7±0.6	-16.6±0.6	-16.0±0.6	-14.8±0.6
$\Delta H^0_{r,298.15}$ (kJ mol^{-1})	25.1±2.4	20.9±1.4	23.7±1.5	31.6±1.8
$\Delta S^0_{r,298.15}$ (J mol^{-1} K^{-1})	133±10	126±7	133±7	153±4
$\Delta H^0_{r,298.15}$ (kJ mol^{-1})[1]	21.0	—	17.2	23.9

[1]Deberdt et al. (1998, 2000)

Table 5. Values of log Q_1 reported in the literature for 20-25°C.

Source	t(°C)	Medium	La	Nd	Gd	Yb
Sonesson (1958a)	20	2.0 mol L^{-1} $NaClO_4$	1.56	1.90	1.84	ND
Sonesson (1958b)	20	2.0 mol L^{-1} $NaClO_4$	ND	ND	ND	1.64
Sonesson (1959)	20	1.0 mol L^{-1} $NaClO_4$	ND	ND	1.82	ND
Kolat & Powell (1962)	20	0.1 mol L^{-1} $NaClO_4$	2.02	2.22	2.16	2.03
Archer & Monk (1964)	25	infinite dilution	2.55	2.67	ND	2.56
Archer & Monk (1966)	25	infinite dilution	ND	ND	ND	2.45
Kovar & Powell (1966)	25	0.1 mol L^{-1} $NaClO_4$	1.82	2.11	2.02	1.84
Migal' & Chebotar (1967)	25	2.0 mol L^{-1} $NaClO_4$	1.54	ND	ND	ND
Migal' & Chebotar (1970)	?	2.0 mol L^{-1} $NaClO_4$	ND	1.90	ND	ND
Migal' et al. (1971)	?	0.5 mol L^{-1} $NaClO_4$	1.35	1.93	ND	ND
Gruzdev & Ermolaev (1974)	20	2.0 mol L^{-1} $NaClO_4$	ND	1.80	ND	ND
Bukietynska et al. (1977)	21	2.0 mol L^{-1} $NaClO_4$	ND	1.83	ND	ND
Bukietynska et al. (1981a)	21	2.0 mol L^{-1} $NaClO_4$	ND	1.81	ND	ND
Bukietynska et al. (1981b)	21	2.0 mol L^{-1} $NaClO_4$	ND	1.93	ND	ND
Bukietynska & Mondry (1984)	21	2.0 mol L^{-1} $NaClO_4$	1.66	1.93	1.96	ND
Daniele et al. (1985)	25	0.25 mol L^{-1}	1.74	ND	ND	ND
Ternovaya (1985)	20	1.0 mol L^{-1} KCl	1.4	1.76	1.70	1.20
Wood (1993)	25	infinite dilution	2.24	2.52	2.43	2.24
Wruck et al. (1997)	20	2.2 mol L^{-1} $NaClO_4$	ND	1.90	ND	ND
Deberdt et al. (1998)	25	infinite dilution	2.20	ND	ND	ND
Deberdt et al. (2000)	25	infinite dilution	ND	ND	2.52	2.32
Wood et al. (2000)	25	0.1 mol L^{-1} NaCl	ND	2.1	ND	ND
This study	25	infinite dilution	2.58	2.90	2.81	2.60
This study[1]	25	0.09 mol L^{-1} NaCl	1.90	2.23	2.14	1.93
This study[1]	25	0.19 mol L^{-1} NaCl	1.73	2.06	1.98	1.76
This study[1]	25	0.98 mol L^{-1} NaCl	1.38	1.76	1.73	1.48
This study[1]	25	2.03 mol L^{-1} NaCl	1.26	1.70	1.70	1.50

[1]The conditional stability constants obtained in this study have been converted to molar units for the purpose of comparison with values obtained in the literature. ND= not determined.

Thermodynamic quantities ($\Delta G^0_{r,298}$, $\Delta H^0_{r,298}$, and $\Delta S^0_{r,298}$) for reaction (1) at 25 °C and infinite dilution are given in Table 4. The values of $\Delta G^0_{r,298}$ were obtained from the formula:

$$\Delta G^o_{r,298} = \frac{-\log \beta_{1,298}}{2.3025R(298.15)} \qquad (11)$$

Where R is the ideal gas constant (8.314 J mol^{-1} K^{-1}) and $\beta_{1,298}$ is the stability constant given in Table 3 for reaction (1) at 25 °C and infinite dilution. The enthalpy of reaction was obtained by differentiating equations (7-10) according to the relation:

$$\left(\frac{\partial \log \beta_1}{\partial T}\right) = \frac{\Delta H^o_{r,298}}{2.3025RT^2} \qquad (12)$$

To apply equation (12), the ionic strength was set equal to zero in equations (7-10) and the resulting expressions were differentiated with respect to temperature. Finally, $\Delta S^0_{r,298}$ was obtained from the expression:

$$\Delta S^o_{r,298} = \frac{\Delta H^o_{r,298} - \Delta G^o_{r,298}}{T} \qquad (13)$$

4. DISCUSSION

4.1. Comparison to Previous Work

The conditional stability constants obtained in this study at 25 °C are compared with those reported in the literature at 20-25 °C in Figure 3. Details of

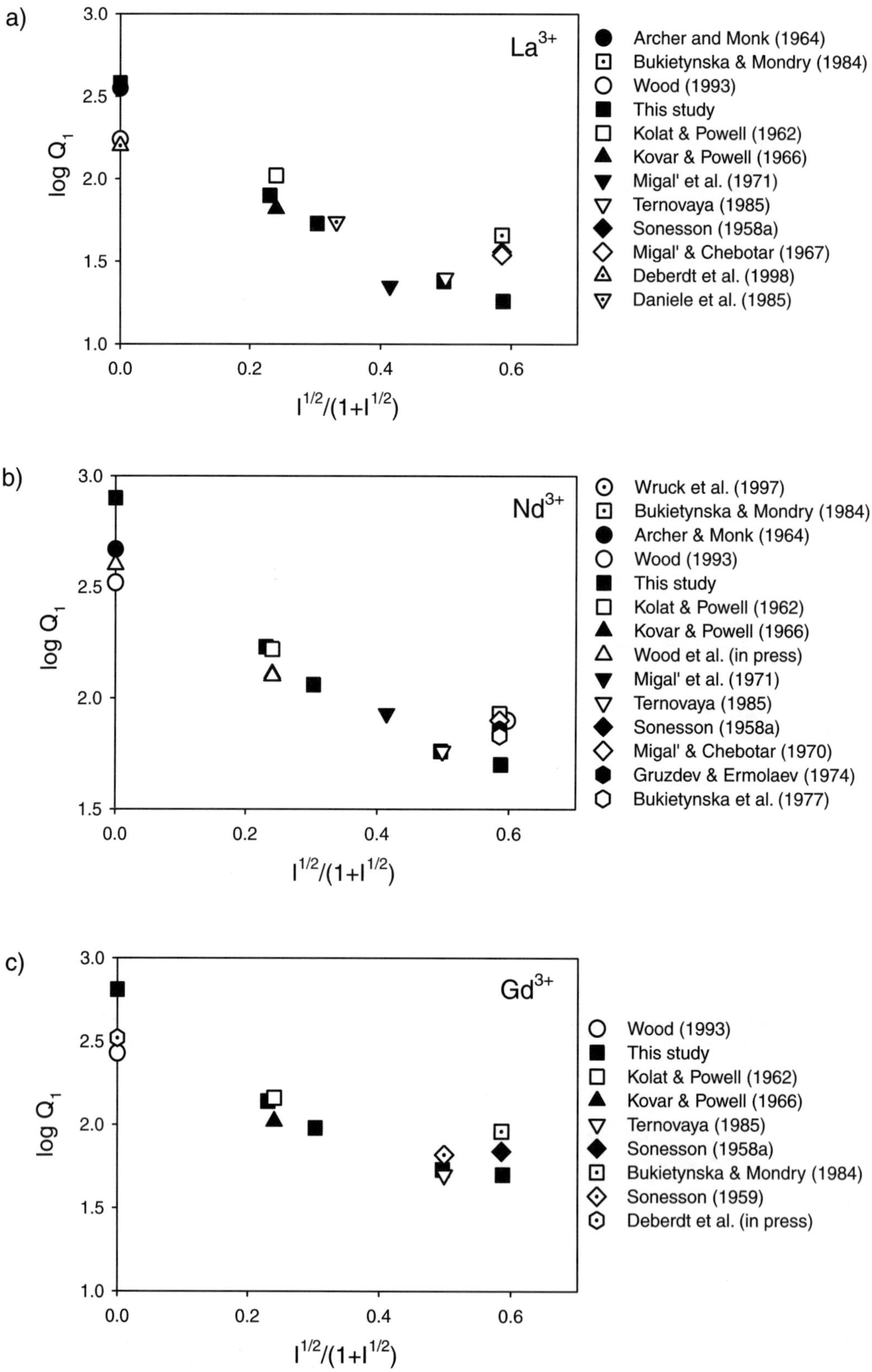
a)
La3+
log Q1
I1/2/(1+I1/2)
3.0
2.5
2.0
1.5
1.0
0.0
0.2
0.4
0.6
Archer and Monk (1964)
Bukietynska & Mondry (1984)
Wood (1993)
This study
Kolat & Powell (1962)
Kovar & Powell (1966)
Migal' et al. (1971)
Ternovaya (1985)
Sonesson (1958a)
Migal' & Chebotar (1967)
Deberdt et al. (1998)
Daniele et al. (1985)
b)
Nd3+
log Q1
I1/2/(1+I1/2)
3.0
2.5
2.0
1.5
0.0
0.2
0.4
0.6
Wruck et al. (1997)
Bukietynska & Mondry (1984)
Archer & Monk (1964)
Wood (1993)
This study
Kolat & Powell (1962)
Kovar & Powell (1966)
Wood et al. (in press)
Migal' et al. (1971)
Ternovaya (1985)
Sonesson (1958a)
Migal' & Chebotar (1970)
Gruzdev & Ermolaev (1974)
Bukietynska et al. (1977)
c)
Gd3+
log Q1
I1/2/(1+I1/2)
3.0
2.5
2.0
1.5
1.0
0.0
0.2
0.4
0.6
Wood (1993)
This study
Kolat & Powell (1962)
Kovar & Powell (1966)
Ternovaya (1985)
Sonesson (1958a)
Bukietynska & Mondry (1984)
Sonesson (1959)
Deberdt et al. (in press)

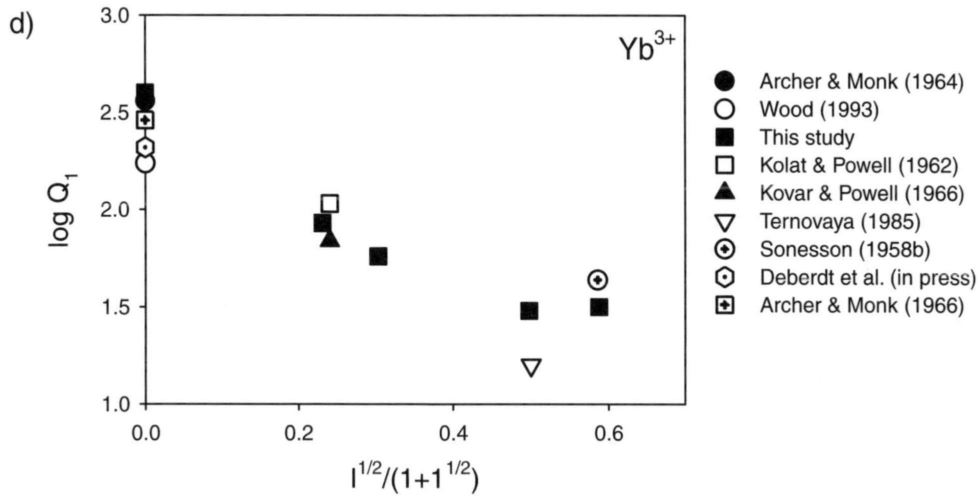

Figure 3: Comparison of log Q_1 values for Ln^{3+}-acetate complexes determined in this study with those reported previously in the literature at 20-25 °C, plotted as a function of ionic strength ($I^{1/2}/(1+I^{1/2})$): a) La^{3+}; b) Nd^{3+}; c) Gd^{3+}; d) Yb^{3+}. Values of conditional stability constants and ionic strengths from this study have been converted from mol kg^{-1} to mol L^{-1} units for the purpose of comparison. The average uncertainty for most of the data depicted is less than or equal to ±0.1 log units.

the values from the literature are given in Table 5. In both Figure 3 and Table 5, we have converted our concentrations and stability constants to molar (mol L^{-1}) units for comparison with the literature values which are given almost exclusively in these units. Note that we did not include previously published data in the fitting process described above, primarily because our experiments were conducted in NaCl solutions, whereas most experiments reported in the literature to date were conducted in $NaClO_4$ solutions. Independent comparison of our stability constants with those determined in non-complexing perchlorate media should permit assessment of the degree to which chloride complexation of the REE may have occurred.

Figure 3 shows that, in general, our results at finite ionic strengths agree very well with those in the literature when comparison is made at the same ionic strength. In particular, our data for all four REE at 0.1 mol L^{-1} ionic strength agree to within 0.1 log unit with those of Kolat and Powell (1962) and Kovar and Powell (1966), with our conditional stability constants falling between those reported by the latter two authors. Moreover, our data at 1.0 mol L^{-1} ionic strength are nearly identical to those of Ternovaya (1985) at the same ionic strength, with the exception of the datum for Yb. The agreement among our results and those reported in the literature is particularly notable because the majority of the results reported previously have been determined in non-complexing $NaClO_4$ media, whereas ours were determined in NaCl media. The agreement among the data obtained in these two media suggest that chloride complexation is quite weak at room temperature and the lower ionic strengths (NaCl concentrations) considered, in concordance with the known weak complexation of the trivalent REE with chloride under these conditions (cf. Wood, 1990a,b; Gammons et al., 1996 and this volume). However, at 2 mol kg^{-1} ionic strength, there is a discrepancy between our results and those obtained in $NaClO_4$ media. This discrepancy is greatest for La^{3+}, but decreases with atomic number. For Yb^{3+}, our stability quotient at 2 mol kg^{-1} is in good agreement with that determined in the landmark study by Sonesson (1958b). There are several reasons why we believe that the observed discrepancy is a result of the competing effect of chloride complexation at the highest ionic strength: 1) the discrepancy is only apparent at the highest concentrations of chloride investigated; 2) the direction of the discrepancy is compatible with the effect of chloride complexation. As shown by Wood et al. (2000), if chloride complexation occurs, stability quotients for REE-acetate complexes calculated neglecting chloride complexation will be underestimated; and 3) the discrepancy decreases on going from La^{3+} to Yb^{3+}, consistent with the apparent decrease in the stability of chloride complexes along the REE series (Wood, 1990a). Furthermore, the relatively weak temperature dependence of the stability quotients for La- and Nd-acetate complexes at I = 2.0

mol kg^{-1} may be explained by the fact that the temperature dependencies of REE–acetate and –chloride complexes are quite similar (Wood et al., 2000), such that the two offset one another.

The agreement among the results at infinite dilution is more variable. For La^{3+} at infinite dilution (Fig. 3a), the stability constant of this study agrees well with the value determined by Archer and Monk (1964). On the other hand, the result of Deberdt et al. (1998) agrees more closely with the value extracted by Wood (1993) via a Pitzer treatment of all the available literature data. However, the value determined in this study is more than 0.3 log units greater than that of Deberdt et al. (1998). For Nd^{3+} (Fig. 3b), the value determined in this study is considerably larger than any of the values determined by Archer and Monk (1964), Wood (1993), or Wood et al. (2000), all of which are relatively close. For Gd^{3+} (Fig. 3c) the values from Wood (1993) and Deberdt et al. (2000) are relatively close, but our value is more than 0.2 log units greater. Finally, for Yb^{3+} (Fig. 3d) the values from this study and Archer and Monk (1964, 1966) are similar, falling within 0.20 log units of one another, whereas the values of Wood (1993) and Deberdt et al. (2000) are lower.

The discrepancies noted among values of log β_1 are most likely due to uncertainties in the activity coefficients of highly charged species such as the trivalent REE. Figure 3 demonstrates that the agreement among log Q_1 values determined at finite ionic strengths is generally much better than that among log β_1 values at infinite dilution. Moreover, quite different methods of extrapolation to infinite dilution were employed in each of these studies. For example, Archer and Monk (1964; 1966) employed the Davies equation to correct their constants to infinite dilution. On the other hand, Wood (1993) used the data of Kovar and Powell (1966) at 0.1 mol kg^{-1} ionic strength and those of Sonesson (1958a,b) at 2.0 mol kg^{-1} ionic strength combined with a modification of the Pitzer equations proposed for REE complexes by Millero (1992). Deberdt et al. (1998; 2000) obtained log β_1 values at infinite dilution by applying the extended Debye-Hückel equation to calculate activity coefficients for their low, but variable ionic strength solutions. Wood et al. (2000) estimated the value of log β_1 for Nd using the extended Debye-Hückel equation recommended by Helgeson (1969). Finally, our study is the only one in which the constants were determined over a range of ionic strengths and extrapolated semi-empirically to infinite dilution.

As mentioned in the Introduction, there are relatively few data available in the literature on the stability constants of REE-acetate complexes at elevated temperature. However, in the next few paragraphs we compare our data to those from the relatively few studies reported in the literature conducted at temperatures greater than 25 °C. Our data for stability constants at elevated temperatures are compared to those determined by Deberdt et al. (1998, 2000) in Figures 4a,b,c. In general, the data exhibit similar trends with respect to temperature, but the stability constants of Deberdt et al. (1998, 2000) are consistently lower than ours by 0.2-0.4 log units. As discussed in the preceding paragraph, the discrepancies shown in Figure 4 most likely arise from differences in the manner in which the results were extrapolated to zero ionic strength.

The conditional stability constants measured by Wruck et al. (1997) for Nd^{3+} as a function of temperature (20-70 °C) in 2.2 mol kg^{-1} $NaClO_4$ are more than 0.2 log units higher than those measured by us in 2.0 mol kg^{-1} NaCl. Similarly, the results of Bukietynska et al. (1981a) for 2.0 mol kg^{-1} $NaClO_4$ at temperatures from 0 to 70 °C are 0.3-0.4 log units higher than those of this study for 2.0 mol kg^{-1} NaCl. These differences between our results in 2.0 mol kg^{-1} NaCl solution and previous measurements in ~2.0 mol kg^{-1} $NaClO_4$ solutions are most likely attributable to the effect of chloride complexation.

The recently reported conditional stability constants for Nd^{3+} in 0.1 mol kg^{-1} NaCl solutions from 25-225 °C of Wood et al. (2000) are compared with conditional stability constants derived in this study from 25-70 °C in Figure 5. It is apparent that the stability constant determined in this study based on a one-species fit of the experimental data at 25 °C is in good agreement with the corresponding value measured by Wood et al. (2000), but only fair agreement with the data at the higher temperatures. There is a discrepancy of 0.2-0.3 log unit between these two data sets at temperatures greater than 25 °C. The uncertainty in the log Q_1 values of Wood et al. (2000) below 100 °C is on the order of ±0.1 log unit, and therefore the error limits of the two sets of data just overlap or almost overlap. It is interesting to note that, although there was no statistical justification for the introduction of a second Nd-acetate complex to fit our data, Figure 5 shows that the value of log Q_1 is sensitive to whether or not a second complex is included in the speciation model. Log Q_1 values calculated with a two-species model are lower than those calculated using a one-species model, and these values are closer to those of Wood et al. (2000). Nevertheless, because there is no statistical justification for including a second complex in the fit to the titration data, and because the conditional stability constants derived for the second complex have unreasonably large uncertainties and exhibit highly irregular trends with respect to temperature and ionic strength, we report only log Q_1 and log β_1 values obtained from the one-species model in

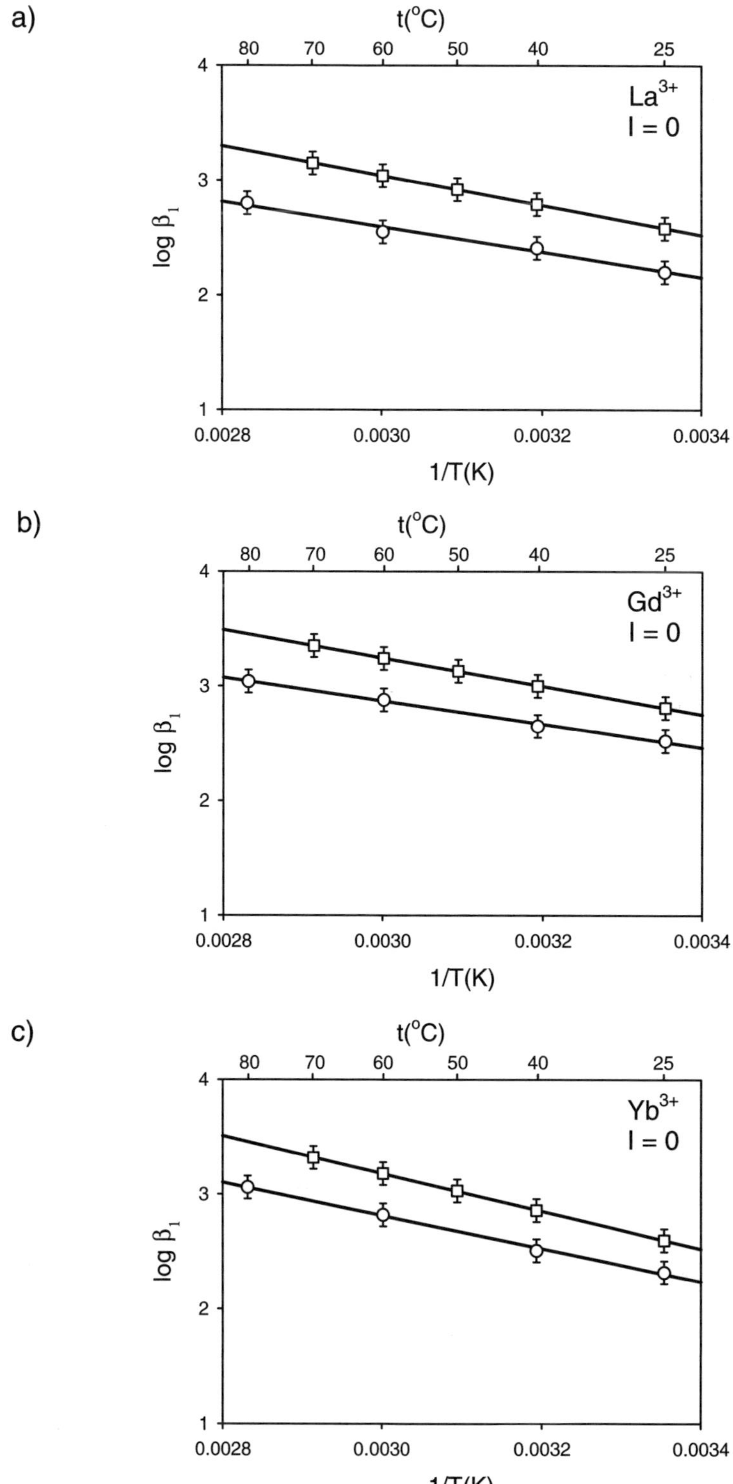

Figure 4: Comparison of log β_1 values obtained in this study with those determined by Deberdt et al. (1998, 2000) at various temperatures for: a) La^{3+}; b) Gd^{3+}; and c) Yb^{3+}. The open squares are the results of this study and the open circles are the results of Deberdt et al. (1998, 2000). The solid lines are linear least squares fits to the data and are provided as visual aids only.

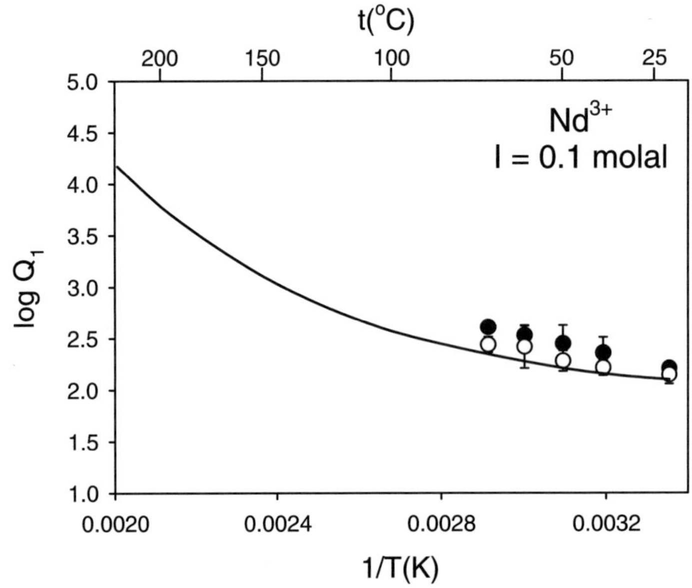

Figure 5: Comparison of log Q_1 values determined in this study for Nd^{3+} in 0.1 mol kg^{-1} NaCl solution with the smoothed values obtained by Wood et al. (2000) in the same medium at various temperatures up to 250 °C.

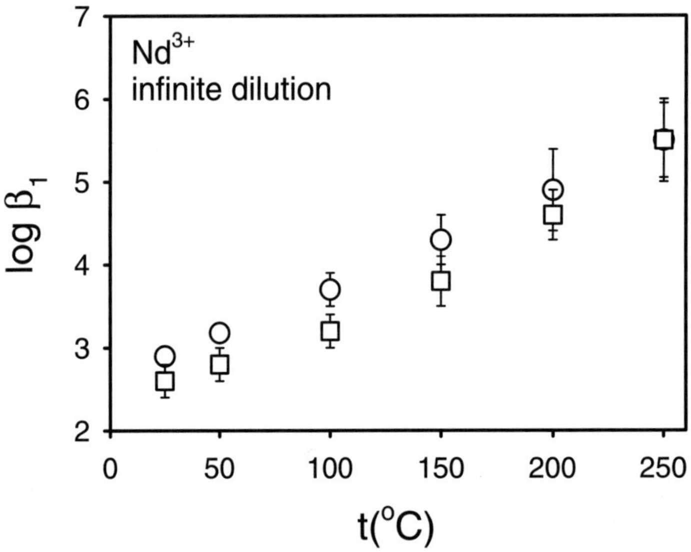

Figure 6: Comparison of stability constants (log β_1) at infinite dilution estimated in this study (filled circles) for the first Nd^{3+} - acetate complex at elevated temperatures with those determined by Wood et al. (2000).

this paper. Results of the two-species model fits to the titration data can be found in Ding (1999).

As shown in the preceding paragraphs, our values of log Q_1 and log β_1 agree to within the combined experimental errors (±0.1-0.3 log units) with the data of Deberdt et al. (1998, 2000) and Wood et al. (2000). The agreement among these studies is encouraging because, although all three studies employed a potentiometric method, there are significant differences in the experimental methodology. In this study, we em-

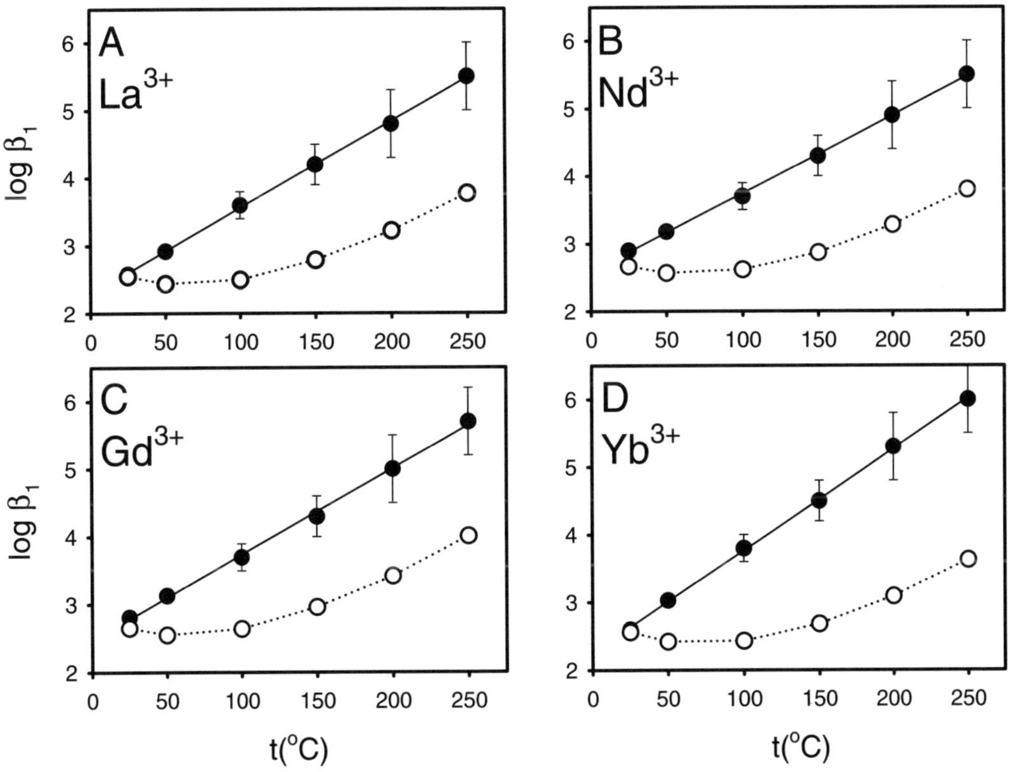

Figure 7: Comparison of stability constants (log β_1) at infinite dilution estimated in this study using the isocoulombic approximation (solid circles, solid line) with those estimated by Shock and Koretsky (1993) (open circles, dotted line) for the first acetate complex with: a) La^{3+}; b) Nd^{3+}; c) Gd^{3+}; d) Yb^{3+}.

ployed glass electrodes, and titrated NaOH into a solution containing $LnCl_3$, NaCl, HCl and acetate, holding ionic strength constant throughout the titration. Deberdt et al. (1998, 2000) also used glass electrodes, but titrated a $LnCl_3$ solution into a sodium acetate/sodium hydroxide buffer solution and ionic strength was low but not constant. Finally, Wood et al. (2000) employed a hydrogen-concentration cell, and titrated a sodium acetate solution into a solution containing $LnCl_3$, HCl and NaCl, maintaining ionic strength constant throughout.

In Table 4 we compare the enthalpies of complexation calculated by us with those determined by Deberdt et al. (1998, 2000). For La the agreement between the two studies is quite reasonable (i.e., the two results agree within their combined uncertainty). However, for Gd and Yb there is a significant discrepancy, with our values being 5-8 kJ mol^{-1} higher than those of Deberdt et al. (2000). Although the temperature trends of the stability constants of Gd- and Yb-acetate complexes determined in this study and by Deberdt et al. (2000) are subparallel (Fig. 4), our results exhibit a temperature trend with a slightly more positive slope than those of Deberdt et al. (2000). It is well known that enthalpies derived from van't Hoff type plots are very sensitive to uncertainties in the log K values, so these slight differences in slope probably account for the observed differences in enthalpies of complexation.

4.2. Extrapolation of Stability Constants to Higher Temperature

Although we have not derived heat capacity changes for the REE-acetate complexation reactions studied here, examination of Figure 4 suggests that, at infinite dilution over the temperature range investigated, $\Delta C^o_{p,r,298} \approx 0$. This would suggest that a simple van't Hoff treatment might be used to extrapolate the measured stability constants to higher temperatures. However, there is no reason to believe that $\Delta C^o_{p,r,298}$ for reaction (1), a non-isocoulombic reaction, would remain near zero over a temperature range greater than the approximately 50 °C range covered in this study. Therefore, in order to extrapolate the stability constants for REE-acetate complexes to higher tem-

Table 6. Stability constants (log β_1) for REE-acetate complexes at elevated temperatures calculated in this study using the isocoulombic method.

t(°C)	La	Nd	Gd	Yb
100	3.6±0.2	3.7±0.2	3.7±0.2	3.8±0.2
150	4.2±0.3	4.3±0.3	4.3±0.3	4.5±0.3
200	4.8±0.5	4.9±0.5	5.0±0.5	5.3±0.5
250	5.5±0.5	5.5±0.5	5.7±0.5	6.0±0.5

perature, the isocoulombic approximation (Lindsay, 1980; Cobble et al., 1982) was used. The complex formation reaction:

$$Ln^{3+} + Ac^- \leftrightarrow LnAc^{2+} \quad (14)$$

is not isocoulombic, but it can be converted to an isocoulombic reaction by addition of other appropriate reactions. In our case, we added reactions (15) and (16) to reaction (14) to obtain reaction (17), which is isocoulombic.

$$Al(OH)^{2+} + H^+ \leftrightarrow Al^{3+} + H_2O \quad (15)$$

$$H^+ + OH^- \leftrightarrow H_2O \quad (16)$$

$$Ln^{3+} + Al(OH)^{2+} + Ac^- \leftrightarrow Al^{3+} + LnAc^{2+} + OH^- \quad (17)$$

The change of heat capacity of reaction (17) is more likely than that of reaction (14) to remain approximately zero over a wide temperature range. Log K of reaction (17) then can be extrapolated to elevated temperatures using a plot of log K vs. 1/T. Values of log K required for reactions (15) and (16) at elevated temperatures were taken from Palmer and Wesolowski (1993) and Busey and Mesmer (1978), respectively. The uncertainty in the extrapolated values of the stability constants is considerably greater than in the raw experimental data determined here. This is a result of the accumulation of errors resulting from the combination of reactions (14), (15) and (16), as well as potential errors introduced by the isocoulombic approximation. The results of the extrapolation, with estimated uncertainties, are given in Table 6 and compared with the values reported by Wood et al. (2000) in Figure 6. The extrapolated stability constants (this study) are in reasonable agreement with the experimentally derived data (Wood et al., 2000) considering their combined uncertainties. It is particularly interesting that the stability constants from the two studies are nearly identical at 250 °C.

Our extrapolated stability constants are compared with those predicted by Shock and Koretsky (1993) in Figure 7. The values estimated by Shock and Koretsky (1993) are lower than ours by several orders of magnitude at temperatures above 50 °C. Deberdt et al. (1998; 2000) and Wood et al. (2000) also found that their experimentally determined stability constants for REE-acetate complexes were higher than the predictions by Shock and Koretsky (1993). Furthermore, our measured and estimated stability constants are in much closer agreement with the experimentally determined values of Deberdt et al. (1998; 2000) and Wood et al. (2000), than are the estimates of Shock and Koretsky (1993). Thus, we believe that the theoretical estimates of Shock and Koretsky (1993) require revision in the light of these new experimental data.

5. CONCLUSIONS

The stability constants of La^{3+}, Nd^{3+}, Gd^{3+}, and Yb^{3+} complexes with acetate have been determined potentiometrically from 25 to 70 °C at ionic strengths of 0.1, 0.2, 1.0 and 2.0 mol kg^{-1}.

Under the conditions of our experiments, complexes higher than the 1:1 complex $LnAc^{2+}$ were not required to fit the data.

The measured conditional stability constants show a mild, positive temperature dependence over the range investigated. The constants decrease with increasing ionic strength.

The measured conditional stability constants for each REE can be represented adequately by an equation of the type: $\log Q_1 = A + B/T + C\cdot I/T + D\cdot T\cdot \exp(I) + \Delta Z_1^2 \cdot A_\gamma \cdot I^{1/2}/(1+I^{1/2})$.

Our conditional stability constants at 25 °C and ionic strengths between 0.1 and 1.0 mol kg^{-1} are in excellent agreement with the conditional stability constants reported in the literature for $NaClO_4$ media. The similarity of REE-acetate stability constants in chloride and perchlorate media suggests that REE-chloride complexes are very weak under the conditions investigated. However, at 2 mol kg^{-1} ionic strength, the difference between the data obtained here and previous work suggest that chloride complexation has become more important at higher NaCl concentrations.

The experimentally derived stability constants were extrapolated to 250 °C using the isocoulombic approximation. These extrapolated constants are in fair agreement with previously measured values (Wood et al., 2000), but suggest that previous theoretical predictions by Shock and Koretsky (1993) are in need of revision.

Acknowledgments—This paper is dedicated to the memory of Dr. David A. Crerar, who was one of the first geochemists to recognize the importance of dissolved organic matter in the complexation of metals in natural aqueous environments and was a pioneer in the study of the thermodynamics of metal-organic complexation at elevated temperatures. SAW is indebted to Professor Crerar for the positive influence on his career. Acknowledgment is made to the Donors of the Petroleum Research Fund, administered by the American Chemical Society, and the State Board of Education of Idaho for financial support of this research. Some of the data reduction and preparation of this manuscript were conducted while SAW was on sabbatical leave at Oak Ridge National Laboratory. This sabbatical leave was financially supported by the Geosciences Program of the Office of Basic Energy Research, U.S. Department of Energy under contract DE-AC05-96OR22464 with Lockheed Martin Energy Research Corporation. Reviews by Moira Ridley, Dick Kettler and Tom Giordano resulted in significant improvement of the manuscript and are gratefully acknowledged.

Editorial handling: T.H. Giordano

REFERENCES

Ananthaswamy J. and Atkinson G. (1984) Thermodynamics of concentrated electrolyte mixtures. 4. Pitzer-Debye-Hückel limiting slopes for water from 0 to 100 °C and from 1 atm to 1 kbar. *Jour. Chem. Eng. Data* **29**, 81-87.

Antipenko B. M., Batyaev N. M., and Privalova T. A. (1973) Relaxation spectroscopy as a method of investigating complex formation reactions in solution. *Russ. J. Inorg. Chem.* **18**, 318-321.

Archer D. W. and Monk C. B. (1964) Ion-association constants of some acetates by pH (glass electrode) measurements. *J. Chem. Soc. (A)*, 3117-3122.

Archer D. W. and Monk C. B. (1966) Dissociation constants of some ytterbium ion-pairs from cation-exchange resin studies. *J. Chem. Soc. (A)*, 1374-1376.

Baes C. F. Jr. and Mesmer R.E. (1976) *The Hydrolysis of Cations*. J. Wiley & Sons, New York.

Banner J. L., Hanson G. N., and Meyers W. J. (1988a) Rare earth element and Nd isotopic variations in regionally extensive dolomites from the Burlington-Keokuk formation (Mississippian): Implications for REE mobility during carbonate diagenesis. *J. Sed. Petrol.* **58**, 415-432.

Banner J. L., Hanson G. N., Meyers W. J. (1988b) Water/rock interaction history of regionally extensive dolomites of the Burlington-Keokuk formation (Mississippian): isotope evidence. In Sedimentology and Geochemistry of Dolostones (eds. V. Shulka and P. A. Baker). *Soc. Econ. Paleontolog. Mineral. Spec. Publ. No.* **43**, 97-113.

Bell J. S., Palmer D. A., Barnes H. L., and Drummond S. E. (1993) Thermal decomposition of acetate: III. Catalysis by mineral surfaces. *Geochim. Cosmochim. Acta* **58**, 4155-4177.

Bénézeth P., Castet S., Dandurand J. L., Gout R., and Schott J. (1994) Experimental study of aluminum-acetate complexing between 60 and 200 °C. *Geochim. Cosmochim. Acta* **58**, 4561-4571.

Bock B., McLennan S. M., and Hanson G. N. (1994) Rare earth element redistribution and its effects on the neodymium isotope system in the Austin Glen Member of the Normanskill Formation, New York, USA. *Geochim. Cosmochim. Acta* **58**, 5245-5253.

Bruckert S. (1970) Effect of soluble organic compounds on pedogenesis in an acid environment. I. Field studies. *Ann. Agron.* **21**, 421-4521.

Bukietynska K. and Mondry A. (1984) Spectroscopic studies on the mixed Nd^{3+}-La^{3+} and Ho^{3+}-Gd^{3+} acetate systems as an example of the determination of stability constants from electronic spectroscopic data. *Polyhedron* **3**, 31-38.

Bukietynska K., Mondry A., and Osmeda E. (1977) Determination of stability constants of inner-sphere lanthanide complexes from the oscillator strengths data. *Jour. Inorg. Nucl. Chem.* **39**, 483-487.

Bukietynska K., Mondry A., and Osmeda E. (1981a) Application of the oscillator strength of "hypersensitive" transitions to the investigation of complex equilibria of lanthanide ions-I. Method of the evaluation of stability constants. *Jour. Inorg. Nucl. Chem.* **43**, 1311-1319.

Bukietynska K., Mondry A., and Osmeda E. (1981b) Application of the oscillator strength of "hypersensitive" transitions to the investigation of complex equilibria of lanthanide ions-II. Spectroscopy, stability and thermodynamics of lanthanide complexes with monocarboxylic and α-hydroxyacids. *Jour. Inorg. Nucl. Chem.* **43**, 1321-1330.

Busey R. H. and Mesmer R. E. (1978) Thermodynamic quantities for the ionization of water in sodium chloride media to 300 °C. *Jour. Chem. Eng. Data* **23**, 175-176.

Carothers W. W. and Kharaka Y. K. (1978) Aliphatic acid anions in oil-field waters - Implications for the origin of natural gas. *Amer. Assoc. Petrol. Geol. Bull.* **62**, 2441-2453.

Choppin G. R. (1983) Comparison of the solution chemistry of the actinides and lanthanides. *Jour. Less-Common Metals* **93**, 323-330.

Choppin G. R. and Schneider J. K. (1970) The acetate complexing by trivalent actinide ions. *J. Inorg. Nucl. Chem.* **32**, 3283-3288.

Cleveland J. M. and Rees T. F. (1981) Characterization of plutonium in Maxey Flats radioactive trench leachates. *Science* **212**, 1506-1509.

Cobble J. W., Murray R. C. Jr., Turner P. J. and Chen K. (1982) High-temperature thermodynamic data for species in aqueous solution. Electr. Power Res. Inst., Palo Alto, Calif., Rep. No. NP-2400, 200 p.

Daniele P. G., De Robertis A., Rigano C., and Sammartino S. (1985) Ionic strength dependence of formation constants. VI. La^{3+} - acetate, -malonate and -citrate complexes in aqueous solution. *Ann. Chim.* **75**, 115-120.

Deberdt S., Castet S., Dandurand J.-L., Harrichoury J.-C., and Louiset I. (1998) Experimental study of $La(OH)_3$ and $Gd(OH)_3$ solubilities (25 to 150 °C), and La-acetate complexing (25 to 80 °C). *Chem. Geol.* **151**, 349-372.

Deberdt S., Castet S., Dandurand J.-L., and Harrichoury J.-C. (in press) Potentiometric study of Gd and Yb acetate complexing in the temperature range 25-80 °C. *Chem. Geol.* **167**, 75-88.

Ding R. (1999) Potentiometric studies of REE complexation in aqueous acetate solutions from 25 ° to 70 °C. Ph.D. thesis, University of Idaho, 177 p.

Ding R., Gammons C.H., and Wood S.A. (1998) Solubility and potentiometric studies of REE complexation with simple carboxylate (acetate, oxalate) ligands from 25 ° to 80 °C. In Arehart, G.B., Hulston J.R. (Eds.) Water-Rock Interaction, Proceedings of the 9th International Symposium on Water-Rock Interaction – WRI-9, Taupo, New Zealand, Mar. 30-Apr. 3, 1998, 831-834.

Dorobek S. L. and Filby R. H. (1988) Origin of dolomites in a downslope biostrome, Jefferson formation (Devonian), central Idaho: evidence from REE patterns, stable isotopes, and petrography. *Bull. Can. Petrol. Geol.* **36**, 202-215.

Drummond S. E. and Palmer D. A. (1986) Thermal decarboxylation of acetate. Part II. Boundary conditions for the role of acetate in the primary migration of natural gas and the transportation of metals in hydrothermal systems. *Geochim. Cosmochim. Acta* **50**, 825-833.

Fabuss B. M., Korosi A. and Shamsul Huq A. K. M. (1966) Densities of binary and terary aqueous solutions of NaCl, $NaSO_4$, and $MgSO_4$, of sea water, and sea water concentrates. *Jour. Chem. Eng. Data* **11**, 325-331.

Fein J. B. (1991) Experimental study of Al-, Ca- and Mg-acetate complexing at 80 °C. *Geochim. Cosmochim. Acta* **55**, 955-964.

Gammons C. H. and Wood S. A. (2000) The aqueous geochemistry of the rare earth elements. Part 8: Solubility of ytterbium oxalate and the stability of Yb(III)-oxalate complexes in water at 25 ° to 80 °C. *Chem. Geol.* **166**, 103-124.

Gammons C. H., Wood S. A., and Williams-Jones A. E. (1996) The aqueous geochemistry of the rare earth elements and yttrium: VI. Stability of neodymium chloride complexes at 25 °C to 300 °C. *Geochim. Cosmochim. Acta* **60,** 4615-4630.

Giordano T. H. (1989) Anglesite ($PbSO_4$) solubility in acetate solutions: The determination of stability constants for lead acetate complexes to 85 °C. *Geochim. Cosmochim. Acta* **53**, 359-366.

Giordano T. H. and Drummond S. E. (1991) The potentiometric determination of stability constants for zinc acetate complexes in aqueous solutions to 295 °C. *Geochim. Cosmochim. Acta* **55**, 2401-2415.

Graf J. L. (1984) Effects of Mississippi Valley-type mineralization on REE patterns of carbonate rocks and minerals, Viburnum Trend, southeast Missouri. *J. Geol.* **92**, 307-324.

Grenthe I. (1962) On the stability of the acetate, glycolate and thioglycolate complexes of tervalent europium and americium. *Acta Chem. Scand.* **16**, 1695-1712.

Grenthe I. (1964) Thermodynamic properties of rare earth complexes. III. Free energy, enthalpy, and entropy changes for the formation of rare earth acetate, glycolate and thioglycolate complexes at 25 °C. *Acta Chem. Scand.* **18**, 283-292.

Gruzdev V. P. and Ermolaev V. L. (1974) Determination of the stability constants of complexes in solution by the method of radiationless energy transfer. *Russ. J. Inorg. Chem.* **19**, 1446-1449.

Helgeson H. C. (1969) Thermodynamics of hydrothermal systems at elevated temperatures and pressures. *Am. Jour. Sci.* **63**, 622-635.

Hennet R. J. C., Crerar D. A. and Schwartz J. (1988) Organic complexes in hydrothermal systems. *Econ. Geol.* **83**, 742-764.

Ilnitskaya E. L. and Ternovaya T. V. (1988) Structure and stability of rare earth element acetates in solution. *Ukr. Khim. Zh.* **54**, 678-685.

Kawamura K., Tannenbaum E., Huzinga B. J., and Kaplan I. R. (1986) Volatile organic acids generated from kerogen during laboratory heating. *Geochem. J.* **20**, 51-59.

Kharaka Y. K., Carothers W. W., and Rosenbauer R. J. (1983) Thermal decarboxylation of acetic acid: Implications for origin of natural gas. *Geochim. Cosmochim. Acta* **47**, 397-402.

Kolat R. S. and Powell J. E. (1962) Acetate complexes of the rare earth and several transition metal ions. *Inorg. Chem.* **1**, 293-296.

Kontak D. J. and Jackson S. (1995) Laser-ablation ICP-MS micro-analysis of calcite cement from a Mississippi-Valley-type Zn-Pb deposit, Nova Scotia: dramatic variability in REE content on macro- and micro-scales. *Can. Min.* **33**, 445-468.

Kovar L. E. and Powell J. E. (1966) Stability constants of rare earths with some weak carboxylic acids. Report TID-4500, Ames Laboratory, Iowa State University, 54 p.

Lindsay W. T. Jr. (1980) Estimation of concentration quotients for ionic equilibria in high temperature water: The model substance approach. Proc. 41st Ann. Meeting Intl. Water Conf., Pittsburgh, PA, Oct. 20, 1980, p. 284-294.

Manning P. G. and Monk C. B. (1962) Solvent extraction studies of ion association. Some europium complexes. *Trans. Faraday Soc.* **58**, 938-941.

Martell A. E., Motekaitis R. J. (1992) Determination and use of stability constants, 2nd.ed., VCH Publishers, New York.

Martens C. S. (1990) Generation of short chain organic acid anions in hydrothermally altered sediments of the Guayamas Basin, Gulf of California. *Appl. Geochem.* **5**, 71-76.

McDaniel D. K., Hemming S. R., McLennan S. M., and Hanson G. N. (1994) Resetting of neodymium isotopes and redistribution of REE's during sedimentary processes: The early Proterozoic Chelmsford Formation, Sudbury Basin, Ontario, Canada. *Geochim. Cosmochim. Acta* **58**, 931-941.

McLennan S. M. (1989) Rare earth elements in sedimentary rocks: Influence of provenance and sedimentary processes. In: Lipin, B.R., McKay, G.A. (Eds.), Geochemistry and Mineralogy of the Rare Earth Elements. *Rev. Mineral.* **21**, 169-200.

Means J. L., Crerar D. A., and Duguid J. O. (1978) Migration of radioactive wastes: Radionuclide mobilization by complexing agents. *Science* **200**, 1477-1481.

Mesmer R. E., Patterson C. S., Busey R. H. and Holmes H. F. (1989) Ionization of acetic acid in NaCl(aq) media: A potentiometric study to 573 K and 130 bar. *J. Phys. Chem.* **93**, 7483-7490.

Migal' P. K. and Chebotar N. G. (1967) Acetato-complexes of lanthanum in water-ethanol solutions. *Russ. Jour. Inorg. Chem.* **12**, 630-632.

Migal' P. K. and Chebotar N. G. (1970) Cerium, praseodymium, and neodymium acetate in aqueous ethanol solutions. *Russ. Jour. Inorg. Chem.* **15**, 625.

Migal' P. K., Chebotar N. G., and Sorochinskaya A. M. (1971) Acetate complexes of lanthanum, cerium, praseodymium, and neodymium in aqueous dioxane solutions. *Russ. Jour. Inorg. Chem.* **16**, 53-55.

Millero F. J. (1992) Stability constants for the formation of rare earth inorganic complexes as a function of ionic strength. *Geochim. Cosmochim. Acta.* **56**, 3123-3132.

Moeller T., Martin D. F., Thompson L. C., Ferrus R., Feistel R., and Randall W. J. (1965) The coordination chemistry of yttrium and the rare earth metal ions. *Chem. Rev.* **65**, 1-50.

Ohr M., Halliday A. N., and Peacor D. R. (1994) Mobility and fractionation of rare earth elements in argillaceous sediments: Implications for dating diagenesis and low-grade metamorphism. *Geochim. Cosmochim. Acta* **58**, 289-312.

Palmer D. R. and Bell J. L. S. (1994) Aluminum speciation and equilibria in aqueous solution: IV. A potentiometric study of aluminum acetate complexation in acidic NaCl brines to 150 °C. *Geochim. Cosmochim. Acta* **58**, 651-659.

Palmer D. A. and Drummond S. E. (1986) Thermal decarboxylation of acetate. Part I. The kinetics and mechanism of reaction in aqueous solution. *Geochim. Cosmochim. Acta* **50**, 813-823.

Palmer D. A. and Drummond S. E. (1988) Potentiometric determination of the molal formation constants of ferrous acetate complexes in aqueous solutions to high tempera-

tures. *J. Phys. Chem.* **92**, 6795-6800.
Palmer D. A. and Hyde K. E. (1993) An experimental determination of ferrous chloride and acetate complexation in aqueous solutions to 300 °C. *Geochim. Cosmochim. Acta* **57**, 1393-1408.
Palmer D. A. and Wesolowski D. J. (1993) Aluminum speciation and equilibria in aqueous solution: III. Potentiometric determination of the first hydrolysis constant of aluminum (III) in sodium chloride solutions to 125 °C. *Geochim. Cosmochim. Acta.* **57**, 2929-2938.
Rard J. A. (1988) Aqueous solubilities of praseodymium, europium and lutetium sulfates. *Jour. Sol. Chem.* **17**, 499-517.
Scherer M. and Seitz H. (1980) Rare-earth element distribution in Holocene and Pleistocene corals and their redistribution during diagenesis. *Chem. Geol.* **28**, 279-289.
Shock E. L. and Koretsky C. M. (1993) Metal-organic complexes in geochemical processes: Calculation of standard partial molal thermodynamic properties of aqueous acetate complexes at high pressures and temperatures. *Geochim. Cosmochim. Acta* **57**, 4899-4922.
Seewald J. S. and Seyfried W. E. Jr. (1991) Experimental determination of portlandite solubility in H_2O and acetate solutions at 100-350 °C and 500 bars: Constraints on calcium hydroxide and calcium acetate complex stability. *Geochim. Cosmochim. Acta* **55**, 659-669.
Semmler J., Irish D. E., and Ozeki T. (1990) Vibrational spectral studies of solutions at elevated temperatures and pressures. 12. Magnesium acetate. *Geochim. Cosmochim. Acta* **54**, 947-954.
Sonesson A. (1958a) On the complex chemistry of the tervalent rare earth ions. I. The acetate systems of lanthanum, cerium, neodymium, and gadolinium. *Acta Chem. Scand.* **12**, 165-181.
Sonesson A. (1958b) On the complex chemistry of the tervalent rare earth ions. II. The acetate systems of praseodymium, samarium, dysprosium, holmium, erbium, and ytterbium. *Acta Chem. Scand.* **12**, 1937-1954.
Sonesson A. (1959) On the complex chemistry of the tervalent rare earth ions. IV. Ion-exchange studies of the gadolinium acetate and glycolate systems. *Acta Chem. Scand.* **13**, 1437-1452.
Sonesson A. (1960) On the complex chemistry of the tervalent rare earth ions. V. The acetate and glycolate systems of yttrium. *Acta Chem. Scand.* **14**, 1495-1498.
Sposito G. (1989) The Chemistry of Soils. Oxford University Press, New York, 277 p.
Surdam R. C., Boese S. W., and Crossey L. J. (1984) The chemistry of secondary porosity. In MacDonald, D.A. Surdam, R.C., eds., Clastic Diagenesis. *Amer. Assoc. Petrol. Geol. Mem.* **37**, 127-149.
Ternoyava T. V. (1985) Spectrographic method for the investigation of complex formation equilibria for the whole series of rare earth elements and several other elements. *Teor. Eksper. Khim.* **20**, 461-467.
van Middlesworth P. E. and Wood S. A. (1998) The aqueous geochemistry of the rare earth elements and yttrium: Part 7: REE, Th and U contents in thermal springs associated with the Idaho Batholith. *Appl. Geochem.* **13**, 861-884.
Willey L. M., Kharaka Y. K., Presser T. S., Rapp J. B., and Barnes I. (1975) Short chain aliphatic acid anions in oil field waters and their contribution to the measured alkalinity. *Geochim. Cosmochim. Acta* **39**, 1707-1711.
Wood S. A. (1990a) The aqueous geochemistry of the rare earth elements and yttrium. Part I. Review of available low temperature data for inorganic complexes and the inorganic REE speciation of natural waters. *Chem. Geol.* **82**, 159-186.
Wood S. A. (1990b) The aqueous geochemistry of the rare earth elements and yttrium. Part II. Theoretical predictions of speciation in hydrothermal solutions to 350 °C at saturated water vapor pressure. *Chem. Geol.* **88**, 99-125.
Wood S. A. (1993) The aqueous geochemistry of the rare earth elements: critical stability constants for complexes with simple carboxylic acids at 25 °C and 1 bar and their application to nuclear waste management. *J. Geol. Engineering* **34**, 229-259.
Wood S. A. (1996) The role of humic substances in the transport and fixation of metals of economic interest (Au, Pt, Pd, U, V). *Ore Geol. Rev.* **11**, 1-31.
Wood S. A. (2000) Organic matter: Supergene enrichment and dispersion. In T.H. Giordano, R.M. Kettler and S.A. Wood, eds., Ore Genesis and Exploration: The Roles of Organic Matter. *Reviews in Economic Geology* **9**, 157-192.
Wood S. A. and Williams-Jones A. E. (1994) The aqueous geochemistry of the rare earth elements and yttrium. Part IV. Monazite solubility and REE mobility in exhalative massive sulfide-forming environments. *Chem. Geol.* **115**, 47-60.
Wood S. A., Tait C. D., Janecky D. R., and Constantopoulos T. L. (1995) The aqueous geochemistry of rare earth elements. V. Application of photoacoustic spectroscopy to speciation at low REE concentrations. *Geochim. Cosmochim. Acta* **59**, 5219-5222.
Wood S. A., Wesolowski D. J., and Palmer D. A. (2000) The aqueous geochemistry of the rare earth elements: IX. A potentiometric study of Nd^{3+} complexation with acetate in 0.1 molal NaCl solution from 25-250 °C. *Chem. Geol.* **167**, 231-253.
Wruck D. A., Zhao P., Palmer C. E. A., and Silva R. J. (1997) Stability quotients of neodymium acetate complexes from 20 to 70 °C by laser-induced photoacoustic spectroscopy. *J. Sol. Chem.* **26**, 267-275.
Yang M., Crerar D. A., and Irish D. E. (1989) A Raman spectroscopic study of lead and zinc acetate complexes in hydrothermal solutions. *Geochim. Cosmochim. Acta* **53**, 319-326.

Water-Rock Interactions, Ore Deposits, and Environmental Geochemistry: A Tribute to David A. Crerar
© The Geochemical Society, Special Publication No. 7, 2002
Editors: Roland Hellmann and Scott A. Wood

The aqueous geochemistry of the rare earth elements and yttrium. Part XI. The solubility of $Nd(OH)_3$ and hydrolysis of Nd^{3+} from 30 to 290 °C at saturated water vapor pressure with in-situ pH_m measurement

SCOTT A. WOOD[1*], DONALD A. PALMER[2], DAVID J. WESOLOWSKI[2], PASCALE BÉNÉZETH[2]

[1]Department of Geological Sciences, Box 443022, University of Idaho, Moscow, ID 83844-3022
[2]Chemical and Analytical Sciences Division, Oak Ridge National Laboratory, P.O. Box 2008, Oak Ridge, TN 37831-6110

* corresponding author (swood@iron.mines.uidaho.edu)

Abstract—The solubility of well-characterized, crystalline $Nd(OH)_3(cr)$ was determined in non-complexing, 0.03-1.0 mol kg^{-1} sodium trifluoromethanesulfonate (sodium triflate, NaTr) solutions from 30° to 290 °C at saturated water vapor pressure (SWVP) over a range of pH_m (pH_m is the negative logarithm of the molal hydrogen-ion concentration). The pH_m was measured *in situ* during the solubility experiments using a hydrogen-electrode concentration cell (50°-290 °C) or a combination glass-electrode (30 °C). From 30° to 200 °C the predominant dissolved Nd species over a wide range of pH_m was found to be the aquo ion, Nd^{3+}. Over the pH_m range where Nd^{3+} predominates, the solubility of $Nd(OH)_3(cr)$ decreased strongly with increasing temperature. Stoichiometric molal equilibrium quotients (Q_{s0}) for the reaction:

$$Nd(OH)_3(cr) + 3H^+ \leftrightarrow Nd^{3+} + 3H_2O \qquad (1)$$

were computed for 0.03, 0.1 and 1.0 mol kg^{-1} sodium triflate. The following equation adequately describes the measured equilibrium quotients as a function of temperature and ionic strength:

$$\log Q_{s0} = -6.662 + 7300.0/T + I(-91.51/T - 7.182\cdot10^{-6}T^2) - 6f^{\gamma}/\ln(10) - 3\log a_w$$

where T is the temperature in Kelvin, I is the ionic strength in mol kg^{-1}, a_w is the activity of water and f^{γ} is a Debye-Hückel term given by

$$f^{\gamma} = -A_{\phi}[x/(1 + 1.2x) + (2/1.2)\ln(1 + 1.2x)] \text{ and } x = I^{1/2}$$

From these equations we calculate $\log K_{s0} = 17.9 \pm 0.3$ at 25 °C, which compares well with a value recently derived from a critical evaluation of the pre-existing literature. We also obtain a value of $\Delta H_f^o(Nd(OH)_3(cr))$ = -1414 ± 5 kJ mol^{-1} at 25 °C and 1 bar which is in good agreement with calorimetrically-derived values reported in the literature.

Our data also suggest that, at low temperature (30° to 100 °C), hydrolysis proceeds from Nd^{3+} to $Nd(OH)_3^0$ over a very narrow pH_m range, and intermediate species are relatively unimportant. However, at 250° and 290 °C, the $Nd(OH)^{2+}$ and $Nd(OH)_2^+$ species are significant, with the relative importance of these two species being sensitive to the ionic strength. We were able to extract stoichiometric molal equilibrium quotients at 250 and 290 °C for the reactions:

$$Nd^{3+} + H_2O(l) \leftrightarrow Nd(OH)^{2+} + H^+$$

$$Nd^{3+} + 2H_2O(l) \leftrightarrow Nd(OH)_2^+ + 2H^+$$

Extrapolation of these equilibrium quotients to zero ionic strength using the Debye-Hückel equation indicates that recent theoretical predictions of the first hydrolysis constant at 250° and 290 °C overestimate the degree of hydrolysis by several orders of magnitude. The solubility minimum, i.e., the region of the solubility curve where $Nd(OH)_3^0$ predominates in solution, decreases with temperature from 30° to at least 100 °C, but increases again at 290 °C. Equilibrium constants for the reaction:

$$Nd^{3+} + 3H_2O(l) \leftrightarrow Nd(OH)_3^0 + 3H^+$$

were also obtained and found to be lower than recent theoretical predictions by increasing amounts as temperature increases (more than 3 orders of magnitude lower at 290 °C).

1. INTRODUCTION

Over the last half century there have been a number of measurements of the solubility of neodymium hydroxide (Table 1) and the hydrolysis of Nd^{3+} (Table 2). Much of this work has been stimulated by the chemical similarity of Nd^{3+} to trivalent actinides (Moeller et al., 1965; Choppin, 1983, 1986, 1989; Brush, 1990; Nitsche, 1990; Millero, 1992), and the need to understand the aqueous geochemistry of the latter in order to predict their fate in the environment.

Also, this work has been driven by interest in the hydrothermal growth of single crystals of REE hydroxides owing to their magnetic, thermal and optical properties (Mroczkowski et al., 1970 and references therein). Finally, a better understanding of the thermodynamics of REE phases and aqueous species would also benefit a number of areas of geochemistry summarized by Ding and Wood (this volume) and references therein. In this paper, we are primarily concerned with the solubility product of $Nd(OH)_3(cr)$, i.e., the equilibrium constant of the reaction:

$$Nd(OH)_3(cr) + 3H^+ \leftrightarrow Nd^{3+} + 3H_2O(l) \quad (1)$$

The stoichiometric molal equilibrium quotient (Q_{s0}) of the above reaction is defined as:

$$Q_{s0} = \frac{[Nd^{3+}]}{[H^+]^3} \quad (2)$$

The solubility product at infinite dilution may be written:

$$K_{s0} = \frac{a_{Nd^{3+}} a^3_{H_2O}}{a^3_{H^+}} = Q_{s0} \cdot \frac{\gamma_{Nd^{3+}} a^3_{H_2O}}{\gamma^3_{H^+}} \quad (3)$$

Table 1 shows that reported values of K_{s0} and Q_{s0} at room temperature for well-crystalline/well-aged $Nd(OH)_3$ vary over about four orders of magnitude. The solubility products for freshly precipitated/slightly aged $Nd(OH)_3$ vary over a similar range. However, as expected, the solubility of poorly crystalline $Nd(OH)_3$ appears to be higher than that for well-crystalline $Nd(OH)_3$. Possible reasons for the discrepancies observed in Table 1 include (see also Diakonov et al., 1998): 1) differences in the crystallinity of the initial $Nd(OH)_3$; 2) partial conversion of $Nd(OH)_3$ to much less soluble hydroxycarbonates or carbonates owing to contamination of the experimental system with CO_2; 3) lack of attainment of equilibrium; 4) inaccurate pH measurement, to which K_{s0} is very sensitive owing to the fact that a_{H^+} is raised to the 3rd power (eq. 3); and 5) sampling and/or analytical uncertainties. There are very few reported values of K_{s0} for $Nd(OH)_3$ at elevated temperatures, and none above 90 °C. However, recently Deberdt et al. (1998) have determined the analogous quantities for $La(OH)_3(cr)$ and $Gd(OH)_3(cr)$ at temperatures up to 150 °C.

Hydrolysis of Nd^{3+} is described by the generalized reaction:

$$mNd^{3+} + nH_2O(l) \leftrightarrow Nd_m(OH)_n^{3m-n} + nH^+ \quad (4)$$

and the hydrolysis constant is given by:

$$K_{hn,m} = \frac{a_{Nd_m(OH)_n^{3m-n}} a^n_{H^+}}{a^m_{Nd^{3+}} a^n_{H_2O}} \quad (5)$$

The expression for $Q_{hn,m}$ is analogous, but is defined in terms of concentrations instead of activities, ignoring the activity of water (cf. eq. 2). Note that,

Table 1. Summary of literature data on the solubility of $Nd(OH)_3(s)$ at room temperature.

Author	T(°C)	Ionic strength and medium	log Q_{s0}
	Crystalline or well aged		
Akselrud (1963)	25	infinite dilution	18.11[1]
Spivakovskii & Moisa (1972, 1977)	25	infinite dilution	16.35[1]
Silva (1982)	25	0.1 mol L^{-1} $NaClO_4$ (extr. to 0 with Davies eq.)	16.0±0.2
Baes & Mesmer (1986)	25	infinite dilution	18.6[3]
Morss et al. (1989)	25	infinite dilution	19.7±0.7[2]
Makino et al. (1993)	22	0.01 mol L^{-1} (extr. to 0)	16.00[4]
Rao et al. (1996)	25	0.1 mol L^{-1} NaCl (extr. to 0 with Pitzer eq.)	14.96
"	90	"	14.18
Merli et al. (1997)	25	infinite dilution	17.5±0.5[2]
Diakonov et al. (1998)	25	infinite dilution	17.61[2]
This study	25	0.03 mol kg^{-1} NaTr (empirically extr. to 0)	17.9±0.3
	Fresh precipitate or slightly aged		
Moeller & Kremers (1944)	25	0.1 mol L^{-1} $Nd(NO_3)_3$	21.3
Moeller & Fogel (1951)	25	0.1 mol L^{-1} $Nd(ClO_4)_3$	20.5
Tobias & Garrett (1958)	25	0.1 mol L^{-1} HCl (extr. to 0 with Davies eq.)	18.94
Meloche & Vrátný (1959)	25	3.72×10^{-2} mol kg^{-1} $Nd(ClO_4)_3$	22.1
Orhanovic et al. (1966)	20	≤0.1 mol kg^{-1} $Nd(NO_3)_3$	18.1
Azhipa et al. (1967)	20	dilute HCl, LiOH	18.57[1]
Kragten & Decnop-Weever (1984)	21.5	1.0 mol L^{-1} $NaClO_4$	19.4±0.2

NOTE: Q_{s0} refers to the reaction: $Nd(OH)_3(s) + 3H^+ \leftrightarrow Nd^{3+} + 3H_2O(l)$ and is given in the same units as the ionic strength for each study. [1]Recalculated by Diakonov et al. (1998); [2]Calculated from calorimetrically-derived thermodynamic data; [3]Calculated based on a correlation with lattice parameters (see text); [4]Cited by Diakonov et al. (1998).

in equation (4), the value of m is limited to unity when the total Nd concentration is maintained at low levels by $Nd(OH)_3$(cr) solubility (see section 4.5). It is difficult to compare the $Q_{hn,m}$ values given in Table 2 because of differences in ionic strength, etc. However, inspection of this table shows that, although the spread in measured $Q_{hn,m}$ values is generally less than that for Q_{s0}, there is some disagreement among the various measurements, and refinements are possible (see also Wood, 1990a). The clearest example is the data at 3 mol L^{-1} $NaClO_4$, for which three separate measurements of $Q_{h1,1}$ vary over an order of magnitude.

The main goal of the study reported in this paper was to determine the solubility product for $Nd(OH)_3$(cr) as a function of temperature up to 290 °C, using *in situ* pH_m measurements, in the hope of resolving the discrepancies in previously reported solubility products. In addition to supplying data previously unavailable above 90 °C, the expected faster reaction kinetics at elevated temperatures should minimize uncertainties owing to a slow approach to equilibrium, which is a decided advantage in attempting to resolve the discrepancies noted above. Secondary goals were to obtain some information on the hydrolysis of Nd^{3+} at elevated temperatures, and to lay the groundwork for using measurements of the solubility of $Nd(OH)_3$(cr) as a means of obtaining stability constants of complexes of Nd^{3+} with other ligands such as acetate, chloride or sulfate (cf. Wesolowski et al., 1998).

2. METHODS

2.1. Synthesis and Characterization of Crystalline $Nd(OH)_3$

Previous studies have shown that $Nd(OH)_3$ is the stable phase in the Nd-O-H system from room temperature up to roughly 800 °C (nearly independent of pressure from 200 to 2000 bars), above which it transforms to NdOOH (Viswanathiah et al., 1976; 1980). Earlier work by Shafer and Roy (1959), which suggested that $Nd(OH)_3$ was stable only to less than 450 °C, was apparently affected by contamination by CO_2, causing the formation of Nd-carbonate phases (Viswanathiah et al., 1976). The partial pressure of CO_2 at which $Nd(OH)_3$ converts to the hydroxycarbonate or carbonate phase may be as low as 10^{-6} bars (Diakonov et al., 1998), requiring that extreme caution be exercised in excluding CO_2 during synthesis and solubility measurements of the hydroxide.

Following Christensen (1966), Merli et al. (1997), and Deberdt et al. (1998), crystalline $Nd(OH)_3$ was synthesized by reacting 99.99% pure Nd_2O_3, supplied by Alfa Aesar (Lot Number P3129A), with deionized (18 MΩ-cm) water at 200-250 °C and sat-

Table 2. Summary of literature data on the hydrolysis of Nd^{3+} at room temperature.

Author	Method	T(°C)	Ionic strength and medium	Hydrolysis constants
Kragten & Decnop-Weever (1984)	Solubility	21.5	1 mol L^{-1} $NaClO_4$	log $Q_{h1,1}$ = -8.1±0.2
				log $Q_{h2,1}$ = -16.2±0.2
				log $Q_{h3,1}$ = -24.3±0.2
				log $Q_{h2,2}$ = -11.6±0.2
Silva (1982)	Solubility	25	0.1 mol L^{-1} $NaClO_4$	log $K_{h2,1}$ = -15.8±0.5
			(extr. to 0 with Davies eq.)	log $K_{h3,1}$ = -23.9±0.2
				log $K_{h4,1}$ < -34
Burkov et al. (1988)	Potentiometry	25	3 mol L^{-1} $NaClO_4$	log $Q_{h1,1}$ = -8.94±0.04
				log $Q_{h2,2}$ = -14.03±0.01
Burkov et al. (1973)	Potentiometry	25	3 mol L^{-1} $NaClO_4$	log $Q_{h1,1}$ = -9.4±0.4
				log $Q_{h2,2}$ = -13.93±0.03
Guillaumont et al. (1971)	Solvent extraction	25	0.1 mol L^{-1} $NaClO_4$	log $Q_{h1,1}$ = -7.0
Kostromina & Badaev (1986)	Spectrophotometry	25?	variable?	log $\beta_{1,1}$ = 7.4±0.4
				log $\beta_{2,2}$ = 13.0±0.7
Moeller (1946)	Potentiometry	25	0.005 mol L^{-1} $Nd_2(SO_4)_3$	log $Q_{h1,1}$ = -9.0
			0.01 mol L^{-1} $Nd_2(SO_4)_3$	log $Q_{h1,1}$ = -8.7
			0.05 mol L^{-1} $Nd_2(SO_4)_3$	log $Q_{h1,1}$ = -7.5
Tobias & Garrett (1958)	Potentiometry	25	3 mol L^{-1} $NaClO_4$	log $Q_{h1,1}$ = -8.5±0.4
Baes & Mesmer (1986)	Critical compilation	25	Infinite dilution	log $K_{h1,1}$ = -8.0
				log $K_{h2,1}$ = -16.9
				log $K_{h3,1}$ = -26.5
				log $K_{h4,1}$ = -37.1
This study	Solubility	30	0.03 mol kg^{-1} NaTr	log $Q_{h3,1}$ = -24.7±0.3

Note: $Q_{hn,m}$ ($K_{hn,m}$) refers to the reaction: $mNd^{3+} + nH_2O(l) \leftrightarrow Nd_m(OH)_n^{3m-n} + nH^+$ and is given in the same units as the ionic strength for each study; β_{nm} refers to the reaction: $mNd^{3+} + nOH^- \leftrightarrow Nd_m(OH)_n^{3m-n}$.

Table 3. Powder X-ray data for the $Nd(OH)_3$(cr) synthesized in this study.

d(Å)	I/I_0	d(Å)	I/I_0
12.45	4	2.22	70
5.48	58	1.86	14
3.19	60	1.85	26
3.08	100	1.83	54
2.76	19	1.61	19
2.42	8	1.42	15

urated water vapor pressure in a Teflon-lined stainless steel reaction vessel for two weeks. The water was first deaerated by vigorously bubbling high-purity Ar gas through it for at least 15 minutes. Other authors (e.g., Rao et al., 1996) have synthesized the hydroxide by reacting solutions of Nd-chloride salts with aqueous hydroxide solutions, followed by aging. However, it is difficult to exclude CO_2 from aqueous hydroxide solutions. In addition, the initial products obtained by precipitating Nd from solutions of chloride salts are known to contain chloride, e.g., $Nd(OH)_2Cl$ (Spivakovskii and Moisa, 1972), and may be amorphous, requiring aging. The method of synthesis employed here eliminates interference by chloride, minimizes interference by CO_2 and results in crystalline $Nd(OH)_3$.

After synthesis, the $Nd(OH)_3$ was washed several times with deionized, deaerated water and separated from the water by centrifugation. The solid was then dried in a vacuum oven at 50 °C and stored under Ar in a polyethylene container with a screw top. The initial Nd_2O_3 was a light blue color; this color changed to lilac upon conversion to $Nd(OH)_3$. No further changes in color were noted upon storage or at the conclusion of any solubility run. However, the grains of $Nd(OH)_3$(cr) were observed to coarsen after experiments at ≥250 °C.

The solid was characterized before the solubility runs using powder X-ray diffraction (XRD), Fourier-transform infrared spectroscopy (FTIR) and thermal gravimetric analysis (TGA). After the runs the solid was characterized by XRD and FTIR. XRD analyses were conducted both at ORNL and the University of Idaho, with similar results. FTIR was conducted at ORNL using KBr pellets and a Bio-Rad FTS-60 instrument.

The results of a typical XRD analysis are given in Table 3. The d-spacings given in Table 3 are in excellent agreement with those given by Roy and McKinstry (1953) and Silva (1982) for $Nd(OH)_3$(cr). The peaks of all XRD patterns were sharp and the background was uniformly low, consistent with well-crystallized $Nd(OH)_3$.

The application of TGA showed that weight loss upon heating occurred in three steps. A weight loss of roughly 9% occurred over the temperature range 292°-442 °C, followed by a weight loss of roughly 4.2% over the range 442°-575 °C. A much smaller, more

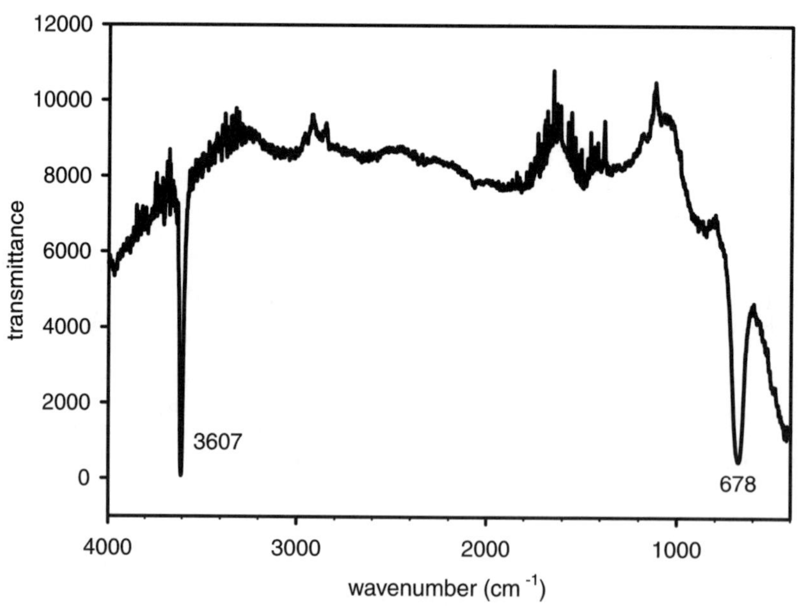

Figure 1. FTIR spectrum of $Nd(OH)_3$(cr) synthesized in this study.

gradual weight loss of ~0.6% occurred over the range 575°-950 °C. The TGA results can be understood in terms of the following reaction:

$$2Nd(OH)_3(cr) \rightarrow Nd_2O_3(cr) + 3H_2O \qquad (6)$$

The first major step in the TGA corresponds to the loss of 2 water molecules (9.22% theoretical), and the second and third steps combined correspond to the loss of 1 water molecule (4.61% theoretical). The total measured weight loss in two separate analyses amounted to 13.80% and 14.15%, compared to the theoretical weight loss of 13.83%.

The FTIR spectrum of a sample of freshly prepared $Nd(OH)_3(cr)$ in a KBr pellet is shown in Figure 1. For this sample, all the steps in the preparation of the KBr pellet were carried out either under an Ar atmosphere in a glove box, or under vacuum. The major features of this spectrum are a sharp, intense band at 3607 cm^{-1} and a broader but intense band at 678 cm^{-1}. These features are characteristic of $Nd(OH)_3$ (Klevtsov et al., 1967), with the former and latter bands attributable to the O-H stretch and O-H bend, respectively, of hydroxide. The spectrum in Figure 1 is also notable for the absence of any significant bands between 1400 and 1500 cm^{-1}, or between 1060 and 1090 cm^{-1}, which are diagnostic of the presence of carbonate (Caro et al., 1972). However, $Nd(OH)_3$ exposed to the atmosphere for more than a few hours invariably exhibited weak bands attributable to Nd-carbonate, indicative of the difficulty in preventing the formation of these carbonates during synthesis, characterization and solubility measurements if extreme precautions are not taken to exclude CO_2.

2.2. Preparation of Solutions

All solutions were prepared by weight, employing reagent-grade chemicals and 18-MΩ-cm deionized water that had been deaerated by vigorously sparging with ultrapure argon for 15 minutes or more. The solutions were stored in polyethylene containers under positive Ar pressure. Concentrated stock solutions of NaTr, NaOH and HTr were prepared and used to make up the desired experimental solutions. The NaTr stock solution was prepared by neutralization of triflic acid with a CO_2-free, concentrated NaOH solution as described by Palmer and Hyde (1993). The compositions of the initial test, reference and titrant solutions employed in this study are given in Table 4.

2.3. Experimental Procedure

2.3.1. The hydrogen-electrode concentration cell (HECC)

Experiments from 50° to 290 °C were conducted using an HECC of the type described in several previous publications (Palmer and Wesolowski, 1993; Bénézeth et al., 1997; Wesolowski et al., 1998; Bénézeth et al., 1999; Palmer et al., 2001). The cell comprised a 1-L Hastelloy reaction vessel fitted with two concentric Teflon cups, separated by a porous Teflon liquid junction. Teflon-coated platinum wires, the ends of which were free of Teflon but coated with platinum black, were inserted into each cup and served as the electrodes. The solutions in each cup were stirred magnetically via Teflon-coated stir bars. The inner cup contained the reference solution with a

Table 4. Initial compositions (mol kg^{-1} H_2O) of test, reference and titrant solutions for solubility experiments.

Series	Test	Reference	Acid titrant	Basic titrant
Nd-1	$m_{H^+} = 3.085x10^{-4}$	$m_{H^+} = 1.285x10^{-3}$	$m_{H^+} = 1.012x10^{-2}$	$m_{OH^-} = 9.490x10^{-3}$
	$m_{Na^+} = 2.964x10^{-2}$	$m_{Na^+} = 2.858x10^{-2}$	$m_{Na^+} = 2.480x10^{-2}$	$m_{Na^+} = 3.321x10^{-2}$
	$m_{Tr^-} = 2.975x10^{-2}$	$m_{Tr^-} = 2.869x10^{-2}$	$m_{Tr^-} = 3.492x10^{-2}$	$m_{Tr^-} = 2.373x10^{-2}$
		$m_{Cl^-} = 9.871x10^{-4}$		
Nd-2	$m_{H^+} = 3.084x10^{-4}$	$m_{H^+} = 1.142x10^{-3}$	$m_{H^+} = 3.030x10^{-2}$	$m_{OH^-} = 3.003x10^{-2}$
	$m_{Na^+} = 2.943x10^{-2}$	$m_{Na^+} = 2.829x10^{-2}$	$m_{Na^+} = 5.103x10^{-3}$	$m_{Na^+} = 3.620x10^{-2}$
	$m_{Tr^-} = 2.974x10^{-2}$	$m_{Tr^-} = 2.943x10^{-2}$	$m_{Tr^-} = 3.540x10^{-2}$	$m_{Tr^-} = 6.168x10^{-3}$
Nd-3	$m_{H^+} = 1.142x10^{-3}$	$m_{H^+} = 1.142x10^{-3}$	$m_{H^+} = 3.030x10^{-2}$	$m_{OH^-} = 3.003x10^{-2}$
	$m_{Na^+} = 2.829x10^{-2}$	$m_{Na^+} = 2.829x10^{-2}$	$m_{Na^+} = 5.103x10^{-3}$	$m_{Na^+} = 3.620x10^{-2}$
	$m_{Tr^-} = 2.943x10^{-2}$	$m_{Tr^-} = 2.943x10^{-2}$	$m_{Tr^-} = 3.540x10^{-2}$	$m_{Tr^-} = 6.168x10^{-3}$
Nd-4	$m_{H^+} = 1.040x10^{-3}$	$m_{H^+} = 1.040x10^{-3}$	None	None
	$m_{Na^+} = 9.925x10^{-2}$	$m_{Na^+} = 9.925x10^{-2}$		
	$m_{Tr^-} = 1.003x10^{-1}$	$m_{Tr^-} = 1.003x10^{-1}$		
Nd-5	$m_{H^+} = 1.022x10^{-3}$	$m_{H^+} = 1.040x10^{-3}$	None	$m_{OH^-} = 3.085x10^{-2}$
	$m_{Na^+} = 9.757x10^{-2}$	$m_{Na^+} = 9.925x10^{-2}$		$m_{Na^+} = 1.642x10^{-1}$
	$m_{Tr^-} = 9.859x10^{-2}$	$m_{Tr^-} = 1.003x10^{-1}$		$m_{Tr^-} = 1.333x10^{-1}$
Nd-6	$m_{H^+} = 1.044x10^{-3}$	$m_{H^+} = 1.044x10^{-3}$	None	None
	$m_{Na^+} = 1.0063$	$m_{Na^+} = 1.0063$		
	$m_{Tr^-} = 1.0074$	$m_{Tr^-} = 1.0074$		

precisely known stoichiometric hydrogen-ion molality. The outer cup contained a suspension of $Nd(OH)_3$(cr) in the test solution (initially ~10 g of solid in ~400 mL of solution). The entire HECC was purged 5-10 times with hydrogen at room temperature before being placed in the aluminum block furnace for equilibration at temperature. Upon completion of the hydrogen purge, the hydrogen pressure was set at ~10 bars at 25 °C. The initial cell configuration in a typical experiment was:

$$H_2, Pt \,|\, HTr\,(m_1), NaTr\,(m_2) \,||\, HTr\,(m_3), NaTr\,(m_4), Nd(OH)_3(cr) \,|\, Pt, H_2$$

with $m_2 = m_4$ at the start of the experiment, and the ratios m_1:m_2 and m_3:m_4 maintained at less than 0.1 to minimize both liquid junction contributions to the measured potential and activity coefficient differences between the two solutions.

Samples of the test solution were withdrawn via a platinum dip tube that had a porous platinum frit gold-welded to its immersed end to prevent particles of the solid from entering the Pt tube during sampling. The portion of the Pt dip tube outside the cell was fitted with a cooling jacket, and samples were withdrawn through a PEEK® valve and a 0.2-μm fluoropolymer (PVDF) syringe filter into pre-weighed polypropylene syringes containing a known mass of high-purity 1 mol L^{-1} HNO_3 (JT Baker Ultrex Reagent). The samples were then analyzed for Nd using inductively coupled plasma-atomic emission spectroscopy (ICP-AES; Thermo Jarell Ash IRIS at ORNL) or inductively coupled plasma-mass spectrometry (ICP-MS; Perkin Elmer Elan 250 at Washington State University) depending on the Nd concentration. In initial experiments, Nd concentrations were also determined by ion chromatography. Precision of repeat analyses was normally in the range 1-10%, depending on how close the Nd concentration was to the detection limit of the analytical technique employed. There was excellent agreement among analytical techniques where analytical working ranges overlapped.

Upon attainment of equilibrium (usually within a day or two; see below for the criteria applied to define attainment of equilibrium), a small amount of either NaOH + NaTr or HTr + NaTr titrant solution at the same ionic strength as the reference and test solutions was injected into the test cell using one of two calibrated positive displacement pumps (Zircadyne 705) to change the pH_m. After equilibrium was reestablished, another sample was taken for total Nd determination. This stepwise process was repeated until a portion of the Nd-hydroxide solubility curve as a function of pH_m was mapped out. Use of two inert, positive-displacement pumps permitted us to reverse the solution titration process and approach equilibrium from both under- and over-saturation.

The working definition of pH_m throughout this paper is $pH_m = -\log [H^+]$, where $[H^+]$ is the hydrogen-ion concentration in mol kg^{-1} units. The convention employed is that triflic acid and NaOH are completely dissociated, and NaTr ion-pairing is treated implicitly by the activity coefficient model used.

Each platinum-hydrogen electrode responds to the half-cell reaction:

$$H_2(g) \leftrightarrow 2H^+ + 2e^- \tag{7}$$

and the difference in potential between the two electrodes is given by the Nernst equation:

$$\Delta E = -\frac{RT}{F}\ln([H^+]_t/[H^+]_r) - E_{LJ} \tag{8}$$

where $[H^+]_t$ and $[H^+]_r$ represent the stoichiometric molalities of hydrogen ion in the test and reference solutions, respectively. The ideal gas and Faraday constants are R and F, respectively, and T is the temperature in Kelvin. Finally, E_{LJ} represents the liquid junction potential calculated using the Henderson equation (Baes and Mesmer, 1986). The limiting equivalent conductances of ions required by the Henderson equation were taken from the following sources: Na^+, H^+ and OH^- (Quist and Marshall, 1965) and Tr^- (Ho and Palmer, 1995). The limiting equivalent conductances of Nd^{3+} were approximated with those of La^{3+}, which are available only from 0 to 100 °C (Robinson and Stokes, 1959), and therefore were extrapolated to higher temperatures. The maximum liquid junction potential calculated in this study was 20 mV, which contributes an uncertainty of <±0.1 in the measured pH_m, assuming that the Henderson equation predicts the value of E_{LJ} to within 25% (Mesmer, 1991). However, the only conditions for which such high values of E_{LJ} were calculated were those where pH_m was sufficiently low at any given temperature that Nd^{3+} accounted for a significant fraction of the ions present in the test compartment. This only occurred at the lowest ionic strength investigated, i.e., 0.03 mol kg^{-1} NaTr, and then only for the most acidic conditions. For most of the data points at 0.03 mol kg^{-1} NaTr, the maximum calculated E_{LJ} was <6 mV, contributing an uncertainty in the measured pH_m of <±0.025. At 0.1 and 1.0 mol kg^{-1} NaTr, the maximum calculated values of E_{LJ} were 2 mV and 0.1 mV, respectively. These contribute uncertainties of <±0.01 and <±0.0005 in the measured pH_m.

2.3.2. Glass-electrode cell

Solubility experiments at 30 °C were conducted in a 500-mL polystyrene bottle. The temperature was maintained constant to within ±0.1 °C using a water bath. The bottle was fitted with a combination glass pH electrode and tubing to maintain a positive-pressure Ar atmosphere. The electrode was calibrated at 30 °C against three to five solutions containing known, stoichiometric amounts of HTr or NaOH and sufficient NaTr to make the total ionic strength identical to that of the solubility experiment (i.e., 0.03 molal). Samples were withdrawn via a syringe and immediately filtered through a 0.2-µm fluoropolymer membrane into pre-weighed polypropylene syringes containing a known mass of high-purity 1 mol L^{-1} HNO_3 (JT Baker Ultrex Reagent). The samples were analyzed as described above. After equilibrium was attained at a given pH_m value, a known quantity of acid or base was titrated into the bottle to shift the pH_m and the entire process was repeated.

2.4. Data Reduction

2.4.1. Determination of stoichiometric molal equilibrium quotients

If polynuclear species are neglected, then the solubility of $Nd(OH)_3$(cr) as a function of pH_m can be expressed in terms of the generalized dissolution reaction:

$$Nd(OH)_3(cr) + (3-n)H^+ \leftrightarrow Nd(OH)_n^{3-n} + (3-n)H_2O \quad (9)$$

for which the conditional equilibrium constant can be written as:

$$Q_{sn} = \frac{[Nd(OH)_n^{3-n}]}{[H^+]^{3-n}} \quad (10)$$

Taking the log of both sides of equation (10) and rearranging yields:

$$\log[Nd(OH)_n^{3-n}] = \log Q_{sn} - (3-n)pH_m \quad (11)$$

Now, differentiating the concentration of the Nd species with respect to pH_m we obtain:

$$\frac{\partial \log[\Sigma Nd]}{\partial pH_m} = n-3 \quad (12)$$

Thus, from equations (11) and (12), we see that, the slope of a plot of log [ΣNd] vs. pH_m at any point yields the value of n-3. If any one species is predominant over a wide range of pH_m, then the solubility curve over that pH_m range should approximate a straight line with slope n-3, where n is the number of hydroxide ions bound to the Nd^{3+} ion in the predominant species.

The total measured Nd concentration in solution can be expressed as:

$$[\Sigma Nd] = [Nd^{3+}] + [NdOH^{2+}] + [Nd(OH)_2^+] + [Nd(OH)_3^0] + \ldots + [Nd(OH)_N^{3-N}] \quad (13)$$

where N is the maximum hydroxide ligand number. Expressing the concentrations of each of the Nd species in terms of its corresponding Q_{sn} value we obtain:

$$[\Sigma Nd] = Q_{s0}\cdot[H^+]^3 + Q_{s1}\cdot[H^+]^2 + Q_{s2}\cdot[H^+]^1 + Q_{s3} + \ldots Q_{sN}\cdot[H^+]^{3-N} \quad (14)$$

This equation can be fit to a set of experimental data for the solubility of $Nd(OH)_3$(cr) as a function of pH_m, at a constant ionic strength and temperature, using non-linear regression, to obtain the values of Q_{sn} as fit coefficients. The best fit model is the one containing the fewest species while minimizing the function:

$$\frac{\sqrt{\sum_{i=1}^{p}\left(\log[\Sigma Nd]_{obs} - \log[\Sigma Nd]_{calc}\right)^2}}{p-q} \quad (15)$$

where p is the total number of solubility measurements and q (= N + 1) is the number of species in the model. In this study, non-linear regression was performed using the least-squares routine in the software package SigmaPlot® (Jandel Scientific), which employs the Marquardt-Levenberg algorithm, and also using a version of the generalized regression program ORGLS (Busing and Levy, 1962). Results obtained using the two regression algorithm were essentially identical.

As will be shown below, for most combinations of temperature and ionic strength, only a single species, the hydrated Nd^{3+} ion, predominated over a wide range of pH_m. In these cases, regression was unnecessary because only one parameter, Q_{s0}, was required to fit the data. In these cases, an estimate of Q_{s0} was calculated for each data point as:

$$Q_{s0} = \frac{[\Sigma Nd]}{[H^+]^3} \quad (16)$$

The accepted value of Q_{s0} was then the average of all the Q_{s0} values calculated according to equation (16) at a given temperature and ionic strength.

Once values of Q_{sn} were determined, then the stoichiometric molal equilibrium quotients ($Q_{hn,1}$) for the hydrolysis reaction (eq. 4) were calculated according to the relation:

$$\log Q_{hn,1} = \log Q_{sn} - \log Q_{s0} \quad (17)$$

2.4.2. Extrapolation to infinite dilution

Values of Q_{s0} were obtained over a sufficient range of ionic strength at each temperature to employ an empirical method of extrapolation to zero

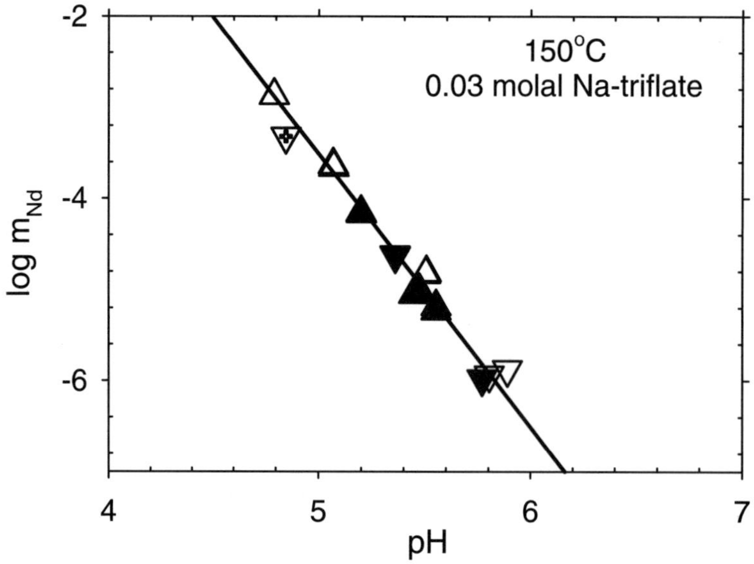

Figure 2. Plot of log $m_{\Sigma Nd}$ vs. pH_m at 150 °C and 0.03 mol kg^{-1} ionic strength demonstrating the reversibility of the solubility equilibrium. Upward-pointing and downward-pointing triangles correspond to approach from undersaturation and oversaturation, respectively. The three different styles of triangles (open, closed, open with cross) represent three separate experimental runs and illustrate the excellent reproducibility of the experimental data. The solid line represents the best fit line through the data with slope constrained at 3.0.

ionic strength. The following function of ionic strength and temperature was fit to the entire set of values of Q_{s0}:

$$\log Q_{s0} = a + b/T + I(c/T + dT^2) - 6f^{\gamma}/\ln(10) - 3\log a_w \quad (18)$$

The fourth term in the equation is the Debye-Hückel expression used in the Pitzer ion-interaction model (Pitzer et al., 1977) with

$$f^{\gamma} = -A_{\phi}[x/(1 + 1.2x) + (2/1.2)\ln(1 + 1.2x)] \quad \text{where } x = I^{1/2} \quad (19)$$

In the fourth term, the coefficient –6 is equal to the change in the charges squared for reaction (1). The parameter A_{ϕ} is the Debye-Hückel osmotic coefficient parameter for water, obtained from the temperature function given by Dickson et al. (1990). The activity of water (a_w) was calculated from the osmotic coefficients for water in NaCl solutions given by Liu and Lindsay (1972), assuming that the activity of water is the same for NaCl and NaTr solutions of equal ionic strength. In equation (18), a,b,c and d are fit parameters determined by regression using ORGLS.

Most other equilibrium quotients were obtained only at a single ionic strength (0.03 or 0.1 mol kg^{-1}). Therefore, these constants were extrapolated to infinite dilution using the Debye-Hückel equation:

$$\log \gamma_i = \frac{-A_{\gamma}\left|Z_i^2\right| I^{1/2}}{1 + B_{\gamma}\mathring{a}I^{1/2}} \quad (20)$$

The A_{γ} and B_{γ} values employed in conjunction with equation (20) were taken from Helgeson and Kirkham (1974).

3. RESULTS

The results of the solubility experiments are given in Appendices I (30 °C) and II (50°-290 °C). Gaps in the sequence of sample numbers correspond to samples obtained when the magnetic stirrer was not functioning properly and such data were not included in the data reduction.

3.1. *Attainment of Equilibrium*

The approach to equilibrium could be monitored by observing the rate of change of pH_m of the test solution (which is proportional to the continuously displayed cell potential). In general, samples were not withdrawn from the cell until the cell potential reached a plateau as a function of time. A lack of change of pH_m with time is a necessary, but not sufficient, criterion for the attainment of equilibrium. However, as mentioned above, the use of two positive

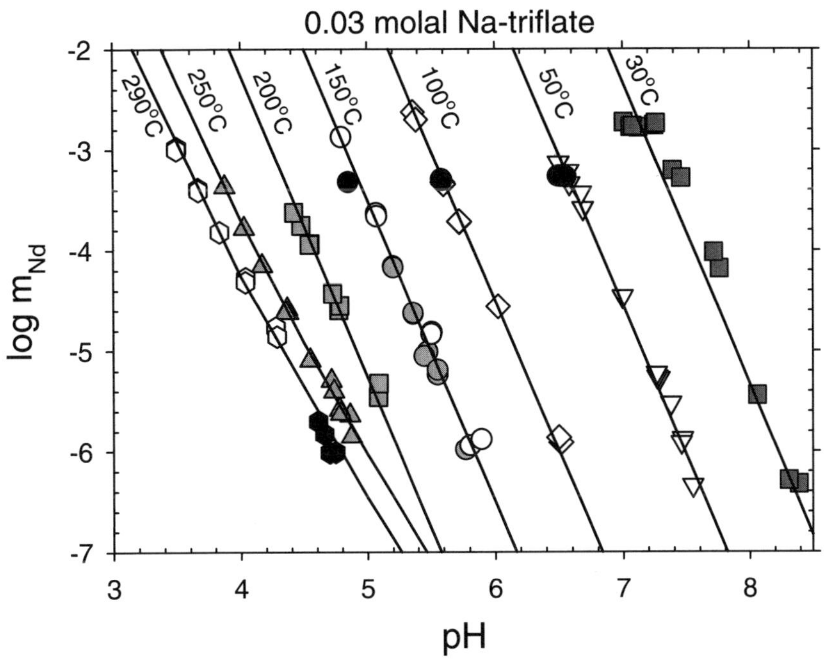

Figure 3. Plot of log $m_{\Sigma Nd}$ vs. pH_m at 0.03 mol kg^{-1} ionic and various temperatures. The four different symbol shadings (none, light gray, dark gray, black) correspond to four separate experimental runs and illustrate the excellent reproducibility of the experimental data. For temperatures 30°-200 °C, the solid lines correspond to a best fit assuming Nd^{3+} is the only Nd species present and have a constant slope of 3.0. At 250° and 290 °C, the solid curves correspond to a best fit assuming Nd^{3+} and $Nd(OH)^{2+}$ are the only species present.

displacement pumps, one to deliver acid and one to deliver base, permitted us to reverse the solubility equilibrium. The data in Figure 2 demonstrate this reversal procedure for 150 °C and 0.03 molal Na-triflate. In the figure, upward-pointing triangles denote approach to equilibrium from undersaturation, whereas downward-pointing triangles denote approach from oversaturation. The three different types of triangles represent three separate runs with three separate batches of $Nd(OH)_3(cr)$. Successive samples during each run were taken at approximately half-day to day intervals (see Appendix II). Figure 2 demonstrates the excellent reproducibility of the data from run to run, and the solid triangles in particular demonstrate that the solubility reaction is readily reversible. These data suggest that equilibrium was closely attained within a day at 150 °C. Similar experiments showed that it took up to 1-3 days to equilibrate at 50-100 °C, but only a few hours at temperatures near 250-290 °C.

It is clear that the efficient agitation provided by the Teflon-coated magnetic stir bar was essential to the relatively rapid approach to equilibrium. On a number of occasions during this study, the magnetic stirrer failed, either because the motor stopped (the brushes required replacement) or the coupling between the motor and the external magnet broke. Data obtained during brief periods in which the stirrer was not functioning were neither reproducible nor reversible, and were discarded. These findings are in agreement with those of Bénézeth et al. (1997, 2001) and Palmer et al. (2001), who noted that equilibration rates were also rapid for the dissolution of boehmite with the efficient stirring in the HECC.

3.2. *Temperature and Ionic Strength Dependence of Q_{s0}*

The data obtained in 0.03 mol kg^{-1} Na-triflate as a function of temperature are shown in Figure 3. This figure demonstrates a number of features. First, the data at each temperature are quite reproducible from run to run. Second, the data from 30° to 200 °C, inclusive, plot along straight lines of slope equal to -3.0, indicating that, over the pH_m ranges shown, the predominant species in solution is the hydrated cation, Nd^{3+}. The data at 250° and 290 °C can be explained by a two-species model, the details of which are discussed below. Third, the solubility of $Nd(OH)_3(cr)$ is markedly retrograde, i.e., decreases strongly with temperature. Finally, it can be seen that the data for 30 °C scatter more strongly about a line with slope -3.0 than the data for other temperatures. This behavior is probably explained by: 1) the greater drift and difficulty in calibrating glass electrodes compared to hydrogen electrodes; 2) slower rates of equilibration at the lower temperatures.

Values of log Q_{s0} for 0.03 mol kg^{-1} Na-triflate were derived from the data from 30° to 250 °C as described in section 2.4.1. From 30° to 200 °C, inclusive, only one species appears to be present over the entire pH_m range shown in Figure 3, and Q_{s0} is reported as the average of values calculated from equation (16). At 250 °C and 0.03 mol kg^{-1} NaTr, two species were apparently present over the range of pH_m shown in Figure 3, and Q_{s0} was obtained using non-linear regression as described in section 2.4.1. The values of Q_{s0} at 250 °C and 0.1 and 1.0 mol kg^{-1} ionic strengths were calculated assuming that Nd^{3+} was the predominant species at the relatively low pH_m (~4) of the solubility measurements made. This is a reasonable assumption because at 0.03 mol kg^{-1} ionic strength and 250 °C Nd^{3+} is predominant at similar values of pH_m, and higher ionic strength should favor the most highly charged species. At 290 °C, Nd^{3+} was not predominant over a sufficiently large range of investigated pH_m at either 0.03 or 0.1 mol kg^{-1} ionic strength, and so no Q_{s0} values were derived from the data at 290 °C. No solubility measurements were obtained at 290 °C and 1.0 mol kg^{-1} ionic strength.

Table 5. Stoichiometric molal equilibrium quotients (log Q_{s0}) determined in this study as a function of temperature and ionic strength.

T(°C)	I (mol kg^{-1})	log Q_{s0}	T(°C)	I (mol kg^{-1})	log Q_{s0}
30	0.03	18.6±0.6	50	0.10	16.68±0.14
50	0.03	16.49±0.16	100	0.10	13.79±0.10
50	0.03	16.32±0.16	150	0.10	11.59±0.10
100	0.03	13.54±0.24	200	0.10	9.88±0.10
100	0.03	13.46±0.10	250	0.10	8.53±0.10
150	0.03	11.42±0.12	50	1.0	16.82±0.10
150	0.03	11.22±0.10	100	1.0	13.93±0.10
150	0.03	11.65±0.13	150	1.0	11.72±0.10
200	0.03	9.74±0.18	200	1.0	9.98±0.10
250	0.03	8.04±0.16	250	1.0	8.60±0.10

The raw values of Q_{s0} obtained in this study are compiled in Table 5. The values of Q_{s0} at 0.03 mol kg^{-1} ionic strength are plotted as a function of inverse temperature in Figure 4. The plot of log Q_{s0} vs. 1/T is highly linear from 30° to 250 °C. Thus, it appears that the change in the heat capacity of reaction (1) is near zero. Equation (18) was fit to the entire data set given in Table 5 to obtain the following expression:

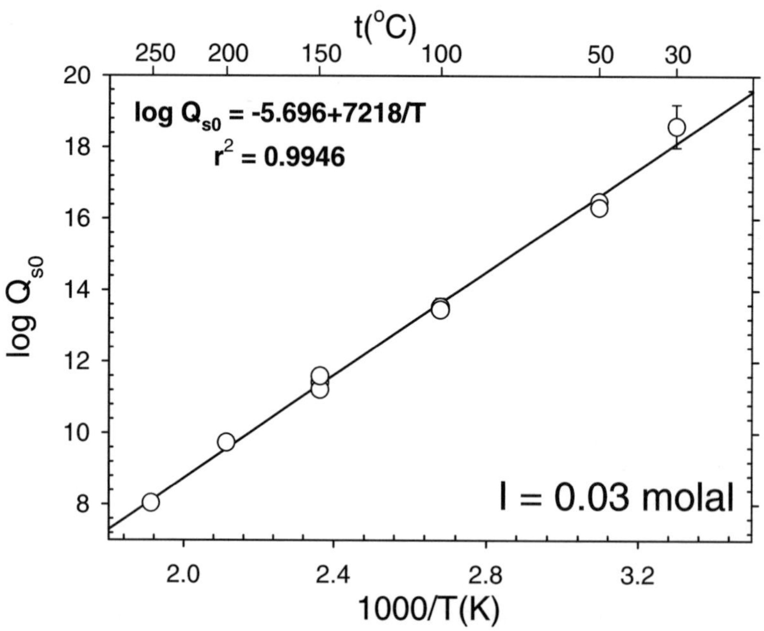

Figure 4. Plot of log Q_{s0} vs. 1/T(K) for reaction (1) at 0.03 mol kg^{-1} ionic strength.

Table 6. Smoothed stoichiometric molal equilibrium quotients (log Q_{s0}) and thermodynamic properties for reaction (1) as a function of temperature and ionic strength.

T(°C)	log Q_{s0}	$\Delta G_r°$ (kJ mol^{-1})	$\Delta H_r°$ (kJ mol^{-1})	$\Delta S_r°$ (kJ mol^{-1} K^{-1})	$\Delta C_{p,r}°$ (kJ mol^{-1} K^{-1})
			I = 0 mol kg^{1}		
0	20.17±0.35	-105.5±1.8	-140.3±5.2	-0.13±0.01	0.00±0.00
25	17.92±0.27	-102.3±1.5	-140.3±5.2	-0.13±0.01	0.00±0.00
50	16.02±0.20	-99.1±1.3	-140.3±5.2	-0.13±0.01	0.00±0.00
75	14.39±0.16	-95.9±1.0	-140.3±5.2	-0.13±0.01	0.00±0.00
100	12.98±0.12	-92.7±0.9	-140.3±5.2	-0.13±0.01	0.00±0.00
125	11.75±0.10	-89.6±0.8	-140.3±5.2	-0.13±0.01	0.00±0.00
150	10.66±0.10	-86.4±0.8	-140.3±5.2	-0.13±0.01	0.00±0.00
175	9.70±0.12	-83.2±1.0	-140.3±5.2	-0.13±0.01	0.00±0.00
200	8.83±0.14	-80.0±1.2	-140.3±5.2	-0.13±0.01	0.00±0.00
225	8.05±0.16	-76.8±1.5	-140.3±5.2	-0.13±0.01	0.00±0.00
250	7.35±0.18	-73.6±1.8	-140.3±5.2	-0.13±0.01	0.00±0.00
275	6.71±0.20	-70.4±2.1	-140.3±5.2	-0.13±0.01	0.00±0.00
300	6.13±0.22	-67.2±2.4	-140.3±5.2	-0.13±0.01	0.00±0.00
			I = 0.03 mol kg^{1}		
0	20.60±0.34	-107.7±1.8	-140.5±5.1	-0.12±0.01	0.00±0.00
25	18.36±0.26	-104.8±1.5	-140.5±5.1	-0.12±0.01	0.00±0.00
50	16.48±0.20	-102.0±1.2	-140.6±5.1	-0.12±0.01	0.00±0.00
75	14.88±0.15	-99.2±1.0	-140.6±5.1	-0.12±0.01	0.00±0.00
100	13.50±0.12	-96.4±0.8	-140.7±5.1	-0.12±0.01	0.00±0.00
125	12.30±0.10	-93.7±0.8	-140.8±5.1	-0.12±0.01	0.00±0.00
150	11.25±0.10	-91.1±0.8	-140.9±5.1	-0.12±0.01	0.00±0.00
175	10.33±0.11	-88.6±1.0	-141.0±5.0	-0.12±0.01	-0.01±0.00
200	9.52±0.13	-86.2±1.2	-141.2±5.0	-0.12±0.01	-0.01±0.00
225	8.81±0.15	-84.0±1.4	-141.3±5.0	-0.12±0.01	-0.01±0.00
250	8.19±0.17	-82.0±1.7	-141.5±5.0	-0.11±0.01	-0.01±0.00
275	7.65±0.19	-80.3±2.0	-141.6±4.9	-0.11±0.01	-0.01±0.00
300	7.20±0.21	-79.0±2.3	-141.8±4.9	-0.11±0.01	-0.01±0.00
			I = 0.1 mol kg^{1}		
0	20.84±0.32	-109.0±1.7	-140.7±4.9	-0.12±0.01	-0.01±0.00
25	18.61±0.25	-106.2±1.4	-140.9±4.9	-0.12±0.01	-0.01±0.00
50	16.74±0.19	-103.6±1.2	-141.1±4.9	-0.12±0.01	-0.01±0.00
75	15.15±0.14	-101.0±0.9	-141.3±4.8	-0.12±0.01	-0.01±0.00
100	13.78±0.11	-98.4±0.8	-141.6±4.8	-0.12±0.01	-0.01±0.00
125	12.60±0.09	-96.0±0.7	-141.9±4.7	-0.12±0.01	-0.01±0.00
150	11.57±0.09	-93.7±0.8	-142.2±4.7	-0.12±0.01	-0.02±0.00
175	10.68±0.11	-91.6±0.9	-142.6±4.6	-0.11±0.01	-0.02±0.00
200	9.90±0.12	-89.6±1.1	-143.1±4.6	-0.11±0.01	-0.02±0.01
225	9.22±0.14	-87.9±1.3	-143.6±4.5	-0.11±0.01	-0.02±0.01
250	8.64±0.16	-86.6±1.6	-144.1±4.4	-0.11±0.01	-0.02±0.01
275	8.16±0.17	-85.7±1.8	-144.7±4.3	-0.11±0.01	-0.03±0.01
300	7.79±0.19	-85.5±2.1	-145.4±4.2	-0.10±0.01	-0.03±0.01
			I = 0.3 mol kg^{1}		
0	21.08±0.28	-110.2±1.5	-141.5±4.4	-0.11±0.01	-0.02±0.01
25	18.85±0.21	-107.6±1.2	-142.0±4.4	-0.12±0.01	-0.02±0.01
50	16.98±0.16	-105.0±1.0	-142.6±4.3	-0.12±0.01	-0.03±0.01
75	15.39±0.12	-102.6±0.8	-143.3±4.2	-0.12±0.01	-0.03±0.01
100	14.03±0.09	-100.2±0.7	-144.1±4.1	-0.12±0.01	-0.04±0.01
125	12.86±0.08	-98.0±0.6	-145.0±4.0	-0.12±0.01	-0.04±0.01
150	11.84±0.08	-95.9±0.7	-146.1±3.9	-0.12±0.01	-0.05±0.01
175	10.96±0.09	-94.1±0.8	-147.3±3.8	-0.12±0.01	-0.050.01
200	10.21±0.10	-92.4±0.9	-148.6±3.8	-0.12±0.01	-0.06±0.02
225	9.56±0.12	-91.2±1.1	-150.1±3.8	-0.12±0.01	-0.06±0.02

(continues)

(Table 6. continued)

250	9.02±0.13	-90.4±1.3	-151.7±3.8	-0.12±0.01	-0.07±0.02
275	8.60±0.15	-90.2±1.5	-153.5±3.8	-0.12±0.01	-0.08±0.02
300	8.31±0.16	-91.2±1.7	-155.5±4.0	-0.11±0.01	-0.08±0.02
			I = 1.0 mol kg^{1}		
0	21.08±0.32	-110.2±1.7	-144.1±3.9	-0.12±0.01	-0.06±0.02
25	18.83±0.27	-107.5±1.5	-145.8±3.9	-0.13±0.01	-0.07±0.02
50	16.92±0.22	-104.7±1.4	-147.8±4.1	-0.13±0.01	-0.09±0.02
75	15.30±0.18	-102.0±1.2	-150.2±4.4	-0.14±0.01	-0.10±0.03
100	13.90±0.15	-99.3±1.1	-152.9±4.8	-0.14±0.01	-0.12±0.03
125	12.69±0.13	-96.8±1.0	-156.0±5.3	-0.15±0.01	-0.13±0.04
150	11.64±0.12	-94.3±1.0	-159.5±6.0	-0.15±0.01	-0.15±0.04
175	10.73±0.13	-92.0±1.1	-163.5±6.8	-0.16±0.02	-0.17±0.04
200	9.94±0.15	-90.0±1.4	-167.9±7.8	-0.17±0.02	-0.19±0.05
225	9.27±0.18	-88.4±1.7	-172.8±8.9	-0.17±0.02	-0.21±0.06
250	8.72±0.22	-87.3±2.2	-178.3±10.3	-0.17±0.02	-0.23±0.06
275	8.30±0.26	-87.1±2.7	-184.4±11.7	-0.18±0.03	-0.26±0.07
300	8.06±0.31	-88.5±3.4	-191.1±13.4	-0.18±0.03	-0.28±0.07

$$\log Q_{s0} = -6.662 + 7300.0/T + I(-91.51/T - 7.182\cdot10^{-6}T^2) - 6f^{\gamma}/\ln(10) - 3\log a_w \qquad (21)$$

Figure 5 shows that equation (21) provides an excellent fit to all the data except that for 30 °C and 0.03 mol kg^{-1} (see section 4.1). Smoothed values of Q_{s0}, together with thermodynamic quantities for reaction (1), i.e., ΔG_r^o, ΔH_r^o, ΔS_r^o, and $\Delta C_{p,r}^o$, were calculated based on equation (21) and are given at infinite dilution and several discrete ionic strengths in Table 6.

3.3. Hydrolysis of Nd^{3+}

Our data for Nd^{3+} hydrolysis are more limited than our data for Q_{s0}. Nevertheless, we can draw a number of preliminary conclusions. Figure 6 shows

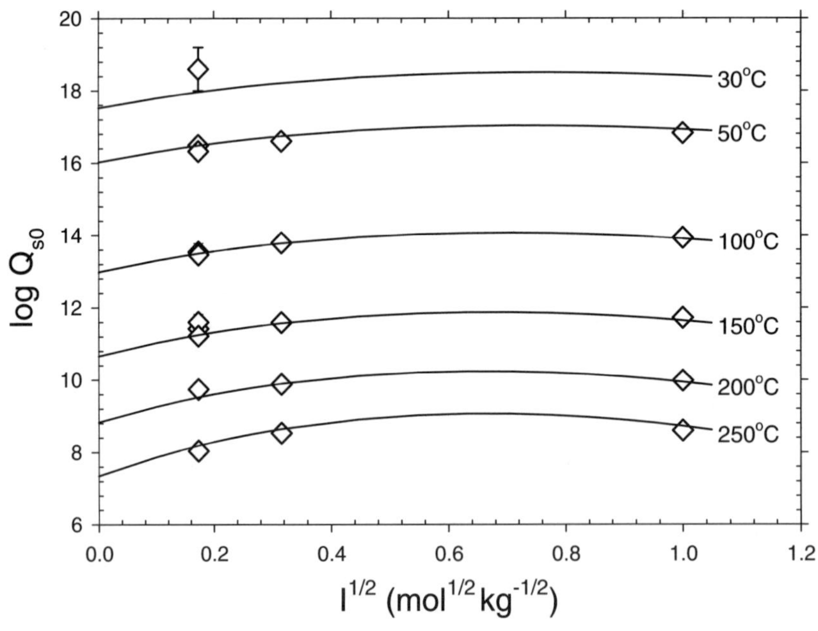

Figure 5. Plot of log Q_{s0} vs. $I^{1/2}$ for reaction (1) at various temperatures. Curves represent best fit of equation (21) in text to the data.

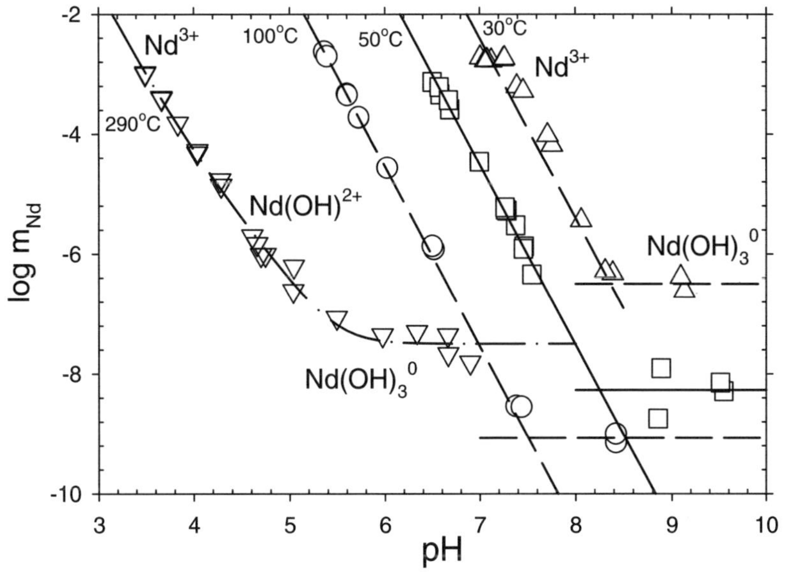

Figure 6. Plot of log $m_{\Sigma Nd}$ vs. pH_m at 0.03 mol kg^{-1} ionic and various temperatures showing the solubility minima. Note that each set of data at 30, 50 and 100 °C can be fit adequately by two straight lines of slopes –3 and 0, suggesting that Nd^{3+} and $Nd(OH)_3^0$ are the predominant species. Three species are required to fit the data at 290 °C.

that, at 30°, 50° and 100 °C, hydrolysis appears to proceed from Nd^{3+} to $Nd(OH)_3^0$ over a very narrow interval of pH_m. It therefore appears that intermediate hydrolysis products such as $Nd(OH)^{2+}$ and $Nd(OH)_2^+$ are not of great importance at the lower temperatures of these experiments. This is consistent with the conclusion of Baes and Mesmer (1986), based on a review of the literature, that the transition from aqueous REE^{3+} to $REE(OH)_3^0$ occurs over a very narrow pH range, and with the behavior of aluminum (Wesolowski and Palmer, 1994). The lack of a significant region of predominance at low temperatures for $Nd(OH)^{2+}$ and $Nd(OH)_2^+$ over the range of ionic strengths investigated in this study also is shown clearly in the work of Silva (1982) (his Fig. 9) at 0.1 mol L^{-1} $NaClO_4$ and Kragten and Decnop-Weever (1984) (their Fig. 1) at 1 mol L^{-1} $NaClO_4$. Values of Q_{s3}, i.e., the equilibrium constant for the reaction:

$$Nd(OH)_3(cr) \leftrightarrow Nd(OH)_3^0 \qquad (22)$$

at 30°, 50° and 100 °C, were determined by averaging the data at the solubility minimum at each temperature.

However, at higher temperatures, intermediate hydrolysis products appear to become more important. As shown in Figures 3 and 6, at 250° and 290 °C, there appears to be a substantial pH_m range where the slope of the solubility curve is neither –3 nor zero. Figure 7 shows that, at 290 °C and 0.03 mol kg^{-1} NaTr, a model involving the species: Nd^{3+}, $Nd(OH)^{2+}$ and $Nd(OH)_3^0$ provides a better fit to the data than a model involving the species: Nd^{3+}, $Nd(OH)^{2+}$ and $Nd(OH)_2^+$. In fact, $Nd(OH)^{2+}$ is the predominant species over a substantial pH_m range. Note that in both fits, the contribution of Nd^{3+} was estimated from the value of Q_{s0} extrapolated assuming that equation (21) is valid to at least 290 °C (an extrapolation of 40 °C); the contribution of Nd^{3+} to the measured solubility data is insufficient to permit the direct derivation of Q_{s0} from a regression of the data. The equilibrium constants for $Nd(OH)^{2+}$ and $Nd(OH)_3^0$, i.e., Q_{s1} and Q_{s3}, respectively, were obtained from non-linear regression of the data shown in Figure 7 after subtracting the minor contribution of Nd^{3+} to the total solubility. The curve fitted to the solubility data at 250 °C and 0.03 mol kg^{-1} ionic strength given in Figure 3 was obtained by regressing the data simultaneously with a model including both Nd^{3+} and $Nd(OH)^{2+}$, and it is obvious that such a model fits the data well. The experiments at 250 °C did not extend to sufficiently high pH_m to be able to detect further hydrolysis products, i.e., $Nd(OH)_3^0$.

The data at 290 °C and 0.1 molal ionic strength also demonstrate the existence of species in addition to Nd^{3+} and $Nd(OH)_3^0$ (Figure 8). In order to model the solubility data at these conditions, the contribution of Nd^{3+} was estimated from log Q_{s0} = 7.92±0.2 obtained from equation (21). The contribution of $Nd(OH)_3^0$ was estimated by noting that reaction (22) does not in-

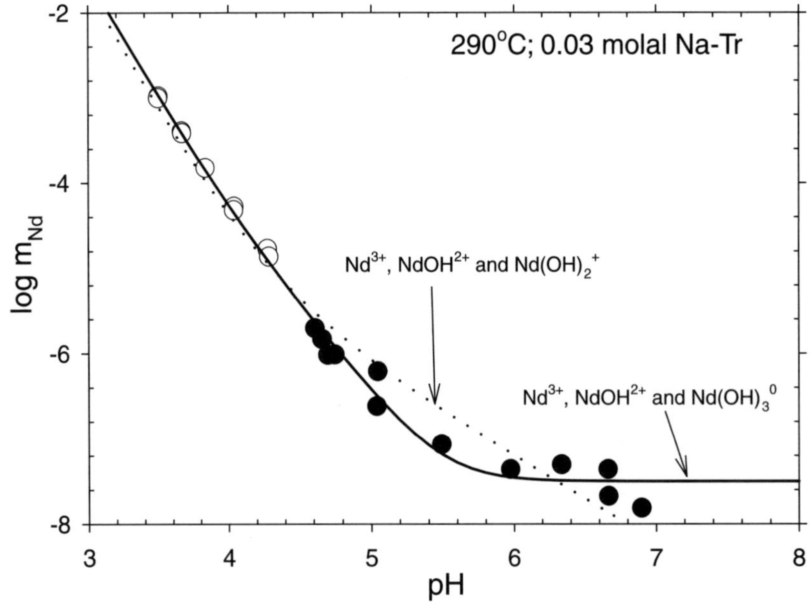

Figure 7. Plot of log $m_{\Sigma Nd}$ vs. pH_m at 0.03 mol kg^{-1} ionic strength and 290 °C showing the fit of two different speciation models. The solid line corresponds to a model including Nd^{3+}, $NdOH^{2+}$ and $Nd(OH)_3^0$; the dotted line corresponds to a model including Nd^{3+}, $NdOH^{2+}$ and $Nd(OH)_2^+$. The open and closed circles represent data obtained using ICP-AES and ICP-MS, respectively.

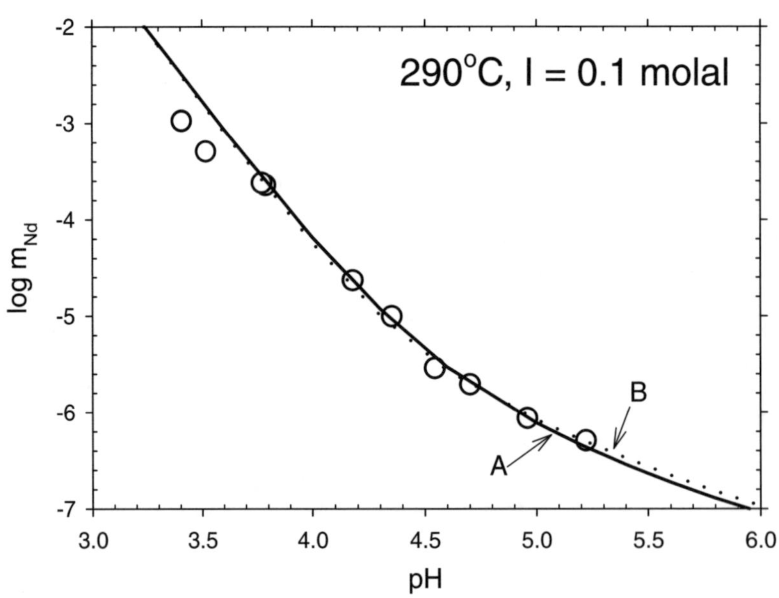

Figure 8. Plot of log $m_{\Sigma Nd}$ vs. pH_m at 0.1 mol kg^{-1} ionic strength and 290 °C showing two speciation models including Nd^{3+}, $NdOH^{2+}$, $Nd(OH)_2^+$ and $Nd(OH)_3^0$. The solid line is the fit provided by model A: log Q_{s0} = 7.7; log Q_{s1} = 3.0; log Q_{s2} = -1.23; log Q_{s3} = -7.50. The dotted line is the model B fit: log Q_{s0} = 7.7; log Q_{s1} = -4.997; log Q_{s2} = -1.12; log Q_{s3} = -7.50.

Table 7. Stoichiometric molal equilibrium quotients for the formation of hydrolyzed species.

T(°C)	I (mol kg^{-1})	n	log Q_{sn}	T(°C)	I (mol kg^{-1})	n	log Q_{sn}
30	0.03	3	-6.5±0.2	290	0.1	2	-1.3±0.2
50	0.03	3	-8.3±0.4	250	0.03	1	3.92±0.04
100	0.03	3	-9.1±0.1	290	0.03	1	3.5±0.1
290	0.03	3	-7.5±0.1	290	0.1	1	≤3.0±0.3

volve any charged species, and therefore Q_{s3} should not be highly dependent on ionic strength. Thus, the value of Q_{s3} determined at 0.03 mol kg^{-1} ionic strength, i.e., log Q_{s3} = -7.50±0.1 was assumed to be valid also at 0.1 mol kg^{-1}. It was determined that use of log Q_{s0} = 7.92±0.2 resulted in the overestimation of the solubility data at low pH, so this value was adjusted to 7.7, which provides a better fit to the data in Figure 8 and is within the uncertainty range given. Thus, log Q_{s0} and log Q_{s3} were fixed at 7.7 and –7.5, respectively, and Q_{s1} and Q_{s2} were obtained by regression. This procedure resulted in curve B in Figure 8, and the values log Q_{s1} = -5.0 and log Q_{s2} = -1.12±0.2. Moreover, the error in the estimate of Q_{s1} given by SigmaPlot was many orders of magnitude larger than Q_{s1}, which suggests that this equilibrium quotient cannot be resolved from the data shown in Figure 8. Next, Q_{s1} was fixed at various values and Q_{s2} was obtained by regression. The value of Q_{s2} obtained was not very sensitive to the value of Q_{s1} selected. Curve A in Figure 8 shows the fit of a model where log Q_{s1} was fixed at 3.0 and log Q_{s2} was refined to a value of –1.23. It is evident that there is very little difference between the fits provided by curves A and B. However, if log Q_{s1} was fixed at a value significantly above 3.0, the fit was degraded significantly. In light of these results, we can really only place an upper limit of 3.0 on the value of log Q_{s1}. The results of the calculations of Q_{s1}, Q_{s2} and Q_{s3} are given in Table 7.

Our results suggest that the relative importance of $Nd(OH)^{2+}$ and $Nd(OH)_2^+$ is quite sensitive to ionic strength at 290 °C. At 0.03 mol kg^{-1} NaTr, $Nd(OH)_2^+$ never appears to be predominant, but $Nd(OH)^{2+}$ exhibits a substantial pH range of predominance. The situation is reversed at 0.1 mol kg^{-1} NaTr. The greater importance of the lower charge species at higher ionic strength is perhaps somewhat unexpected, but the results of the modeling of the solubility data appear unequivocal. Additional solubility experiments over a wider range of ionic strengths and temperatures are required to more fully resolve the relative importance of $Nd(OH)^{2+}$ and $Nd(OH)_2^+$.

Figure 6 shows that the solubility minimum, where $Nd(OH)_3^0$ is the predominant species, decreases with increasing temperature from 30° to at least 100 °C, but apparently increases again at 290 °C. This trend is also reflected in the temperature dependence of log Q_{s3} values given in Table 7.

Stoichiometric molal equilibrium quotients for the various hydrolysis reactions, calculated according to equation (17), are given in Tables 8, 9 and 10. The smoothed values of Q_{s0} obtained from equation (21) and given in Table 6, and the values of Q_{sn} given in Table 7 were used to derive the values of $Q_{hn,1}$. The

Table 8. Comparison of predicted and measured first hydrolysis constants of Nd^{3+}.

T(°C)	log $Q_{h1,1}$	log $K_{h1,1}$ (this study)	log $K_{h1,1}$ (Haas et al., 1995)
	I = 0.03 mol kg^{-1}		
250	-4.3±0.2	-3.8±0.3	-2.02
290	-3.9±0.2	-3.3±0.3	-1.42

NOTE: the constants tabulated refer to the equation:
$Nd^{3+} + H_2O(l) \leftrightarrow Nd(OH)^{2+} + H^+$

Table 9. Comparison of predicted and measured second hydrolysis constants of Nd^{3+}.

T(°C)	log $Q_{h1,1}$	log $K_{h1,1}$ (this study)	log $K_{h1,1}$ (Haas et al., 1995)
	I = 0.1 mol kg^{-1}		
290	-9.2±0.3	-8.4±0.4	-5.41

NOTE: the constants tabulated refer to the equation:
$Nd^{3+} + 2H_2O(l) \leftrightarrow Nd(OH)_2^+ + 2H^+$

Table 10. Comparison of predicted and measured third hydrolysis constants of Nd^{3+}.

T(°C)	log $Q_{h3,1}$	log $K_{h3,1}$ (this study)	log $K_{h3,1}$ (Haas et al., 1995)
	I = 0.03 mol kg^{-1}		
30	-24.5±0.3	-24.0±0.4	-25.71
50	-24.8±0.5	-24.3±0.6	-23.33
100	-22.6±0.3	-22.1±0.4	-18.82
290	-14.9±0.3	-13.9±0.4	-10.51

NOTE: the constants tabulated refer to the equation:
$Nd^{3+} + 3H_2O(l) \leftrightarrow Nd(OH)_3^0 + 3H^+$

latter equilibrium quotients were extrapolated to infinite dilution using the Debye-Hückel equation (eq. 20). In addition, the reaction:

$$Nd^{3+} + 3H_2O \leftrightarrow Nd(OH)_3^0 + 3H^+ \quad (23)$$

has the same ionic species, number of water molecules and charge change as does the reverse of equation (1). If it is assumed that the activity coefficient of the neutral species $Nd(OH)_3^0$ does not change much from 0 to 0.03 mol kg^{-1} ionic strength, then the value of $K_{h3,1}$ also can be obtained from the expression:

$$\log K_{h3,1} - \log Q_{h3,1} = \log Q_{s0} - \log K_{s0}$$
$$= I(91.51/T - 7.182\cdot10^{-6}T^2) - 6f^{\gamma}/\ln(10) - 3\log a_w \quad (24)$$

The values of $\log K_{h3,1}$ obtained using equation (24) differed from those obtained using the Debye-Hückel equation (eq. 20) by 0.1 log units or less. An additional ±0.1 log unit has been added to the total uncertainties in the values of $\log K_{hn,1}$ reported in Tables 8-10 to reflect the additional uncertainty inherent in extrapolation to infinite dilution.

4. DISCUSSION

4.1. Discrepancy Between Model Fit and Experimental Q_{s0} Value at 30 °C

As noted in section 3.2, equation (21) provides an excellent fit to almost all the raw Q_{s0} values given in Table 5 (Fig. 5). However, equation (21) underestimates our experimentally determined value of Q_{s0} at 30 °C and 0.03 mol kg^{-1} ionic strength by 0.64 log unit. This discrepancy may be related to a number of factors, including the lower stability of glass electrodes compared to hydrogen electrodes and less efficient stirring in the case of the 30 °C experiments (the Teflon stir bar frequently became hindered in the bottle causing stirring to stop). However, the most likely explanation is that most of the data obtained at 30 °C relate to amorphous, or at least less well-crystalline $Nd(OH)_3$(s). Note that, the first solubility data points at 30 °C were obtained by approach from undersaturation with respect to $Nd(OH)_3$(cr), and these lie well below the line defined by our experimentally determined value of $\log Q_{s0}$ = 18.6). All other data points were obtained after NaOH had been added to raise pH_m, thus supersaturating the solution, and most of these data points lie above the line in Figure 3. Therefore, we suggest that, in the 30 °C experiments, our initial solubility measurements may relate to less soluble $Nd(OH)_3$(cr), but after supersaturation upon addition of base, a less crystalline, more soluble form of the solid precipitated and exerted control on the measured Nd concentration. Because of the lower temperature and the less efficient stirring in the 30 °C experiments, the solubility equilibrium was not reversible as it was in the higher temperature experiments in the HECC. In this regard it is interesting to note that the first four data points in Appendix I yield $\log Q_{s0}$ = 18.4±0.1, whereas the remainder, obtained after the first addition of base, yield $\log Q_{s0}$ = 19.0±0.2. The first value is in much better agreement with $\log Q_{s0}$ = 17.96±0.25 obtained from equation (21). The solubility results obtained at higher temperatures are considered more reliable because reaction kinetics were more favorable, owing to the higher temperature and more efficient stirring, and hydrogen electrodes are intrinsically more stable than glass electrodes. Moreover, as shown in Figure 2 and discussed above, we were able to reverse the solubility equilibrium at the higher temperatures. Finally, because the relationship between $\log Q_{s0}$ and 1/T(K) is highly linear (Fig. 4), and calculation of $\log Q_{s0}$ at 30 °C from equation (21) involves an extrapolation of only 20 °C, we believe that this value of $\log Q_{s0}$ is more reliable than that obtained directly from the 30 °C experiments.

4.2. Comparison to Previous Work: Solubility Product

The value we obtain for $\log K_{s0}$ for $Nd(OH)_3$ at 25 °C and infinite dilution, via use of equation (21), is 17.9±0.3, which is more than a log unit greater than the values derived from the work of Spivakovskii and Moisa (1972, 1977), Silva (1982), Makino et al. (1993) and Rao et al. (1996), and is more than a log unit less than the value derived from the work of Morss et al. (1989). However, our value agrees to within ±0.5 log units with the estimates derived from the work of Akselrud (1963), Merli et al. (1997), and Diakonov et al. (1998), and is only 0.7 log units less than the estimate obtained by Baes and Mesmer (1986) based on the following equation relating $\log K_{s0}$ and the lattice parameter *a*:

$$\log K_{s0} = -103.7 + 19.04a \quad (25)$$

The value recommended by Diakonov et al. (1998) is based on a critical evaluation of all the solubility and other thermodynamic data available at the time, and is in very good agreement with ours, being only 0.3 log units lower, which is within our experimental uncertainty. This agreement, and the extreme care we have taken in synthesizing $Nd(OH)_3$(cr), establishing the attainment of equilibrium and measuring pH_m, lead us to believe that our value at standard conditions is the most reliable experimentally measured value available. It seems likely that, reported values of $\log K_{s0}$ that are significantly lower than ours, e.g., those of Spivakovskii and Moisa (1972, 1977), Silva

(1982), Makino et al. (1993) and Rao et al. (1996), are a result of the solubility being controlled by the formation of a surface layer of less soluble $NdOHCO_3(s)$ or $Nd_2(CO_3)_3$. As pointed out by Diakonov et al. (1998), standard X-ray diffraction may not detect such a surface precipitate, and IR spectroscopy or another more sensitive technique is required to insure its absence. Ours is the only study to have used IR spectroscopy and TGA, in addition to X-ray diffraction, to characterize the solid starting material. Diakonov et al. (1998) suggest that the NaOH solution used by Rao et al. (1996) to synthesize $Nd(OH)_3(cr)$ may have been contaminated, leading to the inadvertent formation of Nd-hydroxycarbonate or –carbonate. On the other hand, Diakonov et al. (1998) thought it unlikely that the study of Silva (1982) was affected by the same problem because Silva (1982) employed an ultrapure NaOH solution. However, we suggest that synthesis of $Nd(OH)_3(cr)$ by reacting aqueous solutions of Nd^{3+} and NaOH, is particularly susceptible to the formation of less soluble carbonate phases because of the difficulty in completely excluding CO_2 from NaOH solutions.

The only previously reported value of log K_{s0} for crystalline $Nd(OH)_3$ that is significantly greater than ours, is the value given by Morss et al. (1989). This value was calculated from calorimetric data for $Nd(OH)_3(cr)$ and literature thermodynamic data for $H_2O(l)$ and Nd^{3+}. It is well known that comparatively small relative errors in calorimetrically-derived Gibbs free energies of formation can result in much larger relative errors in the resulting Gibbs free energy of reaction, and hence the log K value, for a given chemical reaction. Such errors may be the cause of the discrepancy between our measured log K_{s0} value and the calculated value of Morss et al. (1989). On the other hand, the log K_{s0} value calculated by Merli et al. (1997) from independent, calorimetically-derived thermodynamic data for $Nd(OH)_3(cr)$ is in much closer agreement with our value.

As mentioned in the Introduction, there have been no studies of the solubility of $Nd(OH)_3$ at temperatures greater than 90 °C. Rao et al. (1996) determined K_{s0} at 25° and 90 °C for what they reported to be crystalline $Nd(OH)_3$. Their value of log K_{s0} at 25 °C (14.96) is three orders of magnitude smaller than ours (17.9), but their value at 90 °C (14.18) is more than a half an order of magnitude greater than ours (13.5). Thus, although Rao et al. (1996) observed a decrease in solubility with increased temperature, the magnitude of this decrease is considerably less than that observed in this study. Meloche and Vrátný (1959) also

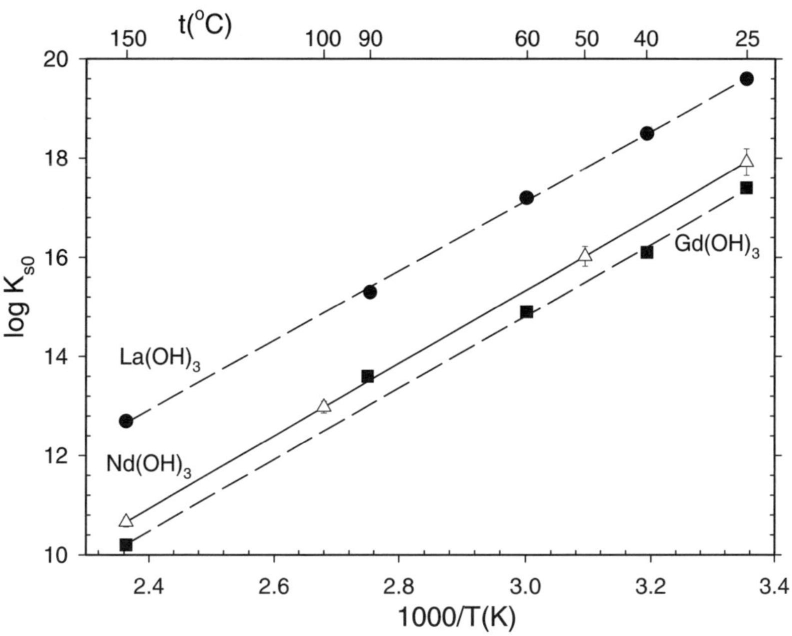

Figure 9. Plot of log K_{s0} vs. 1/T(K) for $La(OH)_3(cr)$, $Nd(OH)_3(cr)$ and $Gd(OH)_3(cr)$. The open triangles and solid line are the data and a regression line, respectively, obtained in this study. The solid symbols are the experimental data reported by Deberdt et al. (1998) and the dashed lines are regression fits to their data. For Gd, the datum at 90 °C appears to be an outlier and was omitted from the regression.

observed a decrease in the solubility of freshly precipitated $Nd(OH)_3$ from 10° to 40 °C, but their values are not directly comparable to ours because their solid was likely to have been poorly crystalline.

The only reported investigation of the solubility of REE-hydroxides at temperatures above 90 °C is that of Deberdt et al. (1998), which were conducted on well-crystallized $La(OH)_3$ and $Gd(OH)_3$ up to 150 °C. The values of K_{s0} obtained by Deberdt et al. (1998) for these hydroxides are compared to the values obtained by us for $Nd(OH)_3$ in this study in Figure 9. If Deberdt et al.'s value of K_{s0} for $Gd(OH)_3$ at 90 °C, which appears to be an outlier, is neglected, then the trends for K_{s0} for all three REE hydroxides as a function of temperature are parallel. Moreover, the solubilities follow a regular pattern, i.e., La >> Nd > Gd. Thus, the two studies appear to be in very good agreement. This agreement is particularly impressive when one considers that pH was determined by quite different means in the two studies. In the study of Deberdt et al. (1998), pH was measured using glass electrodes at temperatures up to and including 90 °C, but was calculated at 150 °C.

4.3. Comparison to Previous Work: Enthalpies of Formation

From equation (21), a value of ΔH_r^o = -140.3±5.2 kJ mol^{-1} was derived for reaction (1) at infinite dilution and 25 °C (Table 6). Using values of $\Delta H_f^o(Nd^{3+})$ = -696.6 kJ mol^{-1} (Morss, 1976) and $\Delta H_f^o(H_2O)$ = -285.8 kJ mol^{-1} (SUPCRT92; Johnson et al., 1992), we obtain a value of $\Delta H_f^o(Nd(OH)_3(cr))$ = -1414±5 kJ mol^{-1}. The calorimetrically-derived values of $\Delta H_f^o(Nd(OH)_3(cr))$ reported by Merli et al. (1997), Morss et al. (1989), and Merli and Fuger (1994), respectively, are -1415.6±2.3 kJ mol^{-1}, -1403.7±1.0 kJ mol^{-1} and -1404.2±3.3 kJ mol^{-1}. The value reported by Merli et al. (1997) is in much better agreement with our enthalpy value than are the values reported by Morss et al. (1989) and Merli and Fuger (1994), which is consistent with the fact that the calculated value of log K_{s0} reported by Merli et al. (1997) is in much better agreement with our value than is the value calculated by Morss et al. (1989). Note that, in their critical evaluation, Diakonov et al. (1998) chose the value of $\Delta H_f^o(Nd(OH)_3(cr))$ reported by Merli et al. (1997) as their recommended value.

It is also of interest to compare our value of $\Delta H_f^o(Nd(OH)_3(cr))$ with the corresponding values for $La(OH)_3(cr)$, $Sm(OH)_3(cr)$ and $Am(OH)_3(cr)$. Morss and Williams (1994) report a value of $\Delta H_f^o(Am(OH)_3(cr))$ = -1371.2±7.9 kJ mol^{-1}, and Cordfunke et al. (1990) determined $\Delta H_f^o(La(OH)_3(cr))$ = -1415.5±1.4 kJ mol^{-1}. Merli et al. (1997) obtained values of $\Delta H_f^o(Am(OH)_3(cr))$ = -1343.6±1.8, $\Delta H_f^o(La(OH)_3(cr))$ = -1416.7±1.3, and $\Delta H_f^o(Sm(OH)_3(cr))$ = -1406.6±2.2 kJ mol^{-1}. The two values of $\Delta H_f^o(La(OH)_3(cr))$ are indistinguishable, within the limits of error, from the value of $\Delta H_f^o(Nd(OH)_3(cr))$ suggested by this study and that of Merli et al. (1997). The value of $\Delta H_f^o(Sm(OH)_3(cr))$ is also similar to that for $Nd(OH)_3(s)$ (only 10 kJ mol^{-1} less exothermic). However, $\Delta H_f^o(Am(OH)_3(cr))$ is more than 40 kJ mol^{-1} less exothermic than $\Delta H_f^o(Nd(OH)_3(cr))$. Note also that the enthalpies of formation of $Am(OH)_3$ determined by Morss and Williams (1994) and Merli et al. (1997) differ by almost 30 kJ mol^{-1}.

4.4. Comparison to Previous Work: Hydrolysis Constants

The only hydrolysis constant that we have been able to determine near room temperature is log $Q_{h3,1}$ at 30 °C and 0.03 molal NaTr. Comparison of this value with previous work is complicated by the fact that no other low-temperature measurements were carried out at the same temperature or ionic strength (Table 2). However, our value of –24.5±0.3 at 30 °C and 0.03 mol kg^{-1} NaTr is of a similar order of magnitude to those reported by Kragten and Decnop-Weever (1984) for 21.5 °C and 1 mol L^{-1} $NaClO_4$ (log $Q_{h3,1}$ = -24.3±0.2), by Silva (1982) for 25 °C and infinite dilution (log $K_{h3,1}$ = -23.9±0.2) and Baes and Mesmer (1986) for 25 °C and infinite dilution (log $K_{h3,1}$ = -26.5). Extrapolation of our 30 °C value to infinite dilution yields a value of –24.0±0.4, which is identical within error limits to the value obtained for infinite dilution at 25 °C by Silva (1982). It is interesting to note that this difference is much less than the 1.9 log unit difference in values of log K_{s0} obtained in the two studies.

We are unaware of any experimentally-determined hydrolysis constants at elevated temperatures with which to compare our measured constants. However, Haas et al. (1995) have provided HKF parameters by which these constants may be estimated. Comparisons of values of log $K_{h1,1}$, log $K_{h2,1}$ and log $K_{h3,1}$ are given in Tables 8, 9, and 10, respectively. From Table 8 it can be seen that, the values of log $K_{h1,1}$ calculated from the HKF parameters of Haas et al. (1995) are higher than those derived from our experimental data by 1.8 and 1.9 log units at 250° and 290 °C, respectively. Our value for log $K_{h2,1}$ is 3 orders of magnitude smaller than that estimated by Haas et al. (1995) (Table 9). A similar comparison of log $K_{h3,1}$ values (Table 9) shows that the estimates of Haas et al. (1995) are more than an order of magnitude smaller than our value at low temperature, but as temperature increases, the estimates of Haas et al. (1995) become larger than our values. At 290 °C, the values of Haas et al. (1995) are larger than ours by more than three orders of magnitude. These findings suggest that use of the HKF parameters of Haas et al. (1995) will result in considerable overestimation of the degree of hydrolysis of the trivalent REE at elevated temperatures.

Other attempts to predict the hydrolysis behavior of the REE at elevated temperature have also been carried out. Wood (1990b) estimated hydrolysis constants of La^{3+}, Sm^{3+}, Lu^{3+} and Y^{3+}. Although he did not provide estimates for Nd^{3+}, comparison of our experimental results for this ion with the similar Sm^{3+} suggests that Wood (1990b) also greatly overestimated the degree of hydrolysis of the REE at elevated temperatures. Kumar (1987) estimated the degree of hydrolysis of Nd^{3+} up to 500 °C. Unfortunately, Kumar (1987) only presented his estimates in graphical form (his Fig. 1), and it is not entirely clear

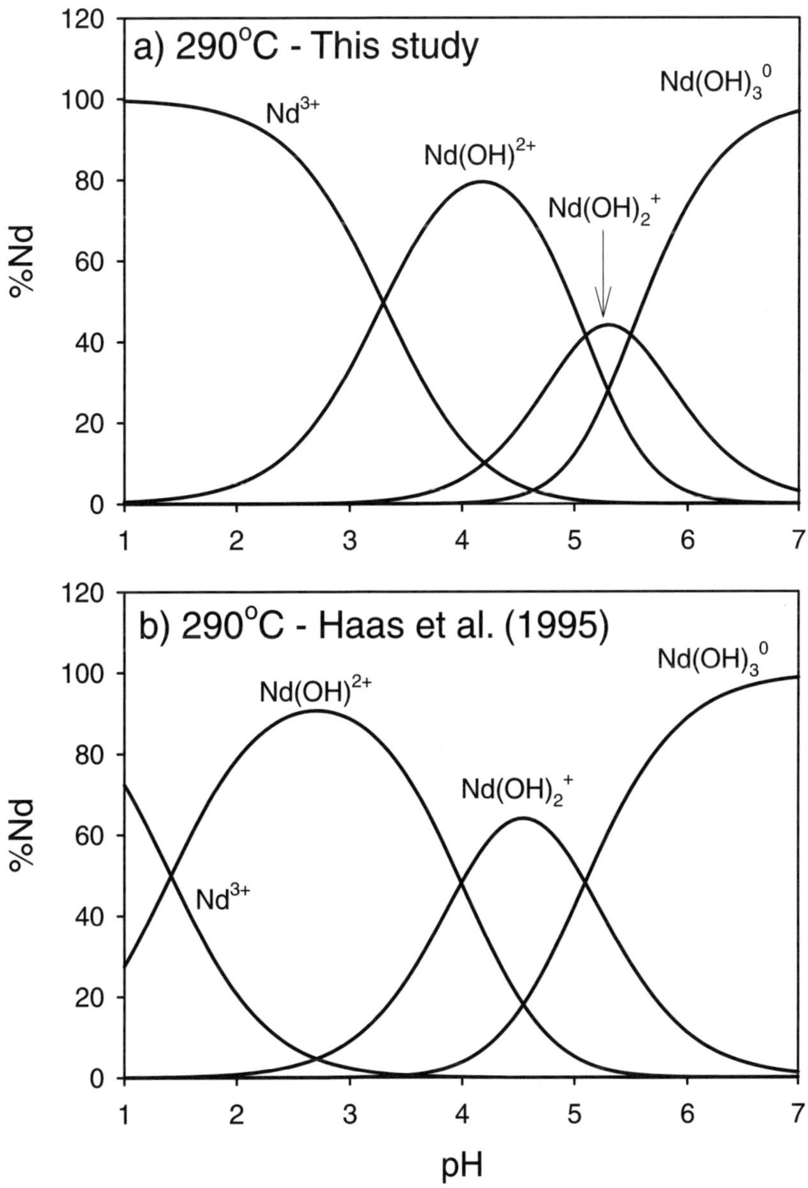

Figure 10. Plot of %Nd vs. pH at 290 °C showing the distribution of Nd^{3+} species among the various hydroxide complexes based on thermodynamic data from: a) this study; and b) Haas et al. (1995).

whether these estimates refer to infinite dilution or the ionic strength of seawater. However, his estimates of roughly log $K_{h1,1}$ = -2.5 and log $K_{h3,1}$ = -12.2 for 300 °C are much closer to our values at infinite dilution than the estimates of either Haas et al. (1995) or Wood (1990b).

Our results for the hydrolysis of Nd^{3+} are in qualitative agreement with those of Deberdt et al. (1998) for La^{3+} and Gd^{3+}. As can be seen from Figure 6, no significant hydrolysis occurs until approximately pH 8.4, 8.3 and 7.7 at 30°, 50° and 100 °C, respectively. The solubility data of Deberdt et al. (1998) reveal no evidence of hydrolysis of La^{3+} up to pH 9.5, 9.0, 8.2 and 7.0, or of Gd^{3+} up to pH 8.0, 7.4, 7.3 and 6.0 at 40°, 60°, 90° and 150 °C, respectively. The pH values noted are the highest investigated by Deberdt et al. (1998), and place upper limits on the degree of hydrolysis. In contrast, the HKF parameters of Haas et al. (1995) would predict a species change from La^{3+} to $La(OH)_3^0$ at pH 8.7, 7.9, 7.0 and 5.6 and from Gd^{3+} to $Gd(OH)_3^0$ at pH 7.8, 7.1, 6.2, and 5.0 at 40°, 60°, 90° and 150 °C, respectively. Thus, the results of Deberdt et al. (1998) confirm our conclusion that the HKF parameters of Haas et al. (1995) increasingly overestimate the degree of hydrolysis of the trivalent REE as temperature increases.

The speciation diagrams depicted in Figure 10 compare the predicted distribution at 290 °C, as a function of pH, of Nd^{3+} species that arises using the theoretical estimates of Nd^{3+} hydrolysis constants calculated from the HKF parameters of Haas et al. (1995), with that arising from the use of our experimentally-determined constants given in Tables 8-10. The Haas et al. (1995) constants predict that hydroxide complexes predominate over Nd^{3+} at a pH of less than 1.5 at this temperature, whereas our data predict that hydroxide complexes do not predominate until pH = 3. The Haas et al. (1995) constants also predict a larger predominance field for $Nd(OH)_2^+$ than do our data. Finally, in the Haas et al. (1995) speciation model, $Nd(OH)_3^0$ becomes predominant at a pH almost one unit lower than in our model. These differences will result in a substantial overestimation of the solubility of Nd-bearing solid phases in the absence of other complexation agents. Moreover, in previous work we have demonstrated that the HKF parameters of Haas et al. (1995) and Shock and Koretsky (1993) underestimate the stability constants of Nd-chloride (Gammons et al., 1996 and this volume) and Nd-acetate complexes (Wood et al., 2000; Ding and Wood, this volume). Thus, use of stability constants for Nd complexes from Haas et al. (1995) and Shock and Koretsky (1993) will result in serious overestimation of the importance of hydroxide complexes relative to chloride and/or acetate complexes in hydrothermal solutions.

Figure 11 illustrates how calculated solubility of $Nd(OH)_3(cr)$ as a function of pH depends sensitively on the precise values of $Q_{hn,1}$. In both Figures 11a and 11b, the value of Q_{s0} obtained in this study at 290 °C was used to calculate the concentration of Nd^{3+} in equilibrium with $Nd(OH)_3(cr)$ as a function of pH. In Figure 11a, the values of $K_{hn,1}$ obtained in this study were employed to calculate the concentrations of the hydroxide complexes, whereas in Figure 11b, the estimates of Haas et al. (1995) were used. It is obvious that the calculated total solubility of Nd^{3+} is much higher in Figure 11b than in Figure 11a. Thus, use of the HKF parameters of Haas et al. (1995) for hydroxide complexes of the REE in calculations of the solubility of REE minerals in the absence of other ligands is likely to result in significant overestimation of the total equilibrium REE concentration at saturation.

4.5. Polynuclear Species

A number of workers have identified polynuclear hydrolysis products of Nd^{3+}, e.g., $Nd_2(OH)_2^{4+}$ and $Nd_3(OH)_5^{4+}$ (cf. Baes and Mesmer, 1986 and references therein; Kragten and Decnop-Weever, 1984). We have detected no evidence for such products in our solubility data. According to the following reactions:

$$2Nd(OH)_3(cr) + 4H^+ \leftrightarrow Nd_2(OH)_3^{4+} + 4H_2O(l) \quad (26)$$

$$3Nd(OH)_3(cr) + 4H^+ \leftrightarrow Nd_3(OH)_5^{4+} + 4H_2O(l) \quad (27)$$

the slope of a curve of log ΣNd vs. pH should yield a value of –4 if either $Nd_2(OH)_2^{4+}$ or $Nd_3(OH)_5^{4+}$ were a predominant species over any range of pH investigated. None of our data correspond to such a slope. In fact, as shown in Figure 3, the data tend to exhibit slopes slightly more shallow than –3. Thus, we conclude that polynuclear species of the type previously proposed, are present in concentrations too low to contribute significantly to our measured solubilities. This conclusion is supported by the fact that calculations carried out using the thermodynamic data given in Baes and Mesmer (1986) for the solubility of $Nd(OH)_3(cr)$ and the hydrolysis of Nd^{3+} at 25 °C also indicate that $Nd_2(OH)_2^{4+}$ is never a significant species over the pH range we investigated, when the total Nd concentration is constrained to a low level by the solubility of $Nd(OH)_3(cr)$.

We cannot rule out entirely the possibility that polynuclear species of stoichiometries different from the two mentioned above were present in significant concentrations in our experiments. However, there are a number of reasons to suspect that polynuclear species were not present in significant quantity, including: 1) we have obtained excellent fits of our models to the solubility data without invoking any polynuclear species; 2) the stability of polynuclear species is strongly dependent on the total concentration of Nd, which is comparatively low in equilibrium with $Nd(OH)_3(s)$ in our experiments; and 3) the stability of polynuclear species tends to decrease with increasing temperature (cf. Wesolowski et al., 1984; Plyasunov and Grenthe, 1994).

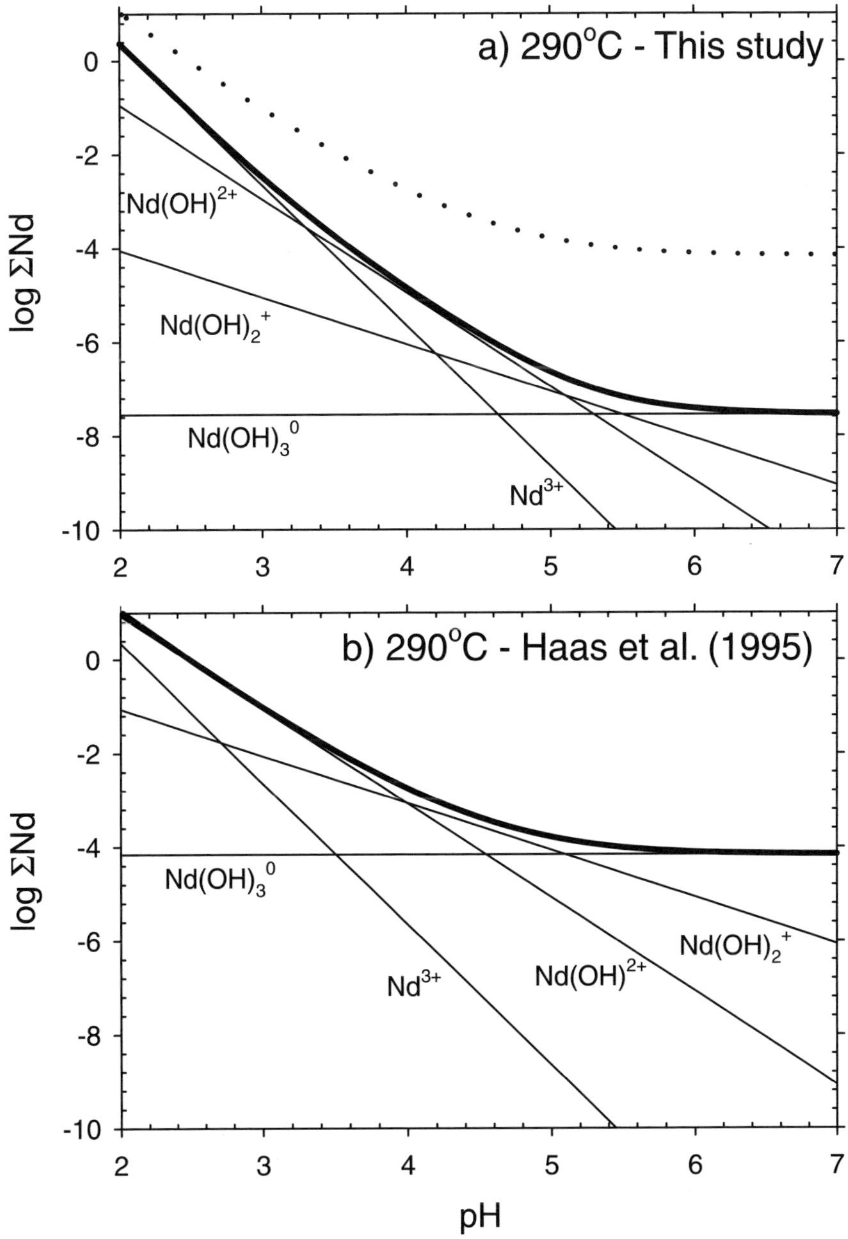

Figure 11. Calculated solubilities of $Nd(OH)_3$(cr) at 290 °C and infinite dilution presented in the form of plots of log $m_{\Sigma Nd}$ vs. pH. Both diagrams were calculated using the value of K_{s0} obtained in this study and the values of hydrolysis constants: a) determined in this study; and b) estimated by Haas et al. (1995). The heavy line and the light lines in each diagram represent the calculated total Nd concentration in solution and the calculated concentrations of individual species, respectively. To aid comparison, the calculated total solubility from b) is reproduced in a) as the dotted line.

4.6. Comparison of Solubilities of $Nd(OH)_3$, $Am(OH)_3$ and $Pu(OH)_3$

Owing to the use of trivalent REE as analogues of the trivalent actinides, it is of interest to compare the solubility of $Nd(OH)_3$ to those of the corresponding hydroxides of some trivalent actinides. Table 11 summarizes reported values of the solubility product (log Q_{s0}) for $Am(OH)_3$ and $Pu(OH)_3$. In many cases, the crystallinity of the initial solid phase is not clear. However, it is evident that the reported values of log Q_{s0} for the hydroxides of Am and Pu scatter over similar ranges as that given in Table 1 for Nd-hydroxide.

The only author to have measured the solubilities of both $Nd(OH)_3$ and $Am(OH)_3$ using the same conditions and techniques was Silva (1982), and his reported solubility products for these two hydroxides are identical within experimental error. Because there is some uncertainty as to the quality of the solubility data obtained by Silva (1982) for Nd, the similarity in his measured solubilities of Nd- and Am-hydroxides cannot be taken as definitive. However, it is worthwhile noting that the recommended solubility products for these two hydroxides given by Baes and Mesmer (1986), based on equation (25) are also nearly identical, owing to their nearly identical cell parameters. It is also interesting to note that the solubility product obtained by Rai et al. (1983) for $Am(OH)_3$ is very close to the value we have obtained for $Nd(OH)_3$, although values for $Am(OH)_3$ determined subsequently by other authors are not in good agreement with our value for $Nd(OH)_3$. One might actually expect that the solubility of the actinide hydroxides should be somewhat higher owing to the effects of radioactivity. However, because of the large spread in measured values of the solubility product for all three of the hydroxides in question, it would appear that the only way to definitively compare their solubilities would be to determine the solubilities of the actinide hydroxides using the same techniques employed in this paper for $Nd(OH)_3$.

Table 11. Summary of literature data on the solubility of $Am(OH)_3(s)$ and $Pu(OH)_3(s)$ at room temperature.

Author	T(°C)	Ionic strength and medium	log Q_{s0}
	Am^{3+}		
Allard et al. (1978)	25	infinite dilution	14[1]
Phillips (1982)	25	infinite dilution	18.7[1]
Silva (1982)	25	0.1 mol L^{-1} $NaClO_4$ (extr. to 0 with Davies eq.)	15.9±0.4[2]
Rai et al. (1983)	25	0.003 mol L^{-1} $CaCl_2$ (extr. to 0 with Pitzer eq.)	17.5±0.3[2]
Bernkopf & Kim (1984)	25	0.1 mol L^{-1} $NaClO_4$	13.9[2]
Baes & Mesmer (1986)	25	infinite dilution	18.7[1]
Stadler & Kim (1988)	25	0.1 mol L^{-1} $NaClO_4$	16.3±0.3[2]
Morss & Williams (1994)	25 25	infinite dilution "	15.5±2[3,5] 12.5±1.6[3,6]
Silva et al. (1995)	25	"	15.2±0.6[4]
Merli et al. (1997)	25	"	16.8[3]
	Pu^{3+}		
Busey & Cowan (1950)	25	infinite dilution	22.4
Latimer (1952)	25	infinite dilution	22.3[1]
Felmy et al. (1989)	25	infinite dilution	15.8±0.4[2]
Morss & Williams (1994)	25	infinite dilution	14.5±2[3]

NOTE: K_{s0} and Q_{s0} refer to the reactions: $Am(OH)_3(s) + 3H^+ \leftrightarrow Am^{3+} + 3H_2O(l)$ or $Pu(OH)_3(s) + 3H^+ \leftrightarrow Pu^{3+} + 3H_2O(l)$ and are given in the same units as the ionic strength.
[1]Estimated values;
[2]Directly measured experimentally;
[3]Calculated from calorimetrically-derived thermodynamic data;
[4]Based on critical evaluation of the literature;
[5]Calculated based on calorimetrically-determined value of $\Delta H_f^o(Am(OH)_3(cr))$ by Morss and Williams (1994);
[6]Recommend value of Morss and Williams (1994) based on critical evaluation.

5. CONCLUSIONS

We have determined the solubility of $Nd(OH)_3(cr)$ as a function of pH, ionic strength (0.03-1.0 m) and temperature (30°-290 °C) in non-complexing Na-triflate media using *in-situ* pH measurements in a well-stirred, hydrogen-electrode concentration cell. The following conclusions can be drawn from this work:

1. In our well-stirred reaction cell, equilibrium is attained within 0.5-1 days at 150 °C, and within a few hours at 290 °C.
2. The un-hydrolyzed, un-complexed Nd^{3+} ion is the predominant species over the majority of the pH range investigated.
3. Log Q_{s0} values obtained in this study decrease with increasing temperature, i.e., $Nd(OH)_3(cr)$ exhibits retrograde solubility, and are a linear function of 1/T(K).
4. The log Q_{s0} values obtained in this study are well described by the simple expression: $\log Q_{s0} = -6.662 + 7300.0/T + I(-91.51/T - 7.182\cdot10^{-6}T^2) - 6f^{\gamma}/\ln(10) - 3\log a_w$.

5. Extrapolation of our results to infinite dilution and 25 °C yields a value of log K_{s0} = 17.9±0.1, in excellent agreement with a recent critical evaluation by Diakonov et al. (1998).
6. Our solubility results for $Nd(OH)_3(cr)$ as a function of temperature are consistent with the results of Deberdt et al. (1998) for $La(OH)_3(cr)$ and $Gd(OH)_3(cr)$.
7. At low temperature (30°-150 °C), hydrolysis proceeds very rapidly over a narrow pH range from Nd^{3+} to $Nd(OH)_3^0$, with little evidence of intermediate species such as $Nd(OH)^{2+}$ or $Nd(OH)_2^+$.
8. At high temperature (250°-290 °C), the intermediate species $Nd(OH)^{2+}$ and $Nd(OH)_2^+$ become more important, but the relative importance of these two species appears to be strongly dependent on ionic strength.
9. Hydrolysis constants derived from our data at elevated temperatures are much smaller than predicted by recent theoretical estimates.
10. The value of $\Delta H_f^o(Nd(OH)_3(cr))$ = -1414±5 kJ mol^{-1}, derived from the temperature dependence of log K_{s0}, compares well with calorimetrically-derived values.

Our results clearly demonstrate that the techniques employed in this study are capable of producing high-quality solubility data for REE hydroxides. It should be possible to use the same techniques to obtain data of comparable quality on a range of REE hydroxides, as well as hydroxides of actinide analogues. Finally, it should be possible to use the study reported here as a foundation from which the complexation of Nd^{3+} with a variety of ligands, such as chloride, sulfate, acetate, EDTA, etc. can be investigated at elevated temperatures. Some preliminary experiments with acetate in our laboratories have indeed shown promise in this regard.

Acknowledgments—We gratefully acknowledge the help provided by Bill Shannon (ICP-MS analyses) and Leslie Baker (X-ray diffraction analyses). Leon Maya and Larry Anovitz assisted in the preparation of samples for analysis by FTIR spectroscopy. This work was conducted while SAW was on sabbatical leave at ORNL and was financially supported by the Geosciences Program of the Office of Basic Energy Research, U.S. Department of Energy under contract DE-AC05-96OR22464 with Lockheed Martin Energy Research Corporation. Additional financial support was provided by the Environmental Management Science Program, U.S. Department of Energy, contract 07-98ER14935.000. The manuscript benefitted from the perceptive reviews of Tom Giordano, Bill Seyfried and Kang Ding.

Editorial handling: T.H. Giordano

REFERENCES

Akselrud N. V. (1963) Basic chlorides and hydroxides of elements of the scandium and lanthanoid group. *Usp. Khim.* **32**, 800-822 (in Russian).

Allard B., Kipatsi H. and Torstenfelt B. (1978) Sorption an Langlivade Radionuklider. Karnbranslesakerhet, Stockholm, Lera Och Berg Del 2, KBS-98.

Azhipa L. T., Kovalenko P. N., and Evstifeev M. M. (1967) Oscillographic determination of the pH of precipitation and activity products of ytterbium and neodymium hydroxides. *Russ. Jour. Inorg. Chem.* **12**, 601-603.

Baes C. F. Jr. and Mesmer R. E. (1986) The Hydrolysis of Cations (Reprint with corrections of the 1976 edition by John Wiley & Sons), Robert E. Krieger. Malabar, Florida.

Bénézeth P., Palmer D. A., and Wesolowski D. J. (1997) The aqueous chemistry of aluminum. A new approach to high temperature solubility measurements. *Geothermics* **26**, 465-481.

Bénézeth P., Palmer D. A., and Wesolowski D. J. (1999) The solubility of zinc oxide in 0.03 m NaTr as a function of temperature, with in situ pH measurement. *Geochim. Cosmochim. Acta* **63**, 1571-1586.

Bénézeth P., Palmer D. A., and Wesolowski D. J. (2001) Aqueous high-temperature solubility studies: II. The solubility of boehmite at 0.03 m ionic strength as a function of temperature and pH as determined by "in situ" measurements. *Geochim. Cosmochim. Acta* **65**, 2097-2111.

Bernkopf M. and Kim J. I. (1984) Hydrolysis reactions and carbonate complexation of americium(III) in natural aquatic systems. Report RCM-02884.

Brush L. H. (1990) Test plan for laboratory and modeling studies of repository and radionuclide for the waste isolation pilot program. Sandia Report.

Burkov K. A., Lilich L. S., Ngo N. D., and Smirnov A. Yu. (1973) Potentiometric study of the hydrolysis of neodymium ions (Nd^{3+}) in 3 M $(Na)ClO_4$ solution. *Russ. Jour. Inorg. Chem.* **18**, 797-800.

Burkov K. A., Busko E. A., and Lilich L. C. (1988) On the hydrolysis of rare earth element metal ions. *Zh. Neorg. Khim.* **33**, 339-342.

Busey H. M. and Cowan H. D. (1950) Behavior of plutonium(III) chloride in titrations with acid and base. LAMS-1105, Los Alamos National Laboratory, Los Alamos, NM.

Busing W. R., Levy H. A. (1962) A General Fortran Least Squares Program, Oak Ridge Natl. Lab. Rep. ORNL-TM-271.

Caro P. E., Sawyer J. O., and Eyring L. (1972) The infrared spectra of rare earth carbonates. *Spectrochim. Acta* **28A**, 1167-1173.

Choppin G. R. (1983) Comparison of the solution chemistry of the actinides and lanthanides. *Jour. Less-Common Metals* **93**, 323-330.

Choppin G. R. (1986) Speciation of the trivalent *f*-elements in natural waters. *J. Less-Common Met.* **126**, 307-313.

Choppin G. R. (1989) Soluble rare earth and actinide species in seawater. *Mar. Chem.* **28**, 19-26.

Christensen A. N. (1966) Hydrothermal preparation of some rare earth trihydroxides and rare earth oxide hydroxides. *Acta Chem. Scand.* **20**, 896-899.

Cordfunke E. H. P., Konings R. J. M., and Ouweltjes W. (1990) The standard enthalpies of formation of hydroxides. IV. $La(OH)_3$ and LaOOH. *Jour. Chem. Thermo.* **22**, 449-452.

Deberdt S., Castet S., Dandurand J.-L., Harrichoury J.-C., and Louiset, I. (1998) Experimental study of $La(OH)_3$

and $Gd(OH)_3$ solubilities (25 to 150 °C), and La-acetate complexing (25 to 80 °C). *Chem. Geol.* **151**, 349-372.

Diakonov I. I., Tagirov B. R., and Ragnarsdottir K. V. (1998) Standard thermodynamic properties and heat capacity equations for rare earth element hydroxides. I. $La(OH)_3$ and $Nd(OH)_3$. Comparison of thermochemical and solubility data. *Radiochim. Acta* **81**, 107-116.

Dickson A. G., Wesolowski D. J., Palmer D. A., and Mesmer R. E. (1990) Dissociation constant of bisulfate ion in aqueous sodium chloride solutions to 250 °C. *Jour. Phys. Chem.* **94**, 7978-7985.

Ding R. and Wood S. A. (this volume) The aqueous geochemistry of the rare earth elements and yttrium. Part X. Potentiometric determination of stability constants of acetate complexes of La^{3+}, Nd^{3+}, Gd^{3+} and Yb^{3+} at 25-70 °C and 1 bar.

Felmy A. R., Rai D., Schramke J. A., and Ryan J. L. (1989) The solubility of plutonium hydroxide in dilute solution and in high-ionic-strength chloride brines. *Radiochim. Acta* **48**, 29-35.

Gammons C. H., Wood S. A., and Williams-Jones A. E. (1996) The aqueous geochemistry of the rare earth elements and yttrium: VI. Stability of neodymium chloride complexes at 25 °C to 300 °C. *Geochim. Cosmochim. Acta* **60,** 4615-4630.

Guillaumont R., Désiré B., and Galin M. (1971) Première constante d'hydrolyse des lanthanides. *Radiochem. Radioanal. Let.* **8**, 189-198.

Haas J. R., Shock E. L., and Sassani D. C. (1995) Rare earth elements in hydrothermal systems: Estimates of standard partial molal thermodynamic properties of aqueous complexes or the rare earth elements at high pressures and temperatures. *Geochim. Cosmochim. Acta.* **59**, 4329-4350.

Helgeson H. C. and Kirkham D. H. (1974) Theoretical prediction of the thermodynamic properties of aqueous electrolytes at high pressures and temperatures: II. Debye-Hückel parameters for activity. *American Journal of Science* **274**, 1199-1261.

Ho P.C. and Palmer D. A. (1995) Electrical conductivity measurements of dilute aqueous sodium trifluoromethanesulfonate solutions at temperatures 0-450 °C and pressures up to 250 MPa. *Jour. Sol. Chem.* **24**, 753-769.

Johnson J. W., Oelkers E. H., and Helgeson H. C. (1992) SUPCRT92: a software package for calculating the standard molal thermodynamic properties of minerals, gases, aqueous species, and reactions from 1-5000 bars and 0-1000 °C. *Comp. Geosci.* **18**, 899-947.

Klevtsov P. V., Klevtsova R. F., and Sheina L. P. (1967) Relationship between the infrared absorption spectra and crystal structure of the hydroxides of the rare earth elements and yttrium. *Jour. Struct. Chem.* **8**, 229-233.

Kostromina N. A., and Badaev Yu. B. (1986) Spectrophotometric study of hydrolysis in the $NdCl_3$-KOH system. *Ukrain. Khim. Zhur.* **52**, 793-796.

Kragten J. and Decnop-Weever L. G. (1984) Hydroxide complexes of lanthanides - VII. Neodymium(III) in perchlorate medium. *Talanta* **31**, 731-733.

Kumar M. D. (1987) Cation hydrolysis and the regulation of trace metal composition in seawater. *Geochim. Cosmochim. Acta* **51**, 2137-2145.

Latimer W. M. (1952) Oxidation Potentials, 2nd ed., Prentice-Hall, New York.

Liu C. and Lindsay W.T. Jr. (1972) Thermodynamics of sodium chloride solutions at high temperatures. *Jour. Sol. Chem.* **1**, 45-69.

Makino H., Yajima T., Yoshikawa H., Yui M. and Sasaki N. (1993) Neodymium(III) hydrolysis constants and solubilities of Nd(III) hydroxide. *Nippon Kagaku Kaishi* **5**, 445-450 (in Japanese; Chem. Abstr. 119: 81217b).

Meloche C. C. and Vrátný F. (1959) Solubility product relations in the rare earth hydrous hydroxides. *Anal. Chim. Acta* **20**, 415-418.

Merli L. and Fuger J. (1994) Thermochemistry of a few neptunium and neodymium oxides and hydroxides. *Radiochim. Acta* **66/67**, 109-113.

Merli L., Lambert B., and Fuger J. (1997) Thermochemistry of lanthanum, neodymium, samarium and americium trihydroxides and their relation to the corresponding hydroxycarbonates. *Jour. Nucl. Mater.* **247**, 172-176.

Mesmer R. E. (1991) Comments on "A new approach to measuring pH in brines and other concentrated electrolytes" by K.G. Knauss, T.J. Wolery, and K.J. Jackson. *Geochim. Cosmochim. Acta* **55**, 1175-1176.

Millero F. J. (1992) Stability constants for the formation of rare earth inorganic complexes as a function of ionic strength. *Geochim. Cosmochim. Acta* **56**, 3123-3132.

Moeller T. (1946) Observations on the rare earths. LV. Hydrolysis studies upon yttrium and certain rare earth(III) sulfate solutions at 25 °C. *Jour. Phys. Chem.* **50**, 242-250.

Moeller T. and Fogel N. (1951) Observations on the rare earths. LXI. Precipitation of hydrous oxides or hydroxides from perchlorate solutions. *Jour. Amer. Chem. Soc.* **73**, 4481.

Moeller T. and Kremers H. E. (1944) Observations on the rare earths. LI. An electrometric study of the precipitation of trivalent hydrous rare earth oxides or hydroxides. *Jour. Phys. Chem.* **48**, 395-406.

Moeller T., Martin D. F., Thompson L. C., Ferrus R., Feistel R., and Randall W.J. (1965) The coordination chemistry of yttrium and the rare earth metal ions. *Chem. Rev.* **65**, 1-50.

Morss L. R. (1976) Thermochemical properties of yttrium, lanthanum, and the lanthanide elements and ions. *Chem. Rev.* **76**, 827-841.

Morss L. R. and Williams C. W. (1994) Synthesis of crystalline americium hydroxide, $Am(OH)_3$, and determination of the solubility-product constants of actinide(III) hydroxides. *Radiochim. Acta* **66/67**, 89-93.

Morss L. R., Haar C. M., and Mroczkowski S. (1989) Standard molar enthalpy of formation of neodymium hydroxide. *Jour. Chem. Thermo.* **21**, 1079-1083.

Mroczkowski S., Eckert J., Meissner H., and Doran J. C. (1970) Hydrothermal growth of single crystals of rare earth hydroxides. *Jour. Cryst. Growth* **7**, 333-342.

Nitsche H. (1990) Basic research for assessment of geologic nuclear waste repositories: What solubility and speciation studies of transuranium elements can tell us. Paper presented at the Intl. Symp. on Sci. Basis for Nuclear Waste Management, Boston, MA Conf. Proceedings.

Orhanovic Z., Pokric B., Füredi H., and Branica M. (1966) Precipitation and hydrolysis of metallic ions. III. Studies on the solubility of yttrium and some rare earth hydroxides. *Croat. Chem. Acta* **38**, 269-276.

Palmer D. A. and Hyde K. E. (1993) Ferrous chloride and acetate complexation in aqueous solutions at high temperatures. *Geochim. Cosmochim. Acta* **57**, 1393-1408.

Palmer D. A. and Wesolowski D. J. (1993) Aluminum speciation and equilibria in aqueous solution: Potentiometric determination of the first hydrolysis constant of aluminum(III) in sodium chloride solutions to 125 °C. *Geochim. Cosmochim. Acta* **57**, 2929-2938.

Palmer D. A., Bénézeth P., and Wesolowski D. J. (2001) Aqueous high-temperature solubility studies. I. The solubility of boehmite as functions of ionic strength (to 5 molal NaCl), temperature (100-250 °C) and pH as determined by in situ measurements. *Geochim. Cosmochim. Acta* **65**, 2081-2095.

Phillips S. L. (1982) Hydrolysis and formation constants at 25 °C. U.S. DOE Report LBL-14313.

Pitzer K. S., Roy R. N., and Silvester L. F. (1977) Thermodynamics of electrolytes. 7. Sulfuric acid. *Jour. Amer. Chem. Soc.* **99**, 4930-4936.

Plyasunov A. V. and Grenthe I. (1994) The temperature dependence of stability constants for formation of polynuclear cationic complexes. *Geochim. Cosmochim. Acta* **58**, 3561-3582.

Quist A. S. and Marshall W. L. (1965) Assignment of limiting equivalent conductances for single ions to 400°C. *Jour. Phys. Chem.* **69**, 2984-2987.

Rai D., Strickert R. G., Moore D. A., and Ryan J. L. (1983) Am(III) hydrolysis constants and solubility of Am(III) hydroxide. *Radiochim. Acta* **33**, 201-206.

Rao L., Rai D., and Felmy A. R. (1996) Solubility of $Nd(OH)_3(c)$ in 0.1 M NaCl aqueous solution at 25°C and 90 °C. *Radiochim. Acta* **72**, 151-155.

Robinson R. A. and Stokes R. H. (1959) Electrolyte Solutions. Butterworths. London, UK.

Roy R. and McKinstry H. A. (1953) So-called $Y(OH)_3$-type structure, and structure of $La(OH)_3$. *Acta Cryst.* **6**, 365-366.

Shafer M. W. and Roy R. (1959) Rare-earth polymorphism and phase equilibria in rare-earth oxide-water systems. *Jour. Amer. Ceram. Soc.* **42**, 563-570.

Shock E. L. and Koretsky C. M. (1993) Metal-organic complexes in geochemical processes: Calculation of standard partial molal thermodynamic properties of aqueous acetate complexes at high pressures and temperatures. *Geochim. Cosmochim. Acta* **57**, 4899-4922.

Silva R. J. (1982) The solubilities of crystalline neodymium and americium trihydroxides. Berkeley, California, Lawrence Berkeley Laboratory, University of California. Thermodynamic Properties of Chemical Species in Nuclear Waste. Topical Report, 57 p.

Silva R. J., Bidoglio G., Rand M. H., Robouch P. B., Wanner H., and Puigdomenech I. (1995) Chemical Thermodynamics of Americium. NEA-OECD Publication, Paris/North-Holland Elsevier, Amsterdam.

Spivakovskii V. B. and Moisa L. P. (1972) Conditions of precipitation of neodymium hydroxide and hydroxide chlorides. *Russ. Jour. Inorg. Chem.* **17**, 34-37.

Spivakovskii V. B. and Moisa L. P. (1977) Composition, activity products, and free energies of formation of scandium, yttrium, lanthanum, and lanthanide hydroxide chlorides and hydroxides. *Russ. Jour. Inorg. Chem.* **22**, 643-646.

Stadler S. and Kim J. I. (1988) Hydrolysis reactions of Am(III) and Am(V). *Radiochim. Acta* **44/45**, 39-44.

Tobias R. S. and Garrett A. B. (1958) The thermodynamic properties of neodymium hydroxide $Nd(OH)_3$, in acid, neutral and alkaline solutions at 25°; the hydrolysis of the neodymium and praseodymium ions, Nd^{3+}, Pr^{3+}. *Jour. Amer. Chem. Soc.* **80**, 3532-3537.

Viswanathiah M. N., Tareen J. A. K., and Kutty T. R. (1976) Hydrothermal equilibria in Ln_2O_3-H_2O systems for neodymium, samarium, ytterbium and yttrium. *Ind. Mineral.* **17**, 54-59.

Viswanathiah M. N., Tareen J. A. K., and Kutty T. R. (1980) Hydrothermal phase equilibria studies in Ln_2O_3-H_2O systems and synthesis of cubic lanthanide oxides. *Mat. Res. Bull.* **15**, 855-859.

Wesolowski D. J. and Palmer D. A. (1994) Aluminum speciation and equilibria in aqueous solution: V. Gibbsite solubility at 50 °C and pH 3-9 in 0.1 molal NaCl solutions (a general model for aluminum speciation; analytical methods). *Geochim. Cosmochim. Acta* **58**, 2947-2969.

Wesolowski D. J., Bénézeth P., and Palmer D. A. (1998) ZnO solubility and Zn^{2+} complexation by chloride and sulfate in acidic solutions to 290 °C with in-situ pH measurement. *Geochim. Cosmochim. Acta* **62**, 971-984.

Wesolowski D. J., Drummond S. E., Mesmer R. E., and Ohmoto H. (1984) Hydrolysis equilibria of tungsten(VI) in aqueous sodium chloride solutions to 300 °C. *Inorg. Chem.* **23**, 1120-1132.

Wood S. A. (1990a) The aqueous geochemistry of the rare-earth elements and yttrium. 1. Review of available low-temperature data for inorganic complexes and the inorganic REE speciation of natural waters. *Chem. Geol.* **82**, 159-186.

Wood S. A. (1990b) The aqueous geochemistry of the rare-earth elements and yttrium. 2. Theoretical prediction of speciation in hydrothermal solutions to 350 °C at saturation water vapor pressure. *Chem. Geol.* **88**, 99-125.

Wood S. A., Wesolowski D. J., and Palmer D. A. (2000) The aqueous geochemistry of the rare earth elements: IX. A potentiometric study of Nd^{3+} complexation with acetate in 0.1 molal NaCl solution from 25-250 °C. *Chem. Geol.* **167**, 231-253.

APPENDIX

APPENDIX I. Raw data for $Nd(OH)_3(cr)$ solubility experiments at 30.1±0.1°C and 1 bar in 0.03 molal Na-triflate.

Sample No.	Time (days)	pH	log m_{Nd}
Nd-1-30-1	2	7.01	-2.73
Nd-1-30-2	6	7.06	-2.77
Nd-1-30-3	7	7.12	-2.78
Nd-1-30-4	8	7.08	-2.77
Nd-1-30-5	9	7.39	-3.21
Nd-1-30-6	10	7.46	-3.28
Nd-1-30-11	28	8.39	-6.32
Nd-1-30-12	30	8.31	-6.29
Nd-1-30-13	35	8.06	-5.44
Nd-1-30-14	38	7.76	-4.18
Nd-1-30-15	40	7.71	-4.02
Nd-1-30-16	42	7.25	-2.76
Nd-1-30-17	45	7.26	-2.74
Nd-1-30-24	73	9.13	-6.61
Nd-1-30-25	78	9.09	-6.38

APPENDIX II. Raw data for $Nd(OH)_3$(cr) solubility experiments at elevated temperatures in Na-triflate medium.

Sample No.	Time (hours)	T(°C)	P(bars)	I (mol kg^{-1})	pH	log $m_{\Sigma Nd}$
Nd-1-150-1	68	150.1	21.5	0.0298	5.473	-5.01
Nd-1-150-2	92	150.1	21.1	0.0298	5.447	-5.05
Nd-1-150-3	116	150.1	22.2	0.0301	5.198	-4.15
Nd-1-150-4	140	150.1	21.7	0.0301	5.200	-4.17
Nd-1-150-5	164	150.1	22.1	0.0298	5.360	-4.63
Nd-1-150-6	172	150.1	22.6	0.0298	5.359	-4.62
Nd-1-150-7	191	150.1	23.1	0.0298	5.359	-4.62
Nd-1-150-11	244	150.1	25.2	0.0297	5.770	-5.98
Nd-1-150-12	259	150.1	24.1	0.0297	5.552	-5.23
Nd-1-150-13	268	150.1	23.2	0.0297	5.552	-5.19
Nd-1-200-1	284	200.1	34.7	0.0298	5.084	-5.46
Nd-1-200-2	292	200.1	35.8	0.0298	5.088	-5.32
Nd-1-200-3	308	200.1	40.3	0.0299	4.772	-4.59
Nd-1-200-4	317	200.1	43.5	0.0300	4.780	-4.55
Nd-1-200-5	332	200.1	52.1	0.0304	4.546	-3.94
Nd-1-200-6	340	200.1	52.0	0.0304	4.541	-3.94
Nd-1-200-7	363	200.1	68.5	0.0307	4.475	-3.75
Nd-1-200-8	385	200.1	71.7	0.0310	4.419	-3.62
Nd-1-200-9	404	200.1	74.0	0.0299	4.724	-4.43
Nd-1-250-1	428	250.1	105.1	0.0301	4.365	-4.58
Nd-1-250-2	436	250.1	106.6	0.0301	4.373	-4.62
Nd-1-250-3	452	250.1	110.2	0.0300	4.713	-5.30
Nd-1-250-4	460	250.1	124.4	0.0300	4.733	-5.40
Nd-1-250-5	476	250.1	124.3	0.0300	4.861	-5.64
Nd-1-250-6	484	250.2	124.4	0.0300	4.873	-5.85
Nd-1-250-7	501	250.1	124.5	0.0301	4.778	-5.58
Nd-1-250-8	506	250.1	124.1	0.0302	4.787	-5.62
Nd-1-250-9	529	250.1	129.6	0.0303	4.546	-5.09
Nd-1-250-10	548	250.1	130.2	0.0304	4.352	-4.62
Nd-1-250-11	573	250.1	135.5	0.0306	4.168	-4.15
Nd-1-250-12	580	250.1	138.1	0.0310	4.026	-3.78
Nd-1-250-13	596	250.1	144.4	0.0322	3.873	-3.37
Nd-2-50-1	28	50.1	10.5	0.0308	6.685	-3.59
Nd-2-50-2	52	50.1	10.1	0.0308	6.686	-3.59
Nd-2-50-3	72	50.1	9.6	0.0308	6.684	-3.59
Nd-2-50-4	92	50.1	10.4	0.0315	6.581	-3.34
Nd-2-50-5	100	50.1	10.3	0.0325	6.498	-3.13
Nd-2-50-6	116	50.2	10.1	0.0318	6.573	-3.22
Nd-2-50-7	164	50.2	9.6	0.0306	6.671	-3.43
Nd-2-50-10	196	50.2	8.7	0.0286	7.376	-5.52
Nd-2-50-11	215	50.2	12.8	0.0285	7.546	-6.34
Nd-2-50-12	239	50.2	11.3	0.0286	6.995	-4.46
Nd-2-50-13	260	50.1	11.3	0.0286	6.997	-4.46
Nd-2-50-14	268	50.2	11.1	0.0284	7.285	-5.28
Nd-2-50-15	284	50.1	10.5	0.0284	7.269	-5.26
Nd-2-50-16	292	50.2	10.2	0.0284	7.264	-5.23
Nd-2-50-17	308	50.1	10.0	0.0284	7.466	-5.87
Nd-2-50-18	313	50.2	10.9	0.0284	7.457	-5.91
Nd-2-50-19	339	50.2	10.5	0.0284	8.888	-7.91
Nd-2-50-20	363	50.2	10.1	0.0284	8.856	-8.75
Nd-2-50-21	383	50.2	10.8	0.0284	9.548	-8.29
Nd-2-50-22	428	50.1	10.1	0.0284	9.513	-8.15
Nd-2-100-1	436	100.3	12.1	0.0284	8.421	-9.14
Nd-2-100-2	452	100.2	11.7	0.0284	8.420	-9.00
Nd-2-100-3	460	100.2	16.6	0.0281	7.379	-8.53
Nd-2-100-4	476	100.2	16.3	0.0281	7.435	-8.55
Nd-2-100-5	484	100.3	15.9	0.0298	5.596	-3.33
Nd-2-100-6	500	100.3	15.6	0.0297	5.602	-3.34
Nd-2-100-7	508	100.3	15.8	0.0369	5.362	-2.63
Nd-2-100-8	524	100.3	15.3	0.0352	5.387	-2.70

Nd-2-100-9	533	100.3	15.8	0.0263	5.724	-3.72
Nd-2-100-10	554	100.3	14.8	0.0263	5.723	-3.71
Nd-2-100-11	574	100.3	14.7	0.0255	6.023	-4.56
Nd-2-100-12	596	100.2	13.1	0.0254	6.513	-5.91
Nd-2-100-13	604	100.3	12.7	0.0254	6.497	-5.76
Nd-2-150-1	619	150.1	18.2	0.0255	5.804	-5.94
Nd-2-150-2	643	150.0	16.6	0.0255	5.890	-5.88
Nd-2-150-3	651	150.1	17.1	0.0255	5.506	-4.81
Nd-2-150-4	668	150.1	15.8	0.0255	5.503	-4.83
Nd-2-150-5	676	150.1	15.6	0.0264	5.068	-3.63
Nd-2-150-6	691	150.1	15.2	0.0263	5.070	-3.66
Nd-2-150-7	722	150.1	15.1	0.0309	4.791	-2.87
Nd-2-150-8	743	150.1	14.6	0.0309	4.792	-2.87
Nd-2-290-1	764	290.1	89.6	0.0316	3.496	-2.98
Nd-2-290-2	787	290.0	89.2	0.0315	3.493	-3.00
Nd-2-290-3	796	290.1	92.2	0.0286	3.663	-3.39
Nd-2-290-4	812	290.1	93.3	0.0286	3.665	-3.41
Nd-2-290-5	820	290.1	94.1	0.0274	3.833	-3.82
Nd-2-290-6	836	290.0	93.6	0.0270	4.039	-4.28
Nd-2-290-7	839	290.0	94.6	0.0271	4.037	-4.32
Nd-2-290-8	844	290.0	95.5	0.0271	4.275	-4.77
Nd-2-290-9	860	290.1	98.7	0.0273	4.285	-4.86
Nd-3-50-1	20	50.2	10.5	0.0313	6.494	-3.26
Nd-3-50-2	27	50.1	10.4	0.0312	6.511	-3.27
Nd-3-50-3	68	50.2	11.0	0.0312	6.540	-3.27
Nd-3-50-4	77	50.1	10.8	0.0312	6.544	-3.27
Nd-3-50-5	92	50.1	14.8	0.0312	6.553	-3.27
Nd-3-100-1	100	100.3	17.6	0.0311	5.583	-3.29
Nd-3-100-2	122	100.2	15.4	0.0311	5.588	-3.30
Nd-3-150-1	164	150.1	18.1	0.0310	4.847	-3.32
Nd-3-150-2	168	150.0	17.8	0.0311	4.847	-3.32
Nd-3-290-2	196	290.1	93.9	0.0295	6.898	-7.81
Nd-3-290-3	212	290.0	95.6	0.0295	6.665	-7.67
Nd-3-290-4	220	290.1	95.7	0.0296	6.661	-7.36
Nd-3-290-5	236	290.1	98.5	0.0297	6.335	-7.30
Nd-3-290-6	244	290.1	99.5	0.0297	5.974	-7.36
Nd-3-290-8	292	290.0	105.3	0.0297	5.490	-7.06
Nd-3-290-9	316	290.0	113.0	0.0296	5.043	-6.21
Nd-3-290-10	332	290.1	114.9	0.0297	5.037	-6.61
Nd-3-290-12	365	290.0	120.8	0.0298	4.609	-5.70
Nd-3-290-13	380	290.1	124.6	0.0299	4.658	-5.83
Nd-3-290-14	388	290.1	124.2	0.0300	4.745	-6.01
Nd-3-290-15	410	290.1	126.8	0.0301	4.699	-6.01
Nd-4-50-1	28	50.2	12.7	0.1017	6.653	-3.37
Nd-4-50-2	51	50.1	12.4	0.1017	6.687	-3.38
Nd-4-50-3	124	50.1	14.4	0.1016	6.709	-3.40
Nd-4-50-4	140	50.2	14.3	0.1016	6.711	-3.40
Nd-4-100-5	148	100.3	17.6	0.1015	5.739	-3.44
Nd-4-100-6	164	100.2	17.3	0.1015	5.744	-3.44
Nd-4-100-7	172	100.2	17.1	0.1015	5.745	-3.43
Nd-4-150-8	188	150.1	22.9	0.1017	5.017	-3.46
Nd-4-150-9	196	150.0	22.6	0.1017	5.018	-3.47
Nd-4-150-10	215	150.1	22.2	0.1016	5.021	-3.48
Nd-4-200-1	222	199.8	59.7	0.1021	4.477	-3.53
Nd-4-200-2	237	200.2	59.4	0.1020	4.473	-3.55
Nd-4-200-3	245	200.1	58.5	0.1021	4.472	-3.54
Nd-4-250-1	260	250.1	86.0	0.1034	4.050	-3.61
Nd-4-250-2	284	250.1	84.1	0.1033	4.066	-3.67
Nd-4-250-3	291	250.1	83.1	0.1034	4.068	-3.67
Nd-5-290-1	22	290.0	100.9	0.1015	3.789	-3.64
Nd-5-290-2	29	290.1	100.5	0.1016	3.772	-3.62
Nd-5-290-3	46	290.0	103.1	0.1012	4.178	-4.63
Nd-5-290-4	54	290.0	102.5	0.1013	4.352	-5.00

Nd-5-290-5	70	290.1	102.2	0.1014	4.544	-5.51
Nd-5-290-6	77	290.0	101.4	0.1015	4.701	-5.69
Nd-5-290-7	105	290.0	102.3	0.1017	4.958	-6.08
Nd-5-290-8	128	290.0	106.3	0.1019	5.221	-6.39
Nd-5-290-9	166	290.0	124.0	0.0974	3.406	-2.97
Nd-5-290-10	173	290.0	124.1	0.0975	3.406	-2.97
Nd-5-290-11	190	289.0	130.3	0.0970	3.517	-3.29
Nd-6-50-1	21	50.2	14.0	1.0085	6.681	-3.55
Nd-6-50-2	27	50.1	13.7	1.0084	6.734	-3.56
Nd-6-50-3	45	50.2	13.6	1.0084	6.793	-3.58
Nd-6-50-4	72	50.2	13.3	1.0083	6.809	-3.59
Nd-6-100-1	93	100.3	16.3	1.0085	5.857	-3.64
Nd-6-100-2	99	100.2	16.1	1.0085	5.859	-3.64
Nd-6-150-1	116	150.1	21.8	1.0112	5.139	-3.70
Nd-6-150-2	124	150.0	21.5	1.0113	5.141	-3.71
Nd-6-200-1	140	200.1	33.7	1.0184	4.558	-3.73
Nd-6-200-2	146	200.1	33.3	1.0189	4.560	-3.75
Nd-6-250-1	164	250.1	58.2	1.0349	4.180	-3.94
Nd-6-250-2	172	250.1	57.8	1.0363	4.182	-3.95

Water-Rock Interactions, Ore Deposits, and Environmental Geochemistry: A Tribute to David A. Crerar
© The Geochemical Society, Special Publication No. 7, 2002
Editors: Roland Hellmann and Scott A. Wood

Experimental calibration of metastable plagioclase-epidote-fluid equilibria at elevated temperatures and pressures: Applications to the chemistry of hydrothermal fluids at mid-ocean ridges

WILLIAM E. SEYFRIED[1], JR., KANG DING[1], AND BIN RAO[2]

[1]Department of Geology and Geophysics, University of Minnesota, Minneapolis, MN, 55455 USA
[2]Department of Earth Science, Nanjing University, Nanjing, P.R. China
e-mail: wes@maroon.tc.umn.edu

Abstract—Mineral solubility experiments were conducted to assess phase relations in the CaO-Na_2O-Al_2O_3-SiO_2-Fe_2O_3-FeO-CuO-Au-H_2O-H_2S-HCl system at elevated temperatures, pressures and dissolved Cl^- concentrations. An important objective of this study was to compare measured concentrations of aqueous species with those predicted from constraints imposed by fluid-mineral equilibria. Mineral phases including plagioclase and epidote solid solutions, together with anhydrite, magnetite, pyrite chalcopyrite and quartz were used to buffer activities of key components of the system. These phases were chosen because they can be directly or indirectly linked to phase relations in subseafloor reaction zones at mid-ocean ridges.

Temperature and dissolved Cl^- concentrations were observed to play important roles in the direction and magnitude of chemical exchange. Increasing temperature from 400 to 425 °C caused most species to increase in concentration, although dissolved Ca decreased. At a given temperature and pressure, an increase in dissolved Cl^- resulted in sharp increases in dissolved Ca, Fe, and Cu, while dissolved gases (H_2, H_2S) decreased as a consequence of salting-out effects. Decreasing pressure from 500 to 300 bars again resulted in a decrease in dissolved Ca, whereas concentration of other major dissolved species increased.

Experimental results are in good agreement with theoretical predictions of fluid chemistry when compositional constraints imposed by metastable plagioclase and epidote solid solutions are explicitly taken into account. Indeed, temperature-dependent changes in dissolved Ca and Na are consistent with metastable plagioclase (An_{70})-fluid equilibria at quartz-saturation, while pH (in-situ) values, are consistent with constraints imposed by metastable epidote-fluid equilibria. Metastable mineral-fluid equilibria is also indicated by dissolved metal concentrations, in particular Fe, Cu and Au. At a dissolved Cl^- concentration of 1.0 molal, pH (in-situ) values of 5.07 and 4.77 are estimated for fluids from experiments at 425 and 400 °C, 500 bars, respectively. As shown here and elsewhere, slight changes in pH (in-situ) result in relatively large changes in dissolved Fe and Cu concentrations at experimental conditions. Measured concentrations of dissolved Al are considerably in excess of theoretical predictions, although in good agreement with recent data from hot spring vents.

Whereas pH and dissolved metal concentrations are highly sensitive to constraints imposed by the composition of plagioclase and epidote solid solutions, dissolved H_2S is affected less by these constraints. Thus, experimentally calibrated vent fluid H_2S(aq) concentrations can be used to constrain the redox evolution of subseafloor reaction zones at mid-ocean ridges. Relatively low H_2S(aq) concentrations of some vent fluids either indicate low temperatures or the absence of pyrite in subseafloor reaction zones.

The generally good agreement between experimental and theoretical models depicting complex hydrothermal alteration processes provides a measure of the recent progress that has been made in theoretical geochemistry and justifies the use of these models to constrain alteration processes in subseafloor reaction zones at mid-ocean ridges.

1. INTRODUCTION

Over the past decades a substantial body of data has accumulated that indicates the effectiveness of mineral dissolution/precipitation processes in controlling the chemistry of hot spring vent fluids at mid-ocean ridges. In particular, investigations of vent fluids at EPR 21°N (Von Damm et al., 1985; Campbell et al., 1988a), EPR 9-10°N (Von Damm et al., 1997), Juan de Fuca Ridge (Von Damm, 1988; Butterfield et al., 1994; Butterfield and Massoth, 1994), MARK and TAG (Campbell et al., 1988b; Edmond et al., 1995), and at the Lucky Strike field, Mid-Atlantic Ridge (Von Damm et al., 1998), reveal dissolved concentrations of aqueous species that are largely consistent with phase equilibria control in subseafloor reaction zones at elevated temperatures and pressures. That many of these studies also involve time-series observations, which often indicate steady-state conditions of dissolved species (e.g., see Campbell et al., 1988a; Edmond et al. 1995), provides added support for subseafloor processes involving mineral-fluid equilibria at reaction zone conditions. Even vent fluids that reveal relatively recent changes in temperature, pressure and salinity owing to subseafloor mixing and phase separation effects (e.g., EPR 9-10°N, Von Damm et al., 1997; Fornari et al., 1998), reveal corresponding changes in fluid chemistry largely in keeping with mineral-fluid phase equilibria controls. For example, "F-vent" at EPR 9°N, which was sampled in 1991 and 1994, where large scale changes in temperature and dissolved Cl^- were observed, also reveals systematic and predictable changes in dissolved metals, H_2 and H_2S, suggesting rock-buffering of vent fluid chemistry. Not only are phase equilibria constraints consistent with dissolved concentrations of the major- and trace-element chemistry of the vent fluid, but also with the temporal variability of hydrogen and oxygen isotopes (Shanks et al., 1995; Berndt et al., 1996; Von Damm et al., 1997).

Experimental and theoretical investigations of fluid-mineral equilibria at elevated temperatures and pressures are obviously needed to constrain reactions that control pH, redox and metal mobility in subseafloor hydrothermal systems (Bowers et al., 1988; Berndt et al., 1989; Seyfried et al., 1991). For example, Berndt and Seyfried (1993) showed that metastable plagioclase feldspar (An60-70)-fluid equilibria at temperatures and pressures of 400 °C and 400 bars, respectively, are needed to account for the relatively high dissolved Ca concentrations and Ca-Na-SiO_2 systematics of hot spring vent fluids. Moreover, Ding and Seyfried (1992b) and Seyfried and Ding (1993, 1995) performed a series of redox- and pH-buffered experiments in the K_2O-Na_2O-Al_2O_3-SiO_2-Fe_2O_3-FeO-CuO-H_2O-H_2S-HCl system to determine thermodynamic data for Cu-chloro complexes over a wide range of conditions. These results, together with previous experimental data (Ding and Seyfried, 1992a), were used to constrain redox conditions in subseafloor reaction zones. Indeed, results of these studies suggest that pH and redox conditions are often buffered by an assemblage of minerals composed of calcic-plagioclase, epidote, quartz, magnetite, anhydrite and pyrite (Seyfried and Ding, 1995). The likely existence of anhydrite as a reactant in subseafloor reaction zones was a surprising result of these experiments, although in keeping with sulfur isotope data and dissolved SO_4^{2-} concentrations in carefully sampled hot spring fluids (Shanks et al., 1995; Von Damm, 1995 and references therein). Although it is unlikely that this complete assemblage of minerals applies to all reaction zones at mid-ocean ridges, it almost certainly does in systems where the permeability allows seawater-derived sulfate to penetrate deeply into the crust, and where sufficient rock-fluid interaction occurs to cause the elimination of pyrrhotite as a primary reactant in basalt/gabbro.

In addition to simulating hydrothermal alteration processes at mid-ocean ridges, mineral-fluid equilibria experiments, involving anhydrite, and plagioclase and epidote solid solutions, afford a means to assess the role of key variables on metal mobility and dissolved gas concentrations for an intermediate redox state, and thus supplement the large body of experimental data available for more oxidizing and reducing systems. Moreover, owing to recent advances in theoretical geochemistry, results of thermodynamic models can be directly compared with the experimental solubility measurements. In this way, the general accuracy of predictive models can be verified lending confidence to the applicability of these models as a means to assess key aspects of subseafloor hydrothermal alteration processes. Similarly, theoretical data can be used to constrain experimental measurements and point out non-equilibrium effects, such as metastable states, which may control one or more components of the fluid phase. For example, we have already established the role of plagioclase metastability (Berndt and Seyfried, 1993) on Ca^{2+}/Na^+ exchange in Cl^--bearing aqueous fluids at elevated temperatures and pressures, and it is likely that other phases behave similarly. Indeed, the common occurrence of zoned crystals of epidote from subareal geothermal systems and results from epidote-bearing experiments (Liou, 1973; Bird and Helgeson, 1980a,b) suggest this may be the case for this phase as well. Thus, in the present study, fluid chemistry coexisting with plagioclase and epidote solid solutions, quartz, magnetite, anhydrite, pyrite, and chalcopyrite was measured and monitored at 400 to 425 °C and 500 to 300 bars in 0.53 and 1.0 molal total dissolved Cl^- solutions. Additionally, the inherent compositional variability of trace elements in the natural minerals used for the experiments, together with constraints imposed by the composition of the reaction cell (Au), allowed us to assess qualitatively the effect of Cl^-, redox, pH and temperature on the mobility of key trace elements. This in turn helps to constrain further the potential mobility of these species during subseafloor hydrothermal alteration processes.

2. METHODS

2.1. Experimental and Analytical Techniques

Mineral solubility experiments were performed using flexible gold reaction cells (Seyfried et al., 1987). A major advantage of the flexible cell system is that it allows internally filtered fluid samples to be withdrawn at experimental conditions, thus mitigating quench effects for most dissolved species.

Reactants used for the experiments reported here were all natural minerals: coarse crystalline plagioclase and epidote solid solution, quartz, magnetite, anhydrite, pyrite, and chalcopyrite. All minerals were obtained from the mineral collections at the University

Table 1. Chemical composition of plagioclase and epidote solid solutions used for hydrothermal experiments. Analyses performed by ICP-MS following fusion with lithium-metaborate and dissolution in HCl (see text).

Oxide (wt. %)	Plagioclase-ss	Epidote-ss
SiO_2	51.23	38.90
TiO_2	0.00	0.00
Al_2O_3	31.43	29.14
Cr_2O_3	0.00	0.00
Fe_2O_3	0.00	0.00
FeO_t	0.41	6.67
MnO	0.00	0.00
MgO	0.17	0.00
CaO	14.29	23.84
Na_2O	3.29	0.00
K_2O	0.00	0.00
H_2O	0.05	0.80
Totals	100.87	100.0
Mol %	An = 70	Ep = 45

Table 2. Abundances in grams of mineral phases used for hydrothermal experiments performed at 0.53 and 1.0 molal total dissolved chloride. Temperatures and pressures ranged from 400-425 °C and 300-500 bars respectively.

Mineral Phases	0.53 molal Series(g)	1.0 molal Series(g)
Plagioclase (An70)	30.15	30.32
Quartz	7.01	7.11
Epidote (Ep45)	9.33	9.34
Anhydrite	8.00	8.01
Magnetite	12.08	12.35
Pyrite	15.98	16.07
Chalcopyrite	4.13	4.37

of Minnesota or Wards Natural Science Establishment. Prior to use, the minerals were acid-cleaned and ground in an agate ball mill, and then sieved through nylon mesh to recover the 74 to 149 micron size fractions. X-ray diffraction, optical microscope examination, electron microprobe analysis and bulk chemical analysis by ICP-MS were performed on the starting solids to constrain mineral compositions. The mole fraction of the anorthite and epidote components of plagioclase and epidote solid solution were determined to be approximately 0.7 and 0.45 respectively (Table 1), whereas all other minerals were nearly pure phases.

Each experiment consisted of a charge containing approximately 30 g plagioclase solid solution, 9 g of epidote solid solution, 7 g of quartz, 12 g of magnetite, 8 g of anhydrite, 16 g of pyrite, and 4 g of chalcopyrite (Table 2). These minerals were placed in a 150-mL Au reaction cell with approximately 100 g of Na-Ca-Cl fluid. The relative abundance of plagioclase, in particular, was used to enhance the rate of approach to plagioclase-quartz-fluid equilibria (see below).

All experiments were designed in accordance with the Gibbs phase rule. As a consequence of this, mineral phases, components and imposed constraints were balanced so as to buffer completely the distribution of aqueous species, assuming, of course, full thermodynamic equilibrium. For example, in their study of plagioclase-quartz-fluid equilibria, Berndt and Scyfried (1993) showed that at a given temperature, pressure, dissolved Cl^- and plagioclase composition, the chemical system was invariant. Moreover,

Table 3. Chemical composition of fluid samples taken during fluid-mineral equilibria experiments at temperatures of 400° and 425 °C, 300-500 bars and 0.53 molal total dissolved Cl. The designator, na, indicates species not analyzed in a particular fluid sample owing to sample size constraints.

Temp		400	400	400	400	425	425
Pressure		500	500	400	300	500	400
Reaction Time (hrs.)	Start	720	2880	336	350	1200	1000
$H_{2(aq)}$(mm)	0	na	0.18	0.25	0.37	0.53	0.67
$H_2S_{(aq)}$(mm)	0	na	8.9	10	12	15	21
SO_4^{2-}(mm)	0	<0.30	0.13	na	0.20	0.41	0.19
Cl(mm)	532	557	547	559	548	579	555
SiO_2(mm)	0	23	22	20	17	24	19
Na(mm)	472	493	480	496	489	503	531
Ca(mm)	30	29	28	27	25	24	20
Fe(mm)	0	3.9	4.1	4.2	5.6	7.3	9.1
Cu(μm)	0	75	na	77	77	100	na
Au(μm)	0	0.3	na	0.2	na	na	na
Ni(μm)	0	1.7	na	1.2	0.8	na	na
Pb(μm)	0	8.0	na	8.3	na	na	na
Al(μm)	0	94.1	16	13	13	na	6

Table 4. Chemical composition of fluid samples taken during fluid-mineral equilibria experiments at temperatures of 400° and 425 °C, 500 bars and 1.0 molal total dissolved Cl. As noted, the designator, na, indicates species not analyzed in a particular fluid sample owing to sample size constraints.

Temp		**400**	**400**	**425**	**425**
Pressure		**500**	**500**	**500**	**500**
Reaction Time (hrs.)		**300**	**2160**	**336**	**2000**
$H_{2(aq)}$(mm)	0	na	0.17	0.42	0.49
$H_2S_{(aq)}$(mm)	0	na	7.4	14	15
SO_4^{2-}(mm)	0	na	0.41	0.42	0.43
Cl(mm)	1021	1030	1050	996	1075
SiO_2(mm)	0	21.0	21	23	22.0
Na(mm)	809	863	850	821	855
Ca(mm)	102	82	80	77	78
Fe(mm)	0	9.85	11	25	26
Cu(μm)	0	98.9	116	185	190
Au(μm)	0	0.36	0.39	0.24	0.26
Ni(μm)	0	na	12	6	14
Pb(μm)	0	5.8	6.2	4.9	6.3
Al(μm)	0	19	20	29	15

their study showed that an ideal solid solution model for plagioclase accounted best for the chemistry of fluids coexisting with An_{60} and An_{82} plagioclase at 400 °C, 400 bars. For Fe- and S-bearing systems, however, as is the case here, redox constraints must be imposed to achieve this same state of invariance. In effect, for a given plagioclase composition, total dissolved Cl^- (charge-balance constraint), temperature and pressure, inclusion of four other mineral phases (e.g., magnetite, pyrite, anhydrite, chalcopyrite) is needed to buffer completely the ten-component system, (Na_2O-CaO-SiO_2-Al_2O_3-FeO-Fe_2O_3-CuO-HCl-H_2O-H_2S), including the composition of epidote solid solution. That the present series of experiments involves epidote solid solution with a specific composition (Table 1) necessarily creates a metastable condition unless this phase recrystallizes completely to a composition consistent with imposed redox constraints (see Bird and Helgeson, 1980a,b). As noted earlier, it is likely that epidote solid solution, like plagioclase, may not recrystallize sufficiently on experimental time scales to achieve full thermodynamic equilibrium. Whether or not epidote composition achieves equilibrium can be established unambiguously, however, from measured fluid chemistry and by comparison of the composition of epidote solid solution before and after the experiments.

Two series of experiments were performed. The first series was conducted at temperatures of 400 to 425 °C, pressures of 300 to 500 bars and 0.53 molal total dissolved Cl^-, while the second was at 1.0 m Cl^- and temperatures and pressures of 400-425 °C and 500 bars, respectively. Reaction progress was evaluated by time-series measurements, which tended to show that steady-state conditions were achieved (Table 3, 4). Steady state and equilibrium, of course, can be quite different. Here, however, we made use of a sampling strategy established previously by Berndt and Seyfried (1993) for their plagioclase-quartz-fluid experiments at conditions virtually identical to those of the present study. In particular, Berndt and Seyfried (1993) showed that metastable plagioclase-fluid equilibria could be effectively established after relatively short reaction periods when the direction of approach to equilibrium was in the Ca-consuming direction, a finding in keeping with much earlier results by Orville (1972). Thus, all of the experiments performed as part of the present investigation were conducted with a starting fluid having "excess" Ca, reactants with "excess" plagioclase, and with reaction times more than necessary to achieve equilibrium as established by Berndt and Seyfried (1993). Although not directly reversing plagioclase exchange reactions created some ambiguity with respect to establishing an equilibrium condition, this was offset by the need to conserve fluid sample to complete the relatively large number of analyses of aqueous species needed to assess the ten-component system under investigation. Dissolved gas measurements, together with determination of trace metal species, such as Cu, Pb and Au, which required a separate procedure for fluid sampling and analysis, placed severe constraints on the inventory of fluid sample from any given experiment.

Fluid samples were analyzed for Na, Si, Ca, Fe, Cu, Al, Pb, Ni, and Au by ICP-MS, while Cl^-, SO_4^{2-} and H_2S (the latter converted to SO_4^{2-} following oxidation by treatment with heated H_2O_2) were analyzed by ion chromatography. Dissolved H_2 was determined by gas chromatography.

Although the source of many of the major and minor elements in solution can clearly be linked to primary components of the starting solids or aqueous fluid, many of the trace elements in solution are derived from dissolution of trace elements in these same solids. The exception, of course, is gold, which was derived from the gold reaction cell. Even though the concentrations of trace elements in the experimental system are different from those in phases in subseafloor settings, the observed changes in the dissolved concentrations of these species (e.g., Pb, Ni, Au) as a function of temperature, pressure and dissolved Cl^- during the experiments can still provide valuable insight on the mobility of these species during subseafloor hydrothermal alteration processes.

The accuracy of reported concentrations of major dissolved cations and anions (mmolal concentrations) is approximately 2%, and 10-20% for trace element species (μmolal concentrations), at the 95% confidence level. The upper range of uncertainty for trace elements largely results from the nature of the sampling process used for these species, which often required dissolution of a metal-bearing precipitate produced during the sampling process (see Seewald and Seyfried, 1990; Seyfried and Ding, 1993). The accuracy of reported concentrations of dissolved gases (H_2 and H_2S) is approximately 10% at the 95% confidence level. Analytical uncertainties given for all aqueous species are based on processing of replicate samples from the experiments and from results of previous research in our lab involving mineral-fluid equilibria studies at a similar range of conditions.

The pH was not measured on fluid samples from the experiments owing to the effects of sulfide mineral precipitation during sample processing, which renders pH values highly uncertain. The pH, of course, could be calculated for fluids from the experiments by making use of phase equilibria constraints as discussed by Ding and Seyfried (1992a). Dissolved metals and $H_2S(aq)$ are sampled separately into acidified solutions in gas-tight syringes to preclude the effects of mineralization on the measured inventory of these species.

2.2. Theoretical Constraints

Mineral solubility calculations were performed at the same temperatures, pressures and total dissolved Cl^- concentrations as used for the experiments to better assess hydrothermal alteration processes. Not only do these calculations constrain experimental results, but also the experiments in turn provide insight into the relative accuracy of theoretical models of fluid-mineral equilibria at elevated temperatures pressures and dissolved Cl^- concentrations.

The standard state adopted in this study for all mineral phases and liquid water is unit activity of the pure phase at the temperature and pressure of interest. For aqueous ions and complexes, the standard state is unit activity of the species in a hypothetical one-molal solution, referenced to infinite dilution in H_2O at the pressure and temperature of the experiment. Accordingly, the activity of H_2O and activity coefficients for solute species approach unity at infinite dilution at any pressure and temperature of interest. Individual molal activity coefficients for charged species (γ_i) were calculated using the B-dot equation of Helgeson (1969).

For neutral species, activity coefficients were calculated from the Setchénow equation:

$$\log \gamma_n = b_{\gamma,n} I \qquad (1)$$

where $b_{\gamma,n}$ is a temperature- and pressure-dependent coefficient characteristic of the nth neutral species and the electrolyte composition (Chen and Marshall, 1982), and I, the true ionic strength, which takes explicit account of the distribution of aqueous species. Ding and Seyfried (1990), for example, determined Setchénow coefficients for $H_{2(aq)}$ and $H_2S_{(aq)}$ in NaCl fluids at temperatures and pressures from 300 to 450 °C and 300 to 500 bars respectively by determining the dissolved concentration of H_2 and H_2S as a function of NaCl concentration in fluids coexisting with pyrite-pyrrhotite-magnetite. Data show that Setchénow coefficients for H_2 and H_2S increase dramatically with decreasing pressure and increasing temperature. Thus, for the range of conditions applicable to the present study, γ_{H_2} and γ_{H_2S} of between 1.2 and 1.5 are indicated, the higher values corresponding to higher dissolved NaCl and temperature, and lower pressure. The existence of species other than NaCl in the fluid has not been accounted for, however, and it is likely that dissolved salts, such as $CaCl_2$ and $FeCl_2$ join NaCl in contributing to the activity/concentration relations for neutral species. This certainly is a concern, but an even greater concern involves the virtual absence of any data for activity coefficients for HCl°, NaCl° and NaOH° at temperatures and pressures in excess of 300 °C and steam saturation, respectively. Thus, for these and similar species, activity coefficients were assigned a constant value of 1.2, in keeping with previous conventions (Ding and Seyfried, 1992a and references therein).

In addition to activity coefficient data for aqueous species, assessment of fluid-mineral equilibria at elevated temperatures and pressures requires detailed knowledge of hydrolysis constants for minerals, activity-composition relations for minerals, where solid solution effects are important, together with dissociation constant data for all aqueous complexes. These data coupled with appropriate mass-balance, mass-action and charge-balance constraints permit determination of the distribution of aqueous species, which then can be compared with results of experimental studies (Ding and Seyfried, 1992a,b; Seyfried and Ding, 1995). In this study, thermodynamic data used for minerals, ions, and gases are from SUPCRT92 (Johnson et al., 1992), whereas data for most aqueous complexes, which are consistent with the SUPCRT92 database, are from Sverjensky et al. (1991, 1997). These data are based on the revised HKF (Helgeson et al., 1981) equations of state for aqueous species (Tanger and Helgeson, 1988; Shock and Helgeson, 1988; 1990; Shock et al., 1989;

Shock et al., 1997) and correlations among equation of state parameters and standard partial molal properties at 25 °C and 1 bar. In addition to these data, however, experimental mineral solubility-data from Ding and Seyfried (1992a,b) were used to obtain stability constants for $FeCl_2°_{(aq)}$ and $CuCl°_{(aq)}$. In the past (see Saccocia and Seyfried, 1990; Seyfried et al., 1991), a Newton-Raphson iterative technique was used to solve simultaneously the full range of mass-action, mass-balance and charge-balance relations needed to assess mineral-fluid equilibria. In this particular case, however, we made use of the EQ3/EQ6 software package (Wolery and Daveler, 1992), which was modified by us to allow calculations to be performed at the same range of conditions as for the experiments. Solid solutions in EQ3/EQ6 are treated with ideal mixing models. Thus, an ideal molecular mixing model was used to define the activities of components of plagioclase solid solution, in keeping with results of Berndt and Seyfried (1993), while an ideal site mixing model was used to express activity-mole fraction relations for epidote and clinozoisite components of epidote solid solution, in keeping with data and interpretations of Bird and Helgeson (1980a,b). This mixing model accounts for the temperature dependence of substitutional order-disorder of Al^{+3} and Fe^{+3} in the M1 and M3 octahedral sites in epidote solid solutions.

3. RESULTS

3.1. Fluid Chemistry

3.1.1. Major elements (Ca, Na, Fe, $SiO_2(aq)$)

Dissolved Ca concentrations decreased from starting conditions to those in fluids at elevated temperatures and pressures for both 0.53 and 1.0 molal Cl^- experiments (Table 3 and 4). Indeed, Ca removal from solution is largely balanced by gains in dissolved Na and Fe. In general, Ca for Na and Fe exchange tend to increase with decreasing pressure and increasing temperature (Table 3 and 4). Moreover, data show that for the redox condition maintained during the experiment, which is intermediate between PPM (pyrite-pyrrhotite-magnetite) and HMP (hematite-magnetite-pyrite), changes in dissolved Cl^- induce large-scale changes in dissolved Fe (Fig. 1), with lesser changes in dissolved Ca and Na. As will be discussed at greater length below, these effects are entirely consistent with formation of aqueous Fe-bearing chloro-complexes at the range of pH and redox conditions fixed by mineral-fluid equilibria at the temperatures and pressures of the experiments (Seyfried and Ding, 1993).

Owing to the presence of quartz, the dissolved concentration of SiO_2 tended to increase with increasing temperature and to decrease with decreasing pressure, consistent with quartz-fluid equilibria. For example, at 400 °C, dissolved SiO_2 decreased systematically from 23 to 16.5 mmolal with decreasing pressure from 500 to 300 bars respectively (Table 3). A similar trend can be seen for the 425 °C series of experiments (0.53 m Cl^-), but here the pressure drop did not extend below 400 bars.

3.1.2. Dissolved gases (H_2 and H_2S) and SO_4^{2-}

Owing to the well-known effect of temperature, pressure and dissolved Cl^- on mineral solubility and activity-concentration relations in the FeO-Fe_2O_3-H_2S-H_2O-HCl system (Ding and Seyfried, 1989; Ding and Seyfried, 1992a,b; Seyfried and Ding, 1995), it is not surprising that dissolved H_2 and H_2S changed significantly during the experiments. Indeed, the dissolved concentration of both gases increased with increasing temperature and decreasing pressures (Tables 3 and 4). At the elevated temperatures and pressures of the experiments, however, salting-out effects are particularly robust (Ding and Seyfried, 1989). It is not surprising, therefore, that the concentrations of H_2 and H_2S tended to decrease with increasing dissolved Cl^- (Fig. 1). Dissolved SO_4^{2-}, in contrast, failed to reveal a clear trend with temperature and pressure, but appeared to increase with increasing dissolved Na (Table 3 and 4), which may reflect formation of $NaSO_4^-{}_{(aq)}$, but this needs to be established better with additional experiments or modeling calculations.

3.1.3. Minor elements (Cu, Au, Ni, Pb and Al)

Dissolved Cu concentrations reveal large-scale changes during the experiments (Table 3 and 4, Fig. 1). As with most minerals, the solubility of chalcopyrite (primary reactant) is greatly affected by temperature, pressure, redox, Cl^- and pH (Crerar and Barnes, 1976; Hemley et al., 1992; Ding and Seyfried, 1992b; Seyfried and Ding, 1993). At 1.0 molal total dissolved Cl^-, for example, a temperature increase from 400 to 425 °C caused Cu concentration to increase significantly. The effect of pressure on dissolved Cu concentrations, however, was relatively modest at the temperatures investigated. It is not entirely clear whether this is real or apparent, being caused entirely or in part by analytical uncertainties related to Cu measurements at relatively low levels where these effects can be relatively large (Seewald and Seyfried, 1990; Seyfried and Ding, 1995).

As noted previously, we also measured the dissolved concentrations of Au, Ni, Pb and Al during the experiments, although sample-size limitations and

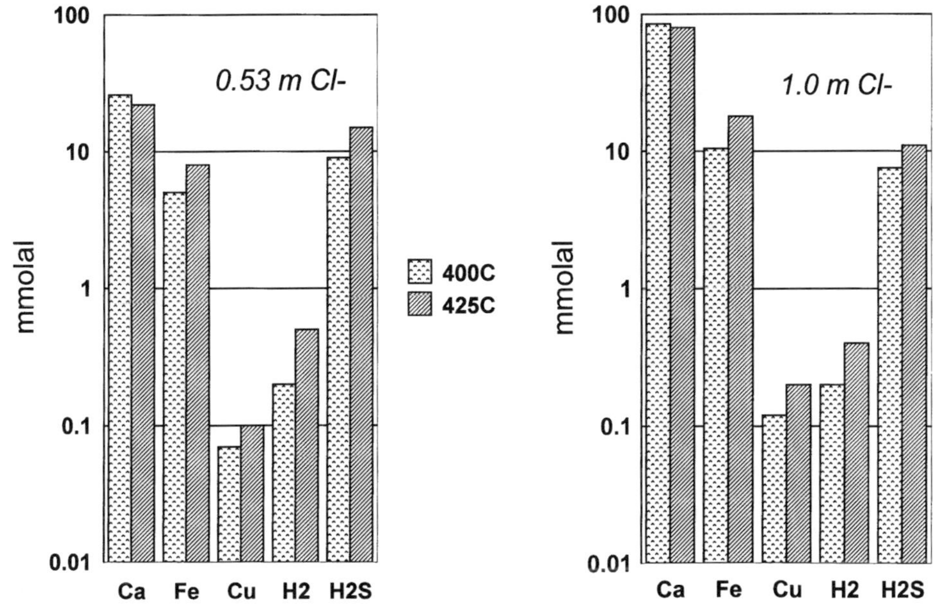

Figure 1. Experimentally determined dissolved Ca, Fe, Cu, H_2, and H_2S in 0.53 m Cl^- (A) and 1.0 m Cl^- (B) fluid coexisting with plagioclase-epidote-quartz-magnetite-anhydrite-pyrite-chalcopyrite assemblage at 400 and 425 °C, 500 bars.

sampling procedures often precluded analysis of all fluid samples for each of these species. Owing to the trace quantities of the source phases (see above), however, we were not able to determine unambiguously mineralogical controls, although results of SEM and EDX analyses of bulk alteration products revealed Ni and Pb associated with both sulfide and oxide. Data which are available for Ni reveal a slight negative pressure effect, something like Cu. Nickel also tended to increase with increasing Cl^- (Table 3 and 4). Lead, in contrast, increased slightly with decreasing pressure (500 to 400 bars only), but failed to record any other unambiguous changes with temperature or dissolved Cl (Table 3, 4), in spite of the existence of well-known chloro-complexes of Pb, which are very much temperature-dependent (Hemley et al., 1992). Possibly, the very small amounts of Pb in the starting solids precluded sufficient mass transfer between solid and aqueous phases to confirm these aqueous speciation effects. In contrast, Au decreased with temperature increase from 400 to 425 °C (1 molal Cl^-), although, again, insufficient data are available to assess unambiguously the role of other key variables, a modest pressure effect notwithstanding.

The dissolved concentration of Al generally falls in the range of approximately 10-20 μmolal, and like the other trace elements (Pb, Au, Ni), it is difficult to demonstrate unambiguous correlations with temperature, pressure and dissolved Cl^-. The dissolved Al concentration of the first sample from the experiment performed at 400 °C, 500 bars and 0.53 m Cl^-, however, is unusually high (94 μm), which may be caused by sample contamination or early-stage non-stoichiometric dissolution effects. That the Al concentration of the subsequent sample from this experiment decreased to a value more typical of other fluid samples from other experiments leads us to question the earlier result. In any event, it is likely that the average dissolved concentration of Al (10-20 μm) reflects control by the bulk assemblage of aluminosilicate phases used as reactants for the experiments.

3.2. Mineral Chemistry

The solid run products from all experiments were examined optically and petrographically, as well as by X-ray diffraction and electron microprobe to evaluate changes in the composition of the starting mineral assemblage. In general, all of the starting minerals were observed at the end of all of the experiments, although there were changes in grain size and crystallinity suggesting mineral dissolution and reprecipitation and crystal growth. In addition, micro-

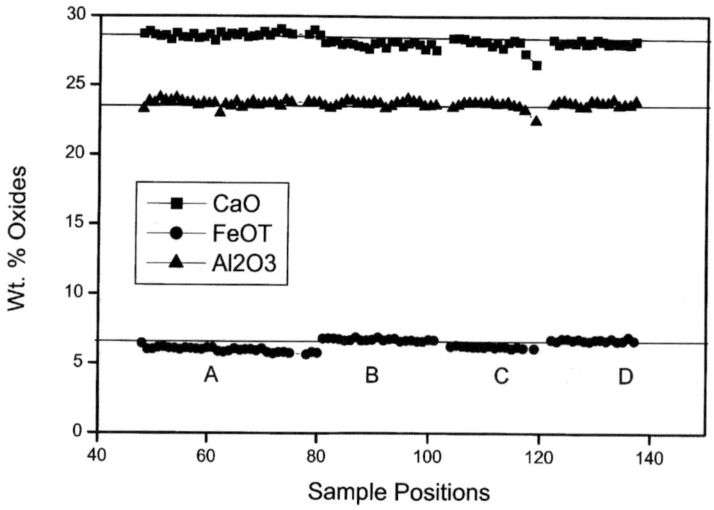

Figure 2. Electron microprobe analyses (CaO, Al_2O_3, FeO) of four (A through D) epidote grains from experiment at 425 °C, 500 bars and 1.0 m Cl^-. Data reveal compositions virtually identical to starting values (solid lines) determined by wet chemical techniques (see text and Table 1). Sample positions refer to transect locations across mineral grains. These data are representative of all samples from all experiments, and indicate existence of metastable epidote-fluid equilibria.

probe analysis revealed very small amounts of Pb and Ni components in sulfide and oxide minerals (as noted above). There can be no question that Pb and Ni in starting oxide, silicate and sulfide phases served as the source of the heterogeneous accumulations of Pb and Ni in alteration phases. Gold crystals were observed in association with the reacted mineral assemblage, which undoubtedly reflects dissolution/re-precipitation of gold from the gold reaction cell, as discussed above.

As expected, aluminosilicate phases failed to show chemical or mineralogical modification. For example, electron microprobe data for epidote and plagioclase solid solutions from the experiment performed at 425 °C, 500 bars and 1.0 molal dissolved Cl^-, reveal little if any differences between starting and reacted compositions, the relatively extreme conditions of the experiment notwithstanding (Table 1, Figs. 2 and 3). Similar analysis of plagioclase and epidote solid solutions from other experiments performed during this study also revealed compositions that do not depart significantly from starting values. That plagioclase solid solution remained An_{70} and indicated virtually no evidence of recyrstallization into a thermodynamically more stable phase is not surprising in light of our previous results with plagioclase solid solutions (Berndt and Seyfried, 1993). That the epidote solid solution also remained unchanged (Ep_{45}) was for us a new finding, but it too is largely in keeping with the well known sluggishness of this phase to equilibrate with aqueous fluids, especially at low to moderate pressures (Liou, 1973). Thus, to accurately assess fluid-mineral equilibria during the experiments, it is essential that constraints imposed by metastable plagioclase and epidote are explicitly taken into account.

4. DISCUSSION

The present study shows that relatively slight changes in temperature, pressure and dissolved Cl^- results in substantial changes in the dissolved concentration of Fe, Cu and other metals in fluids coexisting with a plagioclase-epidote-bearing mineral assemblage. Since plagioclase and epidote solid solutions are abundant phases in natural hydrothermal systems, the results of the experiments can help to constrain pH and redox buffering in these systems. The pH and redox relations of fluids coexisting with these phases, however, depend critically on the relative roles of equilibrium and metastability, which can be evaluated only by assessing mineral solubility simultaneously from both experimental and theoretical perspectives.

In general, results from the present experiments are consistent with data and interpretations from Berndt and Seyfried (1993), in that plagioclase solid solution behaves ideally during fluid-mineral equilibria and

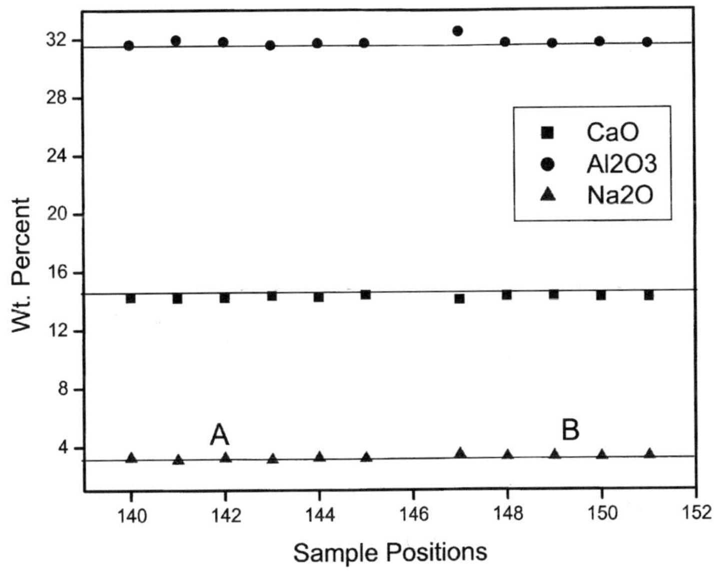

Figure 3. Electron microprobe analyses (CaO, Al_2O_3, Na_2O) of two plagioclase grains (A and B) from a phase equilibria experiment at 425 °C, 500 bars and 1.0 m Cl^-. As with epidote composition, these data reveal compositions (X_{an} = 0.70) identical to starting values (solid lines) determined by wet chemical techniques (see text and Table 1). Sample positions refer to transect locations across mineral grains. The data are representative of plagioclase grains from all experiments, and indicate plagioclase metastability during the experiments.

does not recrystallize into a thermodynamically more stable composition, at least on the time scale of the experiments. Thus, the following reaction can be used to depict metastable plagioclase-fluid equilibria:

Anorthite *Quartz* *Albite*

$$CaAl_2Si_2O_8 + 2Na^+ + 4\ SiO_2 \leftrightarrow 2NaAlSi_3O_8 + Ca^{2+} \quad (2)$$

where the activities of the anorthite and albite components of plagioclase solid solution can be replaced by mole fractions. Even at temperatures and pressures as high as 425 °C and 500 bars, and 1 molal total dissolved Cl^-, dissolved Ca and Na concentrations calculated from equation 2 and appropriate mass-action and mass-balance reactions agree well with experimental data, assuming ideal behavior of anorthite and albite components of An_{70}-plagioclase (Fig. 4a-b). The relative agreement between experimental and theoretical data shown here, however, is, in fact, better than observed by Berndt and Seyfried (1993) for their experiments, which were performed at 400 bars. At 400-425 °C, the higher pressure used for the present experiments (500 bars) minimizes uncertainties in the thermodynamic data for aqueous species estimated using correlation algorithms of Shock and Helgeson (1988), Shock et al. (1989, 1997), and Sverjensky et al. (1997), where pressure effects can be significant, especially at temperatures in the vicinity of the critical temperature of the fluid.

As previously emphasized, the full assemblage of minerals used for the present series of experiments can be shown from theoretical data to buffer completely the dissolved chemistry of the coexisting fluid phase and epidote composition when temperature, pressure, plagioclase composition and total dissolved Cl are fixed. For example, the mole fraction of epidote solid solution predicted, while assuming the ideal site mixing model of Bird and Helgeson (1980a), which explicitly accounts for the temperature dependence of substitutional order-disorder on the octahedral sites, is 0.82 and 0.77 at 425 and 400 °C, respectively. These values, of course, are very different from that of the composition of epidote used for the experiments, which as emphasized, did not reveal evidence of Fe enrichment consistent with theoretical predictions. Thus, for fixed activity of anorthite and epidote components of plagioclase and epidote solid solution respectively, and unit activity for coexisting minerals and water, it can be shown that $a_{Ca^{2+}}/(a_{H^+})^2$ is fixed at any temperature, pressure and dissolved Cl, as follows:

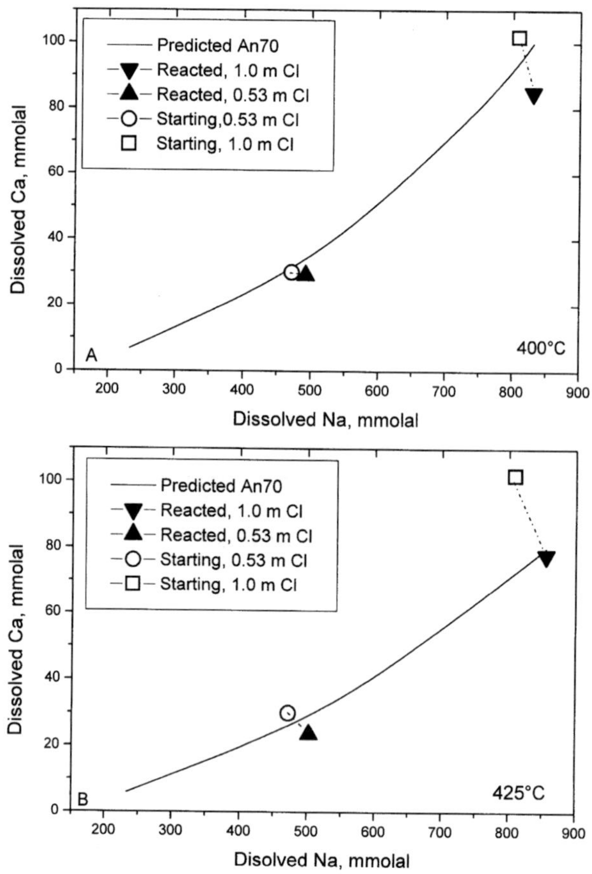

Figure 4. Dissolved Ca concentrations predicted (solid line) and measured (symbols) in fluids coexisting with plagioclase (X_{an} = 0.70)-epidote (X_{ep} = 0.45)-quartz-magnetite-anhydrite-pyrite-chalcopyrite assemblage at 400 (A) and 425 °C (B) 500 bars. Open symbols refer to starting fluid composition, while solid symbols indicate steady-state concentrations. In general, the good agreement between measured and predicted Ca concentrations indicate the important role of metastable mineral-fluid equilibria in controlling fluid chemistry.

$$\underset{\textit{Epidote}}{1.05\ Ca_2Al_2FeSi_3O_{12}(OH)} + 2.0\ H^+ + \underset{\textit{Pyrite}}{0.025\ FeS_2} \leftrightarrow \underset{\textit{Anorthite}}{1.05\ CaAl_2Si_2O_8} + 1.52\ H_2O + Ca^{2+} + \underset{\textit{Quartz}}{1.05\ SiO_2} + \underset{\textit{Magnetite}}{0.3573\ Fe_3O_4} + \underset{\textit{Anhydrite}}{0.05\ CaSO_4} \quad (3)$$

In effect, reaction (3), together with imposed constraints (in particular, plagioclase composition, ΣCl^-, and the charge balance equation), buffers fluid pH. Changing the epidote or plagioclase composition, or total dissolved Cl^-, temperature or pressure, will result in a correspondingly different pH value. For example, increases in the anorthite component of plagioclase solid solution (see Berndt et al. 1989; Berndt and Seyfried, 1993), and dissolved Cl^-, decrease pH, while an increase in the epidote component of epidote solid solution (see Bird and Helgeson, 1980a) increases pH. Results of previous experiments (Ding and Seyfried, 1992a) have shown that for a specific redox condition, dissolved Fe concentration is an exceedingly sensitive indicator of pH, and thus can be used here to test the activity of epidote coexisting with plagioclase solid solution at the conditions of the experiments.

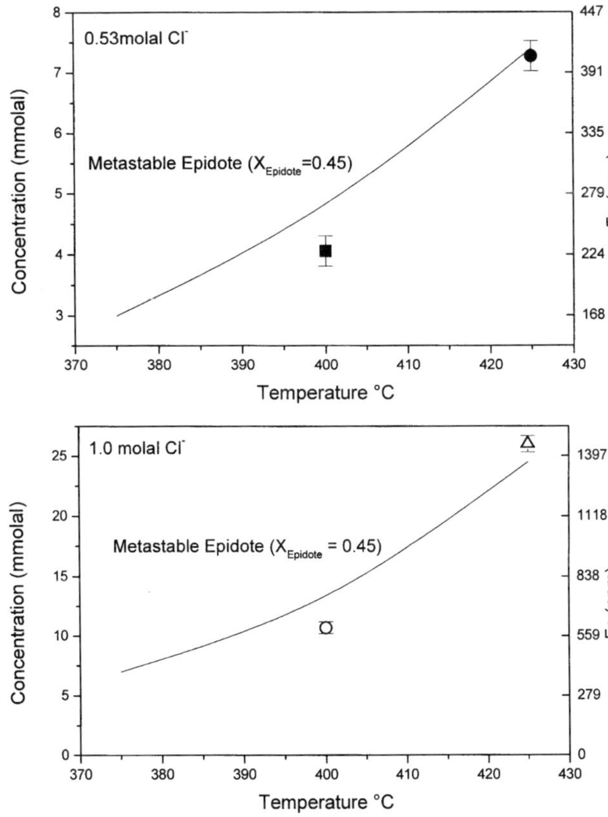

Figure 5. Dissolved Fe concentrations predicted (solid line) and measured (symbols) in 0.53 m total Cl^- (A) and 1.0 m Cl^- (B) coexisting with plagioclase (X_{an} = 0.70)-epidote (X_{ep} = 0.45)-quartz-magnetite-anhydrite-pyrite-chalcopyrite assemblage at 400 and 425 °C, 500 bars. Symbols refer to steady state concentrations in solution. Data reveal generally good agreement with constraints imposed by metastable epidote (X_{ep} = 0.45), rather than an epidote composition (X_{ep} = 0.77 to 0.82) formed in full equilibrium with the assemblage of minerals. As developed in the text, relatively low X_{ep} values give rise to relatively low pH values and correspondingly higher metal concentrations (see Fig. 6).

Measured and predicted concentrations of dissolved Fe are depicted in Figure 5, assuming epidote composition equivalent to that of the starting phase as a function of temperature and dissolved Cl^- concentration. It is immediately clear that the measured dissolved Fe, and therefore fluid pH, is entirely consistent with the metastable epidote composition, for both 0.53 m Cl (A) and 1.0 m Cl^- (B) conditions. At 425 °C, 500 bars and 1.0 molal total dissolved Cl^-, the pH values predicted for metastable epidote (X_{ep} = 0.45) and predicted (X_{ep} = 0.82) compositions are 5.07 and 5.20, respectively (Fig. 6). Although the difference in pH values may appear slight, the corresponding differences in dissolved Fe concentrations are large (Fig. 6). Indeed, a difference in pH (in-situ) of approximately 0.13 results in a 2x variation in dissolved Fe. At temperatures less than 425 °C, model results show that the offset in pH caused by changes in epidote composition (stable vs metastable) decreases largely because the composition of the epidote solid solution predicted to form is less Fe-rich and more like that of the starting metastable phase.

Another test for the pH estimated from fluid-mineral equilibria comes from other dissolved metal concentrations measured during the experiments. For example, Figure 7 shows dissolved Cu concentrations measured and predicted assuming metastable phase equilibria (X_{ep} and X_{an} = 0.45 and 0.70 respectively), together with constraints imposed by chalcopyrite-fluid equilibria at temperatures from 400° to 425 °C, 500 bars, and dissolved Cl^- from 0.53 to 1.0 molal. In general, measured and predicted Cu concentrations are in good agreement, although the measured values exceed those predicted at 425

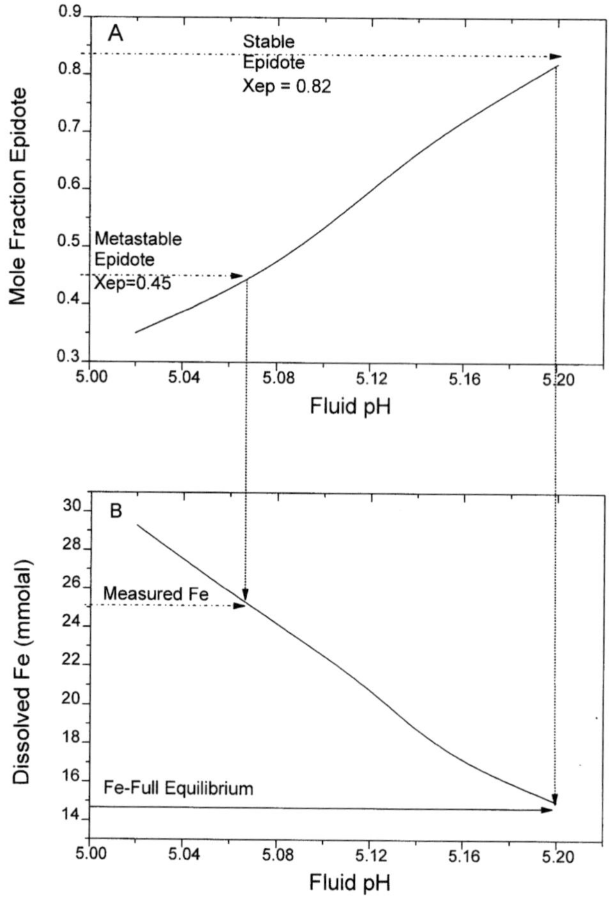

Figure 6. The role of epidote composition on dissolved Fe concentrations and pH (in-situ) of Cl^--bearing aqueous fluid (1.0 m Cl^-) coexisting with plagioclase (X_{an} = 0.70)-epidote-quartz-magnetite-anhydrite-pyrite-chalcopyrite assemblage at 425 °C, 500 bars. The composition of epidote predicted to exist in full equilibrium at experimental conditions departs significantly from that used for the experiment and measured at the termination of the experiment (metastable epidote, X_{ep} = 0.45) (A). The thermodynamically stable epidote composition would give rise to considerably less dissolved Fe (predicted, Fe-Full Equilibrium) than that actually measured (B). In effect, changes in epidote composition result in differences in pH (in-situ), which are responsible for changes in dissolved Fe concentrations.

°C at both dissolved Cl^- conditions. This is likely caused either by uncertainties in the stability constant for $CuCl°_{(aq)}$, which was determined from mineral solubility experiments at conditions very similar to those of the present experiments (Seyfried and Ding, 1993; 1995), or by the existence of another aqueous species of Cu, for which data are presently lacking. It is important to consider, however, that at 425 °C (1.0 molal Cl^-), measured Fe concentration also exceed theoretical predictions, suggesting that at least for this relatively high-temperature-high-Cl^- condition, fluid pH may actually be slightly lower than indicated by metastable mineral-fluid equilibria constraints. A lower pH could be caused by any of a number of factors, but uncertainties in the solid solution models assumed for plagioclase and epidote solid solutions seem most likely. This, coupled with the inherent sensitivity of Cu- and Fe-bearing aqueous species to pH at 400 and 425 °C and 0.53 and 1.0 molal Cl^-, could account for the observed variability. Assuming all of these uncertainties act together to affect pH, a pH variation of as little as 0.02 at 425°C could resolve the apparent offset between predicted and observed metal concentrations. That such a slight adjustment can account for the dissolved concentrations of aqueous species coexisting with such a

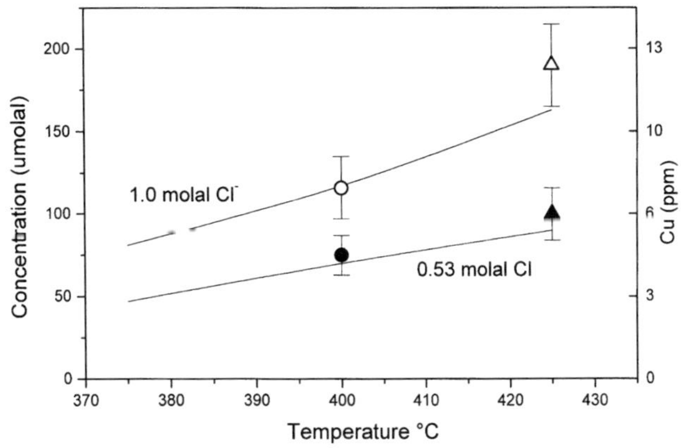

Figure 7. Dissolved Cu concentrations predicted (solid line) and measured (symbols) in fluids (1.0 and 0.53 m total Cl^-) coexisting with plagioclase (X_{an} = 0.70)-epidote (X_{ep} = 0.45)-quartz-magnetite-anhydrite-pyrite-chalcopyrite assemblage at 400 and 425 °C, 500 bars. The slight offset between predicted and measured dissolved Cu concentrations at 425 °C may be due to uncertainties in the distribution of aqueous Cu bearing species or, more likely, slight uncertainties in fluid pH (see text).

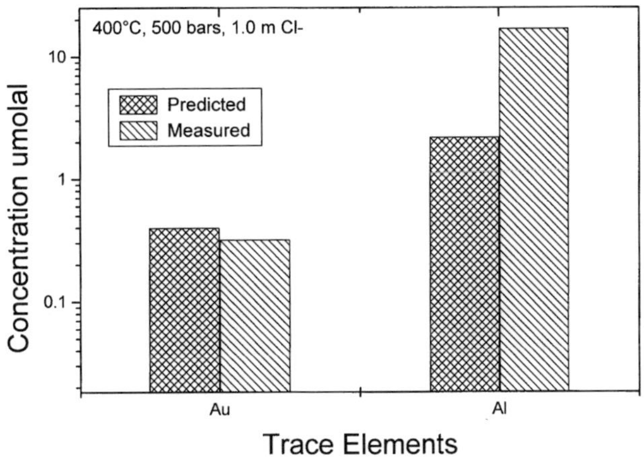

Figure 8. Predicted and measured dissolved concentrations of Au and Al in fluid (umolal) coexisting with plagioclase (X_{an} = 0.70)-epidote (X_{ep} = 0.45)-quartz-magnetite-anhydrite-pyrite-chalcopyrite assemblage at 400 °C, 500 bars and 1.0 molal total dissolved Cl^-. The relative agreement between predicted and measured concentrations for Au lends confidence to pH (in-situ) estimated from constraints imposed by the existence of metastable plagioclase and epidote. The order of magnitude difference between measured and predicted Al concentrations is likely due to uncertainties in Al-speciation at elevated temperatures, pressures and dissolved Cl concentrations.

complex assemblage of minerals at temperatures and dissolved Cl^- concentrations as high as 425 °C and approximately 1.0 molal respectively, is significant indeed, and suggests that, for this chemical system, experimental and theoretical models can be used with a high level of accuracy to constrain a wide range of subseafloor hydrothermal alteration processes. Previously, Seyfried and Ding (1993) used results of experiments in the K_2O-Na_2O-Al_2O_3-SiO_2-Fe_2O_3-FeO-CuO-H_2O-HCl-H_2S system to show that pH values near neutrality at appropriate temperatures and pressure were applicable to hydrothermal reaction zones at mid-ocean ridges. These results are in good agreement with the present

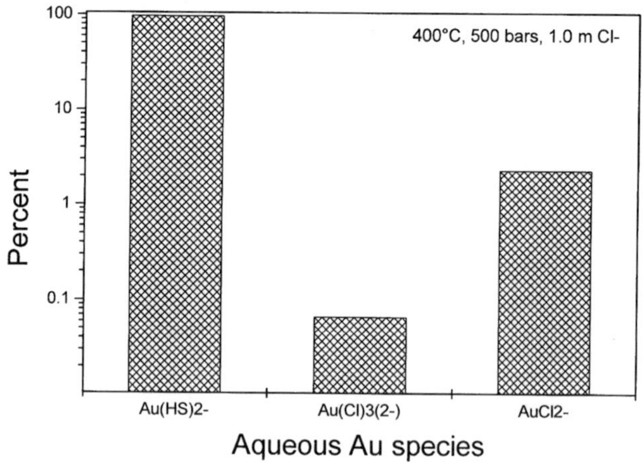

Figure 9. Relative distribution of Au-bearing aqueous species in Cl^- (1.0 molal) and H_2S bearing fluid coexisting with plagioclase (X_{an} = 0.70)-epidote (X_{ep} = 0.45)-quartz-magnetite-anhydrite-pyrite-chalcopyrite assemblage at 400 °C, 500 bar and 1.0 m Cl^-. Relatively high Cl^- concentrations notwithstanding, Au solubility is dominated by bisulfide species. Thus, changes in dissolved Cl^- concentration in hot spring vent systems are unlikely to affect greatly dissolved Au, although $H_2S_{(aq)}$ effects could be significant, assuming the presence of Au-bearing phases. Thermodynamic data used for the calculations are from SUPCRT92 (Johnson et al., 1992) and Sverjensky et al. (1997)

study using phase relations more applicable to basaltic systems when explicit account is taken of the effect of mineral metastability on the pH of coexisting aqueous fluids.

Taking this same approach one step further, we can evaluate the general agreement between theoretical and experimental data for Au, Al and Pb at a particular temperature, pressure and dissolved Cl^- concentration, which may also help to constrain in-situ pH values, again assuming metastable fluid-mineral equilibria. For example, we have already established that gold was among the alteration products observed at the end of the experiments, and certainly existed throughout the experiment, since the experimental reactants were held within the gold reaction cell. Thus, it is likely that pH-dependent gold-fluid equilibria controlled dissolved Au during the experiments. Similarly, Al activity was likely fixed by alumino-silicate phases (plagioclase/epidote solid solutions), and thus, assuming knowledge of the distribution of dissolved Al species at experimental conditions, Al concentration can also be used to constrain pH.

At a temperature and dissolved Cl^- concentration of 400 °C and 1.0 molal, where data are most complete (Table 4), there is in fact good agreement between predicted and measured concentrations of dissolved gold, although the same is not true for dissolved Al (Fig. 8). Assuming pH is not the problem, as indicated by Fe and Cu data, together with dissolved Au, then the relatively high measured concentrations of Al suggest the existence of aqueous Al complexes other than simple hydroxyl species - all that is available in the recently revised SUPCRT database (Shock et al., 1997; Sverjensky et al., 1997). Indeed, it may be that at the conditions of the experiments, alkali bearing Al-complexes (Anderson, 1995), or other Al-bearing aqueous species (Pokrovski et al. 1996) enhance the solubility of the Al-bearing minerals. Other than the relatively high Al concentration measured for the first sample from the experiment at 400 °C, 500 bars (0.53 m Cl), the experimental data are also in generally good agreement with the range of Al concentrations reported for hot spring vent fluids at mid-ocean ridges. For example, end-member Al concentrations recently measured for vents at EPR 9-10°N (Von Damm, personal communication), range from approximately 5 to 18 μmolal. These data also reveal a slight positive correlation with dissolved Cl, suggesting some form of chloro-complexing. Although still preliminary, it is significant that the highest dissolved Al concentration (18 μmolal) is associated with the more Cl-rich vent fluid (850 mmolal), a value which is also in good agreement with results from our 1.0 molal Cl-experiments, suggesting the existence of a Cl-complex of Al in the experiments. Without more data from experiments at lower dissolved Cl^- concentrations, it is not possible to test unambiguously the role of Cl^- and temperature on Al-speciation effects, although this may be one the few occasions where field data are used to constrain

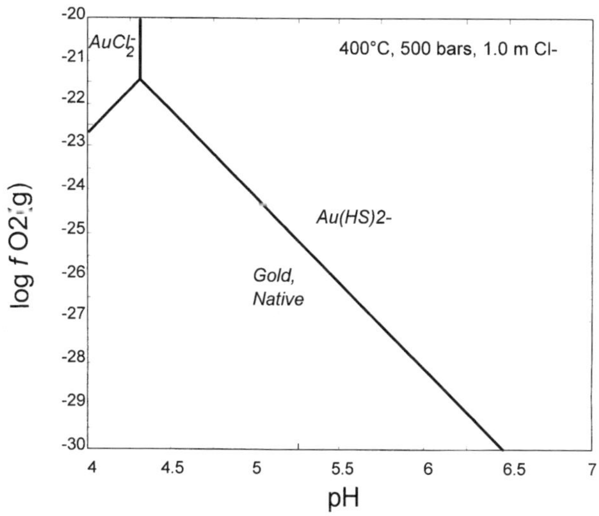

Figure 10. pH-log $f_{O_{2(g)}}$ diagram for the Au-HCl-H_2S-H_2O system at 400 °C, 500 bars and 1.0 molal total dissolved Cl depicting the predominance of Au-bearing species and native gold. H_2S and Au activity are fixed at values consistent with measured concentrations of these species at experimental conditions (Table 4). Calculations were performed with EQ3/EQ6 (Wolery and Daveler, 1992) taking explicit account of activity/concentration relations, and appropriate mass balance, charge balance and mass action expressions. Thermodynamic data used for the calculations are from SUPCRT92 (Johnson et al., 1992) and Sverjensky et al. (1997).

experimental results, and generally shows the benefits of an integrated approach to the study of mineral-fluid interaction at elevated temperatures and pressures.

In apparent contrast with Al, bisulfide complexes are predicted to dominate the distribution of Au-bearing species in fluids coexisting with the metastable plagioclase-epidote-bearing mineral assemblage (Fig. 9). This is important, because it suggests that pH increase and increases in total dissolved H_2S may be more critical than dissolved Cl in contributing to the mobility of gold in subseafloor hydrothermal systems. Of course, it is unlikely that the activity of Au in the submarine hydrothermal system is buffered by native gold, but whichever phase does control gold, this phase would likely respond to changes in fluid chemistry in ways very different than Cu- and Fe-bearing minerals (or even Al-phases), where chloro-complexing is so important in the mobility of these species (Fe and Cu) at elevated temperatures and pressures (Crerar and Barnes, 1976; Hemley et al., 1992; Ding and Seyfried, 1992a,b). To illustrate this better, we can make use of an activity ($f_{O_{2(g)}}$ vs pH) diagram for the Au-H_2S-HCl-H_2O-NaCl system, which shows that at 400 °C, 500 bars and 1.0 molal Cl^-, together with an activity of $H_2S_{(aq)}$ equivalent to that of the experiments at the described conditions, all but the most oxidizing and acidic conditions, $Au(HS)_2^-{}_{(aq)}$ is the predominate aqueous species of gold (Fig. 10).

For the same range of conditions (400 °C, 500 bars, 1.0 molal total dissolved Cl^-), Pb solubility can be calculated to be dominated by chloro-complexes, in particular, $PbCl_{2(aq)}$, which represents over 75% of the total dissolved Pb. Thus, assuming the presence of Pb-bearing phases in subseafloor hydrothermal systems, it is likely that dissolved Pb will correlate with dissolved Cl, like dissolved Fe and Cu, and anti-correlate with pH. Although dissolved Pb data are generally lacking for vent fluids from unsedimented portions of ridges, data that are available for fluids from EPR 9-10°N (Von Damm, personal communication), reveal concentrations on the order of 2 μmolal, in reasonable agreement with values observed here, but nearly a factor of 500 times lower than galena-controlled Pb concentrations.

Metastable plagioclase-epidote-fluid equilibria with imposed constraints not only fix pH, and constrain the mobility of dissolved metals, but when taken together with coexisting oxide and sulfide phases, also fix $f_{O_{2(g)}}$, as follows:

Anorthite *Anhydrite* *Quartz* *Magnetite*

$$CaAl_2Si_2O_8 + 0.5H_2O + CaSO_4 + SiO_2 + 0.5Fe_3O_4$$

Epidote *Pyrite*

$$\leftrightarrow CaAl_2FeO_{12}(OH) + 1.75O_{2(g)} + 0.5FeS_2 \quad (4)$$

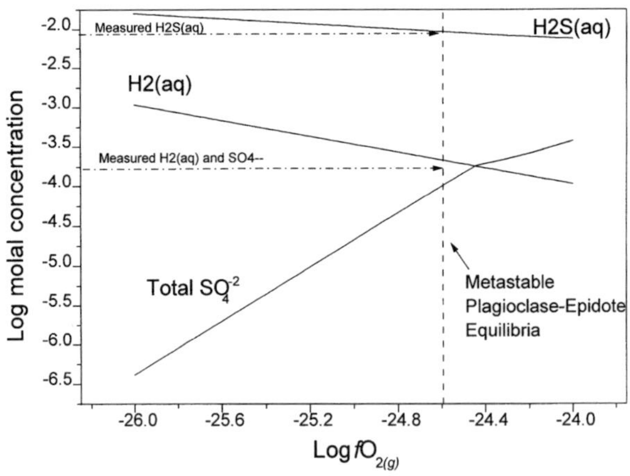

Figure 11. Predicted and measured dissolved sulfur species and $H_{2(aq)}$ in fluid coexisting with plagioclase (X_{an} = 0.70)-epidote (X_{ep} = 0.45)-quartz-magnetite-anhydrite-pyrite-chalcopyrite assemblage at 400 °C, 500 bar and 0.53 total dissolved Cl^- as a function of log $f_{O_2(g)}$. $m_{H_2S(aq)}$ is in very good agreement with theoretical predictions taking account of salting-out effects in NaCl fluid (Ding and Seyfried, 1990) and data from Kishima (1989). Measured H_2 and SO_4^{2-} are slightly below and above theoretical predictions, respectively. Overall agreement is still good, and indicates the potential usefulness of dissolved sulfur species and H_2 in vent fluids as a means of constraining redox in subseafloor reaction zones at mid-ocean ridges.

For example, at 400 °C, 500 bars and 0.53 m Cl^-, a log $f_{O_2(g)}$ of –24.59 can be calculated assuming activities of anorthite and epidote components of plagioclase and epidote solid solutions consistent with measured constraints (Fig. 11). Also plotted on Figure 11 are changes in the dissolved concentrations of total SO_4^{2-}, $H_{2(aq)}$ and $H_2S_{(aq)}$ predicted as a function of changes in log $f_{O_2(g)}$, which also can be compared with values of these species actually measured during the experiment. Data indicate that predicted concentrations of $H_{2(aq)}$ and $H_2S_{(aq)}$ increase with decreasing $f_{O_2(g)}$, whereas the opposite is true for total dissolved SO_4^{2-}. Furthermore, a decrease of log $f_{O_2(g)}$ of a single unit below experimental conditions causes SO_4^{2-} to decrease by approximately two log units. Measured concentrations of $H_{2(aq)}$ and $H_2S_{(aq)}$ are in good agreement with theoretical predictions, although measured SO_4^{2-} exceeds the value predicted by approximately a factor of 2. The relatively slight off-set in predicted and measured concentrations of dissolved SO_4^{2-} notwithstanding, the data reveal that $H_{2(aq)}$ and the $H_2S_{(aq)}/SO_4^{2-}$ ratio can provide internally consistent indicators of redox relations in fluid coexisting with mineral assemblages at temperatures from 400 to 425 °C, and provide an alternative to dissolved H_2 or m_{Fe}/m_{Cu} ratio (see Seyfried and Ding, 1993) as proxies for redox in subseafloor reaction zones, assuming seawater entrainment during vent fluid sampling can be effectively accounted for. Over-determination of redox by combining approaches, however, provides valuable insight on reaction zone conditions, and characteristics such as temperature, pressure, mineralogy, and mineral composition may be determined with much greater accuracy and confidence than presently available by any single approach, providing even greater insight on the temporal evolution of hydrothermal reactions at mid-ocean ridges. This approach could be especially valuable at a hydrothermal system such as TAG, where subseafloor mixing between seawater and vent fluid can result in local redox gradients, which can be discerned better by application of multiple redox proxies.

Another way to visualize the effect of mineral equilibria on the distribution of redox-dependent aqueous species at experimental conditions (e.g., 400 °C, 500 bars, 0.53 m Cl^-) is by use of an activity-activity diagram for the FeO-Fe_2O_3-Al_2O_3-SiO_2-CaO-H_2O-HCl-H_2S system (Fig. 12). Indeed, it can be seen here that for an $a_{Ca^{2+}}/(a_{H^+})^2$ ratio consistent with metastable plagioclase (X_{an} = 0.7)-epidote (X_{ep} = 0.45) equilibria (5.34, see previous discussion), $a_{H_2S_{(aq)}}$-$a_{H_{2(aq)}}$ for fluid coexisting with anhydrite, pyrite and magnetite, plots at a point on the pyrite-magnetite boundary that is relatively close to the HMP (Hematite-Magnetite-Pyrite) invariant point. In general, for all conditions to the right of the dashed line in Figure 12, anhydrite is unstable, and SO_4^{2-}

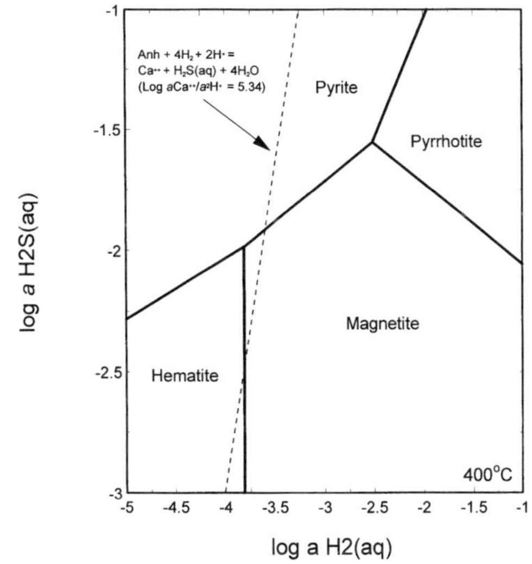

Figure 12. Log $a_{H_2S_{(aq)}}$-log $a_{H_{2(aq)}}$ diagram depicting the anhydrite stability field for fluid coexisting with plagioclase (X_{an} = 0.70)-epidote (X_{ep} = 0.45)-quartz-magnetite-pyrite-chalcopyrite assemblage at 400 °C, 500 and 0.53 molal total dissolved Cl^-. For imposed constraints, the "anhydrite-in" boundary is located near the hematite-magnetite-pyrite (HMP) invariant, but considerably removed from the pyrite-pyrrhotite-magnetite (PPM) invariant. Whether or not the anhydrite-bearing assemblage buffers redox in vent fluids depends largely on the extent of fluid/rock interaction. Hydrothermal systems with very low fluid/rock mass ratios might not be able to eliminate igneous pyrrhotite from controlling redox, while high fluid/rock mass ratios may render pyrite unstable, which would be manifest by a decrease in H_2S, although H_2 may still be relatively high (see text). Thermodynamic data used for the calculations are from SUPCRT92 (Johnson et al., 1992).

converts to $H_2S_{(aq)}$. Changes in $a_{Ca^{2+}}/(a_{H^+})^2$ values, which can result from changes in the activity of anorthite or epidote components of plagioclase and epidote solid solution respectively, can, however, cause some movement of the "anhydrite-in" boundary. For example, an anorthite activity lower than 0.7 would cause this boundary to move to the right in Figure 12, while the opposite would be true for a decrease in the activity of epidote solid solution (e.g., see reaction 3). Owing to constraints imposed by anhydrite stability, however, relatively large changes in the composition/activity of the alumino-silicate assemblage result in only minor changes in the position of the anhydrite stability field (all else being held constant). Thus, stable or metastable equilibria involving plagioclase and epidote solid solutions coexisting with anhydrite, pyrite and magnetite and a Cl—bearing aqueous fluid will likely define a relatively restricted region in $H_2S_{(aq)}$ space, which makes this species in vent fluids a more reliable indicator of redox in subseafloor hydrothermal alteration processes. Of course, changes in temperature, pressure, dissolved Cl^- (salting-out), together with changes in phase assemblage, will exert considerable influence on this, and seafloor applications must explicitly consider any or all of these possibilities to interpret accurately $H_2S_{(aq)}$ concentrations in hot spring vent fluids. Again, it is for this reason, that application of multiple redox constraints are most useful.

4.1. Mineral-fluid Equilibria and Subseafloor Reaction Zones at Mid-Ocean Ridges: Redox Constraints.

Concentrations of dissolved H_2S in vent fluids decrease sharply with decreasing dissolved Cl (Fig. 13). Cl^- variations, of course, are largely the result of phase separation of seawater derived fluid into vapor- and brine-rich components (Bischoff and Rosenbauer, 1987), while $H_2S_{(aq)}$ variations reflect phase equilibria in subseafloor reaction zones, which can also be influenced by Cl^- owing to salting-out effects, as discussed earlier. For example, assuming a reaction zone temperature and pressure of 400 °C and 500 bars, dissolved $H_2S_{(aq)}$ can be predicted from the present study to decrease from approximately 12 to 6 mmolal with increase in Cl from 300 to 1000 mmolal, respectively. Lower temperatures (e.g., 375 °C) result in lower $H_2S_{(aq)}$ concentrations. The effect of Cl^- on $H_2S_{(aq)}$ at the lower temperature is also less owing to temperature-dependent changes in the Setchenow co-

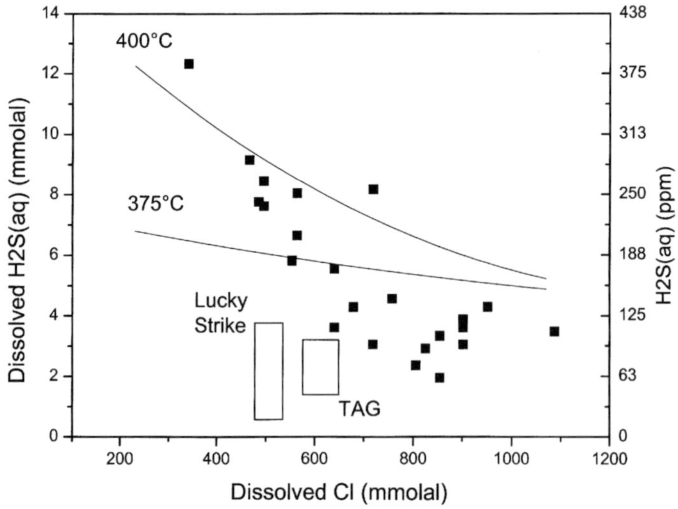

Figure 13. Dissolved $H_2S_{(aq)}$ vs Cl^- concentrations of selected vent fluids at mid-ocean ridges (Von Damm, 1995 and references therein) in comparison with $H_2S_{(aq)}$ concentrations in fluid coexisting with plagioclase (X_{an} = 0.70)-epidote (X_{ep} = 0.45)-quartz-magnetite-anhydrite-pyrite-assemblage at 375-400 °C, 500 bars. The relatively low $H_2S_{(aq)}$ concentrations of many vent fluids may indicate re-equilibration at temperatures less than 375 °C, which is likely the case with the very high-salinity fluids, where temperatures of venting fluid may be as low as 220 °C (see text). In contrast, the conspicuously low H_2S(aq) concentration for vent fluids at Lucky Strike (Von Damm et al., 1998) and TAG (Edmond et al. 1995; Tivey et al., 1995), suggests the loss of pyrite from the reaction zone mineral assemblage (see Fig. 12). This scenario is very much in keeping with the relatively low m_{Fe}/m_{Cu} ratio for the TAG vent fluids (Allen and Seyfried, 1998). Availability of experimentally calibrated theoretical data allows rigorous constraints to be placed on the mineralogic composition of subseafloor reaction zones at mid-ocean ridges.

efficient of $H_2S_{(aq)}$ (Ding and Seyfried, 1990) and the ionic strength of NaCl-fluid. That vent fluids with dissolved Cl^- near seawater values approach H_2S concentrations consistent with phase equilibria constraints involving the anhydrite-bearing system of minerals at 375-400 °C, suggests similar conditions in subseafloor reaction zones. The sharply lower $H_2S_{(aq)}$ concentrations of the high-Cl^- fluids, however, most likely reflect the effects of more extreme cooling, which may be caused by the high density of these fluids precluding rapid ascent to the seafloor. Indeed, the unusually high-Cl^- vent fluids at SJDF ridge have measured temperatures of only 220-290 °C indicating extensive conductive cooling (Von Damm and Bischoff, 1987) with likely attendant loss in $H_2S_{(aq)}$. In cases such as this, it is difficult to constrain unambiguously reaction-zone conditions. Vent fluids at Lucky Strike (Von Damm et al., 1998) and at TAG (Edmond et al. 1995; Tivey et al., 1995), however, are not only characterized by low $H_2S_{(aq)}$ concentrations, but also by low to moderate dissolved Cl^- concentrations (Fig. 13), and relatively high temperatures (Edmond et al., 1995; Tivey et al., 1995; Von Damm et al., 1998), which makes it unlikely that post-reaction zone cooling can account for the relatively low H_2S concentrations for these hydrothermal systems.

An alternative explanation for the relatively low H_2S concentrations at Lucky Strike, TAG, and for similar systems, may involve other chemical and physical factors related to fluid/rock mass ratio and the composition of the substrate through which the hydrothermal fluids circulate. For example, it is well known from the relatively low concentrations of trace incompatible elements in fluids at these sites that fluid/rock mass ratios are relatively high (Von Damm et al., 1998; Campbell et al., 1988b). The reasons for this are likely complex, but could result from tectonic activity that could enhance fluid flow along high-permeability channels, ultimately creating a highly altered substrate. Considering the relatively high solubility of pyrite in aqueous fluids at elevated temperatures, it is likely that the H_2S inventory of the substrate would be effectively depleted in a relatively short period of time, with a corresponding decrease in the concentration of H_2S in the coexisting fluid phase. In effect, the proposed loss of pyrite as a reactant would destroy the H_2S-buffer capacity of the mineral assemblage, and further reaction progress would result in a relatively sharp decrease in H_2S, but only a modest change in H_2 (Fig. 12). That dissolved $H_{2(aq)}$ at TAG (Charlou et al., 1996) is nearly identical to the values measured here at 400 °C, 500 bars (Table 3), is consistent with this interpretation. In addition to accounting for relatively low $H_2S_{(aq)}$, the loss of pyrite in subseafloor reaction zones at TAG, would result in a decrease in m_{Fe}/m_{Cu} ratio, in a manner very much equivalent to what is observed (Allen and Seyfried, 1998).

Whether or not the scenario outlined above to account for the relatively low $H_2S_{(aq)}$ concentrations in hydrothermal fluids at TAG and elsewhere is accurate can not be assessed unambiguously without additional experimental data for an expanded chemical system, at a wider range of temperatures and especially for fluid pressures less than 500 bars. What is clear, however, is that the strategy of conducting experiments to calibrate theoretical models, and then using the models to constrain hydrothermal alteration processes, represents a valuable approach to assess controls on the chemistry of hot spring vent fluids. Indeed, only by use of an integrated approach involving field, experimental and theoretical data can real progress be made in our ability to understand the temporal evolution of exceedingly complex and often metastable chemical systems that characterize subseafloor reaction zones at mid-ocean ridges.

5. SUMMARY AND CONCLUSIONS

Chemical reactions between circulating hydrothermal fluids of seawater origin and rocks of basaltic composition commonly involve metastable mineral assemblages that contain plagioclase solid solutions at temperatures > 400 °C, and low to moderate pressures. The persistence of mineral metastability, however, together with phase separation processes in the NaCl-H_2O system, which can cause substantial changes in total dissolved Cl^- concentrations of coexisting aqueous fluids, can create an exceedingly complex chemical environment, one that may be may be difficult to model thermodynamically. The present investigation, therefore, was conducted to assess the relative agreement between results of experimental and theoretical models depicting phase relations in the CaO-Na_2O-Al_2O_3-SiO_2-Fe_2O_3-FeO-CuO-HCl-H_2O-H_2S system at temperatures from 400-425 °C, 500 bars pressure and dissolved Cl^- concentrations from 0.53 to 1.0 molal. Although experiments were performed at lower pressures (300-500 bars), the dearth of reliable thermodynamic data for aqueous species effectively precludes comparison with experimental results for the relatively low-pressure range of conditions.

In general, for conditions for which comparisons can be made, there is good agreement between experimental and theoretical data when explicit account is made for the presence of metastable phases. Previous experimental data documented this for plagioclase solid solutions, but it is clear from the present investigation that epidote solid solution, a primary component of the starting mineral assemblage, also maintains its composition at initial values, and quite distinct from that which is predicted from phase equilibria constraints. Whereas metastable plagioclase is best indicated by dissolved concentrations of Na and Ca, epidote metastability is revealed by pH effects, which under experimental conditions are manifest by variability of dissolved metal concentrations, especially Fe. For example, at 425 °C, 500 bars and 1.0 molal total dissolved Cl^-, recrystallization of the starting epidote (X_{ep} = 0.45) to a thermodynamically more stable phase (X_{ep} = 0.82), would give rise to a relatively high-pH fluid and correspondingly low dissolved Fe concentrations. That the measured Fe gives a pH nearly equivalent to that in metastable equilibrium with the starting epidote composition (X_{ep} = 0.45), confirms this interpretation, as do electron microprobe data which fail to show compositional modification of the starting epidote.

Dissolved Cu, fixed by the presence of chalcopyrite, and Au, fixed by the gold reaction cell, require a pH (in-situ) that is consistent with the presence of metastable epidote (X_{ep} = 0.45). Dissolved Al concentrations, however, depart sharply from predicted values, suggesting the existence of more complex distribution of Al species than are included in the available thermodynamic database. The experimentally measured Al, however, is in reasonably good agreement with vent fluid data suggesting similar speciation in natural systems. This point notwithstanding, the generally good agreement between predicted and measured dissolved metal concentrations confirms that plagioclase and epidote solid solutions coexisting with Na-Ca-Cl fluids can provide a valuable means of buffering pH of fluids at elevated temperatures and pressures. Although this can be important to pH control in subseafloor hydrothermal systems, it is also important to the design of hydrothermal experiments where pH control is often critically needed to assess unambiguously a wide range of reaction processes. That pH can be changed and maintained simply by changing epidote composition, offers many advantages to alternative approaches.

Dissolved sulfur species and measured $H_{2(aq)}$ concentrations are also consistent with redox constraints imposed in part by metastable epidote-plagioclase-fluid equilibria. In this case, however, it can be shown that for the anhydrite-bearing system, $H_2S_{(aq)}$ is relatively insensitive to compositional variability of the alumino-silicate assemblage, as is the Fe/Cu ratio, two parameters that have been used in vent fluids to constrain redox in subseafloor reaction zones. Although many vent fluids with dissolved Cl^- concentrations near seawater have $H_2S_{(aq)}$ concentrations consistent with a reaction zone mineral assemblage containing metastable plagioclase-epidote, together with quartz, magnetite, anhydrite, and pyrite, this does not appear to be the case for systems having higher dissolved Cl^- concentrations where either distinctly lower temperatures or different phase relations are suggested. It is clear that additional experiments and modeling calculations are needed over a wider range of conditions and with more diverse phase assemblages to distinguish between these two possibilities. The need for additional experiments is underscored further by the fact that the chemistry of many vent fluids is influenced greatly by critical phenomena in the NaCl-H_2O system, which is largely unsupported by theoretical data because this is an exceedingly challenging region of pressure-temperature-composition space. Experiments, however, must also include explicit consideration of mineral kinetics and recrystallization effects, so as to assess better the temporal evolution of minerals and fluids from complex natural systems.

Acknowledgements—We thank Meg Tivey and Scott Wood for their comments and suggestions that greatly improved the final version of the manuscript. We would also like to acknowledge the many helpful discussions with Mike Berndt and Doug Allen, which played an important role in the development of the concepts reflected in the paper. This research was supported by NSF grants: OCE-9605441, OCE-9633132, and OCE-98188908.

Editorial handling: Scott A. Wood

REFERENCES

Allen D. E. and Seyfried W. E. Jr. (1998) Phase equilibria constraints on the chemistry of sub-seafloor reaction zones at 400°C, 500 bars and 1 molal Cl. *Abstracts with Programs - Geological Society of America*, 29, n. 6, p. 25.

Anderson G. M. (1995) Is there alkali-aluminum complexing at high temperatures and pressures? *Geochim. Cosmochim. Acta* **59**, 2155-2161.

Berndt M. E. and Seyfried W. E. Jr. (1993) Calcium and sodium exchange during hydrothermal alteration of calcic plagioclase at 400°C and 400 bars. *Geochim. Cosmochim. Acta* **57**, 4445-4453.

Berndt M. E., Seal R. R. II, Shanks W. C. III, and Seyfried W. E. Jr. (1996) Hydrogen isotope systematics of phase separation in submarine hydrothermal systems: Experimental calibration and theoretical models. *Geochim. Cosmochim. Acta.* **60**, 1595-1604.

Berndt M. E., Seyfried W. E. Jr., and Janecky D. R. (1989) Plagioclase and epidote buffering of cation ratios in mid-ocean ridge hydrothermal fluids: Experimental results in the near supercritical region. *Geochim. Cosmochim. Acta* **53**, 2283-2300.

Bird D. K. and Helgeson H. C. (1980a) Chemical interaction of aqueous solutions with epidote-feldspar mineral assemblages in geologic systems. 1. Thermodynamic analysis of phase relations in the system CaO-FeO-Fe_2O_3-Al_2O_3-SiO_2-H_2O-CO_2. *Am. Jour. Sci.* **280**, 907-941.

Bird D. K. and Helgeson H. C. (1980b) Chemical interaction of aqueous solutions with epidote-feldspar assemblages in geologic systems. II. Equilibrium constraints in the metamorphic/geothermal process. *Amer. J. Sci.* **281**, 576-614.

Bischoff J. L. and Rosenbauer R. J. (1987) Phase separation in seafloor geothermal systems: An experimental study on the effects of metal transport. *Amer. J. Sci.* **287**, 953-978.

Bowers T. S., Campbell A. C., Measures C. I., Spivack A. J., and Edmond J. M. (1988) Chemical controls on the composition of vent fluids at 13°-11°N and 21°N, East Pacific Rise. *J. Geophys. Res.* **93**, 4522-4536.

Butterfield D. A. and Massoth G. J. (1994) Geochemistry of North Cleft Segment Vent Fluids - Temporal Changes in Chlorinity and Their Possible Relation to Recent Volcanism. *J. Geophys. Res-Solid. Earth.* **99**, 4951-4968.

Butterfield D. A., McDuff R. E., Mottl M. J., Lilley M. D., Lupton J. E., and Massoth G. J. (1994) Gradients in the composition of hydrothermal fluids from the Endeavor segment vent field: Phase separation and brine loss. *J. Geophys. Res.* **99**, 9561-9583.

Campbell A. C., Bowers T. S., and Edmond J. M. (1988a) A time-series of vent fluid compositions from 21°N, EPR (1979, 1981, and 1985) and the Guaymas Basin, Gulf of California (1982, 1985). *J. Geophys. Res.* **93**, 4537-4549.

Campbell A. C., Palmer M. R., Klinkhammer G. P., Bowers T. S., Edmond J. M., Lawrence J. R., Casey J. F., Thompson G., Humphris S., Rona P., and Karson J. A. (1988b) Chemistry of hot springs on the Mid-Atlantic Ridge. *Nature* **335**, 514-519.

Charlou J. L., Donval J. P., Jean-Baptiste P., Dapoigny A., and Rona P. A. (1996) Gases and helium isotopes in high temperature solutions sampled before and after ODP Leg 158 drilling at TAG hydrothermal field (26 degrees N, MAR). *Geophysical Research Letters* **23**, 3491-3494.

Chen C.-T. A. and Marshall W. L. (1982) Amorphous silica solubilities; IV, Behaviour in pure water and aqueous sodium chloride, sodium sulfate, magnesium chloride, and magnesium sulfate solutions up to 350 degrees C. *Geochim. Cosmochim. Acta* **46**, 279-287.

Crerar D. A. and Barnes H. L. (1976) Ore solution chemistry V. Solubilities of chalcopyrite and chalcocite assemblages in hydrothermal solution at 200° to 350°C. *Econ. Geol.* **71**, 772-794.

Ding K. and Seyfried W. E. Jr. (1989) H_2 and H_2S concentrations as indicators of temperature in ridge crest hydrothermal systems. *EOS, Trans. Amer. Geophys. Union* **70**, 1397.

Ding K. and Seyfried W. E. Jr. (1990) Activity coefficients of H_2 and H_2S in NaCl solutions at 300-425°C, 300-500 bars with application to ridge crest hydrothermal systems. *EOS, Trans. Amer. Geophys. Union* **71**, 1680.

Ding K. and Seyfried W. E. Jr. (1992a) Determination of Fe-Cl Complexing in the Low Pressure Supercritical Region (NaCl Fluid) - Iron Solubility Constraints on pH of Subseafloor Hydrothermal Fluids. *Geochim. Cosmochim. Acta.* **56**, 3681-3692.

Ding K. and Seyfried W. E. Jr. (1992b) Experimental determination of Cu-Cl speciation at the T-P conditions relevant to ridge crest hydrothermal activity: Implications to log f_{O_2} in hot spring fluids. *EOS, Trans. Amer. Geophys. Union* **73**, 254.

Edmond J. M., Campbell A. C., Palmer M., Klinkhammer G. P., German C. R., Edmonds H. N., Elderfield H., Thompson G., and Rona P. (1995) Time series studies of vent fluids from the TAG and MARK sites (1986, 1990) Mid-Atlantic Ridge: A new solution chemistry model and mechanism for Cu/Zn zonation in massive sulfide ore deposits. In *Hydrothermal Vents and Processes*, Vol. 87 (ed. L. M. Parson, C. L. Walker, and D. R. Dixon), pp. 411. Geological Society Special Publication.

Fornari D. J., Shank T., Von Damm K. L., Gregg T. K. P., Lilley M., Levai G., Bray A., Haymon R. M., Perfit M. R., and Lutz R. (1998) Time-series temperature measurements at high-temperature hydrothermal vents, East Pacific Rise 9 degrees 49'-51'N; evidence for monitoring a crustal cracking event. *Earth Planet. Sci. Lett.* **160**, 419-431.

Helgeson H. C. (1969) Thermodynamics of hydrothermal systems at elevated temperatures and pressures. *Amer. J. Sci.* **267**, 729-804.

Helgeson H. C., Kirkham D. H., and Flowers G. C. (1981) Theoretical prediction of the thermodynamic behavior of aqueous electrolytes at high pressures and temperatures. IV. Calculation of activity coefficients, osmotic coefficients, and apparent molal and standard and relative partial molal properties to 600°C and 5 kb. *Amer. J. Sci.* **281**, 1249-1516.

Hemley J. J., Cygan G. L., Fein J. L., Robinson G. R., and D'Angelo W. M. (1992) Hydrothermal ore-forming processes in light of studies in rock buffered systems: I. Iron, copper, lead, zinc sulfide solubility relations. *Econ. Geol.* **87**, 1-22.

Johnson J. W., Oelkers E. H., and Helgeson H. C. (1992) SUPCRT92 - A Software Package for Calculating the Standard Molal Thermodynamic Properties of Minerals, Gases, Aqueous Species, and Reactions from 1-Bar to 5000-Bar and 0-Degrees-C to 1000-Degrees-C. *Computers & Geosciences.* **18**, 899-947.

Kishima N. (1989) A thermodynamic study on the pyrite-pyrrhotite-magnetite-water system at 300-500 C with relevance to the fugacity/concentration quotient of aqueous H_2S. *Geochim. Cosmochim. Acta* **53**, 2143-2155.

Liou J. R. (1973) Synthesis and stability relations of epidote, $Ca_2Al_2FeSi_3O_{12}(OH)$. *J. Petrol.* **14**, 381-413.

Orville P. M. (1972) Plagioclase cation exchange equilibria with aqueous chloride solution: results at 700°C and 2000 bars in the presence of quartz. *Amer. J. Sci.* **272**, 234-272.

Pokrovski G. S., Schott, J., Harrichoury, J.-C., and Sergeyev A. S. (1996) The stability of aluminum silicate complexes in acidic solutions from 25 to 150°C *Geochim. Cosmochim. Acta* **60**, 2495-2503.

Saccocia P. J. and Seyfried W. E. Jr. (1990) Talc-quartz equilibria and the stability of magnesium chloride complexes in NaCl-$MgCl_2$ solutions at 300, 350, and 400°C, 500 bars. *Geochim. Cosmochim. Acta* **54**, 3283-3295.

Seewald J. S. and Seyfried W. E. Jr. (1990) The effect of temperature on metal mobility in subseafloor hydrothermal systems: Constraints from basalt alteration experiments. *Earth Planet. Sci. Lett.* **101**, 388-403.

Seyfried W. E. Jr., Janecky D. R., and Berndt M. E. (1987) Rocking autoclaves for hydrothermal experiments II. The flexible reaction cell system. In *Hydrothermal Experimental Techniques* (ed. H. L. Barnes and G. C. Ulmer), pp. 216-240. Wiley Interscience.

Seyfried W. E. Jr. and Ding K. (1993) The effect of redox on the relative solubilities of copper and iron in Cl-bearing aqueous fluids at elevated temperatures and pressures: An experimental study with application to subseafloor hydrothermal systems. *Geochim. et Cosmochim. Acta* **57**, 1905-1917.

Seyfried W. E. Jr. and Ding K. (1995) Phase equilibria in subseafloor hydrothermal systems: A review of the role of redox, temperature, pH and dissolved Cl on the chemistry of hot spring fluids at mid-ocean ridges. In *Seafloor Hydrothermal Systems: Physical, Chemical, Biologic and Geological Interactions*, Vol. Geophysical Monograph 91 (ed. S. E. Humphris, R. A. Zierenberg, L. S. Mullineaux, and R. E. Thompson), pp. 248-273. American Geophysical Union.

Seyfried W. E. Jr. Ding K., and Berndt M. E. (1991) Phase equilibria constraints on the chemistry of hot spring fluids at mid-ocean ridges. *Geochim. Cosmochim. Acta* **55**, 3559-3580.

Shanks W. C. III, Boehlke J. K., Seal, R. R. II. (1995) Stable isotopes in mid-ocean ridge hydrothermal systems; interactions between fluids, minerals, and organisms. *Seafloor hydrothermal systems; physical, chemical, biological, and geological interactions.* Vol. Geophysical Monograph 91 (ed. S. E. Humphris, R. A. Zierenberg, L. S. Mullineaux, and R. E. Thompson), pp. 248-273. American Geophysical Union.

Shock E. L. and Helgeson H. C. (1988) Calculation of the thermodynamic and transport properties of aqueous species at high pressures and temperatures: Correlation algorithms for ionic species and equation of state predictions to 5 kb and 1000°C. *Geochim. Cosmochim. Acta* **52**, 2009-2036.

Shock E. L. and Helgeson H. C. (1990) Calculation of the thermodynamic and transport properties of aqueous species at high pressures and temperatures: Standard partial molal properties of organic species. *Geochim. Cosmochim. Acta* **54**, 915-945.

Shock E. L., Helgeson H. C., and Sverjensky D. A. (1989) Calculations of the thermodynamic and transport properties of aqueous species at high pressures and temperatures: Standard partial molal properties of inorganic neutral species. *Geochim. Cosmochim. Acta* **53**, 2157-2183.

Shock E. L., Sassani D. C., Willis M., and Sverjensky D. A. (1997) Inorganic species in geologic fluids: Correlations among standard molal thermodynamic properties of aqueous ions and hydroxide complexes. *Geochim. Cosmochim. Acta* **61**, 907-950.

Sverjensky D. A., Hemley J. J., and D'Angelo W. M. (1991) Thermodynamic assessment of hydrothermal alkali feldspar-mica-aluminosilicate equilibria. *Geochim. Cosmochim. Acta* **55**, 989-1005.

Sverjensky D. A., Shock E. L., and Helgeson H. C. (1997) Prediction of the thermodynamic properties of aqueous metal complexes to 1000 degrees C and 5 kb. *Geochim Cosmochim Acta* **61**(7), 1359-1412.

Tanger J. C. and Helgeson H. C. (1988) Calculation of the thermodynamic properties of aqueous species at high pressure and temperatures: Revised equation of state for the standard partial properties of ions and electrolytes. *Amer. J. Sci.* **288**, 19-98.

Tivey M. K., Humphris S. E., Thompson G., Hannington M. D., and Rona P. A. (1995) Deducing patterns of fluid flow and mixing within the TAG active hydrothermal mound using mineralogical and geochemical data. *J Geophys*

Res Solid Earth **100**(B7), 12527-12555.

Von Damm K. L., Edmond J. L., Grant B., Measures C. I., Walden B., and Weiss R. F. (1985) Chemistry of submarine hydrothermal solutions at 21°N, East Pacific Rise. *Geochim. Cosmochim. Acta* **49**, 2221-2237.

Von Damm K. L. (1988) Systematics of and postulated controls on submarine hydrothermal solution chemistry. *J. Geophys. Res.* **93**, 4551-4562.

Von Damm K. L. (1995) Controls on the chemistry and temporal variability of seafloor hydrothermal fluids. In *Seafloor Hydrothermal Systems: Physical, Chemical, Biologic and Geologic Interactions*, Vol. Geophysical Monograph 91 (ed. S. E. Humphris, R. A. Zierenberg, L. S. Mullineaux, and R. E. Thompson), pp. 222-248. American Geophysical Union.

Von Damm K. L. and Bischoff J. L. (1987) Chemistry of hydrothermal solutions from the southern Juan De Fuca Ridge. *J. Geophys. Res.* **92**, 11,334-11,346.

Von Damm K. L., Bray A. M., Buttermore L. G., and Oosting S. E. (1998) The geochemical controls on vent fluids from the Lucky Strike vent field, Mid-Atlantic Ridge. *Earth Planet. Sci. Lett.* **160**, 521-536.

Von Damm K. L., Buttermore L. G., Oosting S. E., Bray A. M., Fornari D. J., Lilley M. D., and Shanks W. C. I. (1997) Direct observation of the evolution of a seafloor 'black smoker' from vapor to brine. *Earth Planet. Sci. Lett.* **149**, 101-111.

Wolery T. J. and Daveler S. A. (1992) EQ6, A computer program for reaction path modeling of aqueous geochemical systems: Theoretical manual, users guide, and related documentation (version 7.0). **UCRL-MA-110662 PT IV**, 1-337.

Water-Rock Interactions, Ore Deposits, and Environmental Geochemistry: A Tribute to David A. Crerar
© The Geochemical Society, Special Publication No. 7, 2002
Editors: Roland Hellmann and Scott A. Wood

Vapor-transport of ore metals

ANTHONY E. WILLIAMS-JONES, ARTASHES A. MIGDISOV, SANDY M. ARCHIBALD, AND ZHIFENG XIAO

Dept. of Earth and Planetary Sciences, McGill University, 3450 University St., Montréal, Québec, Canada, H3A 2A7
e-mail: willyj@eps.mcgill.ca

Abstract—Recent analyses of the trace-element composition of fumarolic gases, sublimates, and vapor-rich fluid inclusions (rare) suggest that the vapor transport of metals may play a more important role in the formation of hydrothermal ore deposits than is currently recognized. Vapor transport has been modelled thermodynamically by a number of researchers who have generally concluded that it plays a relatively minor role in ore genesis. However, this modelling has been based on the assumption that the transport is controlled almost entirely by the volatility of the corresponding ore minerals, and ignores the potentially important contributions of solvation processes, e.g., interactions between water vapor and the volatilized species.

In this paper, we review the available analytical data for fumarolic gases, sublimates and vapor-rich fluid-inclusions, the experimental data on the volatility of some of the more important ore metals, and some recent experimental data produced mainly in our laboratory on the solubility of Ag, Au, Sn and Cu in chloride-bearing water vapor at temperatures up to 400 °C. Based on this review, and particularly the results of our own experiments, it appears that water vapor is capable of dissolving orders of magnitude higher concentrations of metals than is predicted from volatility calculations. In the case of several metals, notably Cu and Au, (and Hg from volatility alone) the concentrations may be comparable to or even higher than those considered necessary for an ore fluid. We conclude that vapor-transport may play an important role in the formation of a variety of ore deposit types, particularly those forming at higher temperature where vapor is a major or dominant phase in the hydrothermal system.

1. INTRODUCTION

Although there is overwhelming evidence that metallic minerals are concentrated by their precipitation from aqueous liquids in the vast majority of ore-forming hydrothermal systems, the widespread occurrence of boiling has lead some researchers to suggest that the vapor phase may also play an important role in metal transport. This is not a new idea. Indeed, as early as 1644, René Descartes, in his *Principia Philosophae*, proposed that silver-, gold- and lead-bearing veins were produced by the condensation of vapors emanating from the Earth's interior. Nicolas Steno came to the same conclusion in his *De solido intra solidum naturaliter contento* published in 1669. However, the idea was subsequently dismissed by hydrothermalists, notably Pryce (1778) and Charpentier (1778, 1799), who concluded that the main agent of metal transport is not vapor but water in its liquid state.

Pryce, from his observations of veins in the Cornish tin mining district, proposed that water, in its passage through the earth dissolves salts and metals from the surrounding rock and deposits them in fissures to form ores. Charpentier, from an extensive study of veins in the Freiberg silver-lead mines, concluded that the ores were the result of reactions between the country rocks (alteration) and the mineralizing waters. These ideas were expanded upon by Daubrée (1841) and Elie de Beaumont (1847) and by the mid 1800's the hydrothermalists' view of ore formation was firmly entrenched. Interestingly, however, Daubrée (1849) remained open to the possibility that the vapor phase might also deposit metallic minerals and conducted experiments that involved the transport of metals by vapors.

Twentieth century thinking on the genesis of hydrothermal mineral deposits has been very much dominated by the hydrothermalist perspective, i.e., that the medium of metal transport is an aqueous liquid (cf. textbooks by Lindgren, 1933; Bateman, 1950; Park and MacDiarmid, 1975; and Guilbert and Park, 1986; and review articles by Brimhall and Crerar, 1987; Barnes, 1997, Wood and Samson, 1998). In those environments where the vapor phase is volumetrically important, e.g., those of porphyry and epithermal deposits, the vapor is generally considered to be the product of boiling and to be instrumental in producing physicochemical changes favorable for mineralization (Drummond and Ohmoto, 1985).

The first modern papers to seriously attribute a role to vapor in metal transport were those of Brown (1948), who proposed that sulfidic ore deposits could be formed from volatile sulfide species, and Walker and Walker (1956), who proposed that the metals are transported predominantly as volatile chloride species. Krauskopf (1957, 1964) evaluated these proposals quantitatively using thermodynamic data, and concluded that it is feasible to transport most metals by vapor at concentrations sufficient to form economic mineral deposits. He also concluded that, while metals may occur in the vapor as both sulfide and chloride species, only chloride (as HCl) enhances metal solubility. More recently, Wood (1987) used volatility data to determine whether or not vapor transport might have been a significant factor in the formation of the Bushveld PGE deposits, and determined that the amounts of these metals that could be transported at the predicted conditions of ore deposition were insignificant. An important limitation of calculations of vapor-phase solubility based on volatility data, however, and one recognized by Wood (1987), is that they ignore the role of water vapor as a solvent.

One of the first studies to make a convincing geological case for ore formation from an aqueous vapor was by Henley and McNabb (1978). Based on data showing, 1) that intrusions associated with porphyry Cu-Mo deposits are typically emplaced at levels sufficiently shallow for the aqueous fluid to exsolve in the two-phase region, 2) that the mass of vapor exceeds that of liquid and 3) that the two fluids would be stratified by density, these authors developed a model in which the metals are transported to the site of deposition by a plume of vapor. Unfortunately, an important question that was left largely unanswered in their paper was whether the vapor phase was capable of transporting Cu and Mo in the concentrations required to form an ore deposit. The development of new micro-analytical techniques (synchrotron XRF, PIXE, LAM ICP-MS) for measuring concentrations of ore metals in fluid inclusions now offers an opportunity to address this question. Indeed, the first steps in this direction have already been made. Heinrich et al. (1992) showed, using PIXE analyses, that primary vapor-rich inclusions in a granite-hosted quartz-cassiterite vein in the Mole Granite, Australia contain ore level concentrations of Fe, Cu, Pb, and Zn (hundreds to thousands of ppm), and that the concentration of Cu exceeds that of Cu in coexisting brine inclusions. They attributed this preferential partitioning of Cu into the vapor to the formation of strong complexes with sulfur and the relatively high volatility of H_2S, and went on to suggest that, given the strong preference of Au for bisulfide over chloride ligands in hydrothermal liquids, gold might also concentrate in the vapor phase. However, instrumental limitations precluded their analyzing for gold; fortunately this is now possible using recent advances in LAM ICP-MS technology (cf. Heinrich et al., 1999).

In view of arguments by some ore deposit modelers that vapor may play an important role in metal transport, thermodynamic arguments that such transport is feasible, and recent evidence of high concentrations of metals in vapor-rich fluid inclusions, it is now timely to review what is known about the solubility of ore metals in hydrothermal vapors. Most of the data on the capacity of natural vapors to transport significant quantities of metal come from studies of active volcanoes, and particularly those reporting analyses of the condensates of fumarolic gases. Experimental data are limited largely to measurements of the volatility of metals but have been used to estimate metal solubilities in hydrothermal vapors. The few data that are available on the solubility of ore metals in water vapor are those on Au, Ag, Cu, and Sn, which are being collected from experiments performed on chloride-bearing systems at McGill University. In this chapter, we review the data that have been accumulated on the metal contents of volcanic gases and their sublimates, the experimental data that are available on vapor pressures of metals, and our own experimental data on the solubility and speciation of Au, Ag, Sn and Cu chloride-bearing aqueous vapors. Using this information, we make some tentative predictions of the possible concentrations of metals in the vapor phase of chloride-rich hydrothermal systems, and employ an idealized simulation for silver to illustrate the potential capacity of this vapor to transport metals in concentrations sufficient to play an important role in the ore-forming process.

2. METAL CONTENTS OF VOLCANIC GASES

The principal component of volcanic gases is H_2O (typically >90 mole%), which is therefore the dominant solvent for the metals. The next most important constituent is CO_2, which in some cases comprises up to 10 mole % of the total discharge (Table 2). Carbon dioxide is followed in turn by SO_2 and HCl, both of which have been reported to occur in proportions of up to 6 mole %. Other gases, the most important of which are H_2, HF, and H_2S, comprise <1 mole % of the bulk gas. Although water is the main constituent of volcanic gases, it does not necessarily follow that it entirely controls metal solubility. By analogy with liquid water, it is likely that metal solubility is greatly enhanced by complexation reactions involving CO_2, SO_2, HCl, H_2S and HF.

In contrast to the volatile species discussed above, ore metal concentrations in volcanic gases rarely exceed 10^{-5} moles (estimated from analyses of fumarole condensates). Although small by comparison with the major volatile species, these concentrations, as discussed below, are potentially significant in the context of metallic mineralization. In the section that follows we review the available literature on ore metal abundances in fumarolic gases.

Table 1. Proportions of major species in gases released by selected volcanoes. The data are taken from Symonds et al. (1994), and are reported in mole percent.

Volcano	Momotombo	Poas	Mt. Etna	Merapi	St. Augustine	Showa-Shinzan	Mt. St. Helens	Usu
Magma Type	Thol. Basalt		Hawaiite	Andesite			Dacite	
Temp (°C)	658-820	960-1045	1075	767-915	648-870	619-1015	663-802	656-678
log fO_2	-13.5 to -16.5	-9.35 to -10.82	-9.47	-12.49 to -14.86	-12.45 to-17.54	-9.93 to -18.09	-14.25 to -16.76	-16.45 to -17.20
H_2O	97.1 to 97.9	95.28 to 97.08	22.71 to 53.69	88.53 to 95.83	83.91 to 98.31	98.04 to 99.88	91.58 to 98.60	95.80 to 97.30
H_2	0.17 to 0.7	0.39 to 0.67	0.30 to 0.57	0.71 to 1.54	0.38 to 0.69	0.06 to 0.63	0.27 to 0.85	0.27 to 0.34
CO_2	1.44 to 1.47	0.53 to 1.00	17.08 to 33.93	3.26 to 7.56	0.74 to 2.40	0.01 to 0.47	0.89 to 6.94	1.70 to 3.02
CO	0.0015 to 0.0096	0.0031 to 0.01	0.36 to 0.71	0.02 to 0.16	0.002 to 0.016	0.000005 to 0.129	0.0013 to 0.06	0.0029 to 0.0050
SO_2	0.3 to 0.5	1.232 to 2.004	14.69 to 47.7	0.06 to 1.15	0.006-6.18	0.0046 to 0.043	0.067 to 0.208	0.142 to 0.258
H_2S	0.16 to 0.29	0.0055 to 0.4199	0.12 to 0.27	0.12 to 1.16	0.128 to 0.057	0.0004 to 0.109	0.099 to 0.355	0.350 to 0.714
HCl	2.68 to 2.89	0.011 to 0.121	—	0.59	0.157 to 6.04	0.021 to 0.092	0.076 to 0.089	0.0241 to 0.160
HF	0.240 to 0.259	0.011 to 0.121	—	0.4	<0.003 to 0.086	0.01 to 0.031	0.03	0.0116 to 0.0332

Table 2. Concentrations of metal species in fumarole gas condensates . All data are reported in ppm, except gold (ppb). NA: Not analyzed; -: Not detected.

Volcano	Cerro Negro [a]	Momo-tombo [a]	Momo-tombo [b]	Poas [a]	San Cristobal [a]	Tolbachik [c]	Etna [d]	Augustine [e]	Colima [f]	Kudryavy [g]	Merapi [h]	Merapi [i]	White Island [j]	Showa -Shinza [k,l]	Mt. St.[m] Helens	Usu [n,o]	Iwo Jima [p]
Magma Type	Basaltic						Alkali Basalt	Andesitic							Dacite		Rhyolite
Temp (°C)	300-315	456-770	875-886	344-852	562-588	1010	928	870	738-828	535-940	576-796	915	495	800	710	649	165-877
Hg	0.004-0.005	0.006-0.021	–	0.012-0.30	0.012-0.014	–	–	–	NA	0.015-0.340	–	–	–	–	0.020	0.0001	–
As	0.031-0.092	0.23-0.49	2.0-5.5	–	–	23.984	6.146	32.883	0.368-0.525	0.630-1.550	0.28-0.96	0.764	2.8	3.283	1.353	0.015	1.74-4.55
Sb	0.002-0.004	0.003-0.096	1.7	0.003-0.008	0.007-0.008	13.440	0.023	0.205	0.042-0.130	0.02-0.510	≤ 0.16	0.039	0.08	0.133	0.008	0.000	0.02-0.03
Au	1-2	1-24	5	–	–	4	24	–	1-5	NA	NA	–	NA	–	0.03	0.007	1.5-32
Ag	0.003-0.011	0.005-0.014	0.005	0.006-0.25	0.007-0.008	0.006	0.115	–	NA	NA	NA	–	NA	–	0.00001	0.0005	–
Cu	0.2-0.8	0.2-8.4	0.1-1.5	0.1-0.6	0.6	5.962	–	0.194	0.440-0.931	0.032-0.91	0.013-1.0	0.034	0.03	0.021	0.052	0.012	0.006-0.064
Pb	NA	1.9-7.0	0.05-0.6	0.8-5.4	0.4-0.5	0.572	11.999	2.205	0.078-0.480	0.110-9.7	0.32-1.6	1.207	0.18	0.068	0.533	0.006	0.24-1.77
Zn	0.3-0.6	0.4-7.6	0.4-6.0	0.5-8.6	0.9-1.1	9.743	12.620	1.855	4.970-8.140	0.25-13.5	3.6-82	7.935	1.2	0.347	0.019	0.040	0.027-0.24
Sn	0.78-0.98	1.1-2.4	1	1.4-3.9	1.4-1.6	0.092		–	NA	0.1-0.38		–	NA	–	–	7.213	0.2-0.78
W	–	–	–	–	–	–	–	–	NA	0.002-0.060	0.05-0.13	0.054	–	–	0.003	0.003	–
Mo	0.18-0.27	0.15-0.5	2	–	–	–	–	0.766	0.091-0.182	0.002-0.270	0.08-2.8	0.093	0.003	0.049	0.094	0.058	0.005-1.07

a) Gemmell, 1987; b) Quisefit et al., 1989; c) Menyailov and Nikitina, 1980; d) Le Guern, 1988; e) Symonds et al., 1991; f) Taran et al., 2000; g) Taran et al., 1995; h) Symonds et al., 1987; i) Le Guern et al., 1982; j) Tedesco and Totain, 1991; k) Nemeto et al., 1957; l) Symonds, 1992; m) Bernard, 1985; n) Giggenbach and Matsuo,1991; o) Le Guern (unpub.); p) Hedenquist et al., 1994

2.1. Fumarolic Gas Condensates

Analyses of the condensates of volcanic gases indicate that they commonly contain appreciable concentrations of a variety of base and precious metals (Table 1). The distribution of these metals in a given volcano depends on the composition of the associated magma. For example, volcanic gases produced by basaltic magmas will tend to have higher concentrations of copper than those produced by dacitic magmas. By contrast, volcanic gases derived from dacitic magmas will have relatively high concentrations of tin and tungsten. As discussed below, temperature is also an important factor in determining the metal composition of volcanic vapors.

Examples of basaltic volcanoes include Etna (Italy), Momotombo and Cerro Negro (Nicaragua), Poas (Costa Rica), Tolbachik (Russia), and Kilauea (USA). Concentrations of copper analyzed in fumarolic condensates range up to 8.4 ppm, the highest value being from Momotombo (Gemmell, 1987). Significantly, this sample was taken from a fumarole discharging gases at a temperature of 770 °C, the highest temperature recorded by Gemmell (1987) for the fumaroles sampled by him at Momotombo; the other fumaroles had temperatures ranging down to 456 °C and concentrations ranging between 0.2 and 0.4 ppm. Other metals having high concentrations in volcanic gases emanating from basaltic magmas include, Zn, Pb, Sb, As, Ag, and Au. With the exception of As, Sb, and Ag, the highest concentrations reported for these metals are all from Etna (Zn, 13 ppm; Pb, 12 ppm; and Au 24 ppb; Le Guern, 1988). It should be noted, however, that Momotombo is tied with Etna in having the highest reported gold concentration (Gemmell, 1987). The highest concentrations for As (24 ppm) and Sb (13 ppm) are from Tolbachik (Menyailov and Nikitina, 1980) and for Ag (250 ppb) are from Poas (Gemmell, 1987). It is probably significant that the data for Etna and Tolbachik were from fumaroles discharging gases at temperatures of 928 °C and 1010 °C, respectively, two of the highest fumarole temperatures ever recorded.

Studies of andesitic volcanoes, e.g., Merapi in Indonesia (Symonds et al., 1987), Kudryavy in Russia (Taran et al., 1995), White Island in New Zealand (Tedesco and Totain, 1991), Colima in Mexico (Taran et al., 2000) and Augustine in Alaska (Symonds et al., 1992), indicate that the gases are characterized by high concentrations of Zn, Pb, Mo, W, As, Au, and Hg (there are no data for Ag). The highest concentrations of Zn, Pb, and As, are comparable with those of gases from basaltic volcanoes. However, the concentrations of Mo and Hg are substantially higher, up to 3 ppm (Merapi; Symonds et al., 1987) and 340 ppb (Kudryavy; Taran et al., 1995), respectively.

In contrast to those of basaltic and andesitic volcanoes, the gases of felsic volcanoes (rhyolitic and dacitic), e.g., Mount St. Helens in the USA (Bernard and Le Guern, 1986), Usu (Giggenbach and Matsuo, 1991; Le Guern, unpublished data) and Satsuma Iwo-jima, both in Japan (Hedenquist et al., 1994), only have high concentrations of Sn, Mo and Au. Volcanic gases from Usu have the highest concentration of tin (7 ppm; Giggenbach and Matsuo, 1991; Le Guern, unpublished data) and those from Satsuma Iwo-jima have the highest concentrations of Mo (up to 1 ppm) and Au (32 ppb) reported from any volcano (Hedenquist et al., 1994).

2.2. Volcanic Gas Sublimates

Further insights into the role of the vapor phase in transporting metals can be gleaned from fumarole sublimates. The first studies of this type were carried out by Bunsen (1853) who described sublimates from Icelandic volcanoes, and drew attention to the potential importance of HCl and ammonia as possible agents of metal transport. Deville (1862) subsequently provided detailed observations of sublimates from Italian volcanoes and noted that the fumaroles and associated sublimates changed with distance from the volcanic center. The first study that attempted to relate sublimate mineralogy to temperature was that of Lacroix (1907), who reported that fumaroles with temperatures in excess of 650 °C deposited mainly halite, sylvite and aphthitalite ($K_3Na(SO_4)_2$), whereas lower temperature fumaroles (below 650 °C) also deposited sulfides and oxides. At temperatures ≤ 300 °C the variety of minerals decreased, and the sublimates consisted mainly of salammoniac (NH_4Cl), and below 100 °C only gypsum, native sulfur and opal. Somewhat different observations were made by Zeis (1929) in the Valley of 10,000 Smokes. He noted that the principal phases in the high-temperature sublimates were magnetite, ilsemannite ($Mo_3O_8 \cdot nH_2O$), sphalerite, chalcocite, covellite and galena, and that the low-temperature fumaroles consisted primarily of native sulfur and a substance, which he described as "orange arsenic sulphide". The most comprehensive modern studies of volcanic sublimates are those Naboko (1964), on several Kamchatka volcanoes and Stoiber and Rose (1974) on Central American volcanoes. The latter paper also provided an extensive review of sublimate mineralogy for numerous volcanoes in other parts of the world, and noted the occurrence of the following ore minerals in addition to those mentioned above; cassiterite, barite, pyrite, pyrrhotite, orpriment, realgar, and galena-bismuthinite. An important mineral not reported by either Naboko (1964) or Stoiber and Rose (1974) is native gold. However, Fulignati and Sbrana (1998) have since reported the presence of prismatic crystals of this mineral in sublimates of the high-temperature silica zone of a fumarole near the rim of the La Fossa volcano, Vulcano, Italy. These researchers also documented the presence of native tellurium containing minor quantities of selenium. Finally, Korzhinsky et al. (1994) reported the occurrence of rhenium sulfide as 1- to 2-mm thick coatings containing flakes with plate-like morphology on rocks around some high temperature fumaroles at Kudryavy volcano.

Some important limitations of the above studies are that: 1) the fumarole temperatures did not necessarily represent the temperatures at which the sublimates formed; 2) the condensates were rapidly oxidized by the atmosphere, which changed the resulting mineralogy of the sublimates; and 3) interaction of the gases and their condensates with the wall-rock may have produced sublimates substantially different in composition to those initially in equilibrium with the fumarolic gases. In order to overcome these limitations, Le Guern and Bernard (1982) introduced the silica-tube method, which involved inserting a long ($\geq$ 1 m) quartz-glass tube deep into the fumarole and leaving it there for periods ranging from days to months. During this time, sublimates precipitate along the length of the tube, and are zonally distributed in respect to mineralogy due mainly to gradients in temperature. These gradients can be measured accurately, and may correspond to temperatures as great as 1000 °C at the bottom of the tube, and as low as 100 °C at the top. At temperatures <100 °C, condensation of water vapor precludes further sublimate formation. These silica-tube experiments make it possible to reliably relate sublimate minerals to temperature, and thereby avoid the contamination problems mentioned above. In the paragraphs below we document the data available from silica tube experiments conducted on volcanoes of a variety of magma types.

The main high-temperature ore minerals are magnetite, molybdenite, greenockite (CdS) and wolframite, which precipitate only at temperatures >450 °C (Fig. 1). Several other ore minerals, e.g., chalcopyrite, sphalerite, and pyrite, occasionally precipitate at temperature >450 °C but generally form at lower temperature. By contrast, galena, native arsenic, tellurium and selenium, and several Pb-bearing sulfosalts, with rare exception, precipitate at temperatures <450 °C. Some important ore metals, notably gold and silver, are in such low concentrations that, although detected chemically, they are rarely observed as minerals in silica-tube experiments. Important exceptions are the observations of crystals of native gold in silica-tube experiments at Colima (550 and 600 °C; Taran et al., 2000) and Kudryavy (Shmulovich and Churakov, 1998). The Kudryavy volcano is also important as the location of the first discovery of rhenium sulfide (see above). In an experiment of two months duration, rhenium sulfide (ReS_n) was also observed to form in silica tubes together with molybdenite and a solid solution intermediate in composition between the Re sulfide mineral and molybdenite (Korzhinsky et al., 1994).

Among non-ore minerals, the most important at high temperature, and the dominant constituent of many sublimates, is cristobalite. The next most important non-ore minerals precipitated at high temperature are halite and sylvite. However, in contrast to the other minerals discussed above they commonly precipitate along the complete length of a silica tube, i.e., over a very wide interval of temperature. For example, at Momotombo, the range over which these phases precipitated was 860 °C to 200 °C (Quisefit et al., 1989). Other important non-ore minerals precipitating at high temperature are thenardite (Na_2SO_4), aphthitalite ($K_3Na(SO_4)_2$) and salammoniac (NH_4Cl), which like halite and sylvite also persist to low temperature. An observation of possible significance with respect to salammoniac is that the variety of ore minerals is much higher in sublimates containing this mineral than in sublimates from which it is absent (Figure 1). The only non-ore mineral restricted to low temperature is native sulfur, which deposits as a liquid at temperatures <200 °C and as a crystalline solid at temperatures <120 °C.

Important conclusions that can be drawn from the sublimate mineralogy are: 1) the deposition of minerals like molybdenite, chalcopyrite, sphalerite, greenockite, and pyrite at high temperature suggests that only condensates collected from high-temperature fumaroles provide reliable estimates of the concentrations of the corresponding metals in volcanic gases; 2) the predominance of halite and sylvite suggest that chloride complexation may play an important role in metal transport in volcanic gases; 3) the common occurrence of thenardite and salammoniac suggest that sulfate and ammonium may also be important ligands for metal complexation; 4) the variation in mineralogy with magma type and tectonic setting is evidence for corresponding changes in the chemistry of the volcanic gases; and 5) the variety of ore minerals precipitated with salammoniac and the limited variety with halite and sylvite may indicate that ammonium is able to complex a significantly larger number of metals than chloride.

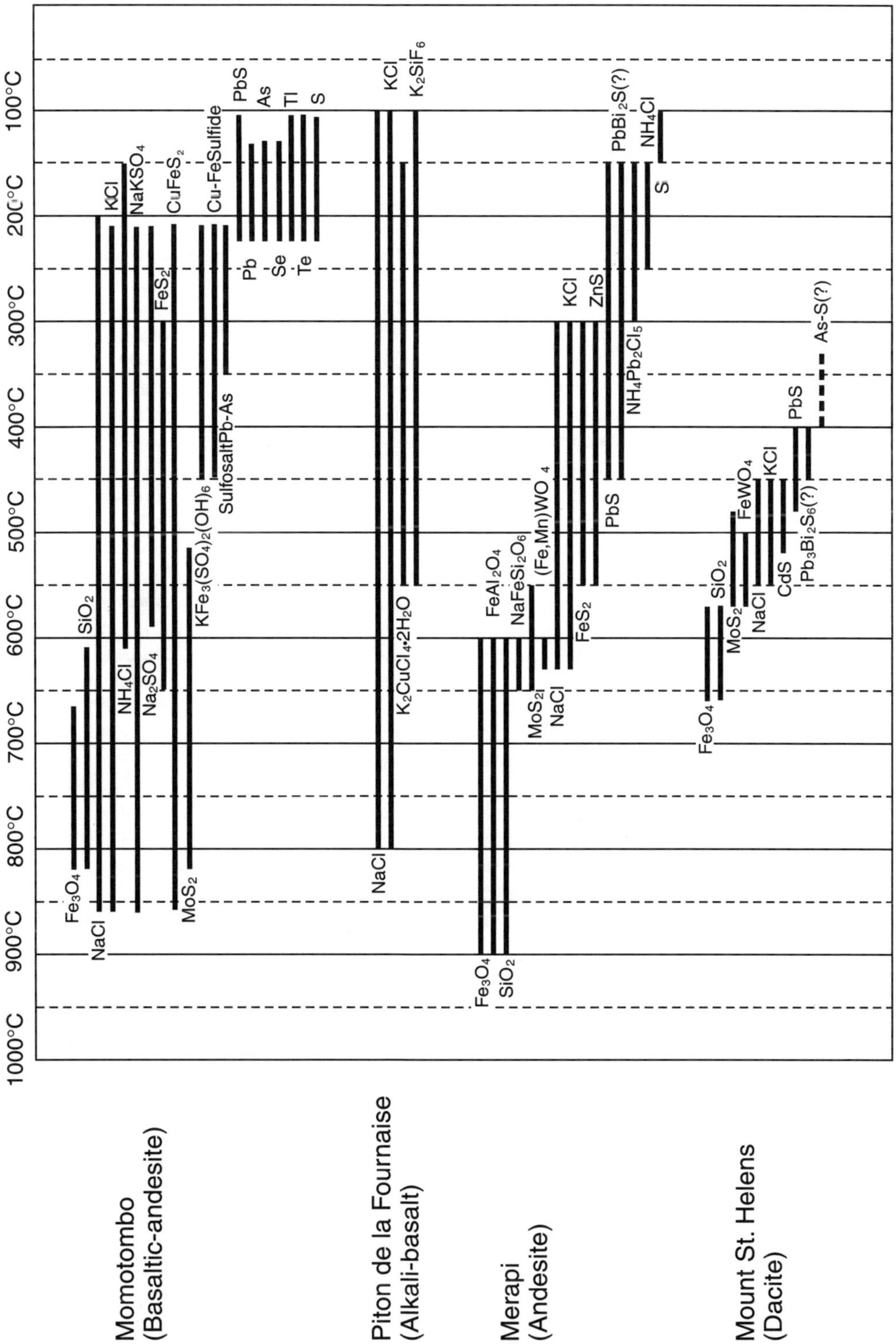

Figure 1. The sublimate mineralogy for silica-tube experiments on fumaroles of four compositionally different volcanoes. Data taken from Quisefit et al., 1989 (Momotombo); Toutain et al., 1990 (Piton de la Fournaise); Le Guern and Bernard, 1982, Bernard, 1985, and Kavalieris, 1994 (Merapi); Bernard and Le Guern, 1986 (Mount St. Helens).

3. THERMODYNAMIC MODELING OF NATURAL GASEOUS SYSTEMS

As discussed in the previous section, measured concentrations of metals in natural volcanic gases vary widely, and depend in large part on the metal considered. Some of this variation is undoubtedly due to differences in natural abundances of the elements. However, a large part of the variation relates to the capacity of a metal to form stable gaseous compounds.

A number of researchers have attempted to estimate metal concentrations in volcanic gases thermodynamically using data on metal volatility and the initial composition of the fluid (e.g., Symonds et al., 1987; Wood, 1987; Spycher and Reed, 1989; Symonds and Reed, 1993; Getahun et al., 1996). These estimates are based on the assumption that for any given temperature the concentration of a metal in the gaseous phase is equal to or less than that required to saturate the vapor with the least volatile mineral containing the metal. They also take into account chemical reactions in the gaseous phase and incongruent evaporation of the minerals. The latter notwithstanding, it follows that metals of highly volatile minerals will saturate at relatively low temperature and those of weakly volatile minerals will saturate at high temperature, leading to a zonation of metals. The concentration of an element in the gas phase of a natural system is thus, to a first approximation, a simple function of the volatility of the corresponding mineral. This volatility can be determined from experimental studies of the partial pressure and composition of the vapor over solid or liquid phases. Data of this type are compiled for many ore and related elements in handbooks of thermochemical data such as the *Gmelin handbook of inorganic chemistry* (periodical), the *Thermodynamic properties tables* of Pankratz (1982, 1984, 1987), *The NBS tables of chemical thermodynamic properties* (Wagman et al., 1982), and *Thermodynamic properties of gases* by Rivkin (1988). In the paragraphs below we review briefly the volatility of a small selection of the more important elements in hydrothermal systems.

3.1. Highly Volatile Elements

Mercury is the most volatile metal in natural systems, and its saturated vapor pressure has been experimentally and theoretically determined over a wide range of temperature using a variety of techniques (e.g., Shpil'rain and Nikanorov, 1971; Cordes et al., 1971; Ambrose and Sprake, 1972; Hubbard and Ross, 1982). Several equations of state have also been developed for mercury vapor (e.g., Alcock et al., 1984). The partial pressure measured over liquid mercury varies from 4×10^{-3} bar at 150 °C to 0.3 bar at 300 °C, which is 3-4 orders of magnitude greater than its vapor pressure over cinnabar (Kulakov, 1975; Ferro et al., 1989).

The next most volatile metal among those commonly occurring in nature is arsenic. Minerals such as orpiment (As_2S_3) and realgar (As_4S_4) are characteristic of low-temperature hydrothermal processes and are commonly found in the sublimates of volcanic gases or post-volcanic fumaroles. The vaporization of realgar has been investigated by mass spectrometry and mass-loss torsion-effusion (Brittain et al., 1983; Munir et al., 1971) as well as by the Knudsen method (Pashinkin et al., 1982; Steblevskii et al., 1986) and Raman spectroscopy (Ozin, 1969; Rogstad, 1972). These studies showed that the primary constituent of the vapor of crystalline As_4S_4 is $As_4S_4^{gas}$, but that minor proportions of $As_4S_5^{gas}$, $As_4S_3^{gas}$, $As_2S_3^{gas}$, $As_2S_2^{gas}$, and S_2^{gas} are also present. The vaporization of orpiment has been investigated using high-temperature mass spectrometry (Lau et al., 1982; Pashinkin et al., 1974), mass spectrometric Knudsen effusion, and thermogravimetric Knudsen effusion (Faure et al., 1973), and measured using a quartz-membrane manometer (Ustyugov et al., 1968). As in the case of realgar, $As_4S_4^{gas}$ was found to be the dominant component of the vapor There were also appreciable proportions of $As_2S_3^{gas}$, and S_2^{gas}, and minor $As_4S_5^{gas}$. The partial pressure of the vapor over realgar varies from 5×10^{-5} to 1 bar at temperatures from 150 to 600 °C, respectively, and over orpiment varies from 1×10^{-5} to 0.5 bar for the same temperature range (Mills, 1974)

Antimony, like As, is also relatively volatile, and this volatility has been investigated extensively through studies of the vaporization of stibnite (Sb_2S_3) (e.g., Hsiao and Schlechten, 1952; Sullivan et al., 1970, Gospodinov et al., 1970). The vapor phase over crystalline and molten stibnite is composed mainly of SbS^{gas}, $Sb_2S_3^{gas}$, and S_2^{gas}. Other components that are present in the vapor include: $Sb_2S_2^{gas}$, $Sb_3S_2^{gas}$, $Sb_3S_3^{gas}$, $Sb_3S_4^{gas}$, $Sb_4S_4^{gas}$, $Sb_4S_5^{gas}$ (Sullivan et al., 1970; Faure et al., 1972; Novoselova and Paschinkin, 1978). The measured vapor pressure over stibnite varies from 10^{-10} to 10^{-3} bar at temperatures from 150 to 600 °C, respectively.

Among the important non-metallic elements occurring in hydrothermal mineral deposits, sulfur is without question the most volatile. The low temperature of melting of sulfur at standard pressure (115 °C) and the low critical temperature of sulfur (444 °C) ensure its high stability in the gas phase. An important characteristic of sulfur vapor is that it displays a high level of polymerization. Sulfur polymers up to S_8 have been reported from experimental studies (Steudel, 1984; 1987), with those of higher order ($S_6 - S_8$) predominating at lower temperature (Rau et al., 1973;

Steudel, 1984; 1987; Lenian et al., 1988). At higher temperatures, S_2^{gas} is the dominant form of sulfur, although there is some disagreement over the temperature at which it becomes dominant. Thus, while Rau et al. (1973) report that the S_2^{gas} polymer predominates at temperatures above 800 K (526.85 °C), Lenian et al. (1988) concluded that the lower temperature limit of S_2^{gas} predominance is 600 K (326 °C). The partial pressure of sulfur vapor over liquid sulfur varies from 10^{-4} bars at 150 °C to 10^{-2} bars at 300 °C.

3.2. *Weakly Volatile Elements*

In comparison with the metals discussed above, minerals containing copper, gold, and silver have extremely low volatility at temperatures commonly encountered in ore-forming hydrothermal systems (<600 °C). Early studies of Cu(I) chloride volatility were conducted by Brewer and Lofgren (1950) at very high temperature (715 °C - 1067 °C) and yielded partial pressures between 0.5×10^{-4} to 6.6×10^{-4} bar. However, the data do not vary systematically with temperature, and total pressure varied significantly. We have estimated the partial vapor pressures of copper (I) chloride ($P_{CuCl^{gas}}$) over the corresponding solid based on data presented in Pankratz (1984) as 2×10^{-15} and 2×10^{-7} bar at 300 °C and 600 °C, respectively. The vaporization of copper (II) has been studied by Dienstbach et al. (1977) using UV/visible and Raman spectroscopy. These authors showed that at elevated temperatures the vapor contains $CuCl^{gas}$, $CuCl_2^{gas}$, and $Cu_2Cl_4^{gas}$.

Owing to the extremely low partial pressure of vapor over native gold, experimental investigations of gold volatility have been performed in a chlorine atmosphere (James and Hager, 1978; Tagirov et al., 1993) and the partial pressures obtained can not be related directly to conditions in natural hydrothermal systems. Based on the results of these studies, gold generally evaporates in the form $Au_2Cl_6^{gas}$ and $AuCl^{gas}$, and at 300 °C and P_{HCl} of 0.4 bars, yields vapor pressures on the order of 10^{-36} bars. Several groups of researchers have reported experimental data on the partial pressures of silver chloride compounds over crystalline and liquid AgCl (e.g., Hildenbrand and Lau, 1996; Tagirov *et al.*, 1993; Gräber and Weil, 1972). Silver chloride evaporation proceeds mainly with the formation of $AgCl^{gas}$ and lesser proportions of $Ag_3Cl_3^{gas}$ (Gräber and Weil, 1972; Visnapuu and Jensen, 1970; Chang and Toguri, 1974). The estimated partial pressures of $AgCl^{gas}$ over liquid (molten) AgCl are 10^{-7} bar at 500 °C and 5×10^{-6} bar at 650 °C.

Owing to the importance of NaCl in natural systems, we decided to also review briefly the volatility of halite, which has been investigated extensively over a wide range of temperatures. These studies have shown that the vapor over halite contains polymers of $Na_nCl_n^{gas}$ dominated by the monomer $NaCl^{gas}$ and the dimer $Na_2Cl_2^{gas}$. (e.g., Kvande, 1979; Ewing and Stern, 1974; Lukashenko et al., 1970; Pitzer, 1996). The measured partial pressures of the vapor over sodium chloride range from 1.9×10^{-15} bar at 300 °C, and 2×10^{-7} bar at 600 °C.

3.3. *Application of Volatility Data*

Volatility data provide an important starting point from which to model vapor-phase transport of elements in nature, and have been used by numerous authors to estimate the trace-element composition of volcanic gases and that of the vapor phase in several active geothermal systems. These models have had some success in predicting the sequences of mineral zonation in the sublimates of volcanic gases and in the case of geothermal systems predicted metals that could potentially have been deposited predominantly from the vapor phase instead of from the liquid. Thus, for example, Quisefit et al. (1989), in simulating the formation of silica-tube sublimates from Momotombo, were able to correctly predict the high-temperature (>650 °C) deposition of native gold, chalcocite and molybdenite, if the corresponding metals were transported as Au-S, Cu-Cl, Cu-Br and Mo-F species. However, they had difficulty in simulating the sequence of mineral deposition at lower temperature. Similarly, although Symonds et al. (1987) were able to correctly predict the high-temperature formation of molybdenite, they overestimated the temperature of sphalerite deposition and failed to predict the deposition of galena despite the fact that this was the principal mineral in sublimates formed at intermediate temperature.

The principal conclusion that can be drawn from the application of volatility data to the behavior of metals in geothermal systems is that, with the exception of mercury, none of the common ore metals fractionate sufficiently into the vapor phase for the latter to constitute an important agent of ore metal transport. Although elements with high volatility like arsenic and antimony are predicted to achieve high concentrations in dry systems, and might therefore be good candidates for vapor-transport in geothermal systems (liquid-vapor), their much higher solubility in the liquid phase (Smith et al., 1987; Spycher and Reed, 1989) precludes their significant fractionation into the vapor phase. Indeed, Spycher and Reed (1989) note that at temperatures ≤300 °C a Broadlands-type geothermal fluid must undergo at least 97% boiling for there to be appreciable concentration of these elements in the vapor phase.

3.4. Vapor-Phase Hydration/Solvation

An important limitation of the modeling discussed in the previous section is that reliance on component fugacities to describe the behavior of gas species ignores potentially important low-energy interactions between the components dissolved in the gas and the gas-solvent. As discussed below, these interactions can significantly enhance the capacity of the vapor to transport metals, and, in some cases, ignoring these interactions may lead to underestimation of the solubility of the metal in the vapor phase by many orders of magnitude.

One of these interactions is predicted to involve the formation of ion-water clusters and has been studied experimentally by several authors (Kebarle, 1975; Davidson and Kebarle, 1976; Kebarle et al., 1977; Sunner et al., 1981; Blades et al., 1990; 1995). The results of these studies suggest that the presence of such clusters may change the properties of dissolved components and solvents sufficiently to allow components with very low volatility to be relatively soluble in water vapor. This explains why the solubility of halite in water vapor, which has been measured by numerous researchers, is several orders of magnitude higher than predicted from volatility data (Styrikovich, 1969; Martynova, 1964; Sourirajan and Kennedy, 1962; Galobardes et al., 1981; Bischoff and Rosenbauer, 1986; Alekhin and Vakulenko, 1987; and Armellini and Tester, 1993). The solubility of NaCl can therefore be described by the following solvation-type reaction:

$$\text{Halite} + n{\cdot}H_2O^{gas} \leftrightarrow NaCl{\cdot}(H_2O)_n{}^{gas} \qquad (1)$$

which involves formation of $NaCl\text{-}H_2O$ clusters; n, the hydration number, varies between 1 and 4. A more complex and potentially more accurate cluster-based statistical mechanical model for NaCl-solvation has been presented by Pitzer and Pabalan (1986).

4. EXPERIMENTAL INVESTIGATIONS OF METAL SOLUBILITY IN WATER VAPOR

In view of the need for experimental data on the solvation of metals by water vapor, we have initiated investigations of a number of systems involving economically important ore-forming metals and elements. As a result of this work, preliminary data are now available for the systems $AgCl\text{-}HCl\text{-}H_2O$, $Au\text{-}HCl\text{-}H_2O$, $SnO_2\text{-}HCl\text{-}H_2O$, and $CuCl\text{-}HCl\text{-}NaCl\text{-}H_2O$.

Experimental studies of the speciation of metals in water vapor can be performed either in dry-vapor systems, i.e., systems undersaturated in respect to liquid water, or in liquid-vapor systems. The former systems involve only solid-vapor interactions, whereas in the latter systems the metals are partitioned among solid, liquid, and vapor. Owing to the overwhelming predominance of molecular species (uncharged) in dry water vapor, investigation of metallic mineral solubility in "dry" systems is associated with fewer technical problems and the number of controlling parameters is smaller. However, it should be noted that the speciation in "dry" systems may differ significantly from that in "liquid-vapor" systems unsaturated with respect to a solid. This results from the fact that in systems unsaturated with respect to solid, the activity of the vapor species is correspondingly lower. We have conducted experimental studies in both kinds of systems. The systems $AgCl\text{-}HCl\text{-}H_2O$, $Au\text{-}HCl\text{-}H_2O$, and $SnO_2\text{-}HCl\text{-}H_2O$ were investigated in the "dry" state, while the system $CuCl\text{-}HCl\text{-}NaCl\text{-}H_2O$ was investigated along the liquid-vapor saturation curve. The latter system was also saturated with respect to a solid. Future experiments are planned to investigate metal partitioning between liquid and vapor in systems unsaturated with respect to a solid.

4.1. Methods

4.1.1. Dry systems

The experimental method is similar to that employed by Migdisov et al. (1998, 1999). Experiments were carried out at temperatures between 300 and 360 °C in an electric furnace (internal diameter 15 cm, height 50 cm) equipped with thick aluminum or copper walls to reduce temperature gradients. A thermal regulator allowed the temperature to be controlled to an accuracy of approximately 0.5 °C. Before runs, the temperature gradient in the experimental system was measured with 3 thermocouples, located at the top, bottom and center of the furnace. The vertical temperature gradient was typically in the range 1.0 to 2.5 °C m^{-1}, but during the first 2 to 3 hours it was between 8 and 20 °C m^{-1} due to the initially cold state of the autoclave. Temperature was measured with two chromel-alumel thermocouples, located in the top and bottom of the furnace.

The experiments were performed in titanium autoclaves, and involved measuring the solubility of synthetic chlorargyrite (AgCl), or cassiterite (SnO_2) or native gold in HCl-bearing water vapor. The autoclaves were constructed from titanium alloy (grade 2 ASTM B348) obtained from Titanium Industries, Inc. (Figure 2). Each autoclave was conditioned initially with nitric acid to produce a protective layer of TiO_2 on the internal surface. Autoclave volumes (100 cm^3) were determined by filling the autoclaves with 25 °C-distilled water from a Teflon flask, and weighing this flask before and after filling. The weighing was performed to an accuracy ±0.1 g. Autoclaves were

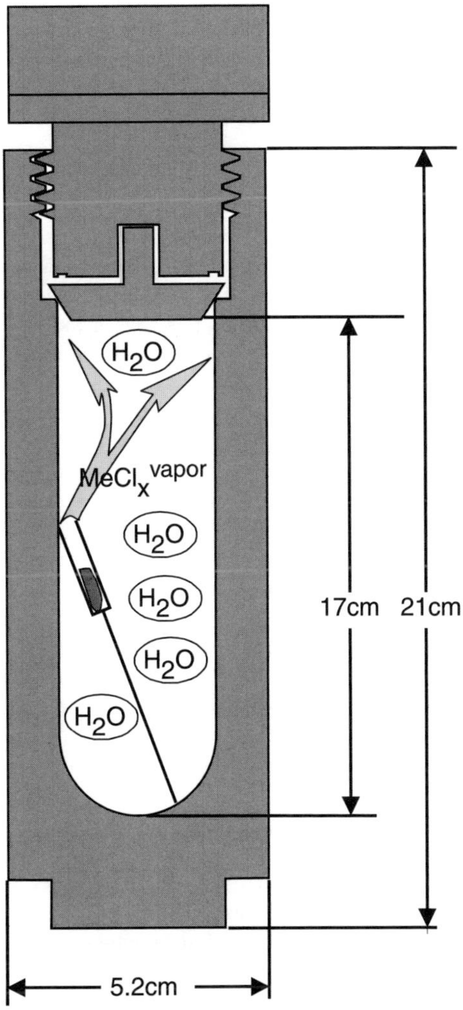

Figure 2. A section through a titanium autoclave of the type used in this study. The autoclaves were constructed from titanium alloy (grade 2 ASTM B348).

loaded with pre-weighed quartz ampoules containing several milligrams of one of the above solids (AgCl, SnO_2, or Au).

The ampoule was suspended near the top of the autoclave using a quartz holder (Fig. 2). A known mass of HCl solution, prepared using distilled nanopure (18-MΩ-cm) water and Optima-grade HCl from Fisher Scientific, was placed in the bottom of the autoclave. The HCl was added in quantities that yielded solutions with pH values of 1.5 to 4.0. Care was taken to ensure that the quantity of HCl solution was insufficient to saturate the system with liquid at the experimental conditions, and that dissolution of the metal was restricted to the vapor. Owing to the extremely low vapor pressures of silver, gold and tin chlorides, compared to that of water vapor, the total pressure in a run was assumed to be equal to the pure water vapor pressure, and was calculated using the measured autoclave volume and the equation of state for water of Kestin et al. (1984).

The redox potentials of the gold and tin systems were buffered by a MoO_2/MoO_3 oxide mixture. One to two grams of this mixture (MoO_2:MoO_3 = 1:1) were placed in a quartz glass tube open at one end, which was suspended in the autoclave. The fO_2 values at the experimental conditions were calculated from data given in Pankratz (1982). This buffer was chosen because of its higher stability in an HCl atmosphere compared to more commonly used Ni/NiO or Fe-containing buffers. Buffer stability calculations were performed using data summarized in Pankratz (1982, 1984). These calculations showed that solid Mo-chlorides are orders of magnitude less likely to form than Fe and Ni chlorides at the conditions investigated. However, the HCl concentration may have varied as a result of interaction between the buffer mixture and gaseous HCl, and consequent removal of HCl from the system. The pH of the condensate was therefore measured after each run. The partial pressure of HCl at the experimental conditions was calculated using the measured pH of the condensate and an assumption that HCl^{gas} behaves ideally.

Before a run, a stream of nitrogen was passed through the autoclave for 20 to 40 minutes to remove atmospheric gases. After the run, the autoclave was air-cooled to room temperature (20 to 30 minutes), and the condensates were collected for analysis. The ampoule was dried at 120°C for 1 to 2 hours to remove water condensed on its walls and on the solid, and was then re-weighed. Chlorargyrite (AgCl) condensed on the walls of the autoclave was dissolved using 3 to 7 ml of nitric acid (pH = 0.5). Gold was dissolved by 5 ml of aqua regia, and tin by a 0.5 HCl solution. The concentrations of dissolved silver and tin in the condensates and washing solutions were determined by graphite furnace-atomic absorption spectroscopy and inductively coupled plasma-mass spectrometry (Activation Laboratories Ltd.), while those of gold were determined by neutron activation at Ecole Polytechnique (Montreal).

A potential source of error in the method is pre-quench condensation of the solid phase on the cold parts of the autoclave during runs. The latter may occur if the sample is at a higher temperature than other internal parts of the autoclave, notably the walls, and could lead to considerable overestimation of metal solubility in the vapor phase. However, since the autoclave was externally heated, it follows that temperatures were highest at the walls and lowest in the center, i.e., the solid reagent was located in the coldest part of the autoclave. Thus, it is very unlikely that any significant condensation occurred during heating and at the experimental temperature. On the other hand, additional quantities of metal could have been released from the solid during quenching, and condensed on the cooler walls of the autoclave together with metal, which had been dissolved in the vapor at the end of the experiment. Another possible source of error is the partitioning of metal from the vapor phase into the liquid (water) during heating or quenching. Owing to the much higher solubility of AgCl, SnO_2 and Au in the liquid phase (by several orders of magnitude higher), this could result in gross overestimation of the solubility of these metals in the vapor phase.

In order to investigate the possibility of significant experimental error due to condensation of metallic species on the walls of the autoclave, we undertook a series of blank experiments in an atmosphere free of water vapor (nitrogen atmosphere, P = 1-80 bar). The mass of solid sublimed in each of these experiments was less than the detection limit of the analytical technique, thereby eliminating the possibility of error due to metallic species condensation.

The issue of experimental error, particularly due to partitioning of components into the liquid water, was further investigated through a series of "kinetic runs". Experiments were carried out under constant water pressure for periods ranging from 1 to 15 days. For all metal species investigated, equilibrium was attained in less than 14 days. Once equilibrium was attained, solubilities were reproducible to approximately ±15-20 %. If there had been significant condensation of metallic species onto the autoclave walls or this species had been partitioned into the liquid water during the experiments, none of the experimental results would have been reproducible, and consequently equilibrium would never have been attained.

4.1.2. Wet system

Experiments measuring the solubility of Cu in water vapor were conducted in a subcritical liquid-saturated CuCl-HCl-NaCl-H_2O (NaCl:HCl = 9:1) system at temperatures of 380 and 400 °C, and total chloride concentrations from 0.01 to 5 *M*. In order to study the partitioning of copper between aqueous vapor and liquid, it is desirable to sample simultaneously the vapor and liquid phases at equilibrium. However, for a variety of reasons, this was not practical in the present study. Owing to the low density of the vapor, a large volume of this phase was needed to obtain enough condensate for analysis. The withdrawal of large volumes of vapor from an autoclave of fixed volume causes the liquid to boil, changing the concentration of both liquid and vapor drastically during the time that a sample is being retrieved. Furthermore, boiling can induce precipitation of $CuCl^{solid}$ in the autoclave or sampling apparatus.

To avoid these problems, experiments were designed to determine copper solubility in the vapor phase by measuring the copper content of condensate collected in isolation of the bulk liquid. Experiments were conducted in the titanium-alloy autoclaves described above. A pressed $CuCl^{solid}$ pellet and copper foil (this buffered the redox potential of the system: CuCl + H_2 ↔ Cu + HCl) were placed in a quartz-glass tube sealed at one end to isolate these solids from the HCl/NaCl solution. The tube and a measured amount of solution were then placed in the autoclave, the tube being oriented vertically with its open end up. The amount of solution, which corresponded to that required to maximize the amount of vapor coexisting with liquid at the conditions of the experiment, was determined from the density data of Bischoff (1991). Three additional quartz-glass tubes (Vycor, 9 mm O.D, 7 mm I.D., volume pre-determined), also sealed at one end, were placed vertically inside the vessels, with their open ends up to collect the vapor condensed at the end of the experiments. At the end of each series of experiments, the autoclaves were quenched rapidly by immersing them vertically in water. This caused vapor occupying the open quartz-glass tubes to be condensed to liquid, and to collect at the bottom of the quartz-glass tubes. The latter were weighed and the solution was then withdrawn and transferred into pre-weighed high-density polyethylene (HDPE) bottles. Any precipitates in the quartz-glass tubes were dissolved with aqua regia, which was then transferred to the HDPE bottles containing the quenched solution. The masses of quenched products were determined by weighing the HDPE bottles and the dried quartz-glass tubes. Other aspects of the experimental procedure were as discussed above for the dry runs.

Theoretically, the mass of solution condensed from vapor in a quartz-glass tube should be the product of the density of the vapor and the internal volume of the tube. However, as quenching is never 100% efficient, some of the high-temperature vapor was lost from the tubes, and the measured masses of condensate were typically lower than the theoretical values. It was therefore necessary to correct for this mass loss to determine the original concentration of copper in the vapor. The correction factor was determined using the following equation:

$$\{Cu\}_{eq} = \{Cu\}_{meas} \times (\text{mass of water vapor})_{calc} / (\text{mass of condensed water})_{meas} \quad (2)$$

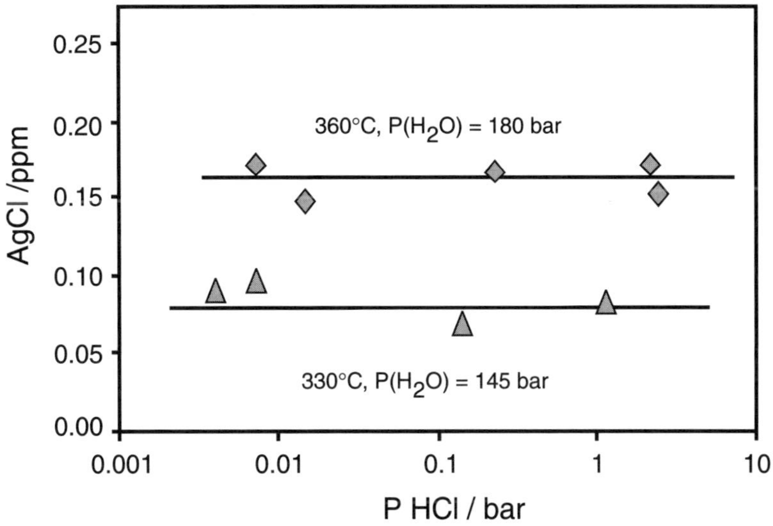

Figure 3. Concentration of silver chloride in water vapor versus the partial pressure of HCl ^{gas}. The zero slope of the data distribution indicates that AgCl solubility was independent of P_{HCl}.

where (mass of water vapor)$_{calc}$ is the mass of water vapor calculated from the internal volume of the quartz glass tube and the density of water vapor at the temperature of the experiment.

The total Cu concentration of each sample was determined either by flame atomic absorption spectrometry (FAAS) or graphite-furnace atomic absorption spectrometry, the latter method being required for samples containing <0.05 mg/kg, Cu, the detection limit for FAAS.

4.2. Results

4.2.1. Dry systems

4.2.1.1. AgCl - HCl - H_2O system. A series of 31 experiments at temperatures of 300, 310, 330, 340, 350 and 360 °C were conducted to evaluate the solubility of AgCl in the vapor phase of the AgCl-H_2O system, and 9 experiments at 330 and 360 °C to evaluate its solubility in the AgCl-HCl-H_2O system. Vapor pressure in the AgCl-H_2O system varied from 20 to 160 bars, and was close to that of the liquid-vapor phase boundary in the system AgCl-HCl-H_2O.

The mole fraction of AgCl in the vapor was calculated from the silver concentration in the quenched condensate. Owing to the low partial pressure of AgCl^{gas}, the total pressure in the autoclave was effectively that of H_2O vapor. Consequently the mole fraction of AgCl;

$$X_{AgCl} = \frac{M_{AgCl}}{M_{H_2O} + M_{AgCl}}$$

in the vapor could be approximated as:

$$X_{AgCl} \cong \frac{M_{AgCl}}{M_{H_2O}},$$

where M is number of moles of the corresponding compound.

The dependence of the concentration of silver chloride in water vapor on the partial pressure of HCl is illustrated in Figure 3. From this diagram, it is evident that AgCl solubility is independent of P_{HCl}, within experimental error, at constant temperature and partial pressure of H_2O. This indicates that there was one dominant gaseous silver species with a stoichiometry Ag:Cl = 1:1.

The concentrations of AgCl in the vapor phase determined from our experiments are shown as a function of P_{H_2O} in Figure 4. From this figure, it can be seen that the isothermal concentration of AgCl in the vapor increases sharply with increasing H_2O pressure.

4.2.1.2. Au-HCl-H_2O system. The solubility of Au in the vapor phase of the system Au-HCl-H_2O was investigated via 23 experiments carried out at temperatures of 300, 340 and 360°C. Water vapor pressures were held constant at each isotherm, from 86 bars (at 300°C) to 144 bars (at 340 and 360°C), and were calculated to be within the vapor-only field at the temperatures of interest.

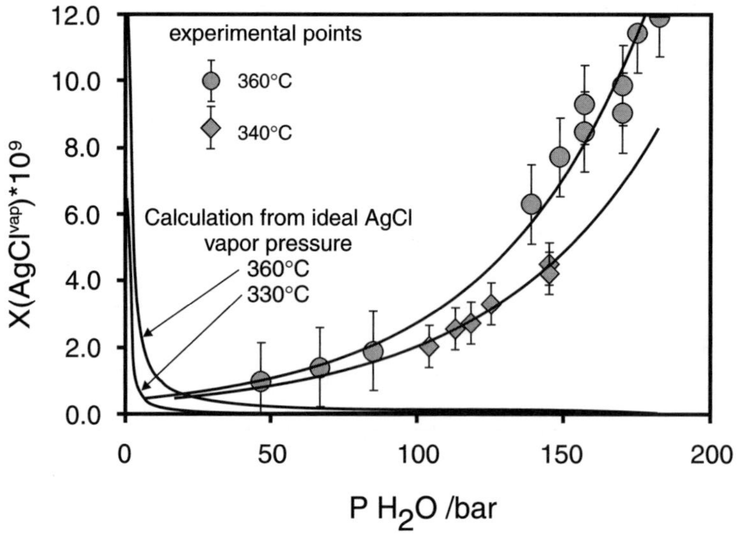

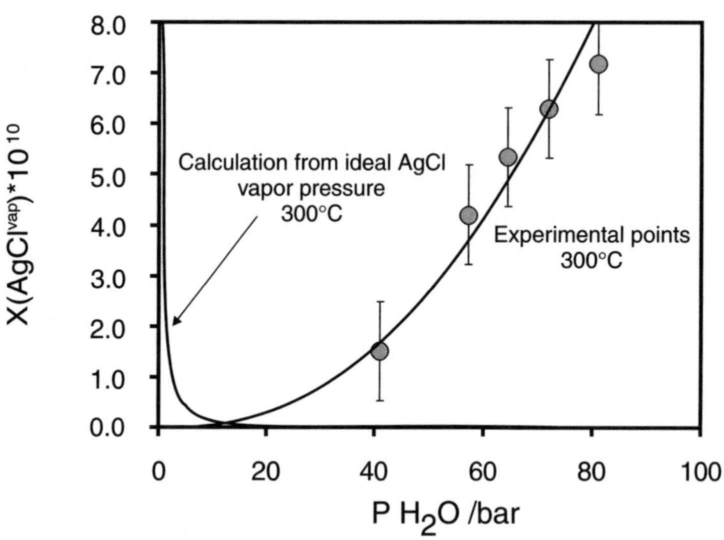

Figure 4. AgCl concentrations in water vapor versus the pressure in the system. Owing to the low partial pressure of $AgCl^{gas}$, the total pressure in the autoclave was effectively that of the water. The heavy solid lines show the silver chloride concentration calculated assuming no hydration of AgCl. The calculations were performed using the data on $AgCl^{gas}$ vapor pressure over crystalline chlorargyrite reported by Tagirov et al. (1993).

The mole fraction of Au in the vapor was calculated from its concentration in quenched condensates using an equation analogous to that employed for Ag:

$$X_{AuCl_m} = \frac{M_{AuCl_m}}{M_{H_2O} + M_{AuCl_m}}$$

and is shown as a function of P_{HCl} in Figure 5. From this figure, it is apparent that the solubility of Au increases with increasing P_{HCl}, but decreases with increasing temperature.

4.2.1.3. SnO_2-HCl-H_2O system. Forty experiments were conducted in the system SnO_2-HCl-H_2O to determine the solubility of SnO_2 in the vapor phase at temperatures of 300 and 320 °C.

The mole fraction of Sn in the vapor was calculated in essentially the same way as that of Ag and Au. Its

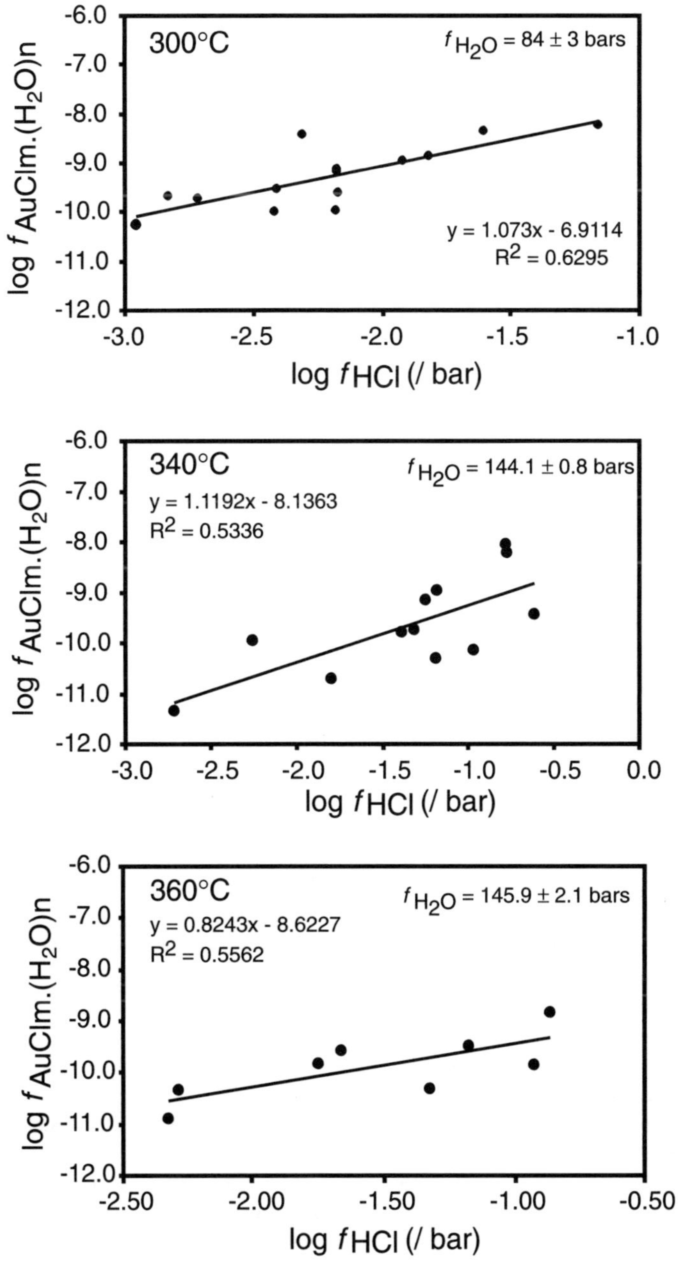

Figure 5. Mole fractions of $AuCl_m{\cdot}H_2O_n$ in water vapor as a function of P_{HCl} in the Au-H_2O-HCl system at 300, 340 and 360 °C. The slope of ~1 for the these temperatures suggests the stoichiometry $AuCl(H_2O)_n$.

dependence on HCl pressure is illustrated in Figure 6. As was the case for the Au-HCl-H_2O system, the solubility of SnO_2 increased with increasing P_{HCl}. Preliminary data for 300 °C suggest that Sn solubility also increases with P_{H_2O}, as in the AgCl-H_2O-HCl system (Fig. 4).

4.2.2. Wet system

4.2.1.1. CuCl - NaCl - HCl - H_2O system. The solubility of CuCl in water-saturated vapor was investigated in a series of 26 experiments conducted at temperatures of 380 and 400 °C. The concentration of

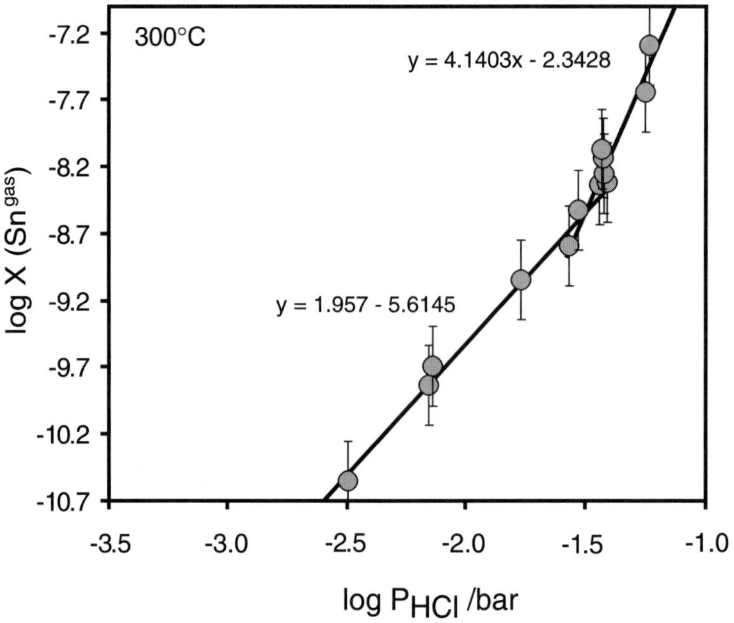

Figure 6. The mole fraction of tin in water vapor as a function of P_{HCl} in the SnO_2-H_2O-HCl system at 300 °C. The slope of the data distribution indicates that tin is present mainly as $SnCl_2 \cdot (H_2O)_n{}^{gas}$, and at high P_{HCl} as $SnCl_4 \cdot (H_2O)_n{}^{gas}$.

copper in the vapor appears to correlate negatively with total chloride concentration (Fig. 7, Table 6). We attribute this correlation to a decrease in vapor density related to a corresponding increase in the salinity of the system (cf. Bischoff, 1991). After corrections were made for vapor density, we were able to conclude to a first approximation that the copper concentration in the vapor is independent of the total Cl^- activity in the coexisting liquid at the temperatures investigated. In view of this, the Cu-Cl stoichiometry of the dominant copper gas species was interpreted to be 1:1. In contrast to its negative correlation with Cl, copper concentration correlates positively with vapor density. The significance of this correlation is discussed below.

5. DISCUSSION

The partitioning of metals into water vapor is the result of two processes that proceed in parallel, namely the dissolution of the metals, i.e., their solvation as $MeL_m{}^{vapor}$ (Me refers to the metal and L to the ligand), and the development of a partial pressure of $MeL_m{}^{gas}$ over the crystalline phase. As the properties of the unsolvated gaseous complexes of the metals investigated experimentally in this study are known (Pankratz 1982, 1984), the concentrations of $MeCl_m{}^{gas}$ can be calculated for various values of P_{H2O} and measured autoclave volumes. The calculations were made assuming inert behavior of components, and predicted maximum concentrations of the four metals (Ag, Au, Sn, and Cu) investigated that are all at least an order of magnitude lower than the analytical detection limit in our experiments. It is also known from the properties of unsolvated gaseous metal complexes that the partial pressure of $MeCl_m{}^{gas}$ over the crystalline phase is constant at constant temperature and constant P_{HCl} but the concentration of $MeCl_m{}^{gas}$ decreases with increasing P_{H2O}. By contrast the metal solubilities in the vapor phase determined experimentally increase exponentially with increasing P_{H2O} at constant temperature and fixed values of P_{HCl}. We therefore conclude that $MeCl_m{}^{gas}$ does not contribute significantly to the total concentration of metals in the vapor phase, and that any description of the system must also take into consideration some interaction among the components of the gaseous mixture.

5.1. Data Treatment

5.1.1. Dry systems

The interactions referred to above could be, firstly, the formation of new stable gaseous compounds with specific stoichiometry (chemical reaction), and secondly, the hydration of metal chloride by water vapor

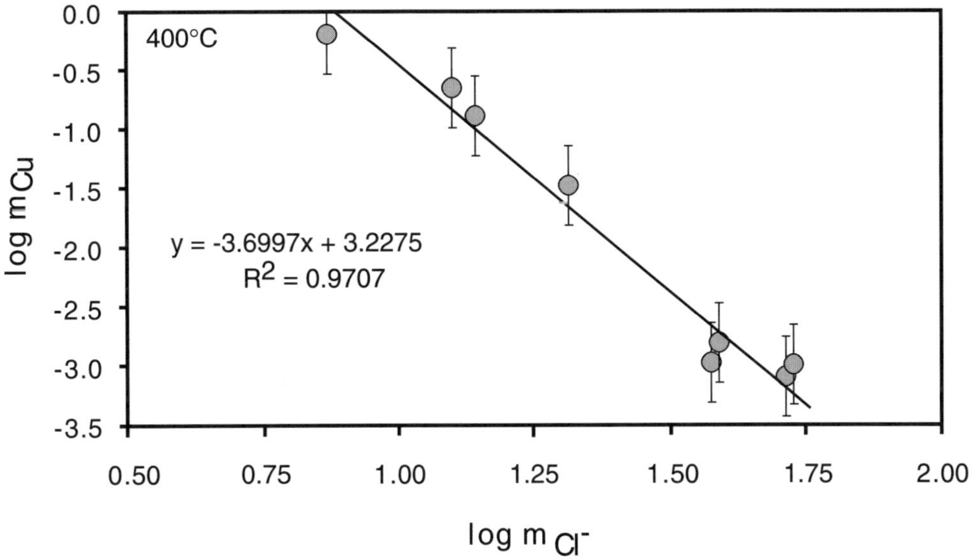

Figure 7. The concentration of copper in the vapor phase versus the total chloride concentration in the system CuCl-HCl-NaCl-H_2O at 400 °C. The negative correlation between these two parameters reflects the decreasing density of vapor due to the salinity increase.

(solubility). In both cases (chemical reaction and solubility) this process can be expressed by the following equations:

$$AgCl^{gas} + m\cdot HCl^{gas} + n\cdot H_2O^{gas} \leftrightarrow AgCl_{m+1}\cdot(H_2O)_n{}^{gas} + m/2\cdot H_2{}^{gas} \quad (3)$$

$$Au^{gas} + m\cdot HCl^{gas} + n\cdot H_2O^{gas} \leftrightarrow AuCl_m\cdot(H_2O)_n{}^{gas} + m/2\cdot H_2{}^{gas} \quad (4)$$

$$SnO_2{}^{gas} + m\cdot HCl^{gas} + n\cdot H_2O^{gas} \leftrightarrow SnCl_m\cdot(H_2O)_{n+0.5m}{}^{gas} + (2\text{-}0.5m)/2\cdot O_2{}^{gas} \quad (5)$$

The Gibbs-Pointing correction was used to account for the change in the pressure (P^o) of the component over the crystalline phase from that for a total pressure of $P = 1$ bar (standard state) to P_2 = pressure for the run as follows:

$$\ln \frac{P_2^o}{P_1^o} = \int_{P_1}^{P_2} \frac{V^o}{R\cdot T}\, dP \quad (6)$$

Given that changes in the molar volume of the crystalline phase with temperature are insignificant for the temperatures investigated, it was assumed that V^o is a constant and

$$\ln \frac{P_2^o}{P_1^o} = \frac{V^o\cdot(P_2 - P_1)}{R\cdot T} \quad (7)$$

(where V^o - is the molar volume and P_1^o, P_2^o are the pressures of this component in states 1 and 2).

It was further assumed from Raoult's Law that:

$$\log P_{MeCl} = \log X_{MeCl} + \log P_{H_2O} \quad (8)$$

Finally, assuming that the partial pressure of HCl is close to its fugacity at the experimental conditions, and taking into account the fact that the equilibrium constant of a homogeneous gaseous reaction is independent of the total pressure in the system, the stoichiometric ratio *Me:Cl* for the hydrated gaseous species was derived from the equilibrium expression for the corresponding reaction as follows:

$$\left(\frac{\partial \log X_{AgCl_{m+1}\cdot(H_2O)_n}}{\partial \log P_{HCl}}\right)_{T,P_{H_2O},P_{H_2}} \cong m+1 \quad (9)$$

$$\left(\frac{\partial \log X_{AuCl_m\cdot(H_2O)_n}}{\partial \log P_{HCl}}\right)_{T,P_{H_2O},P_{H_2}} \cong m \quad (10)$$

$$\left(\frac{\partial \log X_{SnCl_m\cdot(H_2O)_{n+0.5m}}}{\partial \log P_{HCl}}\right)_{T,P_{H_2O},P_{H_2}} \cong m \quad (11)$$

The approach employed here is similar to one developed by Marshall (1972) for treating liquid aqueous solutions and extended to solutions involving vapor by Armellini and Tester (1993). One of the limitations of this approach is that there is commonly considerable difficulty in assigning appropriate standard states to components if the system is heterogeneous (comprises more than one state). Application of the Gibbs-Pointing correction to our two-state (solid-vapor) systems overcame this limitation by making them effectively homogeneous.

The stoichiometric relationship Me:Cl was determined from the slope of the trend of log $X_{MeCl_m \cdot (H_2O)_n}$ versus log P_{HCl}, which was essentially linear for each of the metals investigated; the experiments were conducted at constant partial pressure of water and hydrogen, and constant temperature. Based on equations (9-11) the stoichiometry of the dominant silver and gold species was interpreted to be $AgCl \cdot (H_2O)_n{}^{gas}$ and $AuCl \cdot (H_2O)_n{}^{gas}$, respectively. Tin was interpreted to be present mainly as $SnCl_2 \cdot (H_2O)_n{}^{gas}$, and at high P_{HCl} as $SnCl_4 \cdot (H_2O)_n{}^{gas}$. Significantly, whereas the Me:Cl ratio of the species interpreted to be dominant in the AgCl-HCl-H_2O and SnO_2-HCl-H_2O systems is the same as in the corresponding H_2O-free systems, that for Au is different (1:1 vs 1:3). This indicates that the presence of water vapor increases the stability of gold chloride (I) in gaseous solutions with respect to that of gold chloride (III).

The hydration number (n) of the dominant species in our experiments was determined from the relationship between log $X_{MeCl_m \cdot (H_2O)_n}$ and log P_{H_2O}, which is given by the following equations derived from the expression for the equilibrium constant of the corresponding reactions:

$$\left(\frac{\partial \log X_{AgCl \cdot (H_2O)_n}}{\partial \log P_{H_2O}}\right)_T \cong n-1 \quad (12)$$

$$\left(\frac{\partial \log X_{AuCl \cdot (H_2O)_n}}{\partial \log P_{H_2O}}\right)_{T,P_{HCl},P_{O_2}} \cong n-1-\frac{m}{2} \quad (13)$$

$$\left(\frac{\partial \log X_{SnCl_m \cdot (H_2O)_{n+0.5m}}}{\partial \log P_{H_2O}}\right)_{T,P_{HCl},P_{O_2}} \cong n-1 \quad (14)$$

In the case of Ag, the slope of log $X_{MeCl_m \cdot (H_2O)_n}$ versus log P_{H_2O} is ~2 for each of the temperatures investigated (Fig. 8). As this slope represents the term "*n-1*", we therefore conclude that the above complex has a hydration number of 3. However, it should be noted that the magnitude of the experimental error only permits estimation of the hydration number (n) to a precision of one integer. We interpret this complex to be $AgCl \cdot (H_2O)_3{}^{gas}$

$$AgCl^{cryst} + 3 \cdot H_2O^{gas} = AgCl \cdot (H_2O)_3{}^{gas} \quad (15)$$

Considering that silver is coordinated by three molecules of water and one of chlorine in the species $AgCl \cdot (H_2O)_3{}^{gas}$, it also seems reasonable to predict that the silver is in four-fold coordination.

Unfortunately, the system AgCl-HCl-H_2O was the only dry system investigated, in which the solid dissolved congruently and there was no dependence on f_{O_2}. Investigation of the other systems required that f_{O_2} be buffered, which introduced experimental problems in the runs conducted at constant P_{HCl}. Reaction of HCl with the assemblage MoO_2/MoO_3, which was used to buffer f_{O_2}, although not appreciable, was sufficient to cause some variation in P_{HCl}. In order to estimate the hydration number n for the gold and tin species, it was necessary to replace the graphical method used for silver (Fig. 8) by a numerical approach. Values of n were calculated for each temperature by minimizing the value of the error, using the equation:

$$\text{ERROR} = X_{MeCl}{}^{experimental} - X_{MeCl}{}^{calculated}(n) \quad (16)$$

where $X_{MeCl}{}^{experimental}$ are the values obtained in the runs, and $X_{MeCl}{}^{calculated}(n)$ are the values calculated using the constants obtained for different n. This minimization algorithm is based on a golden section search and parabolic interpolation, and was implemented using a computer program similar to that of Forsythe et al (1976). On the basis of the data presently available from our experiments, gold is tentatively predicted to dissolve in the vapor phase as $AuCl \cdot (H_2O)_3$ and tin as $SnCl_2 \cdot (H_2O)_4$, and possibly as $SnCl_4 \cdot (H_2O)_2$ at low and high partial pressures of HCl, respectively.

5.1.2. Wet system

As discussed above, the Cu:Cl stoichiometry of the dominant copper species in the vapor was 1:1. The solubility of copper in the vapor phase may therefore be described by the reaction:

$$CuCl^{aq} + nH_2O^{vapor} = CuCl \cdot (H_2O)_n{}^{vapor} \quad (17)$$

or

$$CuCl^{solid} + nH_2O^{vapor} = CuCl \cdot (H_2O)_n{}^{vapor} \quad (18)$$

Reaction (18) is identical to reaction (17), except for the phase designation of the reactant copper species.

As discussed earlier, one of the difficulties in treating heterogeneous systems (more than one state) is assigning appropriate standard states to components. In the case of the heterogeneous dry systems (solid-vapor), we were able to overcome this difficulty by applying the Gibbs-Pointing correction, thereby making them effectively homogeneous. Unfortunately this solution is not available to us for our wet system (solid-liquid-vapor). However, several studies of the NaCl-H_2O system have shown that the solubility of NaCl in the vapor phase can be modeled semi-empirically using a relationship derived from classic equilibrium thermodynamics and based on the density of water vapor (Martynova, 1964; Styrikovich, 1969;

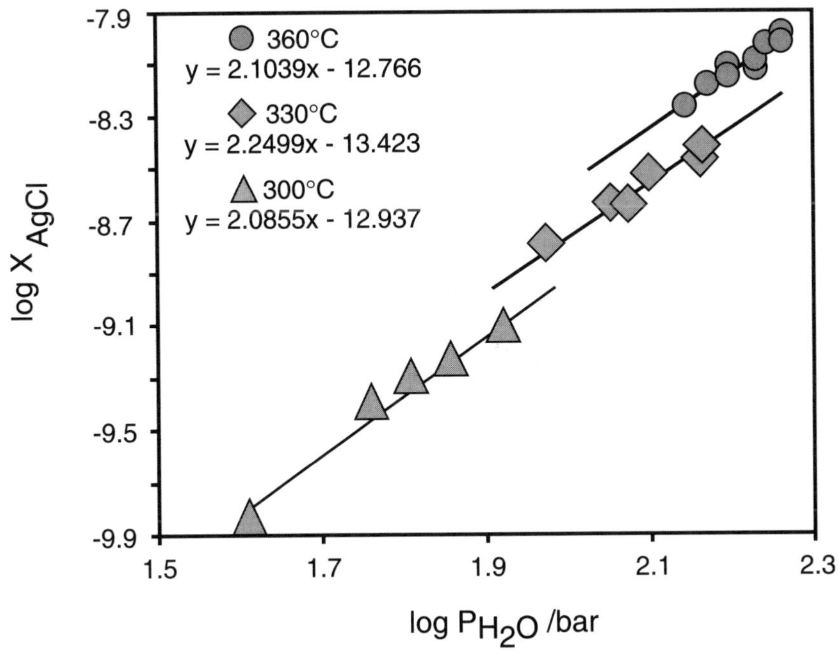

Figure 8. Logarithm of the mole fraction of $AgCl{\cdot}(H_2O)_n$ versus the logarithm of the partial pressure of H_2O for experiments run at 300, 330 and 360 °C. The slope of ~2 for each these temperatures suggests the stoichiometry $AgCl{\cdot}(H_2O)_3$.

Galobardes et al., 1981; Armellini and Tester, 1993). The model assumes that equilibrium between solid NaCl and water vapor is achieved through a solvation-type reaction:

$$NaCl^{solid} + nH_2O^{vapor} = NaCl\,(H_2O)_n{}^{vapor} \quad (19)$$

and that, due to the low density of the water vapor, its solvation power is low and the dissolved NaCl occurs as ion pairs (Mesmer et al., 1988). The equilibrium relationship for reaction (19) is:

$$K_c = \frac{a_{NaCl(H_2O)_n}}{a_{NaCl^{solid}} \cdot a^n_{H_2O^{vapor}}} \approx \frac{C_{NaCl}}{\rho^n_{H_2O^{vapor}}} \quad (20)$$

i.e.,

$$\log C_{NaCl} = n \cdot \log \rho_{H2O} + \log K_c = n \cdot \log \rho_{H2O} - \frac{A}{T} + B \quad (21)$$

where C_{NaCl} is the concentration of NaCl in the vapor phase (ppm or ppb) and A and B are constants. Armellini and Tester (1993) assumed that the stoichiometric coefficient *n* was also a constant based on a review of the literature, which indicated that it varied minimally with temperature and density.

Because, in the present experiments, the pressure change is very small (around 3-5 bars), the fugacity of $CuCl^{gas}$ is directly related to its concentration in the vapor. The equilibrium relationship therefore can be approximated as described above for NaCl:

$$\log K_c = \log m_{CuCl\cdot(H_2O)_n} - n \cdot \log \rho_{H_2O} \quad (22)$$

where K_c is the equilibrium relationship and ρ_{H_2O} is the density of water vapor.

Rewriting equation (22), we obtain:

$$\log m_{CuCl\cdot(H_2O)_n} = n \cdot \log \rho_{H_2O} + \log K_c \quad (23)$$

and differentiating equation (23) with respect to log ρ_{H_2O}, we have :

$$\left(\frac{\partial \log m_{CuCl\cdot(H_2O)_n}}{\partial \log \rho_{H_2O}}\right)_T \cong n \quad (24)$$

As the log molality of copper in the vapor and log density of water vapor were linearly related in our experiments (Fig. 9), log K_c and *n* are given by the corresponding intercept and slope, respectively. The slopes of linear regressions through the data are 3.55, and 3.78 at 380 °C, and 400 °C, respectively, and the hydration number is thus 4 at both temperatures investigated. Reaction (17) therefore becomes:

$$CuCl^{aq} + 4H_2O^{vapor} = CuCl \cdot 4H_2O^{vapor} \quad (25)$$

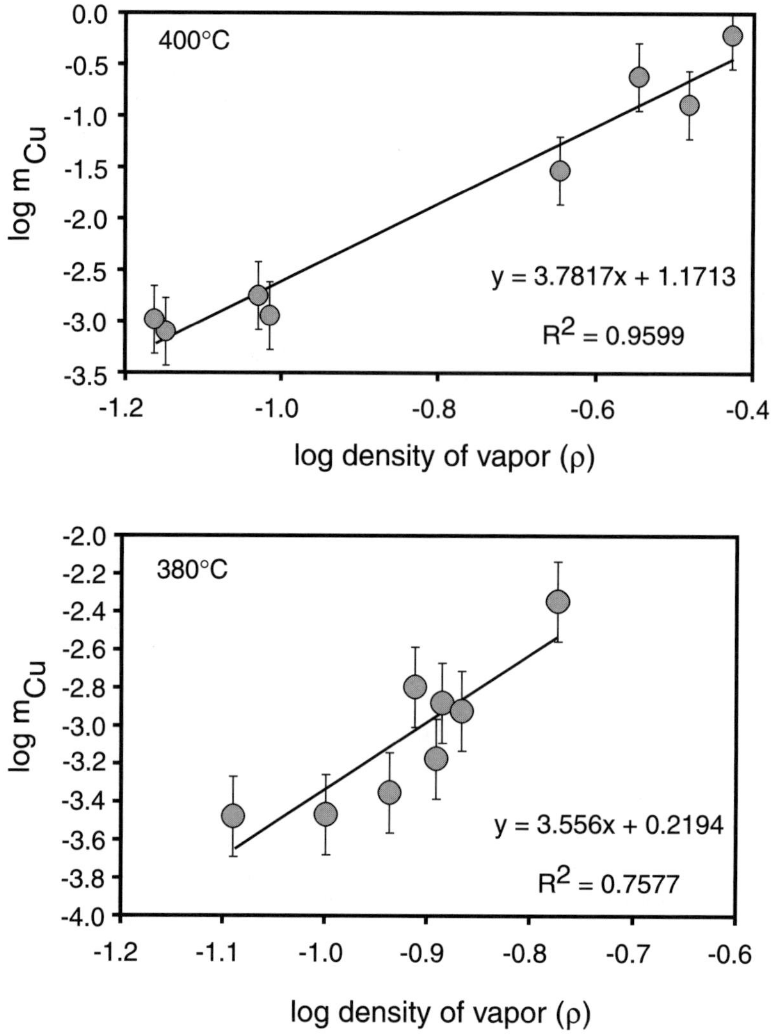

Figure 9. Concentration of copper in water vapor vs.water vapor density at temperatures of 380 and 400 °C. The linear correlation between log molality and log ρ_{H_2O} suggests that copper solubility in the vapor is controlled mainly by vapor density.

and the values of the equilibrium relationship (log K_c) are 0.22 and 1.17 at 380 and 400 °C, respectively.

5.2. Comparison of Experimental Data with Data from Natural Systems

Although the experimental data reported above are for the most part preliminary and restricted to a comparatively limited temperature range (300 and 400 °C), they do at least permit a first-order comparison with data reviewed earlier on the concentrations of the corresponding metals in natural systems. The major limitation in comparing the measured solubilities of the metals in the vapor phase to the apparent solubilities of these metals in volcanic gases is the much higher temperature of the latter. Another limitation is that, in the case of Ag and Cu, metal solubility in volcanic gases is generally controlled by saturation with respect to phases different to those employed in our experiments, e.g., native silver or argentite (Ag_2S) versus chlorargyrite (AgCl) in the case of Ag, and chalcopyrite versus nantokite (CuCl) in the case of Cu. On the other hand, HCl is generally the next most abundant component of volcanic gases after H_2O, CO_2 and SO_2, and commonly comprises several mole percent of the bulk gas, whereas H_2S typically makes up <1 mole percent of the gas.

The maximum concentration of Ag reported for a volcanic gas is 250 ppb, which was measured in a condensate of high-temperature (830 °C) gases collected from a fumarole at Poas (Table 2). However, if we restrict temperature to <400 °C, the highest Ag value drops to 20 ppb and corresponds to condensates of gases released at 170 °C from a fumarole at Cerro Negro (Gemmell, 1987); typically the Ag concentrations of gases at temperatures <400 °C are in the range 3 to 10 ppb (Gemmell, 1987). In our experiments, we obtained mole fractions of Ag of 7.2×10^{-10} and 1.27×10^{-8} at pressures close to water saturation and temperatures of 300 and 360 °C, respectively. These mole fractions correspond to concentrations of 6 ppb and 100 ppb, respectively, and are at least 3–4 orders of magnitude higher than those predicted thermodynamically from volatility data (Tagirov, 1993). However, as can be seen from the data reported above, they compare very favorably with the concentrations measured in condensed gases from volcanic fumaroles.

The highest concentration of Au reported in the literature for the condensate of a volcanic gas is 32 ppb collected from a fumarole on Usu volcano (Giggenbach and Matsuo, 1991). Interestingly, the temperature of the gas was only 165 °C. More typically the concentrations range between 1 and 3 ppb (Gemmell, 1987). By contrast, the concentrations corresponding to the maximum mole fractions of Au obtained in our experiments were 1.1 ppm and 175 ppb at 300 and 360 °C respectively.

Tin concentrations in volcanic gas condensates reported in the literature are significantly higher than those reported for Ag or Au, and range up to 7.2 ppm (for a condensate from a fumarole at Usu volcano degassing at a temperature of 649 °C). At temperature conditions closer to those of our experiments, the concentrations are <2 ppm, e.g., 1.4 ppm for a fumarole degassing at 344 °C (Gemmell, 1987). These concentrations are at least two orders of magnitude higher than those obtained in our experiments, which were conducted at 300 °C, and yielded a maximum mole fraction of Sn of 7.2×10^{-9}, or a concentration of 47 ppb. The reason for this large discrepancy between nature and experiment is unclear. However, it should be emphasized that our investigation of tin solubility is at a very preliminary stage and that it is complicated by the dual valence state of Sn.

Concentrations of Cu in volcanic gases are considerably higher than those of Ag or Au, but similar to that of Sn, the highest reported value being 8.4 ppm from a high-temperature (770 °C) fumarole on Momotombo (Gemmell, 1987). For lower temperature fumaroles (< 400 °C), condensate copper concentrations are typically in the range 200 to 800 ppb. By contrast, concentrations in our experiments (water-saturated) are all substantially higher, ranging from 22 to 277 ppm at 380 °C and 250 ppm to 3.9 wt. % at 400 °C. Although the highest of these values may seem to be unreasonably high, it is interesting to note that Ulrich et al. (1999) report an average copper concentration of 3.3 wt. % for vapor-rich fluid inclusions (analysed by LAM-ICPMS) from the Bajo de la Alumbrera porphyry copper deposit, Argentina; the inclusions coexist with liquid-rich inclusions that homogenize at temperatures between 550 and 650 °C. A possible explanation for the relatively low concentrations of Cu in volcanic gases may be the nature of the saturating phase, i.e., chalcopyrite versus nantokite in our experiments. This does not, however, explain the high concentration of Cu in the porphyry inclusions, which may point to a ligand other than Cl playing an important role in metal transport in porphyry systems.

Given the relatively simple nature of the systems investigated experimentally and the preliminary state of some of these investigations, the results available to date are most encouraging. There is thus every reason to believe that as more complexity is added to these systems, it should be possible to reproduce with considerably greater accuracy the metal concentrations of natural vapors. The next obvious step in experimental investigations of the solubility of metals in aqueous vapors will be to introduce sulfur, a step that is facilitated by ongoing work in our group on the solubility of metals in H_2S (Migdisov et al., 1998, and Zakaznova-Iakovleva et al., 2001).

5.3. Role of Vapor Transport in Ore genesis

A question raised in the introduction to this paper, and one of growing interest to many researchers, in the light of recent reports of high concentrations of metals in vapor-rich fluid inclusions (Ulrich et al, 1999), is whether the concentrations of metals in natural hydrothermal vapors are sufficient to concentrate metals to potentially economic levels. Although it is clearly premature to try and answer this question rigorously, the data made available by our experiments provide an opportunity to roughly evaluate the impact of vapor transport of metals on the ore-forming process. We do this via a relatively simple thermodynamic model to track the removal of silver from rock during progressive interaction with water vapor.

For simplicity, we have restricted the system to the five elements: Ag, Cl, S, O, and H. Chloragyrite was used to represent the Ag phase, and quartz, the inert host-rock. A quartz column containing 5 ppm AgCl was subdivided into 44 segments, or reactors, each having a mass of 1 kg, and differing in temperature

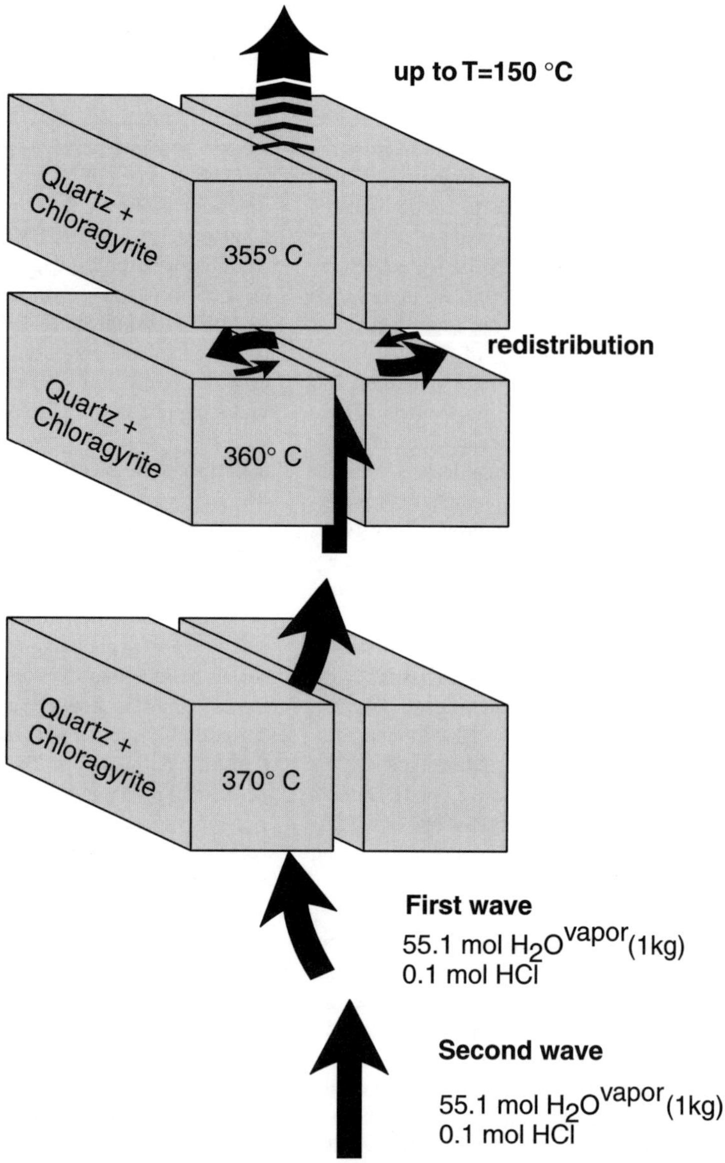

Figure 10. Schematic diagram describing a simplified thermodynamic model in which silver in a rock column composed of chloragyrite and quartz is mobilized during progressive interaction with water vapor. See text for further details.

from the preceding reactor by 5 °C; temperature ranged from 370 °C at the bottom of the column to 150 °C at the top. The pressure in each reactor was that of saturated water vapor (Figure 10). A 1-kg aliquot of water vapor was introduced at the base of the column and allowed to proceed linearly upwards, equilibrating successively with the reactors in the column. This single batch experiment simulated the alteration of the column at a fluid/rock ratio of 1:1. In order to evaluate the effects of higher fluid/rock ratios, additional 1-kg aliquots of fluid were passed through the previously altered column and brought to equilibrium with each reactor along the path.

The results of our simulation are shown in Figure 11. As is evident from this figure, there was appreciable mobilization of silver. After reaction with the first

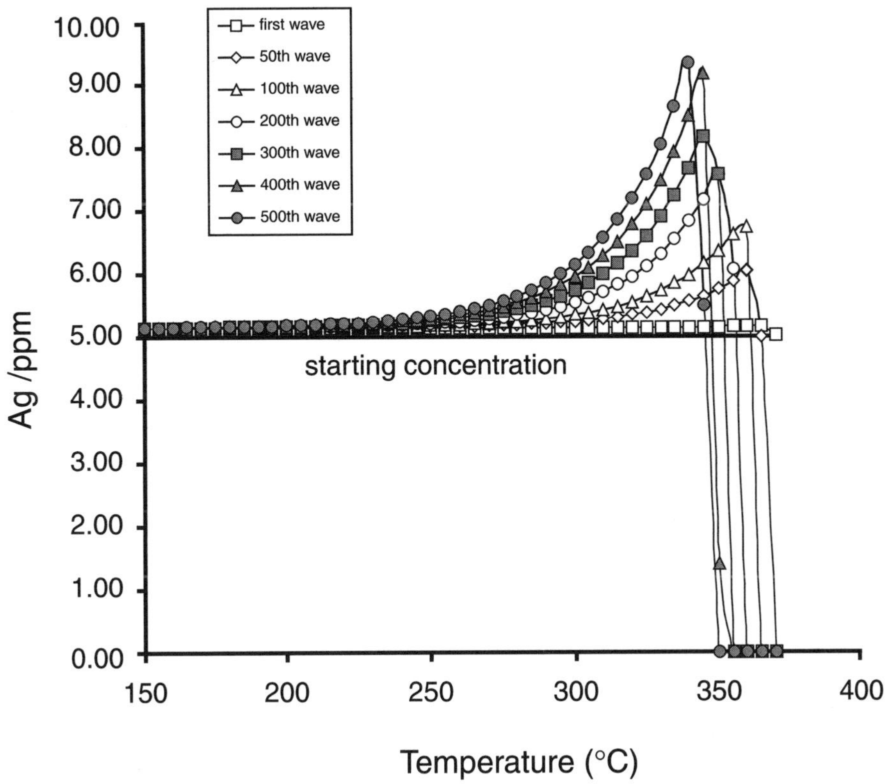

Figure 11. Results of a simulation of the reaction of water vapor with a chloragyrite-bearing quartz vein. The diagram shows the distribution of Ag concentrations in the quartz as a function of temperature at varying vapor-rock ratios (represented by the number of aliquots introduced into the vein).

batch of fluid (fluid/rock ratio, 1:1), the silver content in quartz dropped from 5 ppm to 3 ppm at the base of the column, and gradually increased with decreasing temperature upwards, to a maximum of 5.3 ppm at 365 °C and then gradually decreased to 5.1 ppm at 150 °C. After 50 aliquots of water vapor had been introduced into the column, the maximum silver content had increased to 6 ppm, and the temperature of this maximum had decreased to 355 °C. Most significantly, all the silver formerly present in reactors at 5 °C above this temperature (the first two reactors) had been removed. With further additions of fluid, the peak gradually became better defined, and continued to migrate to lower temperature; after 100 and 500 aliquots of water vapor had been introduced the maxima were 6.8 and 9.3 ppm, respectively, and the corresponding temperatures were 350 and 335 °C. Again, as noted above, all the silver had been removed from reactors at temperatures > 5 °C above those of the maxima.

Although highly idealized, the calculations presented above show that water vapor may be an effective agent for the transport of silver in vapor-dominated hydrothermal systems at temperatures above 300 °C, and an important vehicle for the redistribution of silver in many boiling hydrothermal systems.

6. CONCLUSIONS

The idea that aqueous vapors can transport metals in appreciable concentrations, and may be responsible for the formation of metallic mineral deposits has been around for over 350 years, but in the modern era has been largely ignored. Arguments put forward by Henley and McNabb in the late 1970's, and recent evidence of high concentrations of metals in vapor-rich fluid inclusions, demand that the role of vapor transport in metallic mineral concentration be further evaluated. In this paper we have summarized the data available on the concentrations of metals in volcanic gases (the best examples of possible ore-forming vapors), briefly reviewed metal speciation in gaseous solutions, and presented results of some of the first experiments to be conducted on the solubility of ore-forming metals in aqueous vapors. These experiments, while very preliminary and representing relatively simple analogues of natural systems, do suggest that aqueous vapors may be able to mobilize appreciable quantities of metals in natural hydrothermal systems. It is our hope that this review will encourage further experimentation on the solubility of metals in aqueous vapors, and lead to complementary field-based studies designed to evaluate the potential role of these vapors as agents of transport in the formation of metallic mineral deposits.

Acknowledgments—This research was supported by NSERC research and collaborative grants and an FCAR team grant to AEW-J, S.A. Wood and C.H. Gammons provided important advice on experimental techniques, S. Kecani helped in the design and construction of the analytical equipment and G. Kennedy and S. Lalli assisted in the analytical work. The manuscript was significantly improved by the thoughtful reviews of T.M. Seward, J. Webster, and T. Williams. S.A. Wood furnished important advice on both the preparation and content of the manuscript.

Editorial handling: S.A. Wood

REFERENCES

Alcock C. B., Itkin V. P., and Horrigan M. K. (1984) Vapor pressure equations for the metallic elements: 298-2500 K. *Can. Metall. Q* **23**, 309-313.

Alekhin Yu. V. and Vakulenko A. G. (1987) Solubility and thermodynamic properties of sodium chloride in water vapor at 300-500° and ≤300 bar. *Geokhimiya* **(10)**, 1468-1481.

Ambrose D. and Sprake C. H. S. (1972) Vapor pressure of mercury. *J. Chem. Thermodyn.* **4**, 603-620.

Armellini F. J. and Tester J. W. (1993) Solubility of sodium chloride and sulfate in sub- and supercritical water vapor from 450-550 °C and 100-250 bar. *Fluid Phase Equilib.* **84**, 123-142.

Barnes H. L. (1997) *Geochemistry of Hydrothermal Ore Deposits -3rd edition.* New York, John Wiley and Sons, 972 pp.

Bateman A. M. (1950) *Economic Mineral Deposits*. New York, John Wiley and Sons, 916 pp.

Bernard A. (1985) Les mecanismes de condensation des gaz volcaniques (chimie mineralogie et equilibre des phases condensees majeures et mineures). Ph.D. thesis, Univ. of Brussels, Belgium.

Bernard A. and Le Guern F. (1986) Condensation of volatile elements in high temperature gases of Mt. St. Helens. *J. Volcanol. Geotherm. Res.* **28**, 91-105.

Bischoff J. L. and Rosenbauer R. J. (1988) Liquid-vapor relations in the critical region of the system sodium chloride-water from 380 to 415°C: A refined determination of the critical point and two-phase boundary of seawater. *Geochim. Cosmochim. Acta* **52**, 2121-2126

Bischoff J. L. (1991) Densities of liquids and vapors in boiling sodium chloride-water solutions: A PVTX summary from 300° to 500 °C. *Am. J. Sci.* **291**, 309-338.

Blades A. T., Jayaweera P., Ikonomou M. G., and Kebarle P. (1990) Ion-molecule clusters involving doubly charged metal ions (M^{2+}). *Int. J. Mass Spectrom. Ion Processes* **102,** 251-267.

Blades A. T., Klassen J. S., and Kebarle P. (1995) Free energies of hydration in the gas phase of the anions of some oxo acids of C, N, S, P, Cl, and I. *J. Amer. Chem. Soc.* **117**, 10563-10571.

Brewer L. and Lofgren N. L. (1950) The thermodynamics of gaseous cuprous chloride, monomer and trimer. *J. Amer. Chem. Soc.* **72,** 3038-3045

Brimhall G. H. and Crerar D. A (1987) Ore Fluids: Magmatic to Supergene. In *Thermodynamic Modeling of Geological Materials: Minerals, Fluids and Melts* (eds. I.S.E. Carmichael and H.P. Eugster), Reviews in Mineralogy **17**, 235-321.

Brittain R. D., Lau K. H. and Hildenbrand D. L. (1983) The vaporization of realgar, arsenic sulfide (As_4S_4). *J. Electrochem. Soc.* **130**, 1206-1210.

Brown J. S. (1948) *Ore genesis; a metallurgical interpretation, an alternative to the hydrothermal theory.* Hopewell, N.J., Hopewell Press, 204 pp.

Bunsen M. R. (1853) Recherches sur la formation des roches volcaniques en Islande. *Ann. Chim. Phys. 3rd Series* **38**, 215-289.

Chang S. K. and Toguri J. M. (1974) Effect of temperature and electron energy on mass-spectrum of silver chloride. *J. Less-Common Metals* **38**, 187-192.

Charpentier J. F. W. (1778) Mineralogische Geographie der Chursächsischen Lande. Leipzig.

Charpentier J. F. W. (1799) Beobachtung über die Lagerstätte der Erze, hauptsächlich aus den sächsischen Gebirgen. Leipzig.

Cordes H., Dost L. and Cammenga H. K. (1971) Thermodynamic properties of metal vapors. II. Mercury. *Z. Metallk.* **62**, 915-925.

Daubrée A. (1841) Mémoire sur le gisement, la constitution et l'origine des amas de minerai d'étain. *Annales des mines - 3e série* **XX**, 65-112.

Daubrée A. (1849) Recherches sur la production artificielle de quel-ques espèces minérales cristallines particulièrement de l'oxyde d'étain des filons titanifères des Alpes. *Annales des Mines, 4e série* **XVI**, 129-155.

Davidson W. R. and Kebarle P. (1976) Binding energies and stabilities of potassium ion complexes from studies of the gas phase ion equilibriums $K^+ + M = K^+M$. *J. Am. Chem. Soc.* **98**, 6133-6138.

Descartes R. (1644) *Principia Philosophae,* Amsterdam, I. Rieuwertsz (Publisher), 388 pp.

Deville C. S. (1862) Lettre à M. Elie de Beaumont sur les phénomènes éruptifs de l'Italie méridionale. *Compt. Rend.* **54**, 99-483.

Dienstbach F., Emmenegger F. P. and Schlaepfer C. W. (1977) Vaporization of copper(II) chloride and the structure of the vapor studied using UV/visible and Raman spectroscopy. *Inst. Anorg. Anal. Chem. Helv. Chim. Acta* **60**, 2460-2470.

Drummond S. E. and Ohmoto H. (1985) Chemical evolution and mineral deposition in boiling hydrothermal systems. *Econ. Geol.* **80**, 126-147.

Elie de Beaumont L. (1847) Voyage métallurgique en Angleterre, ou Recueil de mémoires sur le gisement, l'exploitation et le traitement des minerais de fer, étain, plomb, cuivre et zinc, dans la Grande-Bretagne. A. Dufrénoy and L. Élie de Beaumont, Paris, Bachelier, 1837-1839.

Ewing C. T. and Stern K. H. (1974) Equilibrium vaporization rates and vapor pressures of solid and liquid sodium chloride, potassium chloride, potassium bromide, cesium iodide, and lithium fluoride. *J. Phys. Chem.* **78**, 1998-2005.

Faure F. M., Mitchell M. J., and Bartlett R. W. (1972) Vapor pressure study of stibnite (Sb_2S_3). *High Temp. Sci.* **4**, 181-191.

Faure F. M., Mitchell M. J., and Bartlett R. W. (1973) Vapor pressure study of arsenic trisulfide. *High Temp. Sci.* **5**, 128-137.

Ferro D., Piacente V., and Scardala P. (1989) Vaporization enthalpies of black and red mercury sulfides and their heat of transition from vapor pressure measurements. *J. Less-Common Met.* **147**, 1-8.

Forsythe G. E., Malcolm M. A. and Moler C. B. (1976) Computer Methods for Mathematical Computations, Prentice-Hall.

Fulignati P. and Sbrana A. (1998) Presence of native gold and tellurium in the active high-sulfidation hydrothermal system of the La Fossa volcano (Vulcano, Italy). *J. Volcanol. Geotherm. Res.* **86**, 187-198.

Galobardes J. F., Van Hare D. R., and Rogers L. B. (1981) Solubility of sodium chloride in dry steam. *J. Chem. Eng. Data* **26**, 363-366

Gemmell J. B. (1987) Geochemistry of metallic trace elements in fumarole condenstates from Nicaraguan and Costa Rican volcanoes. *J. Volcanol. Geotherm. Res.* **33**, 161-181.

Getahun A., Reed M. H., and Symonds R. (1996) Mount St. Augustine volcano fumarole wall rock alteration: mineralogy, zoning, composition and numerical models of its formation process. *J. Volcanol. Geotherm. Res.* **71**, 73-107

Giggenbach W. F. and Matsuo S. (1991) Evaluation of results from Second and Third IAVCEI Field Workshops on Volcanic Gases, Mt Usu, Japan, and White Island, New Zealand. *Appl. Geochem.* **6**, 125-141.

Gmelin Handbuch der anorganischen Chemie. (Gmelin handbook of inorganic chemistry.) Gmelin - Institut fur Anorganische Chemie und Grenzgebiete in der Max-Planck-Gesellschaft zur Forderung der Wissenschaften. Berlin, Verlag Chemie.

Gospodinov G. G., Pashinkin A. S., Boncheva-Mladenova Z. and Novoselova A. V. (1970) The determination of solid antimony sulphide vapour pressure. *Izvestia Akademii Nauk SSSR, Neorganicheskie Materiali* **4**, 1242 – 1247.

Gräber P. and Weil K. G. (1972) Mass-spectrometric investigations of silver halides. 1. Mass-spectrum appearance potentials, and fragmentation scheme of silver chloride. *Ber. Bunsengeschaft Phys. Chem.* **B 76 (6)**, 417-425.

Guilbert J. M. and Park C. F. Jr. (1986) *The Geology of Ore Deposits*. W. H. Freeman and Company, New York, 985 pp.

Hedenquist J. W., Aoki M., and Shinohara H. (1994) Flux of volatile and ore-forming metals from the magmatic-hydrothermal system of Satsuma Iwojima volcano. *Geology* **22**, 585-588.

Heinrich C. A., Günther D., Audétat A., Ulrich T., and Frischknecht R. (1999). Metal fractionation between magmatic brine and vapor, determined by microanalysis of fluid inclusions. *Geology* **27**, 755-758.

Heinrich C. H., Ryan C. G., Mernagh T. P., and Eadington P. J. (1992) Segregation of Ore Metals between magmatic brine and vapor: A fluid inclusion study using PIXE microanalysis. *Econ. Geol.* **87**, 1566-1583.

Henley R. W. and McNabb A. (1978) Magmatic vapor plumes and ground-water interaction in porphyry copper emplacement. *Econ. Geol.* **73**, 1-20.

Hildenbrand D. L. and Lau K. H. (1996) Thermochemistry of gaseous AgCl, Ag_3Cl_3 and CuCl. *High Temp. Mater. Sci.* **35**, 11-20.

Hsiao C. M. and Schlechten A. W. (1952) Volatility and Stability of Metallic Sulphides, *Journal of Metals* **Jan.**, 65-69

Hubbard S. R. and Ross R. G. (1982) Slope anomaly in the vapor pressure curve of mercury. *Nature* **295**, 682-683.

James S. E. and Hager J. P. (1978) High temperature vaporization chemistry in the gold-chlorine system including formation of vapour complex species of gold and silver with copper and iron. *Metall. Trans.* ***9B***, 501-508.

Kavalieris I. (1994) High Au, Ag, Mo, Pb, V, and W content of fumarole deposits at Merapi volcano, central Java, Indonesia. *J. Geochem. Explor.* **50**, 479-491.

Kebarle P. (1975) Thermochemical information from gas phase ion equilibria. NATO Adv. Study Inst. Ser., Ser. B, B6 (Interact. Ions Mol.), 459-87

Kebarle P., Davidson W. R., French M., Cumming J. B., and McMahon T. B. (1977) Solvation of negative ions by protic and aprotic solvents. Information from gas phase ion equilibria measurements. *Faraday Discuss. Chem. Soc.* **64**, 220-229.

Kestin J., Sengers J. V., Kamgar-Parsi B., and Levelt Sengers J.M.H. (1984) Thermophysical Properties of Fluid H_2O. *J. Phys. Chem. Ref. Data* **13**, 175-183.

Korzhinsky M. A., Tkachenko S. I., Schmulovich K. L., Taran Y. A., and Steinberg G. S. (1994) Discovery of a pure rhenium mineral at Kudriavy volcano. *Nature* **369**, 51-52.

Krauskopf K. B. (1957) The heavy metal content of magmatic vapor at 600°C. *Econ. Geol.* **52**, 786-807.

Krauskopf K. B. (1964). The possible Role of volatile metal compounds in ore genesis. *Econ. Geol.* **59**, 22-45.

Kulakov M. P. (1975) Melting point and vapor pressure of mercury(II) sulfide. *Izv. Akad. Nauk SSSR, Neorg. Mater.* **11**, 553-554.

Kvande H. (1979) Determination of the average molar mass of the vapor above molten sodium chloride by the transpiration method. *Acta Chem. Scand.* **A33**, 407-412.

Lacroix A. (1907) Les mineraux des fumerolles de l'eruption du Vesuve en Avril 1906. *Bull. Soc. Min. Fran.* **30**, 219-266.

Lau K. H., Brittain R. D., and Hildenbrand, D. L. (1982) Vaporization of arsenic trisulfide and the dissociation energy of arsenic monosulfide. *J. Phys. Chem.* **86**, 4429-4432.

Le Guern F. (1988) Ecoulements gazeux reactifs à hautes temperatures, mesures et modelisation. Ph.D. thesis, University of Paris, 314 pp.

Le Guern F. and Bernard A. (1982) A new method for sampling and analyzing sublimates. Application to Merapi Volcano, Java. *J. Volcanol. Geotherm. Res.* **12**, 133-146.

Le Guern F., Gerlach T.M. and Nohl A. (1982) Field gas chromatograph analyses of gases from a glowing dome at Merapi volcano, Java, Indonesia, 1977, 1978, 1979. *J. Volcanol. Geotherm. Res.* **14**, 223-245.

Lenian P., Picquenard E., Corset J., Jensen D.and Stendel R. (1988) Raman spectrometric study of the composition of sulfur vapor at temperatures up to 700° C. *Ber. Bunsenges. Phys. Chem.* **92**, 859-870.

Lindgren W. (1933) *Mineral Deposits, 4th ed.* New York, McGraw-Hill, 930 pp.

Lukashenko E. E., Korobeinikov A. P. and Khomaiko I. A. (1970) Saturated vapor pressure and enthalpies of vaporization of individual sodium and potassium chlorides. *Zh. Fiz. Khim.* **44**, 341-343.

Marshall W. L. (1972) A further description of complete equilibrium constants. *J. Phys. Chem.* **76**, 720-731.

Martynova O. I. (1964) Some problems of the solubility of involatile inorganic compounds in water vapor at high temperatures and pressures. *Russ. J. Phys. Chem.* **38**, 587-592

Menyailov I. A. and Nikitina L. P. (1980) Chemistry and metal contents of magmatic gases: the new Tolbachik vol-

canoes gas (Kamchatka). *Bull. Volcanol.* **43**, 197-207.
Mesmer R. E., Marshall W. L., Palmer D. A., Simonson J. M., and Holmes H. F. (1988) Thermodynamics of aqueous association and ionization reaction at high temperatures and pressures. *J. Sol. Chem.* **17**, 699-718.
Migdisov Art A., Suleimenov O. M., and Alekhin Yu. V. (1998) Experimental study of polysulfane stability in gaseous hydrogen sulfide. *Geochim. Cosmochim. Acta* **62**, 2627-2635.
Migdisov Art. A., Williams-Jones A. E., and Suleimenov O. M. (1999) The solubility of chlorargyrite (AgCl) in water vapor at elevated temperatures and pressures. *Geochim. Cosmochim. Acta* **63**, 3817-3827.
Mills K. C. (1974) *Thermodynamic Data for Inorganic Sulphides, Selenides and Tellurides*. National Physical Laboratory, Teddington, Middlesex, London Butterworths, 844 pp.
Munir Z. A., Street G. B., and Winters, H. F. (1971) Mass-spectrometric and vapor pressure studies on the sublimation of realgar (As_4S_4). *J. Chem. Phys.* **55**, 4520-4527.
Naboko S. I. (1964) Contemporary volcanoes and gas hydrothermal activity. *Geologiya SSSR* **31**, 323-387.
Nemoto, T., Hayakawa, M., Takahashi, K., and Oana, S. (1957) Report on the geology, geophysical, and geochemical studies of of Usu volcano (Showashinzan). *Rept. Geol. Surv. Japan* no. 170, p. 1-149.
Novoselova A. V. and Pashinkin A. S. (1978) *Vapour pressure of volatile metals sulphides*. Moscow, Nauka, 112 pp.
Ozin G.A. (1969) Gas-phase Raman spectroscopy of phosphorus, arsenic, and saturated sulfur vapors. *J. Chem. Soc. D* **(22)**, 1325-1327.
Pankratz L. B. (1982) Thermodynamic properties of elements and oxides. United States Dept. of the Interior, Bureau of Mines: Washington, D.C.
Pankratz L. B. (1984) Thermodynamic properties of halides. United States Dept. of the Interior, Bureau of Mines: Washington, D.C.
Pankratz L. B. (1987) Thermodynamic properties of sulfides. United States Dept. of the Interior, Bureau of Mines: Washington, D.C.
Park C. F. and McDiarmid R. A. (1975) *Ore Deposits-3rd Edition*, New York, McGraw-Hill, 522 pp.
Pashinkin A. S., Izergin A. A., Amirov R. A., Alikhanyan A. S., and Steblevskii, A. V. (1982) Saturated vapor pressure and heat of formation in arsenic sulfide (As_4S_4). *Izv. Akad. Nauk SSSR, Neorg. Mater.* **18**, 1253-1256.
Pashinkin A. S., Molodyk A. D., Belousov V. I., Strel'chenko S. S., and Fedorov, V. A. (1974) Composition of arsenic (III) sulfide vapor. *Izv. Akad. Nauk SSSR, Neorg. Mater.* **10**, 1600-1602.
Pitzer K. S. (1996) Sodium chloride vapor at very high temperatures: linear polymers are important. *J. Chem. Phys.* **104**, 6724-6729.
Pitzer K.S. and Palaban R.T. (1986) Thermodynamics of NaCl in steam. *Geochim. Cosmochim. Acta* **50**, 1445-1454
Pryce W. (1778) *Mineralogia Cornubiensis*. London, J. Phillips (Printers), 331 pp.
Quisefit J. P., Toutain J. P., Bergametti G., Javoy M., Cheyet B., and Person A. (1989) Evolution versus cooling of gaseous volcanic emissions from Momotombo Volcano, Nicaragua: Thermochemical model and observations. *Geochim. Cosmochim. Acta* **53**, 2591-2608.
Rau H., Kutty T.R.N., and Guedes de Carvalho J.R.F (1973) High temperature saturated vapor pressure of sulfur and estimation of its critical quantities. *J. Chem. Thermo.* **5**, 833
Rivkin S. L. (1988) *Thermodynamic properties of gases*. (ed, D. D. Wagman) 4th ed., New York: Hemisphere Pub. Corp.
Rogstad A. (1972) High-temperature laser Raman spectroscopic study on mixtures of arsenic and sulfur vapors. *J. Mol. Struct.* **14**, 421-426.
Shmulovich K. I. and Churakov S. V. (1998) Natural fluid phases at high temperatures and low pressures. *J. Geochem. Explor.* **62**, 183-191.
Shpil'rain E. E. and Nikanorov E. V. (1971) Experimental study of mercury vapor pressure by the boiling point method. *Teplofiz. Vys. Temp.* **9(3)**, 640-643.
Smith C.L., Ficklin W.H., and Thompson J.M. (1987) Concentrations of arsenic, antimony, and boron in steam and steam condensate at the Geysers, California. *J. Volcanol. Geotherm. Res.* **32**, 329-341
Sourirajan S. and Kennedy G. C. (1962) The system H_2O-NaCl at elevated temperatures and pressures. *Am. J. Sci.* **260**, 115-141
Spycher N. F. and Reed M. H. (1989) Evolution of a Broadlands-type epithermal fluid along alternative P-T path; Implications for the transport and deposition of base, precious and volatile metals. *Econ. Geol.* **84**, 328-359
Steblevskii A. V., Alikhanyan A. S., Pashinkin A. S. and Malyusov V. A. (1986) Vapor composition in the arsenic-sulfur system. *Zh. Neorg. Khim.* **31**, 2451-2456.
Steno N. S. (1669) *De Solido intra Solidum Naturaliter Contento*. Florence.
Steudel R. (1984) Elemental sulfur and related homocyclic compounds and ions. In *Sulfur - Its significance for chemistry, for the geo-, bio- and cosmosphere and technology,* vol. **3**, A. Müller and B. Keeds (Eds), Elsevier, Amsterdam. 3-37.
Steudel R. (1987) Sulfur homocycles. In *The chemistry of inorganic homo and heterocycles*, vol. **2**, J. Sowerby and I. Haiduc (Eds), Academic Press. London, 737-739.
Stoiber R. E. and Rose W. I. Jr. (1974) Fumarole incrustations at active Central American volcanoes. *Geochim. Cosmochim. Acta* **38**, 495-510.
Styrikovich M.A. (1969) Steam solutions. *Vestnik Akad. Nauk SSSR* **39**, 70-77
Sullivan C. L., Prusacryk J. E., and Carson K. D. (1970) Molecules in the equilibrium vaporisation of antimony sulfide and selenide. *J. Chem. Phys.* **53**, 1289-1290
Sunner J., Magnera T. F., and Kebarle P. (1981) Ion-molecule equilibrium measurements by high-pressure mass spectrometry. Some recent advances in concepts and technique. *Can. J. Chem.* **59**, 1787-1796.
Symonds R. B. (1992) Getting the gold from the gas: How recent advances in volcanic-gas research have provided new insight on metal transport in magmatic fluids. Rept. Geol. Surv. Japan, No. 279, 170-175.
Symonds R. B. and Reed M. H. (1993) Calculation of multicomponent chemical equilibria in gas-solid-liquid systems: calculation methods, thermochemical data, and applications to studies of high-temperature volcanic gases with examples from Mount St. Helens. *Am. J. Sci.* **293**, 758-864.
Symonds R. B., Rose W. I., Reed M. H., Lichte F. E., and Finnegan D. L. (1987) Volatilization, transport and sublimation of metallic and non-metallic elements in high temperature gases at Merapi Volcano, Indonesia. *Geochim. Cosmochim. Acta.* **51**, 2083-2101
Symonds R. B., Reed M. H. and Rose W. I. (1992) Origin, speciation, and fluxes of trace-element gases at Augustine volcano, Alaska: insights into magma degassing and fumarolic processes. *Geochim. Cosmochim. Acta* **56**, 633-657
Symonds R. B., Rose W. I., Gerlach T. M., Briggs P. H., and Harmons, R. S. (1990) Evaluation of gases, condensates

and SO_2 emissions from Augustine Volcano, Alaska: the degassing of a Cl-rich system. *Bull. Volcanol.* **52**, 355-374.

Symonds R. B., Rose W. I., Bluth G. J. S., and Gerlach T. M. (1994) Volcanic-gas studies: Methods, results and applications. In *Volatiles in Magmas* (eds. M. R. Carroll and J. R. Holloway). Reviews in Mineralogy **30**, 1-66.

Tagirov V. K., Makarov V. A., and Brukvin V. A. (1993) Vapor composition and pressure over chlorides of silver and gold. *Metalli* **5**, 67-71.

Taran Y. A., Bernard A., Gavilanes J-C., and Africano F. (2000) Native gold in mineral precipitates from high-temperature volcanic gases of Colima volcano, Mexico. *Appl. Geochem.* **15**, 337-346.

Taran Y. A., Hedenquist J. W., Korzhinsky M. A., Tkachenko S. I., and Shmulovich K.I. (1995) Geochemistry of magmatic gases from Kudryavy volcano, Iturup, Kuril Islands. *Geochim. Cosmochim. Acta* **59**, 1749-1761.

Tedesco D. and Totain J-P. (1991) Chemistry and emission rate of volatiles from White Island volcano (New Zealand). *Geophys. Res. Let.* **18**, 113-116.

Toutain J. P., Aloupogiannis P., Delorme H., Person A., Blanc P., and Robaye G. (1990) Vapor deposition of trace elements from degassed basaltic lava, Piton de la Fournaise volcano, Reunion Island. *J. Volcanol. Geotherm. Res.* **40**, 257-268.

Ulrich T., Gunther D., and Heinrich C. A. (1999) Gold concentrations of magmatic brines and the metal budget of porphyry copper deposits. *Nature* (London) **399**, 676-679.

Ustyugov G. P., Kudryavtsev A. A., and Kuadzhe B. M. (1968) Saturated vapor pressure of arsenic sulfides. *Izv. Akad. Nauk SSSR, Neorg. Mater.* **4**, 1338-1339.

Visnapuu A. and Jensen, J. V. (1970) Composition and properties of vapors over molten silver chloride. *J. Less-Common Metals.* **76**, 141.

Wagman D. D., Evans W. H., and Parker V. B. (1982) *The NBS tables of chemical thermodynamic properties. Selected values for inorganic and C_1 and C_2 organic substances in SI units.* Washington, Amer. Chem. Soc.; New York: Amer. Inst. of Physics, 392 pp.

Walker A. L. and Walker W. J. (1956) *The origin and nature of ore deposits.* Colorado Springs, Walker Associates.

Wood S. A. (1987) Thermodynamic calculations of the volatility of the platinum group elements (PGE): The PGE content of fluids at magmatic temperatures. *Geochim. Cosmochim. Acta* **51**, 3041-3050.

Wood S. A. and Samson I. M. (1998) Solubility of ore minerals and complexation of ore metals in hydrothermal solutions. In *Techniques in Hydrothermal Ore Deposits* (eds. J. P. Richards and P. B. Larson). Reviews in Economic Geology **10**, 33-80.

Zakaznova-Iakovleva V. P., Migdisov Art. A., Suleimenov O. M., Williams-Jones A. E. and Alekhin Yu. V. (in press) Experimental study of stibnite solubility in gaseous hydrogen sulphide. *Geochim. Cosmochim. Acta*, **65**, 289-298.

Zeis E. G. (1929) The Valley of 10,000 Smokes, the fumarolic incrustations and their bearing on ore deposition. *National Geographic Society Technical Paper* **1**, 61-79.

Water-Rock Interactions, Ore Deposits, and Environmental Geochemistry: A Tribute to David A. Crerar
© The Geochemical Society, Special Publication No. 7, 2002
Editors: Roland Hellmann and Scott A. Wood

Volatile trace-element transport in high-temperature gases from Kudriavy volcano (Iturup, Kurile Islands, Russia)

CHRISTOPH WAHRENBERGER, TERRY M. SEWARD[1] AND VOLKER DIETRICH

Institute for Mineralogy and Petrology, Swiss Federal Institute of Technology, ETH-Zentrum, 8092 Zürich, Switzerland
[1]e-mail: tseward@erdw.ethz.ch

Abstract—Gases, condensates, sublimates and silica-tube deposits were sampled at Kudriavy volcano in 1995 within the temperature range from 160 to 920 °C. The main trace elements transported in the gas phase at 920 °C are Pb, Mo, Bi, Sn, In, As, Se and Te as well as elements partly derived from the wall rock (e.g., Mg, Si, Al). Dominant transporting species as calculated in a heterogeneous closed-system cooling model are chloride complexes (e.g., $CuCl_{(g)}$, $TiCl_{4(g)}$, $FeCl_{2(g)}$). However the Mo-group elements are transported as oxoacids whereas the Sn-group elements and the highly volatile elements of group V and VI of the periodic table are transported dominantly as the sulfide complexes. Bromide and iodide also play an important role in complexing trace elements at temperatures below 600 °C. An open-system equilibrium model where condensed phases are fractionated from the coexisting gas is introduced to simulate the deposition of natural sublimates and is directly compared to a silica-tube sublimation experiment. While the thermochemical models reflect the zonality of sublimates in the tube and some aspects of the sublimate mineral assemblages found in nature, they fail to reproduce the observed temperature distribution. It is concluded that the thermochemical equilibrium models used so far in the investigation of high-temperature volcanic gases do not sufficiently reproduce the volatility of a number of trace elements (e.g., Cu, Zn, Cd and In), because of the lack of consideration of solid solutions in sublimate formation as well as inadequate (and/or unavailable) thermochemical data and the possibility that the underlying equilibrium assumption is only partly correct.

1. INTRODUCTION

Volcanic gases contribute significantly to the general flux of partly hazardous and toxic substances such as heavy-metal volatiles and aerosols to the troposphere (e.g., Nriagu, 1989; Williams et al., 1992). Furthermore, the composition of volcanic gases provides some insight into the chemical output of degassing magmas into hydrothermal convection cells in the earth's crust. This contribution of heat and chemical components to hydrothermal systems plays a fundamental role in the formation of hydrothermal ore deposits.

Despite the importance of elemental transport by volcanic gases, available data on their composition are limited to not more than about a dozen different volcanoes. A compilation of eight condensates is given by Symonds et al. (1992). Additional data are available from Kilauea volcano (Crowe et al., 1987), Satsuma-Iwojima (Hedenquist et al., 1994), Kudriavy (Taran et al., 1995) and Vulcano (Aeolian Islands) (Wahrenberger et al., 2001). An exceptional long-term data set has been obtained from the Showa-Shinzan dome of Usu volcano (Symonds et al., 1996).

The interpretation and theoretical simulation of trace-element transport in the volcanic gas phase is inhibited by a variety of factors. Apart from the practical problems of sampling high-temperature gases on an active volcano, the measured condensate composition should be accepted as being the result of a variety of processes involving magmatic degassing, gas-rock reactions and in the case of medium- to low-temperature gases, the admixture with hydrothermal fluids.

Equilibrium thermodynamic calculations have been applied to these problems (e.g., Symonds and Reed, 1993; Quisefit et al., 1989) and provide insight into the chemistry of volatile transport and speciation for a variety of elements as well as some estimate of the contribution of reactions of the gas with adjacent wall-rock to the measured condensate composition. Thermodynamic modeling has also been used to predict the deposition of sublimates and wall-rock alteration minerals. Such calculations have been carried out on gases from Mt. St. Helens (Symonds and Reed, 1993), Mt. St. Augustine (Symonds et al., 1992; Getahun et al., 1996), Momotombo (Quisefit et al., 1989) and Merapi (Symonds et al., 1987). However, only the latter two studies draw on consistent data sets where gas compositions and condensates have been obtained nearly simultaneously on the same vent.

The applicability of such computational approaches relies on four basic assumptions comprising the accuracy of the input compositions, the accuracy and completeness of the thermochemical data used in the calculation, the assumption of ideality for all phases considered and, the attainment of equilibrium among all phases involved. As yet, in view of the above-mentioned difficulties, the results of such calculations have not been demonstrated to be in agreement with respect to actual observations (e.g., Getahun et al., 1996).

The purpose of this paper is to provide further insight into the applicability of such thermochemical equilibrium calculations to simulate trace-element transport in volcanic gases by comparing the calculated results to the actual observed sublimate mineralogy and chemistry. This study provides a basis upon which to evaluate and improve the underlying assumptions. The study presented here discusses the transport, speciation and deposition of trace elements in sublimates from a high-temperature gas of Kudriavy volcano. It draws on a consistent data set of gases and condensates obtained simultaneously and on the same vents in September 1995. Two different computational approaches are used for the transport simulation which involve both closed- and open-system cooling. The results are then compared to the measured elemental distribution and phase composition observed in a sublimation tube.

2. GEOLOGICAL SETTING

Kudriavy volcano (996 m elevation) is situated in Medvezhya caldera in the north of the island of Iturup (Kuriles), approximately 300 km to the northeast of Hokkaido (Figure 1). The eruptive products are basaltic to andesitic (Ehrlich, 1986; Avdeiko et al., 1992) with the last eruption having occurred in May/June 1883. Simkin and Siebert (1994) reported on a phreatic eruption in 1956, but the resulting eruptive products are apparently missing at the summit. Today's activity is restricted to quiescent degassing within the immediate crater area and to one hot spring at the foot of the neighboring Menshoi'Brat basaltic cone.

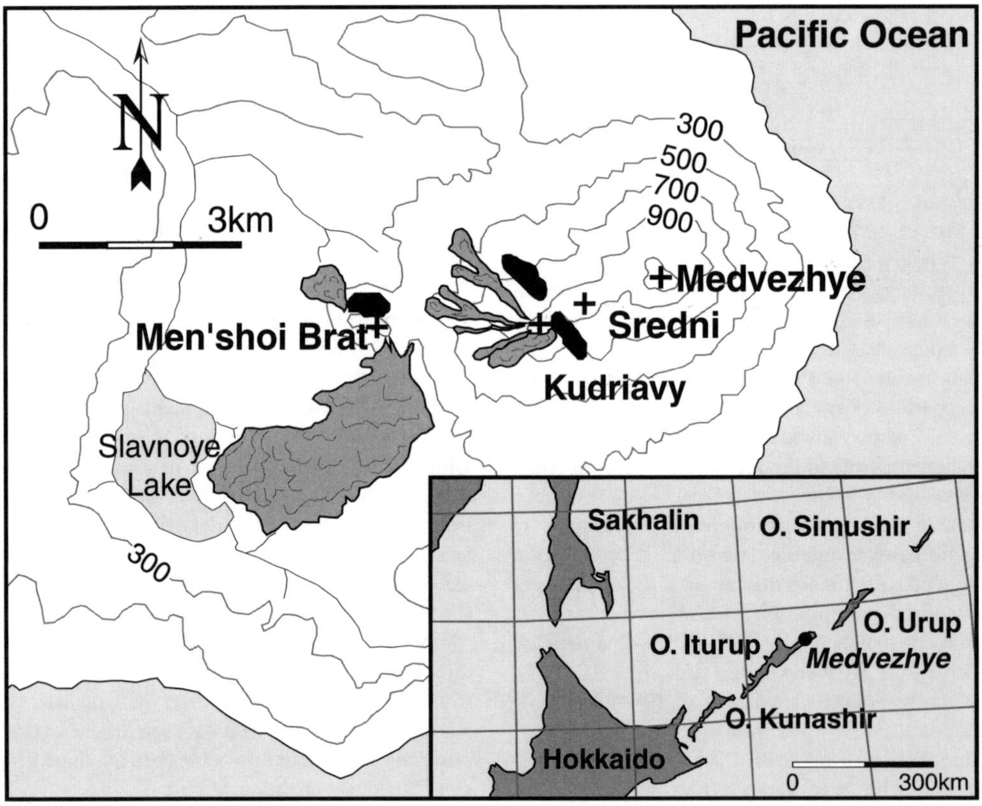

Figure 1: Geographic setting of Kudriavy volcano in Medvezhye caldera. The three major volcanic centers of the caldera Medvezhye, Sredni and Kudriavy lie along a roughly E-W trending line with a younger feature (Men'shoi Brats) to the West. The relative age to these centers is unknown (Erlich, 1986). Contour intervals are in meters above sea level. Dark gray shaded areas show major lava flows, black areas are andesitic domes.

Kudriavy volcano is of particular interest because of the high temperatures (up to 920 °C) and flow rates of up to 80 ms^{-1} of the fumarole discharges. Active deposition of rhenium sulfide occurs (Korzhinsky et al., 1994) in association with a number of other sublimate phases including indium-containing compounds. Of interest, is the occurrence of native metals and other compounds which deposit in sublimation tubes inserted into the high-temperature fumaroles for various periods of time (Korzhinsky et al., 1995, 1996; Tkachenko et al., 1999; Shmulovich and Churakov, 1998). The chemistry and isotopic composition of gases collected between 1990 and 1992 have been discussed by Taran et al. (1995). Total SO_2 emission from Kudriavy in August/September 1995 was estimated to be 75 ± 50 td^{-1} on the basis of COSPEC measurements (Williams et al., 1995).

3. ANALYTICAL METHODS

Gases were sampled using quartz-glass tubing and absorption bottles (Giggenbach, 1975) and subsequently analyzed using the methods (with minor modifications) outlined by Giggenbach and Goguel (1981). Coolant for condensate sampling was an ammonium nitrate/water mixture as well as pure water. Condensates were analyzed by inductively coupled plasma-mass spectrometry (ICP-MS) after dispersion of the sample in an ultrasonic bath and subsequent dilution with high-purity water.

Sublimation tubes (silica glass) were inserted directly into the same vent from which condensates and gas samples were collected (sample K95/2). The tubes were left in the fumaroles for three days and the temperature gradients were measured along the tube axis (396 to 874 °C). Upon arrival in the laboratory, the tube was cut into 5-cm long sectors. Half of each sector was then subjected to analysis by scanning electron microscopy and energy dispersive X-ray analysis using a JEOL 6400 SEM equipped with a Noran Voyager II EDAX system. The sublimates on the remaining part of the sector were dissolved in warm aqua regia, boiled to near dryness and diluted by 1M HCl (Merck "suprapur"). Blanks were prepared by applying the same procedures to an unused silica-glass tube. The solutions were subsequently analyzed by ICP-MS. The temperature distribution of phase composition and chemistry in the tube was thus obtained.

It should be mentioned that the sublimates in the tube were not deposited in perfect axial symmetry and care was taken to analyze material representative of each sector. In addition, the temperatures on the inside tube wall and along the central axis of the tube varied by 30 – 50 °C, depending on the flow velocity. The profile measured along the tube axis is therefore taken as representative of the maximum temperature in the tube. The samples were exposed to air after collection and during transport/storage.

The detection limits for the ICP-MS analysis of silica-tube deposits are worse than that for the condensates. Therefore, not all elements analyzed in the condensate were detected in the sublimation tube deposits. This is caused by the higher blank/background concentrations due to low-level contamination by the acid digestion reagents as well as by the necessarily high dilution factors. Naturally deposited solid sublimate phases collected from active, high-temperature fumaroles were identified using SEM-EDAX in combination with a variety of different X-ray diffraction methods.

4. NUMERICAL PROCEDURES

Trace elements in volcanic gases can be transported in a number of different modes such as,

(i) a true gaseous complex (e.g., SiF_4(g)), (Francis et al., 1996),

(ii) a liquid or solid aerosol suspended in the gas flow (e.g., CuCl(s)), (Amman et al., 1993; Varekamp et al., 1986) and

(iii) as particulate matter produced by abrasion and chemical attack or reaction of the wall-rock or sampling device by the gas (Symonds et al., 1992).

To model this multiphase composition, only heterogeneous equilibria involving solid, liquid (melt) and gaseous components were considered throughout this study. The minimum temperature of all calculations was set to 300 °C. The condensation temperature of hydrated sulfuric acid from Kudriavy gases is calculated to be below 250 °C at 1 bar, in agreement with evaluations of Symonds et al. (1992) in the case of St. Augustine gases. All phases are considered to behave ideally and in the case of gaseous components, to mix ideally. Solid solutions have been neglected. Calculations were restricted to a total pressure of 1 bar.

The thermochemical evaluations were carried out using the CSIRO/SGTE Thermodata program package (Turnbull and Wadsley, 1986). The code is based on the Gibbs free energy minimization procedure of Eriksson (1975). Thermodynamic data for the 519 gaseous inorganic and 404 condensed (solid and liquid) species were compiled from the JANAF tables (Chase et al., 1985) and other sources (Stull et al., 1969; Barin and Knacke, 1973; Barin et al., 1977; Knacke et al., 1991; Robie and Hemingway, 1995). Where necessary, heat capacity functions (e.g. Maier-Kelly) have been used to extrapolate beyond the temperature range for which the data were available as was the case for hematite, $CaMoO_{4(c)}$ (powellite) and $CaWO_{4(c)}$ (scheelite).

Solids and liquids are treated merely as one condensed state designated by the suffix, "(c)", throughout this study. Calculations were carried out in 9 to 12 element systems. The common basis for all thermodynamic systems considered comprises the eight elements C, O, H, S, Cl, Br, F and I. Reference states are the pure elements and substances at 298.15 K and 1 bar. All results of this study are limited by the characteristics of the numerical and computational techniques used, bearing in mind that the thermochemical data carry inherent uncertainties and the compilation from various sources may lead to inconsistencies, although care was taken to ensure internal consistency with respect to different reference states by correcting data to the same convention as necessary. The smallest concentration calculated was 10^{-38} mol mol^{-1} and the terms "stable" and "unstable" used in the following discussions therefore refer to a concentration higher or lower than this number, respectively. Symonds and Reed (1993) have commented on the effect of inconsistencies in the use of the database GASTHERM and estimate the resulting error to be less than the error caused by the uncertainties in $\Delta G°_f$ (± 0.5 log units in terms of volatility and ± 50 °C for deposition temperatures in heterogeneous closed system cooling). As the database of this study partly draws on the same data sources as GASTHERM, a similar estimated uncertainty applies to this study.

The aim of thermodynamic equilibrium modeling of the high-temperature Kudriavy gases was to provide insight into the volatile, trace-element speciation as well as to the nature of the condensed phases with which they are in equilibrium. In addition, the thermodynamic calculation of the temperature range of deposition of various condensed phases may also be compared with that observed in silica-glass sublimation tubes. This comparison permits some evaluation of the validity and application of such equilibrium calculations to the understanding of the volatile phase transport and element speciation in volcanic gas systems.

Two approaches were taken which considered both closed and open systems. For the closed-system calculations, the gas was numerically cooled in 25 °C steps and the element composition was the same in each incremental temperature interval, this approach being similar to those of Symonds et al. (1992) and Getahun et al. (1996). In the open-system calculations, the mass of each element transferred to the condensed phases was removed after each incremental temperature step upon cooling from 1000 to 300 °C.

Input to the computation is the composition of gas sample K95/2 (Tables 1 and 2). All calculations consider heterogeneous systems (i.e., the presence of solids, liquids and a gaseous phase) in which the thermodynamic system is limited to twelve elements. A consequence of the approaches outlined above is that the gas is calculated to cool along an internal gas buffer defined by the $SO_2 + H_2O \leftrightarrow H_2S + {}^3/_2O_2$ equilibrium. The calculations commenced with a starting f_{O_2} slightly below the nickel-nickel oxide buffer at 1000 °C as constrained by the restored composition of sample K95/2 (see below). Such a cooling behavior will only partly reflect reality because many of the gas-phase reactions are quenched at 350 to 400°C due to slower reaction kinetics (Giggenbach, 1987). In addition, the cooling behavior of a naturally flowing gas would be expected to be dependent on the flow rate (Saito et al., 1989) and to have an oxygen fugacity intermediate between that of the internal gas buffer and the corresponding rock buffer.

It has to be kept in mind that the calculations presented here focus on the simulation of an artificial situation, that is, the cooling of a high-temperature gas in the sublimation tube. Naturally, the cooling and ascending gases undergo processes much more complex than modeled here. Ascending, cooling magmatic gases may mix with fluids derived from the hydrothermal system of the volcano and will therefore change their major elemental composition upon cooling (see Taran et al., 1995; Giggenbach, 1987; Wahrenberger et al., 2001). In the case of Kudriavy, it can be shown that such processes are also affecting the halogenide composition and elemental ratios at temperatures as high as 700 °C. As chloride is one of the most important ligands involved in trace-element transport in the gas phase, changes in its concentration will severely influence the characteristics of the trace-element transport.

5. GAS CHEMISTRY AND INPUT COMPOSITION

Gases collected in 1995 are similar in composition to those of the 1990-1992 period (see Table 1 and Taran et al., 1995), although H_2O is somewhat elevated in the 1995 discharges (94.2 – 98.9 vol% in 1995 versus 92.1 – 98.5 vol%) giving rise to the higher H_2O / CO_2 ratio of 79 at largely stable CO_2 / S_t ratios of around 1 for gases in the temperature range of 680 – 920 °C. The gases also contain small amounts of ammonia, even at temperatures as high as 920 °C, indicating the admixture of a hydrothermal component to all of the gases measured. This interpretation is consistent with the equilibration of the NH_3/N_2 system within the hydrothermal envelope at conditions governed by the coexistence of a vapor with a brine and buffered by the host-rock (represented by a $FeO/FeO_{1.5}$ buffer; Giggenbach, 1987). It is possible that the measured carbon monoxide con-

Table 1: Composition of gases from Kudriavy volcano, Iturup, as sampled between 8/27 and 9/1/1995. All values in µmol/mol. Ar and O_2 are below 1 and 2 µmol/mol respectively.

sample No.	t [°C]	H_2O	CO_2	S_t	HCl	HF	NH_3	He	H_2	N_2	CH_4	CO
K95-2	920	942930.141	16765.084	20309.402	1712.085	744.678	2.929	228.22	5677.386	11077.226	b.d.	561.851
K95-3	710	970898.867	10114.12	12865.856	1756.147	592.847	1.605	1.082	3245.972	506.883	b.d.	15.769
K95-4	680	978110.905	10315.327	6463.196	1364.894	538.76	1.963	1.758	2545.095	640.814	b.d.	15.735
K95-5	619.5	974243.571	10645.083	11721.122	467.925	356.893	1.404	1.483	2006.238	541.773	b.d.	12.989
K95-6	470	984030.032	8297.256	3877.897	3113.444	239.345	0.702	0.182	269.822	169.429	b.d.	1.087
K95-7	190	989821.357	5659.778	3965.457	308.955	16.524	0.54	0.048	21.344	201.185	3.576	b.d.
K95-8	160	986575.935	8998.649	3996.99	206.785	12.439	0.704	0.052	20.248	183.843	3.43	b.d.
K95-9	300	987205.185	6237.959	3935.118	2193.711	43.276	1.292	0.119	199.58	182.83	0.299	0.17
K95-10	620	970152.088	14566.745	10402.265	538.143	443.03	0.912	1.533	3233.463	646.013	b.d.	13.781

Table 2: Composition of condensates from Kudriavy volcano, Iturup, as sampled between 8/27 and 9/1/1995. b.d. = below detection limit. Elements measured but below detection limit (<0.5-l ppb) in all condensates are: Li, Be, Sc, Co, Ge, Y, Nb, Pd, Ag, La-Lu, Ta, Re, Os, Ir, Pt, Au, Th, U. All values in ppb with the background/blank values subtracted. Rh and Hg were not measured.

t_{gas} [°C]	920	710	680	470	300	160
sample	K95/2	K95/3	K95/4	K95/6	K95/9	K95/8
			[ng g^{-1}]			
B	1.05E+04	2.01E+04	1.77E+04	1.19E+04	8.73E+03	1.61E+04
Na	1.32E+04	1.08E+03	3.03E+03	n.a.	n.a.	n.a.
Mg	1.19E+03	4.13E+02	3.34E+02	n.a.	4.04E+02	9.12E+01
Al	8.97E+02	1.04E+03	6.65E+02	6.61E+02	8.41E+03	1.78E+02
Si	3.08E+03	4.66E+03	3.80E+03	4.71E+03	4.70E+02	b.d.
K	1.32E+04	n.a.	n.a.	n.a.	n.a.	n.a.
Ca	1.13E+03	n.a.	1.66E+03	n.a.	n.a.	b.d.
Ti	3.45E+02	1.23E+02	b.d.	1.15E+02	2.19E+04	b.d.
V	1.40E+01	4.18E+00	3.69E+00	3.28E+00	1.99E+01	b.d.
Cr	1.82E+02	2.39E+02	1.03E+02	1.27E+02	1.42E+02	2.83E+01
Mn	3.50E+01	7.31E+00	4.60E+00	1.95E+00	2.68E+04	3.47E+00
Fe	4.78E+03	6.01E+02	3.65E+01	b.d.	1.77E+03	3.18E+02
Ni	1.07E+02	8.44E+01	4.19E+02	4.55E+01	4.78E+01	7.35E+01
Cu	5.44E+01	3.33E+01	2.05E+01	1.16E+01	3.11E+01	3.40E+01
Zn	1.28E+03	6.40E+02	5.01E+02	1.15E+02	1.26E+02	3.33E+02
Ga	2.81E+01	1.88E+01	9.46E+00	5.56E+00	4.79E+00	b.d.
As	1.86E+03	2.55E+03	2.40E+03	2.32E+03	1.68E+02	2.64E+02
Se	b.d.	b.d.	7.00E+02	b.d.	b.d.	3.38E+01
Br	6.85E+03	1.04E+04	4.54E+03	4.82E+03	2.92E+03	5.78E+02
Rb	6.21E+01	5.33E+01	2.30E+01	2.66E+00	1.52E+00	b.d.
Sr	b.d.	b.d.	b.d.	4.63E+00	5.07E+01	b.d.
Zr	4.33E+01	2.71E+01	1.59E+01	1.03E+01	2.71E+00	b.d.
Mo	2.04E+02	1.27E+02	1.07E+01	3.04E+00	5.61E-01	1.13E+00
Cd	5.54E+01	b.d.	3.95E+01	1.20E+00	b.d.	b.d.
In	4.37E+01	3.89E+01	4.69E+01	1.08E+01	b.d.	b.d.
Sn	2.87E+02	1.97E+02	4.20E+02	7.53E+01	3.26E+01	9.97E+00
Sb	1.45E+01	2.21E+01	4.06E+01	b.d.	2.58E+00	2.58E+00
Te	2.77E+04	1.96E+01	1.60E+01	2.23E+01	4.94E+00	1.04E+00
I	2.00E+03	3.07E+03	1.72E+03	1.63E+03	1.89E+03	4.69E+02
Cs	9.36E+00	1.39E+01	4.14E+01	2.74E+00	b.d.	b.d.
Ba	8.50E+02	8.92E+02	4.14E+02	1.84E+02	8.09E+01	3.29E+01
W	1.82E+01	1.21E+01	9.80E+00	1.25E+01	6.69E+00	1.10E+01
Tl	7.72E+01	5.46E+01	3.74E+01	2.73E+01	1.43E+01	b.d.
Pb	7.91E+02	6.68E+02	4.70E+02	5.61E+02	1.09E+02	1.38E+01
Bi	6.45E+01	7.29E+01	3.18E+01	6.72E+01	3.38E+00	2.69E+00

tent is underestimated due to slow reaction with NaOH (to form formate) in the sampling bottles during storage as noted previously by Giggenbach and Matsuo (1991).

The data presented in this study (Table 2) indicate that the majority of trace elements are somewhat less concentrated than measured by Taran et al. (1995) in the 1990-1992 period. Of interest is the much higher Br content of the gases measured in this study (approx. 6.85 ppm as compared to 0.75 ppm at 920 – 940°C as reported by Taran et al., 1995). Further significant differences are the lower Si, Al and Zr contents obtained in the 1995 samples. Elements such as Bi, As, B, Tl, Pb, Sn, Mo, Cd and W are highly enriched in comparison to the average wall rock as reported by Avdeiko et al. (1992) (i.e., weight-based, Mg-normalized enrichment factors are $EF_I > 3$). Of low ($EF_I < 3$) or negative EF_I are: Fe, Ti, V, Mn, Si and Al.

To obtain an appropriate input composition for computational modeling, gas sample K95/2 was restored to its last equilibrium composition following the methods outlined by Symonds et al. (1994). The restoration procedure is based on independent f_{O_2} measurements obtained by solid-state electrode where f_{O_2} was found to vary with temperature according to log f_{O_2} = -32.64 + 0.0174 T[K] (Rosen et al., 1993). To restore sample K95/2, the conversion of 6 mmol H_2O to H_2 and of 0.25 mmol CO to CO_2 was necessary to match all "correspondence temperatures" (CT's) within ± 10 °C (see Gerlach and Casadevall, 1986). The last equilibrium temperature is calculated to 984 °C at a log f_{O_2} = -10.989. The presence of N_2, NH_3 and small amounts of Ar (i.e. < 2 μmol^{-1}) in the gas was neglected. As discussed by Taran et al. (1995), Kudriavy exhalations as well as many other volcanic gases do contain "excess" nitrogen. The resulting input composition was obtained by combining the restored gas and condensate analyses of sample K95/2 and this is presented in Table 3.

6. RESULTS AND DISCUSSION

6.1. Silica-Tube Sublimate Mineralogy

Data on the composition of the sublimate phases and their elemental distribution in the silica tube as a function of temperature are given in Figure 2. A number of important rock-forming silicate minerals were found to have sublimed in the higher-temperature regime (i.e., aegerine, $NaFeSi_2O_6$; ferrosilite/enstatite, $(Fe,Mg)SiO_3$; magnetite, Fe_3O_4 and SiO_2; no feldspars were detected). Chloride and sulfide phases were the dominant phases deposited over a wide temperature range. The chloride sublimate phase, (Na,K)Cl contained traces of F, Br and I. Rhenium was detected as a trace element in FeS_2 and in solid solution with MoS_2 but was not detected as a discrete rhenium compound as noted in the natural fumarole sublimate deposits. The rhenium concentration in the condensate collected from the same fumarole into which the sublimation tube had been inserted was below detection limit. A number of phases whose exact stoichiometry could not be determined were also observed. Native iron was detected in a sylvite (KCl) matrix between 770 and 874 °C. Other native elements have been reported in sublimation tube experiments at Kudriavy.

6.2. Natural Fumarole Sublimates

The sublimates which have deposited naturally around various fumaroles at Kudriavy (Table 4) are also distinctly zoned with respect to temperature and distance from the discharging vent. The silicate phases were deposited at the highest temperatures, the oxides and sulfides at lower temperatures and the halides and sulfates at intermediate temperatures. The zonation reflects the changing volatility of trace elements in response to changes in gas chemistry, temperature and air admixture. The extent and significance of the air admixture is as yet unknown.

A number of mineral groupings were recognized in the Kudriavy fumarole sublimates and are as follows:

Pb-Bi sulfides: Cannizzarite ($Pb_4Bi_6S_{13}$), and cosalite ($Pb_2Bi_2S_5$), as well as several other phases with varying (Pb + Bi) / S ratios were identified. These Pb-Bi sulfides concentrate In, Sn, and Zn

Table 3: Composition of sample K95/2 as an input to thermodynamic modeling. All data in mol%. Main composition restored according to Symonds et al. (1994). Below detection limits (< 0.5-1 ppb) are: Li, Be, Sc, Ge, Y, Nb, Pd, Ag, lanthanides (Ce-Lu), Hf, Ta, Re, Os, Ir, Pt and Au.

T_{eq} [°C]	984
log f_{O_2} [bar]	-10.9889
H_2O	94.792
H_2	1.165
CO_2	1.720
CO	0.033
CH_4	0.815E-11
SO_2	1.834
H_2S	0.192
S_2	0.012
COS	0.237E-03
HCl	0.173
HF	0.075
HBr	0.154E-03

B	1.72E-03	As	4.39E-05
Na	1.01E-03	Rb	1.29E-06
Mg	8.70E-05	Zr	8.40E-07
Al	5.89E-05	Mo	3.76E-06
Si	1.94E-04	Cd	8.73E-07
K	5.98E-04	In	6.75E-07
Ca	5.00E-05	Sn	4.29E-06
Ti	1.28E-05	Sb	2.12E-07
V	3.96E-07	Te	3.84E-04
Cr	6.20E-06	I	2.79E-05
Mn	1.12E-06	Cs	1.25E-07
Fe	1.52E-04	Ba	1.10E-05
Co	4.30E-08	La	7.15E-09
Ni	3.22E-06	W	1.75E-07
Cu	1.52E-06	Tl	6.69E-07
Zn	3.46E-05	Pb	6.76E-06
Ga	7.13E-07	Bi	5.46E-07

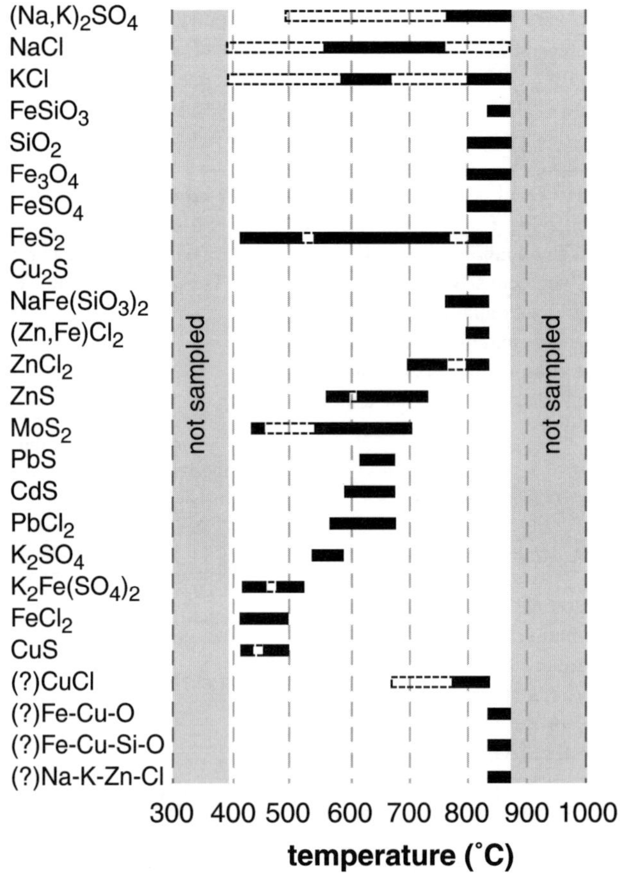

Figure 2: Phases identified in the sublimation tube by semi-quantitative SEM-EDAX spectra. Filled boxes indicate more than five positively identified spectra per 5-cm tube sector and dashed boxes between three and five phases. Question marks designate phases of uncertain stoichiometry. See text for further discussion.

in significant amounts (1 – 10 wt%) and minor Fe and Cu (≈ 1 wt%). In some cases, this incorporation leads to phases with chemistries not attributable to known mineral phases. For example, the In content of Pb-Bi sulfides reaches 13 wt%.

Cd-Zn sulfides: Both end members (i.e., greenockite and wurtzite) were identified. However, solid solutions of intermediate chemistry are more prevalent. Minor amounts up to 1 wt% of Fe, Cu and In apparently substitute for Cd and Zn.

Mo minerals: The Mo-containing minerals detected were powellite at higher temperatures with molybdenite at intermediate temperatures and ilsemannite ($Mo_3O_8 \cdot nH_2O$), forming at the lowest temperatures; tugarinovite (MoO_2) and molybdite (MoO_3) were also present as reported previously by Bykova et al. (1995); a rhenium sulfide mineral also occurs at temperatures below 650 °C (see also Korzhinsky et al., 1994) and apparently, a solid solution occurs between ReS_n and MoS_2 where n varies widely with an average value of ~ 1.5 to 2.0.

Silicates: Mg- and Ca-dominated silicates of the olivine, garnet and (clino-)pyroxene group minerals were detected in the highest temperature zones (i.e., > 750 °C) of the Kudriavy fumarolic field.

Chlorides: Na, Ca and K chlorides have also been identified; a complete solid solution series, (Na,K)Cl has been detected.

These sublimate minerals can be assigned to different temperature assemblages, such as olivine-pyroxenes-magnetite-hematite; powellite-anhydrite-molybdenite ReS_n-garnet (andradite); greenockite-wurtzite-cosalite-cannizzarite and halite-sylvite-hematite. Data on mineral phases presented here confirm and supplement the database already published by Bykova et al. (1995) and Korzhinsky et al. (1996).

6.3. Closed-System Cooling

Closed-system calculations were used to provide insight into the dominant transporting species and investigate the possible coexistence of solid and liquid (i.e., condensed phases). Specifically, the saturation of the gas (with respect to condensed compounds) at near magmatic temperature has to be discussed in the light of possible contamination of the condensate by wall-rock particles. A comparison of predicted solid/liquid phases with the observed sublimate mineralogy provides some test of the validity of the numerical simulation. In total, 27 trace elements were considered in the closed-system equilibrium calculations. All condensed (solid/liquid) species calculated to be stable between 300 and 1000 °C are given in Figure 5 and are discussed in detail below. In the case of gaseous species, representative results for a number of elements are given in Figures 3 and 4.

The results of the closed-system equilibrium calculations can be summarized as follows. Systematic trends in the affinity of trace elements to certain ligands in the gas phase are evident. There is a change from fluoride-dominated, major volatile species to chloride/bromide and in extreme cases iodide complexes with increasing atomic number within the same group, as would be expected. For example, B is mainly volatile as $B(OH)_{3(g)}$, Al as $AlOF_{2(g)}$ and $NaAlF_{4(g)}$, Ga as $GaCl_{(g)}$ and $GaCl_{3(g)}$ and Tl as $TlCl_{(g)}$. At higher temperatures, sulfur (as sulfide) becomes more important as a ligand with first-row transition elements, while O, OH and F decrease in importance relative to chloride as the major transporting ligand, reflecting the well known Irving-Williams stability order. The same order was reported for Mt. St. Helens gas chemistry (Symonds and Reed, 1993). In addition, bromide and iodide complexes become more important in comparison to chloride species with decreasing temperature. The calculations also show that complexes with simple carbonyl and halocarbon ligands are also stable although they are present at low concentrations; for example, at 1000 °C and 1 bar, log f = -28.8 for $Ni(CO)_4$, log f = -34.8 for $SiH_3(CCl_3)_{(g)}$ and log f = - 29.3 for $SiH_3(CF_3)_{(g)}$.

Therefore, the nature of trace-element transport in Kudriavy gases is calculated to be very similar to those of Mt. St. Helens and Merapi (Symonds and Reed, 1993; Symonds et al., 1987). Some differences to St. Augustine (Symonds et al., 1992) are observed by comparing the temperature range between 860 and 1000 °C. For example, $Fe(OH)_{2(g)}$ and $Ni(OH)_{2(g)}$ are important species in Kudriavy and Mt. St. Helens gases but at Mt. St. Augustine, $FeCl_{2(g)}$ and $NiCl_{2(g)}$ dominate by three orders of magnitude over the respective hydroxide species. $PbCl_{2(g)}$ as well as $PbCl_{4(g)}$ dominate strongly over $PbS_{(g)}$ at Augustine which is the opposite to Kudriavy. $MoOCl_{2(g)}$ is the most abundant gaseous Mo-containing species at Augustine below 940 °C, whereas at Kudriavy, this is only true below 375 °C. These differences possibly arise from the higher chloride concentrations (up to 6 vol%) at St. Augustine in comparison to Kudriavy (0.3 vol%, see Table1). We note also the similar stabilities of volatile species in Kudriavy gases calculated by Churakov et al. (2000) with our calculations with minor discrepancies (e.g., copper) arising from differences in the thermodynamic data bases employed.

Forty three condensed phases are calculated to be in equilibrium with Kudriavy gases at various temperatures between 300 to 1000 °C, some of which match the natural sublimation assemblage or the sublimation tube observations as given in Figure 5. However, a number of significant discrepancies between calculated assemblages and those occurring naturally were observed. Iron is calculated to occur only in magnetite and pyrite but occurs both in natural sublimates and in the sublimation tubes, predominantly in silicates and oxide compounds such as olivine ($(Mg,Fe)SiO_4$), augitic-pyroxene ($(Ca,Na)(Mg,Fe,Al)Si_2O_6$), andradite ($Ca_3Fe_2Si_3O_{12}$), hematite (Fe_2O_3), magnetite (Fe_3O_4). Pyrite (FeS_2) and rozenite ($FeSO_4 \cdot 4H_2O$) are found in the natural sublimates. However, they are of limited importance in comparison to the other phases. The thermochemical database comprises data for the pure end members only. In the closed-system calculations, lead is predicted to remain in the gas phase at all temperatures considered. In contrast, lead is found in phases such as cannizzarite ($Pb_4Bi_6S_{13}$), cosalite ($Pb_2Bi_2S_5$) and other Pb-Bi sulfides both in nature and in the sublimation tube. Qualitatively, the actual sublimate phases are correctly predicted by the equilibrium thermodynamic calculations. However, the latter fail to predict the specific paragenesis (e.g., the natural coexistence of powellite and molybdenite with garnet (andradite), as well as the correct temperature ranges of deposition in the sublimation tube (e.g., the deposition of alkali sulfates above 750 °C; the stability range of chalcocite or the dominance of NaCl above 600 °C).

Symonds and Reed (1993) observed the same problem with lead volatility in the case of gases from Mt. St. Helens and invoked inaccuracies in the thermochemical data for the gaseous species to account for such discrepancies. In the case of cannizzarite and

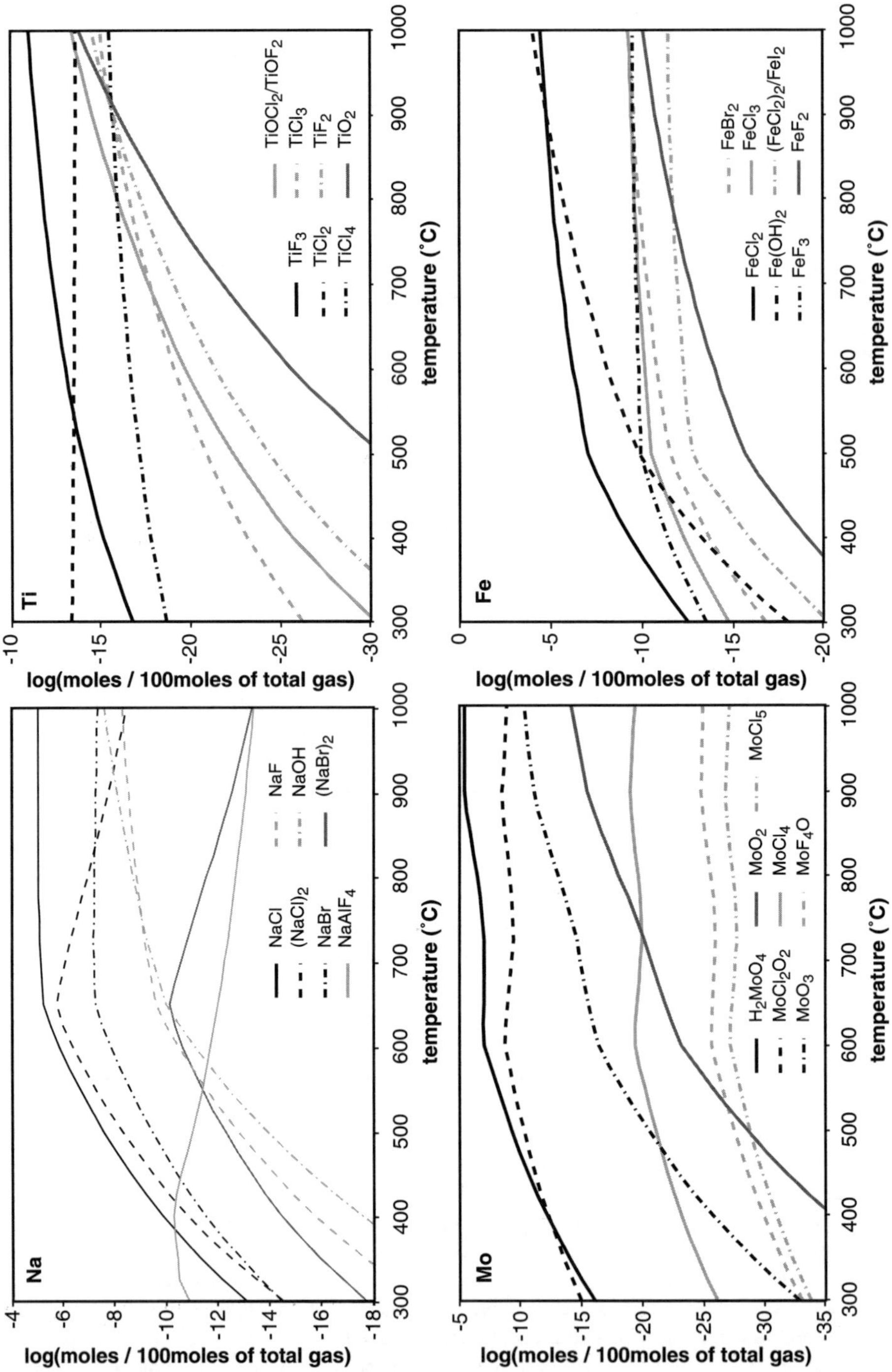

Figure 3: Distribution of (a) sodium (Na, Group IA), (b) titanium (Ti, group IVA, (c) molybdenum (Mo, group VIA) and (d) iron (Fe, group VIIIA) amongst its gaseous species at closed-system equilibrium cooling conditions for the systems: X-Na-Al-Si, X-Ti, X-Ca-Mo-Si and X-Fe-Mg-Si. For solids stable in equilibrium with the gas phase see Fig. 5. Not shown are species which are calculated to be stable but at concentrations below 10^{-25} mol/mol at all temperatures considered, i.e., $(TiCl_3)_{2(g)}$, $TiI_{4(g)}$, $Ti_{(g)}$, $TiI_{(g)}$, $TiBr_{(g)}$, $TiI_{3(g)}$, $TiCl_{(g)}$, $TiI_{2(g)}$, $TiBr_{2(g)}$, and $TiS_{(g)}$; $MoI_{3(g)}$, $Mo_{(g)}$, $MoF_{(g)}$, $MoBr_{(g)}$, $MoI_{(g)}$, $MoO_{(g)}$, $MoF_{2(g)}$, $MoBr_{2(g)}$, $MoF_{4(g)}$, and $MoBr_{3(g)}$; $(FeI_2)_{(g)}$.

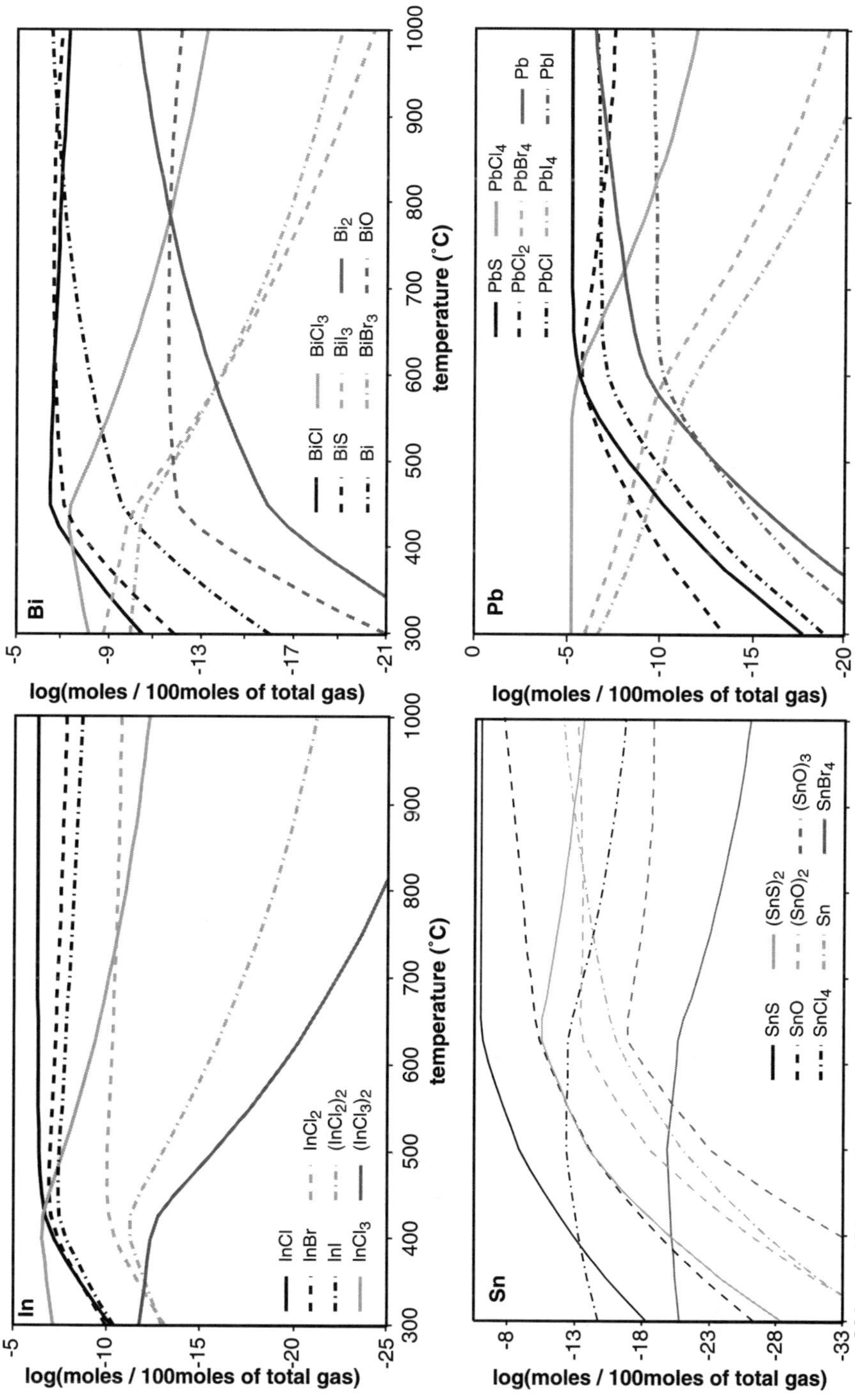

Figure 4: Distribution of (a) indium (In, group IIIB), (b) bismuth (Bi, group VB), (c) tin (Sn, group IVA) and (d) lead (Pb, group IVB) amongst its gaseous species at closed-system equilibrium cooling conditions (for stable solids see Fig. 5). For the systems X-In, X-Bi, X-Sn and X-Pb-Si, all stable gaseous species considered in the calculations are shown.

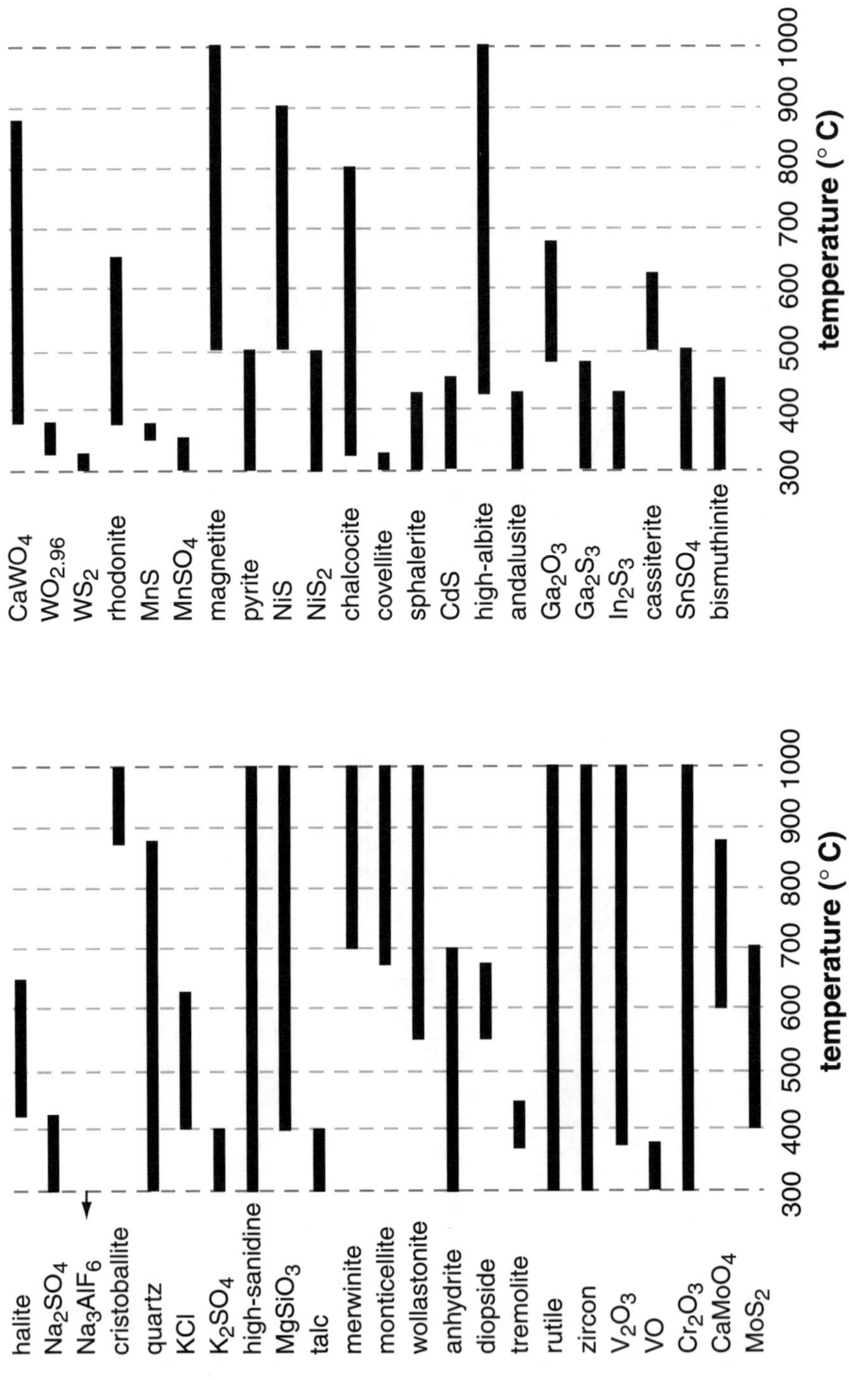

Figure 5: Condensed phases calculated to be stable within the temperature range and the elemental systems considered in the text. The condensed phases of calcium molybdate and tungstate ($CaMoO_{4(c)}$/$CaWO_{4(c)}$) as well as magnetite ($Fe_3O_{4(c)}$) are calculated to be stable on the basis of an extrapolated C°_p function. See text for further discussion.

cosalite, adequate thermodynamic data do not exist and their sublimation could not be considered in the calculations. Alternatively, one could assume that the content of Pb in the condensate was underestimated. Korzhinsky et al. (1996) have shown that siderophile elements are precipitated in the sulfur-containing particles forming in the condensate during sampling and that current analytical techniques do not fully account for the multiphase composition of the condensate. Consequently, the concentrations of elements such as Ag and Pb might be underestimated by an order of magnitude. In the case of sample K95/2, this explanation seems to be insufficient. $PbS_{(c)}$ is only predicted to be stable under the assumption of an underestimation of Pb by at least two orders of magnitude. The failure of the model to predict Fe-pyroxene partly arises because solid solutions are not considered in the calculations. Pure ferrosilite is stable only at pressures above 10 kbars (at 600 °C; Lindsley, 1980). Consequently, the assemblage fayalite + quartz would be expected as a result of these calculations. One of the most important results of the closed-system calculations is the saturation of the gas with respect to major rock-forming minerals at 1000 °C. These phases are: enstatite ($MgSiO_3$), rutile (TiO_2), zircon ($ZrSiO_4$), V_2O_3, Cr_2O_3, magnetite (Fe_3O_4), high-albite ($NaAlSi_3O_8$), monticellite ($CaMgSiO_4$), wollastonite ($CaSiO_3$) and merwinite ($Ca_3MgSi_2O_8$).

No thermochemical data are available for silicates of V, Ti and Cr. These elements are thought to be distributed amongst Fe-Ti oxides and pyroxenitic phases. Similar results have been obtained for St. Augustine, Merapi and Mt. St. Helens gases (Symonds and Reed, 1983; Symonds et al., 1992, 1987). However, it should be noted for example, that at Izalco volcano (Costa Rica), a variety of different vanadium oxides have been observed as sublimation products (e.g., Stoiber and Rose, 1974). Theoretically, several interpretations could account for the calculated supersaturation of the gas phase with respect to rock-forming minerals such as the presence of the phases cited above as aerosols in the gas stream, non-equilibrium behavior (i.e., supersaturation of the gas because of kinetic effects), abrasion, entrainment and dissolution of wall-rock particles in the condensates and incorrect or missing thermochemical data on gaseous and condensed species.

In the case of gases from St. Augustine, Symonds et al. (1992) favored wall-rock contamination as the most plausible explanation for the calculated supersaturation. The basis for this conclusion is the low and similar enrichment factors for wall-rock elements and the absence of silicate phases in the sublimation tubes. However, in the case of Kudriavy, olivine, pyroxenes, iron oxides and SiO_2 are found as natural fumarole sublimates as well as in the sublimation tube. A much stronger argument in favor of wall-rock contamination rather than the use of unreliable and frequently misleading enrichment factors is the fact that the gas is calculated to be saturated with respect to silicate minerals (i.e., SiO_2, Al_2SiO_5) at temperatures above 1300 °C, surpassing a reasonable temperature of the host melt. Therefore, the measured elements are thought to be derived to some extent from wall-rock contamination. It is difficult to judge the extent of this contamination. Symonds et al. (1992) have considered this problem by computationally reacting wall-rock with the measured gas at fixed temperature and pressure.

Generally, the assumption of closed-system cooling seems to oversimplify various aspects of the highly dynamic systems investigated. The observation that solids form not only in the cooler peripheral parts of the vents where processes of air admixture occur (e.g., Getahun et al., 1996) but also on the actual vent walls in contact with the moving gas stream, indicates that the system is open with respect to the retrieval of certain trace elements.

6.4. Open-System Cooling

Twenty elements were considered in open-system calculations of which twelve can be directly compared to the elemental distribution data from the silica tube. The results are presented in Figures 6 to 8 and are also directly compared to sublimation tube mineralogy as outlined above and presented in Figure 2. As the results cannot be compared on a quantitative basis, the amount of each phase has been normalized to its maximum abundance. Note that each condensed phase is considered and scaled separately and therefore, additional features of the results such as the calculated dominance of $NaCl_{(c)}$ over $Na_2SO_{4(c)}$ are not represented in the graphs. In the discussion below, the elemental systems (where X=C-S-O-H-S-Cl-Br-F-I) used for the calculation are discussed separately.
The systems X-Na-Al-Si; X-K-Al-Si(Na); X-Ca-Al-Si(Na) and X-Ca-Mg-Fe-Si (Figures 6a-c) are summarized as follows:

feldspars: anorthite, high-sanidine and high-albite are calculated to be stable in very narrow temperature intervals of 1000 – 960 °C and 1000 – 840 °C; all feldspar condensed phases decrease in abundance by at least seven orders of magnitude in the first 20 °C cooling step.

orthosilicates: $CaSiO_{3(c)}$ (wollastonite), $MgSiO_{3(c)}$ (enstatite), $Ca_3MgSi_2O_{8(c)}$ (merwinite) are all only stable at 1000 °C; $CaMgSiO_{4(c)}$ (monticellite) is stable above 800 °C and $Mg_2SiO_{4(c)}$ is

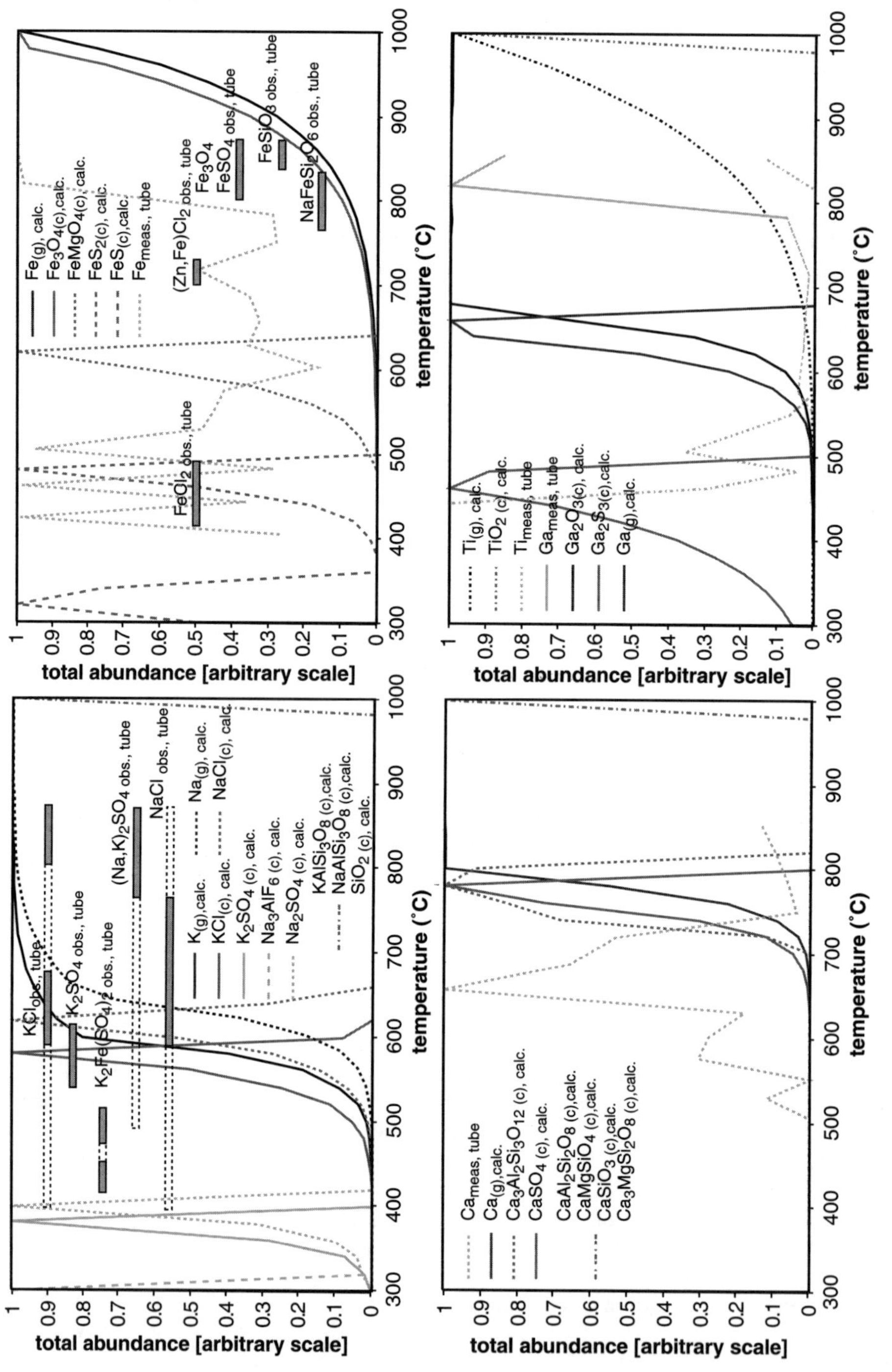

Figure 6: (a) Na/K, (b) Fe, (c) Ca and (d) Ti/Ga phase behavior in open-system calculations as compared to sublimation tube mineralogy. The boxes indicate the presence of solid phases positively identified in the sublimate tubes. Note that $Fe_3O_{4(c)}$ is calculated to predominate over $FeS_{2(c)}$ by two orders of magnitude whereas the sublimation tube chemistry indicates similar strong deposition of iron between 400 and 500 °C as is detected at 800 °C. Titanium is found as a minor constituent in salts of the sublimation tube. See text for further discussion.

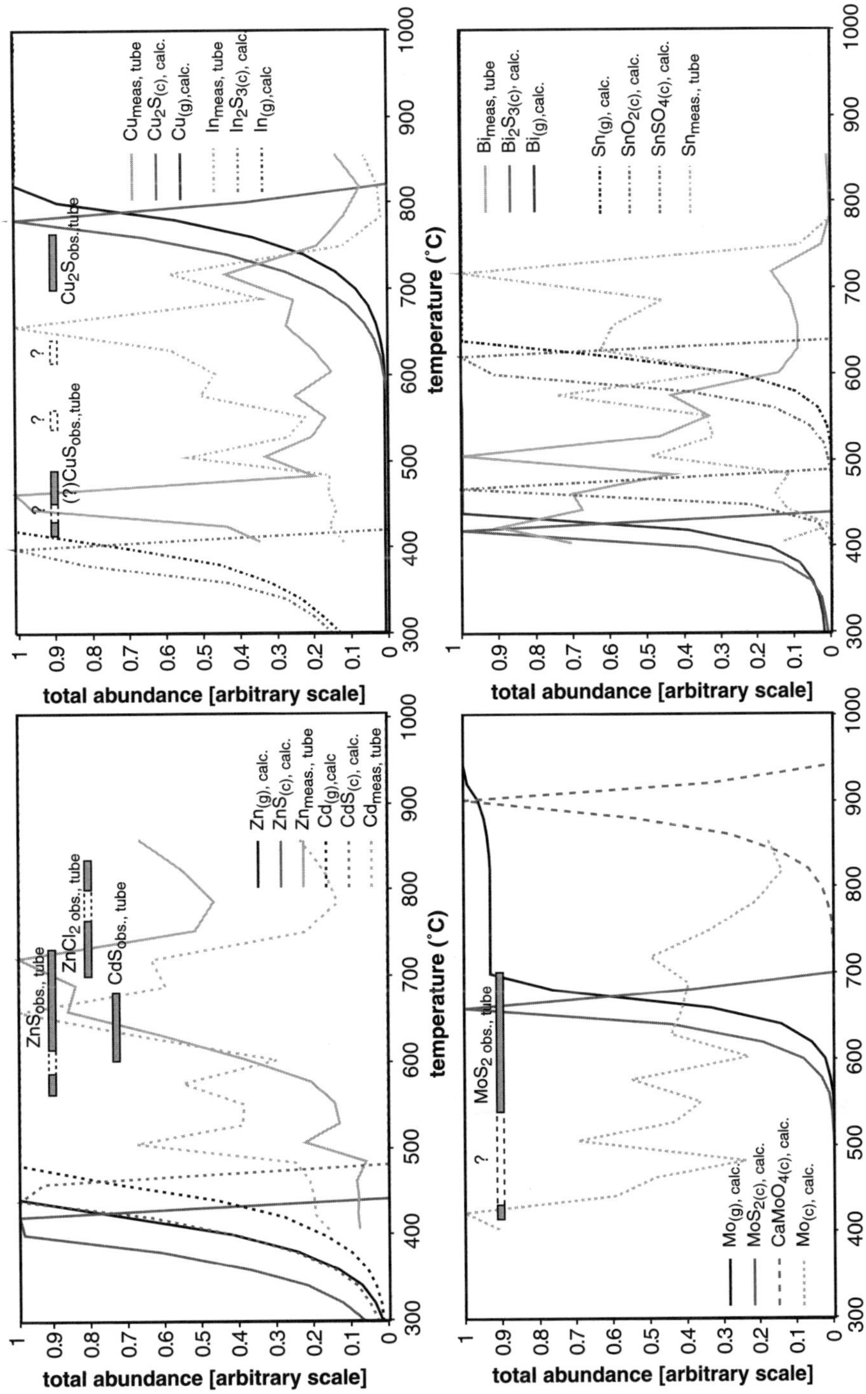

Figure 7: (a) Zn/Cd, (b) Cu/In, (c) Mo and (d) Bi/Sn phase behavior in open-system calculation as compared to sublimation tube chemistry and mineralogy. Indium is found as a constituent of Zn-Cd phases. Tin is found as a constituent of Zn-Cd and Pb-Bi phases in the sublimation tube. Bismuth is found as a constituent in Pb-phases in the sublimation tube.

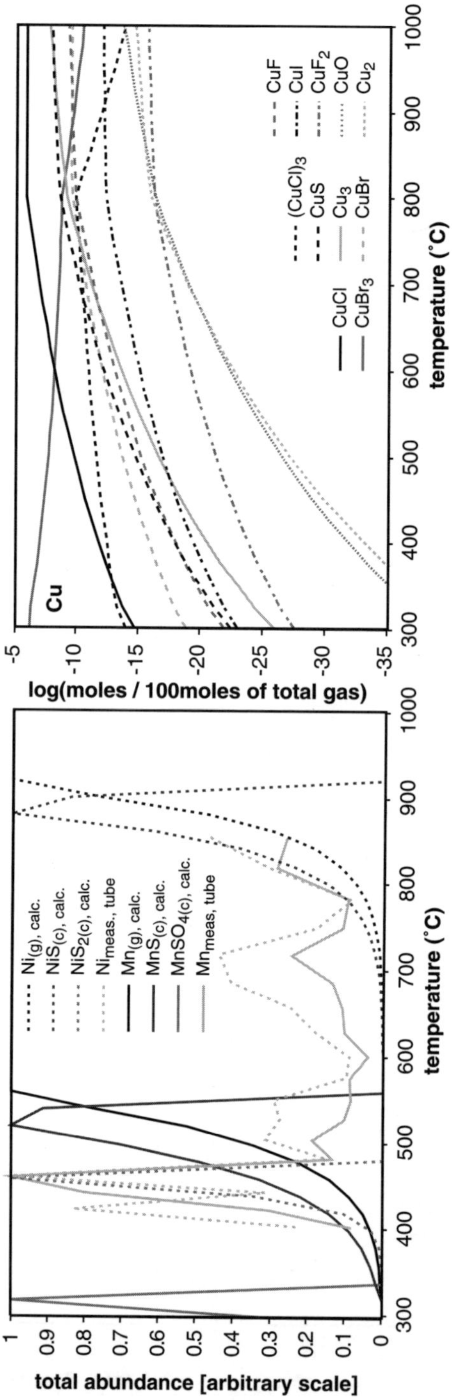

Figure 8: (a) Mn/Ni phase behavior in open-system calculations as compared to sublimation tube chemistry. Again, as discussed in the case of iron (Fig. 6) $NiS_{(c)}$ is calculated to predominate over $NiS_{2(c)}$ which is deposited at lower temperatures whereas the sublimation tube chemistry indicates the maximum deposition at 400 to 500 °C. (b) Distribution of copper (Cu, group IB) amongst its gaseous species at closed-system equilibrium cooling conditions (for stable solids see Fig. 5) in the system X-Cu. All stable gaseous species considered are shown. Note that Cu is predicted to be transported as the $CuBr_{3(g)}$ species below 600 °C. See text and Fig. 7 for further discussion.

calculated to be stable between 980 and 800 °C; grossular ($Ca_3Al_2Si_3O_{12(c)}$), becomes stable upon cooling to 820 – 700 °C; $Al_2SiO_{5(c)}$ (andalusite) is stable between 960 and 920 °C.

chlorides and sulfates: $NaCl_{(c)}$ (halite) and $KCl_{(c)}$ (sylvite) are stable between 640 – 460 °C and 600 – 460 °C respectively, $Na_2SO_{4(c)}$ (thenardite) and $K_2SO_{4(c)}$ (arcanite) between 420 – 300 °C and 380 – 300 °C respectively; $Na_3AlF_{6(c)}$ (cryolite) occurs only at 300 °C. $CaSO_{4(c)}$ (anhydrite) has a wide stability ranging from 780 to 300 °C. However, this phase decreases in abundance by four orders of magnitude above 600 °C.

sulfides: Pyrite ($FeS_{2(c)}$) and troilite ($FeS_{(c)}$) are stable from 560 – 400 °C and from 340 to 300 °C, respectively.

oxides: Magnetite ($Fe_3O_{4(c)}$) decreases from its maximum abundance at 1000 °C and is stable within a 500 °C interval (i.e., 1000 – 500 °C) with the amount being deposited decreasing by two orders of magnitude in the first 200 °C; $MgFe_2O_{4(c)}$ (magnesioferrite) is stable between 620 and 480 °C and predominates over $MgO_{(c)}$; it is the only phase that is stable over two temperature intervals, one from 840 to 640 °C and a second one from 440 to 300 °C (having an abundance of four orders of magnitude lower than the first one); cristobalite, $SiO_{2(c)}$, shows the same features as the orthosilicates: it is only stable at 1000 °C; corundum, $Al_2O_{3(c)}$, is stable over a wide temperature interval ranging from 940 to 300 °C while only decreasing in abundance by three orders of magnitude.

Other systems

X-Ti (Figure 6d): $TiO_{2(c)}$ (rutile) is theoretically stable from 1000 to 440 °C; its abundance changes by seven orders of magnitude in the first 20 °C.

X-Zr-Si, -Cr and -V (not shown): The trends are similar to those for Ti, in contrast to Cr and Zr where only one stable phase exists at 1000 °C (i.e., $Cr_2O_{3(c)}$ and $ZrSiO_{4(c)}$ (zircon)), V has two stable phases: $V_2O_{3(c)}$ at 1000 °C and $VO_{2(c)}$ at temperatures below 340 °C.

X-Zn, -Cd and -In (Figures 7a,b): $ZnS_{(c)}$ (sphalerite), $CdS_{(c)}$ and $In_2S_{3(c)}$ are deposited over similar temperature intervals (440 - 300 °C, 480 - 300 °C and 420 - 300 °C) and are calculated to be the only stable condensed phase for each element.

X-Cu and -Mn-Si (Figures 7b, 8a): $Cu_2S_{(c)}$ has a stability interval of 800 to 580 °C; the phase covellite, calculated to be stable in the closed system (see above), is not predicted to be stable; $MnSiO_{3(c)}$ (rhodonite) is not calculated to be stable in the open system; the reason for this is the high-temperature deposition of Si as cristobalite; $MnS_{(c)}$ is stable from 540 - 340 °C and $MnSO_{4(c)}$ between 320 - 300 °C.

X-Bi, -Sn and –Mo-Ca-Si (Figures 7c,d): $SnSO_{4(c)}$ is deposited between 480 and 300 °C; cassiterite ($SnO_{2(c)}$) deposits from 600 to 500 °C. $Bi_2S_{3(c)}$ (bismuthinite), the only stable Bi-containing phase, is calculated to be stable between 400 and 300 °C; therefore, both minerals coexist theoretically within an interval of 60 °C.

X-Ga and –Ni-Si (Figures 6d, 8a): Each of these elements forms two stable phases. $Ga_2O_{3(c)}$ and $Ga_2S_{3(c)}$ are deposited in the intervals from 920 - 620 °C and 480 -300 °C respectively; $NiS_{2(c)}$ and $NiS_{(c)}$ are stable between 900 and 480 °C.

6.5. Discussion

The features of the open-system calculations can be summarized as follows:

- The results reflect the observed deposition of minerals over restricted temperature intervals in sublimation tubes
- The separation of mineral phases occurs with respect to temperature, such as for example,

$$T\left(\frac{\max.subl.}{NaCl}\right)=620°C\ ,\ T\left(\frac{\max.subl.}{Na_2SO_4}\right)=400°C$$

$$;\ T\left(\frac{\max.subl.}{KCl}\right)=580°C\ ,\ T\left(\frac{\max.subl.}{K_2SO_4}\right)=380°C\ .$$

  However, the corresponding spatial separation in the tube will be a function of the temperature gradient and hence depend directly on the flow rate.
- In some cases, open-system calculations reproduce distinct mineral assemblages found in sublimation tubes and in the natural environment, e.g., garnet-molybdenite-anhydrite or Bi-sulfide (represented by bismuthinite, (($Bi_2S_{3(c)}$) in the calculations) –ZnS-CdS.
- Open-system calculations indicate the deposition of silicate minerals whereas they were calculated to be supersaturated in the closed system, over small temperature intervals of only 20 °C; an exception is TiO_2.
- The results of the calculations differ for open and closed systems with respect to the condensed phases found to be stable; in most cases, the open-system calculations fail to reproduce the correct temperature range of maximum deposition found in the sublimation tubes.

Nevertheless, the open-system calculations are to some extent able to reproduce the processes of subli-

Table 4: Sublimate minerals collected at Kudriavy in 1995.
Temperatures indicate the approximate temperature range of the gas in which the sample was collected. Note that flow conditions will vary across the surface of a rock specimen as will the temperature. Additional information for phase identification beside SEM-EDAX spectra is indicated. n.d. = not detected.

mineral	**theoretical formula**	**additional information**	**observed temperature [°C]**
olivine	$(Mg,Fe)SiO_4$	single X-tal X-ray	>750
enstatite	$Mg_2Si_2O_6$	single X-tal X-ray	>750
augitic pyroxene	$(Ca,Na)(Mg,Fe,Al)Si_2O_6$	shape	650 – 750
wollastonite	$Ca_2Si_3O_6$	shape	650 – 750
andradite	$Ca_3Fe_2Si_3O_{12}$	single X-tal X-ray	650 – 750
powellite	$CaMoO_4$	single X-tal X-ray	650 – 750
anhydrite	$CaSO_4$	single X-tal X-ray	550 – 650
haematite	Fe_2O_3	single X-tal X-ray	550 – 650
magnetite	Fe_3O_4	single X-tal X-ray	550 – 650
unnamed	ReS_n $n \approx 1\text{-}4$ (?)	shape, reflectivity	550 - 650
molybdenite	MoS_2	single X-tal X-ray, reflectivity	550 – 650
greenockite	CdS	single X-tal X-ray, powder X-ray	450 – 550
wurtzite	ZnS $(Zn/Cd)_m \approx .4\text{-}.8$ and $((Zn+Cd)/S)_m \approx 1.0\text{-}.75$	shape	450 – 550
cannizzarite	$Pb_4Bi_5S_{11}$	single X-tal X-ray	450 – 550
cosalite	$Pb_2Bi_2S_5$	shape	450 – 550
pyrite	FeS_2	shape	450 – 550
cristoballite	SiO_2	powder X-ray	450 – 550
halite	$NaCl$	shape, single X-tal X-ray	450 – 550
rozenite	$FeSO_4 \cdot 4H_2O$	single X-tal X-ray	350 – 450
ilsemannite	Mo_3O_8	colour, shape	350 – 450
sinjarite	$CaCl_2 \cdot 2H_2O$	powder X-ray	n.d.
sylvite	KCl	shape	n.d.

mation occurring in a moving gas stream. The differences in the calculated sublimation assemblages as compared to the results of the closed-system approach presented above are partly a consequence of the retrieval of the major elements, Si and Al, in the first 20 °C step of calculation. However, the failure of the model to reproduce the actual observed temperature range of sublimate deposition deserves a more detailed discussion. Two cases may be distinguished:

(i) elements where the calculated volatility is underestimated, i.e.,

$$T\left(\frac{\max.subl.}{calc.}\right) > T\left(\frac{\max.subl.}{obs.}\right),$$

(e.g., Ni, Mo, Mn, Cu and Fe).

(ii) elements where the calculated volatility is overestimated, i.e.,

$$T\left(\frac{\max .subl.}{calc.}\right) < T\left(\frac{\max .subl.}{obs.}\right),$$

(e.g., Ga, Sn, Bi, In, Cd and Zn).

The temperature differences between calculated and observed maximum sublimation discussed are >100 °C (with the exception of Mn where the difference is around 75 °C). While it is correct that the temperature measured along the tube's axis is a maximum estimate of the actual sublimation temperature, the difference in the temperature between the inner tube wall and that along the tube's axis is less than ± 50 °C as determined by direct measurements during the experiment. Therefore, the temperature differences discussed here are significant.

In the case of underestimation of the volatility of trace elements, incorrect or missing thermochemical data as well as dynamic or kinetic effects will be important. Indeed, it would be expected that actual sublimation processes taking place in the tube are intermediate between a closed and an open system. Sublimates forming in the gas stream initially as aerosols might be retained and only deposited at lower temperatures. The behavior of molybdenum may be such an example. In the case of copper, such an argument may be premature. As exemplified in the closed-system approach discussed above and shown in Figure 8b, Cu is predicted to be dominantly volatile as the $CuBr_{3(g)}$ complex below 600 °C. The change in the oxidation state of copper at this temperature is a consequence of the internal gas buffer dominating the redox environment. However, there are no thermochemical data for the hypothetical $CuCl_{3(g)}$ complex. Hence, the volatility of copper is underestimated at temperatures below 600 °C due to a lack of data for a dominant gaseous complex of copper.

It should also be mentioned that other, potentially important volatile species have not been considered in the calculations because thermodynamic data pertaining to their stability are currently unavailable. For example, many volatile species forming in a volcanic gas will be hydrated complexes such as $AgCl(H_2O)_{3(g)}$, $CuCl(H_2O)_{n(g)}$, $AgCl(H_2O)_{n(g)}$, $AuCl(H_2O)_{n(g)}$, and $SnCl_2(H_2O)_{n(g)}$ (Migdisov et al., 1999; Archibald et al., 2000; Williams-Jones et al., this volume) but the stability of these in high-temperature Kudriavy gases has not been considered. In addition, the formation and high-temperature stability of sulfur-containing species such as polysulphanes (Migdisov et al., 1998) and molecular complexes such as $Sb_2S_3(H_2S)_{(g)}$ (Zakaznova-Iakovleva et al., 2001) may be important in volcanic gas systems but have not been considered in our calculations. We have also not considered ionic clustering reactions within the vapor phase (Likholyot, 2000) although these are probably unimportant in volcanic gas systems.

Similarly, erroneous thermochemical data might explain the overestimation of the volatility of certain trace elements. It should also be noted, that most elements do not sublime as pure substances. As demonstrated in Figure 7, elemental distribution of Cd, Zn and In on the one hand, and Pb and Bi on the other, correlate strongly. Therefore, it may be possible to stabilize certain elements in solid solutions at temperatures higher than would be expected for the respective pure substances.

7. CONCLUSIONS AND COMMENTS

The comparison of open-system equilibrium calculations with the results from a silica-tube experiment allows some assessment of the validity of current thermochemical equilibrium modeling as applied to trace-element transport in volcanic gases. At this stage, it has to be concluded that equilibrium models only qualitatively reproduce the sublimation tube chemistry and hence trace-element volatility in the Kudriavy gas. It was shown above that this may be partly explained by unavailable or incorrect thermochemical data as well as the lack of consideration of solid solutions. The latter are believed to stabilize certain sublimates to temperatures higher than predicted by the models, thus causing an overestimation of the volatility of specific elements by the calculation. The models used here are themselves expected to only approximate the real processes in the sublimation tubes which are considered to behave in an intermediate way between a closed and open system, depending on the flow rate.

It was already pointed out above that many of the gaseous equilibria are quenched below 350 – 400 °C. The occurrence of diffusion processes (see also Korzhinsky et al., 1996) and the dependence of gaseous equilibria on the flow rate of a fumarole (Saito et al., 1989) indicate that disequilibrium processes must also occur at much higher temperatures. Therefore, the use of thermochemical equilibrium computations to model heterogeneous equilibria in dynamic volcanic gas systems such as Kudriavy must be treated with caution and regarded as a crude approximation of reality.

Finally, the data reported in this work further emphasize the importance of volcanic gases in transporting heavy-metals and other compounds to the earth's surface and atmosphere. The importance of magma degassing in supplying a metal flux to ore depositing hydrothermal convectional cells is further noted.

Acknowledgments—This study was supported by the Schweizerische Nationalfonds. We are grateful to the Institut für Nichtmetallische Werkstoffe (ETH), for providing access to their SEM-EDAX systems, as well as to I. Farrenkothen and Ch. Moor (EMPA) for help with analyses. We also thank our colleagues at Chernogolovka (K.Shmulovich, M. Korzhinsky and S. Tkachenko) for stimulating discussions and logistical support.

REFERENCES

Amman M., Hauert R., Burtscher H. and Siegmann H.C. (1993) Photoelectric charging of ultrafine volcanic aerosoles, detection of Cu(I) as a tracer of chlorides in magmatic gases. *J. Geophys. Res.* **99**, 551-556.

Archibald S.M., Migdisov A.A. and Williams-Jones A.E. (2000) The stability of Ag and Au chlorides in water vapour at elevated temperatures and pressures. In: Steam, Water and Hydrothermal Systems (P.R. Tremaine, P.G. Hill, D.E. Irish and P.V. Balakrishnan, eds.), NRC Press, Ottawa, 638-644.

Avdeiko G.P., Volynets O.N., Antonov A.Y. and Tsvetkov A.A. (1992) Kurile Island arc volcanism: structural and petrological aspects. *Tectonophysics* **199**, 271-287.

Barin I. and Knacke O. (eds.), (1973) Thermochemical properties of inorganic substances. Springer-Verlag, Berlin.

Barin I., Knacke O. and Kubaschewski O. (eds.), (1977) Thermochemical properties of inorganic substances. Springer-Verlag, Berlin.

Bykova E.Y., Znamenskii V.S., Kovalenker V., Marsii I.M. and Baturin S.V. (1995) Association and deposition conditions of molybdenum minerals in exhalation products of the Kudryavyi volcano, Iturup, Kurile Islands. *Geol. Ore Deposits* **3**, 227-235.

Chase Jr. M.W., Davies C.A., Downey Jr. J.R., Frurip D.J., McDonald R.A. and Syverud A.N. (1985) JANAF thermochemical tables, 3rd ed., Part I and II, J. Phys. Chem., Ref. Data 14, Suppl. No.1, 1-1856.

Churakov S.V., Tkachenko S.I., Korzhinskii M.A., Bocharnikov R.E. and Shmulovich K.I. (2000) Evolution of composition of high-temperature fumarolic gases from Kudriavy Volcano, Iturup, Kuril Islands: the thermodynamic modeling. *Geochem. Internat.* **38**, 436-451.

Crowe B.M., Finnegan D.L., Zoller W.H. and Boynton W.V. (1987) Trace element geochemistry of volcanic gases and particles from 1983-1984 eruptive episodes of Kilauea volcano. *J. Geophys. Res.* **92**, 13708-13714.

Eriksson G. (1975) Thermodynamic studies of high temperature equilibria. XII. SOLGAS-MIX, a computer program for the calculation of equilibrium compositions in multiphase systems. *Chemica Scripta* **8**, 100-103.

Ehrlich E. (1986) Geology of the calderas of Kamtchatka and Kurile Islands with comparison to calderas of Japan and the Aleutians, Alaska. Tech. Rep., USGS, Open File Report 86-291.

Francis P., Chaffin C., Maciejewski A. and Oppenheimer C. (1996) Remote determination of SiF_4 in volcanic plumes: A new tool for volcano monitoring. *Geophys. Res. Letters* **23**, 249-252.

Gerlach T.M. and Casadevall T.J. (1986). Evaluation of gas data from high-temperature fumaroles at Mount St. Helens. *J. Volcan. Geotherm. Res.* **28**, 107-140.

Getahun A., Reed M.H. and Symonds R. (1996) Mount St. Augustine volcano fumarole wall-rock alteration: mineralogy, zoning, composition and numerical models of its formation process. *J. Volcan. Geotherm. Res.* **71**, 73-107.

Giggenbach W.F. (1975) A simple method for the collection and analysis of volcanic gas samples. *Bull. Volcan.* **45**, 235-255.

Giggenbach W.F. (1987) Redox processes governing the chemistry of fumarolic gas discharges from White Island, New Zealand. *Appl. Geochem.* **2**, 143-161.

Giggenbach W.F. and Goguel R.L. (1981) Collection and anlysis of geothermal and volcanic water and gas discharges, New Zealand DSIR Chem. Div. Report, No. CD2401.

Giggenbach W.F. and Matsuo S. (1991) Evaluation of results from the second and third IAVCEI Field Workshop on volcanic gases, Mt. Usu, Japan, and White Island, New Zealand. *Appl. Geochem.* **6**, 125-141.

Hedenquist J.W., Aoki M. and Shinohara H. (1994) Flux of volatiles and ore-forming metals from the magmatic-hydrothermal system of Satsuma-Iwoijma volcano. *Geology* **22**, 585-588.

Knacke O., Kubaschewski O. and Hanselmann K. (eds.), (1991) Thermochemical properties of inorganic substances. Springer-Verlag, Berlin.

Korzhinskiy M.A., Tkachenko S.I., Bulgarov R.F. and Shmulovich K.I. (1996) Condensate composition and native metals in sublimates of high temperature gas streams of Kudryavyi volcano, Iturup, Kurile Islands. *Geochem. Internat.* **34**, 1057-1064.

Korzhinskiy M.A., Tkachenko S.I. and Shmulovich K.I. (1995) How can native Si and Al precipitate from natural volcanic gases with f_{O_2} near Ni-NiO? *Terra Nova Abstr. Supp.* **7**, 17.

Korzhinskiy M.A., Tkachenko S.I. and Shmulovich K.I., Taran Y.A. and Steinberg G.S. (1994) Discovery of pure rhenium sulfide mineral at the Kudriavy volcano, Iturup, Kurile Islands. *Nature* **369**, 51-52.

Likholyot A. (2000) Experimental and theoretical study of ionic clustering reactions involving water. Ph.D. thesis, Inst. für Mineralogie und Petrographie, ETH-Zürich, 142 p.

Lindsley D.H. (1980) Phase equilibria of pyroxenes at pressures < 1 atm. In: C.T. Prewitt (ed.); Pyroxenes. *Min. Soc. Amer. Rev. Mineralogy* **7**, 289-307.

Migdisov A.A., Williams-Jones A.E. and Alekhin Yu.V. (1998) Experimental study of polysulfane stability in gaseous hydrogen sulfide. *Geochim. Cosmochim. Acta* **62**, 2627-2635.

Migdisov A.A., William-Jones A.E. and Suleimenov O.M. (1999) Solubility of chlorargyrite (AgCl) in water vapor at elevated temperatures and pressures. *Geochim. Cosmochim. Acta* **63**, 3817-3827.

Nriagu J.O. (1989) A global assessment of natural sources of atmospheric trace metals. *Nature* **338**, 47-49.

Quisefit J.P., Toutian J.P., Bergametti G., Javoy M., Cheynet B. and Person A. (1989) Evolution vs. cooling of gaseous volcanic emissions from Momotombo Volcano, Nicaragua; Thermochemical model and observations. *Geochim. Cosmochim. Acta* **53**, 2591-2608.

Robie R.A. and Hemingway B.S. (1995) Thermodynamic properties of minerals and related substances at 298.15 K and 1 bar (105 pascals) pressure and at higher temperatures. *USGS Bull.* **2131**, 461 p.

Rosen E., Osadchii E. and Tkachenko S. (1993) Oxygen fugacity directly measured in fumaroles of the volcano Kudriavy (Kurile Islands). *Chem. Erde* **53**, 219-226.

Saito G., Shinohara H. and Matsuo S. (1989) Effect of flow velocity on temperature and chemical composition of fumarolic gases. In: Continental magmatism, JAVCEI, General Assembly Santa Fe, New Mexico, USA;

vol. 131 of Bull. New Mexico Bureau of Mines and Min. Res., 232.

Shmulovich K.I. and Churakov S.V. (1998) Natural fluid phases at high temperatures and low pressures. *J. Geochem. Exploration* **62**, 183-191.

Simkin T. and Siebert L. (1994) Volcanoes of the world. Smithonian Institution, Geoscience Press, 2nd ed.

Stoiber R.E. and Rose Jr. W.I. (1974) Fumarole incrustations at Central American volcanoes. *Geochim. Cosmochim. Acta* **38**, 495-516.

Stull D., Westrum E.F. and Sinke G.C. (1969) The chemical thermodynamics of organic compounds, J.Wiley and Son, New York.

Symonds R.B., Rose W.I., Reed M.H., Lichte F.E. and Finnegan D.L. (1987) Volatilization, transport and sublimation of metallic and non-metallic elements in high temperature gases at Merapi volcano, Indonesia. *Geochim. Cosmochim. Acta* **51**, 2083-2101.

Symonds R.B., Reed M.H. and Rose W.I. (1992) Origin, speciation and fluxes of trace-element gases at Augustine volcano, Alaska: Insights into magma degassing and fumarolic processes. *Geochim. Cosmochim. Acta* **56**, 633-657.

Symonds R.B. and Reed M.H. (1993) Calculation of multicomponent chemical equilibria in gas-solid-liquid systems: calculation methods, thermochemical data and applications to studies of high-temperature volcanic gases with examples from Mount St. Helens. *Am. J. Sci.* **293**, 758-864.

Symonds R., Rose W.I., Bluth G.J.S. and Gerlach T.M. (1994) Volcanic-gas studies; methods, results and applications. In: Volatiles in magmas, *Min. Soc. Amer. Rev. Mineralogy* **30**, 1-66.

Symonds R.B., Mizutani Y. and Briggs P.H. (1996) Long-term geochemical surveillance of fumaroles at Showa-Shinzan Dome, Usu Volcano, Japan. *J. Volcan. Geotherm. Res.* **73**, 177-211.

Taran Y.A., Hedenquist J.W., Korzhinsky M.A., Tkachenko S.I. and Shmulovich K.I. (1995) Geochemistry of magmatic gases from Kudriavy volcano, Iturup, Kurile Islands. *Geochim. Cosmochim. Acta* **59**, 1749-1761.

Tkachenko S.I., Porter R.P., Korzhinsky M.A., van Bergen M.D., Shmulovich K.I. and Steinberg G.S. (1999) Mineral- and ore-forming processes in high-temperature fumarolic gases at Kudryavyi volcano, Iturup, Kuril archipelago. *Geochem. Internat.* **37**, 410-422.

Turnbull A.G. and Wadsley M.W. (1986) The CSIRO-SGTE Thermodata System (Version V). CSIRO, Div. Min. Chemistry Commun. **413**, 1-7.

Varekamp J.C., Thomas E., Germani M. and Buseck P.R. (1986) Particle geochemistry of Etna and Mt. St. Helens. *J. Geophys. Res.* **91**, 12233-12248.

Wahrenberger C., Seward T.M. and Dietrich V. (2001) Gas and metal volatile chemistry at La Fossa crater, Vulcano, Aeolian Islands, Italy. *Geochim. Cosmochim. Acta* (to be submitted).

Williams S.N., Fischer T.P., Shmulovich K.I. and Korzhinsky M.A. (1995) Medvezhya Bull. Global Volcan. Project 20.

Williams S.N., Schäfer S.J., Calavache V.M.L. and Lopez D. (1992) Global carbon dioxide emission to the atmosphere by volcanoes. *Geochim. Cosmochim. Acta* **56**, 1765-1770.

Williams-Jones A.E., Migdisov A.A., Archibald S.M. and Xiao Z. (this volume) Vapor-transport of ore metals.

Zakaznova-Iakovleva V.P., Migdisov A.A., Suleimenov O.M., Williams-Jones A.E. and Alekhin Yu.V. (2001) An experimental study of stibnite solubility in gaseous sulphide from 200 to 320 °C. *Geochim. Cosmochim. Acta* **65**, 289-298.

Water-Rock Interactions, Ore Deposits, and Environmental Geochemistry: A Tribute to David A. Crerar

Editors: Roland Hellmann and Scott A. Wood

A fluid inclusion study of the drusy quartz of the Potosi Dolostone, Southeast Missouri.

BRENDA A. MACMURRAY AND GREG M. ANDERSON*

Department of Geology, University of Toronto, 22 Russell Street, Toronto, ON, M5S 3B1, Canada

*e-mail: greg@quartz.utoronto.ca

Abstract-A fluid inclusion study of drusy quartz in the Potosi formation in the Southeast Missouri barite district (Washington County), north of the Viburnum Trend lead-zinc district, shows that the fluids which deposited the quartz were both less saline and hotter than the fluids which deposited the sulfides in either the Viburnum Trend or the barite district. The drusy quartz is thus apparently unrelated to either mineralization event.

1. INTRODUCTION

Delineating the pathways followed by the fluids responsible for the extensive hydrothermal mineralization in southeastern Missouri has long proved to be a difficult problem. One aspect of this problem not often commented on is that, although there is much discussion of the sources of the fluids and their pathways *towards* the ore deposits, there is little discussion of their pathways *after* ore deposition, i.e., their *exit* paths.

A number of imaginative methods have been used to infer fluid directions towards the ores (Figure 1), and the idea that there may be more than one source and one direction has definitely not been ruled out, so the directions indicated are not necessarily mutually exclusive. There is widespread agreement that at least one source, perhaps the major source, first proposed by Leach and Rowan (1986), lies to the south, where the Ouachita/Allegheny Orogeny at the end of the Paleozoic provided a hydraulic gradient which could account for the regional flow. This is supported by several lines of evidence (Leach et al., 1991; Kaiser et al., 1987, after Leach et al., 1984; Farr, 1986; Leach and Rowan, 1986; Horrall et al., 1997), most recently a modeling study by Appold and Garven (1999). Arrow 1 in Figure 1 represents straight south-to-north flow as implied by two-dimensional flow models along the Viburnum Trend, such as that of Appold and Garven (1999). However, there is evidence that flow was channeled through the Reelfoot rift, Bloomfield Lineament Zone, and/or northwesterly-trending faults (Horrall et al., 1997; see also He et al., 1997), represented in Figure 1 by arrow 4. Another suggestion made here is that, because model uplift contours in the Ouachitas (Beaumont et al., 1987, Figure 5) have a NW – SE direction, there would be a northeasterly component to the northward flow, which would be turned eastward in the vicinity of the Viburnum Trend because of the high permeability contrast across the Trend, the back-reef dolomites being highly permeable. This idea is represented by arrow 5 in Figure 1. An attractive feature of this idea is that it more easily accounts for the predominantly north-south mineral zoning in the Trend. Another suggestion (Gregg and Shelton, 1989a,b; Wagner, 1995) is that at least some flow is from north to south, represented by arrow 3, possibly from the Illinois basin to the northeast.

No matter where the fluids come from, they must exit the mineralized zones and go somewhere. In the case of the Viburnum Trend, there seem to be only two main possibilities, again not mutually exclusive. One is that they exited along the major fault zones to the north of the Trend, perhaps even being responsible for the barite-lead district in Washington County (arrow 2 in Figure 1). Kaiser and Ohmoto (1988) indicate that some of the structures in the region were active during mineralization. The other is that, as mentioned above, they exited through the back-reef facies east of the Viburnum Trend. There are some chemical problems to be resolved in proposing that the barite-lead district mineralization was produced by exiting Viburnum fluids, but it is not impossible.

A very common feature of the dolostone host rocks in the barite-lead district is drusy quartz, the fluid inclusions in which are the focus of this study. According to Wagner (1973) the quartz is pre-barite, and unrelated to the barite-sphalerite-galena ores, but in the context of the exiting Viburnum fluid hypothesis, it could represent early-stage fluids, in which silica derived from the Lamotte sandstone was precipitat-

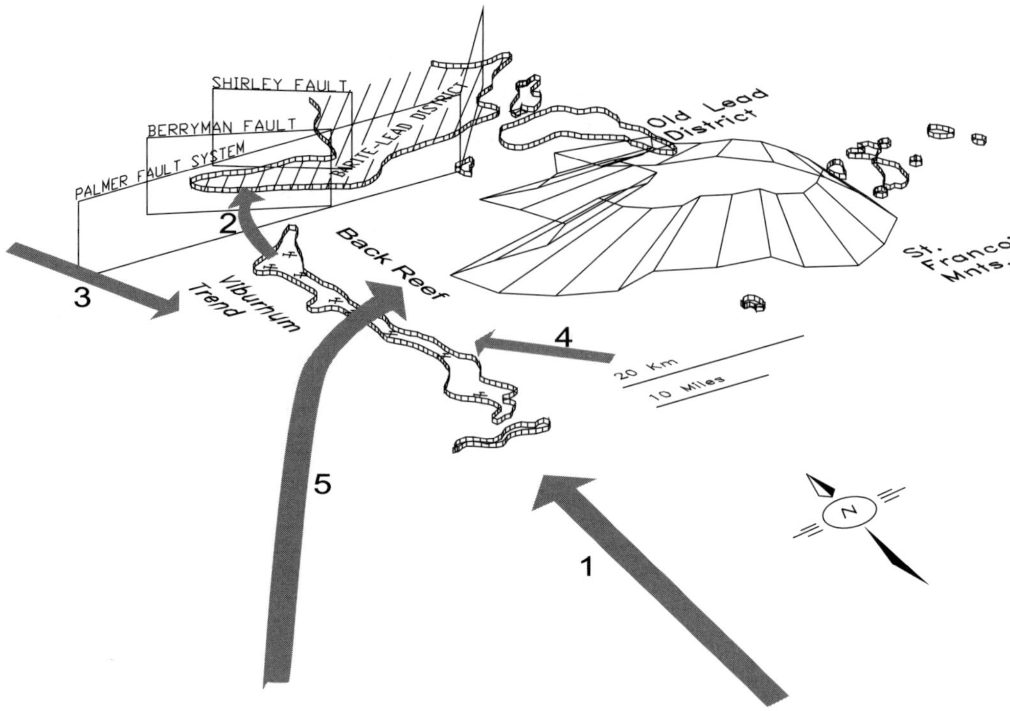

Figure 1: A schematic perspective view of the southeast Missouri area, showing possible fluid pathways discussed in the literature. Numbered arrows are discussed in the text.

ed due to cooling and mixing. In this sense, the drusy quartz represents a very small piece of a much larger puzzle. Fluid inclusions within the drusy quartz of the Potosi Dolostone were examined to determine what kind of fluids were responsible for its deposition.

2. GEOLOGY

The area studied is located within a large structural block bounded on three sides by major fault systems (Kaiser et al., 1987). These include the Palmer, Shirley, Big River and Vineland Fault Systems. In this area, the Potosi Dolostone is a brown, medium- to fine-grained dolostone of Upper Cambrian age. Wagner (1973) describes three primary textural lithologies. These include the more abundant calcarenites and algal stromatolitic biostromes and less common carbonate muds. Dolomitization of these lithologies occurred before the introduction of the silica-bearing fluids which formed chalcedony and drusy quartz mainly in vugs or fractures. The drusy quartz is found most commonly in the digitate algal stromatolitic beds, particularly on the outer edge of the algal columns. This commonly produces a texture of hollow tubes of quartz crystals known as "honeycomb" or "pipe" druse.

3. PROCEDURES

Sample locations were chosen to obtain a representative cross section of the Southeast Missouri Barite District (Figure 2). Fluid inclusion sections (0.2 to 0.5 mm thick) of Potosi Dolostone containing drusy quartz were cut and polished on both sides. The sections were then examined under a transmitted light microscope to locate inclusions, and a rough map of the inclusion locations was made. The sections were then soaked in methyl alcohol for 1.5 to 2 days to remove the section from the glass slide. The sample was carefully broken into small chips, approximately 3-7 mm in size, and placed in the fluid inclusion stage. The equipment used included a Linkam THMS 600 stage with TMS 92 temperature programmer. Temperature calibration in the range –56.8 °C to +28.7 °C was carried out using a natural standard H_2O-CO_2-NaCl inclusion, itself calibrated at $T_{m\ CO_2}$, $T_{m\ clath}$, and $T_{h\ CO_2}$ against organic compounds of known melting points. Most phase changes

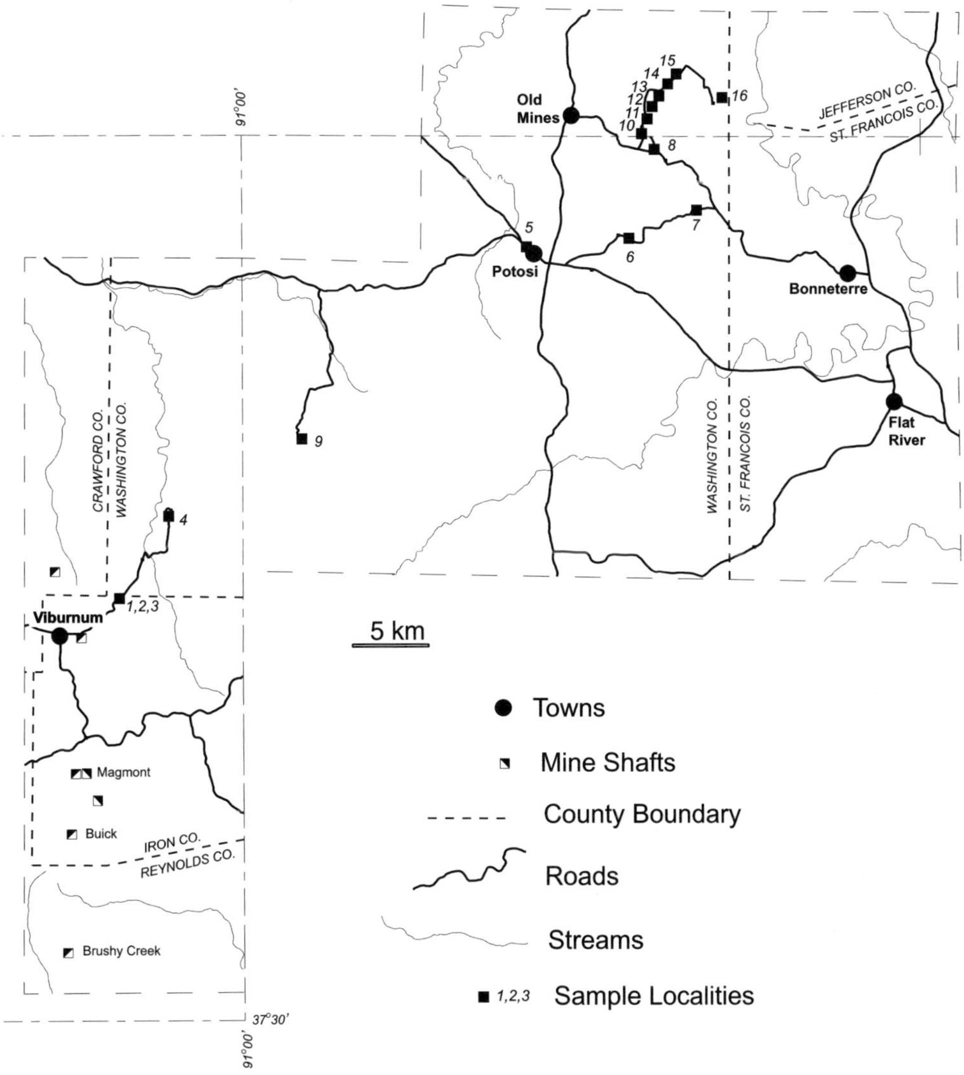

Figure 2: Map of the sampling area, Washington County, Southeast Missouri.

in this temperature range, with the exception of initial melting temperatures, can be determined to ± 0.1 °C. Calibration at other temperatures was carried out using ten of the standard chemicals listed by Macdonald and Spooner (1981), covering a temperature range of –22.8 °C to 306.8 °C. At elevated temperatures, the measurement precision is ±0.5 °C. Due to the small size of the inclusions (average size about 3 by 5 microns), the temperature of first melting could not be reliably determined.

A sketch was drawn of the field of view containing the inclusions to be analyzed. The temperature was reduced to between –95 and –100 °C or until the fluid in the inclusions froze. The temperature was increased by 25°/minute until approximately 10° below the expected temperature of melting of ice ($T_{m\ ice}$). The temperature was then increased by 5°/minute until approximately 2° below the expected temperature of melting of ice. Then the rate of temperature increase was reduced to 0.5°/minute until the ice melted. The melting temperature of ice was recorded and the procedure was repeated twice. If the melting temperatures did not vary by more than 0.5 °C, the average of the three trials was recorded as the melting temperature ($T_{m\ ice}$) of the inclusion. Melting temperature results collected were used to calculate

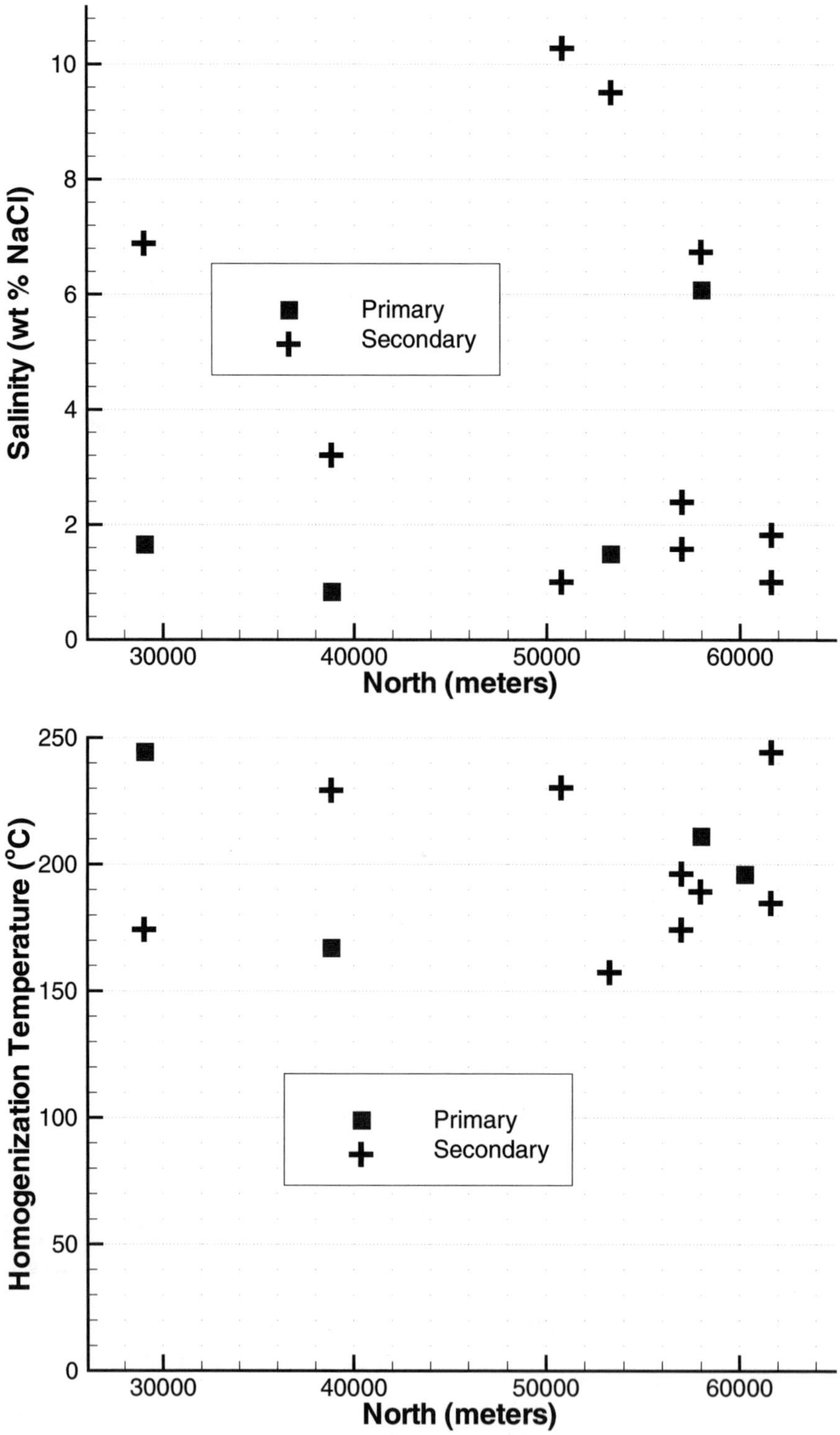

Figure 3: Salinity and homogenization temperatures from the sampling localities in Figure 2, projected onto a N-S plane. No pattern is evident. Projection onto E-W and NE-SW planes are equally uninformative.

salinities expressed as equivalent weight % NaCl (Brown and Hagemann, 1994).

The chips were then heated to determine the homogenization temperature ($T_{h\ L\text{-}V,L}$). The temperature was increased by 25°/minute until the gas bubble appeared to be much smaller. The temperature was then increased by 5°/minute until the gas bubble was seen to bounce quickly around the inclusion. Then the rate of temperature increase was reduced to 0.5°/minute until the gas bubble disappeared. The temperature was reduced until the bubble reappeared and the procedure repeated twice. If the homogenization temperatures did not vary by more than 0.5 °C, the average of the three trials was recorded as the homogenization temperature ($T_{h\ L\text{-}V,L}$) of the inclusion.

Because there were so few inclusions and they were very small, it was difficult to distinguish between primary and secondary fluid inclusions. Primary inclusions tended to be isolated whereas secondary inclusions occurred in groups, along a plane with other tiny inclusions. Both primary and secondary inclusion data were collected and used for this study.

4. RESULTS

The drusy quartz crystals of the Potosi Dolostone are remarkably clean and mostly free of inclusions. Fluid inclusions found within the drusy quartz are very small, simple vapor + liquid inclusions. The inclusions were subrounded to cigar-shaped with occasional subangular triangle-shaped inclusions. The vapor bubble appeared to occupy up to 30% of the inclusion (average approximately 15%). The volume % vapor was fairly constant. When the sample was cooled, no evidence was seen for a separate CO_2 phase. There were, on average, only 4-5 inclusions

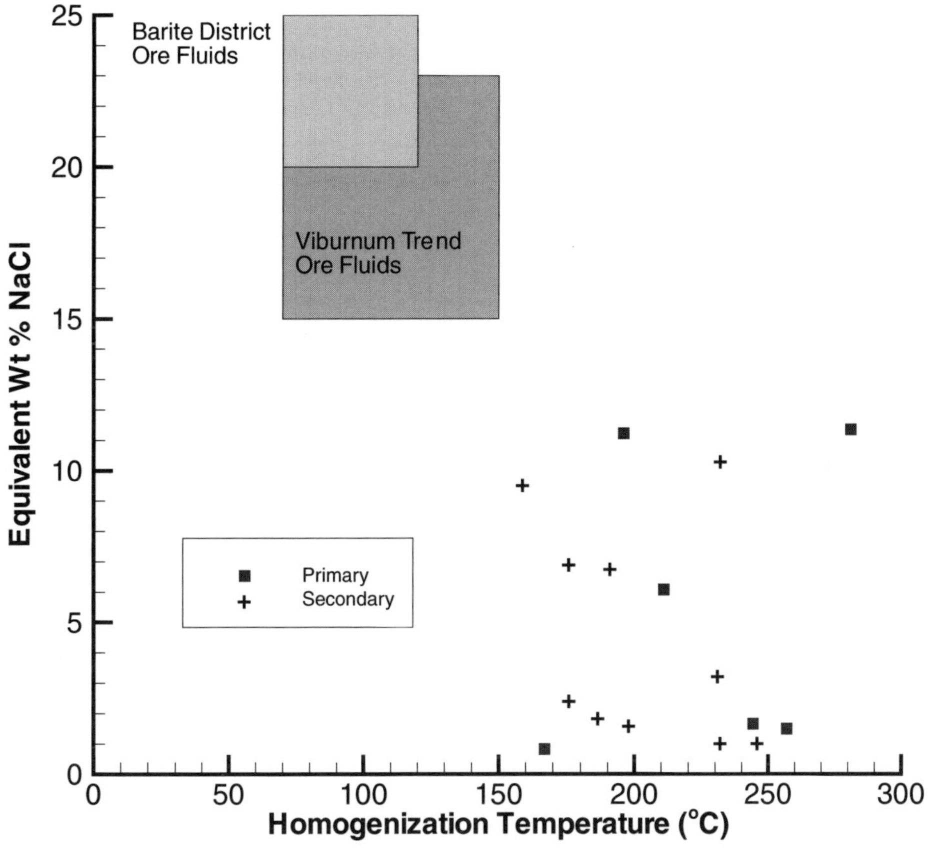

Figure 4: Fluid inclusion measurements of salinity and homogenization temperatures from the sampling localities in Figure 2, compared to the general ranges of fluid inclusion data from the Viburnum Trend and the Barite District.

Table 1. Fluid inclusion data.

Sample Locality (Fig. 1)	Primary T_h	Primary Wt. % NaCl	Secondary T_h	Secondary Wt. % NaCl
3	244.3	1.651		
7	257	1.487		
9	167	0.827		
13	211	6.076		
16	196	11.222		
16	281	11.343		
3			176	6.961
5			232	10.355
5			232	1.075
7			159	9.583
8			176	2.469
8			198	1.651
9			231	3.278
13			191	6.815
15			246	1.075
15			186.5	1.898

found per sample that were large enough to obtain melting and homogenization temperatures. Of these 4-5 inclusions, only 2 or 3 provided useful data (i.e., did not leak).

The data obtained are listed in Table 1. Melting temperatures ($T_{m\ ice}$) ranged from –0.5 °C to –7.7 °C with an average of –3.2 °C. The salinity values calculated from the melting temperatures ranged from 0.8 to 11.3 equivalent weight % NaCl. Homogenization temperatures ($T_{h\ L\text{-}V,L}$) ranged from 159 to 281 °C. The geographic distribution of temperature of homogenization and salinity of primary and secondary fluid inclusions was plotted in three-dimensions with projections on North-South and East-West planes. There did not appear to be any pattern to the geographic distribution for either the temperature of homogenization or salinity (Figure 3).

The fluids found in the inclusions in quartz from the Potosi Dolostone are very different from fluids of the Viburnum Trend and the Southeast Missouri Barite District. The temperatures of homogenization for Viburnum Trend fluids range from approximately 80 to 150 °C (Roedder, 1977; Hagni, 1983; Sverjensky, 1986; Viets and Leach, 1990; Leach et al., 1991, after Leach and Rowan, 1986 and Rowan and Leach, 1990). Leach et al. (1991) reported that salinities of Viburnum Trend fluids are greater than 15 and frequently greater than 20 equivalent weight % NaCl (very saline). The MVT ore-forming fluids of the Viburnum Trend are lower temperature and much more saline than the fluids from the Potosi quartz inclusions (Figure 4). Fluid inclusion data of the barite district suggest that the mineralizing solutions were saline brines with temperatures ranging from approximately 70 to 110 °C and salinities of 20 to 25 equivalent weight % NaCl (Kaiser et al., 1987; Leach, 1979). These fluids are also much more saline and lower temperature than those of the Potosi Formation drusy quartz.

It is of interest to compare these temperatures with the maximum possible burial temperature. Henry (1988) used a stratigraphic reconstruction to estimate the maximum burial depth of the Bonneterre Dolostone in the Viburnum Trend. He determined that the maximum possible burial of the Bonneterre Formation was approximately 1100 m and based on anomalous organic maturities, temperatures from fluid inclusions and fission track annealing studies, concluded that the geothermal gradient in the Ozark region was approximately 100 °C/km (Henry, 1988). Even with this very high geothermal gradient, the temperature of the Bonneterre Formation would not have exceeded 125 °C at maximum burial. Because the Potosi Dolostone is stratigraphically above the Bonneterre Formation and the sequence has not been overturned, its maximum burial temperature would thus be less than 125 °C. Therefore, the fluid inclusion data from this study demonstrate that the silica-bearing fluids that deposited the drusy quartz were higher temperature (159 to 281 °C) than any temperature the Potosi Dolostone host rock could have reached by burial alone.

5. CONCLUSIONS

The fluids found in inclusions in quartz within the Potosi Dolostone are very different from the ore fluids of the Viburnum Trend and the fluids of the Southeast Missouri Barite District. The Mississippi Valley-Type ore-forming fluids of the Viburnum Trend and the mineralizing fluids of the Barite District are lower temperature and much more saline than the fluids in the quartz-hosted inclusions of this study. It would appear then that the drusy quartz is not related in any simple way to the mineralization in either area.

One final note. According to Dake (1930), a conglomerate with rounded, waterworn fragments of apparent Potosi drusy quartz is present at the base of the Lower Ordovician Gasconade formation. However, inquiries at the University of Missouri (Rolla) and the Geological Survey of Missouri found no one who knew of a locality where drusy quartz fragments are exposed in the Gasconade Formation. This observation, if substantiated, and if the fragments were indeed shown to be from the Potosi formation, would imply an age of the drusy quartz from Upper Cambrian to Lower Ordovician. Given the fairly high temperatures of the fluid inclusions in the Potosi drusy quartz, the origin of the fluids responsible would be even more mysterious.

Acknowledgments—We would like to thank Dr. Joe Wagner, chief geologist of the Doe Run Co. in Viburnum, for his interest and assistance in this project. Dr. Wagner personally conducted us on the sampling trip, and provided a wealth of information on the barite district. Dr. Colin Bray instructed in the use of the fluid inclusion apparatus, and Prof. Ed Spooner provided much useful information and criticism. Reviews by Profs. Iain Samson and Anthony Williams-Jones were very useful in improving the manuscript.

Editorial handling: S.A. Wood

REFERENCES

Appold M. S. and Garven G. (1999) The hydrology of ore formation in the southeast Missouri district: Numerical models of topography-driven flow during the Ouachita orogeny. *Econ. Geol.* **94**, 913-935.

Beaumont C., Quinlan G. M. and Hamilton J. (1987) The Alleghenian orogeny and its relationship to the evolution of the eastern interior, North America. *Can. Soc. Petrol. Geol. Memoir.* **12**, 425-445.

Brown P. E. and Hagemann S. G. (1994) MacFlinCor: A computer program for fluid inclusion data reduction and manipulation. In *Fluid inclusions in minerals: Methods and applications* (ed. B. De Vivo and M.L. Frezzotti), Short course of the working group (IMA) "Inclusions in Minerals", Pontigano-Siena, 1-4 September, 1994, pp. 231-250.

Dake C. L. (1930) The geology of the Potosi and Edgehill Quadrangles, v. 23, second series. Missouri Bureau of Geology and Mines.

Farr M. R. (1986) Regional isotopic variation in Bonneterre Formation cements: Implication for brine migration pathways and sources. In *Symposium on the Bonneterre Formation (Cambrian), southeastern Missouri: Stratigraphy, Sedimentology, Diagenesis, Geochemistry, and Economic Geology* (ed. J. M. Gregg and R. D. Hagni), pp. 8. University of Missouri-Rolla, Rolla, Missouri. (abstr.).

Gregg J. M. and Shelton J. L. (1989a) Minor- and trace-element distributions in the Bonneterre dolomite (Cambrian, southeast Missouri): Evidence for possible multi-basin fluid sources and pathways during lead-zinc mineralization. *Geol. Soc. America Bull.* **101**, 221-230.

Gregg J. M. and Shelton J. L. (1989b) Geochemical and petrographic evidence for fluid sources and pathways during dolomitization and lead-zinc mineralization in southeast Missouri: a review. *Carbonates and Evaporites* **4**, 153-176.

Hagni R. D. (1983) Ore microscopy, paragenetic sequence, trace element content, and fluid inclusion studies of the copper-lead-zinc deposits of the southeast Missouri lead district. In *International Conference on Mississippi Valley-Type Lead-Zinc Deposits, Proceedings Volume* (ed. G. Kisvarsanyi et al.), pp. 243-256. University Missouri-Rolla Press.

He Z., Gregg J. M., Shelton K. L., and Palmer J. R. (1997) Sedimentary facies control of fluid flow and mineralization in Cambro-Ordovician strata, southern Missouri. In *Basin-Wide Diagenetic patterns: Integrated Petrologic, Geochemical, and Hydrologic Considerations*, SEPM (Society for Sedimentary Geology) Special Publication No. 57, 81-99.

Henry A. L. (1988) Alteration of organic matter in the Viburnum Trend lead/zinc district of southeastern Missouri. M.Sc. thesis, University of Toronto.

Horrall K. B., Farr M. R., and Hagni R. D. (1997) Evidence for focusing of Mississippi Valley-type ore fluids along the Bloomfield Lineament Zone, Southeast Missouri. Society of Economic Geologists, Special Publication No. 4, 75th Anniversary Volume, D. F. Sangster, ed., pp. 400-412.

Kaiser C. J. and Ohmoto H. (1988) Ore-controlling structures of Mississippi Valley-Type mineralization on the North American midcontinent as products of late Paleozoic convergent plate tectonism. In *North American Conference on Tectonic Control of Ore Deposits and the Vertical and Horizontal Extent of Ore Systems Proceedings Volume* (eds. G. Kisvarsanyi and S. Grant), pp. 424-430. University of Missouri – Rolla Press.

Kaiser C. J., Kelly W. C., Wagner R. J., and Shanks W. C. III (1987) Geologic and geochemical controls of mineralization in the Southeast Missouri Barite District. *Econ. Geol.* **82**, 719-734.

Leach D. L. (1979) Temperature and salinity of the fluids responsible for minor occurrences of sphalerite in the Ozark region of Missouri. *Econ. Geol.* **74**, 931-937.

Leach D. L., Plumlee G. S., Hofstra A. H., Landis G. P., Rowan E. L., and Viets J. G. (1991) Origin of late dolomite cement by CO_2-saturated deep basin brines: Evidence from the Ozark region, central United States. *Geology* **19**, 348-351.

Leach D. L. and Rowan E. L. (1986) Genetic link between Ouachita fold belt tectonism and the Mississippi Valley-type lead-zinc deposits of the Ozarks. *Geology* **4**, 931-935.

Leach D. L., Viets J. G., and Rowan L. (1984) Appalachian-Ouachita orogeny and Mississippi Valley-type deposits. *Geol. Soc. Am. Abstr. Programs* **16**, 572. (abstr.).

Macdonald A. J., and Spooner E. T. C. (1981) Calibration of a Linkham TH 600 programmable heating-cooling stage for thermometric examination of fluid inclusions. *Econ. Geol.* **76**, 1248 - 1258.

Roedder E. (1977) Fluid inclusion studies of ore deposits in the Viburnum Trend, southeast Missouri. *Econ. Geol.* **72**, 474-479.

Rowan E. L. and Leach D. L. (1990) Constraints from fluid inclusions on sulfide precipitation mechanisms and ore fluid migration in the Viburnum Trend lead district, Missouri. *Econ. Geol.* **84**, 1948-1965.

Sverjensky D. A. (1986) Genesis of Mississippi Valley-type lead-zinc deposits. *Ann. Rev. Earth Planet. Sci.* **14**, 177-199.

Viets J. G. and Leach D. L. (1990) Genetic implications of regional and temporal trends in ore fluid geochemistry of Mississippi Valley-type deposits in the Ozark region. *Econ. Geol.* **85**, 842-861.

Wagner R. J. (1973) Stratigraphic and structural controls and genesis of Barite deposits in Washington County, Missouri. Ph.D. thesis, University of Michigan.

Wagner, R. J. (1995) Viburnum Trend considerations, an industry perspective. In: Extended Abstracts: International field Conference on Carbonate Hosted Lead-Zinc Deposits, D. L. Leach and M. L. Goldhaber, eds., p. 331-332, Society of Economic Geologists.

Part C.
Environmental Geochemistry

Water-Rock Interactions, Ore Deposits, and Environmental Geochemistry: A Tribute to David A. Crerar

Editors: Roland Hellmann and Scott A. Wood

Microbial acquisition of nutrients from mineral surfaces

CHRISTOPHER J. TADANIER[1,2,3*], JOHN C. LITTLE[1], DUANE F. BERRY[2], AND MICHAEL F. HOCHELLA, JR.[3]

[1]Charles E. Via Department of Civil and Environmental Engineering,
Virginia Polytechnic Institute and State University, Blacksburg, Virginia 24061 USA
[2]Department of Crop and Soil Environmental Sciences,
Virginia Polytechnic Institute and State University, Blacksburg, Virginia 24061 USA
[3]Department of Geological Sciences,
Virginia Polytechnic Institute and State University, Blacksburg, Virginia 24061 USA

* Author to whom correspondence should be addressed:

Christopher J. Tadanier
Department of Geological Sciences
Virginia Polytechnic Institute and State University
Blacksburg, Virginia 24061 U.S.A.
e-mail: ctadanie@vt.edu

Abstract—Mineral surfaces act as concentrators for both organic and inorganic microbial nutrients and environmental contaminants. The bioavailability of mineral surface-associated nutrients and contaminants is controlled by a complicated set of both biotically- and abioticially-mediated processes which occur at microbe-solution, mineral-solution, and microbe-mineral interfaces. This study was conducted to 1) develop a generalized transport model which can be used to describe nutrient uptake by bacteria, 2) conduct continuous growth experiments using a model system in which the growth-limiting nutrient is adsorbed on a colloidal mineral, and 3) use uptake rate calculations and model-system data to evaluate various physiological responses bacteria may use to obtain nutrients from mineral surfaces. Microorganisms from an undisturbed natural soil were maintained in chemostat (continuous) culture for 45 d, using P_i (orthophosphate) adsorbed on colloidal goethite as a sole phosphorous source. After a short initial acclimation period microbial utilization of adsorbed P_i averaged 82±6 %. The predominant Gram-negative bacterium isolated from chemostat effluent was *Burkholderia cepacia*. Free energy (ΔG) considerations demonstrate that sufficient energy was available for complete bacterial P_i uptake; therefore, trans-membrane P_i transport was not thermodynamically-limited. Rate calculations, which model P_i uptake as a series of discrete transport processes in each cell wall and near-cell compartment, suggest that the microbial community must alter the chemostat environment in order to obtain sufficient P_i for continuous steady-state growth. The nutrient uptake model outlined here, used in conjunction with steady-state chemostat data, may ultimately provide a means to address the broader environmental implications of microbial nutrient acquisition from mineral surfaces on the fate and transport of mineral surface-associated contaminants.

1. INTRODUCTION

Microorganisms display a wide range of nutritional requirements, some being adaptable to varied nutrient conditions, while others are fastidious, occupying only a narrow environmental niche due to strict nutritional needs. However, all microorganisms share the same fundamental biomolecular building blocks — proteins, lipids, nucleic acids, and carbohydrates — which require carbon, oxygen, hydrogen, nitrogen, phosphorous, sulfur, and transition metals in varying amounts. Alkali and alkaline earth metals (typically Na, K, Mg, and Ca) are also required for membrane transport processes, charge balance, structural conformation, signaling, and a variety of other physiological processes. One or more nutrients ultimately limit microbial growth in natural environments, most often carbon, nitrogen, or phosphorous. Oligotrophic (nutrient-limited) conditions generally prevail in natural environments; solution phase nutrients at μM to pM concentrations are commonly encountered due to depletion by microbial growth and sorption onto mineral surfaces.

Mineral surfaces act as concentrators for both organic and inorganic microbial nutrients, with the strength of nutrient-surface association dependent on the physicochemical properties of each. The affinity of microbial nutrients for mineral surfaces, from weakly to strongly associated, is based on one or more fundamental sorptive interactions. Organic molecules may be sorbed to mineral surfaces due to hydrophobic interaction or electrostatic, covalent, or hydrogen bonding mechanisms. Natural organic compounds known to interact with mineral surfaces to varying degrees include biopolymers of plant, animal, and microbial origin such as proteins, lipids, nucleic acids, and polysaccharides; and their associated degradation products such as amino and low molecular-weight fatty acids, nucleosides, and a variety of phosphorylated and aminated sugars (Kummert and Stumm, 1980; Schwertmann et al., 1986; Stevenson, 1994; Chenu, 1995; Cornell and Schwertmann, 1996). Synthetic organic compounds (SOCs) including solvents, manufacturing by-products, hydrocarbon fuels, and pesticides, may also serve as microbial nutrients and have been widely observed in association with mineral surfaces (Kung and McBride, 1991; Rao et al., 1995; Ristori and Fusi, 1995; Brady and Weil, 1999). Inorganic nutrients may be sorbed to mineral surfaces through surface complexation reactions, or they may be contained in surface precipitates, the bulk mineral, or occluded phases. Surface-complexed inorganics may be strongly associated through covalent bonds formed by ligand exchange (inner-sphere-complexed) or relatively weakly associated due to electrostatic interaction through an intervening water molecule (outer-sphere-complexed), whereas inorganic nutrients contained in a mineral phase are characterized as strongly associated (Brown, 1990). Examples of inorganic nutrients observed in association with mineral surfaces include, but are not limited to, nitrogen (NH_4^+, NO_3^-, NO_2^-), phosphorous ($H_2PO_4^-$, HPO_4^{2-}), sulfur (SO_4^{2-}, HS^-), and transition metals (particularly $Fe^{2+/3+}$, and $Mn^{2+/4+}$) (Cornell and Schwertmann, 1996).

The bioavailability of mineral surface-associated contaminants has received substantial attention from both biotic and abiotic points of view, with the impact of environmental factors such as pH, ionic strength, and redox potential on contaminant phase distribution studied extensively (Stotzky, 1986; Lonergan and Lovley, 1991; Ristori and Fusi, 1995; Stone and Torrents, 1995; Bollag et al., 1995; Andreux et al., 1995). However, the influence of microbially mediated processes on nutrient availability has been infrequently studied at environmentally relevant nutrient concentrations. Furthermore, phenomenological mechanisms responsible for macroscopically observed nutrient bioavailability have invariably been speculative in nature, with little or no supporting experimental data available. Therefore, we feel it is imperative to study nutrient bioavailability in natural environments from a biogeochemical perspective, in which microbially-mediated mechanisms responsible for macroscopically observable nutrient bioavailability are considered in terms of interactions at microbe-solution, mineral-solution, and microbe-mineral interfaces. Figure 1 illustrates the conceptual approach we have developed and adopted for studying biogeochemical processes associated with microbial acquisition of nutrients from mineral surfaces.

Chemostat (continuous flow, steady-state) culture offers many possibilities for studying environmental influences on microbial community structure not available using batch culture techniques (Veldkamp, 1977; Smith and Waltman, 1995). Unlike batch culture experiments, in which continuously changing environmental and progressively depleted nutrient conditions bring about a continuous succession of species distributions and phenotypic traits within a microbial community, maintenance of limited but sufficient nutrient conditions in a chemostat ultimately results in development of a stable or steady-state community. Moreover, the initially high nutrient concentrations required for batch culture confer a competitive advantage to microbial species which rely on high growth rate as a survival strategy (*r* strategists), and preferentially inhibit species which optimize utilization of scarce resources (*K* strategists) (Atlas and Bartha, 1998). Steady-state cultures (growth rate of each species is balanced by dilution rate – see Appendix A.1) can be exposed to environmental changes in a systematic manner, allowing controlled study of the physiological response of microorganisms to nutrient-limited conditions more representative of natural environments.

For the research described herein, sorption of orthophosphate (P_i) on colloidal goethite (α-FeOOH) was used as a model nutrient mineral-surface system. Use of this model system is based on its importance from both microbiological and mineralogical points of view. Phosphorous is a major nutrient for all microorganisms, regardless of physiological and metabolic character or environmental niche occupied. Because of its high affinity for mineral surfaces and lack of volatile forms, phosphorous is often the least available of the biologically essential elements, and in global terms may be the single most important factor limiting the extent of life on Earth (Cox, 1995). Indeed, dissipation of P_i in the environment due to current land use practices threatens to adversely impact the food chain at its foundation (i.e., microbial activity and production) and may ultimately limit food production in the future (Runge-Metzger, 1995; Abelson, 1999). Goethite is the most abundant iron-oxide in the environment, and in soils and sediments it may represent a majority of the total surface area exposed (Cornell and Schwertmann, 1996). Phosphate complexes strongly with goethite surfaces, and although not completely characterized, the sorptive behavior of phosphate on goethite is perhaps the most widely studied nutrient mineral-surface system.

Our ultimate goal is to study the actively and passively mediated processes which microorganisms may use to enhance bioavailability of nutrients sorbed on mineral surfaces. Actively mediated processes are based on physiological responses that either enhance nutrient release at the mineral-solu-

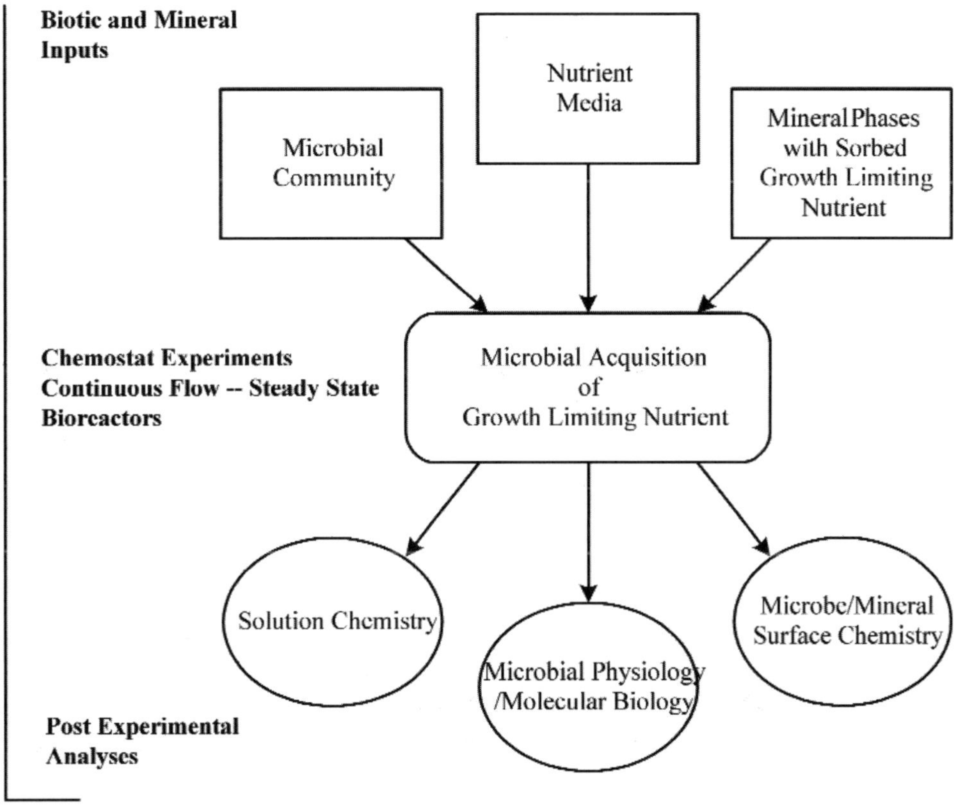

Fig. 1. Conceptual biogeochemical approach for studying microbial acquisition of nutrients from mineral surfaces.

tion interface or minimize resistance to nutrient mass transport in the bulk solution, whereas passively mediated processes involve physiological responses that either enhance nutrient transport across the microbial cell envelope or reduce the cellular nutrient requirement. In this sense, the distinction between actively and passively mediated responses is based on whether or not microbes alter their surrounding environment as a means of increasing sorbed nutrient bioavailability. Specific objectives of the research described here are to 1) develop a generalized nutrient transport/uptake model which allows P_i mass transfer in near-mineral surface, bulk solution, and near-microbe environmental compartments to be described in terms of mineral, solution, and bacterial cell envelop properties, 2) conduct a set of chemostat experiments to obtain basic microbiological, kinetic (growth rate), and solution chemistry data for a natural bacterial community able to effectively utilize P_i sorbed on colloidal goethite as the sole source of phosphorous, and 3) evaluate various active and passive physiological responses which bacteria may use to enhance P_i acquisition from mineral surfaces by comparing chemostat data with previous observations of microbial P_i bioavailability.

2. BIOGEOCHEMISTRY OF P_i-GOETHITE ASSOCIATION

2.1. Phosphate-Goethite as a Model Nutrient-Mineral System

2.1.1. Occurrence of goethite in the environment

Goethite is one of the most widespread minerals in the earth's near-surface environment, and is by far the most abundant iron (hydr)oxide in soils worldwide (Cornell and Schwertmann, 1996; and references therein). It predominates especially under humid and cool to temperate climates. Although the absolute abundance of goethite in a soil is typically only a few weight percent (tens of g goethite/kg soil), its surfaces may account for the majority of the total surface area of the soil as a whole (Schwertmann and Taylor, 1990). This is in part due to the relatively small size and high specific surface area of natural goethite crystals, typically acicular or needle-like with surface area of 20 to 200 m^2/g (Schwertmann and Cornell, 1988), and in part due to formation of goethite coatings on larger mineral grains in soil. Furthermore, because goethite has a high affinity for sorption of many cations and anions (see below), it is thought to be a major factor in the retention and cycling of geochemically and environmentally important elements and compounds.

2.1.2. Chemical forms of phosphorus in the environment

Phosphorus is not an abundant element in the environment, comprising on average less than 0.1 weight percent (1000 mg/kg) of the earth's crust and a negligible percentage of the atmosphere. Total phosphorus content in soils varies from < 20 to 6800 mg P/kg soil (< 0.002 to 0.68 %), with a median value of approximately 500 mg P/kg soil (0.05 %) (Bowen, 1979; Pierzynski et al., 1994; Barber, 1995). Due to strong retention by mineral phase constituents, soil-solution phosphorus concentrations are quite low, with values in the range of 0.002 to 0.03 mg P/kg soil solution ($\leq 10^{-6}$ M) typical in temperate soils (Bohn et al., 1985). Soluble phosphorus concentrations in fresh and ocean waters are also quite low, with average values of 0.02 and 0.07 mg P/kg solution, respectively (Taylor and McLennan, 1985). Phosphorus comprises a relatively small mass fraction in microbiota and plants; 1 to 3 % in bacteria, 0.3 to 0.4 % in marine algae, 0.4 to 1.8 % in marine phytoplankton, 0.5 % in fungi, 0.074 % in lichens, and 0.012 to 1 % in vascular plants (Bowen, 1979). However, these modest phosphorus contents represent bioconcentration of 2 to 4 orders of magnitude due to metabolic activity of living organisms.

Environmental phosphorus, whether in organic or inorganic form, almost invariably occurs as phosphate (PO_4^{3-}, oxidation state +5) (Corbridge, 1990; Frossard et al., 1995). Reduced forms of phosphorus such as orthophosphite (PO_3^{3-}, oxidation state +3), hypophosphite (PO_2^{3-}, oxidation state +1), and phosphine (PH_3, oxidation state -3) occur only rarely in the earth's surface environment, and are generally rapidly oxidized to phosphate (Cox, 1995). Phosphate is incorporated in a wide variety of inorganic and organic forms in nature, which display a broad range of solubilities. Soluble inorganic forms of phosphorus include orthophosphate (P_i) and polyphosphates – linear polymers with 2 or 3 phosphate molecules linked by phosphoanhydride bonds. Both ortho- and polyphosphates undergo protonation, resulting in pH-dependent distributions of partially to fully protonated species. In the range of pH values encountered in all but the most extreme environments (pH 3–11), singly and doubly protonated forms predominate (i.e., HPO_4^{2-} and $H_2PO_4^-$ for P_i; Stumm and Morgan, 1981).

Phosphorus also occurs commonly in relatively insoluble phosphate mineral forms. More than 200 phosphate minerals are known, but those in the apatite group (calcium phosphates – $Ca_{10}(PO_4)_6X_2$, where X represents an anion such as F^-, Cl^-, OH^-, or CO_3^{2-}) are by far the most abundant igneous and sedimentary forms (Corbridge, 1990). A variety of secondary phosphate minerals are formed in soils due to weathering of primary apatite minerals, especially highly insoluble Al- and Fe-phosphate precipitates. Phosphate is also strongly adsorbed on the surfaces of both crystalline and amorphous Al and Fe (hydr)oxide secondary minerals, as discussed in detail in the following sections. The various inorganic and adsorbed mineral forms of phosphate often account for 50 to 70 % of the total phosphorus in soils (Pierzynski et al., 1994).

Phosphorus in organic form occurs in soils and sediments exclusively as phosphate incorporated in biomolecules, and associated degradation products produced by living organisms. Organic forms of phosphorus have been observed to account from as little as 5 % to as much as 95 % of total soil phosphorus content (Stevenson, 1994; Senesi and Loffredo, 1999), with 29 to 65 % being a more representative range (Harrison, 1987; Pierzynski et al., 1994; Paul and Clark, 1996). Less than 50 % of organic phosphorus in most soils can be accounted for in identifiable forms (Senesi and Loffredo, 1999). Chemically identifiable organic phosphorus compounds in soils include inositol phosphates (IP, 2 to 50 %), phospholipids (1 to 5 %), nucleic acids (0.2 to 2.5 %), and trace amounts of phosphoproteins and metabolic phosphates (e.g., ATP and NAD^+) (Stevenson, 1994). The inability to chemically describe the remaining forms of organic phosphorus may be attributed to the occurrence of polymeric phosphate-containing molecules such as teichoic acids and phosphorylated polysaccharides from bacterial cell walls, incorporation of phosphate in soil organic matter (SOM), such as humic and fulvic acids, and formation of insoluble complexes with metal ions or sorption on mineral surfaces (Anderson et al., 1974; Shvets et al., 1991; Stevenson, 1994).

2.1.3. Phosphorus-goethite sorptive interactions

The phosphate-goethite adsorption envelope is typical of oxyanion adsorption on mineral surfaces: fractional phosphate adsorption is maximum at low pH, decreases with increasing pH, and displays maxima or slope changes at the pK_a(s) of the conjugate acid(s) of the adsorbing anion (Hingston et al., 1968; Hingston et al., 1972). Phosphate adsorption on goethite generally exhibits maximum adsorption at about pH 2 (~ pK_1), with discontinuities in the adsorption envelope slope at pH values of approximately 7 (~ pK_2) and 12 (~ pK_3) (Hingston et al., 1968; Hingston et al., 1972; Sigg and Stumm, 1981; Hawke et al., 1989; Geelhoed et al., 1998). The extent of phosphate adsorption on goethite at any pH increases with rising concentration, and often conforms to the Langmuir isotherm (McBride, 1994). However, results of phosphate desorption experiments demonstrate that substantial hysteresis exists between adsorption and desorption isotherms, with residual sorbed phosphate concentrations substantially greater than the equilibrium adsorption isotherm values (Hingston et al., 1974; Barrow, 1983). Explanations for the observed hysteresis in phosphate adsorption-desorption on goethite include kinetic differences (Hingston et al., 1974), slow diffusion out of micropores (Barrow, 1983), and precipitation of a relatively insoluble phosphate mineral phase on the goethite surface (Martin et al., 1988).

Of the common inorganic and organic oxyanions, phosphate is the most strongly adsorbed by goethite surfaces (McBride, 1994). Based on experimental evidence from a variety of direct and indirect methods, a consensus exists supporting inner-sphere complexation of phosphate on goethite by ligand exchange with reactive surface hydroxyl groups (Goldberg and Sposito, 1985; and references therein). Studies of the phosphate-goethite adsorption envelope indicate little influence of ionic strength, a characteristic associated with inner-sphere complexation of metal and oxyanion adsorption on mineral surfaces (Ryden et al., 1977; Barrow et al., 1980; Hayes et al., 1988; Hawke et al., 1989; Davis and Kent, 1990). Several other lines of evidence support inner-sphere complexation, including the adsorption of phosphate on goethite at pH values greater than the point of zero charge (PZC) - i.e., negatively charged surface - (Hingston et al., 1972; Mott, 1981), differences in the kinetics of phosphate adsorption and desorption on goethite (Mott, 1981) and in phosphate adsorption on Fe(III) and Cr(III) oxides (Yates and Healy, 1975), and spectroscopic studies which suggest phosphate (and its structural analog arsenate) adsorption on goethite occurs by ligand exchange with surface hydroxyls (Parfitt et al., 1975; Tejedor-Tejedor and Anderson, 1986; Martin and Smart, 1987; Sun and Doner, 1996; Fendorf et al., 1997).

2.2. Bioavailability of Phosphate in the Environment

The bioavailability of P_i in soil environments is largely determined by its distribution among various organic and inorganic chemical forms and sorption on or presence within mineral and solution phases. About 0.1 % of the estimated 10^{15} kg of P in the earth's crust is subject to at least some turnover due to cycling of P_i between various forms and phases (Paul and Clark, 1996). Phosphate cycling is critical in soils because the uptake of phosphorus by both microorganisms and plants is generally in the form of soluble P_i from the soil solution (Van Veen et al., 1994; Frossard et al., 1995).

Soil solution P_i concentration, and hence bioavailability, is a complex function of pH and soil solution geochemistry, which may vary considerably between soils and with recent environmental conditions and usage for a given soil. Inputs of P_i that impact soil environments include weathering and dissolution of primary and secondary minerals, desorption from mineral surfaces, hydrolysis of phosphate containing biomolecules, and anthropogenic sources such as fertilizers. Conversely, processes which remove P_i from soil solution include incorporation in biomass by microbes and plants, sorption on mineral oxide surfaces, precipitation of secondary minerals, and dilution by infiltration or advective groundwater flow. In general, neither biotically- nor abiotically-mediated reactions involved in phosphorus cycling alter its oxidation state from +5 (Stewart and McKercher, 1982; Tate, 1995).

2.2.1. Influence of P_i distribution on bioavailability

In principle, soil solution P_i concentration may be controlled by the solubility of Ca, Al, and Fe-containing minerals, or sorption on Al and Fe (hydr)oxide, clay, and calcium carbonate mineral surfaces (Lindsay, 1979; Pierzynski et al., 1994; Barber, 1995; Frossard et al., 1995). In acidic to neutral environments phosphate bioavailability is dominated by sorption/desorption to Al- and Fe-(hydr)oxides (Haynes, 1982; Sanyal and De Datta, 1991; Pierzynski et al., 1994; Geelhoed et al., 1998), whereas in alkaline environments solubility of calcium phosphate minerals is thought to control P_i bioavailability (Goldstein, 1994; Lajtha and Harrison, 1995; Iyamuremye and Dick, 1996). The time-dependence of abiotic P_i transformations in soils also supports the role of sorption reactions in determining P_i bioavailability: the kinetics of P_i chemisorption on mineral surfaces are much faster than for precipitation of P_i-containing secondary minerals (Van Der Zee and Van Riemsdijk, 1991).

2.2.2. Microbial response to environmental stimuli - phosphate uptake by bacteria

Essentially all microorganisms are capable of adapting their physiology/metabolism to some extent in response to environmental changes (Stock et al., 1990). Microbial response to environmental stimuli is accomplished through alternative patterns of gene expression, which result in a wide variety of both membrane-bound and cytoplasmic proteins with functions related to nutrient uptake, respiration, sporulation, virulence, resistance to temperature and osmotic shock, and chemical communication between bacterial cells. When confronted with P_i limitation, expression of as many as 38 phosphate starvation-inducible (*psi*) genes has been observed in *E. coli* and closely related bacteria (Wanner, 1994). The *psi* genes of E. coli are arranged in eight separate transcription units (i.e., operons), collectively known as the Pho regulon (Wanner, 1995). Specific functions of Pho regulon gene products include enhanced P_i uptake (Pst system and associated outer-membrane porin PhoE), uptake of small phosphorylated organic molecules (Ugp system), utilization of phosphonate compounds (Phn system), and extracellular phosphoester hydrolysis (periplasmic phosphatase enzymes) (Wanner, 1996).

Because of the continuous-flow design and use of goethite-sorbed P_i as the sole phosphorous source in this study, a microbial community must adapt to solution P_i limitation to remain viable in the long term. Also, a relatively detailed physical description of the pathways involved in bacterial P_i uptake is required to model transmembrane P_i transport. Therefore, consideration of the physiology and regulation of P_i uptake is particularly relevant to this study. Phosphate transport has been most extensively studied in *E. coli*, which serves as the paradigm for environmentally regulated bacterial P_i uptake (Wanner et al., 1996).

Bacterial P_i uptake occurs by a series of diffusionally-driven and enzymatically-mediated energy dependent processes. Phosphorylated molecules diffuse through the outer membrane of Gram-negative bacteria through protein channels and through the peptidoglycan cell wall of Gram-positive cells to the vicinity of the cytoplasmic membrane. With few exceptions, bacteria preferentially transport P_i across the cytoplasmic membrane. P_i is typically released from small phosphorylated organic molecules by enzymatic hydrolysis prior to trans-membrane transport (Torriani-Gorini, 1994). Transmembrane P_i transport is energy-dependent, as it must be transported into the cell against a very unfavorable concentration gradient (cytoplasmic P_i is typically on the order of 10 mM (Shulman et al., 1979; Wanner, 1994), whereas extracellular P_i in natural environments is typically at $\leq \mu M$ levels).

When P_i is present at ~ μM or greater concentrations, *E. coli* take up P_i through the constitutively (continuously) expressed non-specific phosphate inorganic transport (Pit – low affinity) system, whereas at lower extracellular concentrations uptake is primarily through the inducible P_i specific transport system (Pst – high affinity) of the Pho regulon (Rosenberg, 1987). The Pit system utilizes symport (simultaneous transport) of a divalent metal-phosphate complex ($MeHPO_4$ – Me may be Mg^{2+}, Ca^{2+}, Mn^{2+}, or Co^{2+}) and a proton across the cytoplasmic membrane, and is therefore energetically driven by a transmembrane H^+ gradient (Van Veen et al., 1994). The Pst system relies on direct ATP hydrolysis by a membrane-bound transport enzyme to provide the energy required to overcome the energetically unfavorable P_i concentration gradient (Webb and Cox, 1994).

Microbial uptake of P_i from the environment is a highly coordinated activity, subject to a high level of genetic regulation. The Pho regulon is influenced by multiple controls, but its principle regulator is extracellular P_i concentration (Wanner, 1994). P_i control of the Pho regulon involves a form of transmembrane signal transduction in which a cell surface receptor protein (PhoR) modulates phosphorylation of a cytoplasmic transcription regulator protein (PhoB) (Wanner, 1995). When extracellular P_i levels become limiting, PhoR activates PhoB by phosphorylation, thus triggering transcription of Pho regulon *psi* proteins.

Although P_i uptake has been more extensively characterized for *E. coli* than other bacteria, several significant aspects of P_i uptake by *E. coli* appear common among diverse bacterial species. First, the *E. coli* PhoB and PhoR proteins belong to a superfamily of environmentally responsive, two-component regulatory systems which are prevalent in bacteria (Ronson et al., 1987; Parkinson and Kofoid, 1992). Also, both low- and high-affinity P_i transport systems have been observed in a number of both Gram-positive and Gram-negative bacteria, including *Salmonella typhimurium* (Foster and Spector, 1986), *Pseudomonas aeroginosa* (Poole and Hancock, 1984), *Bacillus cereus* (Guddal et al., 1989) and *Bacillus subtilis* (Liu and Hulett, 1997; Qi et al., 1997), *Synechococcus* sp WH7803 (Scanlan et al.,

1997), *Rhizobium meliloti* (Voegele et al., 1997), *Acinetobacter johnsonii* (Van Veen et al., 1993), and *Enterobacter cloacae* and *Klebsiella pneumoniae* (Van Der Ley et al., 1987).

3. NUTRIENT TRANSPORT AND UPTAKE MODELING

Based on the previous discussion of P_i biogeochemistry, it is clear that the soluble P_i concentration has a central role in the interfacial processes which occur between mineral surfaces and bulk solution, and bulk solution and cell envelope. Therefore, those aspects of P_i bioavailability related to transport may be framed in terms of a generalized nutrient transport/uptake model which considers the impact of resistance to mass transfer in the near-mineral surface, bulk fluid, and near-cell membrane environments. However, in a well-mixed chemostat with two-phase mass transport and rapid equilibrium of sorbed components at the mineral-solution interface, the rate of bacterial P_i uptake is likely to be limited by resistance to P_i transport across the cell envelope. Therefore, expressing the steady-state bacterial P_i uptake rate in terms of cell envelope physiology and properties provides a means of evaluating the potential contributions of active and passive microbial responses to P_i limitation.

3.1. Generalized Nutrient Transport/Uptake Model

The acquisition of mineral surface-associated nutrients by microorganisms can be viewed as a sequence of discrete mass transfer steps in which a nutrient molecule must sequentially traverse each of the environmental compartments illustrated in Fig. 2. Each environmental compartment exerts a resistance to nutrient transport, the relative magnitude of which

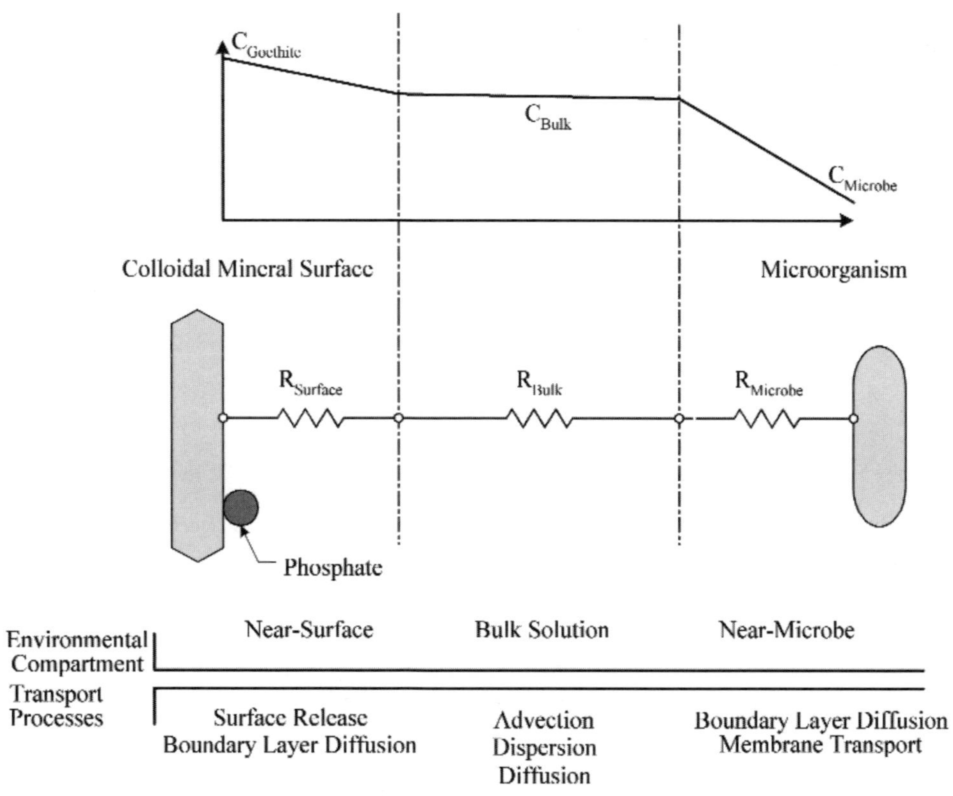

Fig. 2. Graphical representation of the Resistances in Series Model for mass transport in a well-mixed chemostat. Mass transfer in each environmental compartment may be composed of several transport processes acting in series and/or parallel, which combine to yield a composite resistance to mass transfer.

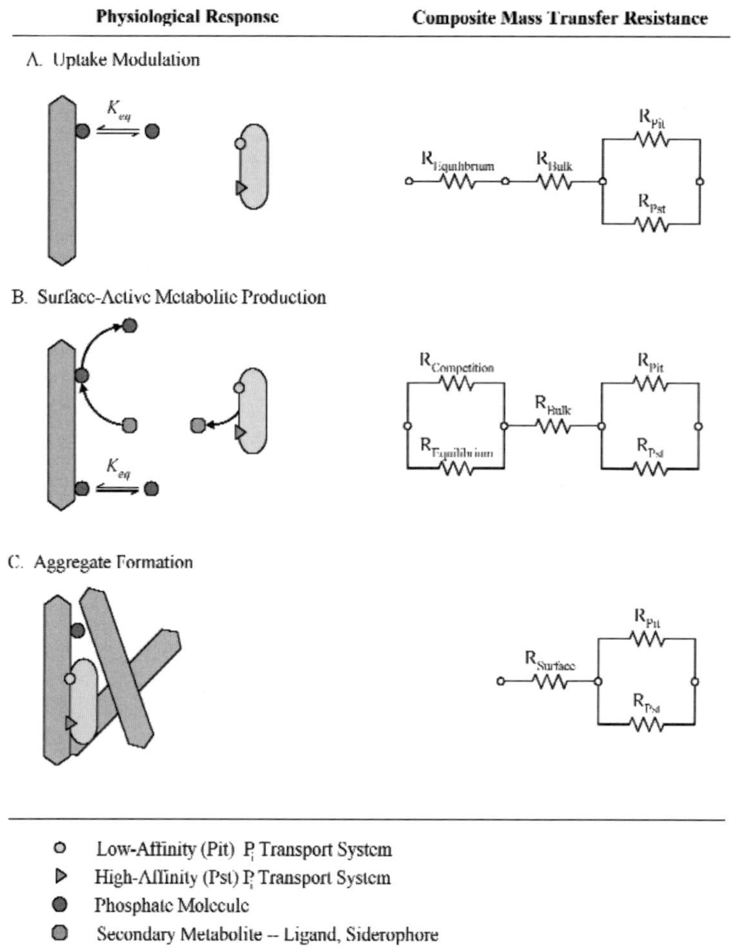

Fig. 3. Conceptual model of several physiological responses microbes may use to lower the overall resistance to phosphate transport or increase its bioavailability.

depends on the biophysicochemical characteristics of the specific mass transfer processes involved. The overall resistance to nutrient transport is given by the sum of resistances of compartments traversed, or in the nomenclature of Fig. 2:

$$R_{Total} = R_{Surface} + R_{Bulk} + R_{Microbe} \quad (1)$$

Furthermore, nutrient transport through each environmental compartment may be accomplished by several processes acting in parallel, resulting in a decreased resistance to mass transfer. For two transport processes acting in parallel the composite resistance to mass transfer is given by:

$$\frac{1}{R_{Composite}} = \frac{1}{R_1} + \frac{1}{R_2} = \frac{R_1 R_2}{R_1 + R_2} \quad (2)$$

Mass transport through the various environments defined in Fig. 2 may be accomplished by several mechanisms. In the near mineral surface environment solute molecules diffuse to and from the mineral surface, where rapid sorption-desorption reactions occur. Transport in the bulk environment may occur by molecular diffusion, dispersion, or advection, with the dominant mechanism being determined by the characteristic dimension or scale of interest. Nutrients diffuse to the microbial surface though a fluid boundary layer after which one or more of several possible diffusive or energy dependent membrane transport processes occur.

Transport of P_i to or across the microbial membrane may in principle limit growth. However, the extent to which microorganisms physiologically respond to imposed environmental conditions, either as single species or in consortia, to lower the overall resistance to P_i transport is not well known. Three possibilities are immediately apparent: (1) microorganisms may alter their membrane character to enhance uptake rate from solution (passive response), (2) microorganisms may excrete surface-active metabolites which alter the equilibrium position of P_i-goethite surface association by competitive exchange or reductive dissolution (active response), or (3) microorganisms may increase their proximity to phosphate-covered mineral surfaces by chemotaxis or inducing microbe-aggregate formation (active response). Each of these responses is conceptually illustrated in Fig. 3. In complex natural environments these responses may occur singly or in combination, with the operative response(s) at any point in time dependent on imposed environmental conditions.

3.2. Modeling P_i Transport Across the Bacterial Cell Envelope

Phosphate transport in the near-microbe environment generally involves a series of three discrete steps: 1. diffusion through the cell boundary layer (i.e., from bulk solution to cell envelope outer surface), 2. diffusion through the outer membrane and periplasm (Gram-negative cells) or S-layer and cell wall (Gram-positive cells), 3. energy-dependent transport across the cytoplasmic membrane (Fig. 4). A composite expression for P_i transport in the near-microbe environment in terms of component mass transfer resistances can be developed by mathematically expressing each mass transfer process as diffusive in nature.

The P_i mass transfer rate J_i through each compartment of the near-microbe environment can be expressed in terms of the resistance to mass transport R_i through the compartment and the concentration difference ΔC_i across the compartment. The general expression for mass transfer rate through any compartment is written as:

$$J_i = \frac{\Delta C_i}{R_i} \tag{3}$$

Physicochemical properties of outer-membrane transport channel proteins and kinetic parameters for energy-dependent cytoplasmic membrane P_i

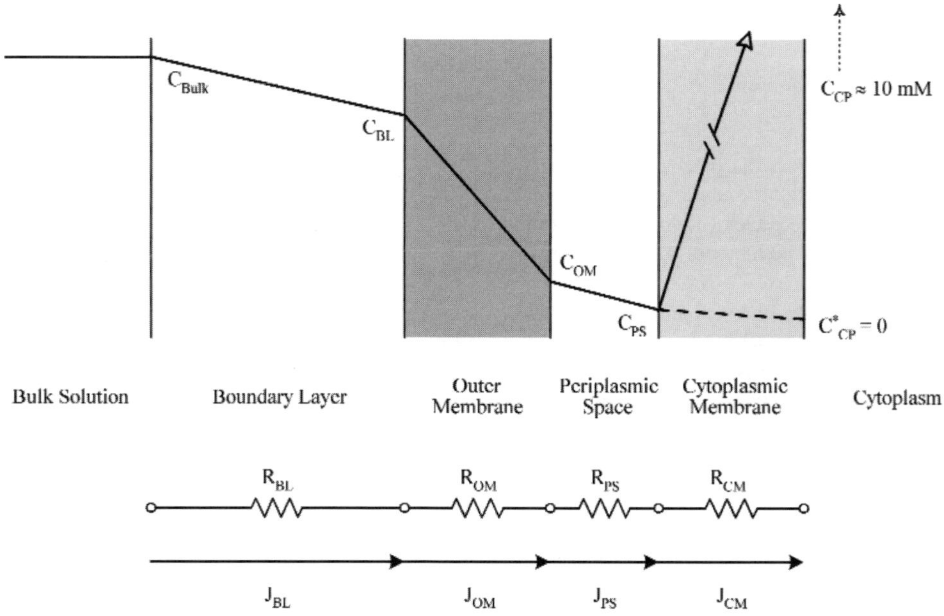

Fig. 4. Phosphate uptake by Gram-negative bacteria, with representative cell envelope P_i concentration profile. P_i transport is by diffusion through the aqueous boundary layer (BL), outer membrane (OM), and periplasmic space (PS), followed by an energy-dependent enzymatically-mediated process in the cytoplasmic membrane (CM). At steady state the rates of P_i transport through all near-cell environmental compartments are equal. Note that the actual cytoplasmic P_i concentration C_{CP} is approximately 10 mM, whereas the apparent P_i concentration C^*_{CP} with respect to the energy-dependent enzymatically-mediated trans-membrane transport process is set to 0.

transport proteins are listed in Table A.2.1 and Table A.2.2, respectively. In the individual near-microbe P_i transport rate expression derivations which follow, parameters in parentheses are either physical constants or representative values used in subsequent numerical calculations.

3.2.1. Diffusion rate through microbial boundary layer (BL)

The P_i diffusion rate through the bacterial cell fluid boundary layer (J_{BL}) is calculated for steady-state diffusion of a solute from solution to the surface of an absorbing sphere, with the concentration difference across the boundary layer as shown in Fig. 4. Based on the cell morphology of bacteria observed in this study (discussed later), a spherical cell shape with diameter of 1.6 μm was used. The diffusion rate J_{BL} and mass transport resistance R_{BL} are expressed as:

$$J_{BL} = 4\pi a D\left(C_{Bulk} - C_{OM}\right);\ \frac{1}{R_{BL}} = 4\pi a D \quad (4)$$

where
a = cell radius (800 nm)
D = diffusivity of P_i in water (0.8×10^{-5} cm^2 sec^{-1} – as tabulated in CRC, 1992).

3.2.2. Diffusion rate through outer membrane (OM)

Diffusion of P_i across the outer membrane occurs through anion-specific channels defined by outer membrane protein molecules (porins). The diffusion rate across the outer membrane (J_{OM}) is calculated from the equation for steady-state diffusion through pores with total area A_P in a wall of thickness x, and a concentration difference across the wall, as shown in Fig. 4. A trans-membrane pore diameter of 1.0 nm is used, based on representative values for Gram negative bacteria listed in Table A.2.1. The diffusion rate J_{OM} and mass transport resistance R_{OM} are expressed as:

$$J_{OM} = \frac{A_p f D}{x}\left(C_{OM} - C_{PS}\right);\ \frac{1}{R_{OM}} = \frac{A_p f D}{x} \quad (5)$$

where
A_p = area for diffusion through outer membrane porin channels (3.0×10^4 P_i transport porins/cell (Hancock et al., 1982; Nikaido, 1996), porin channel diameter of 1.0 nm; $A_p = 2.36\times10^4$ nm^2)
f = Renkin correction factor (Koch, 1990) for porin channel inlet geometry and Stokes drag for diffusion through the channel $f = [1-b/r_p]^2\ [1 - 2.104(b/r_p) + 2.09(b/r_p)^3 - 0.95(b/r_p)^5] = 2.21\times10^{-2}$ (HPO_4^{2-} ionic radius b = 0.29 nm (Hancock et al., 1987) and porin channel radius r_p = 0.5 nm)
D = diffusivity of P_i in water (0.8×10^{-5} cm^2 sec^{-1} – as tabulated in CRC, 1992)
x = thickness of outer membrane (7.5 nm, Koch and Wang, 1982).

3.2.3. Diffusion rate through periplasmic space (PS)

The P_i diffusion rate through the periplasmic space between the outer and cytoplasmic membranes (J_{PS}) is calculated from the equation for steady-state diffusion through a spherical shell of thickness y and radius a (Crank, 1975; Middleman, 1998), with a concentration difference as shown in Fig. 4. Thus, the diffusion rate J_{PS} and mass transport resistance R_{PS} are expressed as:

$$J_{PS} = \frac{4\pi a^2 D}{y}\left(C_{PS} - C_{CM}\right);\ \frac{1}{R_{PS}} = \frac{4\pi a^2 D}{y} \quad (6)$$

where
y = thickness of periplasmic space (7.5 nm, Koch and Wang, 1982)
D = diffusivity of P_i in periplasmic space is conservatively assumed to be the same as in water
a = cell radius (800 nm), as in section *3.2.1* because outer membrane and periplasmic space thicknesses x and y, respectively, are << than cell radius a.

3.2.4. Enzymatically mediated uptake rate through cytoplasmic membrane (CM)

The energy-dependent P_i transport across the cytoplasmic membrane is accomplished by the Pit and Pst enzymatic systems, as discussed in section *2.2.2*. Because P_i transport across the cytoplasmic membrane is an enzymatic process it is subject to substrate saturation, and Michaelis-Menton enzyme kinetics can therefore be used to describe the transport rate across the cell membrane. In terms of Michaelis-Menton enzyme kinetics, the P_i transport rate across the cell membrane (J_{CM}) is a function of the substrate concentration at the surface of the cytoplasmic membrane (C_{CM}):

$$J_{CM} = V = \frac{V_{\max}\left[C_{CM}\right]}{K_M + \left[C_{CM}\right]} \quad (7)$$

where
V_{max} = maximum transport rate when enzyme is saturated by P_i (30 nmol P_i/mg DW·min – Pit system, 20 nmol P_i/mg DW·min – Pst system), 2.8×10^{-13} g DW/cell)
K_M = Michaelis-Menton enzyme half-saturation con-

stant (25×10^{-6} mol P_i/L – Pit system, 0.2×10^{-6} mol P_i/L – Pst system).

Because it is difficult to directly measure the concentration of cytoplasmic membrane-bound enzymes, the Michaelis-Menton constants V_{max} and K_M are expressed on a dry weight of constituents per cell basis (DW/cell).

The energy-dependent enzymatically-mediated P_i transport process described by Eqn. 7, although not physically diffusive in nature, can be modeled using the same mathematical form as a purely diffusive process. Because $[C_{CM}] << K_M$ for the chemostat conditions encountered in this work (see section 4.7), the denominator of Eqn. 7 reduces to K_M. Furthermore, an *apparent* diffusive concentration difference ($C_{CM} - C^*_{CP}$) can be defined across the cytoplasmic membrane as shown in Fig. 4, where C^*_{CP} is set to zero to reflect the independence of the *actual* physical transport process on cytoplasmic P_i concentration; note: the actual C_{CP} is ≈ 10 mM. With these thoughts in mind, Eqn. 7 simplifies to these expressions for J_{CM}, and mass transport resistance R_{CM}:

$$J_{CM} = \frac{V_{\max}}{K_M}\left(C_{CM} - C^*_{CP}\right) \; ; \; \frac{1}{R_{CM}} = \frac{V_{\max}}{K_M} \qquad (8)$$

As depicted in Fig. 3, the Pit and Pst transport systems operate in parallel across the cytoplasmic membrane. However, the resistance to P_i transport of the Pit system is much greater than that of the Pst system ($R_{Pit} \approx 20$ to $200R_{Pst}$), as determined using representative kinetic properties for the two transport systems in several bacterial species (Table A.2.2) and Eqn. 8. Then by Eqn. 2, $R_{composite} \approx R_{Pst}$ for parallel transport, and the Pst system controls P_i transport across the cytoplasmic membrane under P_i -limiting conditions. Therefore, P_i transport by the Pit system will not be considered further.

3.2.5. Microbial cell envelope composite P_i mass transfer resistance

Because at steady state the P_i transport rate through all near-microbe environmental compartments is equal, the transport rate *J* through the composite near-microbe environment can be found by rearranging Eqns. 4, 5, 6, and 8, respectively, as follows:

$$J_{BL}R_{BL} = \left(C_{Bulk} - C_{OM}\right) \qquad (9)$$

$$J_{OM}R_{OM} = \left(C_{OM} - C_{PS}\right) \qquad (10)$$

$$J_{PS}R_{PS} = \left(C_{PS} - C_{CM}\right) \qquad (11)$$

$$J_{CM}R_{CM} = \left(C_{CM} - C^*_{CP}\right) \qquad (12)$$

Summing Eqns. 9–12 yields the following expression:

$$J\left(R_{BL} + R_{OM} + R_{PS} + R_{CM}\right) = \left(C_{Bulk} - C^*_{CP}\right) \qquad (13)$$

Rearranging Eqn. 13 and substituting $C^*_{CP} = 0$ yields:

$$J = \frac{C_{Bulk}}{R_{BL} + R_{OM} + R_{PS} + R_{CM}} \qquad (14)$$

where *J* (mol/s) is the transport rate of P_i into the cell, C_{Bulk} (mol/L) is bulk solution P_i concentration, and R_{BL}, R_{OM}, R_{PS}, and R_{CM} (s/L) are the resistances to P_i transport through the cell boundary layer, outer membrane, periplasmic space, and cytoplasmic membrane, respectively.

4. CHEMOSTAT EXPERIMENTS

A set of chemostat experiments was conducted using a natural soil microbial community and P_i sorbed on goethite as the sole phosphorous source. Basic microbiological, kinetic, and solution chemistry data were collected to facilitate evaluation of various active and passive responses bacteria may use to enhance bioavailability of mineral surface associated P_i.

4.1. Goethite Synthesis and Phosphate Adsorption Characterization

Colloidal goethite was synthesized by precipitation from $FeSO_4$ in an aerated solution buffered to pH 7 with 0.1 M $NaHCO_3$ (Schwertmann and Cornell, 1991). The identity and purity of the synthetic goethite sample was verified by powder X-ray diffraction analysis. No other crystalline mineral phases were detected. Broad weak features, indicative of amorphous material, were also absent in the diffraction pattern. Scanning electron microscopy (SEM) indicated that the size and shape of goethite crystals in our synthetic sample were very similar to those found in natural environments (Schwertmann and Cornell, 1991); colloids were relatively uniform euhedral, acicular crystals, approximately 700 nm long by 70 nm wide. Specific surface area was 85.4 m^2/g as measured by N_2 BET Analysis. The PZC of the prepared goethite was estimated to be 7.9 using the intersection of pH titration curves in 0.001, 0.01, and 0.1 M KCl.

The sorptive behavior of P_i on the goethite synthesized here was characterized by generating adsorption edges at several surface coverages and adsorption isotherms at several pH values. Standard wet-chemical techniques were used to generate these adsorption data, and both adsorption edge and isotherm data were typical of strongly adsorbed oxyanions (Fig. 5). Desorption of phosphate by repeated rinsing with 0.01 N KCl showed strong hysteresis. Phosphate adsorption on goethite was also examined with X-ray photoelectron spectroscopy (XPS) (Fig. 5). The phosphorus signal measured directly from the surface of dried samples was well correlated with the isotherm developed from wet chemical experiments.

4.2. Chemostat Design and Operation

A chemostat was fabricated from a 2.2-L cylindrical borosilicate glass environmental sample bottle. Openings through the teflon-lined screw top lid were provided for glucose-mineral salts (GMS) microbial nutrient media and goethite suspension, pH probe, acid and base for pH control, and low-pressure (0.03 MPa, 5 psig) filtered air. Outflow from the chemostat was by gravity through a horizontal 0.64-cm (0.25-in) diameter teflon tube which penetrated the chemostat wall at a level corresponding to a volume of 2 L. Circumferential and vertical mixing were achieved using a floating teflon stir-bar pressure fit to the chemostat wall and an air-driven lift pump mounted atop the floating stir-bar assembly.

GMS media and phosphate loaded goethite suspension were admitted to the chemostat at equal volumetric rates from separate reservoirs using a dual channel cassette pump. Both GMS media and goethite suspension were prepared as 2X solutions, which resulted in a 1X concentration for all components in the chemostat. Both chemostat influent flow paths were exchanged with autoclaved components at each replenishment of GMS media and goethite suspension. Constant pH was maintained using a pH/ORP controller monitoring an epoxy body pH Triode. Two single-channel peristaltic tubing pumps, actuated by high or low pH setpoint solenoids in the controller, periodically added small volumes of 0.1 N HCl or 0.1 N NaOH to the chemostat in response to minor pH fluctuations. The mathematics of chemostat operation are described in Appendix A.1.

4.3. Glucose-Mineral-Salts Media

Glucose was selected as a carbon and energy source because it can be utilized by most aerobic microorganisms. In addition, glucose does not interact with mineral surfaces to an appreciable extent, and therefore does not compete with P_i for surface adsorption sites. A pH of 4.5 was selected to approximate natural soil pH (~4.2), and to inhibit use of NH_4^+ as an energy source through nitrification (inhibited at pH < 5.5). The mineral salts composition of GMS media is listed in Table 1. In order to replace the natural organic and inorganic nutrient components of an actual soil matrix, vitamin and SL-8 trace-mineral solutions (Atlas, 1995) were added to support growth of fastidious microbial species.

4.4. Analytical Methods

Orthophosphate was measured spectrophotometrically using the molybdate-blue method (APHA et al., 1995). Microbial P_i utilization was determined as the difference between influent P_i concentration and acid hydrolyzable P_i in chemostat effluent. Goethite in the chemostat effluent was reductively dissolved by mild heating at acidic pH with added ascorbic acid prior to determination of acid hydrolyzable P_i. XPS analysis of phosphate loaded goethite surfaces was performed using a Perkin-Elmer 5400 X-ray photoelectron spectrophotometer. Glucose concentration was measured using a commercially available Glucose HK (Hexokinase) enzymatic assay kit (Sigma® Kit GAHK-20). Dissolved oxygen concentration was measured and found to be at saturation in the bulk chemostat effluent. Soluble iron was measured using both graphite furnace atomic emission spectrometry (AES) and ferrozine spectrophotometry (Gibbs, 1978). The concentrations of inorganic cations (Na^+, NH_4^+,K^+, Mg^{2+}, Ca^{2+}) and anions (F^-, Cl^-, NO_2^-, Br^-, NO_3^-, PO_4^{3-}, SO_4^{2-}) were determined by high-pressure ion chromatography (HPIC). The concentrations of low molecular-weight organic acids were measured using high-pressure ion exclusion chromatography (HPIEC). Microbe-mineral aggregates in chemostat effluent were imaged using scanning laser confocal microscopy (SLCM) with acridine orange as a fluorescent biological probe.

4.5. Chemostat Inputs, Operation, and Monitoring

The inoculum used to seed the chemostat consisted of the microbial community present in an undisturbed acidic mineral soil. A dilute slurry was prepared by placing 1 g of well dispersed soil in 100 mL of mineral salts media and gently agitating the mixture for 1 h. Ten mL of the dilute slurry were added to the chemostat, and the selective pressure of the nutrient and physical environments imposed by chemostat operation applied for 45 d. GMS media and goethite slurry were added at equal rates to give a 4-d hydraulic retention time. Growth in chemostat culture under these conditions was expected to result in selective enrichment of those microbial species that were most competitive under the low-pH, low-carbon, mineral-sorbed P_i conditions characteristic of the acidic mineral soil used to prepare the initial

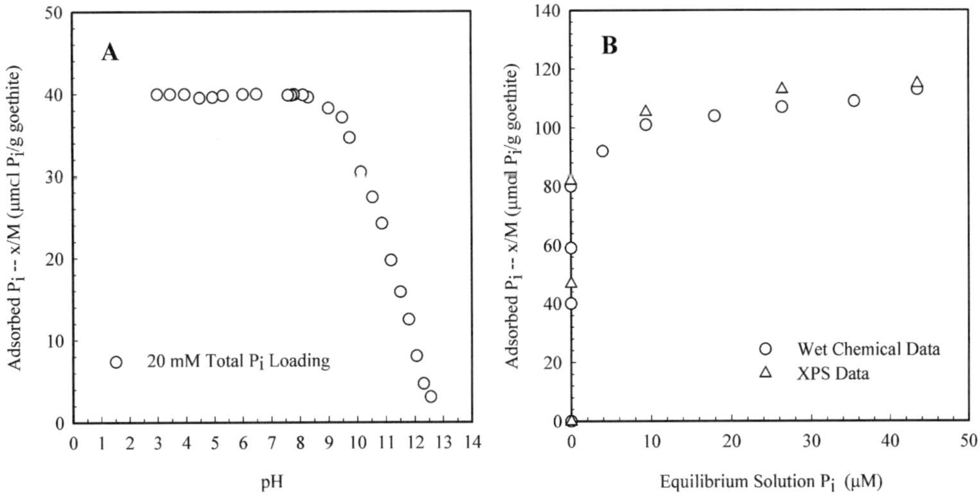

Fig. 5. Equilibrium adsorption of P_i on goethite described by adsorption edge and adsorption isotherm behavior (0-100 μM total P_i, 0.5 g/L goethite, and 0.01 M KCl as background electrolyte). Panel A: Typical adsorption edge for 20 μM total P_i concentration. Sorbed P_i (x/M – μM P_i/g goethite) as a function of pH was determined by both indirect wet-chemical and direct XPS techniques. Indirect wet-chemical measurement of sorbed P_i was made by difference between the total initial P_i concentration of the goethite suspension and the residual P_i concentrations of filtered sub-samples collected at several pH values. XPS peak areas measured from dried goethite suspensions of varying initial pH were directly correlated to sorbed P_i (data not shown). The effect of slight variations in source-sample-detector geometry on measured P_i peak areas was accounted for by using the Fe_{3S} peak area of each sample as an instrumental correction factor. Panel B: Adsorption isotherm data generated from adsorption edge data, pH 6.3. Sorbed P_i (x/M) plotted as a function of equilibrium solution P_i (residual concentration in filtered sample) from both wet-chemical and XPS-based measurements.

inoculum. Phosphorus loading was 40 μmol P_i/g goethite (~ 6 to 12 % surface area coverage for monodentate and bidentate complexes, respectively) and influent glucose concentration was 0.56 mM. After operation for 14 d, influent glucose concentration was increased to 2.78 mM. Effluent samples were collected daily and analyzed for glucose, residual P_i, anions and cations, soluble iron (< 0.2 μm), and organic ligand production.

4.6. Chemostat Effluent Parameters

After a short acclimation period (~ 2 d) in which residual sorbed P_i dropped rapidly, microbial utilization of goethite sorbed P_i averaged 82 ± 6 % throughout the remainder of the experiment. Measurements of inorganic ion concentrations showed that flow rates from the GMS media and goethite reservoirs were balanced and mixing in the chemostat was adequate to preclude hydraulic short-circuiting. A low molecular-weight organic acid (LMWOA – tentatively identified by HPIEC retention time as either citric or gluconic acid) was detected in filtered chemostat effluent. LMWOA production appeared related to carbon availability, increasing from ~ 4 μM at an influent glucose concentration of 0.56 mM to ~ 100 μM at an influent glucose concentration of 2.78 mM. Soluble iron was detected for the duration of the experiment, ranging between 1 and 3 times the solubility limit of amorphous $Fe(OH)_3$ (0.11 μM at pH 4.5).

Formation of colloidal microbe-goethite aggregates was observed, and also appeared related to carbon availability. Aggregates predominated in the chemostat at the low influent glucose level, while at the high-glucose level microbes and goethite were characteristically well dispersed. Soluble polysaccharide levels as high as 20 mg/L were noted under aggregate-forming conditions. The aggregates appeared loosely associated, as brief sonication (30 s) dispersed the microbes and goethite colloids. However, re-association of microbes and goethite was observed within minutes following sonification. A typical microbe-goethite aggregate is illustrated in Fig. 6.

Microscopic examination and Gram staining of chemostat effluent showed that bacteria with motile Gram-negative rod-shaped morphology predominat-

Table 1. GMS media mineral salts composition (1X).

Compound	Concentration (M)
NaCl	4.28×10^{-3}
KCl	4.02×10^{-3}
NH_4Cl	1.08×10^{-3}
$MgSO_4\cdot7H_2O$	6.49×10^{-5}
$CaCl_2\cdot2H_2O$	6.80×10^{-5}

ed, although small numbers of bacteria with other morphologies were observed. Fatty acid profile and 16S rDNA analyses (Midi Labs, Newark, DE) both indicated that the dominant Gram-negative species was *Burkholderia cepacia*. No yeast cells or filamentous fungi were observed. Evidence of approach to steady-state chemostat operation was seen at both influent glucose concentrations used in this study. At the initial influent glucose level of 0.56 mM the culturable cell population was $6.0 \pm 0.8\times10^5$ CFU/mL, LMWOA concentration was 4 μM, and effluent glucose concentration was 1.5 μM; whereas at the subsequent influent glucose level of 2.78 mM, these effluent parameters were $1.2 \pm 0.2\times10^8$ CFU/mL, 105 μM, and 140 μM, respectively.

4.7. Thermodynamics of Microbial P_i Uptake

In accordance with Le Châtelier's principle, P_i uptake from solution by microbes results in desorption from the mineral surface until a new equilibrium condition is established. In a chemostat operating at steady state, and assuming rapid exchange kinetics at the surface-solution interface, the net effect of microbial phosphate uptake is to shift the position of surface complexation equilibrium toward the solution phase. The extent to which phosphate desorbs from goethite surfaces in response to microbial uptake from solution may ultimately depend on the energetics of transmembrane phosphate transport against a very unfavorable concentration gradient or the availability of phosphate at the membrane surface. Simply stated, either the thermodynamics or kinetics of membrane transport may control the influence of solution phase uptake on P_i-mineral surface equilibrium.

Due to the very high equilibrium constant for P_i complexation on goethite surfaces (log $K_{int} \geq 22$, Sigg and Stumm, 1981), when a goethite suspension is loaded with phosphate at less than the total surface sorptive capacity, the solution phase phosphate concentration is extremely low. For example, at the phosphate loading of 40 μmol P_i/g goethite used in our bioavailability study, the observed solution phase P_i concentration was below the 0.1 μM limit of spectrophotometric detection. In fact, the equilibrium solution concentration for the total P_i loading used in chemostat experiments (40 μmol P_i/g goethite in GMS media at pH 4.5) is predicted to be approximately 0.1 to 1 nM, based on numerical simulation using MINTEQA2 (Allison et al., 1991) and K_{int} values available in the literature (Sigg and Stumm, 1981). Direct measurement of P_i concentrations ≤ 1 nM are problematic, even using sensitive radiolabeled ^{32}P methods. Therefore, we must at present rely on predictions made using surface complexation modeling of sorption data to estimate solution phase P_i.

Microorganisms must expend energy to take up P_i against the unfavorable concentration gradient across the cytoplasmic membrane typical of most natural environments (see section *2.2.2*). Energy is also required to transport negatively charged P_i molecules into microbial cells against a negative electrical potential difference ($\Delta\Psi = \Psi_{intercellular} - \Psi_{extracellular}$) across the cytoplasmic membrane. The membrane potential $\Delta\Psi$ varies between about −60 mV and −200 mV depending on the bacterial species and growth conditions (Kashket, 1985), with an average value of about −100 mV for slowly growing or resting cells (Futai and Tsuchiya, 1987). The total electrochemical energy, expressed in terms of free energy ΔG (kJ/mol), required to transport P_i into bacterial cells is:

$$\Delta G = RT \ln \frac{C_{inter}}{C_{extra}} + ZF\Delta\Psi \qquad (15)$$

where C_{inter} and C_{extra} are intercellular and extracellular P_i concentrations, Z is anion charge, F is Faraday's constant, and $\Delta\Psi$ is membrane potential. Under conditions imposed in our bioavailability experiments (T = 298 K, $C_{inter} \approx 10$ mM, $C_{extra} \approx 1$ nM, $Z = -1$ for $H_2PO_4^-$, and $\Delta\Psi \approx -0.1$ V), from Eqn. 15 the ΔG required for P_i uptake is 49.6 kJ/mol, or 9.9×10^{-4} kJ/L for complete uptake of 20 μM P_i. Because of uncertainty in the chemical composition and ionization state of solutes in the cytoplasm, calculation of ionic strength and hence use of activities rather than concentrations in Eqn. 15 is impractical. However, due to the logarithmic nature of the concentration dependence of Eqn. 15 and the large difference in P_i concentration across the cytoplasmic membrane typical of natural systems, the error induced in ΔG calculated using concentrations is minor. For example, assuming cytoplasmic and external ionic strengths of 0.3 and 0.01 M, and using the other values in Eqn. 15 as above, changes the value of ΔG by less than 1 %.

Chemoorganotrophic microorganisms obtain energy for cell maintenance and growth by oxida-

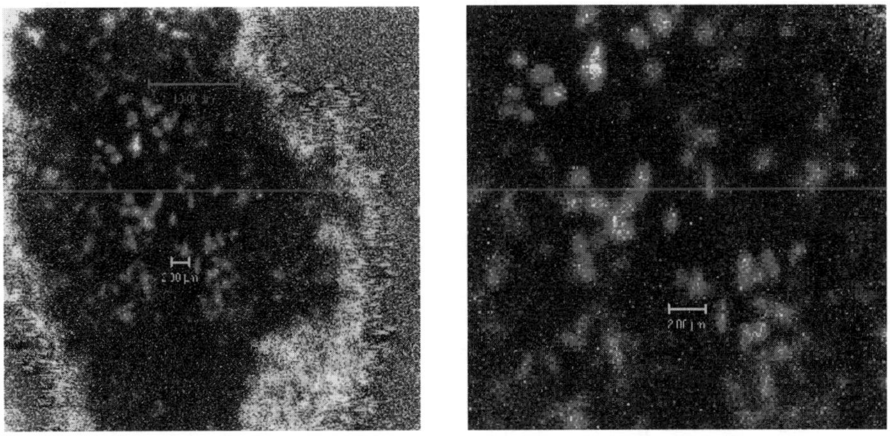

Fig. 6. Scanning laser confocal microscopy (SLCM) images of bacteria-goethite aggregates from chemostat effluent (large scale bar = 10 μm, small scale bar = 2 μm). Bacteria stained with acridine orange dye (lightly contrasted ~ 1-2 μm rod-shaped objects) can be seen in close proximity to colloidal goethite particles (dark regions) within aggregates. These images were collected using a Zeiss LSM-510 SLCM with a 40X, 1.2 n.a. water immersion lens.

tion of organic carbon compounds coupled to formation of high energy phosphoanhydride bonds in ATP. For aerobically respiring bacteria, complete oxidation of 1 mole of glucose to CO_2 and H_2O through the glycolytic and Krebs cycle pathways yields 38 moles of ATP. Given that hydrolysis of 1 mole of phosphoanhydride bonds in ATP gives 31 kJ of energy, the total free energy available due to microbial glucose oxidation in our experiments (0.56 to 2.78 mM glucose) is 0.65 to 3.3 kJ/L. Thus, if 9.9×10^{-4} kJ/L of free energy is thermodynamically required for complete P_i uptake as calculated above, ATP-dependent enzymatically-mediated P_i uptake by the Pst system would require less than 1 % of the energy potentially available by glucose oxidation. Because microorganisms typically spend a large fraction of their energy resources to support ATP-driven solute transport (Voet and Voet, 1995) (30 to 50 % is not uncommon), it appears that transmembrane P_i transport was not thermodynamically-limited in our experiments.

5. MICROBIAL RESPONSES TO P_i LIMITATION

Microbes respond to environmental stimuli such as P_i limitation through alternative patterns of gene expression, which regulate cellular physiology and metabolism. These physiological responses may be grouped into two fundamental categories: passive responses in which the effects of modulated cellular physiology responsible for adaptation to low solution P_i concentrations are confined to the cell itself, or active responses in which cell surface physiology or an exported metabolic product physicochemically alters the surrounding environment to enhance bioavailability of adsorbed P_i. A number of potential passive and active microbial responses to P_i limitation are described in the following sections, and evaluated with respect to both the chemostat experiments presented here and previous observations in the literature.

5.1. Passive Responses

5.1.1. Alteration of cell envelope to enhance P_i uptake

Under P_i-limiting conditions, i.e., several μM or less depending on the species, an enhanced P_i uptake system is commonly induced. In Gram-negative bacteria, these enhanced transport systems are composed of a porin trimer protein channel that spans the outer membrane and a P_i specific transporter protein complex that spans the cytoplasmic membrane. The physicochemical and kinetic properties of P_i transport-related porins and cytoplasmic membrane protein channels for several bacterial species are listed in Tables A.2.1 and A.2.2.

The feasibility of transport system modulation (induction of a high-affinity Pst-like P_i transport system – Fig. 3a) as the sole strategy for enhancing microbial uptake of P_i adsorbed on colloidal goethite surfaces was evaluated by comparing the cell generation time g_{Pi} (h) estimated for P_i uptake limited by cell envelope transport with the chemostat MCRT: if g_{Pi} is greater than MCRT (96 h) then modulated uptake is not sufficient to support steady-state chemostat culture. The resistance to P_i transport through each cell envelope compartment was calculated using representative cell envelope data for Gram-negative bacteria (Tables A.2.1 and A.2.2, and Sections *3.2.1* through *3.2.4*), and values for the individual compartment and total cell envelope P_i transport resistances given in Table 2. Assuming a bulk solution phase P_i concentration of 1 nM, from Eqn. 14 the growth-limiting P_i uptake rate J is 4.9×10^{-23} mol/cell·sec. The amount of P_i required by a single representative Gram-negative cell, estimated using an average dry weight of 2.8×10^{-13} g/cell (Neidhardt et al., 1990) and an average phosphorus content of 2 percent (w/w) (Bowen, 1979), is 1.8×10^{-16} mol/cell. Dividing the phosphate content per cell by the limiting P_i uptake rate J gives a cell generation time g_{Pi} of 1020 h.

Although representative cell envelope parameters were selected in a conservative manner, i.e., to lower the calculated P_i uptake-limited cell generation time g_{Pi} , the time required for steady-state cell growth based on P_i uptake rate limitation exceeded the available chemostat residence time by an order of magnitude ($g_{Pi} \cong 10\times$MCRT). Thus, based on the kinetics of P_i uptake, it appears likely that in addition to inducing a high-affinity Pst-like transport system the microbial culture in this study utilized other as yet unspecified strategies to obtain P_i at a rate sufficient to maintain steady-state chemostat growth. Also, even though the resistance to P_i transport across the outer membrane dominates the overall cell envelope resistance (Table 2), the resistance to P_i transport across the cytoplasmic membrane is not negligible, as has been previously assumed (Koch and Wang, 1982; Koch, 1990).

5.1.2. Reduction in P_i required

In addition to the passively mediated transport strategy outlined in the previous section, which addresses P_i limitation by altering cell envelope structure to enhance P_i uptake through the cell envelope, microorganisms may also reduce their requirement for P_i by substituting non-phosphorylated molecules where possible. For instance, some bacterial species may substitute glycolipids for phospholipids in membranes when exposed to P_i limited environments (Tempest and Neijssel, 1978; Brown and Gilbert, 1993). However, the majority of cellular macromolecules require P_i as an integral component, for which there is no suitable alternative. Using *E. coli* B/r cells grown aerobically on glucose minimal media as an example, complete substitution of non-phosphorylated membrane lipids would only decrease the total cellular P_i requirement by about 12% (Table 3).

5.1.3. Reduction in cell size

Reduction in cell size has also been widely postulated as a general physiological survival response of microbes exposed to nutrient limitation in oligotrophic natural environments. It has been proposed and widely accepted that, in addition to reducing the requirement for limiting nutrients associated with less cell mass, smaller size increases the "efficiency" of microbial nutrient transport. However, we contend that smaller cell size, in and of itself, is not an effective strategy for dealing with nutrient limitation, and use P_i uptake by bacteria to illustrate this point.

Considering only the effect of size on membrane transport by bacteria, smaller size actually decreases P_i uptake rate J. Examination of Eqns. 5-8 shows that cell size (as measured by cell radius a) explicitly appears only in the expressions for diffusion to the outer-membrane surface and diffusion through the periplasmic space. Decreasing a increases the resistance to diffusive mass transport in both the cell boundary layer and periplasmic space (R_{BL} and R_{PS} terms), which by Eqn. 14 actually decreases rather than increases the overall P_i uptake rate J. However, as demonstrated in Table 2, the collective resistance to mass transport in the cell boundary layer and periplasmic space represents < 1% of the total resistance to P_i uptake, and any decrease in overall uptake rate J due to decreased cell size a should be minimal. Conservatively assuming that decreasing cell size does not affect transport across either cell membrane (R_{OM} and R_{CM} terms), the influence of cell size on P_i uptake rate only becomes noticeable at extremely small cell sizes, on the order of 0.1 μm, as illustrated in Fig. 7.

When bacteria encounter nutrient limitation, growth rates slow, and smaller cell sizes are commonly observed. However, reduction in cell size is not a direct physiological response to limitation of any specific nutrient, but rather, a coordinated global response to less favorable growth conditions. Nutritional conditions control the growth rate of bacteria, which in turn dictates the macromolecular composition (amounts of DNA, RNA, ribosomes, etc.) required to support that growth rate (Neidhardt et al., 1990). In fact, cellular components, including total cell mass, protein, DNA, and RNA, are individually varying exponential functions of growth rate, which

Table 2. Representative resistance to P_i transport in cell envelope compartments of Gram-negative bacteria.[a]

Near Microbe Compartment	Resistance to P_i Transport (s/nm^3)	Proportion of Total Resistance (%)
Cell Boundary Layer (R_{BL})	1.24×10^{-13}	0.6
Outer Membrane (R_{OM})	1.80×10^{-11}	88.8
Periplasmic Space (R_{PS})	1.17×10^{-15}	0.006
Cytoplasmic Membrane (R_{CM})	2.14×10^{-12}	10.6
Cell Envelope Total (R_T)	2.03×10^{-11}	100

[a]Values for Resistance to P_i transport in near microbe compartments calculated using Eqns. 4, 5, 6, and 8, respectively, and representative cell envelope data for Gram-negative bacteria listed in Tables A.2.1 and A.2.2.

forms the fundamental relationship between nutrition-imposed growth rate and cellular composition (Schaecter et al., 1958). Obviously, cell size varies in accordance with the amount of cellular machinery required to support any given growth rate. The relationship between cell mass (size) and growth rate is related to the initiation of chromosomal replication: replication is initiated when cell mass reaches a specific value M_i per chromosomal replication origin (*oriC*) (Donachie, 1968).

Reduction in growth rate, and hence cell mass and size, may reduce the requirement for P_i. For the example of *E. coli* B/r cells given in Table 3, 68% of cellular P_i is contained in nucleic acids, the quantities of which are strictly regulated with respect to growth rate. However, there are physical, chemical, and biological constraints which limit the extent to which a cell can down-regulate critical cellular components. In order to remain viable a cell must maintain some minimum set of synthetic machinery, including a complete copy of its genome and some basal level of macromolecules for protein synthesis. The physical size of a bacterial chromosome places a minimum dimension on cell size: the supercoiled nucleoid configuration of bacterial chromosomes has a species-dependent characteristic dimension of hundreds to thousands of nm. Small size also has implications with respect to solution chemistry in the cell cytoplasm (Nealson and Stahl, 1997). For example, at a cytoplasmic concentration typical of many enzymatically mediated reactions (1 μM), a spherical cell of 0.1 μm radius would contain only about 2.5 reactant molecules, whereas a cell of 0.8 mm radius would contain about 1290. The low number of solute molecules which results from small cell size presents a potential impediment to synthesis of macromolecular components and cell growth.

The cells observed in the chemostat experiments described here were characteristically small (~1 μm), but were of a size consistent with the slow growth rate imposed by the oligotrophic nutritional environment and low dilution rate of chemostat operation. The distribution of P_i among cellular components was not examined in this study, and the extent to which substitution of non-phosphorylated lipids were utilized in cell membranes is not known. However, based on the apparent inability of the high affinity Pst uptake system to transport P_i at a rate sufficient to support chemostat growth under the environment applied in our experiments, the example cellular P_i distribution of Table 3, and the insensitivity of P_i uptake rate with cell size shown in Fig. 7, we believe that passive microbial responses to P_i limitation do not adequately explain the high level of P_i bioavailability observed in this study. This suggests that the presence of bacteria altered the physicochemical environment in the chemostat resulting in enhanced P_i release from goethite surfaces.

5.2. Active Microbial Responses to P_i Limitation

5.2.1. Production of surface active biomolecules

Overproduction of several LMWOA metabolites has been observed in chemostat cultures exposed to P_i-limitation (Tempest and Neijssel, 1992). Organic acids may compete directly for sorption sites on mineral (hydr)oxides or inhibit the formation of mineral phase precipitates (Earl et al., 1979; Sibanda and Young, 1986; Schwertmann et al., 1986; Hue, 1991; Violante and Gianfreda, 1993; Geelhoed et al., 1998). Microbial excretion of LMWOA may also accelerate dissolution of (hydr)oxides by lowering pH or by chelating Al or Fe on the mineral surface (Furrer and Stumm, 1986; Zinder et al., 1986; Robert and Berthelin, 1986; Stevenson, 1994; Hersman et al., 1995; Hersman et al., 1996; Holmén and Casey, 1996; Stone, 1997; Barker et al., 1997), thereby releasing P_i from the mineral (Dalton et al., 1952; Banik and Dey, 1983; Bolan et al., 1987; Jayachandran et al., 1989). Reductive dissolution of Fe (hydr)oxides, known to occur in anaerobic environments (Willett, 1982; 1985; 1989; Suter et al., 1991; Ehrlich, 1996), may also occur in aerobic environments when chelating or reducing metabolites are present (Iyamuremye and Dick, 1996). Exopolysaccharides retain water and contain reactive functional groups which may participate in dissolution reactions (Barker et al., 1998). Extracellular enzymes such as Fe(III)-chelate reductase may enhance dissolution of Fe-oxide surfaces (Robinson et al., 1999).

Microbial utilization of P_i and P_i-containing organic compounds sorbed on non-crystalline alu-

Table 3. Distribution of P_i among cellular components of *E. coli*[a] B/r.

Cellular Component	Molecules/Cell	P_i/Molecule	Molecules P_i	% of Total
Cell Envelope				
Phospholipid[b]	2.2×10^7	1	2.2×10^7	12.1
Lipopolysaccharide[c]	1.4×10^6	6	8.6×10^6	4.7
Cytoplasm[d]	1.3×10^7	n.a.	1.3×10^7	7.1
Nucleic Acids				
DNA[e]	2.13 (4.7×10^6 bp)	2/bp	2.0×10^7	11.0
RNA[f]				
rRNA	18700	4460 (1/nucleotide)	8.3×10^7	46.0
tRNA	198000	80 (1/nucleotide)	1.6×10^7	9.1
mRNA	1380	2000 (1/nucleotide)	2.8×10^6	1.5
Miscellaneous[g]			1.6×10^7	8.7

[a] In balanced growth at 37°C in aerobic glucose minimal medium with a doubling time *g* of 40 min (Bremmer and Dennis, 1996; Neidhardt and Umbarger, 1996). Cell mass is 4.33×10^{-13} g/cell.

[b] One P_i molecule per phospholipid molecule based on 76% phosphatidylethanolamine, 20% phosphatidylglycerol, and small amounts of phosphatidylserine and cardiolipin (Neidhardt et al., 1990).

[c] Six P_i molecules per lipopolysacharride molecule (Neidhardt et al., 1990).

[d] Cytoplasmic P_i concentration of 10 mM (Shulman et al., 1979; Wanner, 1994).

[e] Considers chromosomal DNA only, not plasmid DNA.

[f] RNA nomenclature: rRNA - ribsomal RNA, tRNA - transfer RNA, mRNA - messenger RNA.

[g] Miscellaneous includes phosphorylated proteins, coenzymes, free nucleotides (ATP, GTP, etc.), and polyphosphate storage polymers.

minum-oxide precipitates, goethite, kaolinite, and montmorillonite has previously been noted in batch bioavailability experiments (Shang et al., 1996; He and Zhu, 1998). Microbial acquisition of P_i from phosphate-containing minerals has also been observed in several batch bioavailability studies. Several Gram-negative bacterial spp. have been observed to mediate the dissolution of calcium-phosphate minerals through excretion of glycolic or gluconic acid (Goldstein, 1986; 1987; Illmer and Schinner, 1992; Goldstein, 1994). Microbially-mediated release of P_i from feldspar minerals has also been noted (Rogers et al., 1998). However, P_i release from these minerals is often poorly correlated with organic acid production, and the presence of other unidentified mechanisms has been noted (Illmer and Schinner, 1992).

The presence of a LMWOA, exopolysaccharides, and soluble iron substantially in excess of pH-solubility limits in the chemostat effluent of the present study is consistent with excretion of surface active metabolites as a means of enhancing the bioavailability of P_i adsorbed on mineral surfaces. *Burkholderia cepacia* (formerly classified *Pseudomonas cepacia*) was the dominant microbe isolated and identified in the chemostat effluent. *B. cepacia* is closely related to the *Pseudomonads*, many species of which are known to produce secondary metabolites in response to various environmental stimuli (Wubolts and Witholt, 1998; Meyer and Stintzi, 1998). When exposed to P_i limitation, a number of Gram-negative bacteria, including *Klebsiella pneumoniae*, *Erwinia herbicola*, and *Pseudomonas cepacia,* are known to produce gluconic acid by direct oxidation of glucose in the periplasmic space (Tempest and Neijssel, 1992; Babu-Khan et al., 1995). However, the gene which encodes the periplasmic glucose dehydrogenase enzyme responsible for gluconic acid production is not part of the Pho regulon, and cross-regulation between the Pho regulon and gluconic acid formation has not been demonstrated. Exopolysaccharide production is also common among *Pseudomonads*, and has been observed in *P. cepacia* cultures (Sutherland, 1991; Allison and Goldsbrough, 1994).

5.2.2. *Proximity to mineral surfaces*

The pronounced tendency of microbe-mineral aggregate formation observed in this bioavailability study at low glucose levels suggests a possible microbial response to P_i-limitation. In the absence of excess carbon required to produce extracellular surface active metabolites, alteration of cell surface composition or structure in order to induce aggregation with the P_i loaded goethite surfaces may be the least energetically expensive means of enhancing P_i bioavailability. Based on Guoy-Chapman double layer theory, it can be shown that the pH in the electrostatic double layer surrounding a colloidal mineral particle may be as much as several pH units less than that of the surrounding bulk solution (Stumm and Morgan, 1981). The higher $[H^+]$ near the colloidal surface may aid in generation/maintenance of the transmembrane proton gradient which cells use to provide energy for solute transport and ATP formation. Because both the nonspecific (Pit) and specific (Pst) membrane P_i transport systems are endergonic processes which either directly or indirectly depend on a membrane proton gradient, proximity to a colloidal mineral surface with adsorbed P_i not only reduces the mass transfer resistance to the cell surface, but also potentially reduces the energetic expense of transport across the cell envelope.

Aggregation of mineral colloids may also increase the bioavailability of surface-associated nutrients by shifting the position of the solution-surface nutrient equilibrium towards the solution phase in accordance with Le Châtelier's principle. Increased local mineral surface concentration may increase the solution phase P_i concentration within an aggregate by several orders of magnitude. For instance, formation of a goethite aggregate with a characteristic dimension of 50 µm and 90 % free internal volume (solution phase) results in an increase in local goethite surface concentration of approximately 3 orders of magnitude over that in the bulk chemostat solution of this bioavailability study. Furthermore, P_i diffusion within aggregate pore spaces is retarded due to strong association with goethite surfaces, and any increase in solution phase P_i concentration within the aggregate is therefore dissipated extremely slowly by diffusion into the surrounding bulk solution. Analytical solutions for transient mass transport by retarded diffusion

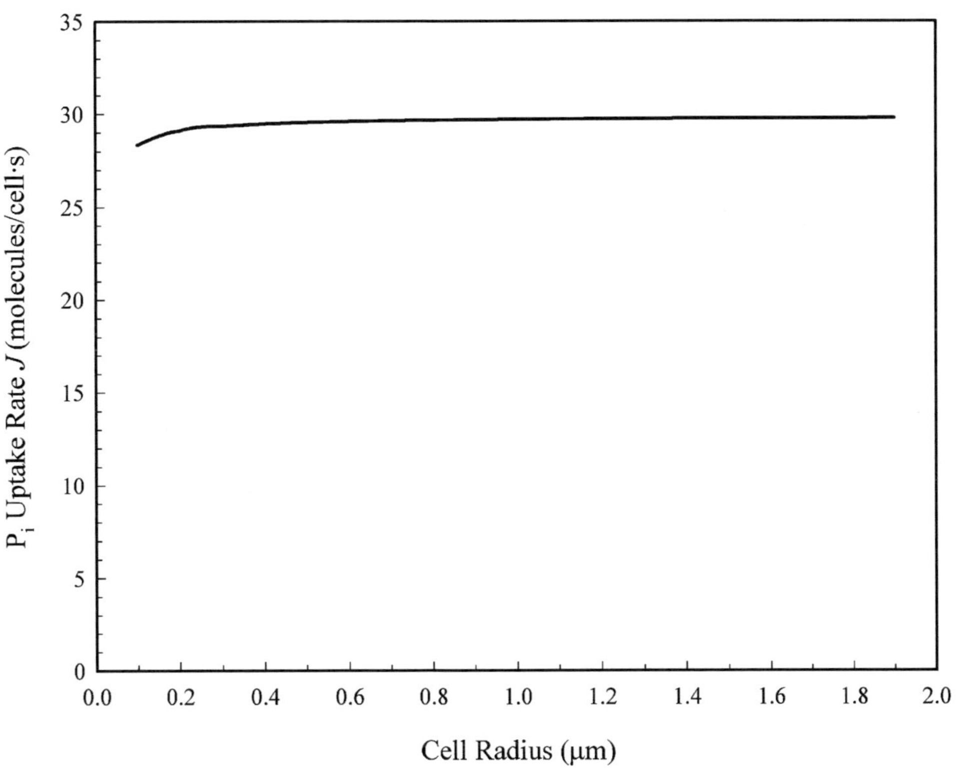

Fig. 7. Effect of cell size on P_i uptake rate by Gram-negative bacteria. P_i uptake rate J calculated using Eqn. 14 and representative P_i transport resistance values for cell envelope compartments listed in Table 2. Note: the numbers of outer membrane Pit system and cytoplasmic membrane Pst system transport protein molecules were conservatively assumed to be constant with cell size; therefore, R_{OM} and R_{CM} terms in Eqn. 14 do not vary with size.

within porous media are found in the literature (Crank, 1975; Middleman, 1998).

Several distinct microbial responses which may be involved in P_i acquisition have been observed in association with microbial adherence/attachment to surfaces, including regulation of extracellular glycoprotein, polysaccharide, and enzyme production (Deretic et al., 1989; Wrangstadh et al., 1990; Dagostino et al., 1991; Giwercman et al., 1991; Davies et al., 1993; Goodman et al., 1993). It has been speculated that microbe-mineral aggregation observed in batch P_i - amorphous Al-oxide bioavailability studies may be involved in release of sorbed P_i (Shang et al., 1996). However, the role and possible mechanisms of exopolysaccharides or other biopolymers in adherence/attachment and subsequent aggregate formation in relation to utilization of mineral-sorbed P_i remain to be elucidated.

6. CONCLUSIONS

Despite the exceptionally strong partitioning of P_i to goethite surfaces, an average of 82 % was bioavailable to a natural soil consortium during the 4-day residence time in chemostat bioreactor studies. Based on extensive experimental and mathematical characterization of the P_i-goethite surface-solution equilibrium in abiotic environments, the release of P_i in our steady-state chemostat experiments appeared clearly related to the presence of a microbial community. This result is also supported by thermodynamic calculations that suggest that sufficient energy was available to these microbes to drive transmembrane P_i transport (i.e., from oxidation of glucose to fuel ATP formation). Further, kinetic calculations of P_i transport from the bulk solution through the cell envelope suggest that the microbes in our chemostat system must actively alter their environment in order to obtain enough P_i to maintain steady chemostat growth. Although, as discussed above, there are a number of possible mechanisms by which this may be accomplished, we have observed formation of microbe-mineral aggregates at low influent glucose concentrations and the production of a competitive ligand (LMWOA) that can enhance P_i desorption at higher influent glucose levels. However, other possible means of microbially-mediated P_i release cannot be excluded based on this study, and the specific roles and interrelationships of these mechanisms remain to be elucidated.

The goethite surface-solution P_i equilibrium which exists in an abiotic setting is disturbed by microbial P_i uptake, and in accordance with Le Châtelier's principle, a different equilibrium condition is established during steady-state chemostat operation. Given the central role of solution concentration in determining the bioavailability of P_i in the two-phase environment of this study, a generalized transport model based on series resistances of the fundamental interfacial and bulk solution transport process can be used to frame microbial acquisition of surface associated nutrients. In combination with steady-state chemostat data, the resistances-in-series model provides a useful tool for examining the presence of and interrelationships among various microbially-mediated nutrient release mechanisms. Ultimately, the generalized resistances in series model may also provide a means to address the broader environmental implications of microbial nutrient acquisition from mineral surfaces on the fate and transport of other mineral surface associated constituents such as heavy metals, agrochemicals, and industrial contaminants.

Acknowledgments—The authors thank Larry Hersman, Jeremy Fein, and an anonymous reviewer for their constructive comments and suggestions.

Editorial handling: R. Hellmann

REFERENCES

Abelson P. H. (1999) A potential phosphate crisis. *Science* **283**, 2015.

Allison D. G. and Goldsbrough M. J. (1994) Polysaccharide production in *Pseudomonas cepacia. J. Basic Microbiol.* **34**, 3-10.

Allison J. D., Brown D. S., and Novo-Gradac K. D. (1991) MINTEQA2/PRODEFA2: A geochemical assessment model for environmental systems [computer program], v. 3.0. DOS. Environmental Research Laboratory, ORD, US EPA, Athens, GA.

Anderson G., Williams E. G., and Moir J. O. (1974) A comparison of the sorption of inorganic orthophosphate and inositol hexaphosphate by six acid soils. *J. Soil Sci.* **25**, 51-62.

Andreux F., Schiavon M., Munier-Lamy C., Mansour M., and Scheunert I. (1995) Factors affecting the movements, reactions, and biotransformations of xenobiotics. In *Environmental Impact of Soil Component Interactions: Natural and Anthropogenic Organics* (eds. P.M. Huang, J. Berthelin, J.-M. Bollag, W.B. McGill, and A.L. Page). CRC Press, Boca Raton. pp. 383-408.

APHA, AWWA, and WEF. (1995) Phosphorus. In *Standard Methods for the Examination of Water and Wastewater*, 19 ed. (eds. A.D. Eaton, L.S. Clesceri, and A.E. Greenberg). American Public Health Association, Washington, D.C. pp. 4-106-4-115.

Atlas R. M. (1995) *Handbook of Media for Environmental Microbiology.* CRC Press, Boca Raton.

Atlas R. M. and Bartha R. (1998) *Microbial Ecology*, 4 ed. Benjamin/Cummings, Menlo Park.

Babu-Khan S., Yeo T. C., Martin W. L., Duron M. R., Rogers R. D., and Goldstein A. H. (1995) Cloning a mineral phosphate-solublizing gene from *Pseudomonas cepacia. Appl. Environ. Microbiol.* **61**, 972-978.

Banik S. and Dey B. K. (1983) Phosphate solubilizing potentiality of the microorganisms capable of utilizing aluminum phosphate as a sole phosphate source. *Zbl. Mikrobiol.* **139**, 17-23.

Barber S. A. (1995) *Soil Nutrient Bioavailability: A Mechanistic Approach*, 2 ed. John Wiley & Sons, New York.

Barker W. W., Welch S. A., and Banfield J. F. (1997) Biogeochemical weathering of silicate minerals. In *Geomicrobiology: Interactions Between Microbes and Minerals* (eds. J.F. Banfield and K.H. Nealson). Reviews in Mineralogy, vol. 35, Mineralogical Society of America, Washington, D.C. pp. 391-428.

Barker W. W., Welch S. A., Chu S., and Banfield J. F. (1998) Experimental observations of the effects of bacteria on aluminosilicate weathering. *Am. Mineralogist* **83**, 1551-1563.

Barrow N. J. (1983) On the reversibility of phosphate sorption by soils. *J. Soil Sci.* **34**, 751-758.

Barrow N. J., Bowden J. W., Posner A. M., and Quirk J. P. (1980) An objective method for fitting models of anion adsorption on variable charge mineral surfaces. *Austral. J. Soil Res.* **18**, 37-47.

Benz R. (1984) Structure and selectivity of porin channels. In *Current Topics in Membranes and Transport Volume 21. Ion Channels: Molecular and Physiological Aspects* (ed. F. Bronner). Academic Press, Orlando. pp. 199-219.

Benz R. and Hancock R. E. W. (1987) Mechanism of ion transport through the anion-selective channel of the *Pseudomonas aeruginosa* outer membrane. *J. Gen. Physiol.* **89**, 275-295.

Benz R., Egli C., and Hancock R. E. W. (1993) Anion transport through the phosphate-specific opr-channel of the *Pseudomonas aeruginosa* outer membrane: effects of phosphate, di- and tribasic anions and of negatively-charged lipids. *Biochim. Biophys. Acta* **1149**, 224-230.

Bohn H. L., McNeal B. L., and O'Connor G. A. (1985) *Soil Chemistry*, 2 ed. Wiley-Interscience, New York.

Bolan N. S., Robson A. D., Barrow N. J., and Aylmore L. A. G. (1987) Effects of vesicular-arbuscular mycorrhiza on the availability of iron phosphates to plants. *Plants and Soil* **99**, 401-410.

Bollag J.-M., Myers C., Pal S., and Huang P. M. (1995) The role of abiotic and biotic catalysts in the transformation of phenolic compounds. In *Environmental Impact of Soil Component Interactions: Natural and Anthropogenic Organics* (eds. P.M. Huang, J. Berthelin, J.-M. Bollag, W.B. McGill, and A.L. Page). CRC Press, Boca Raton. pp. 299-335.

Bowen H. J. M. (1979) *Environmental Chemistry of the Elements.* Academic Press, London.

Brady N. C. and Weil R. R. (1999) Soils and chemical pollution. In *The Nature and Properties of Soils*, 12 ed. Prentice Hall, Upper Saddle River. pp. 723-758.

Bremmer H. and Dennis P. P. (1996) Modulation of chemical composition and other parameters of the cell by growth rate. In *Eschericia coli and Salmonella: Cellular and Molecular Biology*, 2 ed. (eds. F.C. Neidhardt, R. Curtiss III, J.L. Ingraham, C.C.L. Edmund, K.B. Low, B. Magasanik, W.S. Reznikoff, M. Riley, M. Schaechter, and H.E. Umbarger). ASM Press, Washington, D.C. pp. 1553-1569.

Brown G. E. Jr. (1990) Spectroscopic studies of chemisorption reaction mechanisms at oxide-water interfaces. In *Mineral-Water Interface Geochemistry* (eds. M.F. Hochella Jr. and A.F. White). Reviews in Mineralogy, vol. 23, Minerological Society of America, Washington, D.C. pp. 309-363.

Brown M. R. W. and Gilbert P. (1993) Sensitivity of biofilms to antimicrobial agents. In *Microbial Cell Envelopes: Interactions and Biofilms* (eds. L.B. Quesnel, P. Gilbert, and P.S. Handley). Blackwell Scientific, Oxford. pp. 87-97.

Chenu C. (1995) Extracellular polysaccharides: an interface between microorganisms and soil constituents. In *Environmental Impact of Soil Component Interactions: Natural and Anthropogenic Organics* (eds. P.M. Huang, J. Berthelin, J.-M. Bollag, W.B. McGill, and A.L. Page). CRC Press, Boca Raton. pp. 217-233.

Corbridge D. E. C. (1990) *Phosphorous: An Outline of Its Chemistry, Biochemistry, and Technology*, 4 ed. Elsevier Science Publs., Amsterdam.

Cornell R. M. and Schwertmann U. (1996) *The Iron Oxides: Structure, Properties, Reactions, Occurence and Uses.* VCH, Weinheim.

Cox P. A. (1995) *The Elements on Earth: Inorganic Chemistry in the Environment.* Oxford University Press, Oxford.

Crank J. (1975) *The Mathematics of Diffusion*, 2 ed. Clarendon Press, Oxford.

CRC (1992) *Properties of Aqueous Solutions of Electrolytes* (eds. I.D. Zaytsev and G.G. Aseyev). CRC Press, Boca Raton.

Dagostino L., Goodman A. E., and Marshall K. C. (1991) Physiological responses induced in bacteria adhering to surfaces. *Biofouling* **4**, 113-119.

Dalton J. D., Russel G. C., and Sieling D. H. (1952) Effect of organic matter on phosphate availability. *Soil Sci.* **73**, 173-181.

Davies D. G., Chakrabarty A. M., and Geesey G. G. (1993) Exopolysaccharide production in biofilms: substratum activation of alginate gene expression by *Pseudomonas aeruginosa. Appl. Environ. Microbiol.* **59**, 1181-1186.

Davis J. A. and Kent D. B. (1990) Surface complexation modeling in aqueous geochemistry. In *Mineral-Water Interface Geochemistry* (eds. M.F. Hochella Jr. and A.F. White). Reviews in Mineralogy, vol. 23, Minerological Society of America, Washington, D.C. pp. 177-260.

Deretic V., Dikshit R., Konyecsni W. M., Chakrabarty A. M., and Misra T. K. (1989) The *alg*R gene, which regulates mucoidy in *Pseudomonas aeruginosa*, belongs to a class of environmentally responsive genes. *J. Bacteriol.* **171**, 1278-1283.

Donachie W. D. (1968) Relationship between cell size and time of initiation of DNA replication. *Nature* **219**, 1077-1079.

Earl K. D., Syers J. K., and McLaughlin J. R. (1979) Origin of the effects of citrate, tartrate, and acetate on phosphate sorption by soils and synthetic gels. *Soil Sci. Soc. Am. J.* **43**, 674-678.

Ehrlich H. L. (1996) *Geomicrobiology*, 3rd ed. Marcel Dekker, New York.

Fendorf S., Eick M. J., Grossl P., and Sparks D. L. (1997) Arsenate and chromate retention mechanisms on goethite. 1. Surface structure. *Environ. Sci. Technol.* **331**, 315-320.

Foster J. W. and Spector M. P. (1986) Phosphate starvation regulon of *Salmonella typhimurium. J. Bacteriol.* **166**, 666-669.

Friedberg I. (1977) Phosphate transport in *Micrococcus lysodeikticus. Biochim. Biophys. Acta* **466**, 451-460.

Frossard E., Brossard M., Hedley M. J., and Metherell A. (1995) Reactions controlling the cycling of P in soils. In *Phosphorous in the Global Environment: Transfers,*

Cycles, and Management (ed. H. Tiessen). John Wiley & Sons, Chichester. pp. 107-137.

Furrer G. and Stumm W. (1986) The coordination chemistry of weathering: I. Dissolution kinetics of δ-Al_2O_3 and BeO. *Geochim. Cosmochim. Acta* **50**, 1847-1860.

Futai M. and Tsuchiya T. (1987) Proton transport and proton-motive force in prokaryotic cells. In *Ion Transport in Prokaryotes* (eds. B.P. Rosen and S. Silver). Academic Press, San Diego. pp. 3-83.

Geelhoed J. S., Hiemstra T., and Van Riemsdijk W. H. (1998) Competitive interaction between phosphate and citrate on goethite. *Environ. Sci. Technol.* **32**, 2119-2123.

Gibbs M. M. (1978) A simple method for the rapid determination of iron in natural waters. *Wat. Res.* **13**, 295-297.

Giwercman B., Jensen E. T., Hoiby N., Kharazmi A., and Costerton J. W. (1991) Induction of β-lactamase production in *Pseudomonas aeruginosa* biofilms. *Antimicrob. Agents Chemother.* **35**, 1008-1010.

Goldberg S. and Sposito G. (1985) On the mechanism of specific phosphate adsorption by hydroxylated mineral surfaces: A review. *Commun. In Soil Sci. Plant Anal.* **16**, 801-821.

Goldstein A. H. (1986) Bacterial solublization of mineral phosphates: historical perspectives and future prospects. *Am. J. Altern. Ag.* **1**, 51-57.

Goldstein A. H. (1987) Molecular cloning and regulation of a mineral phosphate solublizing gene from *Erwina herbicola. Biotechnology* **5**, 72-74.

Goldstein A. H. (1994) Involvement of the quinoprotein glucose dehydrogenase in the solublization of exogenous mineral phosphates by Gram-negative bacteria. In *Phosphate in Microorganisms: Cellular and Molecular Biology* (eds. A. Torriani-Gorini, E. Yagil, and S. Silver). ASM Press, Washington, D.C. pp. 197-203.

Goodman A. E., Angles M. L., and Marshall K. C. (1993) Genetic response of bacteria in biofilms. In *Trends in Microbial Ecology* (eds. R. Guerrero and C. Pedrós-Alío). Spanish Society for Microbiology, Barcelona. pp. 119-122.

Guddal P. H., Johanson T., Schulstad K., and Little C. (1989) Apparant phosphate retrieval system in *Bacillus cereus. J. Bacteriol.* **171**, 5702-5706.

Hancock R. E. W., Poole K., and Benz R. (1982) Outer membrane protein P of *Pseudomonas aeruginosa*: regulation by phosphate deficiency and formation of small anion-specific channels in lipid bilayer membranes. *J. Bacteriol.* **150**, 730-738.

Hancock R. E., Worobec E. A., Poole K., and Benz R. (1987) Phosphate-binding site of *Pseudomonas aeruginosa* outer membrane protein P. In *Phosphate Metabolism and Cellular Regulation in Microorganisms* (eds. A. Torriani-Gorini, F.G. Rothman, S. Silver, A. Wright, and E. Yagil). American Society for Microbiology, Washington,D.C. pp. 186-190.

Harrison A. F. (1987) *Soil Organic Phosphorous. A Review of World Literature.* CAB International, Oxon.

Hawke D., Carpenter P. D., and Hunter K. A. (1989) Competitive adsorption of phosphate on goethite in marine electrolytes. *Environ. Sci. Technol.* **23**, 187-191.

Hayes K. F., Papelis C., and Leckie J. O. (1988) Modeling ionic strength effects on anion adsorption at hydrous oxide/solution interfaces. *J. Colloid Interface Sci.* **125**, 717-726.

Haynes R. J. (1982) Effects of liming on phosphate availability in acid soils. *Plant and Soil* **68**, 289-308.

He Z. and Zhu J. (1998) Microbial utilization and transformation of phosphate adsorbed by variable charge minerals. *Soil Biol. Biochem.* **30**, 917-923.

Hersman L. E., Lloyd T., and Sposito G. (1995) Siderophore-promoted dissolution of hematite. *Geochim. Cosmochim. Acta* **59**, 3327-3330.

Hersman L. E., Maurice P. A., and Sposito G. (1996) Iron acquisition from hydrous Fe(III)-oxides by an aerobic *Pseudomonas* sp. *Chem. Geol.* **132**, 25-31.

Hingston F. J., Atkinson R. J., Posner A. M., and Quirk J. P. (1968) Specific adsorption of anions on goethite. *Trans. 9th. Int. Cong. Soil Sci.* **1**, 669-678.

Hingston F. J., Posner A. M., and Quirk J. P. (1972) Anion adsorption by goethite and gibbsite: I. The role of the proton in determining adsorption envelopes. *J. Soil Sci.* **23**, 177-192.

Hingston F. J., Posner A. M., and Quirk J. P. (1974) Anion adsorption by goethite and gibbsite II. Desorption of anions from hydrous oxide surfaces. *J. Soil Sci.* **25**, 16-26.

Holmén B. A. and Casey W. H. (1996) Hydroxamate ligands, surface chemistry, and the mechanism of ligand-promoted dissolution of goethite [α-FeOOH(s)]. *Geochim. Cosmochim. Acta* **60**, 4403-4416.

Hue N. V. (1991) Effects of organic acids/anions on P sorption and phytoavailability in soils with different mineralogies. *Soil Sci.* **152**, 463-471.

Illmer P. and Schinner F. (1992) Solublization of inorganic phosphates by microorganisms isolated from forest soils. *Soil Biol. Biochem.* **24**, 389-395.

Iyamuremye F. and Dick R. P. (1996) Organic amendments and phosphorous sorption by soils. In *Advances in Agronomy* (ed. D.L. Sparks). Academic Press, San Diego. pp. 139-185.

Jap B. K. and Walian P. J. (1990) Biophysics of the structure and function of porins. *Quart. Rev. Biophys.* **23**, 367-403.

Jayachandran K., Schwab A. P., and Hetrick B. A. D. (1989) Mycorrhizal mediation of phosphorous availability: synthetic iron chelate effects on phosphorous solubilization. *Soil Sci. Soc. Am. J.* **53**, 1701-1706.

Kashket E. R. (1985) The proton motive force in bacteria: a critical assessment of methods. *Ann. Rev. Microbiol.* **39**, 219-242.

Koch A. L. (1990) Diffusion. The crucial process in many aspects of the biology of bacteria. In *Advances in Microbial Ecology. Vol. 11* (ed. K.C. Marshall). Plenum Press, New York. pp. 37-70.

Koch A. L. and Wang C. H. (1982) How close to the theoretical diffusion limit do bacterial uptake systems function? *Arch. Microbiol.* **131**, 35-42.

Kreusch A., Neubüser A., Schlitz E., Weckesser J., and Schulz G. E. (1994) Structure of the membrane channel porin from *Rhodopseudomonas blastica* at 2 Å resolution. *Protein Sci.* **3**, 58-63.

Krueger R. D., Harper S. H., Campbell J. W., and Fahrney D. E. (1986) Kinetics of phosphate uptake, growth, and accumulation of cyclic diphosphoglycerate in a phosphate-limited continuous culture of *Methanobacterium thermoautotrophicum. J. Bacteriol.* **167**, 49-56.

Kummert R. and Stumm W. (1980) The surface complexation of organic acids on hydrous γ-Al_2O_3. *J. Colloid Interface Sci.* **75**, 373-385.

Kung K. S. and McBride M. B. (1991) Bonding of chlorophenols on iron and aluminum oxides. *Environ. Sci. Technol.* **25**, 702-709.

Lacoste A.-M., Cassaigne A., and Neuzil E. (1981) Transport of inorganic phosphate in *Pseudomonas Aeruginosa. Curr. Microbiol.* **6**, 115-120.

Lajtha K. and Harrison A. F. (1995) Strategies of phosphorous acquisition and conservation by plant species and communities. In *Phosphorous in the Global Environment: Transfers, Cycles, and Management* (ed. H. Tiessen). John Wiley & Sons, Chichester. pp. 139-147.

Lindsay W. L. (1979) Phosphates. In *Chemical Equilibria in Soils.* John Wiley & Sons, New York. pp. 162-209.

Liu W. and Hulett M. (1997) *Bacillus subtilis* PhoP binds to the *phoB* tandem promoter exclusively within the phosphate starvation-inducible promoter. *J. Bacteriol.* **179**, 6302-6310.

Lonergan D. J. and Lovley D. R. (1991) Microbial oxidation of natural and anthropogenic aromatic compounds coupled to Fe(III) reduction. In *Organic Substances and Sediments in Water: Humics and Soil* (ed. R.A. Baker). Lewis Publishers, Inc., Chelsea. pp. 327-338.

Martin R. R. and Smart R. St. C. (1987) X-ray photoelectron studies of anion adsorption on goethite. *Soil Sci. Soc. Am. J.* **51**, 54-56.

Martin R. R., Smart R. St. C., and Tazaki K. (1988) Direct observation of phosphate precipitation in the goethite/phosphate system. *Soil Sci. Soc. Am. J.* **52**, 1492-1500.

McBride M. B. (1994) Chemisorption and precipitation of inorganic ions. In *Environmental Chemistry of Soils.* Oxford University Press, New York. pp. 121-168.

Meyer J.-M. and Stintzi A. (1998) Iron metabolism and siderophores in *Pseudomonas* and related species. In *Pseudomonas* (ed. T.C. Montie). Plenum Press, New York. pp. 201-243.

Middleman S. (1998) *An Introduction to Heat and Mass Transfer: Principles of Analysis and Design.* John Wiley & Sons, New York.

Mott C. J. B. (1981) Anion and ligand exchange. In *Chemistry of Soil Processes* (eds. D.J. Greenland and M.H.B. Hayes). John Wiley & Sons, Inc., New York. pp. 179-219.

Nealson K. H. and Stahl D. A. (1997) Microorganisms and biogeochemical cycles: what can we learn from layered microbial communities. In *Geomicrobiology: Interactions Between Microbes and Minerals* (eds. J.F. Banfield and K.H. Nealson). Reviews in Mineralogy, vol. 35, Mineralogical Society of America, Washington, D.C. pp. 5-34.

Neidhardt F. C., Ingraham J. L., and Schaechter M. (1990) *Physiology of the Bacterial Cell: A Molecular Approach.* Sinauer Associates, Sunderland.

Neidhardt F. C. and Umbarger H. E. (1996) Chemical composition of *Eschericia coli.* In *Eschericia coli and Salmonella: Cellular and Molecular Biology*, 2 ed. (eds. F.C. Neidhardt, R. Curtiss III, J.L. Ingraham, C.C.L. Edmund, K.B. Low, B. Magasanik, W.S. Reznikoff, M. Riley, M. Schaechter, and H.E. Umbarger). ASM Press, Washington, D.C. pp. 13-16.

Nikaido H. (1996) Outer membrane. In *Eschericia coli and Salmonella: Cellular and Molecular Biology*, 2 ed., (eds. F.C. Neidhardt, R. Curtiss III, J.L. Ingraham, C.C.L. Edmund, K.B. Low, B. Magasanik, W.S. Reznikoff, M. Riley, M. Schaechter, and H.E. Umbarger). ASM Press, Washington, D.C. pp. 29-47.

Nikaido H. and Rosenberg E. Y. (1983) Porin channels in *Escherica coli*: studies with liposomes reconstituted from purified proteins. *J. Bacteriol.* **153**, 241-252.

Parfitt R. L., Atkinson R. J., and Smart R. St. C. (1975) The mechanism of phosphate fixation by iron oxides. *Soil Sci. Am. Proc.* **39**, 837-841.

Parkinson J. S. and Kofoid E. C. (1992) Communication modules in bacterial signaling proteins. *Ann. Rev. Genet.* **26** , 71-112.

Paul E. A. and Clark F. E. (1996) *Soil Microbiology and Biochemistry*, 2 ed. Academic Press, San Diego.

Pierzynski G. M., Sims J. T., and Vance G. F. (1994) *Soils and Environmental Quality.* Lewis Publ., Boca Raton.

Poole K. and Hancock R. E. W. (1984) Phosphate transport in *Pseudomonas aeruginosa.* involvement of a periplasmic phosphate-binding protein. *Eur. J. Biochem.* **144**, 607-612.

Qi Y., Kobayashi Y., and Hulett F. M. (1997) The *pst* operon of *Bacillus subtilis* has a phosphate-regulated promoter and is involved in phosphate transport but not in regulation of the pho regulon. *J. Bacteriol.* **179**, 2534-2539.

Rao P. S. C., Bellin C. A., and Lee L. S. (1995) Sorption and biodegradation of organic contaminants in soils: conceptual representations of process coupling. In *Environmental Impact of Soil Component Interactions: Natural and Anthropogenic Organics* (eds. P.M. Huang, J. Berthelin, J.-M. Bollag, W.B. McGill, and A.L. Page). CRC Press, Boca Raton. pp. 263-274.

Ristori G. G. and Fusi P. (1995) Adsorption mechanisms and abiotic catalytic transformations of some agrochemicals by clay minerals. In *Environmental Impact of Soil Component Interactions: Natural and Anthropogenic Organics* (eds. P.M. Huang, J. Berthelin, J.-M. Bollag, W.B. McGill, and A.L. Page). CRC Press, Boca Raton. pp. 337-343.

Robert M. and Berthelin J. (1986) Role of biological and biochemical factors in soil mineral weathering. In *Interactions of Soil Minerals with Natural Organics and Microbes* (eds. P.M. Huang and M. Schnitzer). Soil Science Society of America, Madison. pp. 453-495.

Robinson N. J., Procter C. M., and Guerinot M. L. (1999) A ferric-chelate reductase for iron uptake from soils. *Nature* **397**, 694.

Rogers J. R., Bennett P. C., and Choi W. J. (1998) Feldspar as a source of nutrients for microorganisms. *Am. Miner.* **83**, 1532-1540.

Ronson C. W., Nixon B. T., and Ausbel F. M. (1987) Conserved domains in bacterial regulatory proteins that respond to environmental stimuli. *Cell* **49**, 579-581.

Rosenberg H. (1987) Phosphate transport in prokaryotes. In *Ion Transport in Prokaryotes* (eds. B.P. Rosen and S. Silver). Academic Press, San Diego. pp. 205-248.

Rosenberg H., Gerdes R. G., and Chegwidden K. (1977) Two systems for the uptake of phosphate in *Eschericia coli. J. Bacteriol.* **131**, 505-511.

Runge-Metzger A. (1995) Closing the cycle: obstacles to efficient P management for improved global food security. In *Phosphorous in the Global Environment: Transfers, Cycles, and Management* (ed. H. Tiessen). John Wiley & Sons, Chichester. pp. 27-42.

Ryden J. C., McLaughlin J. R., and Syers J. K. (1977) Mechanisms of phosphate sorption by soils and hydrous ferric oxide gel. *J. Soil Sci.* **8**, 72-92.

Sanyal S. K. and De Datta S. K. (1991) Chemistry of phosphorous transformations in soils. In *Advances in Soil Science* (ed. B.A. Stewart). Springer-Verlag, New York. pp. 1-120.

Scanlan D. J., Silman N. J., Donald K. M., Wilson W. H., Carr N. G., Joint I., and Mann N. H. (1997) An immunological approach to detect phosphate stress in populations and single cells of photosynthetic picoplankton. *Appl. Environ. Microbiol.* **63**, 2411-2420.

Schaecter M., Maaløe O., and Kjeldgaard N. O. (1958) Dependency on medium and temperature of cell size and composition during balanced growth of *Salmonella typhimurium. J. Gen. Microbiol.* **19**, 592-606.

Schwertmann U. and Cornell R. M. (1988) Some properties of soil and synthetic iron oxides. In *Iron in Soils and Clay Minerals* (eds. J.W. Stucki, B.A. Goodman, and U. Schwertmann). Reidel Pub. Co., Dordrecht. pp. 203-244.

Schwertmann U. and Cornell R. M. (1991) Goethite. In *Iron Oxides in the Laboratory: Preparation and Characterization.* VCH, Weinheim. pp. 61-80.

Schwertmann U. and Taylor R. M. (1990) Iron oxides. In *Minerals in Soil Environments* (ed. D.A. Weed). Soil Science Society of America, Madison. pp. 379-439.

Schwertmann U., Kodama H., and Fischer W. R. (1986) Mutual interactions between organics and iron oxides. In *Interactions of Soil Minerals with Natural Organics and Microbes* (eds. P.M. Huang and M. Schnitzer). Soil Science Society of America, Madison. pp. 223-250.

Senesi N. and Loffredo E. (1999) The chemistry of soil organic matter. In *Soil Physical Chemistry*, 2 ed. (ed. D.L. Sparks). CRC Press, Boca Raton. pp. 239-370.

Shang C., Caldwell D. E., Stewart J. W. B., Tiessen H., and Huang P. M. (1996) Bioavailability of organic and inorganic phosphates adsorbed on short-range ordered aluminum precipitate. *Microb. Ecol.* **31**, 29-39.

Shulman R. G., Brown T. R., Ugurbil K., Ogawa S., Cohen S. M., and den Hollander J. A. (1979) Cellular applications of ^{31}P and ^{13}C nuclear magnetic resonance. *Science* **205**, 160-166.

Shvets V. I., Stepanov A. E., Schmitt L., Spiess B., and Schlewer G. (1991) Synthesis and complexation properties of inositol-phosphates. In *Inositol Phosphates and Derivatives: Synthesis, Biochemistry, and Theraputic Potential* (ed. A.B. Reitz). American Chemical Society, Washington, D.C. pp. 155-171.

Sibanda H. M. and Young S. D. (1986) Competitive adsorption of humus acids and phosphate on goethite, gibbsite, and two tropical soils. *J. Soil Sci.* **32**, 67-74.

Sigg L. and Stumm W. (1981) The interactions of anions and weak acids with the hydrous goethite (α-FeOOH) surface. *Colloids and Surfaces* **2**, 101-117.

Smith H. L. and Waltman P. (1995) *The Theory of the Chemostat: Dynamics of Microbial Competition.* Cambridge University Press, Cambridge.

Stevenson F. J. (1994) *Humus Chemistry: Genesis, Composition, and Reactions*, 2 ed. John Wiley & Sons, New York.

Stewart J. W. B. and McKercher R. B. (1982) Phosphorous cycle. In *Experimental Microbial Ecology* (eds. R.G. Burns and J.H. Slater). Blackwell, Oxford. pp. 229-238.

Stock J. B., Stock A. M., and Mottonen J. M. (1990) Signal transduction in bacteria. *Nature* **344**, 395-400.

Stone A. T. (1997) Reactions of extracellular organic ligands with dissolved metal ions and mineral surfaces. In *Geomicrobiology: Interactions Between Microbes and Minerals* (eds. J.F. Banfield and K.H. Nealson). Reviews in Mineralogy, vol. 35, Mineralogical Society of America, Washington, D.C. pp. 309-344.

Stone A. T. and Torrents A. (1995) The role of dissolved metals and metal-containing surfaces in catalyzing the hydrolysis of organic compounds. In *Environmental Impact of Soil Component Interactions: Natural and Anthropogenic Organics* (eds. P.M. Huang, J. Berthelin, J.-M. Bollag, W.B. McGill, and A.L. Page). CRC Press, Boca Raton. pp. 275-298.

Stotzky G. (1986) Influence of soil mineral colloids on metabolic processes, growth, adhesion, and ecology of microbes and viruses. In *Interactions of Soil Minerals with Natural Organics and Microbes* (eds. P.M. Huang and M. Schnitzer). Soil Science Society of America, Madison. pp. 305-428.

Stumm W. and Morgan J. J. (1981) *Aquatic Chemistry*, 2 ed. John Wiley & Sons, New York.

Sugawara E. and Nikaido H. (1992) Pore-forming activity of OmpA protein of *Eschericia coli. J. Biol. Chem.* **267**, 2507-2511.

Sun X. and Doner H. E. (1996) An investigation of arsenate and arsenite bonding structures on goethite by FTIR. *Soil Sci.* **161**, 865-872.

Suter D., Banwart S., and Stumm W. (1991) Dissolution of hydrous iron(III) oxides by reductive mechanisms. *Langmuir* **7**, 809-813.

Sutherland I. W. (1991) *Biotechnology of Microbial Exopolysaccharides.* Cambridge University Press, Cambridge.

Tate R. L. I. (1995) The sulfur and related geochemical cycles. In *Soil Microbiology.* John Wiley & Sons, New York. pp. 359-387.

Taylor S. R. and McLennan S. M. (1985) *The Continental Crust: Its Composition and Evolution.* Blackwell Scientific Publ., Oxford.

Tejedor-Tejedor M. I. and Anderson M. A. (1986) "In situ" attenuated total reflection Fourier transform infrared studies of the goethite (α-FeOOH)-aqueous solution interface. *Langmuir* **2**, 203-210.

Tempest D. W. and Neijssel O. M. (1978) Eco-physiological aspects of microbial growth in aerobic nutrient limited environments. In *Advances in Microbial Ecology* (ed. M. Alexander). Plenum Press, New York. pp. 105-153.

Tempest D. W. and Neijssel O. M. (1992) Physiological and energetic aspects of bacterial metabolite overproduction. *FEMS Microbiol. Lett.* **100**, 169-176.

Torriani-Gorini A. (1994) Introduction: The Pho regulon of *Escherichia coli.* In *Phosphate in Microorganisms: Cellular and Molecular Biology* (eds. A. Torriani-Gorini, E. Yagil, and S. Silver). ASM Press, Washington, D.C. pp. 1-4.

Van Der Ley P., Bekkers A., Van Meersbergen J., and Tommassen J. (1987) A comparative study on the *phoE* genes of three enterobacterial species: implications for structure-function relationships in a pore-forming protein of the outer membrane. *Eur. J. Biochem.* **164**, 469-475.

Van Der Zee S. E. A. T. M. and Van Riemsdijk W. H. (1991) Model for the reaction kinetics of phosphate with oxides and soil. In *Interactions at the Soil Colloid-Soil Solution Interface* (ed. G.H. Bolt). Kluwer, Dordrecht. pp. 205-239.

Van Veen H. W., Abee T., Kortstee J. J., Konings W. N., and Zehnder A. J. B. (1993) Characterization of two phosphate transport systems in *Acinetobacter johnsonii* 210A. *J. Bacteriol.* **175**, 200-206.

Van Veen H. W., Abee T., Kortstee J. J., Konings W. N., and Zehnder A. J. B. (1994) Phosphate inorganic transport (Pit) system in *Escherichia coli* and *Acinetobacter johnsonii.* In *Phosphate in Microorganisms: Cellular and Molecular Biology* (eds. A. Torriani-Gorini, E. Yagil, and S. Silver). ASM Press, Washington, D.C. pp. 43-49.

Veldkamp H. (1977) Ecological studies with the chemostat. In *Advances in Microbial Ecology,* Vol. 1 (ed. M. Alexander). Plenum Press, New York. pp. 59-94.

Violante A. and Gianfreda L. (1993) Competition in adsorption between phosphate and oxalate on an aluminum hydroxide montmorillonite complex. *Soil Sci. Soc. Am. J.* **57**, 1235-1241.

Voegele R. T., Bardin S., and Finan T. M. (1997) Characterization of the *Rhizobium* (*Sinorhizobium*) *meliloti* high- and low-affinity phosphate uptake systems. *J. Bacteriol.* **179**, 7226-7232.

Voet D. and Voet J. G. (1995) *Biochemistry*, 2 ed. John Wiley & Sons, Inc., New York.
Wanner B. L. (1994) Multiple controls of the *Escherichia coli* Pho regulon by the P_i sensor PhoR, the catabolite regulatory sensor CreC, and acetyl phosphate. In *Phosphate in Microorganisms: Cellular and Molecular Biology* (eds. A. Torriani-Gorini, E. Yagil, and S. Silver). ASM Press, Washington, D.C. pp. 13-21.
Wanner B. L. (1995) Signal transduction and cross regulation in the *Escherichia coli* phosphate regulon by PhoR, CreC, and acetyl phosphate. In *Two-Component Signal Transduction* (eds. J.A. Hoch and T.J. Silhavy). ASM Press, Washington, D.C. pp. 203-221.
Wanner B. L. (1996) Phosphorous assimilation and control of the phosphate regulon. In *Eschericia coli and Salmonella: Cellular and Molecular Biology*, 2 ed. (eds. F.C. Neidhardt, R. Curtiss III, J.L. Ingraham, C.C.L. Edmund, K.B. Low, B. Magasanik, W.S. Reznikoff, M. Riley, M. Schaechter, and H.E. Umbarger). ASM Press, Washington, D.C. pp. 1357-1381.
Wanner B. L., Jiang W., Kim S., Yamagata S., Haldimann A., and Daniels L. L. (1996) Are the multiple signal transduction pathways of the Pho regulon due to cross talk or cross regulation? In *Regulation of Gene Expression in Escherichia coli* (eds. E.C.C. Lin and A.S. Lynch). Chapman & Hall, New York. pp. 297-315.
Webb D. C. and Cox G. B. (1994) Proposed mechanism for phosphate translocation by the phosphate-specific transport (Pst) system and the role of the Pst system in phosphate regulation. In *Phosphate in Microorganisms: Cellular and Molecular Biology* (eds. A. Torriani-Gorini, E. Yagil, and S. Silver). ASM Press, Washington, D.C. pp. 37-42.
Weiss M. S. and Schulz G. E. (1992) Structure of porin refined at 1.8 Å resolution. *J. Mol. Biol.* **227**, 493-509.
White D. (1995) Growth. In *The Physiology and Biochemistry of Prokaryotes*. Oxford University Press, Oxford. pp. 34-48.
Willett I. R. (1982) Phosphorous availability in soils subjected to short periods of flooding and drying. *Aust. J. Soil Res.* **20**, 131-138.
Willett I. R. (1985) The reductive dissolution of phosphated ferrihydrate and strengite. *Aust. J. Soil Res.* **23**, 237-244.
Willett I. R. (1989) Causes and prediction of changes in extractable phosphorous during flooding. *Aust. J. Soil Res.* **27**, 45-54.
Willsky G. R. and Malamy M. H. (1980a) Characterization of two genetically separable inorganic phosphate transport systems in *Eschericia coli*. *J. Bacteriol.* **144**, 356-365.
Willsky G. R. and Malamy M. H. (1980b) Effect of arsenate on inorganic phosphate transport in *Eschericia coli*. *J. Bacteriol.* **144**, 366-374.
Wrangstadh M., Szewzyk U., Östling J., and Kjelleberg S. (1990) Starvation-specific formation of a peripheral exopolysaccharide by a marine *Pseudomonas* sp., strain S9. *Appl. Environ. Microbiol.* **56**, 2065-2072.
Wubolts M. G. and Witholt B. (1998) Selected industrial biotransformations. In *Pseudomonas* (ed. T.C. Montie). Plenum Press, New York. pp. 271-329.
Yates D. E. and Healy T. W. (1975) Mechanism of anion adsorption at the ferric and chromic oxide/water interface. *J. Colloid Interface Sci.* **52**, 222-228.
Yoshimura F., Zalman L. S., and Nikaido H. (1983) Purification and properties of *Pseudomonas aeruginosa* porin. *J. Biol. Chem.* **258**, 2308-2314.
Zinder B., Furrer G., and Stumm W. (1986) The coordination chemistry of weathering: II. Dissolution of Fe(III) oxides. *Geochim. Cosmochim. Acta* **50**, 1861-1869.

APPENDIX

A.1. Chemostat (Continuous) Growth

A microbial culture can be maintained in an unvarying physiological condition by operating as an open system in which nutrient input is constant with time and system volume is held constant by balancing the system influent and effluent flows. Under these conditions, the concentrations of all chemical and biological constituents in the culture are constant with time and the system is said to be operating at steady state (White, 1995). The dilution rate D^* (h^{-1}) of the culture due to constant inflow is given by:

$$D^* = \frac{F}{V} \qquad \text{(A.1.1)}$$

where F ($mL{\cdot}h^{-1}$) is the influent flow rate and V (mL) is the system volume. At steady state microbial growth rate is controlled by the dilution rate. Equating the expressions for microbial growth rate and rate of cell loss due to washout in the system effluent at steady state, given by:

$$\frac{dx}{dt} = \mu x \qquad \text{(growth)} \qquad \text{(A.1.2)}$$

and

$$\frac{dx}{dt} = D^* x \qquad \text{(washout)} \qquad \text{(A.1.3)}$$

where x is a measure of microbial population and μ (h^{-1}) is the specific growth rate constant, results in:

$$\mu = D^* \qquad \text{(A.1.4)}$$

A.2. *Physicochemical and Kinetic Properties of Bacterial P_i Transport Proteins*

Table A.2.1. Physicochemical properties of transport related outer membrane proteins (porins) of several Gram-negative bacterial species.

Protein	Organism	Pore Diameter (nm)	Reference
	Non-specific Outer Membrane Transport Proteins		
Omp A	*Escherichia coli*	1.1-1.2	(Sugawara and Nikaido, 1992)
Omp C	*Escherichia coli*	1.08	(Nikaido and Rosenberg, 1983)
Omp F	*Escherichia coli*	1.16	(Nikaido and Rosenberg, 1983)
Orp F	*Pseudomonas aeruginosa*	2.0	(Yoshimura et al., 1983)
Porin	*Rhodobacter capsulatus*	0.8-0.9	(Weiss and Schulz, 1992)
Porin	*Rhodopseudomonas blastica*	1.2	(Kreusch et al., 1994)
	Outer Membrane Transport Proteins Induced by P_i Limitation		
Pho E	*Escherichia coli*	1.0-1.2	(Benz, 1984; Jap and Walian, 1990)
	Salmonella typhimurium	1.4	(Benz, 1984)
Orp P	*Pseudomonas aeruginosa*	0.5-0.7	(Hancock et al., 1982; Benz and Hancock, 1987; Benz et al., 1993)

Table A.2.2. Kinetic parameters for energy dependent bacterial cytoplasmic membrane P_i transport systems.

Bacterial Species	V_{max} (nmol/mg DW·min)†	K_M (μM)	Reference
	Low Affinity Pit System Proteins		
Eschericia coli	55	38.2	(Willsky and Malamy, 1980a)
	60	25	(Rosenberg et al., 1977)
Pseudomonas aeruginosa	6.1‡	19.3	(Poole and Hancock, 1984)
	22.2	10	(Lacoste et al., 1981)
	High Affinity Pst System Proteins		
Eschericia coli	15.9	0.43	(Willsky and Malamy, 1980a)
	10.9	0.6	(Willsky and Malamy, 1980b)
	69	0.16	(Rosenberg et al., 1977)
Pseudomonas aeruginosa	3.0‡	0.46	(Poole and Hancock, 1984)
	12.5	1.1	(Lacoste et al., 1981)
Micrococcus lysodeikticus	73‡	4.3	(Friedberg, 1977)
Methanobacterium thermoaototrophicum	58	0.025	(Krueger et al., 1986)

†DW is dry weight of cell constituents

‡Based on 0.55 mg protein / mg DW (Neidhardt et al., 1990)

V_{max} is the maximum reaction rate when the enzyme is saturated by substrate

K_M is the Michaelis-Menton half-saturation constant (substrate concentration at 0.5 V_{max})

Water-Rock Interactions, Ore Deposits, and Environmental Geochemistry: A Tribute to David A. Crerar
© The Geochemical Society, Special Publication No. 7, 2002
Editors: Roland Hellmann and Scott A. Wood

The effects of ternary surface complexes on the adsorption of metal cations and organic acids onto mineral surfaces

JEREMY B. FEIN

University of Notre Dame, Civil Engineering and Geological Sciences, 156 Fitzpatrick Hall, Notre Dame, IN 46556 USA
e-mail: fein.1@nd.edu

Abstract—Ternary surface complexation, which can be viewed as the co-adsorption of metal cations and organic molecules onto the same mineral surface site, can dramatically enhance both metal cation and organic molecule adsorption under certain pH conditions. In 1990, Schindler (1990) reviewed the experimental studies of ternary complex formation. At that time, only a handful of studies had been conducted, and the only measured thermodynamic stability constants were for Type A (mineral-metal-ligand) complexes. Since 1990, a number of studies have examined both Type A and Type B (mineral-ligand-metal) ternary surface complexation, and in this review I describe the advancement in understanding that has occurred over the past decade. We now have stability constant measurements for both Type A and Type B complexes, but the databases of both are still quite limited. The experimental studies of the past 10 years have reinforced the observation that ternary surface complexation can, under some conditions, control metal and organic mobilities in water-rock systems. In addition, enough data exist to construct predictive techniques for estimating stability constants for ternary surface complexes that have not been directly studied in the laboratory. For Type B complexes, a linear correlation exists between the log stability constant of the ternary surface complex and the sum of the log stability constants of the metal-organic aqueous complex and the organic-surface complex. This relationship can be used to predict whether a mineral-organic-metal surface set is likely to form a ternary surface complex, and it can be used to estimate the extent of adsorption that occurs in such a system. I use this approach to estimate the stability of a goethite-citrate-Co(II) ternary surface complex, demonstrating that the ternary complex can markedly enhance Co(II) adsorption onto goethite over a wide range of pH conditions. Despite having enough data to build predictive models, the accuracy of these models depends strongly on the acquisition of additional data that constrain the thermodynamic stabilities of ternary surface complexes.

1. INTRODUCTION

Metal cations and organic acid ligands co-exist in solution in a wide range of both natural and contaminated water-rock systems. Over the past twenty years, experimental studies have demonstrated that metal cations and organic acid ligands each adsorb to mineral surfaces individually. Furthermore, when present together in contact with mineral surfaces, metal cations and organic ligands can form ternary surface complexes, thereby influencing the adsorptive properties of each solute (e.g., Bowers and Huang, 1986). Prior to 1990, only a small number of studies had documented the formation of ternary metal-organic surface complexes, and even fewer had quantified their thermodynamic stabilities in a surface complexation framework. Schindler (1990) reviewed these early studies, but since 1990 an increasing awareness of metal-organic co-contamination in the subsurface, and an understanding of the influence of natural organic matter on metal cation adsorption, has focussed considerable attention on co-adsorption. The objective of this paper is to review the studies that have been conducted since Schindler's (1990) review, to summarize our current understanding of the stabilities and importance of metal-organic ternary surface complexes, and also to outline what needs to be better understood in order to accurately quantify solute adsorption in metal- and organic-bearing aqueous systems.

2. CHARACTERIZING THE MINERAL SURFACE

Adsorption of solutes onto mineral surfaces can be modeled either with partitioning approaches or with a surface complexation approach. Partitioning approaches describe adsorption in terms of relative amounts in solution and on the mineral surface under a specific set of conditions. Partition coefficients are applicable only to the conditions at which they were determined. Conversely, surface complexation models, which apply the formalism of aqueous ion association reactions to solute adsorption reactions with surfaces, require a detailed understanding not only of the surfaces involved, but also of the adsorption/desorption mechanisms. A surface complexation model treats the adsorbed solute as a distinct thermodynamic entity whose stability can be quantified with an equilibrium constant. The equilibrium constants which describe the extent of adsorption in surface complexation models are *invariant* with respect to most of the parameters which affect partition coefficients. Therefore, one can determine equilibrium constants in systems that isolate specific adsorption reactions, and model complex systems by combining these results into a computational model. Because of the additional components that must be understood and measured, surface complexation modeling is more difficult to apply to adsorption processes than bulk partition modeling. However, the added insight gained from a thorough understanding of the adsorption/desorption mechanisms enables estimation of the extent of adsorption/desorption that occurs under conditions other than those studied in the laboratory.

The pH-dependent surface charge on metal oxides and aluminosilicates can be quantified according to surface complexation theory by the following types of reactions (Hohl and Stumm, 1976; Sposito, 1984):

$$\equiv SOH^{o} \quad + \quad H^{+} \quad \Leftrightarrow \quad \equiv SOH_2^{+} \qquad (1)$$

$$\equiv SO^{-} \quad + \quad H^{+} \quad \Leftrightarrow \quad \equiv SOH^{o} \qquad (2)$$

where $\equiv S$ represents a generic, crystallographically bound metal cation, *S*, at the surface-water interface. This formalism implies that protons are adsorbed or desorbed onto or from specific surface sites, and that the abundance of protons (or other adsorbed cations) at these sites determines surface charge. These sites are considered distinct thermodynamic entities. That is, the equilibria can be quantified with mass-action equations of the form (written for Eqn. 1):

$$K_{(1)} \quad = \quad \frac{[\equiv SOH_2^{+}]}{[\equiv SOH^{0}][H^{+}]} \qquad (3)$$

where *K* represents a conditional (or apparent) equilibrium constant for the subscripted reaction at the ionic strength and surface coverage of interest, and brackets denote concentrations in terms of molality.

Acid/base titrations of mineral powder suspensions constrain the equilibrium constant values for surface protonation or deprotonation reactions. Numerous studies have quantified these values for metal oxides, as well as for more complex aluminosilicates. Sverjensky and Sahai (1996) developed a method for estimating these equilibrium constants for oxides and silicates based on known properties of the solids, in conjunction with electrostatic and solvation theories. A wide range of values are observed for the important rock-forming metal cations. For example, the value of $pK_{(1)}$ (at 0.01-0.03 *m*) ranges from approximately –1.6 for quartz to 7.5 for gibbsite and goethite (Table 3, Sverjensky and Sahai, 1996). Therefore, different mineral surface cation sites can display markedly different surface charges at a given pH.

The advantage of the surface complexation approach lies in the fact that the *K* values are constant at a given pressure and temperature, so system speciation can be directly calculated at any system composition of interest. Because the surface electric field affects the ability of ions to approach and depart from surface sites, these equilibrium constant values must be adjusted to account for this effect, using the following equation (Stumm and Morgan, 1996):

$$K_{Apparent} \quad = \quad K_{Intrinsic} \exp(-\frac{\Delta ZF\Psi}{RT}) \qquad (4)$$

$K_{Apparent}$ represents the equilibrium constant value valid at a particular surface charge (i.e., eq. 3), while $K_{Intrinsic}$ is the fixed equilibrium constant value for a hypothetically uncharged surface. ΔZ, F, Ψ, R, and T represent the change in the charge of the surface species for the reaction, Faraday's constant, the surface potential, the gas constant, and absolute temperature, respectively.

Metal cation adsorption onto mineral surfaces is modeled in the surface complexation approach as attachment of the metal cation M^{m+} onto one of the surface species, for example, the deprotonated negatively charged $\equiv SO^{-}$ site:

$$M^{m+} \quad + \quad \equiv SO^{-} \quad \Leftrightarrow \quad \equiv SOM^{(m-1)+} \qquad (5)$$

The adsorption of metal cations onto metal oxide or silicate surfaces is typically pH-dependent, with little or no adsorption under low-pH conditions (where H^+ ions can effectively compete with the metal cations for the available surface sites). Adsorption increases with increasing pH as the surface sites deprotonate and the surface becomes more negatively charged. Stability constants for metal-surface complexes can be determined by measuring the amount of metal adsorbed onto a mineral surface. The stoichiometry of the surface complex can be constrained by conducting these experiments as a function of pH, ionic strength, and solute:surface concentration ratio (see, e.g., Dzombak and Morel, 1990, for a thorough review of metal adsorption behavior). Furthermore, information concerning the molecular structure of mineral surface complexes can be gleaned from spectroscopic techniques, such as magnetic resonance, x-ray absorption, and Fourier transform infrared spectroscopies (Motschi, 1987; Brown, 1990; Karthein et al., 1991).

Because many types of surface sites are protonated under low-pH conditions, and hence neutrally or positively charged, organic acid anions tend to adsorb most extensively at low pH, such that there is decreasing adsorption of organic anions with increasing pH. Most studies have observed little organic anion adsorption under conditions at which both the anion and the mineral surface are negatively charged (e.g., Stumm et al., 1980). Similar to metal cation adsorption, the adsorption of organic acid anions onto mineral surfaces can be viewed in terms of interactions between the organic molecules and specific sites on the mineral surface. For example, Fein and Brady (1995) model oxalate adsorption onto Al_2O_3 with the following reaction:

$$\equiv AlOH^o \quad + \quad Ox^{2-} \quad + \quad H^+ \quad \Leftrightarrow \quad \equiv AlOx^- \quad + \quad H_2O \tag{6}$$

Adsorption of metal cations or organic acid anions does not typically render mineral surface sites inert. If adsorption results in charge reversal of the surface site, and if the aqueous system contains both dissolved metal cations and organic ligands, then the formation of metal-organic ternary surface complexes can occur. As described by Schindler (1990), two types of ternary surface complexes may form: Type A complexes in which a metal cation bridges the negative charges of the surface and organic acid ligands; and Type B complexes in which an organic poly-functional ligand bridges the positive charges of the surface and a metal cation. In Type A complexes, the coordination sphere of the adsorbed metal cation is only partially filled through its interaction with the surface functional group, creating a positively charged site for further adsorption of ligands. A generic reaction that describes the formation of Type A ternary surface complexes is:

$$\equiv SO^- \quad + \quad M^{m+} \quad + \quad L^{y-} \quad \Leftrightarrow \quad \equiv SOML^{(m-y-1)+} \tag{7}$$

Type B complexes form because adsorption of a poly-functional organic acid anion (such as oxalate illustrated in Eqn. 6) creates a negatively charged surface site, thus promoting co-adsorption of a metal cation. The formation of Type B ternary complexes can be represented as:

$$\equiv SOH^o \quad + \quad M^{m+} \quad + \quad L^{y-} \quad + \quad H^+ \quad \Leftrightarrow \quad \equiv SLM^{(m-y+1)+} \quad + \quad H_2O \tag{8}$$

The formation of either type of ternary surface complex can significantly enhance adsorption of both metal cations and organic acid anions. It is clear that in an organic-dominated system, for example, the formation of Type B ternary complexes would cause adsorption of metal cations under conditions otherwise unfavorable to cation adsorption. Similarly, Type A complexes would enhance adsorption of organic acid anions, but enhanced metal adsorption due to Type A complexation has also been observed (Schindler, 1990).

When the review of Schindler (1990) was published, very few studies had been conducted that had determined thermodynamic stability constants for ternary metal-organic surface complexes, and the only ones that were reported were Type A complexes involving either silica or rutile surfaces. Although some experimental evidence suggested the presence of Type B ternary complexes (e.g., Davis and Leckie, 1978; Davis, 1984), no experimental studies had constrained their stability constants. However, in natural geologic settings and in most metal- and organic-contaminated groundwater systems, the presence of poly-functional organic acid anions (rather than the π-acceptor organic molecule ligands that tend to form Type A complexes) makes it likely that Type B complexes form in these settings. Because of the potential formation of these complexes, and their potential effect on aqueous mass transport, a large number of experimental studies have examined Type B complexation since 1990. The objective of this review is to describe our state of knowledge of both Type A and B ternary surface complexation involving metal cations and organic anions, to summarize the measured thermodynamic stabilities of these complexes, to describe their effects on contaminant transport, and to suggest fruitful lines for future experimentation. While ternary surface complexes exist involving inorganic ligands, such as phosphate and carbonate, this study will focus exclusively on those involving organic molecules.

3. EXPERIMENTAL STUDIES

3.1. Type A Complexation

Since Schindler's review of ternary complexation studies in 1990, a number of studies have been conducted that have examined both Type A and Type B complexation (Table 1). One effect of Type A surface ternary complexes is to increase the extent of metal adsorption onto mineral surfaces. Although the stoichiometry of Eqn. 7 does not indicate the creation of additional binding sites relative to the organic-free systems (Eqn. 5), the presence of organic anions can significantly enhance adsorption over a wide pH range. This is the type of ternary complexation summarized by Schindler (1990). The presence of some organic molecules (in particular π-acceptor ligands) must alter the electron clouds of the sorbing metal cations enough to augment the positive charge of the cation, thereby enhancing the electrostatic attraction with the mineral surface.

Due to the bridging of the cation in Type A ternary surface complexes, this complexation can augment

Table 1. Experimental Evidence for Ternary Surface Complexes – post Schindler (1990)

Reference	Type	Mineral	Metal	Organic	Surface Complexation Interpretation
Lövgren (1991)	A	Goethite	Al	Phthalate	Yes
Stadler & Schindler (1993)	A	Ca-Montmorillonite	Cu	5 & 6 member chelate rings	Yes
Ludwig & Schindler (1995)	A	Anatase	Cu	Chelates with N donor atoms	Yes
Girvin et al. (1996)	A	Gibbsite	Co(II)	NTA	No
Boily & Fein (1998)	A	Corundum	Pb	Benzenecarboxylates	Yes
Nowack & Stone (1999)	A	Goethite	Ca	Phosphonates	No
Boily (1999)	A	Goethite	Cd	Phthalate	Yes
Girvin et al. (1993)	B	δ-Al_2O_3	Co(II)	EDTA	No
Bryce et al. (1994)	B	HFOs	Ni	EDTA	Yes
Nordén et al. (1994)	B	Alumina, Quartz	Eu, Sr	Fulvic Acid	No
Szecsody et al. (1994)	B	Fe-oxide coated Sand	Ca, Co(II) Fe(III)	EDTA	No
Zachara et al. (1994)	B	Assemblage[1]	Co(II)	Humic Acid	No
Zachara et al. (1995)	B	Goethite	Co(II)	EDTA	Yes
Ali & Dzombak (1996)	B	Goethite	Ca, Cu	Phthalate, Chelidamate	Yes
Boily & Fein (1996)	B	Corundum	Cd	Citrate	Yes
Nowack & Sigg (1996)	B	Goethite	Metals[2]	EDTA	Yes
Nowack et al. (1996)	B	Al, Fe oxides[3]	Ni, Pd, Fe(III)	EDTA	Yes
Vohra & Davis (1997)	B[4]	TiO_2	Pb	NTA	Yes
Redden et al. (1998)	B	Goethite	U(VI)	Citrate	No
Vohra & Davis (1998)	B[4]	TiO_2	Pb	EDTA	Yes
Yang & Davis (1999)	B	TiO_2	Cu, Cd	EDTA	Yes

[1] A clay-sized isolate from a natural subsurface sample.
[2] Ca, Zn, Ni, Cu, Pb, Co(II), Co(III), Pd, Fe(III), Al.
[3] δ–Al_2O_3, crystalline and non-crystalline iron-oxides.
[4] Also found evidence for a Type A complex.

the extent of organic adsorption onto mineral surfaces that are otherwise not conducive to anionic adsorption (Lövgren, 1991; Ludwig and Schindler, 1995; Girvin et al., 1996; Boily and Fein, 1998; Nowack and Stone, 1999). It should be noted that enhanced organic adsorption in the presence of aqueous metal cations does not necessarily require the presence of Type A ternary surface complexes. It is possible that Type B complexes could also account for this behavior because the diminished electronegativity associated with a Type B complex (relative to organic-surface complexes themselves) could enable enhanced adsorption of the organic anion due to diminished surface repulsion. For example, Nowack and Stone (1999) demonstrate that the adsorption of some phosphonates, which are used widely as industrial complexing agents, onto goethite is dramatically increased when cations such as Ca^{2+} and Zn^{2+} are present in excess concentrations relative to that of the phosphonates. Fig. 1, from Nowack and Stone (1999), depicts the adsorption of nitrilotris-methylenephosphonic acid (NTMP) onto goethite in the absence of Ca, and with 10 – 1000 μM Ca. The experiments clearly demonstrate that the presence of Ca can significantly increase the adsorption of NTMP in the basic pH range. Nowack and Stone (1999) state that the observed adsorption behavior could be explained by the formation of ternary surface-phosphonate-Ca complexes, but they do not provide thermodynamic stability constants for these complexes. It is likely that the data could also be explained by the formation of a Type A ternary surface complex. Contaminated groundwater systems can possess high concentrations of rock-forming divalent cations (such as Ca^{2+}, Mg^{2+}, Fe^{2+}), as well as contaminant metals (such as Cd^{2+}, Zn^{2+}, etc.). Therefore, especially when concerned with organic contaminant mobility, the formation of Type A surface ternary complexes clearly must be accounted for. However, Type A complexes can also augment the adsorption of metal cations, as observed by Boily (1999). Boily (1999) uses a combination of bulk chemistry analyses with infrared and EXAFS spectroscopies to identify a Type A goethite-Cd-phthalate ternary surface complex and to quantify its thermodynamic stability. However, remarkably few stability constants for Type A complexes have been determined (see Schindler, 1990, and Table 1), so our ability to model these effects is extremely limited.

3.2. Type B Complexation

Type B ternary surface complexation can dramatically affect both metal and organic molecule mobilities in metal- and organic-bearing geologic systems. As the stoichiometry of Eqn. 8 indicates, Type B complexes involve multi-functional organic acid anions. A large number of these types of molecules are found with aqueous metals as contaminants in groundwater systems, and many anthropogenic organic acids are listed as Environmental Protection Agency (E.P.A.) Priority Pollutants (Keith and Telliard, 1979). Organic acids, such as EDTA, citrate, and tributylphosphate, are also common constituents with metals and radionuclides as contaminants in soils (e.g., at U.S. Department of Energy (D.O.E.) sites; Riley et al., 1992), and humic and fulvic acids are naturally-occurring complex organic acids that are virtually ubiquitous in near-surface environments. Therefore, Type B ternary surface complexation is likely to be even more widespread and significant than Type A complexation. For this reason, Type B complexation has received considerable attention over the past 10 years (Table 1).

A number of studies have identified the general effects of organic acid anions on metal adsorption onto mineral surfaces (Fig. 2). As noted by a number of workers in this area (e.g., Davis, 1984; Murphy and Zachara, 1995; Ali and Dzombak, 1996; Boily and Fein, 1996), under low pH conditions the formation of Type B ternary surface complexes enhances the adsorption of aqueous metal cations onto positively- or neutrally-charged mineral surfaces. Conversely, under higher pH conditions, mineral surface sites deprotonate to a greater extent, causing a net decrease in the mineral surface charge, and hence a decrease in the amount of organic anions that can adsorb. The non-adsorbed organic acid anions in solution under these higher pH conditions can form highly stable aqueous metal-organic complexes,

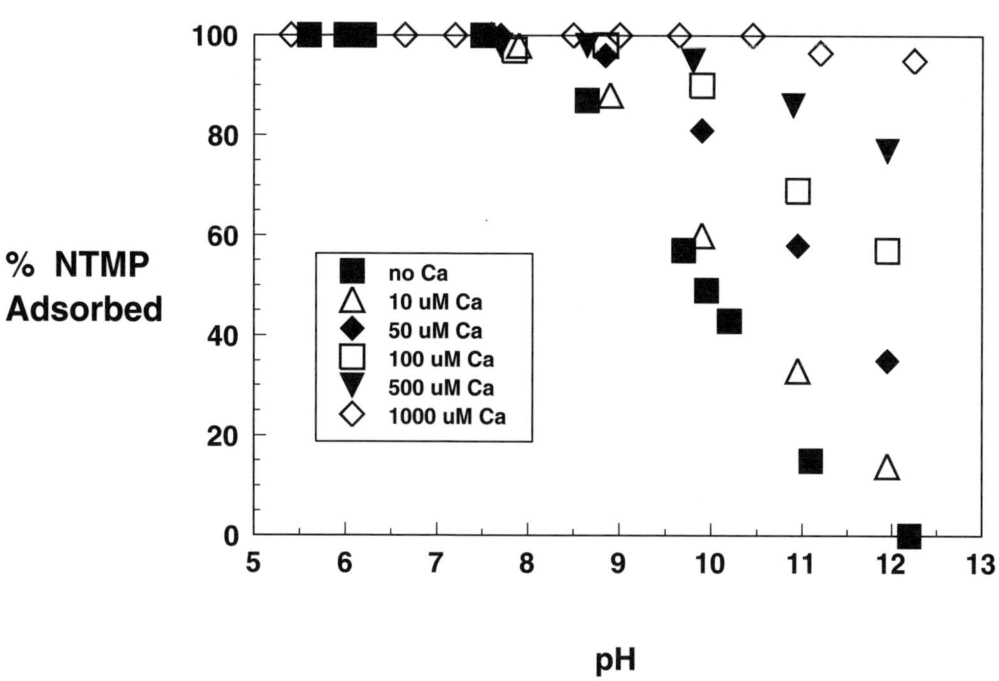

Fig. 1. Experimental data from Nowack and Stone (1999) showing the effect of aqueous Ca^{2+} (from 0 to 1000 μM) on the adsorption of nitrilotris-methylenephosphonic acid (NTMP) onto goethite, with 9.8 μM NTMP and 0.42 g/L goethite in 0.01 M $NaNO_3$. Nowack and Stone (1999) ascribe the adsorption enhancement to a ternary goethite-Ca-NTMP surface complex.

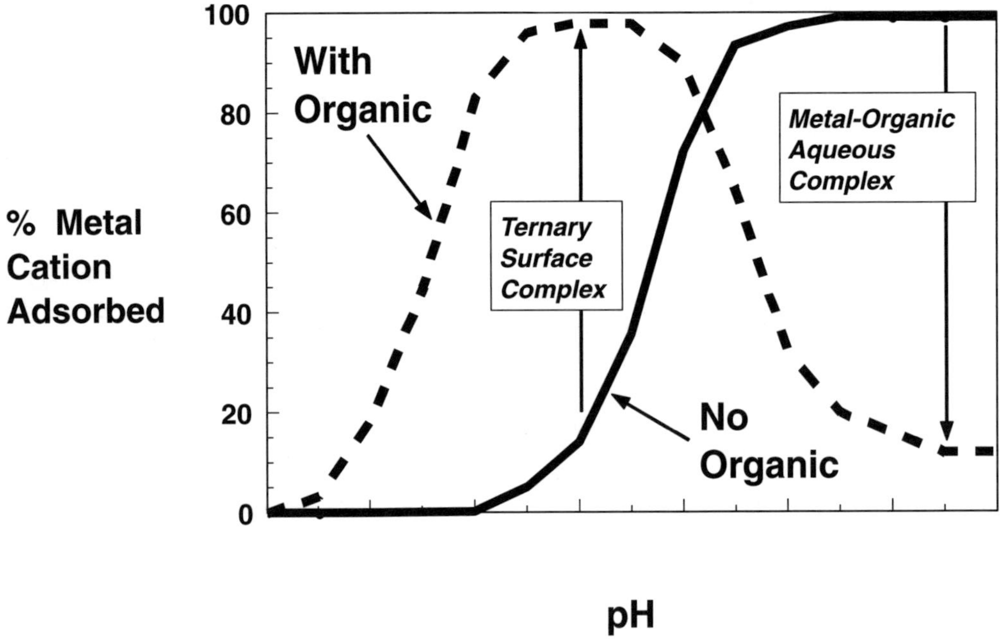

Fig. 2. The generalized effects of dissolved organic acid anions on the adsorption of aqueous metal cations onto an oxide-type mineral surface. The organic anions enhance adsorption under lower pH conditions through the formation of Type B ternary surface complexes, and diminish adsorption at higher pH values by competition involving metal-organic aqueous complexes.

thereby competing with the mineral surface for the available metal and reducing the amount of metal that can adsorb.

A number of field studies have demonstrated that aqueous metals can strongly bind to adsorbed organic matter on mineral surfaces (e.g., Balistrieri et al., 1981; Tessier et al., 1996), qualitatively showing the importance of Type B ternary surface complexes. In addition, experimental studies have been conducted that document the enhancement of metal cation adsorption in the presence of humic and fulvic acids and simpler organic acids (Table 1); however, few provide quantitative constraints on the thermodynamic stabilities of the important ternary surface complexes. Type B ternary surface complexes are formed because adsorption of multi-functional organic acid anions onto positively- or neutrally-charged mineral surfaces creates additional sites for metal cation adsorption on the surface. For example, Zachara et al. (1994) measured over 80% removal of 11 mg/L leonardite humic acid onto 1.3 g/L of subsurface isolates throughout the pH range of their experiments (~4.5 to 8.0), in metal-free, humic acid-mineral systems. In these systems, the adsorption of the humic acid significantly decreased the electrophoretic mobility of the mineral isolates, despite the overall net negative charge of the mineral surfaces. In Co- and humic acid-bearing systems, the extent of Co^{2+} adsorption was significantly higher than that observed in the absence of humic acid, with the maximum amount of adsorption enhancement occurring at approximately pH 6 – 6.5. Zachara et al. (1994) attribute these observations to the adsorption of the humic acid onto positively-charged Al- and Fe-(hydr)oxide sites within the mineral isolates, and subsequent exposure of ionized carboxylate groups to the solution. The net effect of humic acid adsorption is to augment the concentration of negatively-charged surface sites, thereby accounting for both the enhanced electronegativity of the mineral surfaces in humic-only systems, and the elevated Co^{2+} adsorption in the metal and humic experiments. Due to the complex chemistry of both the humic acid and the mineral assemblages used in the experiments, Zachara et al. (1994) could not easily apply a surface complexation approach to quantify the thermodynamic stability of the observed ternary surface complexes. Rather, they applied a linear additivity composite K_d approach to modeling the data.

The effect of EDTA on aqueous metal cation adsorption onto mineral surfaces has received much attention because EDTA has been used to decontaminate nuclear reactors and various nuclear waste processing facilities. At several D.O.E. sites, aqueous radionuclides and metals can exist predominantly as anionic EDTA complexes in groundwater (e.g., Means et al., 1978), suggesting that ternary surface complexation could significantly affect contaminant mobilities. A number of studies have focussed on the effects of EDTA on metal sorption (e.g., Girvin et al., 1993; Szecsody et al., 1994); the structure of EDTA-bearing surface ternary complexes has been investigated using EXAFS (Schlegel et al., 1997), and several studies have quantified the thermodynamic stabilities of the important ternary surface complexes (Bryce et al., 1994; Zachara et al., 1995; Nowack and Sigg, 1996; Nowack et al., 1996; Vohra and Davis, 1998; Yang and Davis, 1999).

The common feature of each of these studies is the general anionic-type of adsorption behavior of the metal or radionuclide involved. Each experiment documents maximum metal adsorption under low-pH conditions, with adsorption decreasing in most cases to negligible amounts at higher pH values. For example, Nowack and Sigg (1996) measured stability constants for EDTA and metal-EDTA (for Ca, Zn, Ni, Cu, Co(II), Pd, Pb, Fe(III), Co(III), and Al) goethite surface complexes. The pH dependence of the adsorption behavior of most of the metals mimics, in general, the adsorption of EDTA alone, indicating that the metal-EDTA complex adsorption is controlled by the same processes that control the adsorption of non-complexed EDTA. The metals that do not follow the typical anion-type trend are ones that form weaker complexes with EDTA, such as Ca, and some of the trivalent metal cations. Ca-EDTA complexes are mostly dissociated under lower pH conditions, and hence Ca adsorption behavior is cation-like under these conditions.

Similar anion-type adsorption behavior for metal-EDTA complexes was observed by Bryce et al. (1994), who measured Ni adsorption onto a hydrous ferric oxide in the presence of EDTA, and by Nowack et al. (1996), who measured Ni- and Pd-EDTA adsorption onto a range of crystalline and non-crystalline metal oxides. Each of these studies modeled the observed adsorption behaviors using the following reaction stoichiometry:

$$\equiv SOH^o \ + \ M^{m+} \ + \ EDTA^{4-} \ + \ H^+ \ \Leftrightarrow \ \equiv S-EDTA-M^{(m-3)-} \ + \ H_2O \qquad (9)$$

using FITEQL (Westall, 1982) to solve for the equilibrium constant for Eqn. 9. Zachara et al. (1995) measured Co(II)-EDTA adsorption onto goethite, as well as onto a number of fluvial sediments. The goethite adsorption behavior observed by Zachara et al. (1995) is consistent with that observed by Nowack and Sigg (1996). That is, between pH values of approximately 5 and 10, adsorption decreases with increasing pH. In addition, Zachara et al. (1995) extended their experiments to lower pH conditions than did Nowack and Sigg (1996), observing a dramatic decrease in metal adsorption with decreasing pH below approximately pH 4 – 5. Zachara et al. (1995) attribute the decreased adsorption below pH 5 to competitive effects of aqueous Fe^{3+}, which, due to goethite dissolution, was present in increasing concentrations with decreasing pH below pH 5 in these experiments. This implies that competition for the available $EDTA^{4-}$ between Co^{2+} and Fe^{3+} caused increasing displacement of Co^{2+} from goethite-EDTA-Co(II) ternary surface complexes, and therefore a decrease in Co adsorption with decreasing pH. Zachara et al. (1995) noted a positive correlation between Co adsorption and iron oxide content of the sediments used in the experiments, but the correlation was weaker than would be predicted based on the goethite experiments alone. It is likely that the diminished adsorption was due again to competitive complexation by Al^{3+} and Fe^{3+} liberated from the minerals by dissolution. This example indicates that accurate use of a surface complexation approach to quantifying solute adsorption requires determination of all important stability constants in the system. The amount of incorporation of Co(II) in ternary surface complexes can only be predicted if the tendency for other system cations to form such complexes is also known.

Type B ternary surface complexation has also been observed involving poly-functional organic acids other than EDTA. Boily and Fein (1996) measured the adsorption of Cd, in the presence and absence of citrate, onto α-Al_2O_3. Enhanced Cd adsorption between approximately pH 4 and 7 in the citrate systems relative to that observed in the organic-free experiments was best accounted for by the formation of the $\equiv$ Al-Cit-Cd^o ternary surface complex. Diminished Cd adsorption was observed in the presence of citrate at higher pH values due to formation of aqueous Cd-citrate complexes, and the inability of citrate to adsorb onto Al-mineral surface sites under high pH conditions. Redden et al. (1998) confirm the existence of a goethite-citrate-U(VI) surface ternary complex using EXAFS and FTIR spectroscopies. Ali and Dzombak (1996) found similar effects of phthalic and chelidamic acids on the adsorption of Cu^{2+} and Ca^{2+} onto goethite, also using a surface complexation approach to quantify the thermodynamic stability of the ternary surface complexes. These studies, and the others involving metal-EDTA systems, demonstrate that the effect of multi-functional organic acids on metal adsorption is strongly pH-dependent, and that it is incorrect to assume that the presence of dissolved chelating agents will always cause increased metal mobility in the subsurface.

Enhanced metal adsorption in the presence of organic acids is not always caused by the formation of ternary surface complexes. Adsorption of organic acid anions onto positively-charged mineral surfaces will offset at least some of the positive charge of the surface, thereby reducing the electrostatic repulsion between the surface and positively-charged aqueous cations (Boily and Fein, 1998; Nowack and Stone, 1999) and enabling cation adsorption to occur on exposed surface sites that are not involved in organic anion adsorption. For example, Nowack and Stone (1999) observed enhanced Cu uptake onto goethite in the presence of a number of different phosphonates from approximately pH 3 to 6. The observed adsorption enhancement was modeled, not with the presence of a distinct ternary surface complex, but rather by accounting for the effects of the adsorbed organic anions on the electric field of the mineral surface. Adsorption of the negatively-charged phosphonates was sufficient to diminish the positive charge of the goethite surface under the low pH conditions. This electric field effect increased the ability of Cu^{2+} to approach and bind to the goethite surface on separate sites from those responsible for phosphonate adsorption. Enhanced Cd adsorption onto goethite in the presence of phosphate was observed by Venema et al. (1997), and was also successfully modeled without the need for a ternary surface complex. Conversely, Boily and Fein (1998) demonstrated that the observed enhancement of Pb adsorption in the presence of benzenecarboxylates was greater than could be accounted for simply due to electric field effects associated with organic anion adsorption, using this as key evidence for the presence of ternary surface complexes involving Pb and the benzenecarboxylates.

4. PREDICTIVE TECHNIQUES

The above review of the experimental studies that have focussed on metal-organic ternary surface complexes illustrates the limitations of our understanding of this adsorption phenomenon. Until a more comprehensive database is assembled, predictive techniques are required to estimate the thermodynamic stabilities of environmentally and geologically important ternary surface species. Schindler (1990) outlines a predictive approach that is based on a statistical treatment of the coordination chemistry of the metal-ligand pair of interest. Although this approach can successfully account for the available experimental data for Type A complexes, it requires a detailed understanding of the coordination chemistry of the adsorbed species.

Linear free energy correlations have been used successfully to relate organic adsorption stability constants to measured stability constants of metal-organic aqueous complexes (e.g., Stumm et al., 1980), and such an approach has also been applied to equilibrium constants for metal-mineral (Langmuir, 1979) and metal-bacteria adsorption reactions (Fein et al., 1997; Fowle and Fein, 2000). In the present study, I concentrate on Type B ternary surface complexes, attempting to relate the overall stability constant of the surface complex to the thermodynamic stabilities of the adsorbed organic species, and the tendency for the aqueous metal cation and the organic to form an aqueous complex. Ternary complexation can be expressed as follows:

$$\equiv SOH^o + M^{m+} + L^{-L} + H^+ \Leftrightarrow \equiv S-L-M^{(m-L+1)} + H_2O \qquad (10)$$

(where L^{-L} represents the fully deprotonated organic acid anion). The organic anion adsorption reaction can be expressed as:

$$\equiv SOH^o + L^{-L} + H^+ \Leftrightarrow \equiv S-L^{(1-L)} + H_2O \qquad (11)$$

and the aqueous metal-organic complexation reaction can be written:

$$M^{m+} + L^{-L} \Leftrightarrow ML_{(aq)}^{(m-L)} \qquad (12)$$

Fig. 3 depicts measured values of the logarithm of the equilibrium constant for Eqn. 10 (log $K_{Ternary}$), plotted as a function of the sum of the log stability constants for Eqns. 11 and 12. The stability constants used in this correlation are listed in Table 2.

As indicated in the present literature review, most measured stability constants for Type B ternary surface complexes are those of metal-EDTA species. The log stability constant value for the Co(II)-EDTA complex from Zachara et al. (1995) was reported as 5.13 for the following reaction:

$$\equiv FeOH_2{}^+ + Co(II)EDTA^{2-} \Leftrightarrow \equiv FeOH_2-EDTA-Co(II)^- \qquad (13)$$

This was altered to the stoichiometry of Eqn. 10 by addition of the surface protonation reaction:

$$\equiv FeOH^o + H^+ \Leftrightarrow \equiv FeOH_2{}^+ \qquad (14)$$

Table 2. Stability Constant Correlation Values

M-L-Solid Set	Log $K_{Ternary}$ Eqn. 10	Log $K_{Solid\text{-}L}$ Eqn. 11	Log $K_{M\text{-}L(aq)}$ [1] Eqn. 12	Reference [2]
Cd-citrate-Al_2O_3	16.8	13.5	5.9	a
Ca-EDTA-FeOOH	22.4	15.0	11.0	b
Zn-EDTA-FeOOH	28.4	15.0	18.2	b
Ni-EDTA-FeOOH	30.7	15.0	20.3	b
Cu-EDTA-FeOOH	31.1	15.0	20.5	b
Pb-EDTA-FeOOH	30.0	15.0	19.7	b
Co(II)-EDTA-FeOOH	28.1	15.0	18.1	b
Pd-EDTA-FeOOH	41.7	15.0	20.7	b
Al-EDTA-FeOOH	26.7	15.0	19.1	b
Fe(III)-EDTA-FeOOH	36.9	15.0	27.6	b
Co(III)-EDTA-FeOOH	52.3	15.0	44.0	b
Pd-EDTA-HFO [3]	32.0	15.0 [4]	20.7	c
Co(II)-EDTA-FeOOH	28.8	15.0 [5]	18.1	d
Ni-EDTA-HFO	28.8	15.0 [6]	20.3	e
Pb-NTA-TiO_2	33.5	14.4	12.7	f

Notes:
1 Values for $K_{(12)}$ adjusted to I=0 where necessary (see text).
2 a=Boily and Fein (1996); b=Nowack and Sigg (1996); c=Nowack et al. (1996); d=Zachara et al. (1995); e=Bryce et al. (1994); f=Vohra and Davis (1997).
3 HFO = hydrous ferric oxide.
4, 5, 6 Values of 15.0 are assumed because these studies do not report EDTA adsorption using the same stoichiometry as in Eqn. 11.

and the reaction describing the formation of the Co(II)EDTA aqueous complex:

$$Co^{2+} + EDTA^{-4} \Leftrightarrow CoEDTA^{-2} \quad (15)$$

assuming that the $\equiv FeOH_2$-EDTA-Co(II) species is thermodynamically equivalent to $\equiv$ Fe-EDTA-Co(II). Ali and Dzombak (1996) model enhanced Cu adsorption onto goethite in the presence of phthalate and chelidamate as ternary surface complexes involving a hydrated surface Fe atom ($\equiv$ FeOH-Cu-ligand), rather than the form used in Eqn. 10. Therefore, inclusion of their data in this treatment is impossible. The correlation that is derived below is valid only for the ternary surface complex stoichiometry described in Eqn. 10. Values of the aqueous metal-organic stability constants are taken from Martell and Smith (1974; 1977; 1982), with values corrected to the reference state of zero ionic strength using the extended Debye-Hückel equation and parameters given by Helgeson et al. (1981). Uncertainties in the relationship arise not only from uncertainties in the measured values of the equilibrium constants for Eqns. 10-12, but also from differences in models used to quantify the electrostatic effects of the mineral surface.

Fig. 3 illustrates that a strong correlation exists between the log of the stability constant of Type B ternary surface complexes (as expressed by Eqn. 10) and the sum of the logs of the equilibrium constants for Eqns. 11 and 12. This sum represents both the tendency for the organic anion to adsorb onto the mineral surface (Eqn. 11) and for the metal and the organic anion to interact (Eqn. 12). The correlation holds for an extremely wide range of stability constant values (over 30 orders of magnitude with respect to the value of $K_{(10)}$). The correlation appears valid for a wide range of metals, for different metal oxide surfaces, and it appears successful for the Cd-citrate data of Boily and Fein (1996), as well (open square in Fig. 3). In addition, although Nowack and Sigg (1996) note significant differences between the ternary complexation behavior involving divalent and trivalent cations, the relationship holds for both types of cations. The circle significantly above the line corresponds to the measured goethite-EDTA-Pd stability constant of Nowack and Sigg (1996). Note that the hydrous ferric oxide-EDTA-Pd stability constant of Nowack et al. (1996) agrees well with the correlation. It is not possible to include the stability constants for Pb-, Cu-, and Cd-EDTA-TiO_2 from Vohra and Davis (1998) and Yang and Davis (1999) in the relationship depicted in Fig. 3 because the stoichiometries these authors use to characterize the surface ternary complexes are different than that shown in Eqn. 10.

Although the correlation depicted in Fig. 3 appears robust over a wide range of stability constant values, it is uncertain whether such a relationship can accurately predict stability constants for

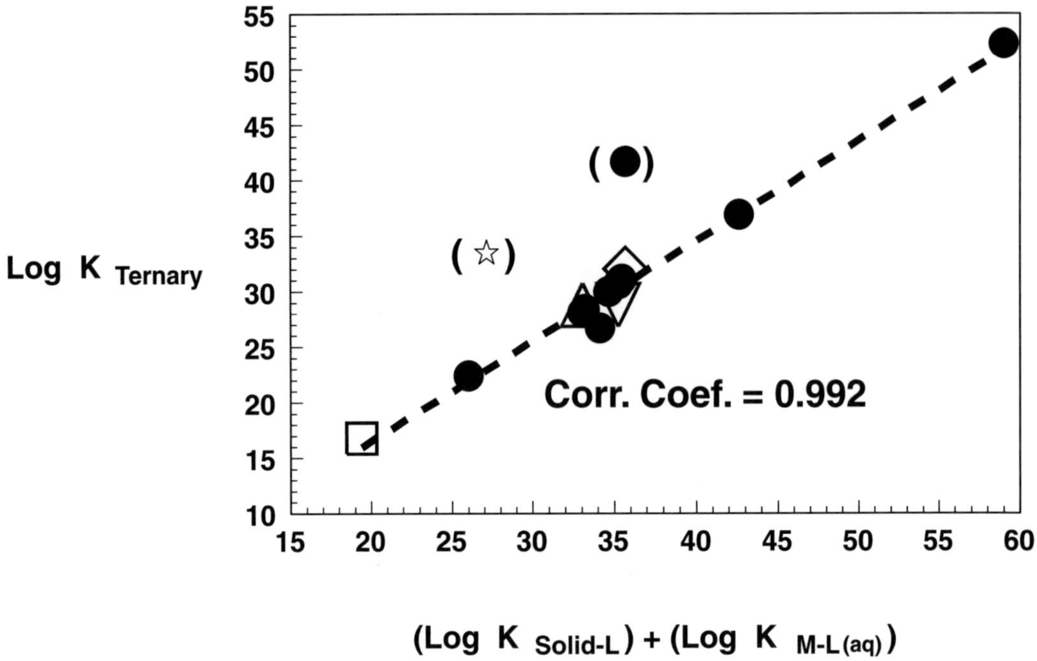

Fig. 3. Correlation plot between the log of the stability constant of ternary surface complexes (Eqn. 10) as a function of the sum of the log of the stability constants of the organic-mineral surface complex (Eqn. 11) and the metal-organic aqueous complex (Eqn. 12). The data are compiled in Table 2: filled circles represent data from Nowack and Sigg (1996); open square from Boily and Fein (1996); open diamond from Nowack et al. (1996); upward pointing triangle from Zachara et al. (1995); and downward pointing triangle from Bryce et al. (1994). The linear regression is calculated neglecting the data (in parentheses) which fall significantly above the correlation line.

Type B ternary surface complexes involving metals and organic acid anions not studied directly in the laboratory. Another system which appears to be inconsistent with this relationship is one involving gibbsite-NTA-Co(II). Girvin et al. (1996) found that NTA does not significantly affect Co(II) adsorption onto gibbsite under low pH conditions, and therefore does not form a Type B ternary surface complex as EDTA does with Co(II) (Girvin et al., 1996, did observe a higher pH effect on NTA adsorption caused by Co(II), and concluded that this was caused by the formation of a Type A gibbsite-Co(II)-NTA complex). Vohra and Davis (1997) find evidence for both Type A and Type B TiO_2–NTA-Pb surface ternary complexes, yet the stability constant (represented by the star in Fig. 3) for the Type B complex does not fit the illustrated correlation. With so few measurements involving organic anions other than EDTA, it is unclear whether citrate or NTA, or both, generally deviate from the correlation in Fig. 3. It is possible that the relationship depicted in Fig. 3 applies only to EDTA, and that the position of the citrate point is coincidental.

In the absence of experimental data, the relationship shown in Fig. 3 can be used to estimate the effects of organic acids on the mobility of aqueous metal cations. For example, Fig. 4 depicts the estimated effects of citric acid on the adsorption of Co(II) onto goethite. The thermodynamic stability of the surface goethite-citrate species has not been determined, so the calculation was conducted assuming that the stability constant value is equivalent to the value determined by Boily and Fein (1996) for the surface species $\equiv AlCit^{-2}$ (with a log stability constant value of 13.5). Stability constant values for ≡ Al-ligand surface complexes are reasonably close to those for ≡ Fe-ligand complexes (Stumm et al., 1980). The stability of the aqueous Co-citrate complex is taken from Martell and Smith (1977), and with an ionic strength correction, the log stability constant is taken as 6.3. The sum of the two aforementioned log stability constant values is 19.8. Using the linear relationship shown in Fig. 3, the sum of 19.8 yields an estimated log stability constant for the ternary surface complex (≡ FeCitCo) of 16.6. This value, along with equilibrium constants for the other important reactions between the goethite surface, dissolved citrate, and dissolved Co(II), can be used to calculate the distribution of Co in the system as a function of pH. The calculation, conducted using FITEQL (Westall, 1982), used stability constant values for aqueous complexes from Martell and Smith (1977; 1982); and for Co(II)-goethite surface species from Zachara et al. (1995).

Figure 4 shows that citrate, which is a co-contaminant with Co at a number of D.O.E. sites (Riley et al., 1992), may significantly affect the mobility of Co(II) through the formation of a ternary surface complex. The presence of citrate dramatically enhances the adsorption of Co over a wide range of pH conditions. For example, at pH 6.0 with no citrate present, less than 10 percent of the total Co in the system is adsorbed onto the goethite surface. In the citrate-bearing system, the formation of the ternary surface complex causes the concentration of adsorbed Co to increase to over 97 percent of the total. Clearly, such a dramatic change in the adsorption behavior of Co(II) would markedly change the estimated mobility of Co(II) in groundwater systems. This example shows that ternary surface complexation can dominate the chemical behavior of trace metals in the environment. However, it should also be noted that under some conditions, ternary surface complexes are not likely to significantly alter metal mobilities. For example, a similar estimation of the thermodynamic stability of the ≡Fe-oxalate-Co ternary surface complex on goethite yields a log stability constant of 12.7. Speciation calculations using this value indicate that oxalate causes only a relatively small enhancement of Co(II) adsorption onto goethite due to ternary surface complex formation (Fig. 4). Even with two orders of magnitude more oxalate than Co(II) in the system, the ≡Fe-oxalate-Co ternary surface complex comprises no more than 20 percent of the total Co budget at any pH value, and shifts the Co adsorption edge to more acid pH values by less than 0.5 pH units.

5. CONCLUSIONS AND FUTURE WORK

Although a number of studies have been conducted since Schindler's 1990 review of ternary surface

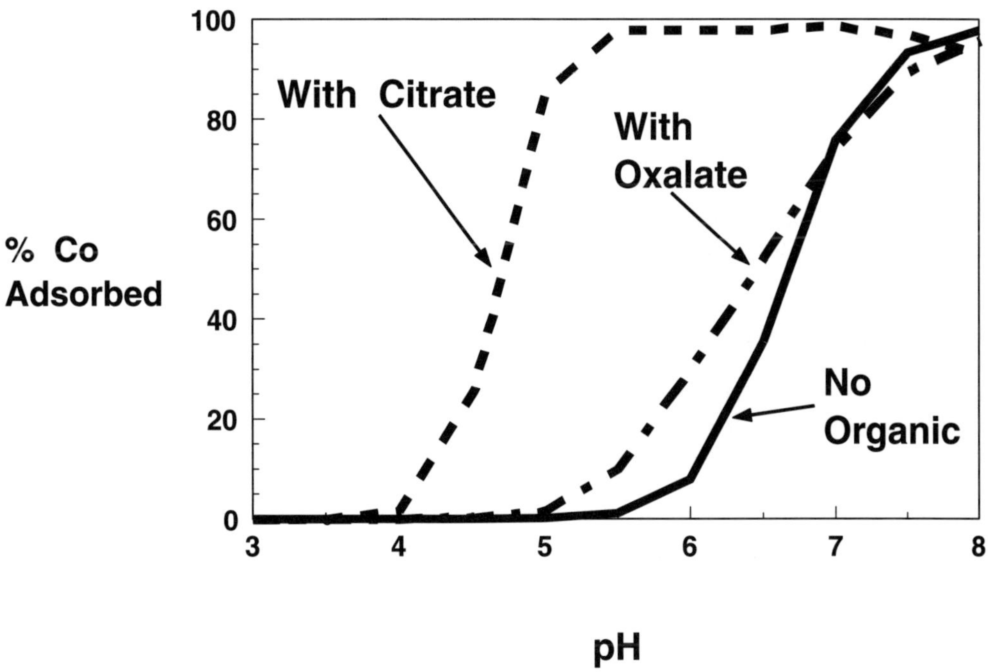

Fig. 4. Calculated adsorption behavior of Co(II) onto goethite in the presence and absence of citric acid. The stability constant for the goethite-citrate-Co(II) ternary surface complex was estimated from the linear relationship depicted in Fig. 3. Calculation conducted for the following conditions: 10^{-5} *m* total Co(II), 2.7 x 10^{-3} *m* total Fe surface sites, 0.01 *m* $NaNO_3$. Solid curve corresponds to a system with no organic acid; dashed curves correspond to those containing either 10^{-3} *m* total citric or oxalic acid.

complexes, our understanding of this adsorption process is still limited. Although Type A complexation (mineral-metal-organic) can significantly enhance the adsorption of aqueous metal cations onto mineral surfaces, these ternary complexes are more likely to affect organic molecule distributions. A number of studies have demonstrated that the presence of aqueous metal cations in organic anion bearing systems can augment the extent of adsorption of the anions through charge reversal due to cation adsorption onto the mineral surface. Despite these potentially large effects, thermodynamic stability measurements for Type A surface complexes are extremely scarce. Because of this lack of data, it is difficult to construct predictive models that enable accurate estimation of the thermodynamic stability of these complexes. Future work quantifying stability constants of Type A complexes would begin to alleviate this problem.

Type B ternary surface complexation (mineral-organic-metal) can significantly enhance metal adsorption under mildly acidic to neutral pH conditions. This enhancement has been demonstrated, and quantified in terms of surface complexation theory, by a number of experimental studies over the past 10 years. Although measurements of the thermodynamic stabilities of Type B ternary surface complexes are still fairly limited, enough exist to enable estimations to be made of the controls on the stabilities of these complexes. A strong linear correlation exists between the log stability constant of Type B ternary surface complexes and the sum of the log stability constants of the metal-organic aqueous complex and the organic-surface complex of the components. The number of experimental studies that constrain this relationship is limited, and the primary goal of future experimental work should be to expand the data set to include metals and organic acid anions that are environmentally or geologically important, and have yet to be studied. The observed relationship enables the thermodynamic stability of Type B ternary surface complexes to be estimated, and the resulting changes in metal adsorption behavior to be determined for various conditions. These changes in adsorption behavior can be large, as in the case illustrated in this study for goethite-citrate-Co(II) complexation, or they can be negligible, as for oxalate (see Fig. 4). At present, due to the small data set that constrains the behavior, the uncertainties associated with the estimated stability constant values are unknown, but are likely to be large. Clearly, in order to more accurately determine whether Type B ternary complex formation is important in a given system, a much larger data set must be used to constrain this relationship, so future work should be focused toward this end.

Ternary surface complexation between mineral surfaces, organic molecules, and metal cations can dramatically affect the fate and transport of both organic molecules and metals in geologic systems. The thermodynamic stabilities of a handful of ternary surface complexes have been measured over the past decade. Our ability to estimate the importance of ternary surface complexation in systems not directly studied in the laboratory has improved since 1990, but it is still limited in scope and accuracy. Due to the common occurrence of metals and organic molecules as co-contaminants, it is crucial that we improve our understanding of ternary surface complexes in order to accurately model the mobility of contaminants in the subsurface.

Acknowledgments—Although I never worked with David Crerar, his work and his students have helped to shape my scientific pursuits. Crerar had been instrumental in quantifying thermodynamic properties of species in water-rock systems, and while he never worked directly with ternary surface complexes, I feel that this contribution is in keeping with the type of work in which he would have engaged. I would like to acknowledge a review of an early version of this article provided by Peter Wightman, who constantly maintains vigilance in keeping me honest with my correlation models. Thorough reviews by Jean-François Boily, Lars Lövgren, Bernd Nowack, and Berhard Wehrli have greatly improved the manuscript.

Editorial handling: Roland Hellmann

REFERENCES

Ali M. A. and Dzombak D. A. (1996) Effects of simple organic acids on sorption of Cu^{2+} and Ca^{2+} on goethite. *Geochim. Cosmochim. Acta* **60**, 291-304.

Balistrieri L., Brewer P. G., and Murray J. W. (1981) Scavenging residence times of trace metals and surface chemistry of sinking particles in the deep ocean. *Deep-Sea Res.* **28A**, 101-121.

Boily J.-F. (1999) The surface complexation of ions at the goethite (α- FeOOH)/Waterinterface: A multisite complexation approach. Ph.D. Thesis, Umeå University, Sweden.

Boily J.-F. and Fein J. B. (1996) Experimental study of cadmium-citrate co-adsorption onto α-Al_2O_3. *Geochim. Cosmochim. Acta* **60**, 2929-2938.

Boily J.-F. and Fein J. B. (1998) Adsorption of Pb(II) and benzenecarboxylates onto corundum. *Chem. Geol.* **148**, 157-175.

Bowers A. R. and Huang C. P. (1986) Adsorption characteristics of metal-EDTA complexes onto hydrous oxides. *J. Coll. Interface Sci.* **110**, 575-590.

Brown G. E. Jr. (1990) Spectroscopic studies of chemisorption reaction mechanisms at oxide-water interfaces. In: *Mineral-Water Interface Geochemistry* (eds. M.F. Hochella, Jr. and A.F. White). Reviews in Mineralogy **23**, 309-363. Mineralogical Society of America, Washington, D.C.

Bryce A. L., Kornicker W. A., and Elzerman A. W. (1994)

Nickel adsorption to hydrous ferric oxide in the presence of EDTA: Effects of component addition sequence. *Environ. Sci. Technol.* **28,** 2353-2359.

Davis J. A. (1984) Complexation of trace metals by adsorbed natural organic matter. *Geochim. Cosmochim. Acta* **48**, 679-691.

Davis J. A. and Leckie J. O. (1978) Effect of adsorbed complexing ligands on trace metal uptake by hydrous oxides. *Environ. Sci. and Technol.* **12**, 1309-1315.

Dzombak D. A. and Morel F. M. M. (1990) *Surface Complexation Modeling: Hydrous Ferric Oxide.* John Wiley and Sons, New York.

Fein J. B. and Brady P. V. (1995) Mineral surface controls on the diagenetic transport of oxalate and aluminum. *Chem. Geol.* **121**, 11-18.

Fein J. B., Daughney C. J., Yee N., and Davis T. (1997) A chemical equilibrium model of metal adsorption onto bacterial surfaces. *Geochim. Cosmochim Acta* **61,** 3319-3328.

Fowle D. and Fein J. B. (2000) Experimental measurements of the reversibility of metal-bacteria adsorption reactions. *Chem. Geol.* **168**, 27-36.

Girvin D. C., Gassman P. L., and Bolton H. Jr. (1993) Adsorption of aqueous cobalt ethylenediaminetetraacetate by δ–Al_2O_3. *Soil Sci. Soc. Am. J.* **57**, 47-57.

Girvin D. C., Gassman P. L., and Bolton H. Jr. (1996) Adsorption of nitrilotriacetate (NTA), Co and CoNTA by gibbsite. *Clays and Clay Minerals* **44**, 757-768.

Helgeson H. C., Kirkham D. H., and Flowers G. C. (1981) Theoretical prediction of the thermodynamic behavior of aqueous electrolytes at high pressures and temperatures: IV. Calculation of activity coefficients, osmotic coefficients, and apparent molal and standard and relative partial molal properties to 600°C and 5 Kb. *Amer. J. Sci.* **281**, 1249-1516.

Hohl H. and Stumm W. (1976) Interaction of Pb^{+2} with hydrous γ–Al_2O_3. *J. Coll. Interface Sci.* **55**, 281-288.

Karthein R., Motschi H., Schweiger A., Ibric S., Sulzberger B., and Stumm W. (1991) Interactions of chromium(III) complexes with hydrous δ-Al_2O_3: Rearrangements in the coordination sphere studied by electron spin resonance and electron spin-echo spectroscopies. *Inorg. Chem.* **30**, 1606-1611.

Keith L. H. and Telliard W. A. (1979) Priority pollutants: I-A perspective view. *Environ. Sci. Technol.* **13**, 416-423.

Langmuir D. (1979) Techniques of estimating thermodynamic properties for some aqueous complexes of geochemical interest. In: *Chemical Modeling in Aqueous Systems* (ed. E.A. Jenne). American Chemical Society, Washington, D.C. pp. 353-387.

Lövgren L. (1991) Complexation reactions of phthalic acid and aluminium(III) with the surface of goethite. *Geochim. Cosmochim. Acta* **55**, 3639-3645.

Ludwig C. and Schindler P. W. (1995) Surface complexation on TiO_2 II. Ternary surface complexes: Coadsorption of Cu(II) and organic ligands (2,2'-bipyridyl, 8-aminoquinoline, and o-phenylenediamine) onto TiO_2 (Anatase). *J. Coll. Interface Sci.* **169**, 291-299.

Martell A. E. and Smith R. M. (1974) *Critical Stability Constants. Volume 1: Amino Acids.* Plenum Press, New York.

Martell A. E. and Smith R. M. (1977) *Critical Stability Constants. Volume 3: Other Organic Ligands.* Plenum Press, New York.

Martell A. E. and Smith R. M. (1982) *Critical Stability Constants. Volume 5: First Supplement.* Plenum Press, New York.

Means J. L., Crerar D. A., and Duguid J. O. (1978) Migration of radionuclide wastes: Radionuclide mobilization by complexing agents. *Science* **200**, 1477-1486.

Motschi H. (1987) Aspects of the molecular structure in surface complexes; Spectroscopic investigations. In: *Aquatic Surface Chemistry* (ed. W. Stumm). John Wiley & Sons, New York.

Murphy E. M. and Zachara J. M. (1995) The role of sorbed humic substances on the distribution of organic and inorganic contaminants in groundwater. *Geoderma* **67**, 103-124.

Nordén M., Ephraim J. H., and Allard B. (1994) The influence of a fulvic acid on the adsorption of europium and strontium by alumina and quartz: Effects of pH and ionic strength. *Radiochim. Acta* **65**, 265-270.

Nowack B., Lützenkirchen J., Behra P., and Sigg L. (1996) Modeling the adsorption of metal-EDTA complexes onto oxides. *Environ. Sci. Technol.* **30**, 2397-2405.

Nowack B. and Sigg L. (1996) Adsorption of EDTA and metal-EDTA complexes onto goethite. *J. Coll. Interface Sci.* **177**, 106-121.

Nowack B. and Stone A. T. (1999) The influence of metal ions on the adsorption of phosphonates onto goethite. *Environ. Sci. Technol.* **33**, 3625-3631.

Redden G., Bargar J., and Bencheikh-Latmani R. (1998) Characterization of U(VI)-goethite adsorption interactions in the presence of citrate using EXAFS spectroscopy. *Stanford Synchrotron Radiation Laboratory Activity Report* #2513.

Riley R. G., Zachara J. M., and Wobber F. J. (1992) Chemical contaminants on DOE lands and selection of contaminant mixtures for subsurface science research. DOE/ER-0547T. Washington, D.C.: U.S. Department of Energy.

Schindler P. W. (1990) Co-adsorption of metal ions and organic ligands: Formation of ternary surface complexes. In: *Mineral-Water Interface Geochemistry* (eds. M.F. Hochella, Jr. and A.F. White). Reviews in Mineralogy **23**, 281-307. Mineralogical Society of America, Washington, D.C.

Schlegel M. L., Manceau A., and Charlet L. (1997) EXAFS study of Zn and ZnEDTA sorption at the Goethite (α-FeOOH)/water interface. *J. Phys. IV France* **7**, Colloque C2, 823-824.

Sposito G. (1984) The surface chemistry of soils. Oxford Univ. Press. New York.

Stadler M. and Schindler P. W. (1993) The effect of dissolved ligands upon the sorption of Cu(II) by Ca-montmorillonite. *Clays and Clay Minerals* **41**, 680-692.

Stumm W., Kummert R., and Sigg L. (1980) A ligand exchange model for the adsorption of inorganic and organic ligands at hydrous oxide interfaces. *Croat. Chem. Acta* **53**, 291-312.

Stumm W. and Morgan J. J. (1996) *Aquatic Chemistry.* 3rd Edition. John Wiley and Sons, Inc. New York.

Sverjensky D. A. and Sahai N. (1996) Theoretical prediction of single-site surface-protonation equilibrium constants for oxides and silicates in water. *Geochim. Cosmochim. Acta* **60**, 3773-3797.

Szecsody J. E., Zachara J. M., and Bruckhart P. L. (1994) Adsorption-dissolution reactions affecting the distribution and stability of Co^{II}EDTA in iron oxide-coated sand. *Environ. Sci. Technol.* **28**, 1706-1716.

Tessier A., Fortin D., Belzile N., DeVitre R. R., and Leppard G. G. (1996) Metal sorption to diagenetic iron and manganese oxyhydroxides and associated organic matter: Narrowing the gap between field and laboratory measurements. *Geochim. Cosmochim. Acta* **60**, 387-404.

Venema P., Hiemstra T., and van Riemsdijk W. H. (1997) Interaction of cadmium with phosphate on goethite. *J. Coll. Interface Sci.* **192**, 94-103.

Vohra M. S. and Davis A. P. (1997) Adsorption of Pb(II), NTA, and Pb(II)-NTA onto TiO_2. *J. Coll. Interface Sci.* **194**, 59-67.

Vohra M. S. and Davis A. P. (1998) Adsorption of Pb(II), EDTA, and Pb(II)-EDTA onto TiO_2. *J. Coll. Interface Sci.* **198**, 18-26.

Westall, J. (1982) FITEQL: A computer program for determination of equilibrium constants from experimental data. Version 2.0. Report 82-02. Corvallis, OR: Department of Chemistry, Oregon State Univ.

Yang J.-K. and Davis A. P. (1999) Competitive adsorption of Cu(II)-EDTA and Cd(II)-EDTA onto TiO_2. *J. Coll. Interface Sci.* **216**, 77-85.

Zachara J. M., Resch C. T., and Smith S. C. (1994) Influence of humic substances on Co^{2+} sorption by a subsurface mineral separate and its mineralogic components. *Geochim. Cosmochim. Acta* **58**, 553-566.

Zachara J. M., Smith S. C., and Kuzel L. S. (1995) Adsorption and dissociation of Co-EDTA complexes in iron oxide-containing subsurface sands. *Geochim. Cosmochim. Acta* **59**, 4825-4844.

Water-Rock Interactions, Ore Deposits, and Environmental Geochemistry: A Tribute to David A. Crerar

Editors: Roland Hellmann and Scott A. Wood

The dynamics of arsenic in saturated porous media: fate and transport modeling for deep aquatic sediments, wetland sediments, and groundwater environments

PETER R. JAFFÉ*, SOOKYUN WANG, PETER L. KALLIN[1], AND SHERWOOD L. SMITH[2]

Department of Civil and Environmental Engineering, Princeton University, Princeton, NJ 08544, USA
[1]Now at OMNI Environmental Corporation, Princeton, NJ, USA
[2]Now at Frontier Research Program for Global Change SEAVANS, Tokyo, Japan

*Corresponding author. E-mail: jaffe@princeton.edu

Abstract—The fate and transport of many trace metals, metalloids, and radionuclides in porous media is closely linked to the biogeochemical reactions that occur as a result of organic carbon being sequentially degraded by different microorganisms using a series of terminal electron acceptors. The spatial distribution of these biogeochemical reactions is affected by processes that are often unique to and/or characteristic of a specific environment. We have developed a generic model formulation and applied it to simulate the fate and transport of arsenic in three distinct hydrologic settings, namely deep aquatic sediments (no rooted plants), wetland sediments, and groundwaters. The key differences in physical processes that have been considered are sedimentation, roots and their effects on biogeochemistry and advective transport, and differences in mixing processes. Steady-state formulations were applied to the sedimentary environments while a time-dependent formulation was applied to the groundwater case. Results of numerical simulations show that these physical processes significantly affect the chemical profiles of different electron acceptors, their reduced species, and arsenate as well as arsenite that will result from the degradation of an organic carbon source in the sediments or groundwater. Even though specific biological transformations are allowed to proceed only in zones where they are thermodynamically favorable, results show that mixing as well as abiotic reactions can make the profiles of individual electron acceptors overlap and/or appear to reverse their expected order.

1. INTRODUCTION

The fate and transport of many trace metals, metalloids, and radionuclides in porous media are closely linked to the biogeochemical reactions that occur as a result of organic carbon being degraded by different microorganisms using a series of terminal electron acceptors. Throughout the redox profile that develops in such environments, trace metals can be mobilized/immobilized via processes such as reduction/oxidation, sorption/desorption, precipitation/dissolution, and/or the formation of complex ions. Although the work presented here is generic and applies to a wide range of trace metals, metalloids, and radionuclides, for this study, we will focus the simulations specifically on arsenic as a model contaminant.

Arsenic is a ubiquitous contaminant that is, among others, associated with the chemical and power industry, as well as agriculture. The predominant forms of arsenic in natural water systems are As(V) as H_3AsO_4 (arsenate), and As(III) as H_3AsO_3 (arsenite). Arsenate is dominant in oxic waters, but under reducing conditions is converted to arsenite (Massacheleyn et al., 1991). Like phosphate, arsenate exists mainly in its deprotonated forms at natural pH levels, and is therefore readily sorbed onto positively charged surfaces of minerals such as Fe(III) oxides (Fendorf et al., 1997; Dzombak and Morel, 1990). Bacteria have been identified which, in environments that are more reducing than nitrate reduction, can use arsenate as a terminal electron acceptor, reducing it to arsenite (Newman et al., 1997 a,b; Dowdle et al. 1996). Arsenite is the more toxic form of arsenic, and it exists primarily as a neutral dissolved species at ambient pH values, in zones that have a more reducing environment. Compared to arsenate, transport of arsenite is therefore not significantly retarded due to sorption (Gulens et al., 1979; Holm et al., 1979; Oscarson et al., 1980), but it has a high affinity to form complexes with sulfides, forming sulfide minerals such as orpiment (As_2S_3).

Given that the transport and bioavailability of many metals and metalloids are affected by similarly complex sets of chemical reactions and physical mechanisms, we have developed a general analytical framework to improve our understanding of how the fate and transport of these metals are affected by various biogeochemical and physical transport processes. Because these processes are highly environmentally dependent, the biogeochemical dynamics of arsenic will be different in various environmental settings. In our simulations, we will focus on three important, permanently saturated, hydrologic settings, where redox conditions can range from highly oxidized to highly reduced, namely: deep aquatic sediments without rooted plants (referred to hereafter as deep sediments), wetland sediments with rooted macrophytes, and groundwaters. Our objective is to assess the main differences in model development as well as arsenic fate and transport in these environments.

Common to the modeling approach is that, distinct redox zones will develop as a result of the biodegradation of organic matter and the sequential utilization of different terminal electron acceptors, as shown in Table 1. The spatial distribution of these zones will determine where arsenic is reduced/oxidized, sorbed to different solid phases, and/or precipitates. Clearly pH is a key parameter in determining the fate of arsenic and other chemical constituents in the environments, and spatial/temporal pH values will depend on the different biotic and abiotic reactions occurring in each redox zone. Complex microbial and chemical redox reactions could change pH conditions in the aquatic system by producing/consuming acid and base constituents during the decomposition of organic matter. For the purpose of this work, however, pH will be considered a model input and will not be simulated explicitly.

1.1. Differences in Model Formulations between Deep Sediments, Wetland Sediments, and Groundwater.

A fundamental difference between saturated sediments near the sediment/water interface (deep and wetland sediments) and in the subsurface (groundwater environments) is that, at the interface, sedimentation provides a source for new and usually oxidized sediments. This is important, in that this sedimentation process normally provides a source of fresh iron and manganese oxides. Thus, in sedimentary environments, a quasi-steady-state spatial profile can develop containing a reacting solid phase of Mn(IV) and/or Fe(III). In a system where the solid phase is stationary, such as groundwaters, steady-state conditions cannot result in the presence of such reacting solid phases. It should be noted that only a fraction of the Mn(IV) and Fe(III) is bioavailable. Amorphous iron oxide may be bioavailable while crystalline forms are not. To simulate biogeochemical dynamics in groundwater environments where the bioavailable fraction of Mn(IV) and Fe(III) are being consumed as terminal electron acceptors, one needs to use non-steady-state models that can simulate the profiles of these species over space and time.

For deep and wetland sediments, an oxidized surface layer of sediment is typically present at the sediment/water interface (Gambrell and Patrick, 1978; Ponnamperuma, 1984; Mitsch and Gosselink, 1993), which is followed by a steep redox gradient, resulting from the different reactions shown in Table 1 proceeding sequentially. Concentration gradients of redox species of more than an order of magnitude over a distance of a centimeter have been observed in different sediments (Brendel and Luther, 1995). Because of much lower total bioavailable organic carbon in groundwater environments, and often higher water-flow velocities, these same redox gradients occur over significantly longer distances in these settings (Hunter et al., 1998).

Arsenic may be immobilized due to sorption onto different organic and inorganic soil constituents, or due to precipitation. Adsorption of arsenate onto iron and manganese oxides is an important process where the oxidation potential is high. As the system becomes more reduced, iron and manganese oxides are depleted, and trace metals, as well as arsenic, can desorb and remobilize as a dissolved species, although in the presence of sulfide ions, are usually immobilized as sulfides.

In addition to the processes described above, redox profiles in wetland sediments are also affected by the presence of higher plants. Wetland plants have evolved specialized adaptations to supply oxygen to the roots, and to transfer oxygen from the roots into the surrounding sediment (Armstrong, 1979; Sand-Jensen et al., 1982; Bedford et al., 1991; Mendelssohn, 1993). This transfer of oxygen maintains an oxidizing potential in the rhizosphere, which is important for root survival. It protects roots from phytotoxins, which are common in swamps at low redox potentials, and prevents transport into the root of highly soluble Fe(II) ions by precipitating them at the root surface following oxidation to Fe(III) (Grosse, 1997). The rates of oxygen transfer into wetland sediments can be substantial and highly variable (Armstrong, 1979; Bedford et al., 1991; Brix, 1993; Sorrel, 1994). These rates are dependent on the type of vegetation and root size, as well as the chemical composition and oxygen demand of the sediments (Brix et al., 1996). Oxygen release rates measured for *E. sphacelata* ranged from non-detection for roots placed in a deoxygenated solution, to 55 μmol h^{-1} g^{-1} dry weight when placed in a solution with a redox potential of -200 mV (Sorrel et al., 1993). The degree of iron plaque accumulation on roots is highly variable, and depends on the concentration of ferrous iron in solution as well as its complexation with organic ligands and the availability of other exchange sites on different soil mineral and organic fractions

Table 1. Key biogeochemical model stoichiometric relationships.

Layer	Stoichiometric relationships for the major model redox reactions[1]
1.	$(CH_2O)_{106}(NH_3)_{16}(H_3PO_4) + 106\ O_2 \rightarrow 106\ CO_2 + 16\ NH_3 + H_3PO_3 + 106\ H_2O$
	a. $NH_4^+ + 2\ O_2 \rightarrow NO_3^- + H_2O + 2\ H^+$
	b. $2\ Mn^{2+} + O_2 + 2\ H_2O \rightarrow 2\ MnO_2 + 4\ H^+$
	c. $4\ Fe^{2+} + O_2 + 10\ H_2O \rightarrow 4\ Fe(OH)_3 + 8\ H^+$
	d. $H_2S + 2\ O_2 \rightarrow SO_4^{2-} + 2\ H^+$
	e. $CH_4 + 2\ O_2 \rightarrow CO_2 + 2\ H_2O$
2.	$(CH_2O)_{106}(NH_3)_{16}(H_3PO_4) + 84.8\ NO_3^- + 84.8\ H^+ \rightarrow 106\ CO_2 + 42.4\ N_2 + 16\ NH_3 + H_3PO_3 + 148.4\ H_2O$
3.	$(CH_2O)_{106}(NH_3)_{16}(H_3PO_4) + 212\ MnO_2 + 424\ H^+ \rightarrow 106\ CO_2 + 212\ Mn^{2+} + 16\ NH_3 + H_3PO_3 + 318\ H_2O$
	a. $MnO_2 + H_2S + 2\ H^+ \rightarrow Mn^{2+} + S^0 + 2\ H_2O$
	b. $MnO_2 + 2\ Fe^{2+} + 4\ H_2O \rightarrow Mn^{2+} + 2\ Fe(OH)_3 + 2\ H^+$
4.	$(CH_2O)_{106}(NH_3)_{16}(H_3PO_4) + 424\ FeOOH + 848\ H^+ \rightarrow 106\ CO_2 + 424\ Fe^{2+} + 16\ NH_3 + H_3PO_3 + 742\ H_2O$
	a. $2\ Fe(OH)_3 + H_2S + 4\ H^+ \rightarrow 2\ Fe^{2+} + S^0 + 6\ H_2O$
5.	$(CH_2O)_{106}(NH_3)_{16}(H_3PO_4) + 53\ SO_4^{2-} + 53\ H^+ \rightarrow 106\ CO_2 + 53\ HS^- + 16\ NH_3 + H_3PO_3 + 106\ H_2O$
	a. $SO_4^{2-} + CH_4 + 2\ H^+ \rightarrow H_2S + CO_2 + 2\ H_2O$
6.	$(CH_2O)_{106}(NH_3)_{16}(H_3PO_4) \rightarrow 53\ CO_2 + 53\ CH_4 + 16\ NH_3 + H_3PO_4$

[1] In these reactions, the organic material is represented by the Redfield stoichiometry, which differs for terrestrial ecosystems (Vitousek et al., 1988), and can be adjusted depending on the characteristics of the sediment organic carbon source, wetland litter, or carbon source injected into the subsurface. To facilitate comparison between the simulations, the same carbon formulation was used in all cases. It should also be noted that sulfur is not included in the Redfield formula and that therefore the sulfur source used here is external from the organic matter.

(Mendelssohn, 1993). Adsorption of arsenic, as well as other trace metals, can be significantly enhanced in the rhizosphere due to the formation of these iron oxyhydroxide plaques on roots (Otte et al., 1995). In addition to oxygen transfer, evapotranspiration also affects the redox profile of wetland sediments, by inducing advection that changes the vertical flow velocity profile.

1.2. Simulation of the Dynamics of Trace Metals and Metalloids in Saturated Porous Media.

Biologically-mediated redox dynamics in fresh- and marine water sediments, as well as groundwaters, have been simulated in terms of the sequential utilization of different electron acceptors during the biodegradation of sediment organic matter (Rabouille and Gaillard, 1991; Sweerts et al., 1991; Matsunaga et al., 1993; Van Cappellen and Wang, 1995; Dhakar and Burdige, 1996; Park and Jaffé, 1996; Wang and Van Cappellen, 1996; Hunter et al., 1998). Based on these biogeochemical processes, trace-metal dynamics in groundwater have been described by several authors (e.g., Egesgraad and Kipp, 1992; Lensing et al., 1994; Yeh and Tripathi, 1991). In many of these studies, it was assumed that the metal is at equilibrium with the surrounding geochemistry. Clearly this is not always the case. Fe(II) for example, can be found in measurable amounts in oxic and nitrate-reducing waters, where it has been transported from zones where Fe(III) is being reduced. Non-equilibrium conditions have therefore been implemented in formulating the cycling of iron and manganese by many of these investigators (e.g., Hunter et al., 1998), and for contaminant trace metals that undergo redox reactions where the metals are kinetically driven towards chemical equilibrium by biological reactions (Smith and Jaffé, 1998) or abiotic reduction (Jardine et al., 1999).

In this work, we will use the same biogeochemical formulations for the three environmental settings being compared. The fundamentals of the model formulation and required modifications for deep sediments, wetland sediments, and groundwaters are described below. The generic equations are given in the Appendix, and a detailed description of the model implementation for sediments is given in Smith and Jaffé, (1998). The spatial (vertical in sediments, and horizontal in groundwater) profiles of fifteen coupled constituents, including solid (organic matter, manganese oxides, iron oxides, and orpiment) and dissolved substances (dissolved organic carbon, oxygen, nitrate, sulfate, ammonia, dissolved manganese, dissolved iron, sulfide, methane, arsenate, and arsenite), and their interactions, including phase changes such as precipitation, dissolution, and sorption, are described.

To model the profile of the redox conditions in saturated porous media, we describe the degradation of organic matter and the sequential reduction of electron acceptors, taking into account where appropriate, transport processes such as molecular diffusion, dispersion, bioturbation, irrigation, and advection. Bioturbation and irrigation are described as a mixing process similar to dispersion and molecular diffusion, as discussed by Berner (1980). Bioturbation and irrigation are processes driven by benthic organisms, which are present in deep sediments and wetland sediments, but are absent in groundwater environments. Based on the electron acceptor that is being used by the microorganisms degrading the organic substrate, the spatial domain is divided into six different zones: aerobic respiration, denitrification, manganese reduction, iron reduction, sulfate reduction, and methanogenesis. The microbial utilization of the electron acceptors in each of these zones is described by the stoichiometric equations shown in Table 1. For wetland sediments, effects of plants are included by adding proper source/sink terms to account for organic carbon release (e.g., litter, exudates, root turnover), oxygen release, nutrient uptake (e.g., nitrogen), and evapotranspiration-induced advection. It is assumed here that sedimentation provides a source of fresh, bioavailable iron and manganese oxides, as well as particulate organic carbon at the sediment/water interface. For the groundwater case, a dissolved organic carbon source at the origin of the domain represents the injection of an organic substrate through a well or the leaching of biodegradable organic carbon from a source such as a landfill.

The microbial degradation of the organic matter is described by a Monod-type formulation. Reactions such as oxygen consumption by reduced compounds (e.g., Mn^{2+}, Fe^{2+}, HS^-) are formulated as second-order reactions, and precipitation of Mn^{2+}, Fe^{2+}, with HS^- and other compounds such as phosphorus and carbonate (or dissolution of the respective mineral), are formulated as being proportional to the degree of over- or undersaturation, as shown in the appendix. Diffusion and/or advection of different chemical species across redox boundaries, followed by mixing, will result in many additional biotic and abiotic redox reactions (e.g., abiotic reduction of MnO_2 by HS^-). The most relevant reactions for the problem at hand and typical rate coefficients have been discussed by several authors (Von Gunten and Zorbist, 1993; Matsunaga et al., 1993; Smith and Jaffé, 1998). Coefficients for the biotic and abiotic reaction rates included in the simulations are given in Table 2. The model formulation allows for an easy incorporation of additional reactions that are identified as being important. Although none of the systems simulated are close to reaching chemical equilibrium conditions, the pE at a specific location is approximated to that of the dominant redox couple (Morel and Hering, 1993). The dominant redox couple in each sediment layer is assumed to be given by the half reaction of the respective electron acceptor for organic matter oxidation.

With the computed concentration profiles of the oxidized and reduced electron acceptors, pE, species such as organic carbon and ammonia, and inputted values for the pH and alkalinity, we use MINTEQA2 (an EPA-supported chemical speciation model) to determine the chemical speciation under equilibrium conditions for each of a set of specified chemical constituents (e.g., trace metals, metalloids, sulfides) at each location in the sediments and/or groundwater. The difference between the actual and these equilibrium concentrations determines how far from equilibrium a reaction is. Although it is not a mechanistically-based formulation, it is convenient to describe the rate at which equilibrium conditions are approached by the product of a rate constant and the gradient given by this concentration difference. This allows a consistent, dynamic formulation for species such as arsenic and all of the other species reacting with it, considering changes in oxidation state, complexation with different ligands, and precipitation. Even though this approach of coupling a transport and chemical speciation model is computationally expensive, it has the advantage of being general and allowing for the easy incorporation/substitution of additional chemical species.

Sorption of arsenate, as well as the other species simulated, onto solid organic carbon is described by a linear sorption isotherm, and sorption onto manganese and iron oxide by the double-layer surface complexation model, accounting for interactions among these species (Dzombak and Morel, 1990). The relevant reactions and respective coefficients for the sorption model are the same as in Smith and Jaffé (1998). Colloids and dissolved organic carbon may play a role in facilitating pollutant transport in

Table 2. Key biotic/abiotic reaction parameters used in the simulation

		sediments	groundwater♦
Maximum specific growth rate when [yr^{-1}]			
O_2 is the electron acceptor		4.0	1.0
NO_3^- is the electron acceptor		1.5	0.04
Mn(IV) is the electron acceptor		0.2	0.01
Fe(III) is the electron acceptor		0.032	0.005
SO_4^{2-} is the electron acceptor		1.8	0.17
Half-saturation constant of [μm]			
organic carbon		100	53.85
O_2		100	20
NO_3^-		20	20
Mn (IV)		1	3.71
Fe (III)		1	3.71
SO_4^{2-}		10	10
Threshold concentrations [μm] for indicator χ_{eA}			
O_2		0.5	
NO_3^-		6	
Mn (IV)		1	
Fe (III)		5	
SO_4^{2-}		15	
Biological As reduction rate [yr^{-1}]		130 ♣	
Abiotic As oxidation rate [M^{-1} yr^{-1}]		6.0×10^6 ♠	
Maximum utilization rate of [yr^{-1}]			
ammonia oxidation (nitrification)		30	
aerobic methane oxidation		10	
Half-saturation constant of [μm]			
ammonia oxidation (nitrification)		30	
aerobic methane oxidation		10	
Second-order rate coefficients for the redox reactions [M^{-1} yr^{-1}]			
Oxidants	Reductants		
MnO_2 (s)	Fe^{2+}	1.0×10^4	
MnO_2 (s)	HS^-	8.0×10^5	
FeOOH (s)	HS^-	1.0×10^3	
O_2 (aq)	XMn^{2+} (ads)	2.1×10^7	
O_2 (aq)	XFe^{2+} (ads)	1.6×10^8	
O_2 (aq)	Fe^{2+}	2.1×10^7	
O_2 (aq)	Mn^{2+}	4.6	
O_2 (aq)	HS^-	2.0×10^5	
NO_3^-	Fe^{2+}	1.6×10^3	
$[SO_4^-]_{total}$	CH_4	1.0×10^4	

♦ based on Kallin (1999); ♣ Newman et al., 1997b; ♠ Scott and Morgan, 1995; all other parameters are based on Smith and Jaffé (1998)

porous media (McCarthy and Zachara, 1989; Dunnivant et al., 1992), which was not accounted for in the simulations shown here, but can be easily included by allowing for a partitioning of the relevant species between the dissolved organic carbon and the aqueous phase.

2. RESULTS

A series of simulations were conducted to compare the dynamics of arsenic in deep sediments, wetland sediments, and in groundwaters. Physical and transport parameters used in these simulations are

Table 3. Physical and transport parameters used in the simulation

	sediments	groundwater
domain length [cm]	17.5	5000
infiltration/groundwater velocity [cm yr^{-1}]	50	2000
dispersivity [cm]	-	25
coefficient of bioturbation [cm^2 yr^{-1}]	150	-
coefficient of irrigation [cm^2 yr^{-1}]	300	-
characteristic depth of bioturbation layer [cm]	10.5	-
characteristic depth of irrigation layer [cm]	10.5	-
porosity	0.5	0.35
dry bulk density of soil [g cm^{-3}]	1.7	2.0
pH	6.75	6.5
Fe(III) deposition rate [moles cm^{-2} yr^{-1}]	1.0×10^{-3}	-
Mn(VI) deposition rate [moles cm^{-2} yr^{-1}]	3.3×10^{-4}	-
carbon deposition rate [moles cm^{-2} yr^{-1}]	5.0×10^{3}	-
evapotranspiration rate [cm yr^{-1}]	50	-
depth of rhizosphere [cm]	8.5	-

given in Table 3. For the sedimentary environments, constant concentration boundary conditions were set for the dissolved constituents at the water/sediment interface, and constant flux conditions for the solid components. Zero gradient conditions were set for the bottom of the domain. Zero gradient boundary conditions do affect the simulated profiles near the bottom of the domain, and changes in the gradient of some profiles near the bottom of the domain can be attributed to these boundary conditions. For the groundwater simulations, upstream constant concentration boundary conditions were set, which were equal to the initial conditions for all constituents except dissolved organic carbon. The initial conditions for dissolved organic carbon were zero throughout the domain, and the boundary conditions were set to 1.25×10^{-4} M, representing a constant injection or leaching of carbon for t > 0. At the outflow of the domain, zero gradient boundary conditions were used for all dissolved species. In sediments as well as wetlands one might have either groundwater recharge or discharge. The results shown here will focus on groundwater recharge, or surface water entering the sediments.

We will first compare the effect of roots on the dynamics of electron acceptors and arsenic in sediments. Figures 1 and 2 show the respective deep sediment and wetland sediment profiles of electron acceptors being consumed during the degradation of organic matter, while Figs. 3 and 4 show the profiles of their corresponding reduced forms, as well as ammonia, in deep sediments and wetland sediments, respectively (methane concentration is not shown). Simulations were conducted for a series of oxygen-release and evapotranspiration rates (Kallin, 1999). For the simulations shown here, when roots are present, oxygen is being released into the sediments at a rate of 3 g O_2 m^{-2} day^{-1}, and evapotranspiration is 50 cm yr^{-1}. Root density was assumed to decrease as a cosine function from the water/sediment interface to a depth of 8 cm. Oxygen release, as well as water uptake as a function of depth, was assumed to be proportional to the root density.

As can be seen by comparing Figs. 1 and 2, the average oxygen concentration in the rhizosphere (upper 8.5 cm) of wetland sediments is higher in comparison to that in deep sediments (no roots present). The direct result of this increased oxygen concentration can be seen in that the oxidized species (NO_3^-, Mn(IV), Fe(III), and SO_4^{2-}) are transported deeper into the wetland sediments before being used as an electron acceptor. Compared to deep sediments where no roots are present, for the wetland case (with roots), this results in a decrease in the concentration gradient at the water/sediment interface and increased depth of depletion. Conversely, the reduced species shown (Mn^{2+}, Fe^{2+}, and HS^-) are being produced deeper in the wetland sediments, resulting in a decreased gradient at the water/sediment interface, and hence a reduced flux of these species into the water column (see Figs. 3 and 4). It is interesting to note the very sharp Fe^{2+} peak in these profiles. Actual Fe(III) reduction occurs only at the specific depths associated with these Fe^{2+} peaks, and the rapid disappearance of the dissolved Fe^{2+} below the Fe(III) reduction zone is due to the precipitation of FeS. These Fe^{2+} profiles are extremely sensitive to the precipitation rate, and measurements of these rates under different environmental conditions are needed. For the simulations shown in this work, a precipitation rate (k_s, as defined in the appendix) of 40 M^{-1} yr^{-1} was selected (M is the molarity.).

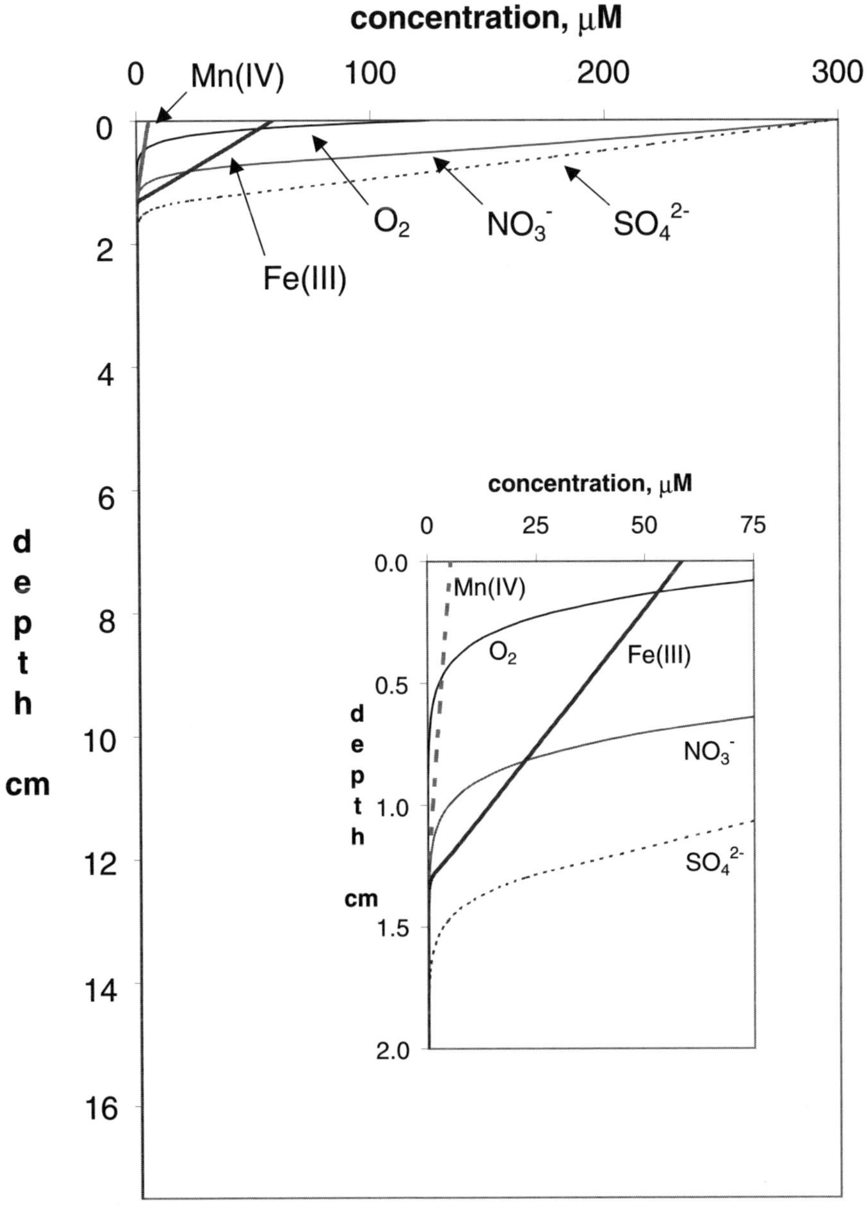

Fig. 1. Simulated concentration profiles of terminal electron acceptors in deep sediments (no roots present).

Because evapotranspiration induces flow in the sediments towards the roots, the effect of evapotranspiration on these profiles is even more marked. Evaporation-induced advection increases the flux of dissolved species (O_2, NO_3^-, SO_4^{2-}) from the water/sediment interface into the sediment. This increases the overall mass of available electron acceptors entering the sediments to the microorganisms degrading the organic matter. Furthermore, since a given mass of water is withdrawn from the rhizosphere due to evapotranspiration, a mass balance for the individual species dictates that their concentrations increase due to this water withdrawal. This effect can be seen clearly for the reduced Mn, Fe, and S, which all increase in concentration throughout the whole rhizosphere, reaching concen-

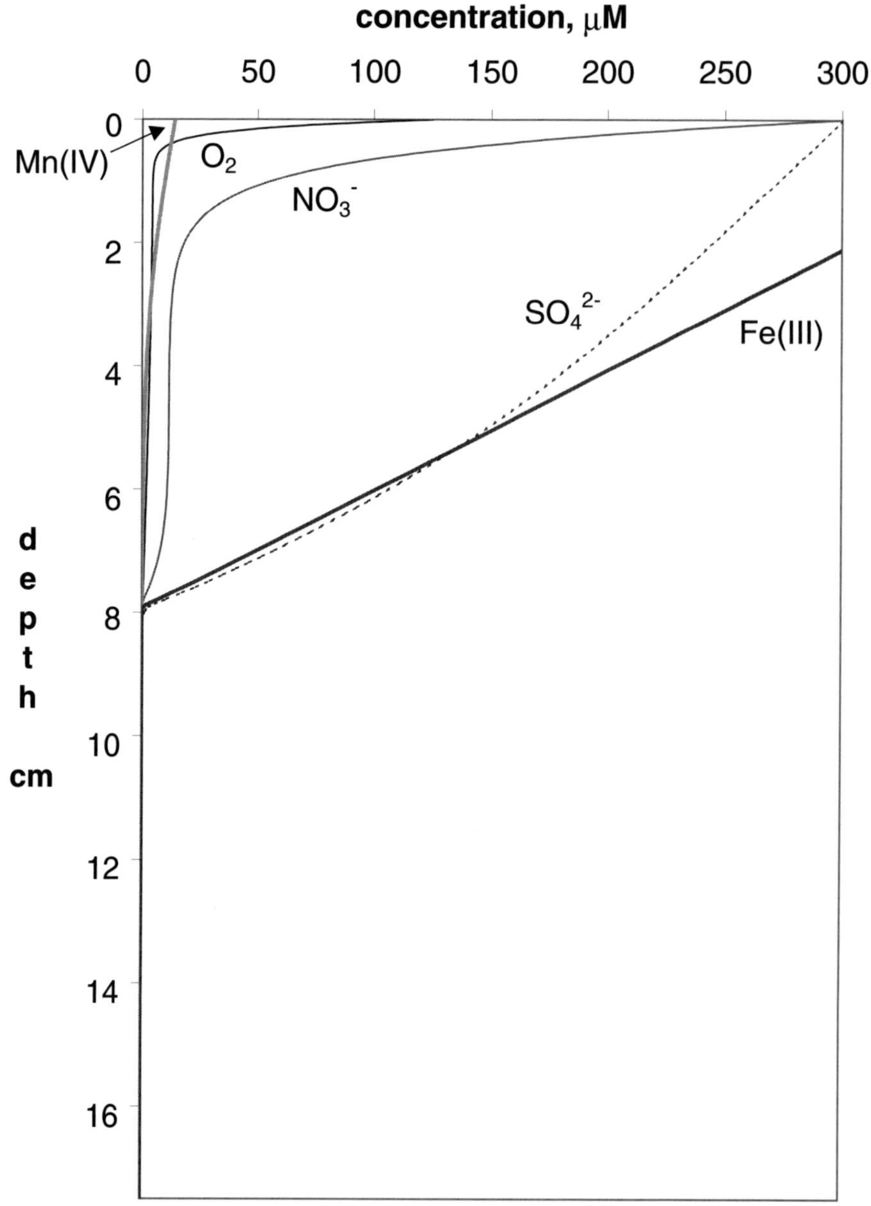

Fig. 2. Simulated concentration profiles of terminal electron acceptors in wetland sediments (roots present).

trations that are significantly higher than for the case with no roots present (compare Figs. 3 and 4).

Nitrate and ammonia uptake by plants was described by a Fickian expression, proportional to the concentration difference between the ambient concentration and a specified cutoff concentration. The nitrate and ammonia uptake rate coefficients were chosen to result in an annual biomass growth of 3,000 g/m^2, assuming 2.5% nitrogen by weight. Additionally, sorption of NH_4^+ to the sediments was assumed to follow a linear sorption isotherm. Therefore, evapotranspiration affects the nitrate profile only in terms of increasing the overall flux into the sediments. Higher oxygen releases into the sediments by roots result in a larger conversion of ammonia into nitrate. The combined effect of increased

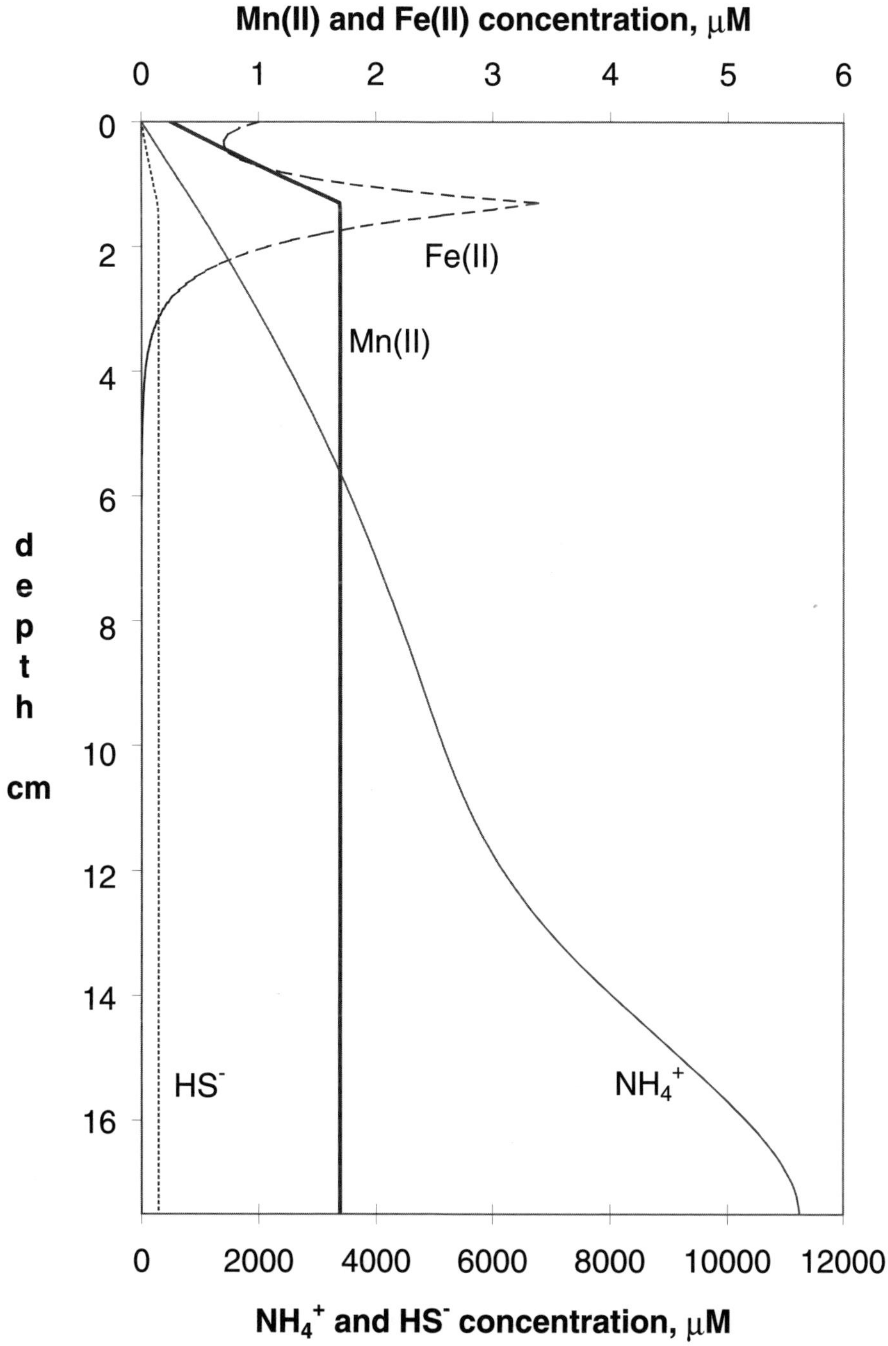

Fig. 3. Simulated concentration profiles of reduced species in deep sediments (no roots present).

nitrification and uptake by plants results in a significant decrease in the ammonia concentration in the rhizosphere of wetland sediments as compared to the deep sediment case.

Figure 5 shows the simulated profiles of dissolved arsenate and dissolved arsenite for deep sediments and wetland sediments. These profiles show that in deep sediments much steeper gradients for arsenate

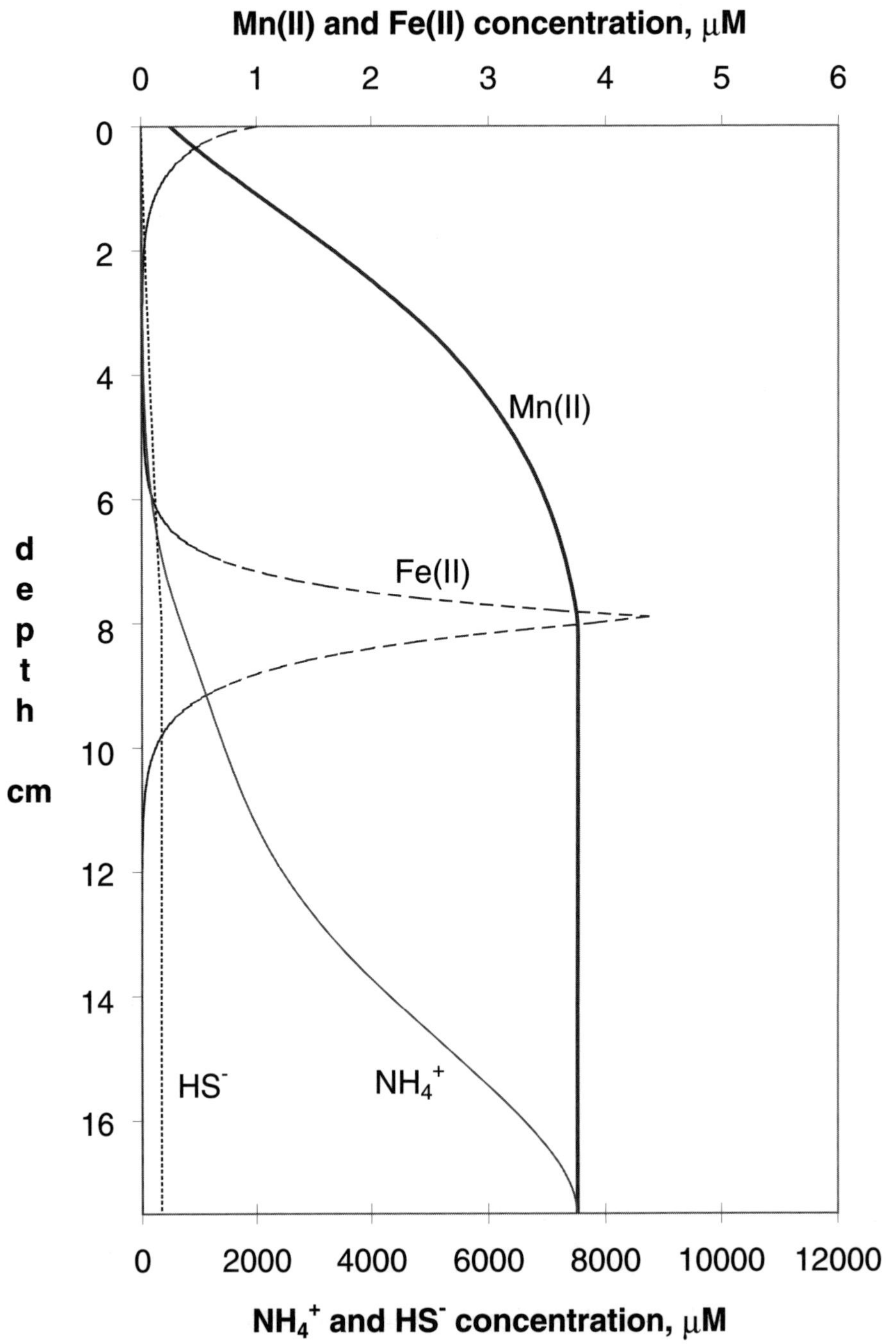

Fig. 4. Simulated concentration profiles of reduced species in wetland sediments (roots present).

and arsenite can develop at the sediment/water interface, when macrofaunal irrigation effects are not intense, as compared to the wetlands case where the profiles are less steep, due to the effects of plant roots. Therefore, a key impact of the root activity is to decrease the flux of dissolved arsenate entering the sediments from the overlaying water column, as well as the flux of the more toxic arsenite that is returned into the water column from the sediments. These fluxes are altered in their magnitude not only

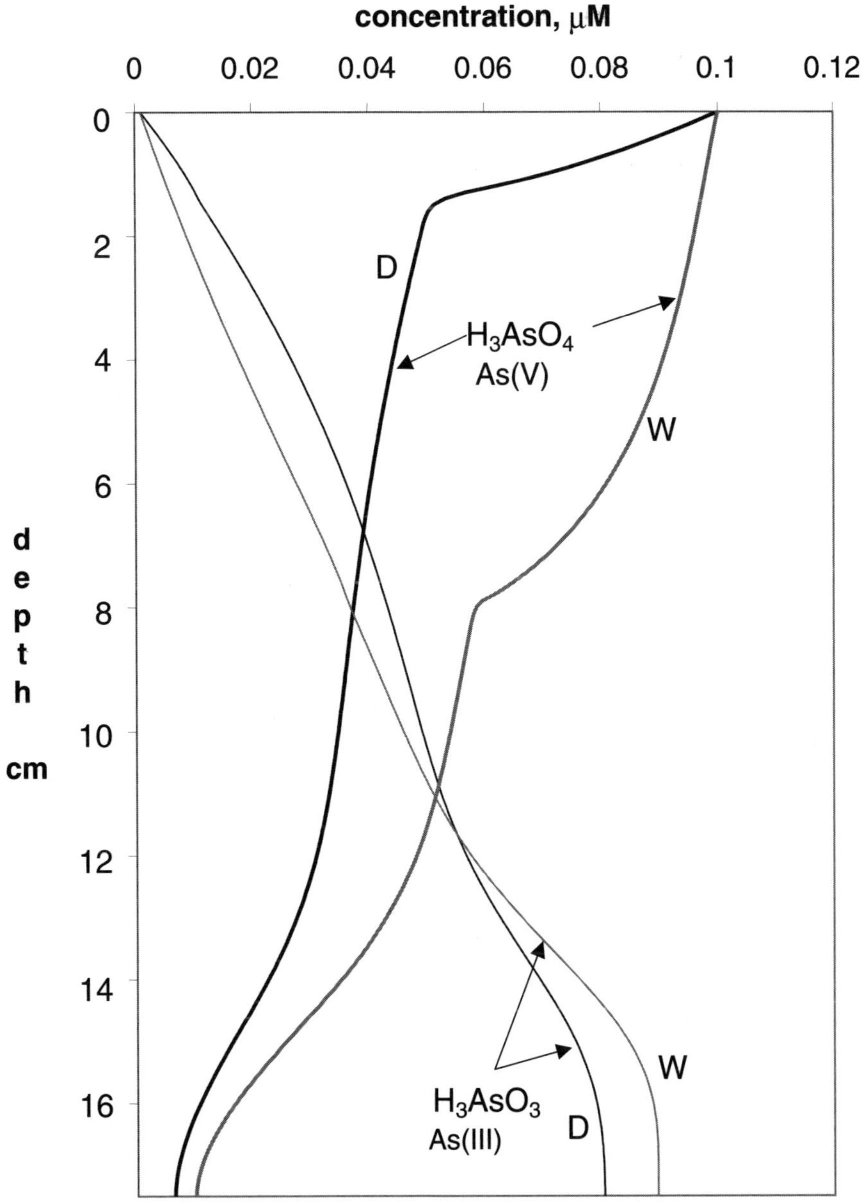

Fig. 5. Simulated concentration profiles of arsenate (As V) and arsenite (As III) in wetland (W) and deep (D) sediments.

because of the change in the concentration gradients, but also due to the increase in the advective flow induced by evapotranspiration. Since these are steady-state simulations, the dissolved-phase fluxes are not affected by the increased sorption of arsenate onto the larger amounts of iron and manganese oxides in wetland sediments. However, when the total flux of arsenic, including the sorbed phase, is considered, the presence of plants tends to increase the amount of arsenic buried in the sediments (Kallin, 1999). If we were to consider temporal spikes in the water column's arsenic concentration, sorption would play a larger role in decreasing the dissolved-phase concentration and the effect of roots on buffering the flux of arsenite into the water column would be even larger. It should be noted that a

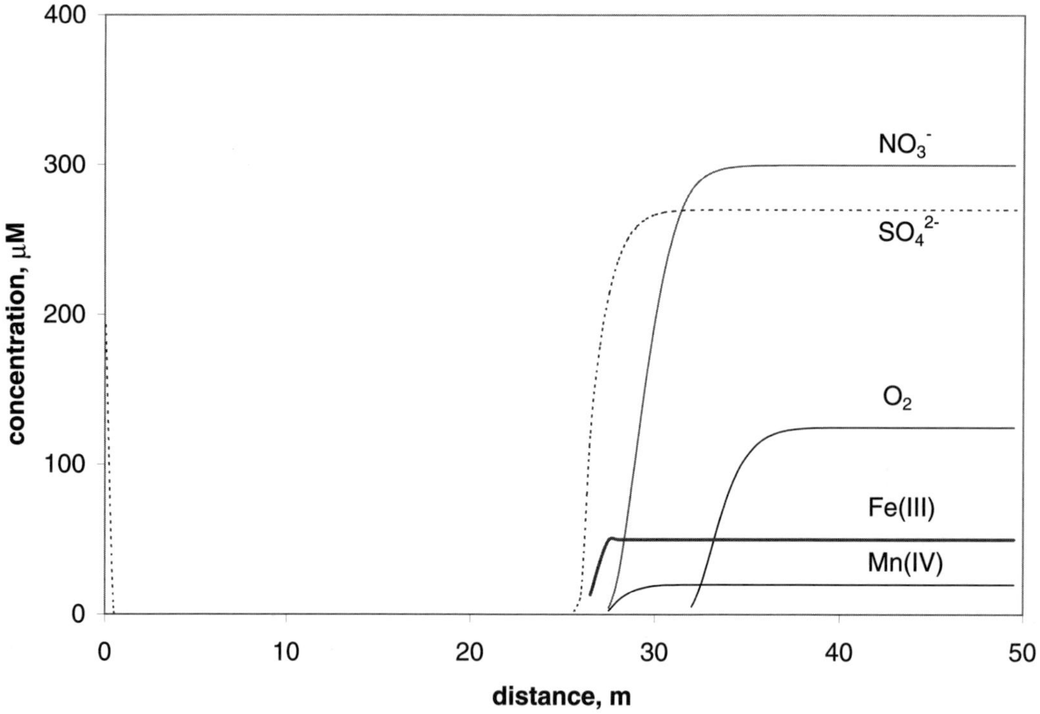

Fig. 6. Simulated concentration profiles of terminal electron acceptors in groundwater.

wide range of root density distribution can be expected, depending upon the type of wetland vegetation, and that different root density distributions will affect the individual concentration profiles.

Because of the very different nature of the groundwater environment, groundwater simulations were conducted for different flow conditions and length scales. Figure 6 shows the simulated one-dimensional profiles of the different electron acceptors, and Fig. 7 that of the reduced species, one year after a carbon source was injected at the origin. Note that in the groundwater environment the solid-phase bioavailable Mn(IV) and Fe(III) are not replenished at the interface as in the sedimentary environment, and are therefore depleted over time.

The simulated dynamics of arsenic in the groundwater environment are illustrated in Fig. 8 for time increments of 0.2 years, up to a one-year period, after a carbon source was injected at the origin. While As(V) is reduced to As(III), the total aqueous phase arsenic concentration increases significantly as time progresses. The reason for this increase is that a large fraction of the As(V) is sorbed while the As(III) sorbs to a much smaller extent. As dissolved As(V) is reduced, sorbed As(V) will desorb to reestablish equilibrium adsorption conditions. Furthermore the ratio of dissolved to adsorbed As(V) increases over time because the bioavailable amorphous Fe(III) has been consumed as an electron acceptor, which decreases the total sites available for sorption. In these as well as in the sediment simulations, it was assumed that 30% of the total Fe(III) is amorphous and available to bacteria. Arsenate sorption to amorphous iron oxide was assumed to be the same as to the crystalline iron oxide.

As discussed earlier, arsenate can act as an electron acceptor for certain bacteria during the degradation of organic matter, and the same generic formulation was used for arsenate reduction as for the other dissolved electron acceptors. As opposed to the major electron acceptors described in the model, arsenate reduction was allowed to proceed in parallel with the reduction of the other electron acceptors after nitrate has been consumed. For this reason arsenate reduction proceeds even when the environment becomes highly reduced and methanogenesis occurs, such as deep in the sediments or in the groundwater zone where sulfate has been con-

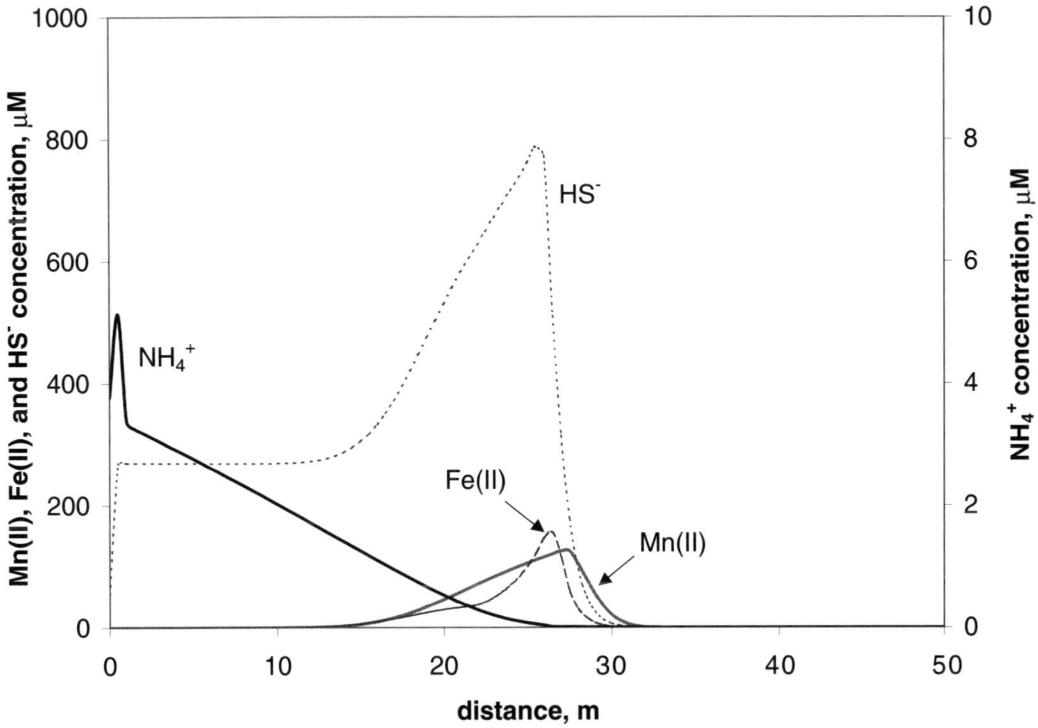

Fig. 7. Simulated concentration profiles of reduced species in groundwater.

sumed. Further research is necessary to establish if there is a lower redox limit for the reduction of arsenate. The effect of the current formulation on arsenic reduction can be seen in Figs. 5 and 8, where arsenate concentration steadily decreases and arsenite concentration increases with depth or time. Even though microbiological arsenate reduction is not initiated until all the nitrate has been consumed, Fig. 5 shows that a concentration gradient for arsenate and arsenite extends from the water/sediment interface through the aerobic and nitrate reducing zones. This gradient develops due to mixing (mainly bioturbation and irrigation) and does not imply that arsenate is being reduced in these zones. For the groundwater simulations, where there is much less longitudinal mixing, one can see clearly from Figs. 6 and 8 that after a one-year period, arsenite production is starting at the location where nitrate is depleted ($x \approx 27$ m).

3. DISCUSSION

The results presented here illustrate a methodology capable of simulating the biogeochemical dynamics of trace metals and metalloids, specifically arsenic, in porous media. Redox profiles, which develop in porous media in response to the utilization of different terminal electron acceptors during the biodegradation of organic carbon, were simulated, and coupled with the dynamics of arsenic as it is transported across the different redox zones. Although there are significant differences in the formulation of physical transport processes among the three environmental settings considered here, the same generic biogeochemical formulations are applicable in all cases. The key differences in physical processes that have been considered here are sedimentation, enhanced oxygen transport by roots, transport-induced advection, and differences in mixing. We have shown that these processes have significant effects on the concentration profiles of all the chemical species simulated and how these profiles may overlap with each other. If a solid phase is being consumed, as is here the case for Mn(IV) and Fe(III), a dynamic formulation needs to be implemented if the solid matrix is stationary. For a non-stationary solid matrix, as is the case for sediments where the sediment/water interface is moving over time due to

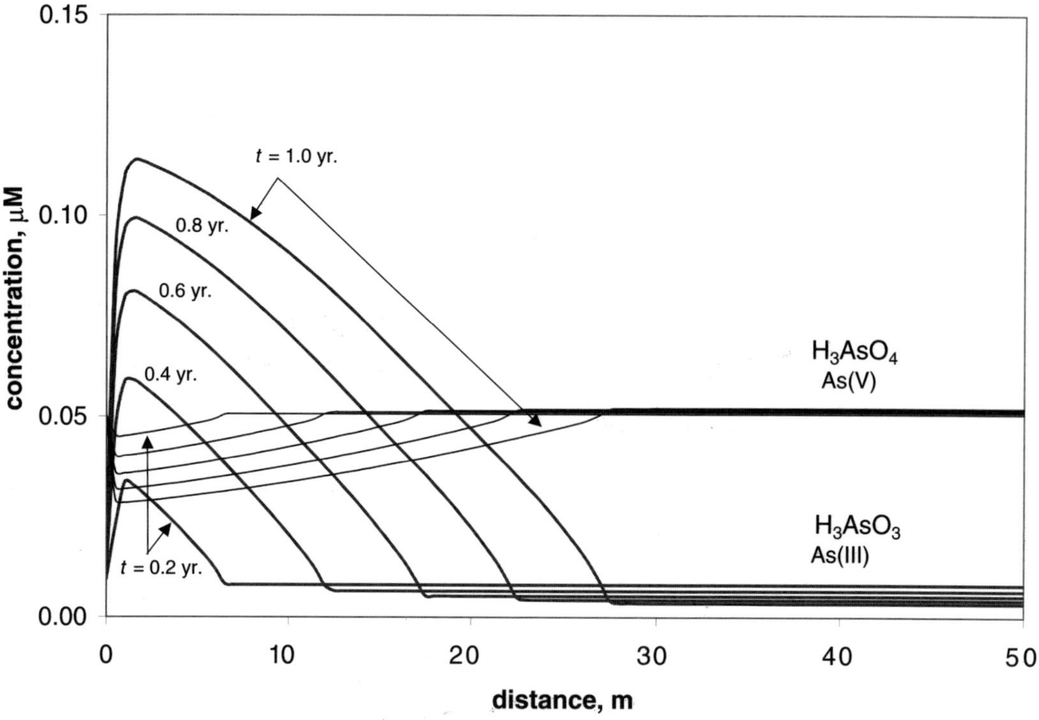

Fig. 8. Time evolution (0.2 < *t* < 1.0 yr) of aresenate and arsenite concentrations with distance in groundwater.

sedimentation, a steady-state formulation will yield zones where manganese and iron reduction are occurring. A steady-state formulation may therefore be appropriate to simulate a long-term, quasi-steady state in sedimentary environments. Given that sorption of arsenate to sorbents such as iron oxides affects the fate of arsenic in different porous media environments, it is extremely important to properly simulate the dynamics of these sorbents. Significant amounts of manganese and iron oxides can build up in the rhizosphere of wetland sediments, which can have an important effect on the overall immobilization of arsenic in wetlands.

A common feature used in the formulations examined here is that individual chemical species are not assumed to be at chemical equilibrium. Chemical equilibrium calculations are only used to determine the deviation from equilibrium conditions, based on which a kinetic formulation is implemented. This formulation allows different species to coexist. For example, as shown in all of our simulations, species such as Fe^{2+} do coexist with other more oxidized species such as NO_3^-. This is commonly observed in sediments as well as groundwaters, and could not be simulated if equilibrium formulations were implemented.

These reaction kinetics can have significant impacts on the individual chemical profiles. As can be seen from the insert in Fig. 1, when the water flow is from the oxidized zone to the reduced zone, the electron acceptors are depleted sequentially in the order shown in Table 1. Conversely, when the water flows from reduced to more oxidized zones (e.g., groundwater for larger distances from the origin or groundwater discharge type wetlands), the various biotic and abiotic reactions occurring simultaneously can alter the spatial sequence of the profiles of the electron acceptors from that described in Table 1. This, as well as the overlapping concentration profiles near the water/sediment interface, illustrates that knowledge of the electron acceptor profiles alone does not provide adequate information to identify the zones where a specific microbiological process is occurring. Additional measurements such as the partial pressure of H_2, which can define the bacteria capable of existing at that partial pressure (e.g.,

Lovley and Goodwin, 1988), might be required to better assess which degradation process is occurring at a given location. Such measurements may also help to more efficiently compare the accuracy of the model simulations to field data.

The simulated concentration profiles of electron acceptors in deep sediments shown here compare well to a series of profiles measured in lake sediments and described in Smith and Jaffé (1998). This was achieved using rate coefficients consistent with those reported in the literature (Smith and Jaffé, 1998; Kallin, 1999), and it therefore increases our confidence in the model. Yet accurate model verification, especially for field conditions, is difficult to achieve due to the complexity of these models and the large number of model coefficients. Even though the groundwater model formulation for the main electron donors/acceptors described here differs from that described by Hunter et al. (1998), simulations conducted for the same flow conditions and equivalent rate coefficients, resulting from either model, show a remarkably close match in the concentration profiles of the electron acceptors (not shown). This indicates that the model outputs are rather robust and not very sensitive to small alterations in the process formulations. For example, this model uses Monod kinetics for the biological reactions, while that by Hunter et al. (1998) relies on first-order kinetics.

Given the limitations in model validation, the utility of biogeochemical models such as those presented here is mainly as a scientific tool to understand variable interactions, and to be used as an aid in interpreting trends in field observations. For example, by comparing the simulations for deep sediments to those for wetland sediments, one can gain a reasonable understanding on how the colonization of aquatic plants will affect the fluxes of trace metals or metalloids in a shallow lake. Alternatively, in the case of groundwater, the simulation shows to what degree a trace metal or metalloid can be immobilized if a carbon source is added either on purpose or accidentally to the subsurface.

Finally, it should be noted that the one-dimensional models presented here do not allow for heterogeneities in the porous media where localized differences in the microbiological processes can develop. This is expected to occur in natural settings and would contribute even more to the smearing of specific concentration profiles.

Acknowledgements—The development of the model to simulate trace-metal dynamics in wetland sediments was funded in part by the U.S. EPA, grant # R827288. The simulations for arsenic in groundwater were conducted with a model developed for trace metal/radionuclide bioremediation, funded in part by the DOE NABIR program, grant # DE-FG02-98ER62705. The authors also wish to thank Drs. Nikolaidis, Van Cappellen, Charlet, and two anonymous reviewers for their constructive reviews of this manuscript.

I met David Crerar in 1982, and our first collaborative effort was on a small project dealing with the geochemistry and transport of arsenic in a shallow groundwater system. From that time, and because of David's love of nature and his wide range of interests, we had many deeply stimulating and enjoyable discussions about a variety of biogeochemical processes in different environmental settings, ranging from wetlands to lakes and groundwaters. It is for this reason that I thought this article would be appropriate in a special issue dedicated to him. I count myself fortunate to have known David as a colleague and close friend.

— Peter Jaffé
Princeton, NJ
September 2000

Editorial handling: R. Hellmann

REFERENCES

Armstrong W. (1979) Aeration in higher plants. *Adv. Bot. Res.* **7**, 226-332.

Bedford B., Bouldin D. R., and Beliveau B. D. (1991) Net oxygen and carbon dioxide balances in solutions bathing roots of wetland plants. *J. Ecol.* **79**, 943-959.

Berner, R.A. (1980) *Early Diagenesis; A Theoretical Approach.* Princeton University Press, Princeton.

Brendel, P. J. and Luther G. W. III (1995) Development of a gold amalgam voltametric microelectrode for the determination of dissolved Fe, Mn, O_2, and S(-II) in porewaters of marine and freshwater sediments. *Environ. Sci. Technol.* **29**, 751-761.

Brix H. (1993) Macrophyte-mediated oxygen transfer in wetlands: Transport mechanisms and rates. In *Constructed Wetlands for Water Quality Improvement* (ed. G. A. Moshiri). Lewis Publishers, Inc., Boca Raton. pp. 391-398.

Brix H., Sorrel B. K., and Schierup H. H. (1996) Gas fluxes by in situ convective flow in *Phragmites australis. Aquatic Botany* **54**, 151-163.

Dhakar S. P. and Burdige D. J. (1996) A coupled, non-linear, steady state model for early diagenetic processes in pelagic sediments, *Amer. J. Sci.* **296**, 296-330.

Dowdle P. R., Laverman A. M., and Oremland R. S. (1996) Bacterial dissimilatory reduction of arsenic(V) to arsenic(III) in anoxic sediments. *Appl. Environ. Microbiol.* **62**, 1164-1669.

Dunnivant F. M., Jardine P. M., Taylor D. L., and McCarthy J. F. (1992) Cotransport of cadmium and hexachlorobiphenyl by dissolved organic-carbon through columns containing aquifer material. *Env. Sci. Technol.* **26**, 360-368.

Dzombak D. A. and Morel F. M. M. (1990) *Surface Complexation Modeling: Hydrous Ferric Oxide.* John Wiley, New York.

Egesgraad P. and Kipp K. L. (1992) A geochemical transport model for redox-controlled movement of mineral fronts in groundwater flow systems: A case of nitrate removal by oxidation of pyrite. *Wat. Resour. Res.* **28**, 2829-2843.

Fendorf S., Eick M. J., Grossl P., and Sparks D. L. (1997) Arsenate and chromate retention mechanisms on geothite, 1. Surface structure. *Env. Sci. Technol.* **31**, 315-320.

Gambrell R. P. and Patrick W. H. Jr. (1978) Chemical and microbiological properties of anaerobic soils and sediments. In *Plant Life in Anaerobic Environments* (eds. D. D. Hook and R. M. M. Crawford). Ann Arbor Science Publishers, Ann Arbor. pp. 375-424

Grosse W. (1997) Gas transport in trees. In *Contributions to Modern Tree Physiology* (eds. H. Renneberg, W. Eschrich, and H. Ziegler). Backhuys Publishers, Leiden. pp. 57-74.

Gulens J., Champ D. R., and Johnson K. E. (1979) Influence of redox environments on the mobility of arsenic in ground water. In *Chemical Modeling in Aqueous Systems*. (ed. E. A. Jenne). Amer. Chem. Soc. Symp. Ser. 93, American Chemical Society, Washington, D.C. pp. 81-95.

Holm T. R., Anderson M. A., Iverson D. G., and Stanforth R. S. (1979) Heterogeneous interactions of arsenic in aquatic systems. In *Chemical Modeling in Aqueous Systems*. (ed. E. A. Jenne). Amer. Chem. Soc. Symp. Ser. 93, American Chemical Society, Washington, D.C. pp. 711-736.

Hunter K. S., Wang Y., and Van Cappellen P. (1998) Kinetic modeling of microbially-driven redox chemistry of subsurface environments: coupling transport, microbial metabolism and geochemistry. *J. Hydrol.* **209**, 53-80.

Jardine P. M., Fendorf S. E., Mayes M. A., Larsen I. L., Brooks S. C., and Bailey W. B. (1999) Fate and transport of hexavalent chromium in unsaturated heterogeneous soil. *Env. Sci. Technol.* **33**, 2939-2944.

Kallin P. L. (1999) Modeling the fate and transport of trace metal contaminants in natural and constructed surface flow wetlands. Ph.D. thesis, Princeton University.

Lensing H. J., Vogt M., and Herrling B. (1994) Modeling of biologically mediated redox processes in the subsurface, *J. Hydrol.* **159**, 125-143.

Lovley, D.R. and Goodwin S. (1988) Hydrogen concentration as an indicator of the presence of the predominant terminal electron-accepting reaction in aquatic sediments. *Geochim. Cosmochim. Acta* **52**, 2993-3003.

Massacheleyn P. H., Delaune R. D., and Patrick W.H. (1991) Effect of redox potential and pH on arsenic speciation and solubility in contaminated soil. *Environ. Sci. Technol.* **25**, 1414-1419.

Matsunaga T., Karametaxas G., Von Gunten H. R., and Lichtner P. C. (1993) Redox chemistry of iron and manganese minerals in river recharged aquifers: A model interpretation of a column experiment. *Geochim. Cosmochim. Acta* **57**, 1691-1704.

McCarthy J. F. and Zachara J. M. (1989) Subsurface transport of contaminants - mobile colloids in the subsurface environment may alter the transport of contaminants. *Environ. Sci. Technol.* **23**, 496-502.

Mendelssohn I. A. (1993) *Factors controlling the formation of oxidizing root channels: a review and annotated bibliography. Tech. Report WRP-DE-5.* U.S. Army Engineer Waterways Experiment Station.

Mitsch W. J. and Gosselink J. G. (1993) *Wetlands*, 2nd ed. Van Nostrand Reinhold Co., New York.

Morel F. M. M. and Hering J. G. (1993) *Principles and Applications of Aquatic Chemistry*. J. Wiley & Sons, New York.

Newman D. K., Beveridge T. J., and Morel F. M. M. (1997a) Precipitation of arsenic trisulfide by *Desulfotomaculum auripigmentum*. *Appl. Env. Microbiol.* **63**, 2022-2028.

Newman D. K., Kennedy E. K., Coates J. D., Ahmann D., Ellis D. J., Lovley D. R. and Morel F. M. M. (1997b) Dissimilatory arsenate and sulfate reduction in *Desulfotomaculum auripigmentum* sp. nov. *Arch. Microbiol.* **168**, 380-388.

Oscarson D. W., Huang P. M., and Liaw W. K. (1980) The oxidation of arsenite by aquatic sediments. *J. Environ. Qual.* **9**, 700-703.

Otte M. L., Kearns C. C., and Doyle M. O. (1995) Accumulation of arsenic and zinc in the rhizosphere of wetland plants. *Bull. Environ. Contam. Toxicol.* **55**, 154-161.

Park S. S. and Jaffé P. R. (1996) Development of a redox potential model for the assessment of postdepositional heavy metal mobility. *Ecological Modelling* **91**, 169-181.

Ponnamperuma F. N. (1984) Effects of flooding on soils. In *Flooding and Plant Growth* (ed. T. T. Kozlowki). Academic Press, New York. pp. 10-46.

Rabouille C. and Gaillard J.-F. (1991) A coupled model representing the deep-sea organic carbon mineralization and oxygen consumption in surficial sediments. *J. Geophys. Res.* **96**, 2761-2776.

Sand-Jensen K., Prahl C., and Stokholm H. (1982) Oxygen release from roots of submerged aquatic macrophytes. *Oikos*. **38**, 349-354.

Scott M. J. and Morgan J. J. (1995) Reactions on oxide surfaces 1. Oxidation of As(III) by synthetic birnessite. *Environ. Sci. Technol.* **29**, 1898-1905.

Smith S. L. and Jaffé P. R. (1998) Modeling the transport and reaction of trace metals in water-saturated soils and sediments. *Water Resour. Res.* **34**, 3135-3147.

Sorrel B. K. (1994) Airspace structure and mathematical modelling of oxygen diffusion, aeration and anoxia in *Eleocharis sphacelata* R. Br. roots. *Aust. J. mar. Freshwater Res.* **45**, 1529-1541.

Sorrel B. K., Brix H., and Orr P. T. (1993) Oxygen exchange by entire root systems of *Cyperus involucratus* and *Eleocharis sphacelata*. *J. Aquatic Plant Management* **31**, 24-28.

Sweerts J. R. A., Bar-Gillissen M-J., Cornelese A. A., and Cappenberg T. E. (1991) Oxygen-consuming processes at the profundal and littoral sediment-water interface of a small meso-eutrophic lake (Lake Vechten, the Netherlands). *Limnol. Oceananography* **36**, 1124-1133.

Van Cappellen P. and Wang Y. (1995) Metal cycling in sediments: Modeling the interplay of reaction and transport. In *Metal Contaminated Aquatic Sediments* (ed. H. E. Allen). Ann Arbor Press, Chelsea. pp. 21-64.

Vitousek P.M., Fahey T., Johnson D.W., and Swift M.J. (1988) Element interactions in forest ecosystems - succession, allometry and input-output budgets. *Biogeochemistry* **5**, 7-34.

Von Gunten U. and Zobrist J. (1993) Biogeochemical changes in groundwater infiltration systems: Column studies. *Geochim. Cosmochim. Acta* **57**, 3895-3906.

Wang Y. and van Cappellen P. (1996) A multicomponent reactive transport model of early diagenesis: Application to redox cycling in coastal marine sediments. *Geochim. Cosmochim. Acta* **60**, 2993-3014.

Yeh G. T. and Tripathi V. S. (1991) A model for simulating transport of reactive multispecies components: Model development and demonstration. *Water Resour. Res.* **27**, 3075-3094.

APPENDIX — FORMULATION OF MODEL EQUATIONS

Mass balance equations for dissolved species

- In deep sediments:

$$\frac{\partial}{\partial t}\left(\phi\left[1+K_i^{eff}\right]\right)C_i^{aq} = -(V-W)\frac{\partial \phi C_i^{aq}}{\partial z} + W\frac{\partial\left(K_i^{eff}\phi C_i^{aq}\right)}{\partial z} + \frac{\partial}{\partial z}\left[\left(D_{h_i}+D_b+D_{irr}\right)\cdot\frac{\partial \phi C_i^{aq}}{\partial z} + D_b\frac{\partial\left(K_i^{eff}\phi C_i^{aq}\right)}{\partial z}\right] + \sum R_i$$

- In wetland sediments:

$$\frac{\partial}{\partial t}\left(\phi\left[1+K_i^{eff}\right]\right)C_i^{aq} = -\frac{\partial}{\partial z}\left[\left(V(z)-W\right)\phi C_i^{aq}\right] + W\frac{\partial\left(K_i^{eff}\phi C_i^{aq}\right)}{\partial z} + \frac{\partial}{\partial z}\left[\left(D_{h_i}+D_b(z)+D_{irr}(z)\right)\cdot\frac{\partial \phi C_i^{aq}}{\partial z} + D_b(z)\frac{\partial\left(K_i^{eff}\phi C_i^{aq}\right)}{\partial z}\right] + \sum R_i$$

- In groundwater:

$$\frac{\partial}{\partial t}\left(\phi\left[1+K_i^{eff}\right]\right)C_i^{aq} = D_h\frac{\partial^2 \phi C_i^{aq}}{\partial x^2} - V\frac{\partial \phi C_i^{aq}}{\partial x} + \sum R_i$$

t : time [T]

ϕ : porosity

K_i^{eff} : effective partition coefficient for the equilibrium adsorption of species i [M/M]

C_i^{aq} : concentration of the dissolved species i [M/V]

V : infiltration velocity of water (positive down) in sediments or groundwater velocity [L/T]

z : distance from the sediment/water interface [L]

W : velocity of the sediment/water interface (from deposition: negative values corresponding to accumulation of sediment) [L/T]

D_{hi} : coefficient of hydrodynamic dispersion (mechanical dispersion + molecular diffusion) of species i [L^2/T]

D_b : coefficient of bioturbation (mixing of sediment by aquatic and benthic organisms) [L^2/T]

D_{irr} : coefficient of irrigation (mixing of pore-water by benthic organisms) [L^2/T]

ΣR_i : sum of consumption/production of species i by biotic/abiotic reactions [M/VT]

x : distance parallel to groundwater flow from the origin [L]

The effective partition coefficient for species i, K_i^{eff}, is defined as:

$$K_i^{eff} = \frac{(1-\phi)\cdot[\text{total adsorbed concentration of component species } i]}{\phi\cdot[\text{total dissolved concentration of component species } i]}$$

Mass balance equations for solid species

- In deep sediments:

$$\frac{\partial}{\partial t}\left[(1-\phi)C_i^s\right] = W\frac{\partial}{\partial z}\left[(1-\phi)C_i^s\right] + \frac{\partial}{\partial z}\left[(1-\phi)D_b\cdot\frac{\partial C_i^s}{\partial z}\right] - \sum R_i$$

- In wetland sediments:

$$\frac{\partial}{\partial t}\left[(1-\phi)C_i^s\right] = W\frac{\partial}{\partial z}\left[(1-\phi)C_i^s\right] + \frac{\partial}{\partial z}\left[(1-\phi)D_b(z)\cdot\frac{\partial C_i^s}{\partial z}\right] - \sum R_i$$

- In groundwater:

$$\frac{\partial}{\partial t}\left[(1-\phi)C_i^s\right] = -\sum R_i$$

C_i^s : concentration of the solid species i [M/V]

The bioturbation and irrigation coefficients were assumed to decrease with depth according to:

$$D_b = \frac{D_b^0}{\left[1+\exp\{z-z_b\}\right]},\quad D_{irr} = \frac{D_{irr}^0}{\left[1+\exp\{z-z_{irr}\}\right]}$$

D_b^0, D_{irr}^0 : coefficients of bioturbation and irrigation at the sediment/water interface in sediments [L^2/T],

z_b, z_{irr} : characteristic depths for bioturbation and irrigation [L]

The rate of oxidation of organic carbon, C_C, by a given terminal electron acceptor of concentration C_{eA}:

$$R_C^{eA} = -\chi_{eA}\mu_{meA}\left(\frac{C_{eA}}{K_{seA}+C_{eA}}\right)\left(\frac{C_C}{K_{seA,C}+C_C}\right)$$

χ_{eA} : indicator coefficient (1 when this electron acceptor is being utilized, and 0 when it is not)

μ_{meA} : maximum rate of carbon oxidation via this electron acceptor [M/VT]

K_{seA} : half-saturation coefficient for the electron acceptor [M/V]

$K_{seA,C}$:half-saturation coefficient for the carbon substrate (for consumption of this electron acceptor) [M/V]

The rate of consumption of a given terminal electron acceptor, which equals the rate of production of its corresponding reduced species:

$$R_{eA} = -\alpha_{eA}\cdot R_C^{eA} =$$

$$-\alpha_{eA}\chi_{eA}\mu_{meA}\left(\frac{C_{eA}}{K_{seA}+C_{eA}}\right)\left(\frac{C_C}{K_{seA,C}+C_C}\right)$$

$$= -R_{\text{corresponding reduced species}}$$

α_{eA}: stoichiometric coefficient representing the ratio of the number of moles of a given electron acceptor required to oxidize one mole of organic carbon

The rate of oxidation of ammonia and corresponding production of nitrate and consumption of oxygen:

$$R_{NH_3}^{O_2} = -\alpha_{NH_3,O_2}\mu_{mNH_3,O_2}$$

$$\times\left(\frac{C_{NH_3}}{K_{s,NH_3}+C_{NH_3}}\right)\left(\frac{C_{O_2}}{K_{s,O_2}^{NH_3}+C_{O_2}}\right) = -R_{NO_3^-}^{O_2}$$

Second-order rate of redox reactions:

- Rates of oxidation of dissolved reduced species i (e.g., Mn^{2+}, Fe^{2+}, HS^-, CH_4) by oxidant j:

$$R_i^j = \mu_{j,i}C_jC_i$$

$\mu_{j,i}$: second order rate coefficient

- Rates of oxidation of adsorbed Fe(II) and Mn(II) (represented as a function of the concentration of the reduced metal adsorbed onto Mn(IV) oxides):

$$R_{M^{2+}}^{ox} = \mu_{M^{2+}}^{ox}\left[XO_2\cdot M^{2+}\right]\cdot C_{O_2}$$

$\mu_{M^{2+}}^{ox}$: rate coefficient for oxidation of adsorbed metal M^{2+}

$\left[XO_2\cdot M^{2+}\right]$: concentration of metal M^{2+} adsorbed onto metal oxides, XO_2(the adsorption is calculated by the equilibrium speciation model)

For a given solid, S, composed of N_s species, the net rate of precipitation, R_S^{net}:

$$R_S^{net} = k_s\cdot\left(\prod_i^{N_s}[FIA_i]^{\alpha_{i,s}} - K_{sp}^S\right)$$

k_s : precipitation rate coefficient specified for solid S [reaction dependent, for $N_s = 2$ and $\alpha_{i,s} = 1$, $M^{-1}VT^{-1}$]

FIA_i : dissolved free ion activity of component i [M/VT]

$\alpha_{i,s}$: stoichiometric coefficient of the i-th component of solid S

K_{sp}^S : equilibrium solubility product for solid S [reaction dependent, for $N_s = 2$ and $\alpha_{i,s} = 1$, M^2/V^2]

Integration of the role of plants in wetland sediment model

- Variable velocity field (described as the superposition of a groundwater infiltration velocity field and an evapotranspiration-induced velocity field)

$$V(z) = V_{Inf} + \alpha ET\cdot w(z)$$

V_{Inf} : groundwater infiltration velocity [L/T]

α : proportionality constant (typically assumed as 0.5)

ET : total annual evapotranspiration [L/T]

$w(z)$: a normalized root density function

$$= \cos\left[\frac{\pi}{2} \cdot \frac{z}{z_{rhiz}}\right]$$

z_{rihz} : depth of the rhizosphere [L]

- Oxygen diffusion in the rhizosphere:

$$Oxysource(z) = \delta \cdot \frac{a}{z_{rhiz}} \cdot Rhizload \cdot w(z)$$

δ : an indicator function (1 when $z < z_{rhiz}$, 0 otherwise)

a : normalization factor

$Rihzload$: the oxygen loading [M/L^{2}T]

- Nitrogen uptake:

$$Nitup(z) = \delta \cdot ET \cdot w(z) NitMT_0 \left(C_0^{NO_3} - C^{NO_3}(z)\right)$$

$Nitup\ (z)$: nitrate uptake rate [M/VT]

$NitMT_0$: baseline mass transfer rate coefficient [1/T]

$C_0^{NO_3}$: specified minimum nitrate concentration below which uptake stops [M/V]

$C^{NO_3}(z)$: aqueous nitrate concentration [M/V]

Water-Rock Interactions, Ore Deposits, and Environmental Geochemistry: A Tribute to David A. Crerar
© The Geochemical Society, Special Publication No. 7, 2002
Editors: Roland Hellmann and Scott A. Wood

Primary drainage of NAPL governed by time-dependent interfacial properties

DAVID M. TUCK

NAPLogic, Inc., P.O. Box 72, Princeton, NJ 08542-0072, USA
Email: dtuck@naplogic.com

Abstract—Primary drainage involving a non-aqueous phase liquid (NAPL) invading a water-saturated, initially water-wet porous medium is a process having important practical interest. It occurs during release of organic liquid contaminants such as gasoline, solvents, or other organic wastes. It also occurs during primary migration of oil following its generation in a source zone. Standard conceptual and numerical pore-scale models of this process do not incorporate time in any explicit manner in their representation of the capillary forces that control this process. Time-dependence of interfacial chemistry, however, is a well-known phenomenon, with some of the earliest interfacial chemical work exhibiting its existence, and quantitative evaluation becoming established in the late 1930's. Time-dependence of interfacial properties is manifested during primary drainage through the sequential processes of interface dilation and compression as a fluid-fluid interface advances through a porous medium. This sequence is illustrated and described schematically to demonstrate its importance in the drainage process.

Log-linear interfacial chemical kinetics of a known system was incorporated into a small pore-scale model to assess its influence on primary drainage. NAPL saturation was a function of time as well as applied capillary pressure. NAPL migration rate became a log-linear function of applied capillary pressure, and the maximum NAPL saturation exhibited a minimum at the entry pressure for pure bulk NAPL without any interfacially active components. This arises from competition between applied capillary pressure and the interfacial kinetics for control over the timing of when and whether NAPL penetrates a given pore or not. The interfacial chemical kinetics dominate at lower applied capillary pressure. More time becomes available for the interfacial chemical kinetics to lower interfacial tension sufficiently to enter pores that would otherwise be by-passed as the applied capillary pressure decreases. These results stand in marked contrast with the "standard" equilibrium view of the capillary pressure-saturation relationship. Comparison of model results, assuming standard Tempe cell dimensions, indicates initial penetration of a medium occurs at lower capillary pressure and final saturation is higher when time-dependent interfacial chemistry is incorporated.

1. INTRODUCTION

Subsurface non-aqueous phase liquid contamination, both light (LNAPL) and dense (DNAPL), poses a major environmental problem for many commercial and industrial sites. For the purposes of this paper, NAPL, LNAPL, and DNAPL are interchangeable; the behavior will be comparable with respect to the interfacial chemistry. Understanding DNAPL behavior in porous media is important for designing effective strategies for locating subsurface DNAPL and for designing effective remediation systems. The displacement of water by a non-aqueous phase liquid (NAPL) is called primary drainage. It is also important for petroleum migration from a source area. Two important interfacial properties play major roles in subsurface NAPL movement during primary drainage: the surface or interfacial tension between fluids and the contact angle between the fluid-fluid interface and the mineral-grain solid surfaces. "Surface tension" is generally applied to gas-liquid systems, while "interfacial tension" is applied to liquid-liquid systems.

Interfacial properties are generally treated as static by assuming equilibrium values when modeling multiphase behavior in porous media, either explicitly in pore-scale models or implicitly in continuum models. This assumption is accurate for three cases: 1) single component NAPLs, 2) systems where the magnitude of the kinetic changes are small with respect to the equilibrium values, or 3) systems where the interfacial chemical kinetics are rapid with respect to the process under consideration. Interfacial chemical kinetics involves the rates of interfacial reactions, such as partitioning of surface active substances into a fluid-fluid interface. Assuming equilibrium, however, is not necessarily accurate for the process of primary drainage (i.e., initial penetration of a water-wet porous medium by a NAPL). The first objective of this paper is to briefly review literature on time-dependent interfacial chemistry to show it is not a new or unknown phenomenon. The second objective is to illustrate why equilibrium is not necessarily a valid *a priori* assumption with respect to primary drainage, and how time-dependent interfacial effects are manifested during primary drainage. The final objective is to show the effects of incorporating time-dependent interfacial chemistry in a small pore-scale model.

1.1. Time-Dependence in Interfacial Chemistry

Early indications that interfacial chemical equilibrium is not established instantaneously are over a hundred years old. Tate (1864) showed that interfacial tension of solutions measured using the drop volume method depends on the rate of formation of the drop. The faster a drop is formed (i.e., the younger its average surface age), the higher is the measured interfacial tension. Dupré (1869) studied the surface tensions of aqueous soap solutions, noting that they decreased as the surface aged. Lord Rayleigh (1879) developed the pulsating jet method for measuring surface tension, and later (Rayleigh, 1890) showed it could be used to resolve surface tension changes over short time-scales (on the order of seconds). Rayleigh also concluded that surface tension of sodium oleate solutions decreased with increasing age of the surface. Bohr (1909) further developed and refined the vibrating or oscillating jet method, facilitating its application by later workers (Rideal and Sutherland, 1952; Addison and Litherland, 1953; Thomas and Potter, 1975; Zhang and Zhao, 1989). Several research groups used other methods to show that surface tension of aqueous soap solutions varied with time (Bigelow and Washburn, 1928; Tartar et al., 1940). Andreas et al. (1938) extended and further developed the pendant drop technique for measuring surface and interfacial tensions. Ward and Tordai (1944) showed that the interfacial tension of solutions is, in general, time-dependent. Later, Ward and Tordai (1946) were the first to explicitly apply a diffusion model to fit kinetic data for interfacial tension. Several authors have since measured time-dependent surface or interfacial tension covering many orders of magnitude in time scales. Time-scales of interfacial or surface tension kinetics range from milliseconds (Thomas and Potter, 1975; Zhang and Zhao, 1989) to seconds and minutes (Lucassen and Giles, 1975; Tuck and Rulison, 1999a), to minutes and hours (Pan et al., 1998), to hours, days, and longer (Tartar et al., 1940; Schroth et al., 1995; Tuck and Rulison, unpublished data; Tuck and Rulison, 1999a).

Time-dependence in interfacial chemistry is thus a well-known phenomenon. This time-dependence can occur over a very large range of time scales, depending on the system under consideration. It is therefore not necessarily appropriate to *assume* equilibrium values for interfacial tension and contact angle. Tuck and co-workers (Tuck et al., 1998) have shown that the *effective* value of interfacial tension between water and a dyed-DNAPL may vary significantly during primary drainage when interfacial chemical kinetics is considered. They showed that the faster a penetrating front of a complex NAPL (exhibiting time-dependent surface chemistry) penetrates a porous medium, the younger will be the average interface age, and therefore the higher will be *effective* interfacial tension. Whether or not equilibrium values will accurately reflect their *effective* values during primary drainage of a complex NAPL in an initially water-wet porous medium depends on the relative time scales of the drainage process and the interfacial chemical kinetics. It is inappropriate to assume equilibrium values measured for pure phases when modeling the behavior of complex phases. Examples of these situations include DNAPLs stained with an azo dye (Tuck and Rulison, 1999b) or field NAPLs such as gasoline or spent solvents (Powers and Tamblin, 1995; Tuck et al., 1998).

The mechanism through which time-dependent interfacial properties are manifested during primary drainage is illustrated in a later section. The potential influence of interfacial chemical kinetics on primary drainage is demonstrated in the subsequent section through the development and implementation of a pore-scale model in which time-dependent interfacial tension is incorporated.

1.2. Theory

1.2.1. Surface chemistry and pore-scale interface movement

Interfacial chemistry is important in multiphase flow in porous media because of its role in capillary forces. Capillary forces determine which pores will be penetrated by an infiltrating NAPL during primary drainage. This role is expressed through the Young-Laplace equation (Dullien, 1992):

$$P_c = P_n - P_w = \gamma_{nw} \cos\theta \left(\frac{1}{r_1} + \frac{1}{r_2} \right) \quad (1)$$

where P_c is the capillary pressure (dynes/cm^2 or N/m^2), which is defined as the difference between the non-wetting phase pressure P_n and the wetting phase pressure P_w, γ_{nw} is the interfacial tension between the nonwetting (n) and wetting (w) fluid phases (dynes/cm or mN/m), θ is the contact angle between the fluid-fluid interface and the solid or mineral surface (degrees or radians), and r_1 and r_2 are the principal radii of curvature of the fluid-fluid interface (cm or m). The two radii of curvature are equal if the capillary geometry is cylindrical, in which case the Young-Laplace equation simplifies to:

$$P_c = P_n - P_w = 2\gamma \cos\theta \left(\frac{1}{r} \right) \quad (2)$$

Typically, pore-scale models of porous media assume circular cross-sectional geometry for the structure of pores and pore-throats (Celia et al.,

1995). The interfacial tension (γ) and contact angle (θ) are represented as single-valued, i.e., equilibrium, parameters in most pore-scale models, or contact angle hysteresis is incorporated by specifying advancing and retreating contact angles (Giordano and Slattery, 1983a; Dullien, 1992). Equation 2 (or appropriate modifications) is used to determine interface stability and thus whether an immiscible fluid phase (gas or NAPL) penetrates a given pore during primary drainage. A fluid-fluid interface may be "trapped" in a given pore throat until conditions change to make it unstable.

A fluid-fluid interface is unstable once it passes the neck of a pore throat (Haines, 1930; Miller and Miller, 1956; Melrose, 1965; Dullien, 1992). The interface advances extremely rapidly through the pore-body into the next pore throats where it reestablishes new stable positions. This rapid interface "jump" is called a Haines jump, named after the person who first described it (Haines, 1930). Interface migration through a porous medium occurs through a sequence of relatively stable locations in pore throats linked by a series of Haines jumps. The applied capillary pressure controls stable interface locations in a model porous medium (and within any given pore if the pore-throat geometry is allowed to change with axial distance through the pore-throat). This is because most models treat the interfacial chemistry as static via equilibrium assumptions for γ and θ. The interface locations within a pore-throat will change, however, if interfacial chemistry is not at equilibrium and capillary pressure is not sufficiently high to overcome the initial capillary forces.

1.2.2. Interfacial chemical kinetics

Equilibrium values of interfacial tension or contact angle are determined by partitioning or adsorption equilibria of interfacially active (surface active) components at the interfaces whenever any of the phases contain such components. This is expressed quantitatively via the Gibbs equation (Gibbs, 1948; Adamson, 1990). The Gibbs equation for a three-component system at constant temperature is (Adamson, 1990):

$$d\gamma = -\Gamma_1^\sigma \, d\mu_1 - \Gamma_2^\sigma \, d\mu_2 - \Gamma_3^\sigma \, d\mu_3 \qquad (3)$$

where Γ_i^σ is the surface excess of component i, μ_i is chemical potential of component i, and the superscript σ designates that the parameter is evaluated at the surface between the fluid phases. In this paper, components *1* and *2* are the respective bulk fluids. Component *3* is assumed to be interfacially active; it may be either predominantly present in the aqueous phase (water soluble) or predominantly present in the NAPL (oil soluble). Positive surface excess means that a greater amount of component i is accumulated in the surface than is present in a volume of bulk solution containing the same amount of solvent. A reduction in interfacial tension occurs in such cases. The first two terms of Eqn. 3 can be eliminated if we assume that a Gibbs surface equation (Gibbs, 1948; Adamson, 1990) can be defined, i.e., if we define a surface such that $\Gamma_1^\sigma = \Gamma_2^\sigma = 0$.

Equation 3 is an equilibrium equation. The surface excess of component *3*, Γ_3^σ (and therefore the interfacial tension between the liquid phases) changes with time if surface equilibrium has not been attained. Isothermal perturbation of a surface equilibrium can be brought about in five ways: 1) addition of pure component *1* to the bulk liquid *1* phase, 2) addition of pure component *2* to the bulk liquid *2* phase, 3) addition of component *3* to either of the liquid phases, 4) compression of the interfacial area, or 5) dilation of the interfacial area. The first three ways change the chemical potentials, μ_i, the fourth way increases the surface excess, and the fifth decreases the surface excess.

Time becomes an important variable for interfacial tension when an interface is expanding or being compressed more rapidly than interfacially-active components in either phase can partition into or out of the interface. Time-dependence is a function of the specific system and varies over several orders of magnitude of time-scale, as noted in the introduction. Time-dependence of interfacial tension during a dilation process is also known as *dynamic surface* or *interfacial tension* (Lucassen and Giles 1975) and as *surface dilational viscosity* (Giordano and Slattery, 1983a, b, 1987). The latter term arises by analogy that interfacial tension resists interface dilation as viscosity resists fluid flow. Interfacial chemical kinetics may become important for any given dynamic process involving compression or dilation of fluid-fluid interfaces. Imbibition, the immiscible displacement of a non-wetting fluid by a wetting fluid in a porous medium, and drainage, the immiscible displacement of a wetting fluid by a non-wetting fluid in a porous medium, are two such processes.

1.2.3. "Haines" jumps and interface dilation

Crawford and Hoover (1966) examined the movement of air-water interfaces through small diameter glass and Teflon tubes (ID ≅ 1 mm) filled with various sized water-wetted or non-wetted beads. They simultaneously monitored injection pressure (in cm of water) and injection volume, which was done in discrete volumetric increments, $\Delta V = 1.7 \times 10^{-4}$ cm^3. Their pressure plots show distinct discontinuities, which

Crawford and Hoover associated with rapid movements of the air-water interfaces as they advanced from pore-to-pore, i.e., the pressure discontinuities correspond to Haines jumps. The discontinuities occurred in less than 0.5 seconds (the approximate limit of time resolution in their measurements) in pores with volumes ranging in size from approximately 0.5 to 14 x 10^{-5} cm^3. These pore dimensions are roughly the scale of medium to very coarse sand. Thus, 0.5 seconds is a measure of the upper limit for the time required for an air-water interface to advance through a single pore-body during a Haines jump in their media. The "creeping" stages occurred over time-spans ranging between approximately 20 and 185 seconds. These results indicate that air-water interfaces spent at least 40 times longer in pore-throats than in "jumping" to the next pore throat, and that ratio may exceed 400. Thus, the assumption that interfaces spend negligible time moving through pore bodies is a reasonable first approximation (Tuck et al., 1998).

Heller (1968) took 16-mm movies (64 frames per second) of drainage experiments of 1,1,1-trichloroethane from two unconsolidated packs of 80-mesh and 140-mesh fluorite grains. Exposure time was half of the 15.6-millisecond frame-repetition-time. He documented Haines jumps covering a few pores (~1 to 10 by visual estimation) and several covering several tens of pores occurring over the time interval between exposures. Therefore these Haines jumps took place in no more than 23.4 milliseconds. Haines jumps covering several tens of pores imply that relatively small regions in the porous medium have slightly larger pore-throat sizes than the surrounding porous regions.

1.2.4. Interfacial viscosity

Boussinesq (1913), in his analysis of free-boundary flows, identified two coefficients of interfacial viscosity, interfacial shear viscosity (ε) and interfacial dilational viscosity (κ). These interfacial viscosity coefficients are incorporated into the standard Newtonian model for fluid interfaces (Scriven, 1960; Slattery, 1964). Interfacial shear viscosity is the two-dimensional equivalent of the ordinary fluid shear viscosity; it is the force required to produce relative motion between adjacent line elements in a surface (Adamson, 1990). Surface or interfacial dilational viscosity is defined by the following equation (Adamson, 1990):

$$\frac{1}{\kappa} = \frac{1}{\Delta\gamma}\frac{1}{A}\frac{dA}{dt} \tag{4}$$

where γ is the surface or interfacial tension, A is the interfacial area, and t is time. He notes that the equilibrium quantity corresponding to κ is the modulus of surface elasticity, E, given by:

$$E = \frac{d\gamma}{d\ln A} \tag{5}$$

He further states that "... κ and E are probably the most important rheological properties of interfaces and films" (Adamson, 1990, p. 127). Of the two coefficients of surface viscosity, interfacial dilational viscosity is generally more important for two reasons: first, κ is generally on the order of 100 times greater than ε; and second, interfaces are more often subject to dilation strains than shear strains (Adamson, 1990).

1.2.5. Interfacial compression and dilation in porous media

Fluid-fluid interfaces continually go through compression and dilation during drainage or imbibition. Drainage in a porous medium is illustrated in Fig. 1, which is a qualitative time sequence showing the advance of a multicomponent fluid containing a surface active component (i = 3) through an idealized pore. The porous medium is visualized as a vertical slice through uniform, spherical grains in a cubic packing (Graton and Fraser, 1935). The appropriate equilibrium interfacial tension equation is Eqn. 3, assuming isothermal conditions and a Gibbs' surface. The surface excess, Γ_3^σ, is illustrated schematically by a surfactant molecule partitioned across the fluid-fluid interface; each surfactant molecule shown in the interface represents a schematic surface excess of one (1). The initial stage, shown in Fig. 1a, is taken to be a "fresh" interface, just established in the top pore-throat, and the schematic surface excess equals 2. The hydrostatic fluid pressure difference is assumed to remain constant.

Component *3* diffuses to and partitions into the interface as the interface ages. The surface excess increases and interfacial tension decreases. The stable position of the interface changes as a result, and the interface begins to creep down into the pore throat. The interfacial area is compressed simultaneously with the creeping interface advance. Both processes, adsorption of component *3* and interface compression, act to increase the surface excess, Γ_3^σ, and thus decrease interfacial tension according to Eqn. 3. Schematic surface excess in Fig. 1b is designated as 4^+, with the superscript "+" indicating the surface compression portion of the increased surface excess. These processes continue until the interface reaches the neck of the pore throat, where, in Fig. 1c, the schematic surface excess is 6^{++}.

A Haines jump results in extremely rapid dilation of the interface as illustrated in the sequence pro-

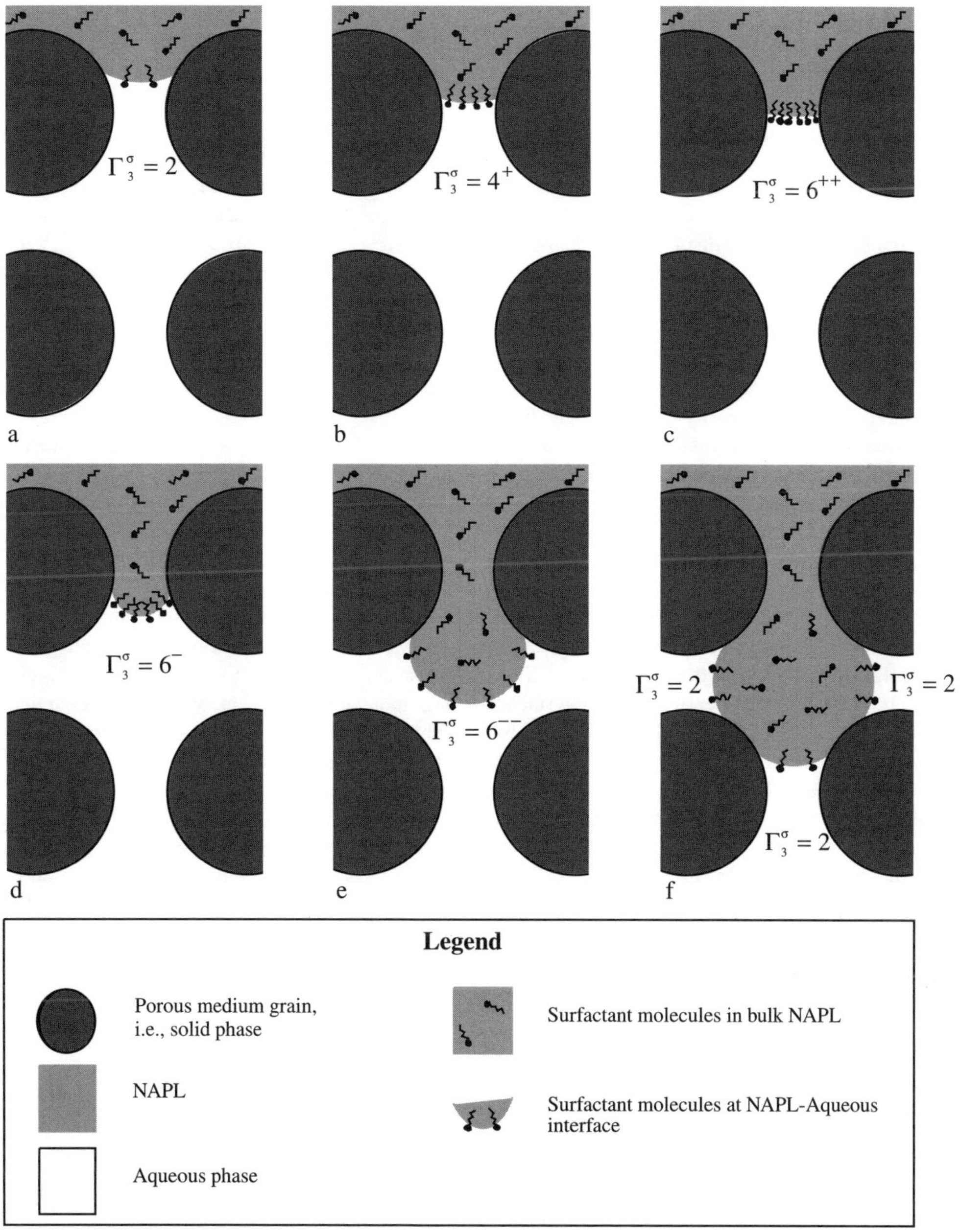

Fig. 1. Schematic illustration of the mechanism by which dynamic interfacial tension effects are manifested during primary drainage in a porous medium. The schematic values of the surface excess, Γ_3^σ, are displayed. See text for discussion.

ceeding from Figs. 1c-f. Following the method above, the schematic surface excess in Fig. 1d is 6^-, and by Fig. 1e it is 6^{---}, with the superscript "-" indicating decreased surface excess due to surface dilation. The surface excess continues to decrease, purely by dilation, and, by the time the "jumping" interface encounters the three new pore throats and establishes three new semi-stable interfaces, it has decreased, in this idealized medium, back to 2, where the sequence started. This schematic sequence illustrates how interfacial chemical kinetics becomes important during multiphase flow in porous media. The sequence assumes that the time required for a Haines jump is significantly less than the time required to attain interfacial equilibrium.

1.2.6. Early modeling work

Giordano and Slattery (1983a, b) extended the qualitative theory proposed by Slattery (1974, 1979) in their model of immiscible displacement of a NAPL blob in a capillary of constant radius incorporating the influence of interfacial viscosity. One of their major explicit assumptions was that the dimensionless interfacial viscosity, $N_{\kappa+\varepsilon}$, is much greater than unity. $N_{\kappa+\varepsilon}$ is the ratio of the sum of the interfacial shear ε and dilational viscosity κ to the bulk fluid shear viscosity-capillary radius product ($\mu_2 R$); here the subscript *2* indicates the bulk viscosity of the discontinuous fluid phase:

$$N_{\kappa+\varepsilon} = \frac{\kappa+\varepsilon}{\mu_2 R} \qquad (6)$$

Giordano and Slattery (1983a, b) concluded that interfacial viscosity could have a significant effect on tertiary oil recovery, and that the interfacial dilational viscosity was of greater importance than interfacial shear viscosity. They also showed that rate of displacement is a function of interfacial viscosity.

In a later paper, Giordano and Slattery (1987) modeled immiscible displacement of a NAPL blob in a capillary whose radius varies in sinusoidal fashion with axial position. Again, they found that interfacial viscosity played a major role in determining the average rate of displacement. Quasi-stable interfaces are established as an interface begins into the narrow portion of the sinusoidal capillary. They creep slowly toward the neck of the pore throat. Beyond this point the interfaces become unstable, and "jump" to the next pore throat where they begin the slow "creeping" stage anew. Their 1983 and 1987 models, however, did not consider the effects involved with the multiply-connected pores of a 2-D or 3-D porous medium network.

The rate at which a non-wetting, immiscible phase penetrates a porous medium is strongly controlled by the amount of time the interfaces spend "creeping" down into the pore throats until they reach their unstable limit at the pore necks (Miller and Miller, 1956; Melrose, 1965; Giordano and Slattery, 1983b, 1987). In the traditional view, this is purely a function of the applied fluid pressure difference, i.e., capillary pressure. Applied capillary pressure is not the only control, however, for complex fluids. The time-dependence of interfacial chemistry can play a major role in such systems. As noted above, Giordano and Slattery (1983a) showed that displacement rate of NAPL blobs trapped in cylindrical and sinusoidal capillaries is a function of dimensionless interfacial viscosity. Also, Tuck et al. (1998) showed that different *effective interfacial tension* values are likely to apply when the physical capillary pressure exceeds the entry pressure for the medium. A small pore-scale model is developed in the next section to examine the effects of time-dependent interfacial tension (i.e., *dynamic interfacial tension* or *interfacial dilational viscosity*) on primary drainage in porous media.

2. PORE-SCALE MODEL

2.1. Model Domain

A small, two-dimensional pore-scale model was formulated consisting of a 7x7 array of spherical pore bodies connected by cylindrical pore throats arranged in a square lattice (see Fig. 2). The pore throats were assumed to be of infinitesimal length, and hence did not contribute to the pore volume. Two populations of pore body and pore throat sizes were designed to mimic fine-to-medium sand, based on empirical results of Pirkle et al. (1997a). Figure 3 is a graphical presentation of the data contained in Tables 1 and 2. Five evenly-spaced pore-body and pore-throat sizes were assigned to each population. The populations shared one overlapping category of pore-body and pore-throat size. The central nine pores of the medium were assigned the smaller pore-body and pore-throat size-categories. Pore-body and pore-throat sizes were assigned randomly by my assistant (my, at that time, three-year old daughter, Gaia) by drawing cards from a shuffled deck consisting of aces through fives. The horizontal boundaries of the medium are no-flow boundaries. The upper boundary of the medium is a constant non-wetting-phase pressure boundary. The lower boundary is a constant wetting-phase pressure boundary. For the sake of convenience, the term "capillary pressure" will be used interchangeably for the "boundary fluid pressure difference". In reality, "capillary pressure" means the fluid-pressure difference existing across a specific NAPL-water interface. This will vary throughout the model, with values below the applied fluid-reservoir pressure difference. The pore throats at the boundaries fall in the coarser pore-throat size population. Tables 1 and 2 contain the pore body and pore throat size specifications, respectively, corresponding to the different size categories. This pore-scale model can be conceived, in physical terms, as a two-dimensional slice through a granular medium containing a random heterogeneity, either due to packing variation of uniform grains (Graton and Fraser, 1935) or variation in

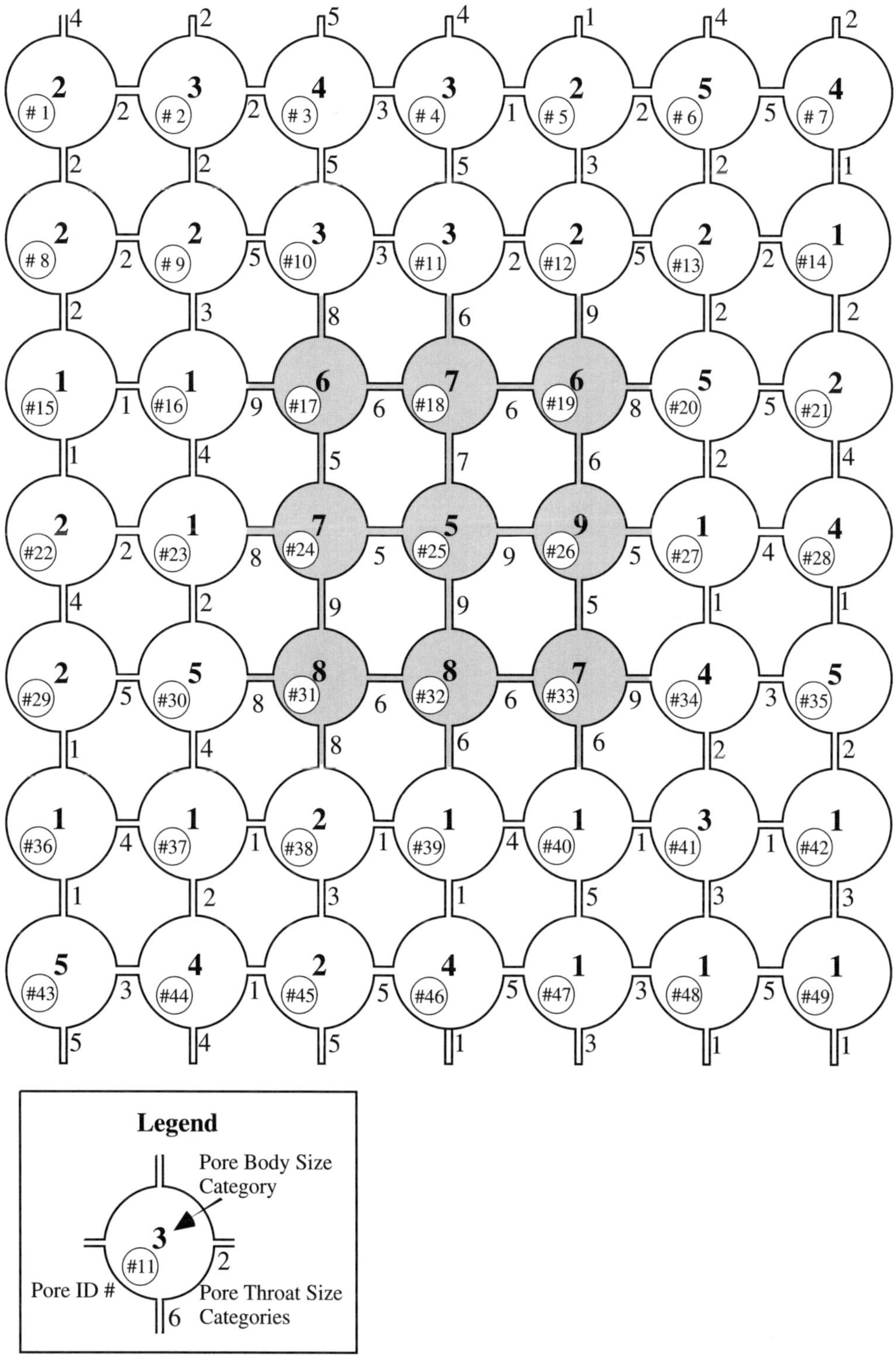

Fig. 2. Time-dependent interfacial chemistry effects were assessed in this 7x7 node pore-scale model. The center nine pores (nodes) and associated pore-throats were assigned to the smaller size range. Pore-body and pore-throat sizes are given in Table 1.

Table 1. Pore-body and pore-throat size categories and pore volumes.

Pore Category Number	Pore Radius (cm^3, $x10^{-2}$)	Volume per Pore (cm^3, $x10^{-4}$)	Number of Pores in Category	Total Volume for Size Category (cm^3, $x10^{-3}$)
1	5.4	6.5958	13	8.58
2	5.2	5.8898	11	6.48
3	5.0	5.2350	5	2.62
4	4.8	4.6325	6	2.78
5	4.6	4.0772	6	2.45
6	4.4	3.5682	2	0.713
7	4.2	3.1034	3	0.931
8	4.0	2.6808	2	0.536
9	3.8	2.2985	1	0.230
			Total Pore Volume	25.3

Table 2. Pore throat size categories and pore entry pressures for PCE ($\gamma = 48.6$ mN/m, $\theta = 26.5°$).

Pore Throat Category	Pore Throat Radius (cm)	Capillary Entry Pressure (mbar)
1	5.4 x 10^{-3}	16.04
2	5.2 x 10^{-3}	16.62
3	5.0 x 10^{-3}	17.26
4	4.8 x 10^{-3}	18.01
5	4.6 x 10^{-3}	18.79
6	4.4 x 10^{-3}	19.64
7	4.2 x 10^{-3}	20.58
8	4.0 x 10^{-3}	21.61
9	3.8 x 10^{-3}	22.74

grain size (Morrow, 1971). Alternatively, it could be conceived as a model of a rough walled fracture with the central region being of smaller average aperture.

The drainage experiment results reported by Heller (1968) provide support for including small-scale heterogeneous regions within a pore-scale model. As mentioned above, Haines jumps covering several tens of pores imply relatively small regions within the medium that have slightly larger, effective pore-throat radii. The qualifying term "effective" distinguishes the irregular geometry of real porous media from the idealized representation of cylindrical and spherical geometry used here and in most pore-scale models.

2.2. *Incorporating Time-Dependent Interfacial Chemistry*

The Young-Laplace equation modified for circular cross-sectional elements (Eqn. 2) governs the capillary behavior. Thus, the standard criterion for movement of an interface past a given pore throat of radius r_{ij} is:

$$P_c = P_n - P_w > \frac{2\gamma cos\theta}{r_{ij}} \tag{7}$$

where the subscripts n and w indicate non-wetting and wetting fluids, respectively, and i and j designate a specific pore-throat. This criterion holds for single-valued interfacial tension and contact angles. Trapping of wetting fluid by isolation from continuous wetting fluid was allowed in the model. Wetting fluid becomes trapped in finer pore-body/pore-throat domains as a NAPL penetrates the coarser domains surrounding them. Ferrand and Celia (1992) have shown this to be an important process in heterogeneous porous media. This model was designed to capture this effect on a smaller scale.

When time-dependence is incorporated into the interfacial parameters, the criterion for interface movement through a pore can be specified as:

$$P_c = P_n - P_w > \frac{2\gamma(t) cos\theta(t)}{r_{ij}} \tag{8}$$

where the interfacial tension, $\gamma(t)$, and contact angles, $\theta(t)$, are each expressed as functions of time.

2.2.1. *Interfacial chemical kinetics*

Tuck and co-workers (Tuck et al., 1998; Tuck and Rulison, 1999a, b; Tuck and Rulison, unpublished data) have characterized the time-dependence of interfacial tension and contact angle in the Sudan IV-dyed-tetrachloroethylene (PCE)-water-glass system. This system is of interest because several investigators have used this dye in visualization experiments involving PCE or other DNAPLs (Kueper and Frind,

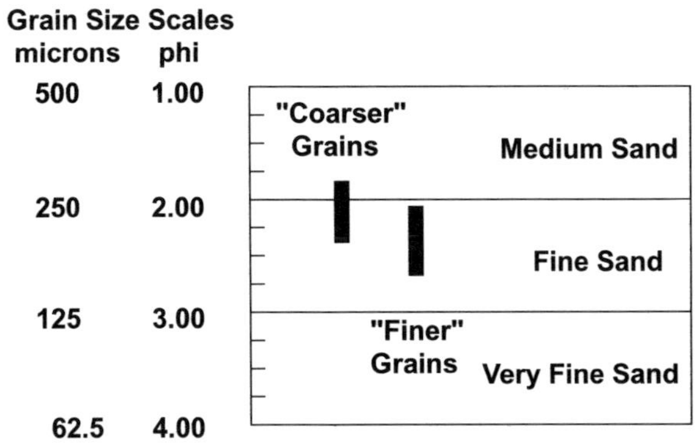

Fig. 3. Comparison of pore-scale model dimensions with approximate dimensions for "equivalent" granular porous media. The two pore-body and pore-throat size populations overlap slightly near the boundary between fine and medium sand.

1991; Poulsen and Kueper, 1992; Brewster et al., 1995; Fortin et al., 1997). Sudan IV belongs to the azo class of dyes (Catino and Farris, 1978). Virtually all visualization studies involving dyed DNAPLs have used this dye or minor variations of it (Tuck, 1999). Tuck and co-workers (Tuck et al. 1998; Tuck and Rulison, 1999a, b) found that both $\gamma(t)$ (expressed in dynes/cm or, equivalently, mN/m) and $\theta(t)$ (expressed in degrees) for a system consisting of 0.5 g/L of Sudan IV in PCE, deionized water, and glass can be described accurately over time spans of hours to days by the following log-linear equations (see Figs. 4 and 5):

$$\gamma(t) = 48.6 - 3.64 \log(t) \qquad r^2 = 0.991 \qquad (9)$$

$$\theta(t) = 21.4 + 5.96 \log(t) \qquad r^2 = 0.933 \qquad (10)$$

where t is the age of the interface in seconds. The model applied here simplified Eqn. 8 by assuming an "equilibrium" value for the contact angle (26.5°), i.e., no contact angle time-dependence was assumed. The focus of the modeling work was solely on the effects of time-dependent interfacial tension. The interfacial tension time-dependence was modeled using Eqn. 9, and the critical ages were calculated to satisfy the inequality of Eqn. 8 for each pore-throat size category as a function of applied capillary pressure (see Fig. 6). The log-linear equation results in higher interfacial tension values than for pure PCE for interface ages less than one second. This is unlikely to be physically realized. Hence, any critical age less than one second was taken to indicate the interface was instantly unstable. In such cases, the interface continues into the next pore (or pores) until it encounters a pore-throat of sufficiently small radius to stop it.

2.2.2. Numerical experiments

Three sets of numerical experiments were performed. The first set included two experiments using Eqn. 7, i.e., assuming capillary equilibrium, as the criterion for interface movement, one using 48.6 mN/m for the interfacial tension and the other using 25.3 mN/m (from Eqn. 9 using an interface age of 30 days). These experiments were run assuming the conventional capillary pressure-saturation experimental procedure of incrementing capillary pressure and assuming equilibrium capillary pressure-saturation conditions are achieved following each pressure increment. Capillary pressure was incremented in 0.2 mbar steps for this first set of experiments.

The second set of numerical experiments was run assuming constant fluid reservoir pressure difference P_c throughout an experiment. The constant value of P_c was changed for each experiment in increments of 2 mbar from 2 to 20 mbars for this series of experiments. Near the entry pressure for pure PCE in this medium (approximately 16 mbar), P_c was changed in increments of 1 mbar. The NAPL (PCE) saturation history was then computed for each value of P_c. Viscous flow control of pore-filling times was approximated by assuming that 0.01 seconds was required for all pore-throat and pore-body sizes. Heller's (1968) experimental documentation of Haines jumps during drainage suggests this time estimate is at least the right order of magnitude. Actual times should vary as a function of the pore-throat radius and length (Blunt and King, 1991). Since infinitesimal length was assumed for each pore-throat, however, a constant value for the fill time is a reasonable first approximation. This only becomes important for pressures above 16 mbar.

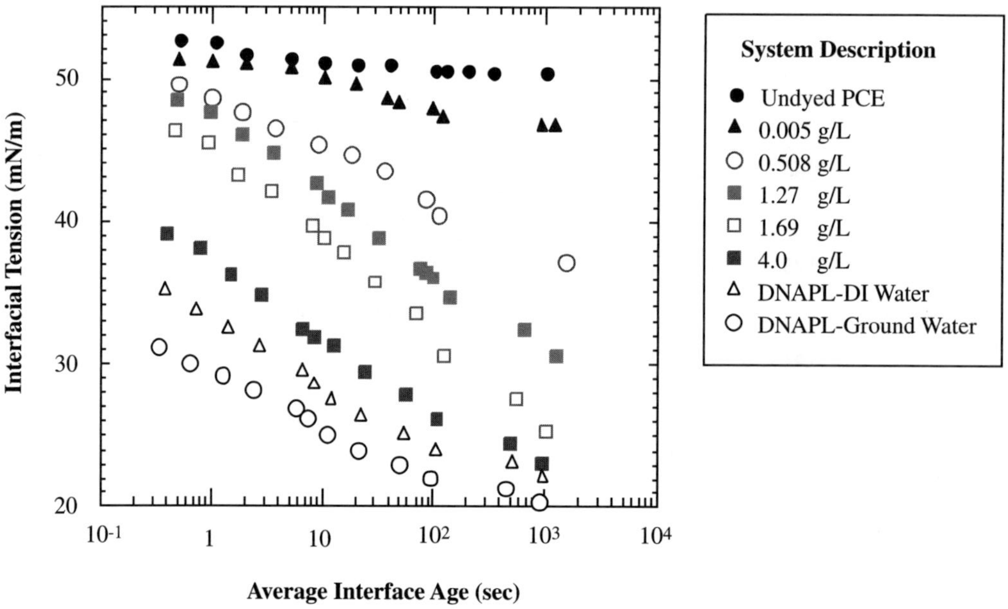

Fig. 4. Time dependence of interfacial tension in the system water-tetrachloroethylene (PCE)-Sudan IV dye. DNAPL-DI water curve is for the surface between a sample of the Savannah River Site A/M Area DNAPL (site DNAPL) and deionized water. DNAPL-Ground Water curve is for the surface between the site DNAPL and an equilibrated sample of the A/M Area ground water.

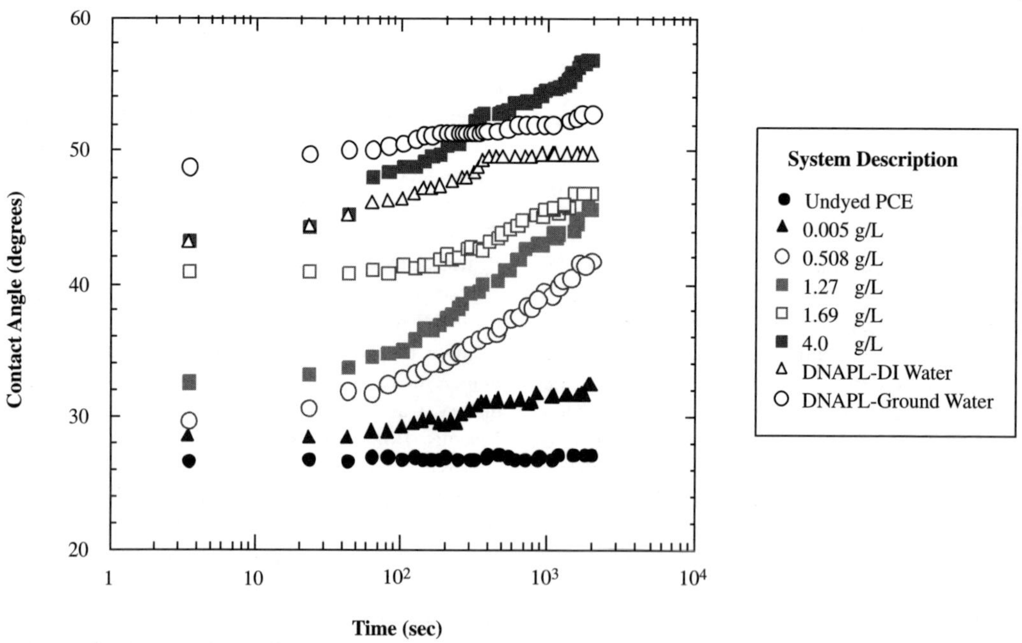

Fig. 5. Time dependence of contact angle in the water-tetrachloroethylene (PCE)-Sudan IV dye-glass system. DNAPL-DI water curve is for the surface between a sample of the Savannah River Site A/M Area DNAPL (site DNAPL) and deionized water. DNAPL-Ground Water curve is for the surface between the site DNAPL and an equilibrated sample of the A/M Area ground water.

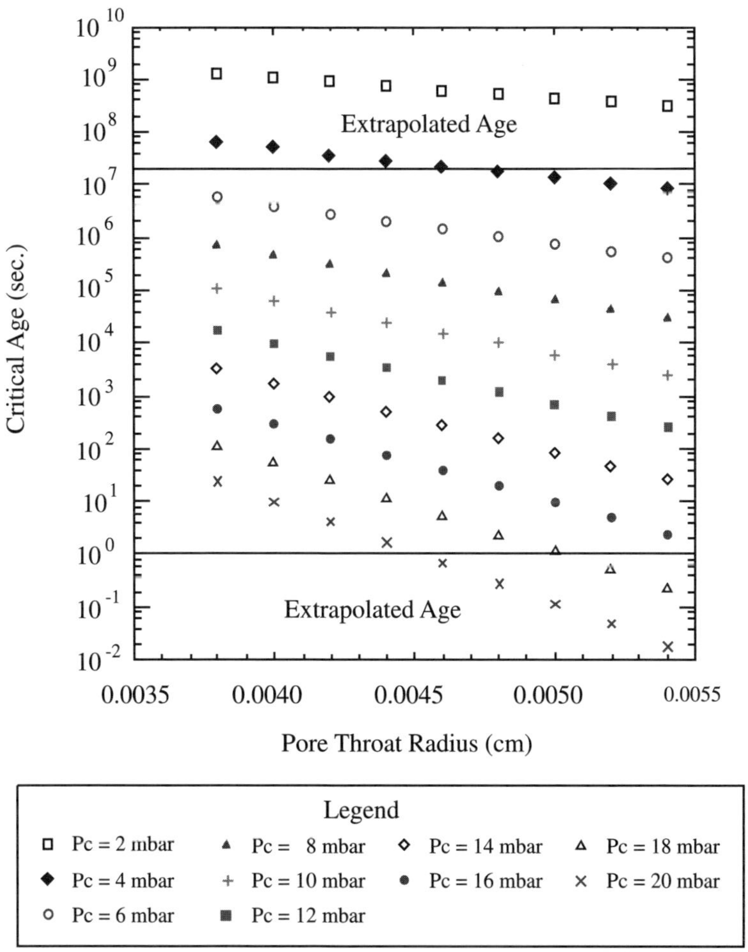

Fig. 6. Critical age for interface instability as a function of applied capillary pressure and pore-throat radius. Critical ages greater than approximately 3 x 10^6 seconds are extrapolations of the empirical data. Critical ages less than one second indicate conditions in which the pore-throat will not support an interface. Symbols are the calculated results for the model pore-throats listed in Table 2.

The third set of numerical experiments was run as a standard capillary pressure-saturation experiment, but assuming time-dependence of interfacial tension was operating and pressure-step equilibration time (the length of time following each pressure increment) was varied. The pressure was incremented in 2-mbar steps for 2 to 20 mbars, and time-dependent penetration of the medium was recorded for the pressure-step-equilibration time for that experiment. In these last experiments, an interface instantly becomes unstable (i.e., jumps to the next pore) if the age of an interface in a pore-throat was greater than or equal to the critical age for a stable interface at the incremented value of P_c.

3. MODEL RESULTS

3.1. Equilibrium Interfacial Tension

Results from the first set of experiments are shown in Fig. 7. Different values for "equilibrium" interfacial tension were assumed in these experiments. Note that the entry pressure for this medium for undyed PCE is between 16.0 and 16.2 mbars. The maximum NAPL saturation attained in both experiments was 0.8542.

3.2. Time-Dependent Interfacial Tension

Model results for the second set of experiments are presented in Fig. 8. Dynamic interfacial chemistry

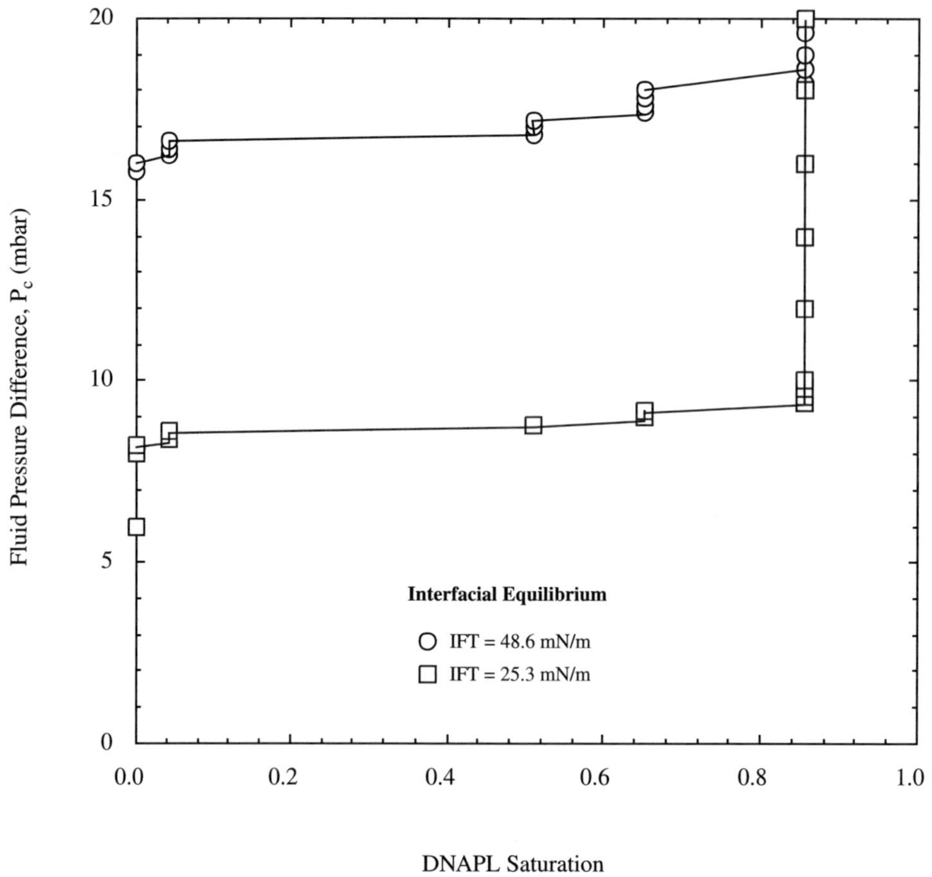

Fig. 7. Pore-scale model results using single-valued ("equilibrium") interfacial tension. The interfacial tension between pure PCE and water is 48.6 mN/m; note the entry pressure at approximately 16 mbars.

was incorporated in these experiments. Maximum saturation varied under the different applied capillary pressures in a non-monotonic fashion (see Fig. 9). Maximum saturation ranged from 0.8542 to 0.9894. A minimum value for the maximum saturation was obtained under the 17-mbar and 18-mbar capillary pressure conditions, i.e., those constant capillary pressure conditions just above the entry pressure for pure PCE. The slopes of the curves illustrate that time was an important variable in reaching the final saturation. The locations of the curves along the time scale indicate that drainage occurs more quickly under the higher applied capillary pressure conditions. Three different NAPL migration rates were calculated for each experiment from the saturation histories in the bottom row of pores. The rates were based on the average distance from the non-wetting phase reservoir to the bottom row of pores and on three different "arrival" times: first arrival of NAPL, the interpolated time when NAPL saturation reached 0.5, and the time when NAPL saturation reached 1.0. These rates are shown as a function of the constant capillary pressure in Fig. 10.

3.3. Discrete Pressure-Step Increments

Model results for the third set of experiments are presented in Fig. 11, where they are presented in comparison to the results in which single, "equilibrium" values of interfacial tension were assumed. Capillary pressure at which NAPL first enters the medium decreased and the maximum saturation attained in an experiment increased as the equilibration time was increased between pressure increments. Maximum saturation is shown as a function of the pressure-step equilibration time in Fig. 12. Maximum saturation in these experiments ranged between 0.8703 and 0.9504.

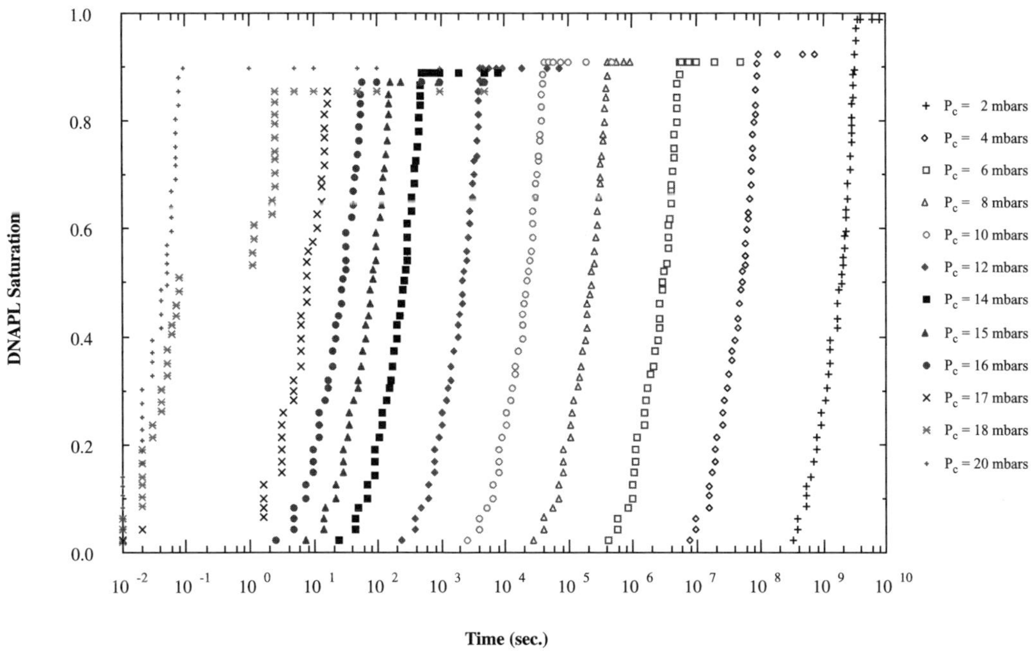

Fig. 8. Pore-scale model results with time-dependent interfacial chemistry incorporated. Saturation becomes a function of time as well as applied capillary pressure.

4. DISCUSSION

4.1. *Equilibrium Interfacial Tension*

Results from the first set of experiments are consistent with typical results from uniform grain-size porous media. The rapid increase in PCE saturation reflects the near homogeneity of this medium. The rapid build-up of PCE saturation occurs between capillary pressures of 16 and 18 mbars for pure PCE (γ_{nw} = 48.6 mN/m) and between 8 and 9 mbars for the lower assumed interfacial tension value (25.3 mN/m). All of the PCE saturation occurs in the coarser pores surrounding the finer pore domain. These results are consistent with those reported by Ferrand and Celia (1992). Irreducible wetting phase saturation is determined by the finer pore-throat heterogeneity embedded within the model. Results for the lower "equilibrium" interfacial tension are consistent with the standard assumption of Leverett J-function scaling of capillary pressure saturation curves based on changing the interfacial tension between the fluids (Leverett, 1941; Bear, 1972). The capillary pressure-saturation curve is merely shifted to lower capillary pressures in response to the lower interfacial tension. These results confirm that the model is a reasonable approximation of a real porous medium, and that it demonstrates behavior comparable to pore-scale simulations in which a much greater number of pores were used and the standard assumptions of equilibrium interfacial chemistry were employed (Ferrand and Celia, 1992).

4.2. *Time-Dependent Interfacial Tension*

There are several things to note in Fig. 8. First, there are discrete discontinuities in the S_{PCE}-time curves. This is especially apparent in the P_c-S_{PCE} curve for 18 mbars. These discontinuities occur when multiple pore throats become unstable simultaneously. To some extent the apparent discontinuities are an artifact of the model since discrete categories of pore throat sizes were used rather than a smooth distribution, and also, because of the small total number of pores. This result is physically reasonable, however, as Haines jumps can occur simultaneously through many different pores in real porous media (Payatakes,

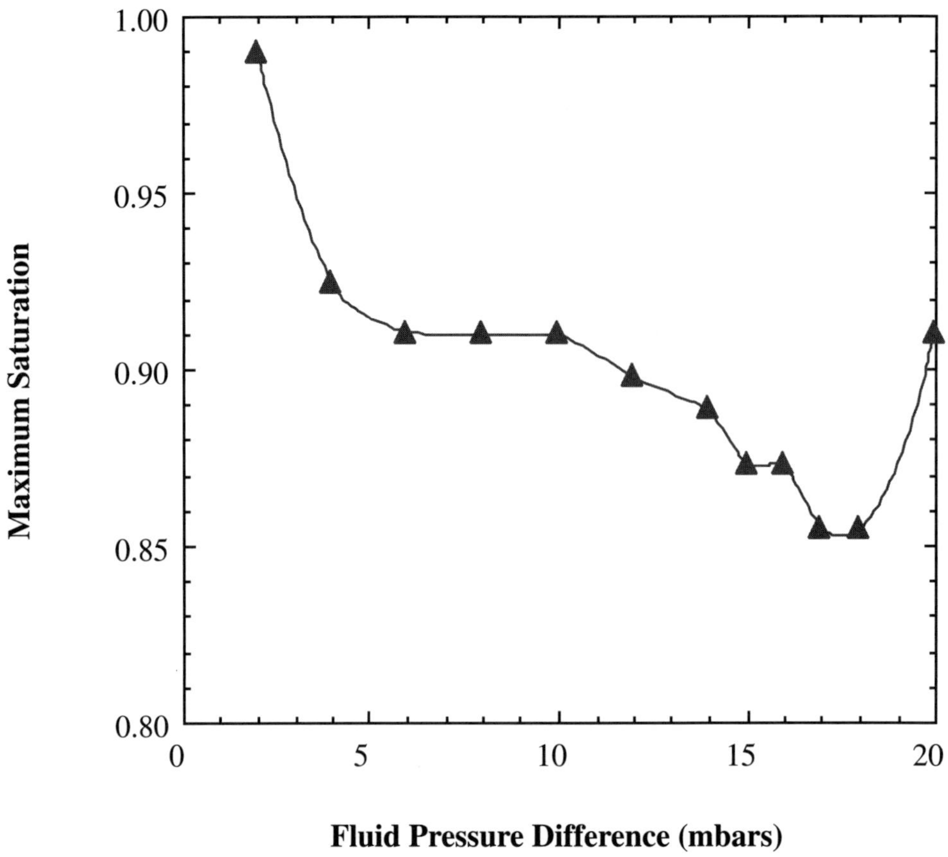

Fig. 9. Maximum NAPL saturation as a function of applied capillary pressure with interfacial chemical kinetics incorporated in the pore-scale model. Maximum saturation occurs at a pressure significantly lower that the entry pressure for the pure bulk NAPL with no interfacially-active component.

1982). It also happens when an interface jumps through multiple pore bodies because the pore throat radius at the end of the first pore body being "jumped" (or subsequent pores "jumped") is not sufficiently small to establish a stable interface and therefore stop its advance (Payatakes, 1982). Heller's (1968) movie-frames of 1,1,1-trichloroethane drainage that show Haines jumps covering several to several tens of pores provide visual experimental documentation that this occurs.

The second thing to note is that the non-wetting phase saturation, S_n, becomes a function of both capillary pressure and time, with the time-dependence arising through the interfacial chemical kinetics:

$$S_n = S_n(P_c, t) \tag{11}$$

The time-dependence enters from the time-dependent surface chemistry at capillary pressures below the entry pressure for pure PCE and from a combination of viscous flow and the time-dependent surface chemistry at capillary pressures above the pure PCE entry pressure. The time scale varies as a function of the capillary pressure boundary conditions. However, non-wetting phase penetration of the medium does not necessarily stop at a threshold capillary pressure, as applied by Leverett (1941). Leverett's work presupposes equilibrium interfacial chemistry, and hence a threshold capillary entry pressure for the medium/fluid combination. The threshold entry pressure becomes somewhat nebulous under the conditions of time-dependent surface chemistry. Penetration of the medium still occurs in the combination modeled here, even at capillary pressures significantly below the "threshold" pressure for undyed PCE. NAPL penetrates the medium at different rates. The penetration or migration rate is a function of the boundary fluid pressure difference. However, penetration or migration of NAPL does not

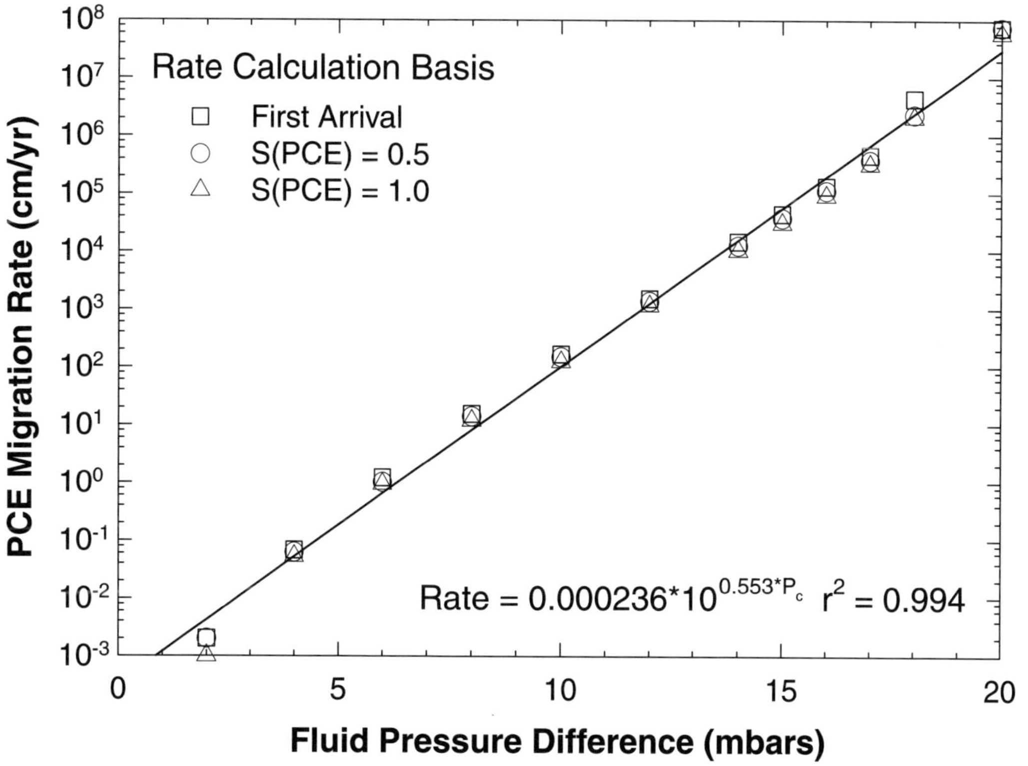

Fig. 10. NAPL migration rate as a function of applied capillary pressure with interfacial chemical kinetics incorporated in the pore-scale model.

stop at fluid pressure differences below the "threshold" capillary pressure, assuming equilibrium interfacial tension. These rates are more a function of the non-linear chemical kinetics than of the capillary pressure. The approximate log-linear curve reflects the log-linear nature of the interfacial chemistry time dependence given by Eqn. 9.

The third and very important, although rather obvious, observation is that penetration of a porous medium occurs at capillary pressure boundary conditions lower than those predicted by the standard equilibrium capillary pressure-saturation model, *if static values are assumed for interfacial tension and contact angle which are higher than the true equilibrium values*. This is important to bear in mind when interpreting multiphase flow visualization experiments in which NAPLs dyed with azo dyes are used. These dyes have generally been assumed to have no effect on the interfacial chemistry. The work of Tuck and coworkers (Tuck et al., 1998; Tuck and Rulison, 1999b), as illustrated in Figs. 4 and 5, demonstrates that this assumption is inappropriate.

The fourth, and perhaps most intriguing observation, is that maximum nonwetting phase saturation, $S_{n,\ max}$, may not increase monotonically with capillary pressure, as the standard, equilibrium capillary pressure-saturation model predicts. This is illustrated in Fig. 9. In fact, lower maximum NAPL saturation occurred at fluid reservoir pressure differences near the entry pressure for the equilibrium capillary pressure-saturation model for the pure solvent, in this specific case approximately 16 mbars. This minimum appears to arise from competition between whether the capillary pressure boundary values or the time-dependent surface chemistry will determine whether a given pore is penetrated or not. Near the entry pressure for the medium, the capillary pressure and time-dependent surface chemistry play nearly equal roles in determining when and which pores become filled with NAPL. The capillary pressure is the dominant control at higher capillary pressure boundary values. Alternatively, the time-dependent chemistry dominates control of when and whether a given pore will become filled at

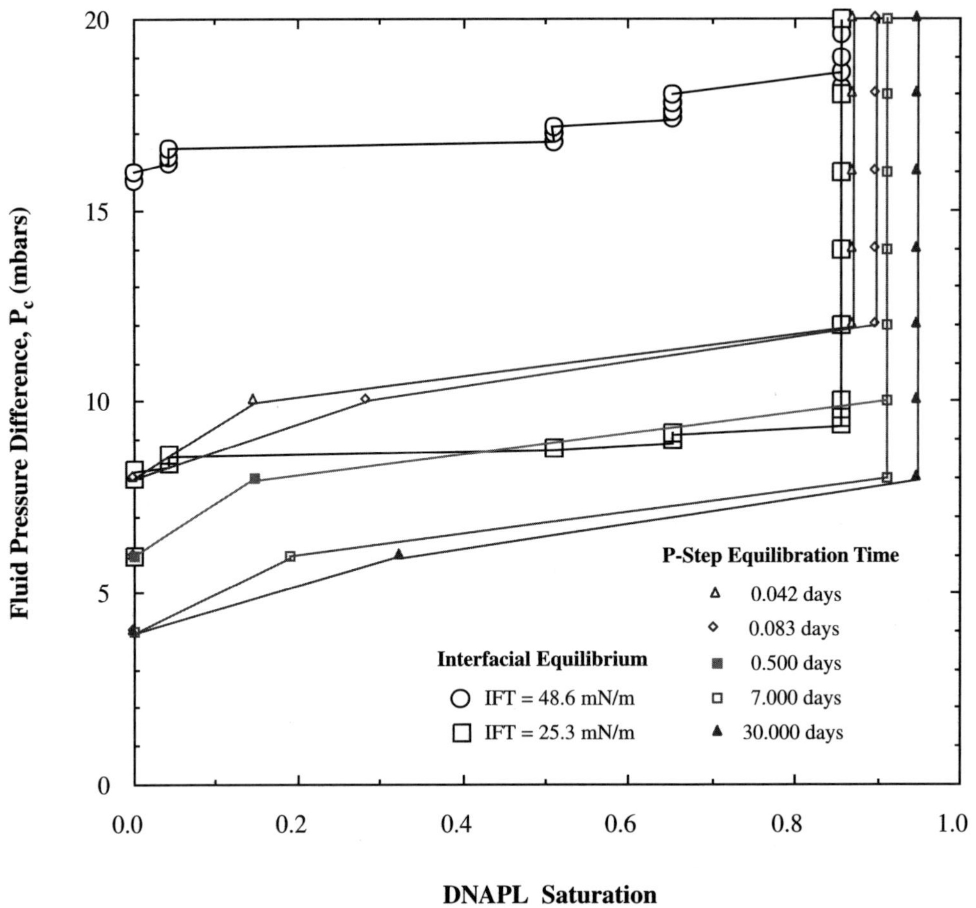

Fig. 11. Pore-scale model results for incremented pressure-step experiments under conditions of time-dependent interfacial tension as a function of pressure-step equilibration time.

capillary pressures below the entry pressure for a pure, single-component NAPL (in this case, PCE) with no surface-active components.

An analysis of variance was performed to test whether or not the minimum saturations were significant. The actual entry pressure for undyed PCE was 16.04 mbar. Four experiments were run in which the difference between this entry pressure and the modeled pressure was less than 2 mbars. Those experiments also yielded the lowest values for $S_{n,max}$. These were thus assigned to one group, having four samples. The remaining experiments, with the exception of the P_c = 2 mbar, were assigned to a second group, resulting in a subtotal of 7 experiments. A model II analysis of variance was performed to test the null hypothesis that the mean saturation values for both groups are equal (Sokal and Rohlf, 1969). The experiment F-statistic was 40.16. The critical value of the F-distribution for p = 0.001 with the Among Groups degrees of freedom equal to one (*df* = 1) and the Within Groups degrees of freedom equal to nine (*df* = 9) is 22.9 ($F_{0.001[1,9]} = 22.9$, Rohlf and Sokal, 1969). Thus, the two groups have statistically different means ($p < 0.001$), i.e., the minimum in the plot of S_{max} vs. P_c is statistically significant. The maximum NAPL saturations for boundary fluid pressure differences within ±2 mbar of the pure-solvent equilibrium entry pressure are statistically significantly lower than maximum NAPL saturations for boundary fluid pressure differences outside this range.

4.3. Discrete Pressure-Step Increments

The results of the third set of experiments (Fig. 11), those in which the fluid reservoir pressure difference was incrementally ramped up, provide a qualitative

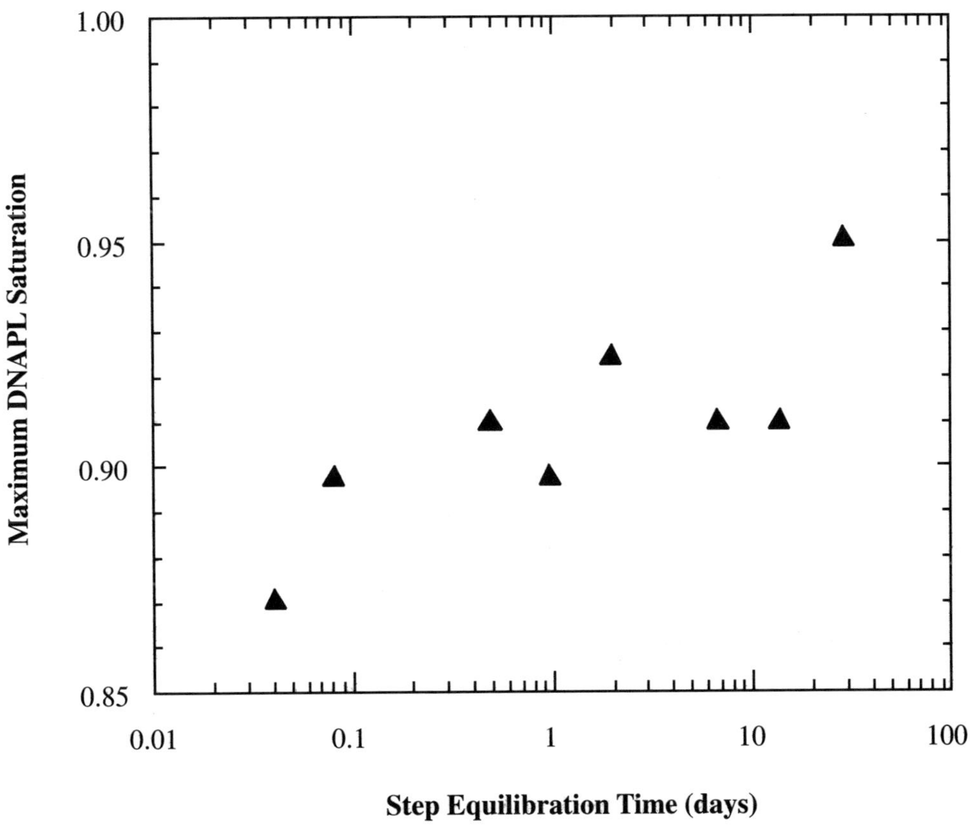

Fig. 12. Maximum saturation as a function of the pressure-step equilibration time.

view of the effects that time-dependent surface chemistry would be expected to have on current experimental techniques for measuring the capillary pressure-saturation relationship. The results suggest, depending on the time used for equilibration between pressure increments, that we could expect measured results to approximate those based on simply assuming the equilibrium interfacial tension for the dyed PCE-water system, with the caveat that we would expect to see higher final PCE saturation. But this is very deceptive. In fact, what we would more likely see is that our current experimental approach using a real porous medium sample would yield a curve somewhere between that of undyed PCE and the equilibrium-interfacial tension curve. The difference between what the current model results suggest and what we would actually measure arises from the extremely small sample of porous medium represented by this model.

We can begin to understand why this is so by thinking about what would happen if the model domain were expanded by simply linking multiple copies of the current 7x7 pore-scale model. We would obtain nearly the same results if we expand the model domain only in the direction transverse to flow. If, on the other hand, we were to expand the model domain in the direction parallel to flow, the results would be much different. The reason for this is that rates of PCE migration are slow at pressures below the entry pressure for undyed PCE (see Fig. 10). A low migration rate results in a relatively higher saturation in the short model because it can still cover an appreciable fraction of the model domain. Low migration rates under the low fluid pressure differences would penetrate a progressively lower fraction of the model as the length of the domain increased in the direction parallel to flow. A "standard" capillary pressure-satu-

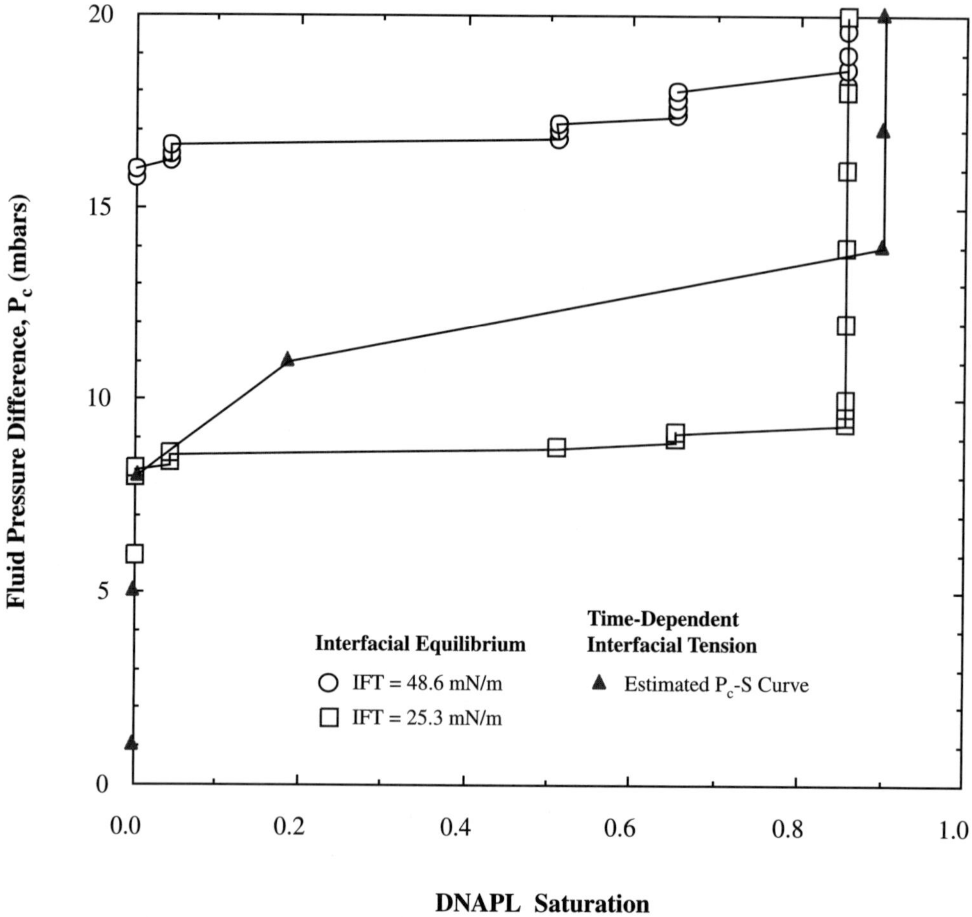

Fig. 13. Estimated capillary pressure-saturation curve for a 7.5 cm long, 2-D medium composed of linked copies of the 7x7 pore-scale model presented here. Saturation was approximated based on the fraction of the medium covered in a pressure-increment/time-step using the migration rates presented in Fig. 10. Bounding "equilibrium" interfacial tension curves for the 7x7 model are shown for comparison.

ration curve, as we would measure it in a Tempe cell, can be estimated using the following approximation:

$$S_{est}(P_c) = \left(\frac{D_{mig}}{D_{cell}}\right) S_{max} \tag{12}$$

where $S_{est}(P_c)$ is the estimated saturation at a fluid reservoir pressure difference of P_c, D_{mig} is the estimated distance PCE would migrate during the pressure-step equilibration time, D_{cell} is the length of the Tempe cell, and S_{max} is the estimated maximum saturation. D_{mig} can be estimated using the following relation:

$$D_{mig} = R_{mig}(P_c)\left(\frac{\Delta t}{365}\right) \tag{13}$$

where $R_{mig}(P_c)$ is the migration rate as a function of the boundary fluid pressure difference estimated from the exponential curve fit in Fig. 10 and Δt is the pressure-step equilibration time in days. This approach yields an approximate capillary pressure-saturation relationship for the 2-D model domain based on the fractional length of porous medium covered during a pressure-increment/time-step, as determined by the migration rates of Fig. 10.

Common practice in geotechnical companies is to measure the capillary pressure-saturation relationship using seven pressure levels over a two-week period, i.e., an average equilibration time of 2 days. Fig. 13 contains the estimated capillary pressure-saturation relationship for this Tempe-cell model domain based on the fraction of the porous medium covered during a pressure-increment/time-step as determined by the migration rates of Fig. 10. I assumed a Tempe cell length of 7.5 cm (3 inches),

0.9 for S_{max}, and the following sequence of pressures: 1 mbar, 5 mbar, 8 mbar, 11 mbar, 14 mbar, 17 mbar, and 20 mbar. The estimated curve falls between the bounding curves generated assuming single-values for the interfacial tension. Even this estimated curve is probably biased toward the curve for the lower interfacial tension. The reason is that the oil-wetted pressure plate used to separate the oil reservoir from the porous medium would likely adsorb a significant amount of any surface active component before it got into the medium. Hence, the current experimental technique is likely to bias results toward higher interfacial tension than what may actually occur in a "real" setting.

4.4. Model Limitations

4.4.1. Model size

Clearly, a major criticism of this model can be leveled at the paucity of its size. I believe the results to be valid, however, for several reasons. First, it demonstrates the qualitative behavior observed in standard capillary pressure-saturation measurements made in real porous media consisting of narrow grain size, hence pore size, distributions (see Fig. 7). Rapid water drainage occurred over a narrow range of capillary pressure once the critical capillary pressure was exceeded. The model also illustrates the concepts of by-passing and stranding of water in the finer pores. Hence residual water saturation was observed. This result is comparable to those of Ferrand and Celia (1992) who used a much larger number of pores in their simulations. In addition, the results with respect to NAPL-front penetration rate are qualitatively similar to the model results of Giordano and Slattery (1983a; 1987) for one-dimensional displacement rates of discontinuous NAPL in capillary tubes in which they explicitly incorporated interfacial viscosity (i.e., time-dependent surface chemistry).

4.4.2. Interfacial chemistry time-dependence

Another criticism of the model can be leveled against the representation of the interfacial tension time-dependence. The logarithmic equation implies that interfacial tension can decrease infinitely, including reaching negative values at sufficiently large time. This is clearly a physical impossibility. The interfacial tension eventually reaches a stable equilibrium or meso-equilibrium value, rather than continuing to decline. A better representation of this time-dependence would be that due to Hua and Rosen (1988). They represented the time dependence of dynamic surface tension using the following equation:

$$\gamma_t - \gamma_m = \frac{\gamma_o - \gamma_m}{1 + \left(\frac{t}{t^*}\right)^n} \tag{14}$$

where γ_t is surface or interfacial tension at time t, γ_o is the surface or interfacial tension of the pure solvent, γ_m is the meso-equilibrium surface or interfacial tension, and t^* and n are constants. This equation captures the approach of surface or interfacial tension to an equilibrium or meso-equilibrium value and the observations that time-dependent interfacial tension rarely exceeds that of the pure, bulk solvent. This form could easily be incorporated into the model if sufficient data were available. Currently there are insufficient time data for the system modeled here to have reached meso-equilibrium.

Tuck and Rulison (unpublished data) monitored the interfacial tension in the system modeled here out beyond eight hours (>28,800 sec). Interfacial tension continued to decline log-linearly for this whole time following Eqn. 9 and showed no trend that the log-linear phase of the decline was nearing. Similarly, they followed the contact angle for more than 30 days (2.592×10^6 sec). Again, the contact angle continued to follow the log-linear trend of Eqn. 10 with no sign of deviation. Comparison of these experimental times with the critical ages illustrated in Fig. 6 reveals that the contact angle data fall within the majority of the modeled time-scales, but the modeled interfacial tension implies an extrapolation of the experimentally measured data. Only pore-throats down to category 6 (see Table 2), however, were ever invaded in the model, with the exception of the experiment with constant capillary pressure of 2 mbar (2 mbar results). Thus, the actual maximum time any NAPL-water interface spent in a specific pore-throat (i.e., the maximum interface age used in the modeling) was 204,700 seconds, again with the exception of the 2 mbar results. Extrapolation of the interfacial tension data presents no problem if we assume that the contact angle behavior is indicative of what the interfacial tension behavior would be, i.e., that the log-linear rate of decrease would continue beyond 30 days. The log-linear equations for the interfacial tension and contact angle still give physically reasonable values well beyond the time frames discussed here. Thus, there is no conflict at all if the interfacial chemical kinetics are viewed as purely arbitrary, and not meant to model a specific system. This might be very reasonable in the geologic time regime characteristic of primary oil migration. In that case, surface-active components will migrate at different rates than the bulk oil due to adsorption loss on mineral surfaces. This could significantly retard the surface chemical kinetics at the leading edge of the migrating oil.

4.4.3. Interface age

Another limitation of this model is that the effective age of an interface gets reset essentially to zero after each Haines jump in order to directly apply the interfacial chemical kinetics of Eqn. 9. The interfaces measured via the drop volume technique start from a "fresh" condition since they arise from the break occurring when a drop falls from the measurement instrument (see Tuck and Rulison, 1999b, for details). Interfaces in a porous medium do not form in this fashion during drainage. Instead, they go through cycles of dilation and contraction, as illustrated in Fig. 1.

An "effective" age of the "fresh" interfaces can be defined in terms of the initial interfacial surface excess, Γ_3^σ, when a quasi-stable interface is first formed following a Haines jump. A rigorous model of the concepts involved in this paper would entail modeling the surface expansion and subsequent change in interfacial excess, Γ_3^σ, as an interface makes a Haines jump and establishes its new "stable" positions in the subsequent pore throats. A relatively simple approximation of this effect can be made by proportionately reducing the interfacial excess in effect when an interface reaches the narrow-point of the pore throat, as illustrated in Fig. 1c. The factor for the reduction will be approximated by the ratio of the interfacial area in the pore-throat just before a Haines jump occurs to the sum of the new interfacial areas established in the subsequent pore-throat after completion of the Haines jump. The justification for this approximation lies in the direct relationship between interfacial tension reduction and the surface excess as expressed in Eqn. 3. This approach leads to the following equation to estimate the initial interfacial tensions of the "fresh" interfaces in this model:

$$\gamma_{t=0} = \gamma_o \left(1 - \frac{\frac{1}{3} \Gamma_3^\sigma \mu_3}{\gamma_o} \right) \tag{15}$$

where γ_o is the interfacial tension of the pure NAPL or solvent, and $\gamma_{t=o}$ is the adjusted interfacial tension for the "fresh" interfaces in the porous medium following a Haines jump. This approach assumes: 1) that the new interfaces are established simultaneously, 2) that the areas of the new interfaces are all identical or at least negligibly different, and 3) that the surface excess just before the Haines jump is distributed equally over the new interfaces.

4.5. Experimental Support

Tuck and co-workers (Pirkle et al. 1997a,b; Tuck et al., 2000) conducted experiments to test the role of time-dependent surface chemical effects on Sudan IV-dyed PCE displacement of water using glass bead porous media. In their early work (Pirkle et al., 1997a), they measured the entry pressures of PCE with various Sudan IV dye concentrations in uniform glass bead porous media. Each experiment from that work lasted less than eight hours. They measured an entry pressure of 15.45 cm of PCE with a standard deviation of 0.66 cm (n=4) for the dye concentration used in the model results reported here. This translates to an expected capillary entry pressure of 24,500 baryes (g/cm-sec^2). The 95% confidence interval for the entry pressure of PCE with this dye concentration in this medium was thus 13.35 to 17.55 cm of PCE (21,200 to 27,900 baryes). In a later experiment (Pirkle et al., 1997b; Tuck et al., 2000), they set a dyed PCE column above a water-saturated glass bead medium of the same grain size to a height of approximately 9 cm, and then monitored the experiment for nearly 70 days. Thus, the surface chemical time-dependence became the primary variable governing penetration of the glass bead medium in the latter experiment since the target fluid pressure difference was less than 60% of the expected entry pressure and less than 70% of the lower 95% confidence limit for entry pressure. They saw visual penetration of the medium to a depth of approximately 2 mm. They also noted that the glass beads in this top 2 mm section of the column had become stained red by the Sudan IV dye. Limitations of their experimental apparatus, however, allowed a spike in capillary pressure that exceeded the lower 95% confidence limit, but did not exceed the mean entry pressure measured previously. Thus, their results, at this time, offer only qualitative evidence supporting the theory and model presented here. Further experimental work is needed.

4.6. Model Implications

The major conceptual advance of this work is to incorporate time explicitly into the capillary pressure-saturation relationship via incorporation of time-dependent surface chemistry. The result is that the rate of advance of a NAPL front during primary drainage can be estimated from this type of model. Following the multi-scale model concept of Celia et al. (1993), embedding such a pore-scale model into a continuum model might thus allow more accurate estimation of NAPL migration rates and be used to help determine time step length for the continuum-scale model component.

An additional connection to continuum-scale behavior arises from the interaction between the interfacial chemical kinetics and the porous medium heterogeneity. The model results imply that this interaction may determine the portion of finer pore-size heterogeneities that are by-passed during primary drainage. This interplay will be a function of the applied capillary pressure and the magnitude of the heterogeneity, as well as the interfacial chemical kinetics.

The use of equilibrium interfacial tensions will lead to overestimating the rate of NAPL migration in models involving complex fluids containing interfacially active components. This will be especially true for situations where the capillary pressure is relatively low.

The interfacial chemistry changes occur due to adsorption and partitioning of interfacially active components to the fluid-solid and fluid-fluid interfaces as discussed above. Equation 3 indicates that the magnitude of the equilibrium change in interfacial tension is directly proportional to the chemical activity of the surface-active component. Thus, the magnitude and rate of interfacial tension decline and contact angle increase will change with long-term migration as the interfacially active component(s) is (are) lost by adsorption onto the porous medium solids. This should have the effect of decreasing the rate of NAPL migration as it moves away from the source area. This effect may have particular relevance to primary migration of oil out of a source rock toward a trap since these distances are on the scale of geologic basins, as opposed to tens to hundreds of meters at a contaminated site.

Current experimental techniques for measuring the capillary pressure-saturation relationship are likely to cause significant inaccuracy for complex NAPLs containing surface active components. This arises from the interplay of the time-dependent chemical kinetics, the applied fluid reservoir pressure difference, the resulting rate of NAPL migration, and the length of the pressure-step equilibration time.

5. CONCLUSIONS

Surface chemical kinetics is a well-established phenomenon in multicomponent fluid systems. Timescales for interfacial tension change range from milliseconds to hours, and even days and longer, while those for contact angles can range up to months and perhaps longer, depending on the system. *A priori* assumption of equilibrium values for interfacial tension and/or contact angle may not be appropriate since most systems of practical interest, whether experimental or field situations, involve such multicomponent systems.

Whether or not dynamic surface chemistry must be considered depends on three issues. The first is whether or not a surface-active component is present in the system of interest. The equilibrium assumption is perfectly valid for simple, two-component systems involving pure bulk fluids. In field situations and in laboratory systems in which dyes are used to aid visualization, the equilibrium assumption needs to be examined. The second issue concerns the relative time scales of the process under consideration and the specific dynamic surface chemical kinetics of the system involved. Any process involving interfacial dilation or compression that occurs on a time-scale significantly shorter than the time-scale of the surface chemical kinetics should be evaluated for the need to incorporate the kinetics. Primary drainage in a porous medium of a multicomponent, non-wetting liquid (for example, a complex NAPL in water-wetted media) containing a surface-active component is a process for which the interfacial chemical kinetics should be evaluated. The third issue is whether the magnitude of the interfacial property changed is great enough to significantly alter the process, i.e., are the concentration and surface chemical behavior of a surface active species sufficient to produce significantly different capillary behavior.

The results of incorporating interfacial kinetics in a pore-scale model illustrate its importance. Contrary to the standard, equilibrium capillary pressure-saturation measurement procedures, the attainment of "equilibrium" may take far longer than ordinary laboratory time-scales. The non-wetting phase saturation becomes a function of both the capillary pressure (the traditional view) *and the surface chemical kinetics* under these conditions. The rate of migration of a NAPL front becomes a function of the interfacial chemical kinetics. Maximum NAPL saturation can no longer be seen as a monotonically increasing function of the capillary pressure boundary conditions alone. Rather, in pore-scale heterogeneous media at least, competition between whether the capillary pressure boundary conditions or the interfacial chemical kinetics are more important for determining when and if an interface advances result in a minimum in a plot of maximum NAPL saturation versus capillary pressure. The minimum occurs near the entry pressure for the pure bulk fluid without surface active components.

Current experimental techniques for measuring the capillary pressure-saturation relationship can mask the effects of time-dependent surface chemistry through both large pressure increments and short equilibration times. In addition, it is likely that oil-wetted pressure-plates could remove significant amounts of surface-active components in a NAPL prior to its encountering the test medium.

The effects of modeling multiphase flow by assuming equilibrium values for interfacial properties will be to over-estimate the rate at which a NAPL front advances, particularly at capillary pressures which just exceed the entry pressure for the medium. Another effect will be to underestimate the time required for NAPL to penetrate finer-grained units. In addition, it is likely that NAPL saturation may be off by a significant amount. More detailed work is needed to verify the results presented here. Both experimental and more elaborate modeling are warranted to test the predictions.

Acknowledgements—It is with great pleasure that I thank Dr. Roland Hellmann and Dr. Scott Wood for arranging this special issue honoring Dave Crerar. Dave was instrumental in bringing me to Princeton and connecting me with my subsequent advisor, mentor, and colleague, Peter Jaffe, Professor of Civil Engineering, Princeton University. It is, of course, impossible to say what my career would have looked like without this connection, but I doubt that it would have been as rich in intellectual stimulation. Dave's rapid decline in health, having occurred during my time at Princeton, was a great loss. Dave's capabilities and knowledge, at what should have been the height of his productivity, would have been an even greater blessing than Peter's alone.

I wish to acknowledge Dr. Roland Hellmann of the Univ. of Grenoble, Dr. Jonathan C. Johnson of the USGS, Dr. Tissa Illangasekare of the Colorado School of Mines, and an anonymous reviewer for their comments on the paper. They greatly improved its content and presentation. I also wish to thank Westinghouse Savannah River Co. and the U.S. Department of Energy, who have both directly and indirectly supported the work which forms the foundation of this paper.

Editorial handling: R. Hellmann

REFERENCES

Adamson A.W. (1990) *Physical Chemistry of Surfaces*, 5th ed. John Wiley & Sons Inc., New York.

Addison C.C. and Litherland D. (1953) The properties of freshly formed surfaces. Part XVIII. Dynamic surface potentials and the adsorption process: Molecular orientation in soluble films of decyl alcohol. *J. Chem. Soc.* 1143-1150.

Andreas J.M., Hauser E.A., and Tucker W.B. (1938) Boundary tension by pendant drops. *J. Phys. Chem.* **42**, 1001-1019.

Bear J. (1972) *Dynamics of Fluids in Porous Media.* American Elsevier Publishing Company, Inc., New York.

Bigelow S.L. and Washburn E.R. (1928) Variations in the surface tensions of solutions. *J. Phys. Chem.* **32**, 321-353.

Blunt M. and King P. (1991) Relative permeabilities from two- and three-dimensional pore-scale network modelling. *Transp. Porous Media* **6**, 407-433.

Bohr N. (1909) Determination of the surface-tension of water by the method of jet vibration. *Philos. Trans. Royal Soc. (London) A* **209**, 281-317.

Boussinesq M.J. (1913) Sur l'existence d'une viscosité superficielle, dans la mince couche de transition séparant un liquide d'un autre fluide contigu. *Ann. Chim. Phys.* **8**[E] **Série, 29**, 349-357.

Brewster M.L., Annan A.P., Greenhouse J.P., Kueper B.H., Olhoeft G.R., Redman J.D., and Sander K.A. (1995) Observed migration of a controlled DNAPL release by geophysical methods. *Ground Water* **33**, 977-987.

Catino S. and Farris R.E. (1978) Azo dyes. In *Kirk-Othmer Encyclopedia of Chemical Technology* (eds. H.F. Mark, D.F. Othmer, C.G. Overberger, and G.T. Seaborg). John Wiley & Sons, New York. pp. 387-433.

Celia M.A., Rajaram H., and Ferrand L.A. (1993) A multiscale computational model for multiphase flow in porous media. *Adv. Water Resour.* **16**, 81-92.

Celia M.A., Reeves P.C., Ferrand L.A. (1995) Recent advances in pore-scale models for multiphase flow in porous media. *Reviews Geophysics*, Supplement, "U.S. National Report to the Union of Geodesy and Geophysics 1991-1994", 1049-1057.

Crawford F.W. and Hoover G.M. (1966) Flow of fluids through porous mediums. *J. Geophys. Res.* **74**, 2911-2917.

Dullien F.A.L. (1992) *Porous Media: Fluid Transport and Pore Structure*, 2nd ed. Academic Press, Inc., San Diego.

Dupré A. (1869) *Théorie Mécanique de la Chaleur.* Gauthier-Villars, Paris.

Ferrand L.A. and Celia M.A. (1992) The effect of heterogeneity on the drainage capillary pressure-saturation relation. *Water Resour. Res.* **28**, 859-870.

Fortin J., Jury W.A., and Anderson M.A. (1997) Enhanced removal of trapped non-aqueous phase liquids from saturated soil using surfactant solutions. *J. Contam. Hydrol.* **24**, 247-267.

Gibbs J.W. (1948) *The Collected Works of J. Willard Gibbs. Vol. I. Thermodynamics.* Yale University Press, New Haven.

Giordano R.M. and Slattery J.C. (1983a) Effect of interfacial viscosities upon displacement in capillaries with special application to tertiary oil recovery. *A.I.Ch.E. J.* **29**, 483-492.

Giordano R.M. and Slattery J.C. (1983b) Stability of static interfaces in a sinusoidal capillary. *J. Colloid Interface Sci.* **92**, 13-24.

Giordano R.M. and Slattery J.C. (1987) Effect of interfacial viscosities upon displacement in sinusoidal capillaries. *A.I.Ch.E. J.* **33**, 1592-1602.

Graton L.C. and Fraser H.J. (1935) Systematic packing of spheres with particular relation to porosity and permeability. *J. Geol.* **43**, 785-909.

Haines W.B. (1930) Studies in the physical properties of soil. V. The hysteresis effect in capillary properties, and the modes of moisture distribution associated therewith. *J. Agric. Sci.* **20**, 7-116.

Heller J.P. (1968) The drying through the top surface of a vertical porous column. *Soil Sci. Soc. Am. Proc.* **32**, 778-786.

Hua X.Y. and Rosen M.J. (1988) Dynamic surface tension of aqueous surfactant solutions I. Basic parameters. *J. Colloid Interface Sci.* **124**, 652-659.

Kueper B.H. and Frind E.O. (1991) Two-phase flow in heterogeneous porous media 2. Model application. *Water Resour. Res.* **27**, 1059-1070.

Leverett M.C. (1941) Capillary behavior in porous solids. *Pet. Trans., A.I.M.E.* **142**, 152-169.

Lucassen J. and Giles D. (1975) Dynamic surface properties of nonionic surfactant solutions. *J. Chem. Soc., Farad. Trans.* **71**, 217-232.

Melrose J.C. (1965) Wettability as related to capillary action in porous media. *Soc. Pet. Eng. J.* **5**, 259-271.

Miller E.E. and Miller R.D. (1956) Physical theory for capillary flow phenomena. *J. Appl. Phys.* **27**, 324-332.

Morrow N.R. (1971) Small-scale packing heterogeneities in porous sedimentary rocks. *A.A.P.G. Bull.* **55**, 514-522.

Pan R., Green J., and Maldarelli C. (1998) Theory and experiment on the measurement of kinetic rate constants for surfactant exchange at an air/water interface. *J. Colloid Interface Sci.* **205**, 213-230.

Payatakes A.C. (1982) Dynamics of oil ganglia during immiscible displacement in water-wet porous media. In *Annual Review of Fluid Mechanics* (eds. M. van Dyke, J.V. Wehausen, and J.L. Lumley). Annual Reviews Inc., Palo Alto, CA. pp. 365-393.

Pirkle W.A., Iversen G.M., and Tuck D.M. (1997a) DNAPL penetration of capillary barriers due to time-dependent wetting relationships. Final Report submitted in completion of SCUREF Project SC-00058, SCUREF Task 232. University of South Carolina-Aiken.

Pirkle W.A., Iversen G.M., and Tuck D.M. (1997b) The effect of an organic dye on DNAPL entry pressure into water saturated porous media. Final report submitted in completion of SCUREF Project SC-0001-EC, SCUREF Task 185, Phase II. University of South Carolina-Aiken.

Poulsen M.M. and Keuper B.H. (1992) A field experiment to study the behavior of tetrachloroethylene in unsaturated porous media. *Environ. Sci. Technol.* **26**, 889-895.

Powers S.E. and Tamblin M.E. (1995) Wettability of porous media after exposure to synthetic gasolines. *J. Contam. Hydrol.* **19**, 105-125.

Rayleigh L. (1879) On the capillary phenomena of jets. *Proc. Royal Soc. (London)* A **29**, 71-97.

Rayleigh L. (1890) On the tension of recently formed liquid surfaces. *Proc. Roy. Soc. (London)* **47**, 281.

Rideal E.K. and Sutherland K.L. (1952) The variation of the surface tension of solutions with time. Solutions of alcohols. *Trans. Farad. Soc.* **48**, 1109-1123.

Rohlf F.J. and Sokal R.R. (1969) *Statistical Tables*. W. H. Freeman and Company, San Francisco.

Schroth M.H., Istok J.D., Ahearn S.J., and Selker J.S. (1995) Geometry and position of light nonaqueous-phase liquid lenses in water-wetted porous media. *J. Contam. Hydrol.* **19**, 269-287.

Scriven L.E. (1960) Dynamics of a fluid interface: Equation of motion for Newtonian surface fluids. *Chem. Eng. Sci.* **12**, 98-108.

Slattery J.C. (1964) Surfaces-I. Momentum and moment of momentum balances for moving surfaces. *Chem. Eng. Sci.* **19**, 379-385.

Slattery J.C. (1974) Interfacial effects in the entrapment and displacement of residual oil. *A.I.Ch.E. J.* **20**, 1145-1154.

Slattery J.C. (1979) Interfacial effects in the displacement of residual oil by foam. *A.I.Ch.E .J.* **25**, 283-289.

Sokal R.R. and Rohlf F.J. (1969) *Biometry: The Principles and Practice of Statistics in Biological Research.* W. H. Freeman and Company, San Francisco.

Tartar H.V., Sivertz V., and Reitmeier R.E. (1940) The surface tension of aqueous solutions of some paraffin chain colloidal electrolytes, including a comparison of the capillary rise and sessile bubble methods of measurement. *J. Am. Chem. Soc.* **62**, 2375-2380.

Tate T. (1864) On the magnitude of a drop of liquid formed under different circumstances. *Philos. Mag.* **27**, 176-180.

Thomas W.D.E. and Potter L. (1975) Solution/air interfaces. I. An oscillating jet relative method for determining dynamic surface tensions. *J. Colloid Interface Sci.* **50**, 397-412.

Tuck D.M. (1999) Discussion of "DNAPL migration through a fractured perching layer" by Daniel B. Stephens, James A. Kelsey, Mark A. Prieksat, Mel G. Piepho, Chao Shan, and Mark D. Ankeny. *Ground Water* **37**, 485-486.

Tuck D.M. and Rulison C. (1999a) Azo dyes and their interfacial activity: Implications for multiphase flow experiments(U). Westinghouse Savannah River Company, WSRC-MS-99-00138. Aiken, SC.

Tuck D.M. and Rulison C. (1999b) Interfacial activity of Sudan IV dye in the tetrachloroethylene-water-glass system. Westinghouse Savannah River Company, WSRC-TR-98-0038. Aiken, SC.

Tuck D.M., Iversen G.M., Pirkle W.A., and Rulison C. (1998) Time-dependent interfacial property effects on DNAPL flow and distribution. In *Nonaqueous-Phase Liquids: Remediation of Chlorinated and Recalcitrant Compounds* (eds. G.B. Wickramanayake and R.E. Hinchee). Battelle Press, Columbus, OH. pp. 73-78.

Tuck D.M., Iversen G.M., and Pirkle W.A. (2000) DNAPL penetration of saturated porous media under conditions of time-dependent surface chemistry. In *Savannah River Site Environmental Remediation Systems in Unconsolidated Upper Coastal Plain Sediments - Stratigraphic and Structural Considerations: Carolina Geological Society 2000 Field Trip Guidebook, Aiken, SC* (eds. D.E. Wyatt and M.K. Harris). Carolina Geological Society. pp. K1-K16.

Ward A.F.H. and Tordai L. (1944) Existence of time-dependence for interfacial tension of solutions. *Nature* **154**, 146-147.

Ward A.F.H. and Tordai L. (1946) Time-dependence of boundary tensions of solutions. I. The role of diffusion in time-effects. *J. Chem. Phys.* **14**, 453-461.

Zhang L.-H. and Zhao G.-X. (1989) Dynamic surface tension of the aqueous solutions of cationic-anionic surfactant mixtures. *J. Colloid Interface Sci.* **127**, 353-361.

Water-Rock Interactions, Ore Deposits, and Environmental Geochemistry: A Tribute to David A. Crerar
© The Geochemical Society, Special Publication No. 7, 2002
Editors: Roland Hellmann and Scott A. Wood

In situ speciation of uranium(VI) at the silica-water interface: A combined TRLIFS and surface complexation study

UTA GABRIEL[1]*, LAURENT CHARLET[1], AND CARL W. SCHLÄPFER[2]

[1]Laboratoire de Géophysique Interne et Tectonophysique, OSUG, UJF - CNRS (UMR 5559)
Grenoble, France

[2]Institut de Chimie Inorganique, Université de Fribourg, Switzerland

*Corresponding author, present address: INRS Géoressources, Université du Québec,
C.P. 7500, Sainte-Foy, QC, G1V 2L2, Canada (e-mail: ugabriel@NRCan.gc.ca)

Abstract—The sorption of uranyl ions ($U^{VI}O_2^{2+}$) to amorphous silica has been studied by laser-induced time-resolved fluorescence spectroscopy (TRLIFS) at trace concentration levels (1 and 0.1 μM) in the presence of atmospheric CO_2. Uranyl is adsorbed on silica between pH 4 and 8.8, and two uranyl surface complexes have been identified in this pH range by their fluorescence properties. Characterized by distinct lifetimes (170 ± 25 μs at low pH and 360 ± 50 μs at high pH) and luminescence emission spectra (250 cm^{-1} red shift between the low pH and high pH spectra), they have been assigned to the $\equiv SiO_2UO_2^{\circ}$ and $\equiv SiO_2UO_2OH^-$ species. When fluorescence data were compared to analytically determined total sorbed concentrations, a third non-luminescent ternary surface complex ($\equiv SiO_2UO_2OHCO_3^{3-}$) had to be postulated to account for adsorption between pH 8.0 and 8.8.

The ability of uranium-rich silica particles to aggregate, and thus not to migrate in soils, was assessed by a series of centrifugation experiments using silica suspensions at pH 4.4 and 8.4. About 20 % of the total uranyl concentration (U_{tot} = 1 μM) is easily removed by centrifugation (1000 rpm, 5 minutes) and can considered to be immobile, as these particles aggregate easily in natural porous media. Under more rigorous centrifugation conditions (10,000 rpm, 40 to 60 minutes) the intermediate mobility fraction is removed. The uranyl removal is 98 % at pH 8.4, but only 70 % at pH 4.4. Of the 30 % of U_{tot} that is mobile at pH 4.4, 17 % fluoresces, and of this amount, 7 and 10 % are assigned to the sorbed and the dissolved uranyl fractions, respectively. Therefore, the risk of silica particle-induced transport of uranyl is more elevated in slightly acidic waters.

1. INTRODUCTION

Silica comprises a major part of igneous and metamorphic rocks of the earth's continental crust, including minerals in sedimentary environments (sand, biogenic silica, etc.). Consequently, adsorption on silica affects many superficial processes such as weathering rates, scavenging of diverse elements in the ocean, and the fate of contaminants in soils and groundwaters (Van Cappellen and Gaillard, 1996; Van Cappellen and Qui, 1997). All of these phenomena are controlled by chemical processes taking place at the interface between the mineral surface and the bulk solution. During the past decade, considerable progress has been made in the study of silica surface chemistry (Giamelo et al., 1990; Costa et al., 1990; Belyak et al. 1993; Dove and Rimstidt, 1994), dissolution kinetics (Dove and Crerar, 1990; Brady and Walther, 1990), and sorption of certain aqueous ions (e.g., thorium: Östhols, 1995; uranyl: Michard et al., 1996). However, the identity and structure of the actual surface species controlling the processes on the silica/water interface remain subject to controversy.

Owing to the development of surface-sensitive spectroscopic techniques such as infra-red spectroscopy (IR), nuclear magnetic resonance spectroscopy (NMR), low-energy electron diffraction (LEED), atomic force microscopy (AFM), and X-ray absorption spectroscopy (XAS) (Burneau and Gallas, 1998; Hommel et al., 1998; Reich et al., 1998; Sylwester et al., 2000; Dent et al., 1992), direct atomic level information on bonding environments at the silica/water interface has become available.

Based on surface titration and electrokinetic measurements (Schindler et al., 1976; James and Healy, 1972), surface complexation models (SCM's) of silica minerals were developed which postulate the formation of inner-sphere $\equiv OMe^+$ and $\equiv OMeOH^{\circ}$ surface complexes between a metal cation (Me^{2+}) and a primary surface hydroxyl $\equiv OH^{\circ}$ site (e.g. a surface silanol group). The various applications of these models for the interpretation of adsorption data and reaction kinetics demonstrated their validity in describing the speciation at the mineral/water interface. In the presence of CO_2 and an excess of metal ions with respect to surface reactive sites, the following species are assumed to form: $\equiv OMe^+$, $\equiv OMeOH^{\circ}$, $\equiv OMe(HCO_3)^{2-}$, and $\equiv OMe(OH)_2^{\circ}$. These surface species have been inferred indirectly from surface titration, electrokinetics, adsorption, and reactive transport data, and from an analogy between the formation of surface and aqueous species (Schindler, 1991).

The present study focuses on the uranyl cation (UO_2^{2+}), which is used as a fluorescent molecular probe of the silica/water interface. Time-resolved laser-induced fluorescence spectroscopy (TRLIFS) allows an in situ investigation of the complexes it forms at the silicate surface. The uranyl ion is the most common and mobile form of uranium in oxidizing aqueous media. It is present in large concentrations in mining waste sites, atomic weaponry installations, and is expected in worst case scenarios to migrate out of spent nuclear fuel disposal sites and penetrate into soils and aquifers. Laser-induced fluorescence spectroscopy has long been a useful technique for characterizing the molecular environment of long fluorescence lifetime compounds (rare earth elements, uranyl ion, organic molecules) and their interactions with the surface of silica (Glinka and Krak, 1995; Kushnirenko et al., 1993; Stanton and Maatman, 1963; Wheeler and Thomas, 1984).

The main advantage of TRLIFS over the widely used XPS and EXAFS methods is that the former can provide detailed information on surface species present at very low surface concentrations. In addition, small changes (e.g., during hydrolysis) in the environment of the monitored molecular probe can be measured. This technique also has the advantage of not requiring any ultrahigh vacuum conditions. Since the laser beam interacts with the investigated suspension in a non-perturbing manner, the surface composition remains practically unchanged during the spectroscopic investigation.

In the present study we report on the first in situ spectroscopically-observed speciation of uranyl ions sorbed on amorphous silica. This TRLIFS-based speciation is compared with that obtained by standard macroscopic surface complexation models. We further study the centrifugation of U-rich particles at various pH's, centrifugation speeds, and centrifugation times. Based on these parameters, the risk of free uranyl and uranyl-silica particle transport is assessed.

2. METHODS

2.1. Experimental Methods

The silica sample used in this study (Aerosil 200) was provided by Degussa (Germany). It is a nonporous, X-ray amorphous material composed of spherical particles with a diameter of about 25 nm and a specific surface area of 169 m^2/g (Östhols, 1995). All chemicals used in this study were of p.a. grade.

Amorphous silica suspensions (particle concentration c_p = 1 g dm^{-3}) were prepared in 50-cm^3 polypropylene bottles in equilibrium with the atmosphere, at room temperature (22 ± 1 °C). For each of the three total uranyl concentrations (0, 0.1, and 1 μM) a series of 11 samples was prepared, with pH values adjusted between 3 and 9 using varying proportions of HNO_3 and Na_2CO_3 solutions at low and high pH values, respectively. $NaNO_3$ was added (mainly at low and intermediate pH values) in order to adjust the ionic strength to 0.01 M. The samples were shaken for at least one week at room temperature. Afterwards, the pH was measured once again and this final pH registered. A combination glass electrode (Metrohm) was calibrated using standard buffer solutions (Merck). Silica concentrations were determined colorimetrically using the molybdate blue method (Strickland and Parsons, 1972).

2.2. Time-resolved Laser-induced Fluorescence Spectroscopy (TRLIFS)

The fluorescence spectra for the surface speciation analysis were recorded at the Radiochemistry Institute, Research Center Rossendorf, Germany, using the following instruments and devices. The suspension samples were introduced into a fluorescence measurement cell (quartz, 4 cm^3, path length 1 cm; stirred) and excited with a pulsed NdYAG laser at 266 nm (Spectra Physics, 4 mJ/pulse, 8 ns pulse duration). The fluorescence was captured with an optic fiber (length: 0.5 m; Spectrophysics). The spectra were registered using a diode array (EG&E, 700 diodes, 0.3 nm resolution). The diode array was controlled by a delay generator (600 V) which was triggered by the laser itself and delayed by a delay generator. The energy delivered by the laser was controlled by an optical meter after passing the measurement cell. Previously, this experimental set-up was used to characterize the fluorescence of the uranyl ion in 0.1 M $HClO_4$. Bernhard et al. (1996) reported a lifetime of 1.55 ± 0.2 μs, which agrees with the values of 1.7 ± 0.2 μs measured by Eliet at al. (1995) and 1.9 ± 0.2 μs by Kato et al. (1994).

The experimental conditions of the present measurements are listed in Table 1. The delayed time gate (40.1 μs) was chosen in order to eliminate the fluorescence of dissolved uranyl hydroxyl, nitrate, and carbonate complexes (speciation in Fig. 1, equilibrium constants in Table 2) which are known to have lifetimes of about 33 μs or less (Eliet et al., 1995; Deniau et al., 1993). Uranyl sorbed on silica is known to have elevated lifetimes (Wheeler and Thomas, 1984: 440 μs). Duff et al. (2000) also reported elevated luminescence lifetimes for uranyl sorbed on lake

Table 1. Measurement conditions for the characterization of the surface complexes

delay	40.1 μs
gate	10 μs
delay step	5 and 2 μs
number of time steps	31
laser shots per spectrum	200
spectra per measurement	3

sediments (lifetimes ≤ 526 μs, measured at liquid N_2 temperature). The fluorescence of the dissolved uranyl silicic acid complex ($UO_2H_3SiO_4^+$) is thought to depend on silicic acid polymerization (Moll et al., 1998). Uranyl luminescence lifetimes of 19 ± 4 μs and 90 to 200 μs are reported in the presence of monomeric and polymeric silicic acid, respectively. Therefore, the dissolved uranyl silicic acid complex has to be considered in the fluorescence data interpretation. The silicic acid concentration in the present study was measured and found to vary from 1.5 to 2.2 mM, depending on pH. Speciation of uranyl ion in the presence of 2 mM silicic acid is given in Fig. 1.

In order to characterize uranyl surface complexes by their emitted time-resolved fluorescence spectra, one must first deconvolute all other fluorescence components. The fluorescence of solution complexes was already discussed. The silica suspension without uranyl (due to possible impurities of the silica or to defects in the lattice of the silica particles) also emitted a fluorescence signal in the studied wave number range. The spectra depict a broad, symmetrical, and rather noisy maximum at about 20,000 cm^{-1} (Fig. 2). The luminescence intensity could be described with a bi-exponential decay function (lifetimes of 8 ± 1 and 53 ± 4 μs). Neither the peak positions in the spectrum nor the lifetimes changed as function of pH (3 ≤ pH ≤ 8). This fluorescence signal should be present in all samples and independent of the uranyl adsorption. This spectrum was therefore subtracted from all of the spectra of the uranyl-silica system, thereby resulting in a zero baseline outside the range of the characteristic peaks of the uranyl fluorescence (for details see Gabriel, 1998). Typical spectra of the water-amorphous silica-uranyl system at a total uranyl concentration of 1 μM are given in Fig. 3.

The removal of particles from solution was assessed by a series of centrifugation experiments, in which both centrifugation time and speed were varied. Two 200-ml samples (pH 4.4 and 8.4) were prepared. After centrifugation (Beckmann Avanti J-20 centrifuge, fixed angle rotor JA-14, polypropylene tubes 40 ml), the fluorescence decay of the original and the supernatant solutions were recorded in the 50- to 450-μs time range using a pulsed nitrogen laser (VSL-337ND, Laser Science, Inc.; 250 μJ/pulse, 15 Hz, wave length 337 nm), a fluorescence measurement cell (quartz, 4 cm^3, path length 1 cm, stirred), a photomultiplier (Hamamatsu 94302, 500 V), a gelatin filter (Kodak 55, maximal transmission 67-69 % at 510-520 nm), a digital oscilloscope (digitizing oscilloscope

Table 2. Equilibrium constants (log K) for solution speciation (Grenthe et al., 1992; Stumm and Morgan, 1996, Moll et al., 1998)

Reaction			**log K (I=0)**	**log K (I=0.01)**
$UO_2^{2+} + H_2O$	↔	$UO_2OH^+ + H^+$	-5.2	-5.3
$UO_2^{2+} + 2\ H_2O$	↔	$UO_2(OH)_2^\circ + 2\ H^+$	-10.3	-10.4
$UO_2^{2+} + 3\ H_2O$	↔	$UO_2(OH)_3^- + 3\ H^+$	-19.2	-19.2
$UO_2^{2+} + H_2CO_3^\circ$	↔	$UO_2CO_3^\circ + 2\ H^+$	-7.0	-7.1
$UO_2^{2+} + 2\ H_2CO_3^\circ$	↔	$UO_2(CO_3)_2^{2-} + 4\ H^+$	-16.4	-16.3
$UO_2^{2+} + 3\ H_2CO_3^\circ$	↔	$UO_2(CO_3)_3^{4-} + 6\ H^+$	-28.5	-27.7
$2\ UO_2^{2+} + 3\ H_2CO_3^\circ$	↔	$UO_2CO_3(OH)_3^{3-} + 5\ H^+$	-17.6	-17.7
$UO_2^{2+} + H_4SiO_4^\circ$	↔	$UO_2H_3SiO_4^+ + H^+$	-1.7	-1.8
$UO_2^{2+} + NO_3^-$	↔	$UO_2NO_3^+$	0.3	0.06
H_2O	↔	$OH^- + H^+$	-14.0	-13.9
$H_4SiO_4^\circ$	↔	$H_3SiO_4^- + H^+$	-10.0	-9.9
$H_4SiO_4^\circ$	↔	$H_2SiO_4^{2-} + 2\ H^+$	-23.0	-22.7
$H_2CO_3^\circ$	↔	$HCO_3^- + H^+$	-6.4	-6.3
$H_2CO_3^\circ$	↔	$CO_3^{2-} + 2\ H^+$	-16.7	-16.4

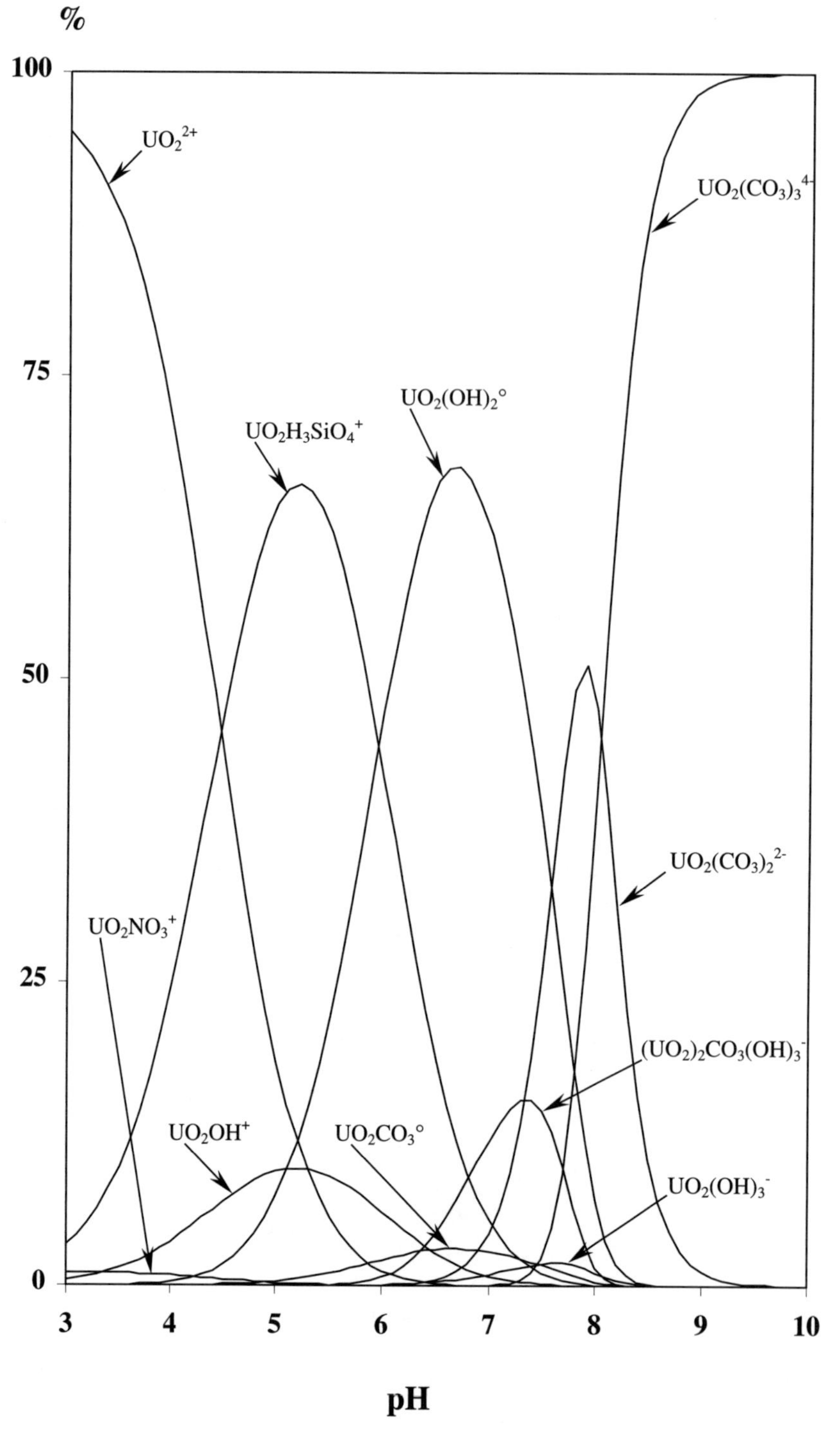

Fig. 1. Uranyl solution speciation based on the constants listed in Table 2 (Grenthe et al., 1992); code: GRFIT, Ludwig (1993); $[UO_2^{2+}]_{tot}$ = 1 μM; $[H_2CO_3°]_{free}$ = 22.4 μM; $[H_4SiO_4°]_{tot}$ = 2 mM; I = 0.01 M; $[NO_3^-]_{tot}$ = 0.01 M.

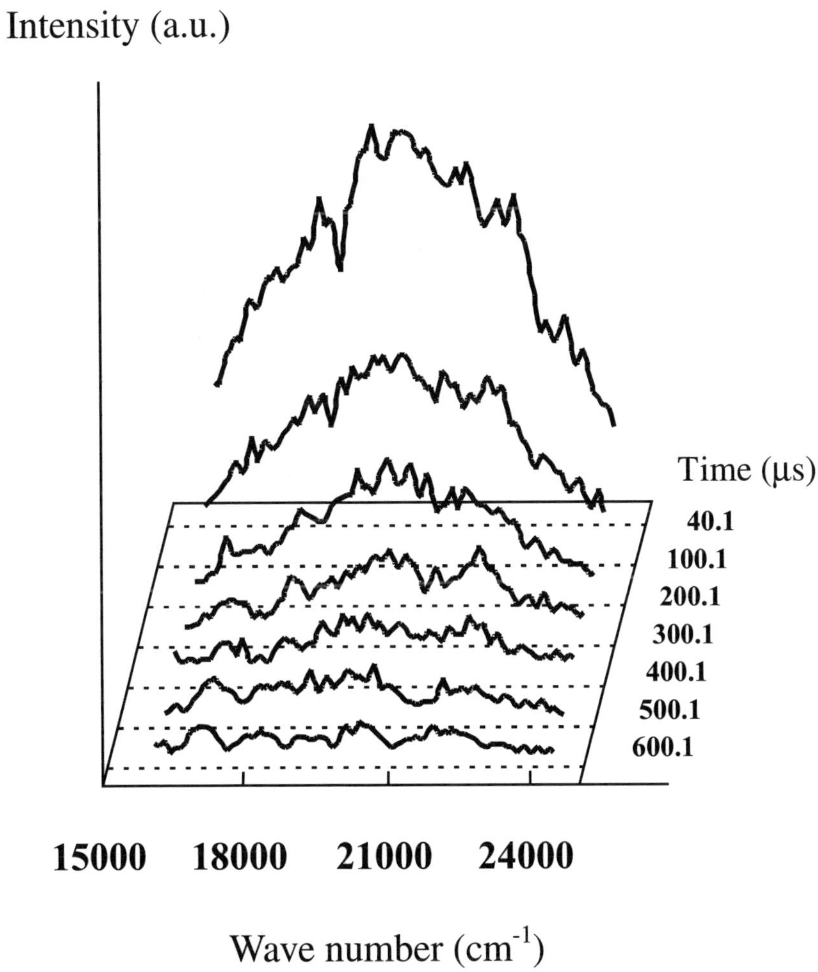

Fig. 2. Time-resolved luminescence spectra of a silica suspension at pH 5.11 (c_p = 1 g dm^{-3}; no uranyl ions).

54501A 100MHz, Hewlett Packard), and a photodiode (PIN S1223-01, Hamamatsu) at the Physical Spectrometry Laboratory, Grenoble, France. The fluorescence intensity is expressed in terms of % of the original suspension fluorescence (before centrifugation).

The sorption isotherms were measured by TRLIFS according to a third method, using the experimental set-up for fluorescence measurements at the Physical Spectrometry Laboratory, Grenoble, France. The suspensions were filtered through 0.45-μm filter membranes. No significant differences, with respect to dissolved uranyl concentrations, were observed when suspensions were filtered through 0.45-, 0.2-, 0.1-, and 0.025-μm filters. After filtration samples were instantly acidified to pH 1 with nitric acid to avoid sorption on the container walls. The dissolved uranyl concentration was measured according to the method given in Brina and Miller (1992). The addition of H_3PO_4 (final concentration 0.75 M) was used to form stable uranyl solution complexes with a typically long fluorescence lifetime of about 200 μs. This value is in good agreement with measurements of 200 μs by Moulin et al. (1995) and 232.1 ± 0.5 μs by Kato et al. (1994) and Meinrath et al. (1993).

The initial fluorescence intensity is proportional to the concentration of the emitting molecules. By extrapolation of the exponential decay to time zero, the initial fluorescence intensity was determined. The method was calibrated using the standard additions method. Corrections were made for base line and

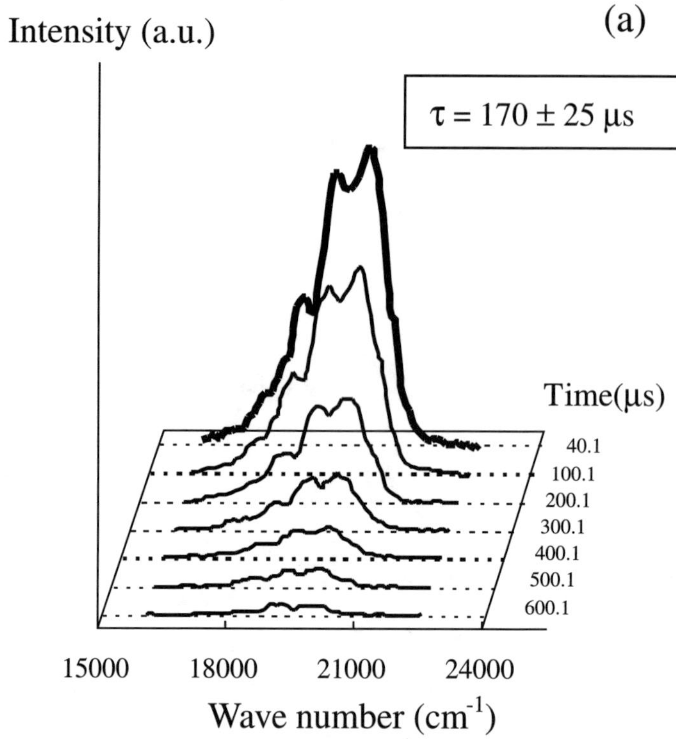

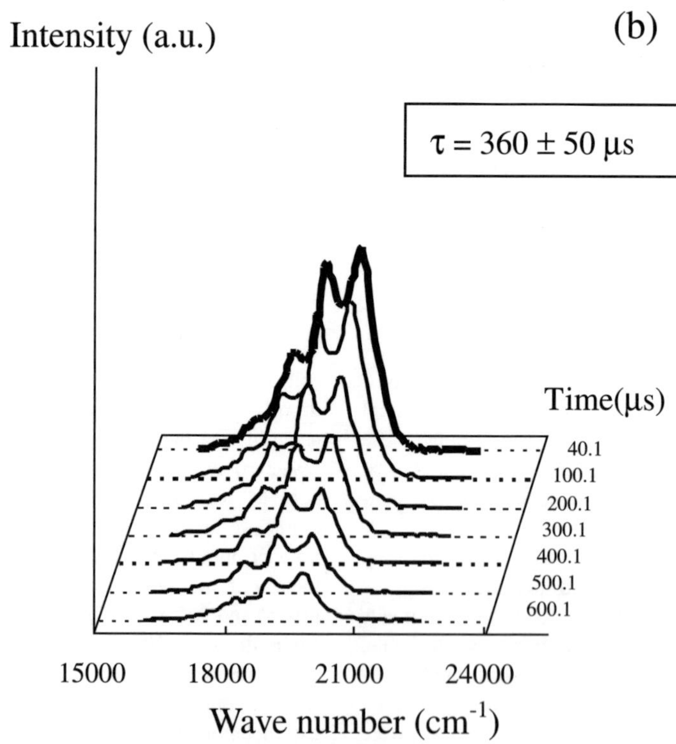

Fig. 3. Experimental time-resolved luminescence spectra of silica suspensions at pH 4.4 (a) and pH 8.2 (b). Intensities are shown in the same arbitrary units; $[UO_2^{2+}]_{total}$ = 1 μM; c_p = 1 g dm^{-3}. The spectra obtained 40.1 μs after the laser pulse (thick lines) have been used in Figs. 4 and 5.

instrumental drift. The detection limit and precision were 3 nM and 5 %, respectively (Gabriel et al., 1998).

2.3. Data Treatment

The fluorescence emitted by complex i is a function of the time t and wave number ν, and its decay follows an apparent first order law (Brina and Miller, 1992):

$$I_i(t,\nu) = I_i(0,\nu)\exp\left(-\frac{t}{\tau_i}\right) \quad (1)$$

where $I_i(t,\nu)$ is the fluorescence intensity of complex i at the wave number ν and time t; τ_i is the characteristic fluorescence lifetime of complex i.

When all parameters of the experiment are kept constant, the observed fluorescence intensity $I_i(t,\nu)$ of species i at time t and wave number ν is related to the concentration S_i of the emitting complex i by:

$$I_i(t,\nu) = \omega_i(t,\nu)\, S_i \quad (2)$$

where $\omega_i(t,\nu)$ is an experimental parameter. It incorporates the effects of absorption, radiant and non-radiant transition probabilities (all characteristic for the emitting complex i), and several parameters characteristic of the experimental set-up, such as pre-filter and post-filter effects. A detailed discussion of this is given by Berthoud et al. (1988) and Eliet (1996). The vector $\omega_i(t,\nu=1\ ..\ l)$ represents the fluorescence spectrum of species i (l is the number of data points in the spectrum).

The fluorescence intensity of n emitting complexes is additive. The overall fluorescence intensity of a composite solution or suspension is then described by

$$I_i(t,\nu) = \sum_{i=1}^{n} \omega_i(t,\nu)\, S_i \quad (3)$$

The sum of concentrations of all emitting surface complexes, S_i, is equal to or smaller than the analytically-determined total concentration of sorbed uranyl ion, S_{tot}:

$$S_{tot} \geq \sum_{i=1}^{n} S_i \quad (4)$$

The difference between S_{tot} and the summation on the right-hand side (Eqn. 4) gives the concentration of possible silent, non-emitting surface species. Concentrations of all emitting complexes are related to the emitted fluorescence light, given by Eqn. 3. Two parameters, $\omega_i(t,\nu)$, the emission at a given time t and a wave number ν, and the emission lifetime τ_i, are characteristic of an adsorbed species. These parameters were used separately in order to obtain quantitative information with respect to the surface speciation.

2.3.1. Wave number-resolved data (spectra at 40.1 μs)

In order to reduce the data from the spectra from 700 to 60 we smoothed the experimental data using the built-in function of the software package Mathcad 6+ (spline).

At a given time t the measured fluorescence intensity $I(\nu)$, given by:

$$I(\nu) = \sum_{i=1}^{n} \omega_i^t(\nu)\, S_i \quad (5)$$

is analogous to the Lambert-Beer law for absorption. The fluorescence of a series of suspensions is given by the following matrix equation (dimensions of the matrices are given in brackets)

$$\mathrm{I}(l,s) = \mathrm{w}(l,n)\,\mathrm{S}(n,s) \quad (6)$$

with

- *I(l,s)* matrix of the measured fluorescence intensities. The columns are the emission spectra for the given solution.
- *w(l,n)* matrix of the fluorescence spectra. The columns give the emission spectra of the significantly different surface complexes.
- *S(n,s)* matrix of the relative concentrations. The columns give the speciation of the surface species.
- *l* number of data points per spectrum
- *s* number of solutions (or measured spectra)
- *n* number of significantly different complexes

The mathematical procedure consists of decomposing the matrix of the fluorescence data *I(l,s)* into a linear product of the matrices of the relative concentrations of the fluorescent complexes *S(n,s)* and their specific fluorescence spectra *w(l,n)*.

Singular value decomposition of *I(l,s)* allows for the determination of the number of emitting species n. Evolving Factor Analysis (EFA, Gampp et al., 1987) can be used for the model-free decomposition of the matrix *I(l,s)* after the number of species have been determined, based on the following conditions:

1. linear relationship between the measured fluorescence intensity at time t and the concentration of the fluorescent complex S_i (Eqn. 2)
2. mass conservation
3. the first and the last spectrum are due to only one species, respectively
4. when stepping from one pH to another only one new emitting species appears
5. the species disappear as pH increases in the same order as they appeared (i.e., "first in - first out").

The condition of the mass conservation is easily replaced by Eqn. 4 if linear programming (Press et al., 1992) is used to solve the optimization problem.

2.3.2. Time-resolved integral values of the fluorescence intensity (decays)

The EFA of the spectra indicated that two complexes dominate the experimental fluorescence spectra of the suspensions containing 1 μM uranyl. Thus, Eqn. 3 can be reduced to a two-term equation. At a given wave number ν (or within a given ν range) the experimental decay is described by:

$$I(t) = \omega_1^0 S_1 \exp\left(-\frac{t}{\tau_1}\right) + \omega_2^0 S_2 \exp\left(-\frac{t}{\tau_2}\right) \quad (7)$$

The experimental intensity integral $I(t)$ between 19000 and 20000 cm^{-1} was used.

2.4. Surface Complexation Modeling

The adsorption chemical equilibrium of uranyl ions onto silica by surface complexation can be described by the general equation:

$$p \equiv Si(OH)_2 + q\, UO_2^{2+} + r\, H^+ + s\, CO_3^{2-} \leftrightarrow$$

$$\left(\equiv Si(OH)_2\right)_p \left(UO_2\right)_q (H)_r \left(CO_3\right)_s^{(2q+r-2s)} \quad \beta_{pqrs}^s \quad (8)$$

where β_{pqrs}^s is the conditional equilibrium constant, and p, q, r and s are the (positive or negative) stoichiometry coefficients. The conditional equilibrium constant includes activity coefficients for solution species which were calculated with the Davies equation (Davies, 1962). Activity coefficients for all surface species are assumed to be equal to 1. The $\equiv Si(OH)_2$ surface species is a hypothetical average surface adsorption site, susceptible to lose one proton according to:

$$\equiv Si(OH)_2 \rightarrow \; \equiv SiO_2H^- + H^+ \qquad \beta_{10-10}^s \quad (9)$$

The total concentration of surface sites $[\equiv Si(OH)_2]_{tot}$ was set to 0.51 mM (after Östhols, 1995) and the total uranyl ion concentration $[UO_2^{2+}]_{tot}$ was defined as constant. The total proton concentration $[H^+]_{tot}$ is calculated using a charge-balance expression. The free carbonic acid concentration $[H_2CO_3]_{free}$ is controlled by CO_2-equilibrium with the atmosphere ($P_{CO_2} = 10^{-3.51}$ atm).

The conditional equilibrium constants β_{pqrs} and $\beta_{10\text{-}10}$ were corrected for the Coulombic energy of the surface charge to obtain the corresponding intrinsic constants:

$$\beta_{pqrs(\text{int})}^s = \beta_{pqrs}^s \exp\left(\frac{(2q+r-2s)F\psi}{RT}\right) \quad (10)$$

where Ψ is the effective surface potential. The total molar charge of the surface σ_{tot} is given by:

$$\sigma_{tot} = \sum_p \sum_q \sum_r \sum_s (2q+r-2s)\beta_{pqrs}^s$$
$$\times \left[\equiv Si(OH)_2\right]^p \left[UO_2^{2+}\right]^q \left[H^+\right]^r \left[CO_3^{2-}\right]^s \quad (11)$$

To relate the total molar surface charge σ_{tot} to the surface potential Ψ the constant-capacitance model (Schindler and Gamsjäger, 1972) was used, since the formation of outer-sphere complexes could be excluded (Dent et al., 1992; Reich et al., 1996; 1998; Sylwester et al., 2000):

$$\Psi = \frac{\sigma_{tot} F}{A_s c_p \kappa} \quad (12)$$

where A_s is the specific surface area ($m^2\, g^{-1}$), c_p is the particle concentration (g dm^{-3}), F is the Faraday constant, and κ is the capacitance (F m^{-2}). The intrinsic stability constants, the total site concentration, and the capacitance were determined from the adjustment of the experimental fluorescence data using the program GRFIT (Ludwig, 1993). The evaluation was based mainly on the goodness-of-fit (Westall, 1982).

3. RESULTS AND DISCUSSION

3.1. Identification of Different Surface Complexes

In Fig. 4 the experimental uranium fluorescence spectra recorded on silica suspensions equilibrated at different pH values and total uranyl concentrations (1 and 0.1 μM) are compared. The spectra recorded for different total uranium concentrations reveal similar behavior. At pH values below 4.4 the fluorescence signal is very weak, but around pH 4.4 the intensity increases strongly. In this pH range (pH ≤ 4.4), macroscopic sorption measurements indicate that the concentration of sorbed uranium shifts from 50 % to 80 % of the total uranyl concentration (discussed further on). The fluorescence spectra in this pH range show the same maximum peaks, whatever the total uranium concentration, and the same uranyl molecular environment is assumed to dominate the fluorescence signal. In the 4.5 to 5.5 pH range the fluorescence intensity remains at the same elevated level, but a second component appears in the spectra, which is characterized by maxima at smaller wave numbers (i.e. with a 250-cm^{-1} red shift). In the pH range 5.5 to 8 the spectra are dominated by this sec-

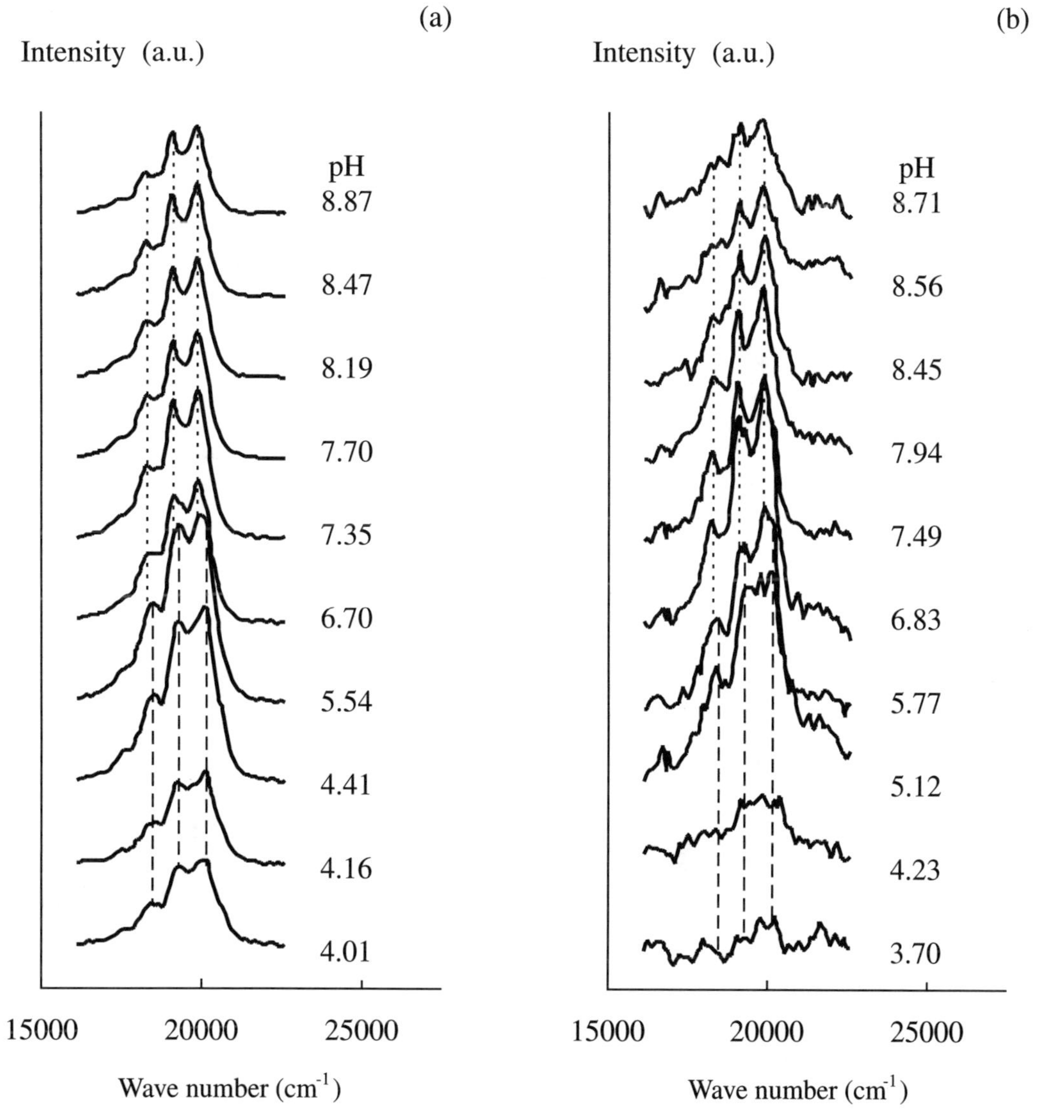

Fig. 4. Experimental spectra of the silica suspensions (c_p = 1 g cm^{-3}) containing 1 µM (a) and 0.1 µM (b) uranyl; intensity values are divided by 10 for (a).

ond component spectrum. Beyond pH 8.2, the overall intensity decreases.

A comparison of both fluorescence components indicates that they also differ as a function of their lifetimes. A longer lifetime is observed for the higher pH fluorescence component (Fig. 3). By an adjustment of a mono-exponential decay function to the experimental data, lifetimes of 170 ± 25 and 360 ± 50 µs were determined for samples prepared at low and high pH values, respectively. The mono-exponential decays at pH 4.4 and 8.2 did not completely exclude the existence of minor fluorescence components at these pH values, but indicated nonetheless the dominance of distinct molecular environments in both spectra.

3.2. *Quantification of the Surface Speciation*

The spectra obtained for a total uranyl concentration of 1 µM (Fig. 4a) were analyzed by singular value decomposition (Gampp et al., 1987). This

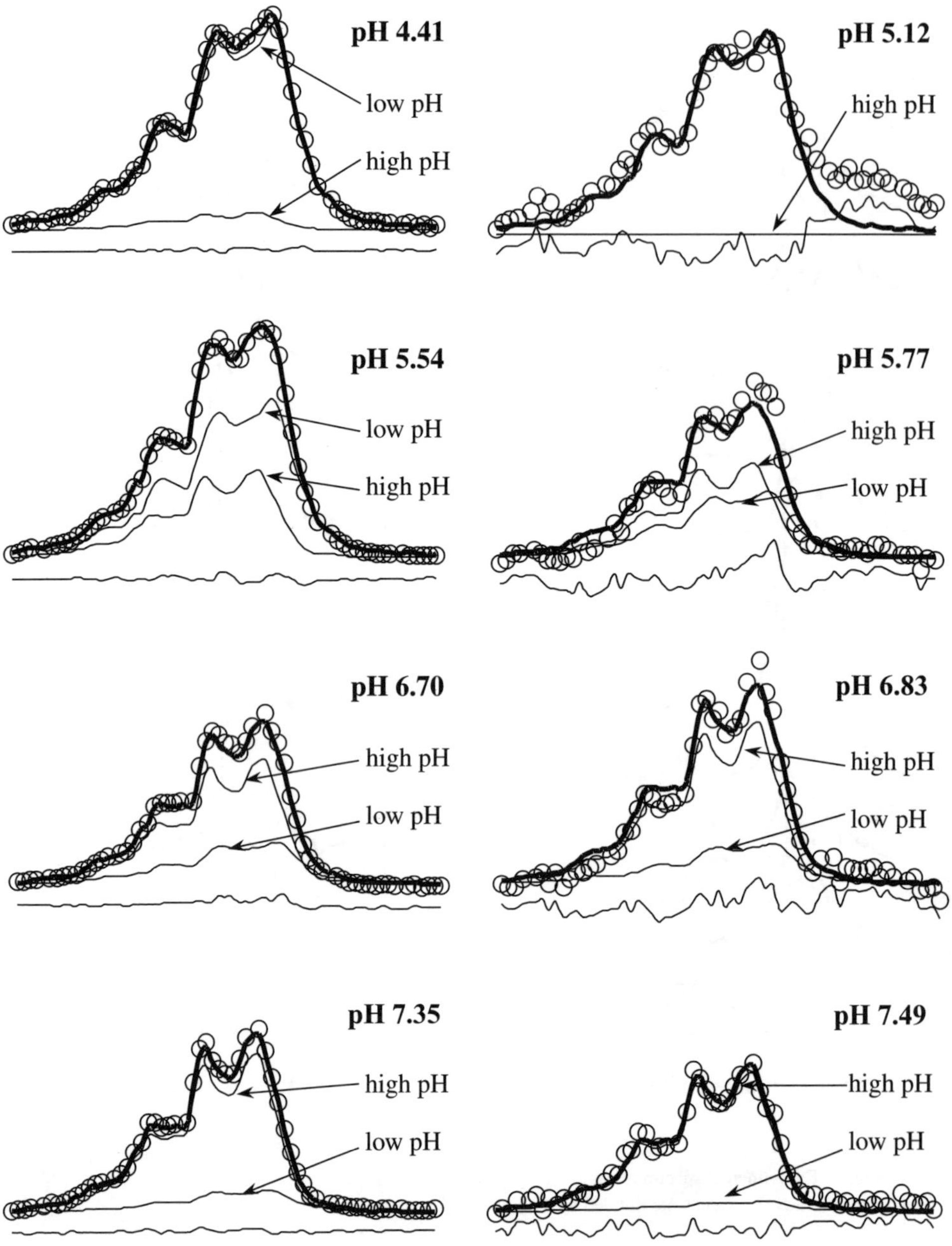

Fig. 5. Experimental spectra (o symbol) at a total uranyl concentration of 1 μM (left column) and 0.1 μM (right column), adjusted using evolving factor analysis (thin lines: prevailing spectra at low/high pH; thick line: sum of both; thin line below: remaining error; c_p = 1 g dm^{-3}; intensity values are divided by 10 for the left column).

analysis confirmed the existence of two distinctively different spectra. For the deconvolution of these spectra the evolving factor analysis (EFA) was used. The spectra of the intermediate pH values, where dominance shifts from the first spectrum to the second, are represented in the left column of Fig. 5. The already mentioned dominance shift between pH 5.5 and 6.7 is confirmed. The experimental data at a total uranyl

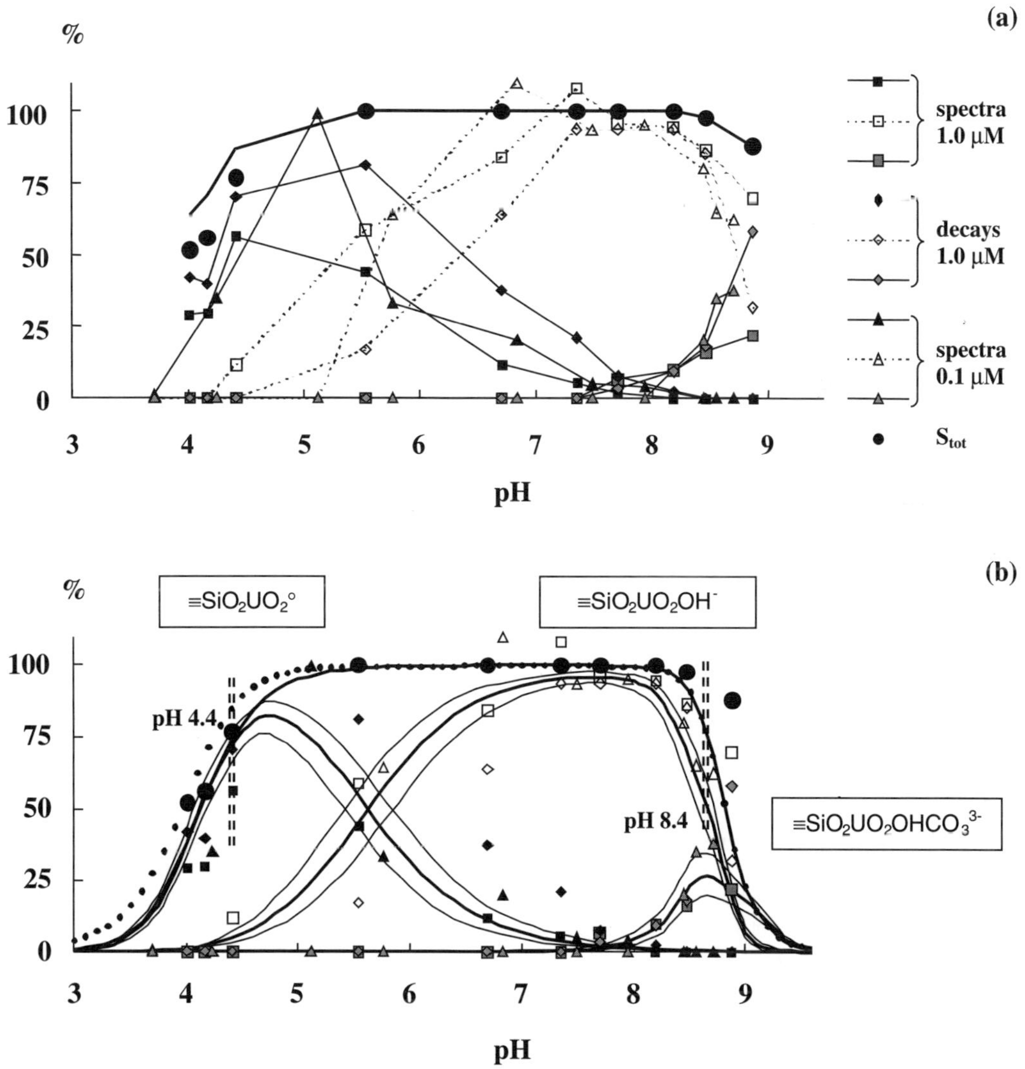

Fig. 6. Results of the experimental spectra and decays (a) and uranyl surface speciation model (b). Stability and intrinsic surface constants are given in Tables 2 and 3; filled and open symbols indicate the lower and higher protonated state; grey symbols show the silent surface complex contribution, i.e., the difference between S_{tot}, the sorbed amount of uranyl (filled circles, in % of total concentration) and the two luminescent surface complexes; thin lines in (b) show the sensitivity analysis; the thick line in (a) and small dots in (b) include the dissolved uranyl silicic acid complex (see text for details.)

concentration of 0.1 μM were adjusted with the same two composite spectra (right column in Fig. 5). The spectrum at pH 5.1 is still completely controlled by the first component spectrum, whereas at pH 5.8 both spectra have about the same importance. At pH 6.8 the second component dominates the spectrum.

In order to be able to deduce surface species from the intensities of the two fluorescence components, the overall sorbed concentration needed to be characterized. Uranyl ions are nearly completely removed from solution between pH 5.5 and 8.5 by adsorption on filterable silica (> 0.45 μm, large dots in Fig. 6). Below pH 4.4, two processes prevent uranyl adsorption: (i) protons compete efficiently with uranyl ions for silanol reactive groups and (ii) dissolved silicic acid, and very marginally also nitrate and hydroxyl ions, complex uranyl in solution (Fig. 1). On the other hand, above pH 8.5 dissolved carbonate ions out-compete surface silanol ligands for uranyl complexation. Stable $UO_2(CO_3)_3^{4-}$ are formed in solution, and uranyl adsorption starts declining (i.e., Fig. 1).

Since the uranyl complexed by (polymeric) dissolved silicic acid also emits a fluorescence signal with a long lifetime (90-200 µs, Moll et al. 1998), the sum of the concentrations of the emitting species may be greater than the total sorbed concentration in the low pH range. In order to evaluate this effect the concentration of the calculated dissolved uranyl silicate ($UO_2H_3SiO_4^+$) was added to the total sorbed concentration (thickest solid line in Fig. 6a), this resulting in positive % changes of 9, 10, and 6 at pH values of 4.0, 4.2, and 4.4, respectively. The large domain of complete adsorption between pH 5.5 and 8.5 is not affected. For this reason it was justifiable to set the measured total concentration of sorbed uranyl, S_{tot}, larger or equal to the sum of the concentrations of all the present uranyl containing surface complexes according to Eqn. 4. The relative concentrations of the two fluorescent surface complexes obtained by EFA are indicated in Fig. 6a by the filled and open squares (1 µM uranyl) and filled and open triangles (0.1 µM), respectively. The second component dominates beyond pH 5.5 ± 0.2. It appeared that the fluorescence yield at time 40.1 µs after the laser shot of the low-pH surface complex was about double the one of the high-pH surface complex (Figs. 3, 4).

In order to test whether the dissolved uranyl silicic acid complex has a different fluorescence spectrum than the surface species, 4 samples were centrifuged at relatively low speed for 10 minutes. In Fig. 7, the fluorescence spectra are compared. No spectral differences can be observed between the original and the centrifuged samples, suggesting that the spectral data, in contrast to the lifetime data, cannot be used to distinguish between surface and solution (polymeric/monomeric) uranyl silica species.

The initial intensities of the exponential decay functions (adjusted to the experimental decays in the 40.1 to 600.1 µs time range, data not shown, Gabriel, 1998) were used as a third indication for the concentrations of the two fluorescence components (diamonds in Fig. 6). The shift from one surface complex to another occurred at a slightly higher pH (between pH 5.5 and 6.7). Thus, the spectra and lifetime values at 1 µM uranyl lead to the same in situ speciation of uranium as in the silica suspension. The experimental decay signals of the 0.1 µM uranyl series were not used for the speciation estimation since the signal after 200 µs had a too low signal/noise ratio. It was therefore not possible to adjust a bi-exponential decay function.

A difference between the experimentally determined sorbed concentration S_{tot} and the sum of the concentrations of the two fluorescent surface complexes S_1 and S_2 was observed between pH 8 and 9. This difference was interpreted as being due to the presence of a non-fluorescent surface complex (relative concentrations are represented by gray-filled symbols in Fig. 6).

3.3. Surface Complexation Modeling

Uranyl adsorption on amorphous silica has been studied extensively by EXAFS (Dent et al., 1992; Reich et al., 1998; Sylwester et al., 2000). Adsorption is reported to take place in the equatorial plane formed around the U^{VI} atom of the linear uranyl ($O{=}U^{2+}{=}O$) cation. Free uranyl in an acid, non-complexing aquatic environment is surrounded by five to six water molecules, these forming the equatorial plane of the uranyl cation. In all complexation or adsorption reactions some of these water molecules are replaced by negatively charged ions, such as nitrate, hydroxyl, or surface groups (inner-sphere complexes). This leads to an uneven charge distribution in the equatorial plane. In some cases this leads to a splitting of the equatorial shell because the nearest atoms are likely to be found at different distances. This was observed by Waite et al. (1994) on ferrihydrite. Farges et al. (1992) confirmed the splitting of the equatorial shell in solids by XRD.

The measured small U-Si distances suggest that uranyl forms bi-dentate surface complexes. Sylwester et al. (2000) determined an U-U closest neighbor distance suggesting the existence of polymers at pH 6.46 and a total uranyl concentration of 41.5 µM. Based on the solubility of soddyite (Moll et al., 1996; Moll, 1997) precipitation is likely at $[UO_2^{2+}]_{tot}$ = 41.5 µM, but may also depend on the reaction rate. At pH 3.14 and a total uranyl concentration of 9.79 mM, no molecular U-U near-order was detected (Sylwester et al., 2000). Reich et al. (1998) studied uranyl adsorption on silica at even higher total uranyl concentrations (≤ 0.05 M in the pH range 3.5 to 4.5) and did not detect uranyl surface polymers. Based on the available evidence it is not possible to completely exclude the existence of uranyl hydroxy-polymers at the silica/water interface. However, the similarities in the present spectral data obtained for total uranyl concentrations that vary over one order of magnitude indicate mononuclear uranyl adsorption (Figs. 4a,b,5).

In the characterization of uranyl solution speciation, TRLIFS has been widely used to distinguish different hydrolysis states and polymers (Eliet et al., 1995; Brendler et al., 1996; Bernhard et al., 1996; Moulin et al., 1995). Hydrolysis and polymerization can be observed by TRLIFS because charge distribution and symmetry in the equatorial plane change. Equatorial and axial oxygen bounding distances are inversely related (Pauling principle: Pauling, 1929; Hudson et al. 1995), this causing the altered fluorescence signal.

Based on all of the above information we propose two fluorescent surface complexes, each with a mono-nuclear bidentate structure. The adsorption reaction takes place in the equatorial plane where two of the five (or six) water molecules of the free uranyl are replaced by silanol surface groups, this resulting in the release of two protons (Table 3). The water molecules of the free uranyl ion have not been included in Table 3. The low-pH complex is assumed to be a neutral uranyl complex $\equiv SiO_2UO_2{}^{\circ}$. The second fluorescent surface complex is thought to include an additional hydroxyl ion (i.e., in its formation three protons are released: $\equiv SiO_2UO_2OH^-$).

One explanation for the non-observed fluorescence of the postulated surface complex at high pH may be the presence of carbonate ions, this leading to a ternary surface site-uranyl-carbonate structure. Bargar et al. (1999, 2000) have observed ternary uranyl carbonate complexes on hematite. The for-

Table 3. Intrinsic constants (log K) for silica surface speciation (constant capacitance model; c_p = 1 g dm^{-3}; A_s = 169 m^2 g^{-1}; κ = 7 F m^{-2}; $[NO_3^-]_{tot}$ = 0.01 M, $[\equiv Si(OH)_2]_{tot}$ = 0.51 mM; $[H_2CO_3{}^{\circ}]_{free}$ = 22.4 μM; $[H_4SiO_4{}^{\circ}]_{tot}$ = 2 mM; $[UO_2{}^{2+}]_{tot}$ = 1 μM)

Reactions			**logK (I=0)**	**logK (I=0.01 M)**
$\equiv Si(OH)_2$	↔	$\equiv SiO_2H^- + H^+$	-6.98	-6.94
$\equiv Si(OH)_2 + UO_2{}^{2+}$	↔	$\equiv SiO_2UO_2{}^{\circ} + 2H^+$	-4.69	-4.78
$\equiv Si(OH)_2 + UO_2{}^{2+} + H_2O$	↔	$\equiv SiO_2UO_2OH^- + 3H^+$	-10.33	-10.37
$\equiv Si(OH)_2 + UO_2{}^{2+} + H_2CO_3{}^{\circ} + H_2O$	↔	$\equiv SiO_2UO_2OHCO_3{}^{3-} + 5H^+$	-22.02	-21.98

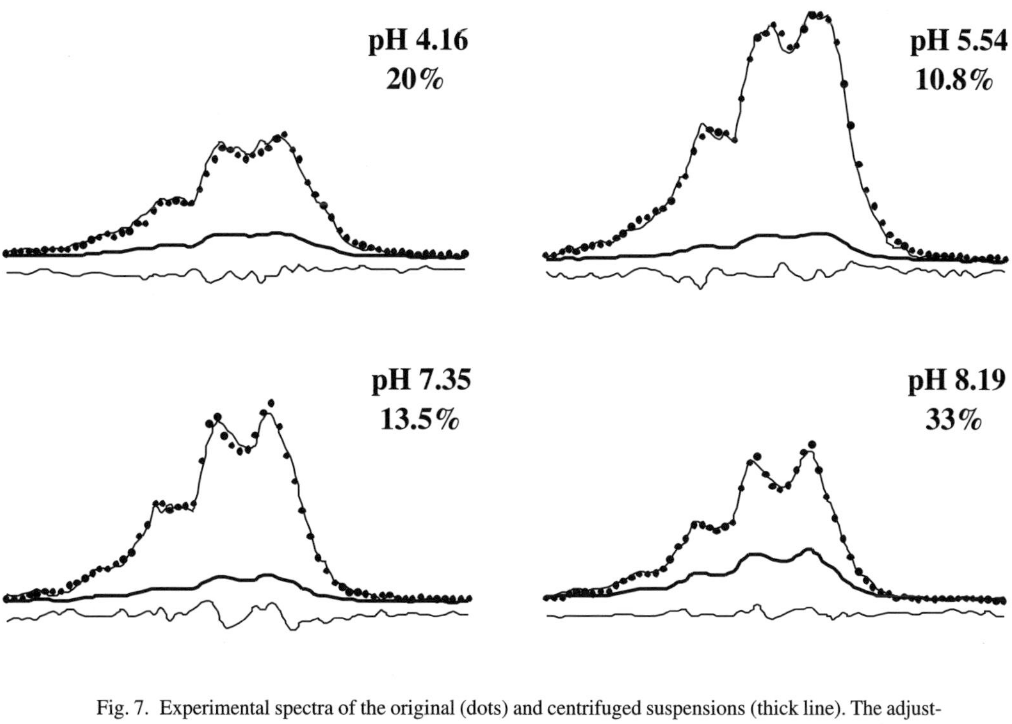

Fig. 7. Experimental spectra of the original (dots) and centrifuged suspensions (thick line). The adjusted thin line permits comparison of the magnified spectrum to the original one; the remaining error is represented by the thin lines below; percentages correspond to remaining fluorescence after centrifugation; total uranyl concentration = 1 μM.

mer study concluded that the entire pH range is dominated by ternary uranyl carbonate complexes. However, iron-bearing oxides and hydroxides are known to have a high affinity for dissolved carbonate, but this behavior is not known for silica surfaces.

Bernhard et al. (1996) reported uranyl complexation with carbonate in solution. This resulted in quenching of the fluorescence and therefore leading to no measurable signal 1 μs after the laser pulse. The same was found by Kato et al. (1994) who did not detect a fluorescence signal due to aqueous uranyl carbonate complexes. On the other hand, Bernhard et al. (1996) report fluorescence lifetimes of 270 and 300 μs for liebigite and zellerite (both uranyl carbonates), respectively. This might be due to a lesser amount of water in the molecular environment of uranyl in the mineralogical lattice.

For the non-fluorescent surface complex we propose the following structure: $\equiv SiO_2UO_2OHCO_3^{3-}$. Its formation involves the release of three protons. This complex has been included in the following surface speciation model, even though both its existence and structure are hypothetical for the moment. The surface speciation is represented by the thick lines in Fig. 6b. The optimum data fit was obtained with a capacitance κ of 7 F m^{-2}. This high capacitance value indicates a very thin double layer. The corresponding logarithmic complexation constants are summarized in Table 3. The logarithmic deprotonation constant of the silica surface was assumed to be 6.94, in accordance with the literature (Charlet et al., 1993; Östhols, 1995).

The most accurate region in the speciation diagram (Fig. 6b) is between pH 5.5 and pH 7. We evaluated the impact of a 0.1-log unit change in the equilibrium constants as a sensitivity analysis. The thin lines in Fig. 6b indicate the results showing (i) increases in the complexation constants (log K + 0.1) of the $\equiv SiO_2UO_2^{\circ}$ and $\equiv SiO_2UO_2OHCO_3^{3-}$ complexes and a decreased formation constant (log K – 0.1) of $\equiv SiO_2UO_2OH^-$, or (ii) the inverse. The sensitivity analysis showed that a 0.1-log unit change of the constants did not change the adsorption or the desorption edge, but shifted the intersection of the dominance fields of the first and second complexes from pH 5.5 to 5.9. The theoretical curves are closer to the experimental data deduced from the spectra than to those deduced from the luminescence decays. The differences between both data sets might be due to different sensitivities of the spectra and the luminescence decays with respect to certain structural parameters, such as bonding distances and symmetry.

3.4. Particle-facilitated Transport of Uranyl

Uranium sorbed on silica may be either mobile or immobile in the environment, depending on the particle mobility, i.e., depending on the tendency to aggregate in the subsoil. As was observed in preliminary experiments, centrifugation does not remove the entire fluorescence signal of the uranyl silica suspensions. This tendency was further studied in a series of centrifugation experiments performed on silica suspensions at equilibrium at two different pH values (4.4 and 8.4) with 1 μM uranium. Fluorescence decays (50 to 450 μs) of the supernatant were recorded after centrifugation at different times and velocities. Results are presented in Fig. 8, where the registered fluorescence intensity at 50 μs has been normalized to the intensity of the suspension before centrifugation (also at 50 μs).

Irrespective of the equilibrium pH (4.4 and 8.4), 20 % of the fluorescence is easily removed after a light centrifugation (5 minutes at 1000 rpm). At pH 4.4 (but not at pH 8.4) a further 45 % of the fluorescence is removed by centrifugation for 1 h at the same centrifugation speed. At pH 8.4 centrifugation at 1,000 rpm for one hour did not reduce the uranyl fluorescence in the supernatant. A high-gravity field (centrifugation at 10,000 rpm for 60 min) is needed to reduce the uranyl fluorescence to zero. However, at pH 4.4, even high-gravity fields (40 minutes at 10,000 rpm) were not able to remove more than another 15 % of the fluorescence signal, such that 20 % of the fluorescence signal remained in suspension and/or solution.

As already discussed in the quantification of the surface speciation, the fluorescence signal cannot simply be assumed to correspond directly to the total sorbed concentration because dissolved uranyl silica complexes potentially emit a similar luminescence signal. Therefore, at pH 4.4, the 20 % fluorescence signal which remains should be compared to the dissolved uranyl fraction which is complexed by silicic acid, which is 10 % of the total uranyl concentration. At this pH, 13 % of the total uranyl concentration is not bound to silicic acid and can therefore be assumed to have no long-life fluorescence. The sum of all the fluorescent fractions (sorbed and dissolved) can be assumed to correspond to 87 % of the total uranyl concentration. On this basis the different fractions at pH 4.4 (Fig. 9) have been calculated. The non-removable fluorescence fraction is attributed partly to the sorbed (7 %) and partly to the dissolved (10 %) uranyl. With respect to these interpretations, it must be kept in mind that experimental artifacts, such as the incomplete removal of silica particles

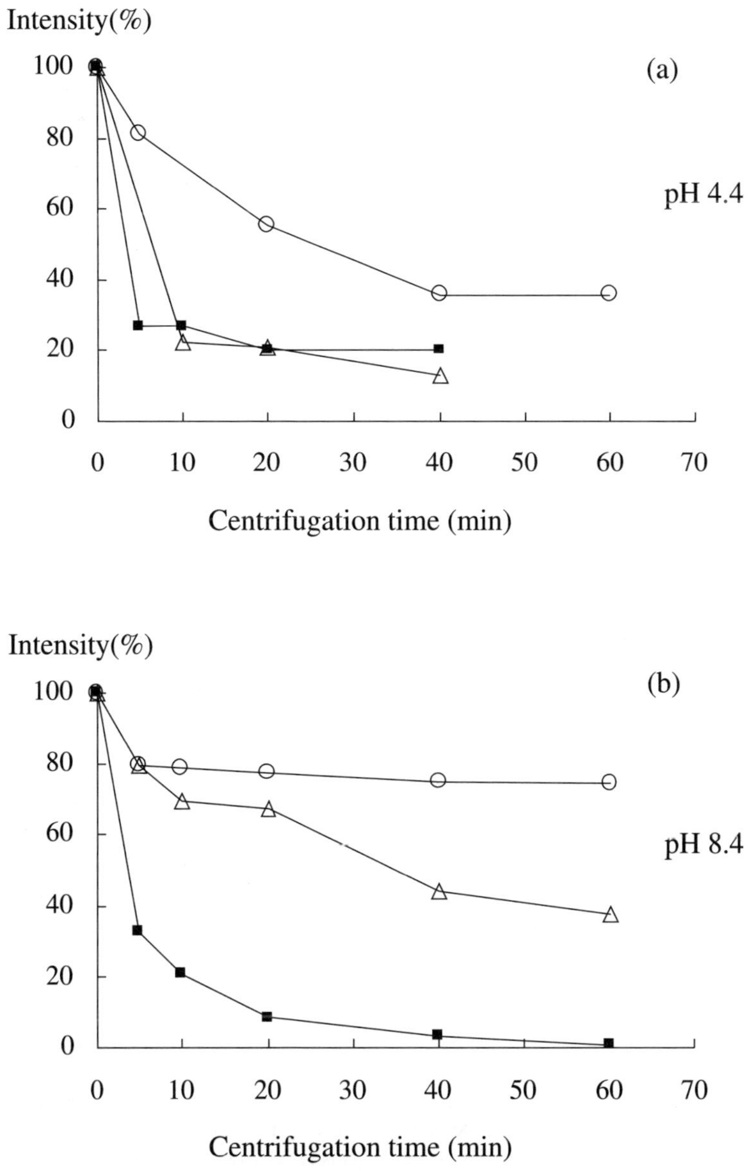

Fig. 8. Reduction of the observed fluorescence intensity (at 50.2 μs after the laser shot) after 5 to 60 minutes centrifugation at 1,000 (circles), 2,000 (triangles), and 10,000 (squares) rpm at pH 4.4 (a) and pH 8.4 (b).

attached to the quartz cell walls, may play a role in influencing the results.

In general, aggregation does occur where expected; the most rapid aggregation is observed in the slightly acidic pH range where silica particles have nearly no net surface charge. At acid pH, the proposed uranyl surface complex is also not charged. Intermediate aggregation is observed at pH 8.4 where silica particles have large negative surface charges and the proposed uranyl surface complex is also charged negatively. The 17 % uranyl at pH 4.4 which does not aggregate at all under the centrifugation conditions presented here is most probably bound to silicic acid, resulting in charged molecules (see Eqn. 8 in Table 2).

These results suggest that the risk of uranyl transfer by silica colloids in groundwater and the hydro-

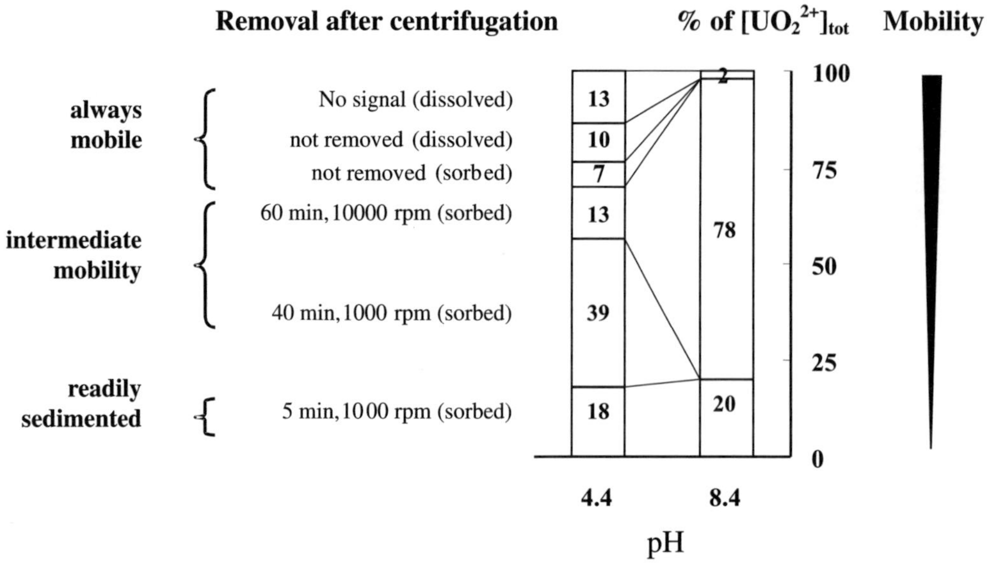

Fig. 9. Removed percentage of uranyl (% of total concentration) at different centrifugation conditions.

sphere is controlled not only by pH conditions, but also by the flow velocity. In slightly acidic waters, uranyl is more mobile as it may be complexed with aqueous silicic acid, and in slightly basic conditions it is potentially transported as a sorbed species on small silica particles which have an intermediate susceptibility to aggregation behavior.

4. CONCLUSIONS

For the first time, to our knowledge, two distinctively different molecular environments of uranyl sorbed on amorphous silica have been observed and quantified by spectroscopic analysis. This has been done by TRLIFS at trace-level concentrations, i.e., at levels not accessible to conventional spectroscopic techniques such as NMR or EXAFS. This is important since trace-level concentrations are perhaps more relevant to natural environments, and in addition sorption mechanisms may change with the sorbate to sorbant ratio, i.e., the structure of the surface complexes may change as a function of surface concentration (Morris et al., 1994; Chrisholm-Brause et al., 1994).

In general, uranyl adsorption to amorphous silica is known to result in the formation of bidentate surface complexes, where two water molecules in the equatorial plane of the uranyl ion are replaced by silanol groups (Dent et al., 1992, Reich et al., 1998, Sylwester et al., 2000). The different observed molecular environments in this study are thought to represent different levels of hydrolysis. Polymerization of uranyl seems unlikely as U-U near-order was observed in this system only at elevated uranyl concentrations likely to cause surface precipitation. The addition of ions such as carbonate or silicic acid may also cause a change in the structure of the surface complex. Carbonate ions are most likely not an integral part of the fluorescent surface species since they are known to drastically reduce fluorescence lifetimes in aqueous environments. At this point it seems very unlikely that sorbed silicic acid molecules are distinguishable from structural units belonging to amorphous silica. Therefore this process was not considered in the speciation calculations in the present study.

The two complexes observed by TRLIFS were assigned to $\equiv SiO_2UO_2^{\circ}$ and $\equiv SiO_2UO_2OH^-$ species. In accordance with literature data, the existence of a non-fluorescent uranyl-silica-carbonate-hydroxo complex $\equiv SiO_2UO_2OHCO_3^{3-}$ was postulated in order to account for the difference between the sum of the concentrations of the fluorescent surface complexes and the experimentally determined total sorbed uranyl concentration (Waite et al., 1994).

In this study silica particles were shown to be a potential candidate for facilitating the transport of uranium in subsoils. The major risk appears to occur in the acidic pH range where dissolved silicic acid and a fraction of the silica particles, despite their nearly neutral surface charge, are least likely to aggregate and thus be removed from solution. Thus, under favorable transport conditions, uranium sorbed to silica may migrate in the subsoil.

Acknowledgements—We are grateful to the radiochemistry department of the Forschungszentrum Rossendorf, Germany, where U.G. spent several weeks for the TRLIFS measurements. We thank Gerhardt Geipel, Axel Brachmann, Henry Moll, Vinzenz Brendler, and Thorsten Stumpf from this department, whose valuable assistance helped immensely in the writing of this manuscript. U.G. thanks the "Pole Européen" (UJF-Grenoble) for a doctoral scholarship. The authors would like to express their gratitude to the group working on porous silica at the Spectrometry Physics Laboratory, Grenoble, for providing material and assistance for the elaboration of the device that was used to quantify the uranyl solution concentrations. We thank W. Runde, G. Bidoglio, E. Giffaut and one anonymous reviewer for their constructive comments which led to the improvement of the manuscript. We also thank Roland Hellmann for assistance in the analyses of the aqueous silica, for comments on the manuscript, and improving the English.

Editorial handling: R. Hellmann

REFERENCES

Bargar J. R., Reitmeyer R., and Davis J. A. (1999) Spectroscopic confirmation of uranium(VI) carbonate adsorption complexes on hematite. *Environ. Sci. Technol.* **33**, 2481-2484.

Bargar J. R., Reitmeyer R., Lenhardt J. J., and Davis J. A. (2000) Characterization of U(VI)-carbonate ternary complexes on hematite: EXAFS and electrophoretic mobility measurements. *Geochim. Cosmochim. Acta* **64**, 2737-2749.

Belyak Yu. N., Glinka Yu. D., Krut A. V., Naumenko S. N., Ogenko V. M., and Chuiko A. A. (1993) IR spectroscopic study of the dynamics of thermal dehydration and dehydroxylation of extremely hydrated samples of dispersed silica. *J. Appl. Spectroscopy* **59**, 515-522.

Bernhard G., Geipel G., Brendler V., and Nitsche H. (1996) Speciation of uranium in seepage waters of a mine tailing pile studied by time-resolved laser-induced fluorescence spectroscopy (TRLFS). *Radiochim. Acta* **74**, 87-91.

Berthoud T., Decambox P., Kirsch B., Maichien P., and Moulin C. (1988) Direct uranium trace analysis in plutonium solutions by time-resolved laser-induced spectrofluorometry. *Anal. Chem.* **60**, 1296-1299.

Brady P. V. and Walther J. W. (1990) Kinetics of quartz dissolution at low temperatures. *Chem. Geol.* **82**, 253-264.

Brendler V., Geipel G., Bernhard G., and Nitsche H. (1996) Complexation in the system $UO_2^{2+}/PO_4^{3-}/OH^-$(aq): potentiometric and spectroscopic investigations at very low ionic strengths. *Radiochim. Acta* **74**, 75-80.

Brina R. and Miller A.G. (1992) Direct determination of trace levels of uranium by laser-induced kinetic phosphorimetry. *Anal. Chem.* **64**, 1413-1418.

Burneau A. and Gallas J.-P. (1998) Hydroxyl groups on silica surfaces - Vibrational spectroscopies. In *The Surface Properties of Silicas* (ed. A.P. Legrand). John Wiley & Sons, Chichester, 147-234.

Charlet L., Schindler P. W., Spadini L., Furrer G., and Zysset M. (1993) Cation adsorption on oxides and clays: The aluminium case. *Aq. Sci.* **55**, 291-303.

Chrisholm-Brause C., Conradson S. D., Buscher C. T., Eller P. G., and Morris D. E. (1994) Speciation of uranyl sorbed at multiple binding sites on montmorillonite. *Geochim. Cosmochim. Acta* **58**, 3625-3631.

Costa D., Fubini B., Giamello E., and Volante M. (1990) A novel type of active site at the surface of crystalline SiO_2 (alpha-quartz) and its possible impact on pathogenicity. *Can. J. Chem.* **69**, 1427-1434.

Davies C. W. (1962) *Ion Association.* Butterworth, Washington, D.C.

Deniau H., Decambox P., Mauchien P., and Moulin C. (1993) Time-resolved laser-induced spectrofluorometry of UO_2^{2+} in nitric acid solutions. Preliminary results for on-line uranium monitoring applications. *Radiochim. Acta* **61**, 23-28.

Dent A. J., Ramsay D. F., and Swanton S. W. (1992) An EXAFS Study of uranyl ion in solution and sorbed onto silica and montmorillonite clay colloids. *J. Colloid Interface Sci.* **150**, 45-60.

Dove P. M. and Crerar D. A. (1990) Kinetics of quartz dissolution in electrolyte solutions using a hydrothermal mixed flow reactor. *Geochim. Cosmochim. Acta*, **54**, 955-969.

Dove P. M. and Rimstidt J. D. (1994) Silica-water-interactions. In *Silica: Physical Behavior, Geochemistry and Materials Applications* (eds. P.J. Heaney, C.T. Prewitt, and G.V. Gibbs). Reviews in Mineralogy **29**, Mineralogical Society of America, Washington, D.C., pp. 259-308.

Duff M. C., Morris D. E., Hunter D. B., and Bertsch P. M. (2000) Spectroscopic characterization of uranium in evaporation basin sediments. *Geochim. Cosmochim. Acta* **64**, 1535-1550.

Eliet V. (1996) Applications des techniques de fluorescence pour l'étude de l'uranium dans des milieux homogènes et hétérogènes: Réactions d'hydrolyse et photoréduction sur le bioxyde de titane. Ph.D. thesis, Université de Paris Sud, France.

Eliet V., Bidoglio G., Omenetto N., Parma L., and Grenthe I. (1995) Characterization of hydroxide complexes of uranium(VI) by time-resolved fluorescence spectroscopy. *J. Chem. Soc. Faraday Trans.* **91**, 2275-2285.

Farges F., Ponader C. W., Calas G., and Brown G. E. Jr. (1992) Structural environments of incompatible elements in silicate glass/melt systems. *Geochim. Cosmochim. Acta*, **56**, 4205-4220.

Gabriel U. (1998) Transport réactif de l'uranyle: Mode de fixation sur la silice et la goethite; expériences en colonne et réacteur fermé; simulations. Ph.D. thesis, Université Joseph Fourier, Grenoble, France.

Gabriel U., Gaudet J.-P., Spadini L., and Charlet L. (1998) Reactive transport of uranyl in a goethite column: An experimental and modelling study. *Chem. Geol.* **151**, 107-128.

Gampp H., Maeder M., Meyer C. J., and Zuberbuehler A. D. (1987) Evolving factor analysis. *Comments Inorg. Chem.* **6**, 41-60.

Giamelo E., Fubini B., Volante M., and Costa D. (1990) Surface oxygen radicals originating via redox reactions during the mechanical activation of crystalline SiO_2 in hydrogen peroxide. *Colloids and Surfaces* **45**, 155-165.

Glinka Y. D. and Krak T. B. (1995) Luminescence spectra of

uranyl ions adsorbed on disperse SiO_2 surfaces. *Physical Review B* **52**, 14985-14995.

Grenthe I., Fuger J., Konings R. J. M., Lemire R. J., Muller A. B., Nguyen Trung C., and Wanner H. (1992) *Chemical Thermodynamics of Uranium.* North Holland, Amsterdam.

Hommel H., Legrand A. P., Doremieux C., and D'espinose de la Caillerie J.-B. (1998) Hydroxyl groups on silica surfaces - NMR spectroscopies. In *The Surface Properties of Silicas.* (ed. A.P. Legrand). John Wiley & Sons, Chichester, 235-284.

Hudson E. A., Terminello L. J., Viani B. E., Reich T., Bucher J. J., Shuh D. K., and Edelstein N. M. (1995) X-ray absorption studies of uranium sorption on mineral substrates. In *Applications of Synchrotron Radiation Techniques to Material Science II. Symposium* (eds. L.J. Terminello, N.D. Shinn, G.E. Ice, K.L. D'Amico, and D.L. Perry). Materials Research Society, pp. 235-240.

James R. O. and Healy T. W. (1972) Adsorption of hydrolyzable ions at the oxide-water interface. I. Co(II) adsorption on SiO_2 and TiO_2 as model systems. *J. Colloid Interface. Sci.* **40**, 42-52.

Kato Y., Meinrath G., Kimura T., and Yoshida Z. (1994) A study of U(VI) hydrolysis and carbonate complexation by time-resolved laser-induced fluorescence spectroscopy (TRLFS). *Radiochim. Acta* **64**, 107-111.

Kushnirenko I. Ya., Glinka Yu. D., Degoda V. Ya., Krak T. B., and Ogenko V. M. (1993) Luminescent properties of uranyl ions adsorbed on the surface of disperse silicon dioxide. *J. Appl. Spectroscopy* **59**, 687-692.

Ludwig C. (1993) Koordinationschemie an der Grenzschicht Oxid/Wasser, I. Ternaere Oberflaechenkomplexe mit Cu(II) und organischen Liganden und TiO_2 (Anatas), II. Ein Modell zur Beschreibung der Deprotonierung von $Al_{13}O_4(OH)_{24}(H_2O)_{12}{}^{7+}$. Ph.D. thesis, University of Bern, Switzerland.

Meinrath G., Kato Y., and Yoshida Z. (1993) Spectroscopic Study of the uranyl hydrolysis species $(UO_2)_2(OH)_2{}^{2+}$. *J. Radioanal. Nucl. Chem.* **174**, 299-314.

Michard P., Guibal E., Vincent T., and Le Cloirec P. (1996) Sorption and desorption of uranyl ions by silica gel: pH, particle size, and porosity effects. *Microp. Mat.* **5**, 309-324.

Moll, H. (1997) Zur Wechselwirkung von Uran mit Silikat in waessrigen Systemen. Ph.D. thesis, TU Dresden, Germany.

Moll H., Geipel G., Matz W., Bernhard G., and Nitsche H. (1996) Solubility and speciation of $(UO_2)_2SiO_4{\cdot}2H_2O$ in aqueous systems. *Radiochim. Acta* **74**, 3-7.

Moll H., Geipel G., Brendler V., Bernhard G., and Nitsche H. (1998) Interaction of uranium(VI) with silicic acid in aqueous solutions by time-resolved laser-induced fluorescence spectroscopy (TRLFS). *J. Alloys & Compounds* **271-273**, 765-768.

Morris D. E., Chrisholm-Brause C. J., Barr M. E., Conradson S. D., and Eller P. G. (1994) Optical spectroscopic studies of the sorption of $UO_2{}^{2+}$ species on a reference smectite. *Geochim. Cosmochim. Acta* **58**, 3613-3623.

Moulin C., Decambox P., Moulin V., and Decaillon J. G. (1995) Uranium speciation in solution by time-resolved laser-induced fluorescence. *Anal. Chem.* **67**, 348-353.

Östhols E. (1995) Thorium sorption on amorphous silica. *Geochim. Cosmochim. Acta* **59**, 1235-1249.

Pauling L. (1929) The principles determining the structure of complex ionic crystals. *J. Am. Chem. Soc.* **51**, 1010-1026.

Press W. H., Vetterling W. T., Teukolsky S. A., and Flammery B. A. (1992) *Numerical Recipes*, 2nd ed. Cambridge University Press, Cambridge.

Reich T., Moll H., Denecke M. A., Geipel G., Bernhard G., Nitsche H., Allen P. G., Bucher J.J., Kaltsoyannis N., Edelstein N. M., and Shuh D. K. (1996) Characterization of hydrous uranyl silicate by EXAFS. *Radiochim. Acta* **74**, 219-223.

Reich T., Moll H., Arnold T., Denecke M.A., Henning C., Geipel G., Bernhard G., Nitsche H., Allen P.G., Bucher J.J., Edelstein N.M., and Shuh D.K. (1998) An EXAFS study of uranium(VI) sorption onto silica gel and ferrihydrite. *J. Spectroscop. Related Phenom.* **96**, 237-243.

Schindler P. W. (1991) A solution chemist's view of surface chemistry. *Pure and Appl. Chem.* **63,** 1697-1704.

Schindler P. W. and Gamsjäger H. (1972) Acid-base reactions of the titanium dioxide (anatase)-water interface and the point of zero charge of titanium dioxide suspensions. *Kolloid -Z. u. Z. Polymere* **250**, 759-763.

Schindler P. W., Fuerst B., Dick R., and Wolf P. U. (1976) Ligand properties of surface silanol groups. I. Surface complex formation with iron(3+), copper(2+), cadmium(2+) and lead(2+). *J. Colloid. Interface Sci.* **55**, 469-475.

Stanton J. and Maatman R. W. (1963) The reaction between aqueous uranyl ion and the surface of silica gel. *J. Colloid Sci.* **18**, 132-146.

Strickland C. and Parsons J. L. (1972) Determination of reactive silicate. In *Practical Book of Sea Water Analysis.* Fisheries Research Board of Canada, pp. 65-70.

Stumm W. and Morgan J. J. (1996) *Aquatic Chemistry. Chemical Equilibria and Rates in Natural Waters.* John Wiley & Sons, Inc., New York.

Sylwester E. R., Hudson E. A., and Allen P. G. (2000) The structure of uranium(VI) sorption complexes on silica, alumina, and montmorillonite. *Geochim. Cosmochim. Acta* **64**, 2431-2438.

Van Cappellen P. and Gaillard J. F. (1996) Biogeochemical dynamics in aquatic sediments. In *Reactive Transport in Porous Media* (eds. P.C. Lichtner, C.I. Steefel, and E.H. Oelkers). Reviews in Mineralogy **34**, Mineralogical Society of America, Washington, D.C., pp. 335-376.

Van Cappellen P. and Qui L. (1997) Biogenic silica dissolution in sediments of the Southern Ocean. I. Solubility. *Deep-Sea Res. II* **44**, 1129-1149.

Waite T. D., Davis J. A., Payne T. E., Waychunas G. A., and Xu N. (1994) Uranium(VI) adsorption to ferrihydrite: Application of a surface complexation model. *Geochim. Cosmochim. Acta* **58**, 5465-5478.

Westall J. C. (1982) A computer program for the determination of chemical equilibrium constants from experimental data. Version 1.2. Oregon State University, Corvallis, Oregon, USA.

Wheeler J. and Thomas J. K. (1984) Photochemistry of the uranyl ion in colloidal silica solution. *J. Phys. Chem.* **88**, 750-754.

Water-Rock Interactions, Ore Deposits, and Environmental Geochemistry: A Tribute to David A. Crerar

Editors: Roland Hellmann and Scott A. Wood

Ecosystem dynamics of zinc and manganese within a mine-waste impacted wetland

COLLEEN M. HANSEL[1], MATTHEW J. LA FORCE[1], STEPHEN SUTTON[2], AND SCOTT FENDORF[*1]

[1]Dept. of Geological and Environmental Sciences, Stanford University, Stanford, CA 94305, USA
[2]Consortium for Advanced Radiation Sources and Department of Geophysical Sciences, University of Chicago, Chicago, IL 60637 USA
[*]fendorf@stanford.edu

Abstract—Wetland ecosystems, which serve as a natural filter for many environmental contaminants, influence the fate and transport of metals. The partitioning and specific binding mechanisms of metals within the soil, water, and plants have important implications for delineating potential bioavailability and/or release of metals upon changing physicochemical conditions within aquatic environments. Here we investigate the mechanisms of Mn and Zn partitioning and sequestration within a natural, mine-waste impacted wetland. The partitioning of Mn and Zn within the Cataldo wetland subsurface is dominantly in the mineral and plant root (*Typha latifolia, Phalaris arundinacea, Scirpus microcarpus, Equisetum arvense*) fractions while solution concentrations are negligible. Manganese and Zn concentrations of some plant species are equivalent to those of the soil and, in some cases, may exceed soil-associated Mn and Zn. Yet, owing to higher soil densities, the mass contributions are ≥10-fold higher in the mineral fraction (Mn: 7,840 kg; Zn: 6,360 kg) relative to total plant roots (Mn: 707 kg; Zn: 319 kg). Manganese is predominantly present as a rhodochrosite-like ($MnCO_3$) phase within the soil and plant roots. Zinc, on the other hand, varies from a smithsonite-like ($ZnCO_3$) phase in the soil to its hydrated derivative, hydrozincite [$Zn_5(OH)_6(CO_3)_2$], on the root surface. X-ray fluorescence microtomography and x-ray microprobe analysis reveal a strong correlation between Mn, Zn, and Ca on the surface of *P. arundinacea* roots. Manganese, Zn, and Ca exist as isolated nodules on the root surface with concentrations diminishing adjacent to these regions as well as in the interior of the root. Thus, carbonate ligands appear to be the dominant means by which Zn and Mn are sequestrated within this ecosystem and may have a pronounced influence on metal sequestration in general.

1. INTRODUCTION

Several metal ions are essential for both plant and animal metabolism yet may have deleterious effects when exceeding threshold concentrations. The introduction of these toxic metals to terrestrial and aquatic environments is largely due to anthropogenic sources. Mining activities, for instance, have resulted in widespread contamination of surface waters, groundwater, soils, sediments, and vegetation. The Coeur d'Alene (CdA) mining district in the Idaho panhandle exemplifies the detrimental effects of mine-wastes on the environment. The CdA mining district is situated within the Lewis and Clark shear zone, composed of numerous faults trending primarily east-west and constituting an intracontinental plate boundary (Bennett et al., 1989). Mining of lode deposits allowed for the prosperity of more than 90 mines in the area and, hence, CdA earned the reputation as one of the most opulent mining districts worldwide. The district has the largest recorded Ag production in the world, as well as abundant quantities of Au, Cd, Cu, Pb, Sn, and Zn (Bennett et al., 1989). The primary economic minerals are galena (PbS), sphalerite (ZnS), and Ag-bearing tetrahedrite ($Cu_{12}Sb_4S_{13}$). The gangue minerals within the region are chiefly quartz (SiO_2), sericite [fine-grained muscovite, $KAl_2(AlSi_3O_{10}(OH)_2)$], ferroan dolomite [$Ca(Mg, Fe^{2+})(CO_3)_2$], and siderite ($FeCO_3$), but the region also hosts an abundance of calcite ($CaCO_3$), pyrite (FeS_2), chalcopyrite ($CuFeS_2$), and pyrrhotite ($Fe_{1-x}S$, x = 0-0.2) (Bennett et al., 1989; Hobbs et al., 1965). Minor gangue minerals also include the carbonates, dolomite [$CaMg(CO_3)_2$] and ankerite [$Ca(Mg, Fe^{2+}, Mn)(CO_3)_2$].

Unfortunately, the CdA mining district is also known as the home to the Bunker Hill Superfund Site (Bennett, 1994). Mining practices employed until the 1960's involved disposal of mine tailings directly into the South Fork of the Coeur d'Alene River (CdAR), which were subsequently transported downstream (Rabe and Bauer, 1977; Reece et al., 1978). Consequently, mine tailings were deposited in the Main Stem of the CdAR as well as Lake Coeur d'Alene (Reece et al., 1978; Horowitz et al., 1992; Harrington et al., 1998). Dredging of the riverbed to allow for navigation routes, as well as flooding of the river, resulted in deposition of these mine tailings on the floodplain of the CdAR.

From nearly a century of mining in the CdA district, an estimated 115 million metric tons of mine tailings were produced with about 60% of them discharged directly into the South Fork of the CdAR (Javorka, 1991). Ellis (1940) noted an absence of biota in the CdAR from the South Fork at Kellogg to the CdA Lake delta. Additionally, Ellis (1940) reported that some contaminants in the river would kill plankton within 36 hours and were lethal to fish within 72 hours. In 1968, federal and state regulations required the installation of tailings ponds within the CdA mining district in an attempt to suppress further pollution of the CdAR system (Rabe and Bauer, 1977). Although settling ponds greatly improved water quality within the South Fork and Main Stem of the CdAR, metals remained within the groundwater and river at levels toxic to aquatic organisms (Mink, 1971; Galbraith et al., 1972; Norbeck, 1974; Rabe and Bauer, 1977). Additionally, erosion of riverbanks along the South Fork of the CdAR contributes high concentrations of suspended metal-laden sediments to the river system (Ioannou, 1979). Sediments of the Main Stem of the CdAR contain elevated concentrations of Cd (22.6 mg kg^{-1}), Mn (1,600 mg kg^{-1}), Pb (4,180 mg kg^{-1}), and Zn (6,650 mg kg^{-1}) (Reece et al., 1978). South Fork sediments are further enriched in Cd (43.3 mg kg^{-1}), Mn (8420 mg kg^{-1}), Pb (11400 mg kg^{-1}), and Zn (7550 mg kg^{-1}) (Reece et al., 1978). Furthermore, sediments of Lake CdA are highly enriched in As (201 mg kg^{-1}), Fe (82,486 mg kg^{-1}), Mn (5,953 mg kg^{-1}), Pb (3,820 mg kg^{-1}), and Zn (2,995 mg kg^{-1}) (Harrington et al., 1998; Horowitz et al., 1992).

Numerous wetlands are situated adjacent to the CdAR and lake providing a habitat for a variety of terrestrial and aquatic biota. High metal concentrations within these wetlands jeopardize the health and safety of the habitants. For instance, high Pb levels within aquatic plants and soils have recently been found responsible for approximately 7000 tundra swan deaths even though they spend only a few weeks in the region during spring migration (Blus et al., 1998). Contaminant dynamics within wetland ecosystems involve a complex continuum between the soil, water, plants, and microbial communities. Aquatic plants influence metal partitioning either directly by way of uptake and incorporation of metals within plant tissue or indirectly by changing the physicochemical conditions at the soil/root interface (Figure 1). During plant growth, radial oxygen diffusion and plant respiration provides enhanced microenvironments of $O_{2(g)}$ and carbonate at root surfaces (Armstrong, 1978). Consequently, upon diffusion toward the roots, metals may undergo redox transformations. Ferrous iron, for example, is readily oxidized forming ferric (hydr)oxide

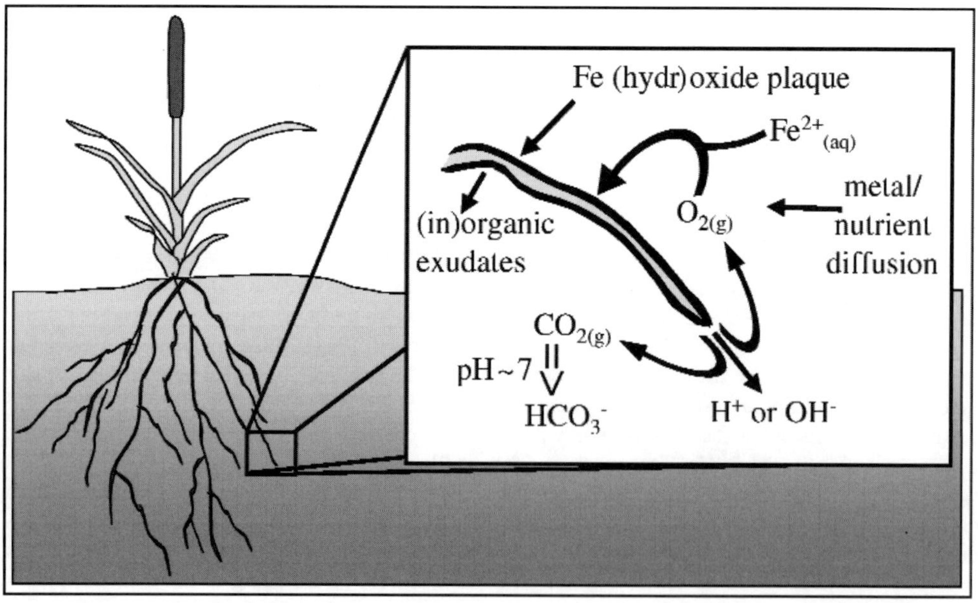

Figure 1. Simplistic schematic representation of processes occurring within the rhizosphere of aquatic plants. Oxidation zone surrounding roots ranges between ~1 and 4 mm (Marschner, 1995). Diagram is not to scale.

coatings (plaque) directly on root surfaces which may then provide a substrate for metal adsorption or nucleation (Bartlett, 1961; Mendelssohn and Postek, 1982; Bacha and Hossner, 1977; Chen et al., 1980; Taylor et al., 1984).

As such, metal ions may undergo a number of sorption paths within wetland soil environments: (1) precipitation or adsorption to soil minerals; (2) adsorption or complexation with soil organic matter; (3) precipitation or adsorption to root surfaces; or (4) complexation with root exudates and subsequent translocation within the plant. The partitioning and specific binding mechanisms of metals have important implications for delineating potential bioavailability and/or release of metals upon changing physicochemical conditions within aquatic environments. Accordingly, here we elucidate the partitioning and mechanisms of Mn and Zn sequestration within the soils of a natural, yet contaminated, wetland.

2. MATERIALS AND METHODS

2.1. Study Site Location and Description

The Cataldo Mission Flats constitute the floodplain of the Main Stem of the CdAR and are located approximately 30 km southeast of the city of Coeur d'Alene and 50 km east of Lake Coeur d'Alene (La Force et al., 1998) (Fig. 2). Fluvially deposited and dredged mine tailings from the Main Stem of the CdAR are situated upon the floodplain (Galbraith et al., 1972), and consequently the flats contain elevated concentrations of As, Fe, Mn, Pb, and Zn. Flooding of the CdAR results in palustrine-emergent wetlands within the contaminated floodplain, one of which is the focus of this study and is hereinafter referred to as the Cataldo wetland (Cowardin et al., 1979). Briefly, the palustrine wetland classification includes all nontidal wetlands situated on river floodplains or shoreward of lakes, river channels, or estuaries, which are dominated by trees, shrubs, persistent emergents, emergent mosses or lichens (Cowardin et al., 1979). In the United States, the majority of wetlands are palustrine-emergent (Cowardin et al., 1979; Fretwell et al., 1996); therefore, investigating Mn and Zn within the Cataldo wetland will have considerable implications for other mine-waste influenced wetlands. Seasonal flooding of the CdAR supplies a yearly influx of sediment and water (possibly metal-laden) to the Cataldo wetland. During winter, however, the groundwater from the wetland discharges into the low-level CdAR.

The Cataldo wetland hosts an abundance of vegetation and wildlife. Wetland inhabitants include a variety of terrestrial mammals and waterfowl. Trees in the area include alder, aspen, birch, cottonwood, willow, ponderosa pine, and white pine. The wetland accommodates a number of aquatic plant species including common cattail (*Typha latifolia*), small-fruited bulrush (*Scirpus microcarpus*), ovate spike-rush (*Eleocharis ovata*), common rush (*Juncus effusus*), common horsetail (*Equisetum arvense*) and reed canarygrass (*Phalaris arundinacea*). Four indigenous plant species vegetate ca. 95% of the wetland area: *T. latifolia* (50%), *P. arundinacea* (20%), *E. arvense* (20%), and *S. microcarpus* (5%). Isolated patches of various plant species and shrubs or unvegetated soil occupy the remaining 5% of the wetland area.

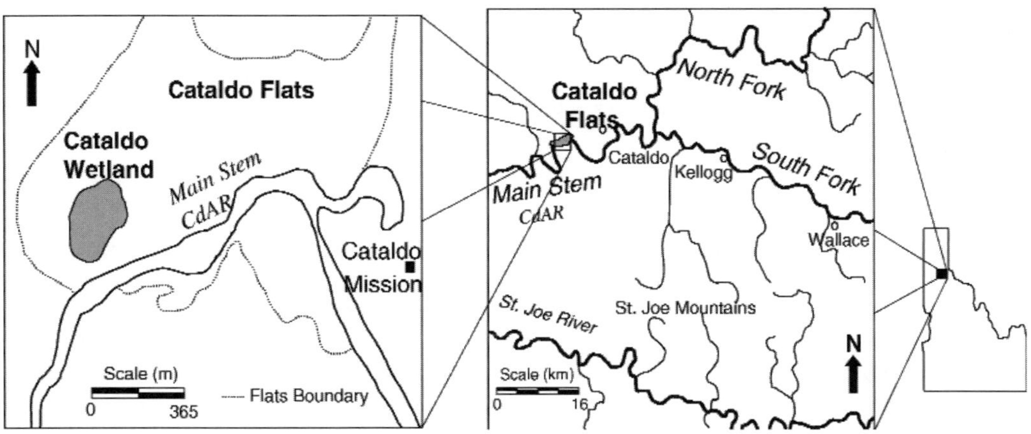

Figure 2. Location of the Cataldo Flats (right) and study site, Cataldo wetland (left).

2.2. Sample Collection and Preparation

Plant, soil, and water samples were collected in May of 1998 from the Cataldo wetland. Three samples of the four dominant plant species (*T. latifolia, P. arundinacea, E. arvense, and S. microcarpus*) were collected and stored on ice for transport to the laboratory. Plants were then divided into root and shoot portions and washed (lightly scrubbed) thoroughly with double-deionized water to remove any attached soil particles from the roots. Samples were dried to a constant weight at 70 °C and ground in a mill to pass a 20-mesh (0.84 mm) sieve. Plant roots used for spectroscopic analyses were collected and immediately washed, freeze-dried, and mounted on an acrylic plate and sealed with Kapton polymide film to prevent moisture loss while minimizing x-ray absorption.

Soil samples were obtained using a 4-cm diameter PVC piston-coring device (La Force et al., 2000a). Cores collected via this procedure were then immediately capped and placed upright into a N_2-purged sample transporter (La Force et al., 2000a). Upon returning to the laboratory, the top 0-15 cm were homogenized in a N_2-purged atmosphere glovebox (Labconco, Kansas City, MO).

Membrane-separated diffusion cells (peepers) have been used extensively to monitor metal ions in solution (Brandl and Hanselmann, 1991; Carignan 1984; Carignan et al., 1985). Peepers used in this experiment were created following the procedures of La Force et al. (2000a). The peepers were placed in the wetland for two weeks, which was a sufficient time to reach equilibrium at this site (La Force et al., 2000a). In the field, aqueous samples were removed from the peepers and Eh and pH were recorded using a portable meter. Samples were then acidified with concentrated HCl and kept on ice until metal analysis 3 h later. A field blank of double deionized water with 2 drops of concentrated HCl was also run with the samples as a quality control; metal concentrations in the blank were negligible.

2.3. Total Metal Analysis

Concentrations of metal(loid)s within plant portions were determined using a wet digestion procedure consisting of nitric acid followed by successive additions of 30% hydrogen peroxide and heating. Specifically, samples (0.2 g) were digested with 4 mL of nitric acid for 24 h. Four mL of hydrogen peroxide were then added to each sample followed by heating in a digestion block (100±2°C) for 30 min. Hydrogen peroxide-heating treatments were repeated until plant residue disappeared.

Soil samples underwent total digestions using a combination of acid additions, shaking, and heating. Samples (1 g dry weight) were first digested with 5 mL of aqua regia (3:1 $HCl:HNO_3$ by volume) followed by shaking for 2 h. Ten mL of hydrofluoric acid were then added followed by shaking for an additional 2 h and heating in a digestion block (100±2 °C) for 1 h. Lastly, 2 g of concentrated boric acid were added to each sample and shaken for 1 h. Samples were then centrifuged and supernatants were passed through a 0.2-μm filter.

Supernatants from plant and soil digestions, as well as water samples, were analyzed for Mn and Zn via inductively coupled plasma-optical emission spectroscopy (Thermo Jarrell Ash IRIS ICP-OES; Franklin, MA) with a 10% accuracy range and quality control was checked every 15 samples. Detection limits were defined as 3σ (where σ is the standard deviation) of 5 blanks (Klesta and Bartz, 1996). Detection limits were: Mn 0.01 mg L^{-1} and Zn 0.02 mg L^{-1}.

2.4. Soil/Root Density Calculations

Soil and root densities were determined for use in Mn and Zn mass calculations. Six cores were extracted within the four dominant plant areas of the wetland. Cores were 15 cm in length and contained a soil volume of 68.7 cm^3. Bulk density of the soil was determined by typical procedures (USDA, 1996). The mineral density was determined after loss on ignition (USDA, 1996). Similarly, root densities were determined by air drying the soil, removing the readily visible root fractions, and then following the above-described method for mineral fraction determinations. It is important to note that the root density may be underestimated due to the difficulty in isolating and removing fine root hairs; hence metal concentrations associated with the root fraction are conservative and may be much larger.

2.5. Microscopic/Spectroscopic Techniques

In order to determine the bulk mineralogy of Mn and Zn in the solid-phase, x-ray absorption spectroscopy [near-edge structure (XANES) spectroscopy and extended x-ray absorption fine structure (EXAFS) spectroscopy] were performed on beamlines IV-1, IV-2, and IV-3 (beamline IV is an 8 pole wiggler) at the Stanford Synchrotron Radiation Laboratory (SSRL), running under dedicated conditions. The ring operated at 3 GeV with a current ranging from ≈100 mA to ≈50 mA. Energy selection was accomplished using a Si (220) monochromator typically with an unfocused beam. Depending on the beamline, higher-order harmonics were either eliminated by a harmonic rejection mirror or by detuning ~ 60% for Mn and 30% for Zn. Absorption spectra were recorded by fluorescent x-ray production using a wide-angle ionization chamber (Lytle et al., 1984) for model compounds or a 13-element Ge semiconductor detector for unknown samples. A Cr (for Mn) and Cu (for Zn) filter along with Soller slits were used to minimize the effects of scattered primary radiation when using the fluorescent ionization chamber. Incident and transmitted intensities were measured with in-line ionization chambers. The energy range studied was -200 to +500 eV about the K-edge of Mn (6539 eV) and –200 to +1000 eV about the K-edge (9659-eV) of Zn. Each scan was calibrated internally by placing elemental foil between the second and third in-line ionization chambers. Between 5 and 10 individual spectra were averaged for each sample. The reported Mn EXAFS spectra were obtained at 11 K to reduce the thermal disorder of the specimen; all other spectra were collected at ambient temperature and pressure.

X-ray absorption spectra of model Mn compounds used for analysis included feitknechtite (β-MnOOH), bixbyite (α-Mn_2O_3), rhodochrosite ($MnCO_3$), manganite (γ-MnOOH), pyrolusite (β-MnO_2), and birnessite (δ-MnO_2). Manganese standards of pyrolusite, bixbyite, and rhodochrosite were purchased from Fischer Scientific, Strem Chemicals, and Alpha Aesar, respectively. Hausmanite, feitknechtite, and manganite were synthesized from procedures modified after Hem and Lind (1983), Hem et al. (1982), and Bricker (1965) as described in Fendorf et al. (1999a); birnessite was made following procedures of Fendorf and Zasoski (1992). The purity of all mineral standards was checked using x-ray diffraction.

Spectra were analyzed using WinXAS (XANES) and EXAFSPAK (EXAFS) software (George, 1993). Fluorescence spectra were background-subtracted and the atomic absorption was normalized to unity. Isolation of backscattering contributions was accomplished by fitting a spline function to the absorption envelope and then subtracted it from the spectra. The isolated function was then transformed from units of eV to $Å^{-1}$ to produce the EXAFS function [$\chi(k)$] where k ($Å^{-1}$) is the photoelectron wave vector, which was then weighted by k^3. The $\chi(k)$ spectrum was Fourier-transformed to produce the radial structure function (RSF) in R-space (Å). First-shell peaks were individually isolated and backtransformed (Fourier-filtered). Fourier-filtered (FF) spectra were then fit with the predicted function in which the coordination number (CN), interatomic distance (R), and disorder (σ^2) were varied to give the best fit between the experimental and predicted spectra.

X-ray microprobe and x-ray fluorescence microtomography (FCMT) were used to determine spatial distributions and associations of Mn and Zn on and within *P. arundinacea* roots. Microtomography scans were collected on beamline (undulator) 13-ID-C at the Advanced Photon Source, Argonne National Laboratory. For FCMT, the intact (unaltered) root was mounted on the rotation axis of a x-y-theta stepping motor stage. Fluorescence data were collected with a solid-state energy dispersive x-ray detector which allows multiple elements to be detected simultaneously. The FCMT data were obtained by translating the root through the x-ray beam and collecting fluorescence at each 2-micron step. At each position, the fluorescence signal from a given element is proportional to the integrated number of atoms of that element along the ray transected by the beam. The sample is then rotated by 0.5 about the vertical axis and the line scan repeated. The process is continued until the particle has been rotated through a total of 180 at the end of which a 2D plane has been sampled.

Microprobe scans were also obtained to reveal a more detailed (higher magnification) cross-sectional view of the root. Manganese and Zn were mapped simultaneously along the same transect such that direct spatial associations and distributions were ascertained. Scans were performed on beamline 10.3.2 at the Advanced Light Source Lawrence, Berkeley National Laboratory using micro-x-ray fluorescence. The ring operated at ~1.9 GeV with a current of approximately 400 mA. The beam energy was set at ~10 keV and a dual-channel semi-conductor Si(Li) detector was used for fluorescent x-ray detection. The technique involved irradiating the sample with a micron-sized beam of light of a fixed energy, then rastering the sample position to obtain an elemental distribution map over the chosen area. One channel of the detector was windowed to collect K_α emission from Mn and the other for Zn. The scan covers a 250 µm x 250 µm section of the root with data collection at 5-µm increments.

2.6. Solution Analysis and Speciation

Water samples obtained from the peepers were analyzed for Mn and Zn via inductively coupled plasma-optical emission spectroscopy (Thermo Jarrell Ash IRIS ICP-OES; Franklin, MA) with a 10% accuracy range and quality control was checked every 15 samples. Detection limits were defined as 3σ (where σ is the standard deviation) of 5 blanks (Klesta and Bartz, 1996). Detection limits were: Mn 0.01 mg L^{-1} and Zn 0.02 mg L^{-1}.

Saturation indices (SI) were obtained to determine if the solution is oversaturated, undersaturated, or in equilibrium with the observed solid phases as follows:

$$SI = \log(IAP/K_{sp}) \quad (1)$$

where K_{sp} is the solubility constant for the solid-phase and the ion activity product (IAP) is the total activity of the ions that comprise the solid-phase. Saturation indices for Mn and Zn phases were calculated using the computer code MINTEQA2. The following equations and solubility constants were used for saturation-index computations (Johnson, 1982; Johnson et al., 1992; Schindler et al., 1969):

$$MnCO_{3(s)} \leftrightarrow Mn^{2+} + CO_3^{2-} \quad \log K_{sp} = -10.58 \quad (2)$$

$$ZnCO_{3(s)} \leftrightarrow Zn^{2+} + CO_3^{2-} \quad \log K_{sp} = -9.87 \quad (3)$$

$$Zn_5(OH)_6(CO_3)_2 + 6H^+ \leftrightarrow 5Zn^{2+} + 2CO_3^{2-} + 6H_2O \quad \log K_{sp} = 9.65 \quad (4)$$

For the soil analysis, a pH value of 7.05 and peak concentrations for Mn (6 mg kg^{-1}) and Zn (0.03 mg kg^{-1}) were utilized in calculations. Although carbonate concentrations were not directly measured in the field, carbonate values in anaerobic soils are typically an order of magnitude higher than atmospheric and were thus set at $P_{CO_2} = 10^{-2.5}$ atm.

3. RESULTS

3.1. Metal Budgets

The bulk wetland soil pH is circumneutral with values decreasing slightly with depth (Table 1). Similarly, Eh values indicate a decreasing soil redox potential with depth and the presence of reducing conditions. Rhizosphere pH and Eh values were not obtained on plant samples due to the difficulty in performing micro-scale field measurements. However, rhizosphere pH values may differ by 2 units from the bulk soil pH depending on plant and soil characteristics (Marschner, 1995).

Table 1. Water chemistry as a function of depth.*

Depth	Eh	pH	Mn	Zn
cm	mV		mg L^{-1}	
-10	410	7.29	0.17	nd
-5	325	7.16	0.10	nd
0	222	7.05	6.0	nd
5	213	6.94	0.47	0.03
10	219	6.81	1.0	nd
15	200	6.94	0.22	nd

* nd = not detected

Within wetland ecosystems, Mn and Zn may accumulate within the soil, solution, and vegetation. The partitioning of Mn and Zn within these fractions is reported in Tables 1 and 2. Solution concentrations of Mn and Zn are minimal within a range of 10 cm above and 15 cm below the soil/water interface (Table 1). Dissolved Mn concentrations peak at 6 mg L^{-1} at the soil/water interface while aqueous Zn is only detected at a depth of 5 cm (0.03 mg L^{-1}). In contrast, Mn and Zn concentrations are elevated in the soil and plant fractions (both root and shoot portions) (Table 2). Total concentrations of Mn and Zn within the soil did not differ significantly throughout the various regions of the wetland and hence soil concentrations are represented as a wetland average. Concentrations of Mn and Zn with plant roots are generally comparable with the soil and, at times, exceed those of the soil. Additionally, total Mn and Zn concentrations are on average higher in the roots relative to the shoot (aerial) portions.

To obtain a better understanding of the relative contribution of metal abundance within the wetland subsurface (excluding plant shoots), Mn and Zn mass calculations were obtained using soil and root densities, metal concentrations, and wetland area (Table 3). The Cataldo wetland's area is ~78,200 m^2, with four plant species (*T. latifolia, P. arundinacea, E. arvense, and S. microcarpus*), accounting for ca. 95% (~74,290 m^2) of the total area. Since the soil and plant samples were extracted to a 0.15-m depth, we consider a wetland volume of 11,730 m^3 and the four dominant plant species stands comprise

Table 2. Manganese and zinc concentrations within the wetland soil and plants.*

	Roots		Shoots		Soil	
	Mn	Zn	Mn	Zn	Mn	Zn
	mg kg^{-1}					
					1600	1300
T. latifolia	790	350	3400	86		
P. arundinacea	1500	950	1200	510		
E. arvense	1900	300	1800	250		
S. microcarpus	1000	1300	1500	250		

* values reported as mean of triplicate analysis, relative standard deviations are within 10%

11,144 m^3. Using the fractional abundance of each plant species, the volume of soil within each plant stand was calculated. Within the top 15 cm, the corresponding soil mineral densities range from 0.42 to 0.50 g cm^{-3} and root densities from 0.05 to 0.06 g cm^{-3}. Soil and root densities were consistent throughout individual plant stand areas and thus reflect a considerable degree of homogeneity within plant species.

Using the aforementioned densities and volumes, the mass of Mn and Zn within the soil and four plant species were obtained (Table 3). Manganese associated with the roots of the different plant species is remarkably similar, excluding *S. microcarpus.* Zinc, on the other hand, is lower in both in *S. microcarpus* and *E. arvense* relative to *T. latifolia* and *P. arundinacea.* Within the top 15 cm of the Cataldo wetland, 7,840 kg of Mn are associated with the soil and 707 kg with plant roots. Similarly, 6,360 kg of Zn are affiliated with the soil fraction and 319 kg with roots. Plant roots thus sequester ca. 8% of Mn and 5% of Zn sequestration with this ecosystem.

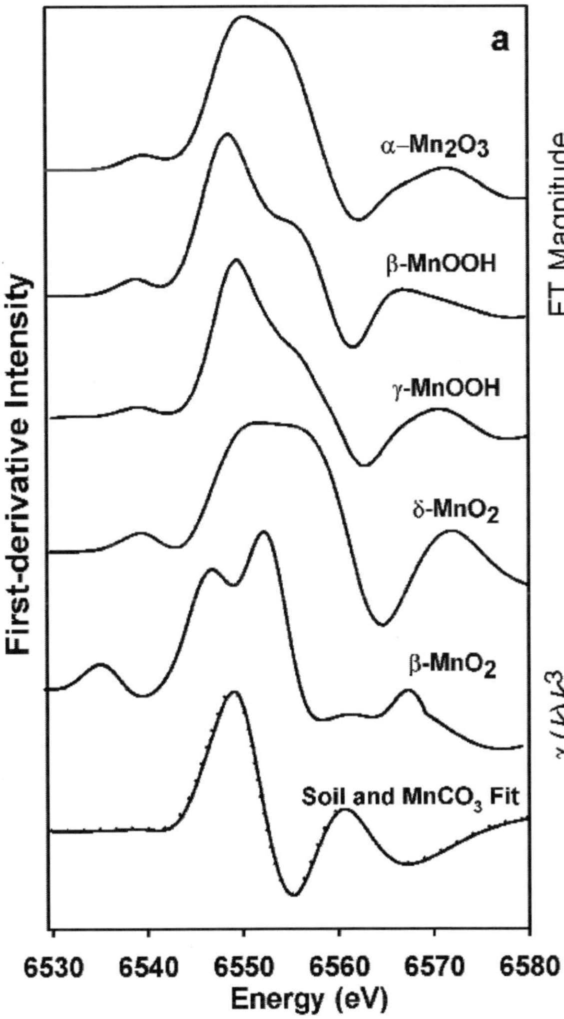

Figure 3. Spectroscopic analyses of soil Mn illustrated by (a) XANES spectra for various Mn standard compounds and wetland soil, (b) Fourier-transformed EXAFS spectrum, and (c) normalized k^3-weighted EXAFS spectra. Theoretical fit is indicated by the dashed line. The Fourier-transform spectrum is not phase-shift corrected and thus peak positions are less than true interatomic distances.

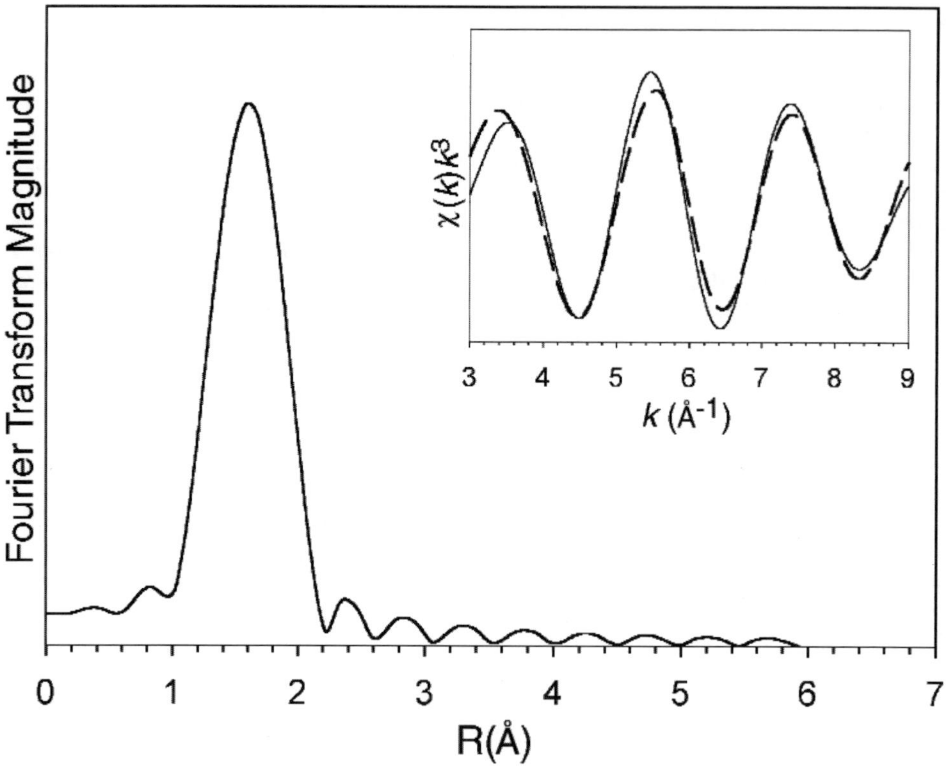

Figure 4. Fourier-transformed (FT) spectrum of Zn in the soils resulting in a radial structure function in real space (Å). Fourier-transform is not phase-shift corrected and thus peak positions are less than true interatomic distances. Inset represents the k^3-weighted, Fourier-filtered first shell (Zn-O) from the FT spectrum illustrating the theoretical fit (dotted line) to the experimental curve (solid line).

Table 3. Manganese and Zn budgets within the wetland.

Plant Species	Abundance* %	Volume¶ m^3	Density $g\ cm^{-3}$		Mass (kg)† Mn		Zn	
			Soil	Roots	Soil	Roots	Soil	Roots
T. latifolia	50	5865	0.42	0.05	3900	230	3200	100
P. arundinacea	20	2346	0.50	0.06	1900	210	1500	140
E. arvense	20	2346	0.42	0.05	1600	230	1300	35
S. microcarpus	5	587	0.47	0.06	440	37	360	44
Total‡	95	11144			7840	707	6360	319

* abundance based on relative proportions of different plants within wetland
¶ volume calculated from relative abundance and total wetland volume (11,730 m^3) to a 15-cm depth
† mass reported as mean of 6 samples; relative standard deviations are within 10%
‡ remaining 5% = unvegetated and mixed plant and shrub regions

3.2. Speciation of Mn and Zn in Soils

The positioning of the main absorption edge in the XANES spectrum and corresponding first-derivative curve can be used to differentiate Mn oxidation states in soil (higher oxidation states occur at greater energies) (Manceau et al., 1992; Schulze et al., 1995; Friedl et al. 1997; Fendorf et al., 1999b; Ressler et al., 2000). For example, Mn(II) and Mn(IV) are separated by ~4 eV in their first-derivative XANES spectra. Manganese carbonate has a lower pre-edge intensity and absorption edge position, owing to its six-fold co-

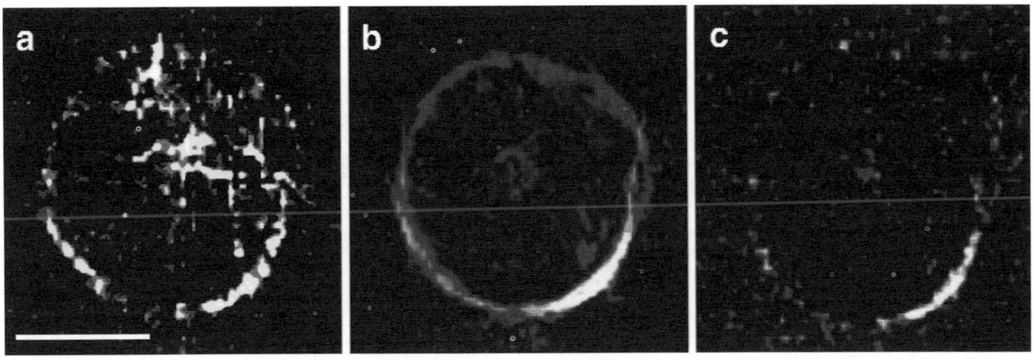

Figure 5. Metal distributions on and within *P. arundinacea* roots. Microtomographic images of (a) Mn (b) Zn and (c) Ca. Scale bar represents 300 μm.

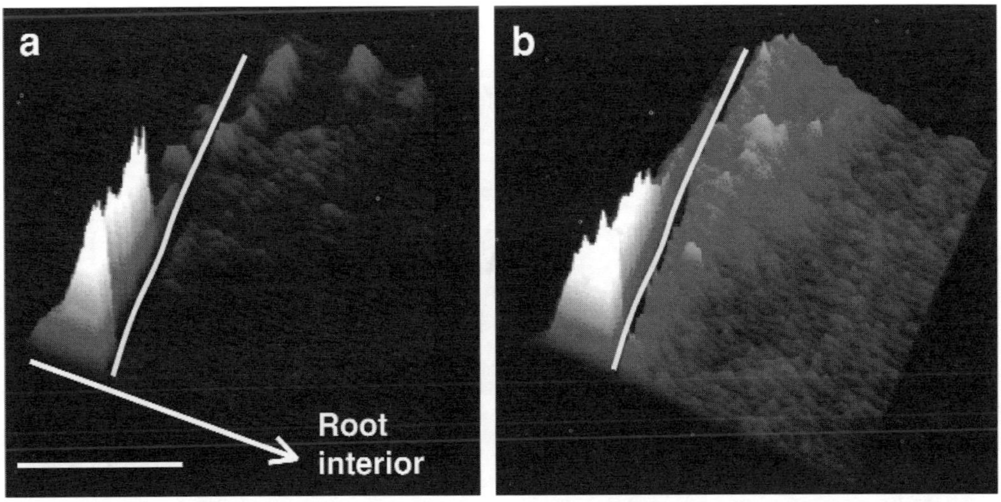

Figure 6. Higher magnification x-ray microprobe image of (a) Mn and (b) Zn along a transect from the root center to the exterior of the plaque (scale bar represents 50 μm). Preserved roots were sliced in half longitudinally to reveal a cross-sectional view. The yellow line represents the root epidermis and relative elemental concentrations increase with color brightness.

ordination and lower oxidation state than higher valent Mn (hydr)oxides (Fig. 2) (Manceau et al., 1992; Schulze et al, 1995). Quantitative analysis of our unknown soil samples collected from the wetland revealed that pre-edge peak amplitudes and main absorption edges were consistent in all Mn samples and are consistent with the rhodochrosite spectrum. Thus, Mn(II) is the dominant solid-phase oxidation state and appears to have a rhodochrosite-like structure. In addition, linear combination fitting of the unknown Mn species were best reproduced using the rhodochrosite spectrum (Fig. 3).

We used EXAFS spectroscopy to confirm the local structure of Mn in the solids as rhodochrosite-like phases. The Fourier transformed and unfiltered k^3-weighted $\chi(k)$ EXAFS function are shown in Fig. 3 with associated spectra predicted for rhodochrosite. Predicted spectra were calculated using published values for the first two Mn shells in rhodochrosite (Effenberger et al., 1981; Friedl et al., 1997). Unfor-

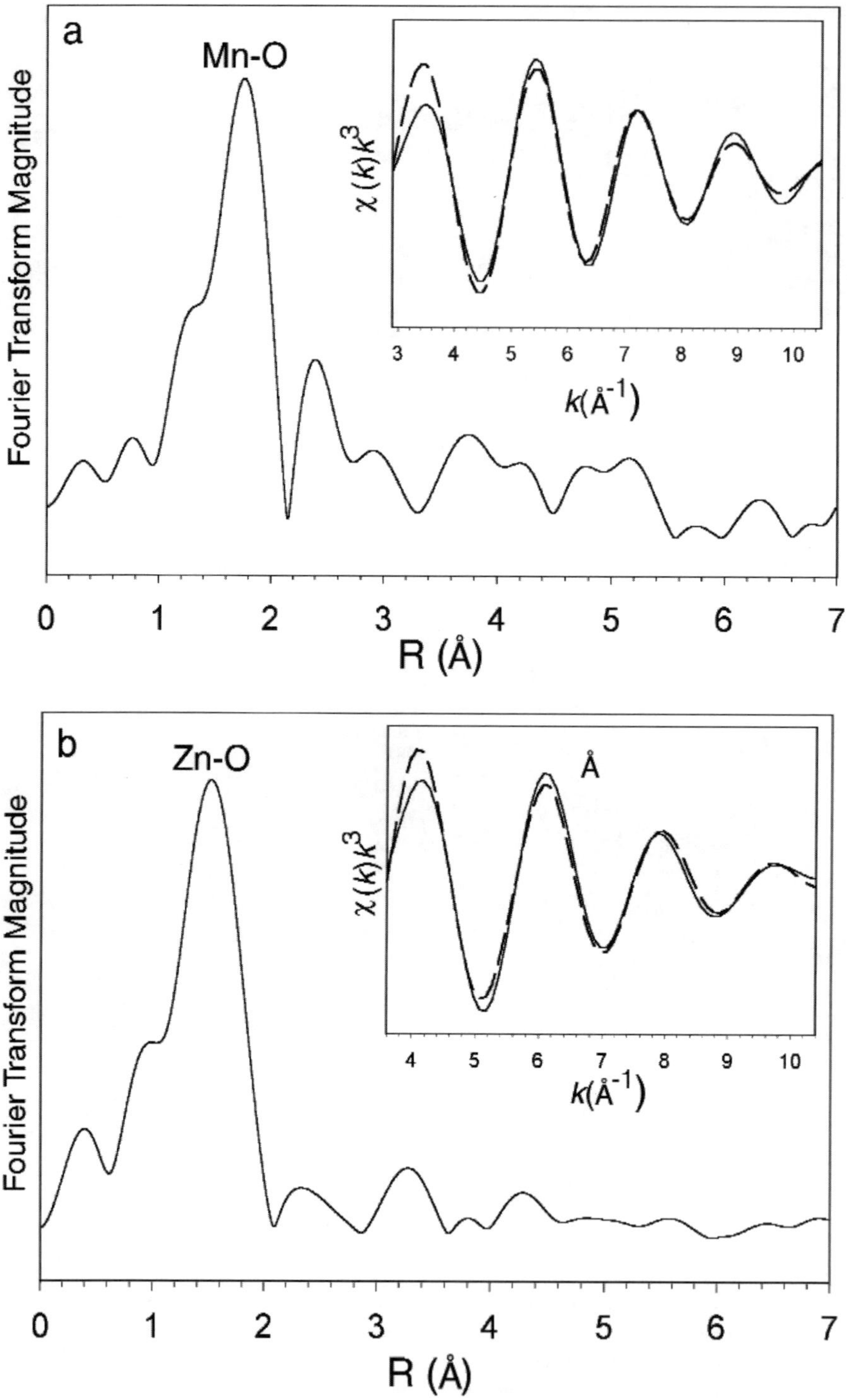

Figure 7. Fourier-transformed (FT) spectra of (a) Mn and (b) Zn resulting in a radial structure function in real space (Å). Fourier-transforms are not phase-shift corrected and thus peak positions are less than true interatomic distances. Insets represent the k^3-weighted, backtransformed first shells [(a) Mn-O and (b) Zn-O] from the FT spectra illustrating the theoretical fits (dotted lines) to the experimental curve (solid lines).

tunately, it is difficult to decipher whether the manganese carbonate is a pure species or a solid-solution.

The coordination environment for Zn within the soil was also determined using EXAFS spectroscopy (Fig. 4). Zinc is octahedrally coordinated by oxygen with a Zn-O radial distance of 2.11 Å corresponding to a smithsonite-like ($ZnCO_3$) phase or Zn-substituted $CaCO_3$ (Effenberger et al., 1981; O'Day et al., 1998). Although second-shell radial distances could be used to differentiate these two possibilities, we could not resolve second shells sufficiently to allow for this distinction.

3.3. Spatial Distribution and Speciation of Mn and Zn on Plant Roots

On the surface of *P. arundinacea* roots, microtomographic images illustrate that Mn and Zn are elevated in discrete zones on the exterior of the epidermis whereas, in contrast, concentrations diminish in areas adjacent to these zones. Zinc and Mn are strongly correlated aside from the root interior where Mn concentrations are considerably higher (Fig. 5a,b). Additionally, Ca levels are elevated at sites of Zn and Mn precipitation (Fig. 5c). A more detailed (higher magnification) depiction of Mn and Zn distributions further illustrates the strong correlation between Mn and Zn (Fig. 6a,b). Moreover, a nodular morphology of the Mn/Zn precipitates is portrayed. Scans were performed on extensive portions of various roots and the results were consistent with those illustrated in Figures 5 and 6.

The local coordination environment of Mn and Zn, determined using EXAFS spectroscopy (Fig. 7), reveals that Mn is octahedrally coordinated by oxygen. Similar to the soil results, the Mn-O interatomic distance is 2.20 Å, which is in close agreement with published values (2.19 Å) for rhodochrosite ($MnCO_3$) (Effenberger et al., 1981; Friedl et al., 1997). In contrast, the first shell for Zn is a composite of three distinct sites (2 octahedral and 1 tetrahedral) having an average of four different interatomic distances (2.08, 2.06, 2.15, and 1.96 Å). The Zn-O(H) distances correspond to published values for hydrozincite [$Zn_5(OH)_6(CO_3)_2$] (Ghose, 1964; O'Day et al., 1998). Accordingly, on the basis of the EXAFS analyses, the discrete nodules observed on *P. arundinacea* roots are composed of Mn and Zn carbonate phases.

3.4. Geochemical Modeling of Mn- and Zn-Carbonate Phases

MINTEQA2 was used to determine if the pore waters were saturated with respect to the observed carbonate phases within the soil and on plant roots. The bulk pore water is near equilibrium (SI ≈ -0.193) with respect to rhodochrosite but undersaturated (SI ≈ -3.02) with respect to smithsonite. It is important to note that the K_{sp} values assume a pure species yet as mentioned previously, we could not differentiate between a pure smithsonite species or a Zn-substituted Ca carbonate phase. Alternatively, considering that wetland water levels are beginning to decline at this point in time, Zn phases may be in a transitional period and it may therefore be feasible that smithsonite is dissolving.

It is well-established that steep gradients in nutrient concentrations, pH, redox potential, and carbonate values occur at the soil-root interface (Marschner, 1995). Within the rhizosphere, carbonate concentrations can approach 5 mmol L^{-1}, while pH values can be 2 units greater and nutrient concentrations several times higher than those in bulk solution (Bedford et al., 1991; Marschner, 1995; Linehan et al., 1989). Although it is difficult to predict the chemical environment within the rhizosphere, if we consider merely a 1 unit pH increase (8.05) and Zn concentrations three times that in the bulk solution, the pore water is in equilibrium with hydrozincite (SI ≈ 0). Furthermore, Zn is an essential nutrient to plants and thus more realistic Zn concentrations are most likely greater than 5 times that in the soil; given such a concentration, saturation with respect to hydrozincite would be exceeded. Similarly, Mn concentrations at the root surface will be several times higher than within the bulk solution. Considering that the pore water within the bulk solution is in equilibrium with rhodochrosite, the more favorable conditions within the rhizosphere will promote rhodochrosite saturation.

4. DISCUSSION

The partitioning of Mn and Zn within the Cataldo wetland is dominantly between the mineral fraction and plant roots while solution concentrations are negligible. Concentrations of Mn and Zn associated with roots vary among plant species; however, at times, root concentrations are roughly equivalent to those of the soil. Yet, owing to higher soil densities, the mass contributions are ≥10-fold higher in the mineral fraction relative to roots of all plant species. Nevertheless, root surfaces are providing an essential substrate for metal nucleation accounting for 8% of Mn and 5% of Zn sequestration with the wetland soils.

Spectroscopic analyses reveal the dominance and significance of carbonate ligands in precipitation, and subsequent sequestration, of Mn and Zn. XAS data reveal that a rhodochrosite-like phase is the dominant Mn phase within the soil and on the surface of *P. arundinacea roots*. Similar to Zn, first-shell data cannot differentiate between a pure rhodochrosite or a Ca-Mn carbonate solid-solution. On root surfaces, however, Ca exhibits a strong correlation with Mn and Zn and may therefore suggest that Ca carbonate ($CaCO_3$) is in some manner associated with Mn and Zn deposition on root surfaces, and perhaps within the soil as well.

Zinc is also bound by carbonate ligands, yet the phase varies from smithsonite-like in the soil to its hydrated derivative hydrozincite [$Zn_5(OH)_6(CO_3)_2$] on the root surface. Smithsonite formation is very slow under surface temperatures and pressures; hydrozincite, on the other hand, is formed over a wide range of $CO_{2(g)}$ partial pressures (Zachara et al., 1989a). Structural dissimilarities among rhodochrosite and hydrozincite impede epitaxial growth and subsequent solid-solution (Zachara et al., 1989a). Thus, the probable association between the rhodochrosite-like and hydrozincite-like phases within the root-borne nodules is as a mixture of discrete precipitate domains.

On the basis of our analyses, Zn is incorporated into a smithsonite-like ($ZnCO_3$) phase and/or as Zn-substituted $CaCO_3$ within the soil. Zinc may initially adsorb on calcite where it may be incorporated into the $CaCO_3$ crystal lattice or nucleate as a smithsonite precipitate (Zachara et al., 1989b; Reeder, 1996; Zachara et al., 1991). Smithsonite formation, albeit slow, may be induced in an elevated carbonate environment or by initial nucleation onto $CaCO_3$ surfaces (surface catalysis). Alternatively, if hydrozincite is metastable, it may initially precipitate either in the presence or absence of $CaCO_3$, followed by its transformation into a smithsonite-like phase; however, the time-dependent feasibility for such conversion is unknown (Zachara et al., 1989a; Bottcher, 1995). In general, Zn precipitation may involve a single phase or a convolution of carbonate phases within the soil.

Although only Spring results are presented, Mn concentrations, speciation, and structural environments are equivalent throughout the year (La Force et al., 2000b). Correspondingly, when wetland water levels decline and the soils become aerated, XAS analyses of the roots and soil still reveal Mn carbonates as the dominant phases (La Force et al., 2000b). However, sequential selective extractions of the wetland soils indicate that the concentration and proportion of Mn-carbonates relative to other Mn phases varies an order of magnitude throughout the year (La Force et al., 2000b). Conversely, the structural environment of Zn within the soil varies significantly throughout a year with carbonate phases dominating during Spring when water levels and microbially respired carbonate levels are elevated (Bostick et al, 1999). As water levels decline, however, Zn oxide phases dominate the partitioning of Zn. The data suggest, therefore, that the presence and dominance of Mn and Zn carbonate phases is primarily a consequence of secondary precipitation rather than residual ore deposition. Zinc on the root surface, however, remains associated with a carbonate phase (hydrozincite) throughout the year and thus denotes a more consistent environment in the rhizosphere for carbonate formation and stability (data not shown).

The strong correlation between Mn and Zn and their morphology are illustrated by the x-ray microprobe and microtomography images of *P. arundinacea* roots. Manganese and Zn exist as discrete carbonate "nodules" on the root surface with concentrations diminishing in areas adjacent to these clusters. Due to the structural dissimilarity of rhodochrosite and hydrozincite, these phases seemingly exist in isolated regions within the nodule. Roots, therefore, provide a substrate for carbonate nucleation, as well as stimulate an enhanced carbonate microenvironment by way of root and/or microbial respiration. Owing to the structural similarity between soil and root-associated carbonate phases, it may be speculated that a similar nodular morphology exists on mineral surfaces with surface defect sites providing the site of nucleation.

5. CONCLUSIONS

Past mining practices have jeopardized environmental quality in many regions, including the Coeur d'Alene district, by introducing toxic metals to the water, soils, and biota. Public health relies on the immobilization of such metals to restrict their introduction into drinking water and food chains. Wetlands provide a "sink" for metals, yet the mechanisms of metal sequestration within these unique ecosystems are not well understood. Accordingly, we have shown in this study the dominant role of carbonates in Mn and Zn sequestration within a natural, mine-waste impacted wetland.

Metal budgets indicate that the soil mineral fraction and the surface of plant roots are the dominant players in Mn and Zn retention, while dissolved concentrations are negligible. Manganese and Zn are highly correlated within the soil and on the surface of plant roots, occurring as carbonate phases. On plant roots, and possibly within the soil, Mn, Zn, and Ca carbonate nodular deposits occur at discrete sites on the surface.

Accordingly, these results have profound consequences on predicting the fate and modeling the transport of metals within anaerobic soils. In this study, we elucidate the behavior of Mn and Zn cycling in wetland ecosystems. Carbonate ligands appear to be the dominant means by which Zn and Mn are sequestrated and immobilized within this wetland ecosystem, and may have a pronounced influence on metal sequestration in general. As such, carbonate precipitation of metals within the soil as well as at the soil/root interface must be considered when implementing remediation strategies, such as phytoremediation and constructed wetlands.

Acknowledgements—We gratefully acknowledge the staff of the Advanced Photon Source, Argonne National Laboratory, specifically Matt Newville for fluorescence microtomography analyses. We also thank Stanford Synchrotron Radiation Laboratory and Lawrence Berkeley National Laboratory for granted beamtime and technical assistance. This research was supported by the Exploratory Research Program of the U.S. EPA.

Editorial handling: Scott A. Wood

REFERENCES

Armstrong W. (1978) Root aeration in the wetland condition. In *Plant Life in Anaerobic Environments* (eds. D.E. Hook and R.M.M. Crawford) Ann Arbor Science, Michigan pp. 269-298.

Bacha R. E. and Hossner L. R. (1977) Characteristics of coatings formed on rice roots as affected by iron and manganese additions. *Soil Sci. Soc. Am. J.* **41**, 931-935.

Bartlett R. J. (1961) Iron oxidation proximate to plant roots. *Soil Science* **92**, 372-379.

Bedford B. L., Bouldin D. R., Beliveau B. D. (1991) Net oxygen and carbon-dioxide balances in solutions bathing roots of wetland plants. *J. Ecology* **79**, 943.

Bennett E. H., Siems P. L., and Constantopoulos J. T. (1989) The geology and history of the Coeur d'Alene Mining District, Idaho. In *Guidebook to the Geology of Northern and Western Idaho and Surrounding Area.* (eds. V. E. Chamberlain, R. M. Breckenridge, and B. Bonnichsen) Idaho Geological Survey Bulletin 28. pp. 137-156.

Bennett E. H. (1994) A history of the Bunker Hill Superfund Site, Kellogg, Idaho. Idaho Geological Survey. Pacific Northwest Conf., Spokane, Washington.

Blus L. J., Henny C. J., Hoffman D. J., Sileo L., and Audet D. J. (1998) Persistence of high lead concentrations and associated effects in tundra swans captured near a mining and smelting complex in northern Idaho. *Ecotoxicology* **8**, 125.

Bostick B. C., Hansel C. M., La Force M. J., and Fendorf S. E. (1998) Zinc local structure and partitioning within a contaminated wetland. *Stanford Synchrotron Radiation Laboratory 1998 Activity Report*. pp. 391-394.

Bottcher M. E. (1995) The formation of rhodochrosite-smithsonite ($MnCO_3$-$ZnCO_3$) solid solutions at 5°C. *Mineral. Mag.* **59**, 481-488.

Brandl H. and Hanselmann K. W. (1991) Evaluation and application of dialysis porewater samplers for microbiological studies at sediment-water interfaces. *Aquatic Sci.* **53**, 54-73.

Bricker O. (1965) Some stability relations in the system Mn-O_2-H_2O at 25° and one atmosphere total pressure. *Amer. Miner.* **50**, 1296-1355.

Carignan R. (1984) Interstitial water sampling by dialysis: Methodological notes. *Limnol. Ocean.* **29**, 667-670.

Carignan R., Rapin F., and Tessier A. (1985) Sediment porewater sampling for metal analysis: A comparison of techniques. *Geochim. Cosmochim. Acta* **49**, 2493-2497.

Chen C. C., Dixon J. B., and Turner F. T. (1980) Iron coatings on rice roots: Mineralogy and quantity influencing factors. *Soil. Sci. Soc. Am. J.* **44**, 635-639.

Cowardin L. M, Carter V., Golet F. C., and LaRoe E. T. (1979) Classification of wetland and deepwater habitats of the United States. *U.S. Fish and Wildlife Service Report* FWS/OBS-79/31, 131 p.

Effenberger H, Mereiterÿ K., and Zemannÿ J. (1981) Crystal structure refinements of magnesite, calcite, rhodochrosite, siderite, smithsonite, and dolomite, with the discussion of some aspects of the stereochemistry of calcite type carbonates. *Zeitschrift für Kristallographie* **156**, 233-243.

Ellis M. M. (1940) Pollution of the Coeur d'Alene River and adjacent waters by mine wastes. *U.S. Bureau of Fisheries,* Special Scientific Report 1, Washington, D.C.

Fendorf S. E. and Zasoski R. J. (1992) Chromium(III) oxidation by delta-MnO_2. 1. Characterization. *Environ. Sci. Technol.* **26**, 79-85.

Fendorf S. E., Jardine P. M., Patterson R. R., Taylor D. L., and Brooks S. C. (1999a) Pyrolusite surface transformations measured in real-time during the reactive transport of Co(II)$EDTA^{2-}$. *Geochim. Cosmochim. Acta* **63**, 3049-3057.

Fendorf S. E., Jardine P. M., Taylor D. L, Brooks S. C., and Rochette E. A. (1999b) Auto inhibition of oxide mineral reductive capacity toward Co(II)EDTA. In *Mineral-water interfacial reactions kinetic and mechanisms.* (eds. D.L Sparks and T.J.Grundl). ACS Symposium Series 715, Washington, DC.

Fretwell J. D., Williams J. S., and Redman P. J. (1996) National water summary on wetlands resources. *USGS Water-Supply Paper* 2425.

Friedl G., Wehrli B., and Manceau A. (1997) Solid phase in the cycling of manganese in eutrophic lakes: New insights from EXAFS spectroscopy. *Geochim. Cosmochim. Acta* **61**, 275-290.

Galbraith J. H., Williams R. E., and Siems P. L. (1972) Migration and leaching of metals from old mine tailings deposits. *Groundwater* **10**, 33-44.

George G. N. (1993) EXAFSPAK. Stanford Synchrotron Radiation Laboratory.

Ghose S. (1964) The crystal structure of hydrozincite, $Zn_6(OH)_6(CO_3)_2$. *Acta Crystall.* **17**, 1051-1057.

Harrington J. M., La Force M. J., Rember W. C., Fendorf S. E., and Rosenzweig R. F. (1998) Phase associations and mobilization of iron and trace elements in Coeur d'Alene Lake, Idaho. *Environ. Sci. Technol.* **32**, 650-656.

Hem J. D. and Lind C. J. (1983) Non-equilibrium models for predicting forms of precipitated manganese oxides. *Geochim. Cosmochim. Acta* **47**, 2037-2046.

Hem J. D., Roberson C. E., and Fournier R. B. (1982) Stability of beta MnOOH and manganese oxide deposition from springwater. *Water Resources Res.* **18**, 563-570.

Hobbs S. W., Griggs A. B., Wallace R. E., and Campbell A. B. (1965) Geology of the Coeur d'Alene district, Shoshone County, Idaho. *U.S. Geological Survey Professional Paper* **478**, 1-28.

Horowitz A. J., Elrick K. A., and Cook R. B. (1992) Effect of mining-related activities on the sediment-trace element geochemistry of Lake Coeur d'Alene, Idaho, USA—Part 1. *U.S. Geological Survey Open-File Report 92-109.*

Ioannou C. (1979) Distribution, transport and reclamation of abandoned mine tailings along the channel of the South

Fork of the Coeur d'Alene River and Tributaries, Idaho. M.S. thesis, University of Idaho.

Javorka E. (1991) Lake water quality assessment Coeur d'Alene Lake, Benewah and Kootenai Counties, Idaho. Coeur d'Alene Basin Interagency Group, Coeur d'Alene Tribe of Idaho, Coeur d'Alene Subagency, Plummer, Idaho. p. 42.

Johnson K. S. (1982) Solubility of rhodochrosite ($MnCO_3$) in water and seawater. *Geochim. Cosmochim. Acta* **46**, 1805-1809.

Johnson J. W., Oelkers E. H., and Helgeson H. C. (1992) SUPCRT92: A software package for calculating the standard molal thermodynamic properties of minerals, gases, aqueous species, and reactions from 1-bar to 5000-bar and 0° to 1000°C. *Comput. Geosci.* **18**, 899-947.

Klesta E. J. Jr. and Bartz J. K. (1996) Quality assurance and quality control. In *Methods of Soil Analysis. Part 3. Chemical Methods* (ed. D.L. Sparks). SSSA ASA, Madison WI. pp. 19-49.

La Force M. J., Hansel C. M., and Fendorf S. E. (2000a) Constructing simple wetland sampling devices. *Soil Sci. Soc. Am. J.* **64**, 809-811.

La Force M. J., Hansel C. M., and Fendorf S. E. (2000b) Seasonal transformations of manganese in a mine-waste impacted wetland. *Soil Sci. Soc. Am. J.* In press.

La Force M. J., Fendorf S. E., Li G. C., Schneider G. M., and Rosenzweig R. F. (1998) A laboratory evaluation of trace element mobility from flooding and nutrient loading of Coeur d'Alene River sediments. *J. Environ. Qual.* **27**, 318-328.

Linehan D. J., Sinclair A. H., and Mitchell M. C. (1989) Seasonal changes in Cu, Mn, Zn and Co concentrations in soil in the root-zone of barley (*Hordeum vulgare L.*) *J. Soil Sci.* **40**, 103-115.

Lytle F. W., Greegor R. B., Sandstone D. R., Marques E. C., Wong J., Spiro C. L., Huffman G. P., and Huggins F. E. (1984) Measurements of soft x-ray absorption spectra with a fluorescent ion chamber. *Nucl. Instrum. Methods Phys. Res. Sect. A.* **226**, 542-548.

Manceau A., Gorshkov A. I., and Drits V. (1992) Structural chemistry of Mn, Fe, Co, and Ni in manganese hydrous oxides: Part I. Information from XANES spectroscopy. *Amer. Miner.* **77**, 1133-1143.

Marschner H. (1995) Mineral Nutrition of Higher Plants. Academic Press, NY.

Mendelssohn I. A., and Postek, M. T. (1982) Elemental analysis of deposits on the roots of *Spartina alterniflora* Loisel. *Amer. J. Bot.* **69**, 904-912.

Mink L. L. (1971) Water quality of the Coeur d'Alene River basin. M.S. thesis, University of Idaho.

Norbeck P. M. (1974) Water table configurations and tailings distribution, Coeur d'Alene Valley, Idaho. M.S. thesis, University of Idaho.

O'Day P. A., Carroll S. A., and Waychunas G. A. (1998) Rock-water interactions controlling zinc, cadmium, and lead concentrations in surface waters and sediments, U.S. Tri-State Mining District. 1. Molecular identification using x-ray absorption spectroscopy. *Environ. Sci. Technol.* **32**, 943-955.

Rabe F. W. and Bauer S. B. (1977) Heavy metals in lakes of the Coeur d'Alene River Valley, Idaho. *Northwest Science* **51**, 183-197.

Reece D. E., Felkey J. R., and Wai C. M. (1978) Heavy metal pollution in the sediments of the Coeur d'Alene River, Idaho. *Environ. Geol.* **2(5)**, 289-293.

Reeder R. J. (1996) Interaction of divalent cobalt, zinc, cadmium, and barium with the calcite surface during layer growth. *Geochim. Cosmochim. Acta* **60**, 1543-1552.

Ressler T., Wong J., Roos J., and Smith I. L. (2000) Quantitative speciation of Mn-bearing particulates emitted from autos burning (methylcyclopentadienyl)manganese tricarbonyl-added gasolines using XANES spectroscopy. *Environ. Sci. Tech.* **34**, 950-958.

Schindler P., Reinert M., and Gamsjäger H. (1969) Löslichkeitskonstanten und free bildungs enthalpien von $ZnCO_3$ and $Zn_5(OH)_6(CO_3)_2$ bei 25ºC. *Helvetica Chim. Acta* **52**, 2327-2332.

Schulze D. G., Sutton S. R., and Bajt S. (1995) Determining manganese oxidation-state in soils using x-ray absorption near-edge structure (XANES) spectroscopy. *Soil Sci. Soc. Amer. J.* **59**,1540-1548.

Taylor G. J., Crowder A. A., and Rodden R. (1984) Formation and morphology of an iron plaque on the roots of *Typha latifolia* L. grown in solution culture. *Amer. J. Bot.* **71**, 666-675.

USDA (1996) Soil Survey Laboratory Methods Manual, Report no. 42, Vol 3., Washington, D.C.

Zachara J. M., Kittrick J. A., Dake L. S., and Harsh J. B. (1989a) Solubility and surface spectroscopy of zinc precipitates on calcite. *Geochim. Cosmochim. Acta* **53**, 9-19.

Zachara J. M., Kittrick J. A., and Harsh J. B. (1989b) The mechanism of Zn^{2+} adsorption on calcite. *Geochim. Cosmochim. Acta* **52**, 2281-2291.

Zachara J. M., Cowan C. E., and Resch C. T. (1991) Sorption of divalent metals on calcite. *Geochim. Cosmochim. Acta* **55**, 1549-1562.

Subject Index

A

acetate 6-8, 10, 12, 13, 15, 18, 19, 20, 21, 22, 23, 24, 209-225, 231, 248, 251
acetic acid 211, 213-214
actinide 210, 229, 250-251
active sites 10, 11, 15, 17, 63-65, 69, 71, 77, 423
adsorption 10, 11, 12, 15, 16, 17, 19, 21, 23-30, 66-69, 71-72, 112-113, 365-376, 423-439, 443
adsorption, contaminants 340, 358
adsorption, enhanced 367-368, 371-373
adsorption, mechanisms 339-340
adsorption, microbial nutrients 339-340
adsorption, mineral surfaces 10, 339-340, 342-343
adsorption, proton 10-11, 16, 27, 69, 71, 111-126
aerosol 307, 309, 319, 325
albite, crystalline 10, 14, 18, 19, 20, 23, 24, 61-62, 75, 77, 83-94, 265, 318-319
albite, glass 83-94
aluminum (Al) 10, 12, 15, 16, 27-29, 47, 75, 80, 259-260, 262-263, 269-271, 275, 312-313, 315-316, 368, 371, 373-374
americium hydroxide 246, 250
ammonia 283, 310, 382-384, 386-387, 396
andalusite 318
andesite 281, 283, 285, 308,
andradite 314, 324
anhydrite 258-260, 263, 266, 269-274, 276, 314, 318, 323-324
ankerite 441
anorthite 18, 19, 21, 24, 90, 259, 265-266, 271-273, 319
antimony (Sb) 282-283, 286-287, 312-313, 325, 441
aphthitalite 283-284
aragonite 182-184, 186
arcanite 323
arsenate 389-380, 382, 387-392
arsenic (As) 282-286, 312-313, 379-397, 442-443
arsenite 379, 382, 387-389, 391-392
atomic force microscopy (AFM) 83-94, 97-98, 102-103, 105-109, 112, 117, 119-122, 124-126
azo dye(s) 400, 407, 413

B

bacteria 3, 4, 6, 7, 18, 22, 23, 339-364, 372
bacteria, active responses 340-341, 345, 347, 349, 353, 355-358
bacteria, cell envelope 341, 345, 347, 349, 353-358
bacteria, environmental stimuli 344, 353, 356
bacteria, nutrient uptake model 345-349,
bacteria, passive responses 341, 345, 347, 349, 353-355
Bajo de la Alumbrera deposit, Argentina 299
barite 97-109, 112, 119-122, 124-125, 283, 329-330, 333-334
basalt 281-283, 285, 308
biodegradation 381, 391
biotic cycling 55-57
biotite 18, 37-57, 61
bismuth (Bi) 312-315, 317, 321, 323-325
bismuthinite 284, 318, 323
bisulfide complexes 270-271, 280
boron 312-313, 315
bromide 287, 315
bulrush 443
Bunker Hill 441

C

cadmium (Cd) 312-314, 321, 324-325, 368-369, 371-373, 441-442
calcite 39, 136, 139-141, 144-145, 155-156, 170, 175, 182-184, 186, 441, 452
calcium (Ca) 3, 258-259, 262-263, 265-266, 272-273, 275, 368-369, 371, 373
cannizzarite 313, 315, 324
capillary pressure 400-401, 404, 406-407, 409-419
carbonate 136, 330, 442, 446, 448, 451-453
cassiterite 280, 283, 288, 318, 323
Cataldo 443-444, 446-447, 451
cation exchange 39, 46-48, 55-57, 211
cattail 443
centrifugation vs. mobility 436-438
cerium (Ce) 192, 194-195, 197, 200, 203-204
Cerro Negro volcano 282-283, 299
chalcopyrite 258-260, 263, 267-269, 272-273, 284, 298-299, 441
chalk 129-149, 153-176
chemical affinity 17, 23, 27, 29, 30, 97, 186
chemical weathering rates 4, 37-57
chemostat growth 340-341, 345, 349-355
chlorargyrite 192, 288-289, 292, 298
chloride complexes 191-206, 211, 219-220, 224, 258, 262-263, 270-271, 279-280, 284, 287, 289, 294-299, 310, 315
chromium (Cr) 312-313, 318-319, 323
citrate 6, 8, 12, 14-15, 18-22, 39, 368-369, 371, 373-375
co-adsorption 365, 367
cobalt (Co) 368, 370-376
Coeur d'Alene (CDA) 441-443
Colima volcano 282-284
condensates, volcanic gas 280, 282-284, 299, 307-309, 312-313, 315, 319
contaminant transport 367-368, 371
copper (Cu) 258-260, 262-264, 267-276, 280, 282-283, 287, 290-291, 294-299, 312-315, 321-323, 325, 368, 371-373, 441
corundum 61, 64, 67, 112-113, 115-117, 323

cosalite 313, 315, 319, 324
COSPEC 309
cristobalite 187, 284, 318, 323-324
critical age 407, 409, 414, 417
critical radius 102, 105, 108
cryolite 323
crystal growth 77, 79, 81, 97-109, 112, 119-120, 122-126
crystal morphology 76-78
crystal twin boundaries 77, 81
crystal twin laws 79
crystallization 79

D

Debye-Huckel 216, 220, 236, 244
diffusion 99, 325, 343-344, 346-348, 354, 357, 382, 400, 442
diopside 61, 318
dissociation constants, acid 213-215
dissolution 3-7, 10, 11, 12, 13, 14, 15, 16, 17-30, 39, 47-48, 54, 57, 61-72, 75-80, 83-94, 112, 119, 235, 257, 260-261, 263
dissolution, rates (see also kinetics of dissolution) 61-72, 77, 83-94, 119-122, 124, 237
dissolution, kinetics of biotite, in natural setting (Panola granite) 37-57
dissolution, kinetics of biotite, experimental, 54
dissolution, caustic solution 75
dissolution, incongruent 75-76
dissolution, mechanism 5, 11
dissolution, non-stoichiometric 263
dissolution, selective 75-77, 79
dissolution-precipitation in porous media-numerical models 129-149, 153-176
dissolved silica 259-260, 262
distribution, mobility of elements 47-49, 53, 57, 97, 258-259, 261, 271
DNAPL (dense non-aqueous phase liquid) 399-400, 406-408, 410-411, 414-416
dolomite 136, 329, 441
drainage 50, 399-404, 406-407, 410, 412, 417-419
duripans 43, 50-52, 57
dysprosium (Dy) 192, 195, 204

E

East Pacific Rise 257-258, 270-271
ecosystem 56, 381, 442, 446-447, 452-453
EDTA 15, 19, 120, 368-374
electron acceptor 6, 379-386, 390-393, 396
element mobility (see also distribution, mobility of elements) 365-376
enstatite 61, 319, 324
enthalpy of complexation 217, 223
enthalpy of formation 239-240, 246
entropy 183, 217, 239-240
environmental remediation 399, 453
epidote 258-260, 262-276
epitaxial growth 107-108
epitaxial replacement 40-41, 54
epithermal deposits 279
equilibrium 55, 62, 67-69, 77, 105-106, 111-112, 115, 125, 181-189, 192-193, 203, 230-231, 234-237, 244, 248-250, 259-260, 264, 267-268, 275, 284, 290, 295-298, 300, 307-308, 310, 312-313, 315-317, 322, 325, 381-382, 390-392
equilibrium, stable 181-189
equilibrium, metastable 181-189, 258, 260, 264-273, 275-276
equilibrium constant 23, 28, 200, 204, 230, 235, 238-241, 243-244, 365-366, 372-373, 425, 430, 435-436
Equisetum arvense 443-444, 446-448
erbium (Er) 192, 194, 196, 198, 200, 203-204
europium (Eu) 192, 194, 195, 197, 200, 203-204, 368
europium anomaly 191, 205-206
EXAFS 369, 371, 424, 434, 438, 444-445, 447, 449, 451
extrapolation 4, 25, 37, 210, 220, 223, 409, 417

F

fayalite 319
fish 442
flexible gold reaction cells 258
flow boundary, constant pressure 404
flow boundary, no flow 404
fluid, saturation 407, 409-420
fluid, trapping 401, 404, 406, 419
fluid inclusions 280, 299, 329-335
fluid percolation 145, 147, 154, 156-157, 171, 173
fractal dimension 91-93
FTIR spectra 232-233, 371
fulvic acid 10, 22, 210, 368-369
fumaroles 280, 282-286, 299, 309, 313, 325

G

gadolinium (Gd) 192, 194-195, 197, 200, 203-204, 210, 215-221, 223-224, 230, 245-246, 248, 251
gadolinium hydroxide 230, 245-246, 251
galena 271, 283-284, 287, 329, 441
gallium (Ga) 312-313, 315, 318, 320, 323, 325
garnet 314-315, 323
geothermal systems 191, 287
Gibbs equation 401
Gibbs-Pointing correction 295-296
glass electrode 212-213, 223, 234, 238
goethite 40, 44, 47, 50, 61-72, 340-344, 347, 349-353, 355-358, 366, 368-369, 371-375
gold (Au) 259-260, 262-264, 269-271, 275, 279-280, 282-284, 287-296, 299, 312-313

granite 49, 186, 280
greenockite 284, 314, 324
grossular 323
groundwater 3, 8, 10, 29, 380-384, 390-392, 408, 441-443

H

Haine's jump 401-402, 404, 406-407, 411-412, 418
halite 283-284, 287-288, 315, 318, 323-324
heat capacity 238-240, 309
Helgeson's B-dot equation 193, 220, 261
hematite 40, 62-72, 262, 272-273, 309, 314-315
heterogeneity 393, 404, 406, 411, 418-419
holmium (Ho) 192, 194, 196, 198, 199-200, 203-205
horsetail 443
humic acid 10, 22, 210, 368-370
hydration 69, 288, 292, 294, 296-297
hydraulic conductivity 39, 43, 50-21
hydrogen 259-263, 272-275
hydrogen sulfide (H_2S) 259-263, 272-275
hydrogen-electrode concentration cell (HECC) 223, 233-234, 237, 244
hydrolysis, of silicates 3, 5, 16, 75, 261
hydrolysis, of cations 69, 71, 192-193, 211, 213, 229-251, 434, 438
hydrothermal 75, 97-98, 111, 191-192, 205, 209-211, 248, 258-259, 261, 264, 269, 271-275, 279-280, 286-287, 299, 301-302, 307, 310, 325
hydrothermal atomic force microscopy (HAFM) 112, 117, 119-122, 124-126
hydroxide complexes 206, 229-251, 270
hydroxycarbonate 230, 245, 446, 451-452
hydrozincite 446, 451-452

I

ICP-AES (ICP-OES) 91, 192, 234, 242, 444-445
ICP-MS 234, 242, 259-260, 309
Idaho 441
ilsemannite 283, 314, 324
imbibition 401-402
indium (In) 309, 312-314, 317-318, 321, 323, 325
infinite dilution 183, 210-211, 216-217, 220-223, 230-231, 235, 244, 246-251
interfacial capillary forces 400-401
interfacial chemistry 400-401, 404-406, 409, 411-413, 417, 419
interfacial compression, dilation 401-404, 418-419
interfacial contact angle 399-401, 406-408, 417, 419
interfacial curvature 400
interfacial energy 102-103, 105
interfacial kinetics 399-402, 406, 412-413, 417-419
interface stability 401, 409
interfacial tension, equilibrium 399-401
interfacial tension, measurement, drop volume method 400, 418
interfacial tension, measurement, oscillating jet method 400
interfacial tension, measurement, pendant drop method 400
interfacial tension, time-dependent 399-401, 404-407, 409-419
interfacial viscosity, shear 402, 404,
interfacial viscosity, dilational 401-402, 404
interstep spacing 97, 105
iodide 315
iron (Fe) 259-260, 262-264, 266-268, 270-272, 274-276, 280, 289, 312-313, 368, 371, 373-375, 381, 383, 385-388, 390-391, 442-443
iron oxide 44, 46-47, 61-72, 368, 370-371, 379-382, 389-392, 442
isocoulombic reaction 205, 223-224
isovolumetric weathering 49, 51, 54, 57
Izalco volcano 319

K

kaolinite 40-41, 43-44, 46-47, 49-50, 54, 61,
Kapton polyimide 444
Kellogg, Idaho 442
kinetics 28, 61-72, 182, 189, 392-393, 423
kinetics of dissolution (see also dissolution rates; dissolution kinetics) 28, 83-94
kink dynamics 112, 119, 121-126
Knudsen method 286
Kudryavy volcano 282-284, 307-325
Kurile Islands 307-308

L

LAM ICP-MS 280, 299
lanthanide (see also rare earth elements) 191-205, 209-225, 229-251, 312-313
lanthanum (La) 192, 194-195, 197, 199-200, 203-205, 210-211, 214, 216-221, 223-224, 230, 234, 245-248, 251
lanthanum hydroxide 230, 245-246, 248, 251
laser ablation 280, 299
lateral spreading rate 100, 102, 104-105, 109
leached layers 4, 21, 25, 29, 84, 89-90, 92-93
leaching rates of Si 52-53
lead (Pb) 259-260, 262-264, 270-271, 279-280, 282-283, 285, 312-315, 317, 319, 321, 325, 329, 368, 371-374, 441-443
Leverett J-function 411-412
ligand numbers 193, 195-196, 199-205, 210-211, 235
linear free energy correlations 372-375
liquid junction potential 213, 234

LNAPL (light non-aqueous phase liquid) 399
lutetium (Lu) 192, 194, 196, 198-200, 203-204

M

magnesioferrite 323
magnetite 258-263, 266-274, 283, 313-315, 318-319, 323-324
manganese (Mn) 312-313, 322, 325, 380-388, 390-391, 442-453
mass spectrometry 286
massive sulfide deposits 206
mass-loss torsion-effusion 286
matric potentials 42-43, 50
Merapi volcano 281-283, 285, 307, 315, 319
mercury (Hg) 282-283, 286-287, 312
merwinite 318-319
metal-organic interactions 209-225, 365-376
microscopy 24, 84-86
mid-ocean ridges 257-258, 269-270, 272-276
migration rate 410, 412-413, 415-416, 418-419
mine tailings 441-443
mine waste 441-443
mineral/water interfacial reactions 83-94, 97-109, 111-126, 339-360, 365-376, 423-439
Mississippi Valley Type Deposit 209, 329-335
mixed-flow reactor 97-99, 106-107
models 181-189
molybdenite 284, 287, 314-315, 323-324
molybdenum (Mo) 280, 282-283, 287, 289, 312-316, 321, 325
molybdite 314
Momotombo volcano 281, 283-285, 307
monticellite 318-319,
morphological forms 78-79
Mount Etna volcano 281-283
Mount St. Helens 281-283, 285, 307, 315, 319

muscovite 39-41, 53, 53-54

Ñ

nantokite 298-299
NAPL (non-aqueous phase liquid) 399-419
native sulfur 283-284
neodymium (Nd) 191-193, 203-205, 209-211, 214, 216-220, 222-224, 229-251
neodymium hydroxide 229-251
neodymium isotopes 209
Nernst equation 212, 234
nickel (Ni) 259-260, 262-264, 312-313, 315, 322, 325, 368, 371, 373
nitrate 379, 381-382, 386, 390-391
nitrogen uptake 382, 397
non-linear susceptibility 112-113, 117
nonwetting phase 400, 413
NTA 368, 373-374
nuclear waste 111, 209-210, 371
nucleation 76, 122-124
numerical simulation 379-397, 407, 409
nutrient bioavailability 340-341, 357

O

oil, primary migration 417, 419
oil, tertiary recovery 404
olivine 314-315, 319, 324
ore genesis 279, 299
organic carbon 46, 49, 379-385, 391
organic ligands 4, 6, 7-13, 18-25, 30, 209-225, 365-376
organic liquid contaminants 399, 419
organic mobility 368, 376,
orpiment 286
oxalate 6-8, 10-26, 367, 375-376
oxygen fugacity , 271-272, 310

P

palladium 368, 371, 373
palustrine 443
Panola Granite 37-57
partitioning 280, 288, 290, 294, 442-443, 446, 451-452
PCE (tetrachloroethylene) 406-416, 418
PEEK 212-213, 234
peepers 444-445
percolation: reconstructed porous media method 154, 156, 160, 166, 170
permeability, synthetic 153-176
permeability, measured in chalk 153-176
pH measurement 62, 211-213, 233-235, 261
Phalaris arundinacea 443-449, 451
pH-jumps 61-72
phosphorus, as a nutrient 339-341
phosphorus, bioavailability 343-345
phosphorus, chemical forms 342
phosphorus, distribution 343-344,
phthalate 368-369, 373
Piton de la Fournaise 285
PIXE 280
plagioclase 16, 17, 23, 24, 29, 39-40, 43, 49, 53-54, 206, 258-260, 262-276
plankton 342
plutonium hydroxide 250
Poas volcano 281-283, 299
point of zero charge 11-12, 66, 115, 117, 119
polynuclear species 211, 235, 248-249
polysaccharides 6, 15, 22, 23
pore diameter distributions 155, 159-160, 163, 166, 171
pore geometry 400-401, 406
pore scale models, numerical 404-409
pore water saturation 43
pore-scale interface movement 400-401
porosimetry methods, mercury 155
porosimetry methods, water 155
porosimetry methods, density 155
porosity 41, 51-52, 129-130, 132, 136, 139, 142, 149, 153-176

porous media 148, 379-397, 399-404, 406-407, 411, 413, 415-418
porphyry Cu-Mo deposits 279, 299
potential-determining ions 66
potentiometry 211-213, 233-235
Potosi Dolostone 330, 333-335
powellite 309, 314-315, 324
power spectral density 84, 86-89, 93
praseodymium (Pr) 192, 194-195, 197, 200, 203-204
precipitation and growth mechanisms 97-109, 111-126
pressure solution, vertical, horizontal stress 129-130, 136-137, 141, 149
pressure solution, pore fluid pressure 133, 142
pressure solution, strain rate 137-138, 140-149
pressure solution, calcite solubility 139-141, 144
pressure solution, thermal activation energy 144-145, 147
pressure solution, strain hardening 132, 143-144, 149
pressure solution, mechanisms, dissolution 134-135
pressure solution, mechanisms, diffusion 130, 133-135
pressure solution, mechanisms, precipitation 130, 133-135
pressure solution, thermodynamics, chemical potential vs. stress 132-134
pressure solution, models, free face 134-135
pressure solution, models, water-film diffusion 134-135
pressure solution, models, island-channel 134-135
progress variable 183-186
pyrite 158-261, 263, 266-276, 283-284, 315, 318, 323-324, 441
pyroxene 314-315, 319, 324
pyrrhotite 258, 261, 273, 284, 441

Q

quartz, drusy 329-335
quartz 12, 18-23, 27, 39-40, 46, 61, 66, 187, 258-260, 262-263, 265-274, 276, 318-319, 441
quartz-glass tubes 284, 290, 309

R

radius of curvature 404, 406-407, 409, 412
Raman spectroscopy 286-287
rare earth elements (see also lanthanides) 191-206, 209-225, 229-251
reaction path model 183-184, 186
reactive sites (see active sites)
realgar 284, 286
redox zones 380, 382, 391-392
reduction 379, 381-384, 390-392
regolith elemental fluxes 37-57
resistances-in-series model 345-347
rhenium (Re) 284, 309, 312-314
rhenium sulfide 284, 309, 314
rhizosphere 7, 9, 380-381, 384-385, 385, 392, 442, 446, 451-452
rhodochrosite 445, 449, 451-452
rhodonite 318, 323,
rhyolite 282-283
rock deformation, dry vs. fluid assisted 129-149, 153-176
roots 3, 6, 7, 39, 56, 442-452
rozenite 315, 324
rutile 318-319, 323, 367

S

salammoniac 283-284
salting out 262, 272-273
samarium (Sm) 192, 194-195, 197, 200, 203-204, 246-247
samarium hydroxide 246
San Cristobal volcano 282
sanidine 21, 318-319
saprolite 37-57
Satsuma Iwojima volcano 282-283, 307
saturation, hydraulic 41-43, 52
saturation state 11, 16, 17, 23, 29, 61, 446
scanning laser confocal microscopy (SLCM) 350, 353
scanning probe microscopy (see also AFM, HAFM) 83-94, 111-126
scheelite 309
Scirpus microcarpus 443-444, 446-448
seafloor hydrothermal fluids 205-206, 210, 257-276
second harmonic generation (SHG) 111-119, 126
secondary ion mass spectrometry (SIMS) 21, 83, 92
sediments 8-10, 380-392, 441-442
selenium (Se) 284-285, 312
sericite 441
Setchénow equation 261
Showa-Shinzan volcano 281-282, 307
siderite 441
silanol group 76, 423, 433-435, 438
silica particles 424-425, 436-439
silica-tube experiments 192
silica-water-interface 423-439
silver (Ag) 279-280, 284, 287, 289, 291-292, 296, 298-301, 312-313, 319, 441
silver chloride (see also chlorargyrite) 191-206, 291-292
sinjarite 324
smithsonite 451-452
soils 37-57, 380, 384, 441-443, 446-449, 451-452
solubility 7, 13, 15, 19-22, 81, 191-206, 229-251, 258, 261-262, 268, 270-271, 274, 280, 287-293, 295-296, 298-299, 302
solubility product 229-251
solute fluxes 37-39, 50, 52, 54-55, 57
solvation 288, 294, 297
sorption, (see also adsorption) 97, 379-382, 386, 389-390, 395-396
Southeast Missouri 329-335
sphalerite 283-284, 287, 318, 323, 329, 441
spiral growth 97, 102, 104-107, 109
St. Augustine volcano 281-283, 307, 309, 315
stability constants 191-206, 209-225, 366-368, 371-376

step kinetics 121
stibnite 286
strontium (Sr) 368
sublimates, volcanic 280, 283-287, 307-309, 313-315, 319-320, 324-325
sulfate 258-260, 262, 284, 313-315, 323, 382, 390
sulfide 279-280, 283-285, 313-315, 323, 379-380, 382
sulfur polymers 286-287
Superfund 441
supersaturation 97-103, 105-109, 244
surface analysis 83-94
surface area 11, 16, 19, 20, 37-39, 44, 46-47, 50-51, 54-55, 62, 68, 71, 77, 83-84, 86, 89, 91-93
surface complexes (surface complexation) 15, 29, 62, 65-68, 365-376, 423-439
surface complexes, fluorescence of 424-425, 427, 429-430, 433-438
surface elasticity 402
surface excess 401-403, 418
surface potential 112-119, 366-367
surface roughness 83-93
surface strain energy 77, 81
sylvite 283-284, 313, 315, 323-324
synchrotron XRF 280

T

TAG 257, 272, 274-275
tellurium (Te) 284-285, 312-313
Tempe cell 416
terbium (Tb) 192, 194, 196, 198, 200, 203-205
ternary surface complexes 365-376, 435-436
tetrachloroethylene (see PCE)
tetrahedrite 441
thallium (Tl) 285, 312-313, 315
thenardite 284, 323
thermal gravimetric analysis 232-233
thermodynamic modeling 258, 261-262, 286-287, 309-310
thermodynamics, microbial uptake of inorganic phosphorus (P_i) 352-353
thermodynamics 181-189
thermodynamics, constraints 181-189
thermodynamics, system 181, 189
thermodynamics, First Law of 182
thermodynamics, degrees of freedom 186-187
thermodynamics, chemical potential 182, 186-187
thulium (Tm) 192, 194, 196, 198, 200, 203-204
time-resolved laser-induced luminescence spectroscopy (TRLIFS) 424-427, 434-435, 438
time-scale 400-401, 417, 419
tin (Sn) 279-280, 282-283, 289, 292-294, 296, 299, 312-313, 317, 321, 325
titanium (Ti) 312-313, 316, 319-320, 323
Tolbachik volcano 282-283
tracer test 39, 50-51
transients, pH 61-72
transport modeling 379-397
triaxial press 137
triflate 236-238, 250
trifluoromethanesulfonate 211
troilite 323
tugarinovite 314
tundra swan 442
tungsten (W) 282-283, 312-313
two-dimensional (heterogeneous) nucleation 97-100, 102-103, 105-109
Typha latifolia 443-444, 446-448

U

unsaturated flow 50-51, 53
uranium (U) 312, 368, 371, 423-439
uranyl, carbonate complexation of 424-426, 434-436
uranyl, solution speciation 425-426
uranyl, surface complexes 423-439
uranyl, equatorial plane of 434-435
uranyl, mobility, , 434, 436
Usu volcano 281-283, 299, 307

V

vanadium (V) 312-313, 318-319, 323
vapor transport 279-302, 307-325
Viburnum Trend 329, 333-334
volatility 280, 286-288, 299, 310, 313, 315, 324-325
volcanic gases 280-285

W

water exchange rates 71
wetlands 380-382, 384, 386-390, 392, 395-396, 441-453
wetting phase 400-402, 406, 411
White Island 282-283
wolframite 284
wollastonite 318-319, 324
wurtzite 314, 324

X

XANES 444-445, 447-448
X-ray computerized tomography (CT) 153-176
X-ray diffraction, powder 232, 245
X-ray fluorescence microtomography (FCMT) 445, 449
X-ray photoelectron spectroscopy (XPS) 83-84, 92, 350-351, 424

Y

Young-Laplace equation 400, 406
ytterbium (Yb) 192, 194, 196, 198, 200, 203-204, 210-211, 215-217, 219-221, 223-224

Z

zeolite ZSM-5, Al zoning, enrichment 75, 80
zeolite ZSM-5, ferroelasticity 75, 77, 79
zeolite ZSM-5, hydrothermal synthesis 75-76

zeolite ZSM-5, morphology 76-78
zeolite ZSM-5, symmetry 77-79, 81
zeolite ZSM-5, twinned crystal 75-81
zinc (Zn) 280, 282-283, 312-314, 320-321, 323-325, 368, 371, 373, 441-453
zircon 39